SLIDING MODE CONTROL

Edited by **Andrzej Bartoszewicz**

Sliding Mode Control
Edited by Andrzej Bartoszewicz

Published by InTech
Janeza Trdine 9, 51000 Rijeka, Croatia

Publishing Process Manager Ivana Lorkovic

Technical Editor Teodora Smiljanic

Cover Designer InTech Design Team

Additional hard copies can be obtained from orders@intechopen.com

Sliding Mode Control, Edited by Andrzej Bartoszewicz
p. cm.
ISBN 978-953-307-162-6

For EU product safety concerns:
IN TECH d.o.o., Prolaz Marije Krucifikse Kozulić 3, 51000 Rijeka, Croatia,
info@intechopen.com or visit our website at intechopen.com.

Meet the editor

Andrzej Bartoszewicz received the M.Sc. degree in 1987 and the Ph.D. degree in 1993 both from Technical University of Łódź, Poland. Then he obtained the degree of doktor habilitowany in control engineering and robotics from Academy of Mining and Metallurgy in Cracow, Poland. He was visiting scholar at Purdue University, West Lafayette, In., USA and at Strathclyde University, Glasgow, UK. Then for one year he was at University of Leicester, UK. Currently he is Professor at Technical University of Łódź, vice dean of Department of Electrical, Electronic, Computer and Control Engineering, head of Electric Drive and Industrial Automation Group and vice-director of Institute of Automatic Control. He has published three monographs and over 250 papers, primarily in the fields of sliding mode control and congestion control in data transmission networks.

Contents

Preface

The theory of variable structure systems with sliding modes is currently one of the most important research topics within the control engineering domain. Moreover, recently a number of important applications of the systems primarily in the field of power electronics, control of electric drives, robotics and position regulation of sophisticated mechanical systems have also been reported. Therefore, the objective of this monograph is to present the most significant latest developments in the theory and engineering applications of the sliding mode control and to stimulate further research in this field.

The monograph consists of 28 chapters. It begins with six contributions devoted to various significant issues in power electronics. In the first chapter, Ben Saad et al. propose, test and compare sliding mode and fuzzy sliding mode controllers for DC-DC converters. In the second chapter, Petkova et al. consider the operation of the single-phase inverter and single-phase active power filter and prove, both in simulations and laboratory experiments, the effectiveness of sliding mode controllers in these two applications. Then, Al-Hosani et al. also consider the design of DC-DC buck and boost converters. They develop the sliding mode approach which implements – very common in industry – proportional integral derivative (PID) controllers. The main idea of that chapter may be summarized as enforcing sliding mode such that the output converter voltage contains proportional, integral and derivative components with the predefined coefficients. Chattering is then reduced through the use of multiphase power converter structure. The proposed design methods are confirmed by means of computer simulations. In the next chapter, Zhu and Kang consider arc welding/cutting power supply and propose a "synthetic" sliding mode and PI controller. They propose to use the PI controller in the current loop and the sliding mode controller in the voltage loop. The results are verified by experiments conducted on a 20 kW arc welding/cutting power source. They show on one hand good dynamic performance of the system, and on the other decreased undesirable voltage overshoot. Another contribution concerned with power electronics is the chapter by Ayad et al. which presents sliding mode control of fuel cells, supercapacitors and battery hybrid sources for vehicle applications. Then, the chapter by Susperregui presents and evaluates first-order and higher-order sensorless sliding mode control algorithms, for a doubly-fed induction generator. The algorithms not only aim at governing active and reactive power exchange between the doubly-fed induction generator stator and the grid, but also ensure the synchronization required for smooth connection of the generator stator to the grid.

Sliding mode systems are a feasible option not only for power converter control but also for electric drive regulation. Therefore an important issue of induction motor control is addressed in the next two chapters. The chapter by Cortes-Romero and Sira-Ramirez presents a combination of two control loops, one employing a discontinuous sliding mode controller and another one based on the combination of generalized proportional integral control and generalized proportional integral disturbance observer. The authors of the chapter demonstrate – by experiments performed on an actual induction motor test bed with a voltage controlled brake – that the proposed combination results in robust position and tracking control of induction motors. In the next chapter, written by Reama et al. a new simple and easy to implement adaptive sliding mode scheme for speed and flux control of induction motor using online estimation of the rotor resistance and load torque are proposed. The two chapters on control of induction motors are followed by a contribution of Patre and Panchade, which is concerned with a unified sliding mode approach to torque, position, current and speed regulation of DC drives. Then the next chapter, by Santo et al., presents the design and implementation of a sliding mode position controller for a linear switched reluctance actuator devoted primarily for robotic applications. The section devoted to the problem of electric drive control ends up with a chapter on friction compensation for a mini voice coil motors. The chapter written by Lin et al., demonstrates that sliding mode control approach may reliably eliminate stick slip oscillations and reduce the steady state error. This conclusion is drawn based on experimental results performed on a mini voice coil motor mounted on a compact camera module.

The next three chapters are concerned with selected issues in robotics. The first of them, written by Becerra and Sagues proposes a robust controller for image-based visual servoing for differential drive mobile robots. The second one, by Rivera et al., is devoted to the application of a higher order, namely super-twisting sliding mode controller for trajectory tracking of an under-actuated manipulator and also for induction motors. Then Herman and Kozłowski consider rigid, serial manipulators and present an extensive survey of selected non-adaptive sliding mode controllers expressed in terms of the inertial quasi-velocities. They also point out a number of advantages offered by sliding mode control schemes using inertial quasi-velocities. The next seven chapters present successful applications of sliding mode control paradigm in other areas than power electronics, electric drives and robotics. The section devoted to those applications begins with the chapter by Fung and Chang on sliding mode force and motion control of three very popular mechanisms, i.e. slider-crank, quick-return and toggle mechanism. Then Tournes et al. propose a higher order sliding mode control scheme for automatic docking of space vehicles. The issue of higher order sliding mode control is also considered in the chapter, by Suares et al. In that contribution higher order sliding mode is successfully used to suppress nonlinear dynamics in physical plants with friction which is inevitable in all mechanical systems. Higher order sliding mode approach is further considered in the chapter by Salgado-Jiménez et al. on control of remotely operated vehicles which are nowadays indispensable in performing the inspection tasks and maintenance of numerous underwater structures, common in the oil industry, especially in deep and not easily accessible to humans waters. That chapter demonstrates that sliding mode control is a viable option for controlling underwater vehicles which operate in a highly dynamic and uncertain environment often affected by waves and strong currents. Another interesting and very well worked out application is described in the next chapter authored by Schindele and Aschemann. They propose three types of sliding mode controllers (conventional, second-order and proxy) for a linear axis driven by four pneumatic muscles and verify performance of these controllers on a laboratory test rig. Then Kim et al. present adaptive sliding mode controller of adhesion force between the rail and the wheel in railway rolling stocks. The section concerned with various applications of sliding mode control concludes with the chapter by Wen on optimal fuzzy sliding mode control of biochips and biochemical reactions.

The last section of this monograph presents selected new trends in the theory of sliding mode control. It begins with a chapter by Nowacka-Leverton and Bartoszewicz pointing out some advantages of sliding mode control systems with time-varying switching surfaces. Then the chapter by Sato et al. discusses a new variable structure design method which results in good transient performance of the controlled system and small steady state error. The next chapter by Chen and Tseng is devoted to the attenuation of an important and fairly undesirable effect of chattering. The authors present a new controller design procedure aimed at chattering reduction by low-pass filtering of the control signal. Also the subsequent chapter, written by Mihoub et al., considers the chattering phenomenon. It efficiently combines multi-model approach to the reaching phase performance improvement with the second order sliding mode controller design for discrete time systems. Another significant theoretical issue is considered by Adloo et al. Those authors propose sliding mode controller for two dimensional (2-D) systems and discuss the switching surface design and the control law derivation. In the penultimate chapter of this monograph, Yasser et al. propose to incorporate some elements of artificial intelligence, namely appropriately trained neural networks, into the sliding mode control framework and demonstrate the advantages of this approach. Finally, the last chapter of this book – written by Nied and de Oliveira – also concentrates on some aspects of combining neural networks with sliding mode control, however their goal is quite different from that of Yasser et al. Indeed Nied and de Oliveira present a sliding mode based algorithm for on-line training of artificial neural networks, rather than exploiting neural networks in variable structure controller construction.

In conclusion, the main objective of this book was to present a broad range of well worked out, recent application studies as well as theoretical contributions in the field of sliding mode control. The editor believes, that thanks to the authors, reviewers and the editorial staff of Intech Open Access Publisher this ambitious objective has been successfully accomplished. It is hoped that the result of this joint effort will be of true interest to the control community working on various aspects of non-linear control systems, and in particular those working in the variable structure systems community.

Andrzej Bartoszewicz
Institute of Automatic Control,
Technical University of Łódź
Poland

Part 1

Sliding Mode Control in Power Electronics

1

Sliding Mode Control and Fuzzy Sliding Mode Control for DC-DC Converters

Kamel Ben Saad, Abdelaziz Sahbani and Mohamed Benrejeb
Research unit LARA,
National engineering school of Tunis (ENIT), Tunis,
Tunisia

1. Introduction

Switched mode DC-DC converters are electronic circuits which convert a voltage from one level to a higher or lower one. They are considered to be the most advantageous supply tools for feeding some electronic systems in comparison with linear power supplies which are simple and have a low cost. However, they are inefficient as they convert the dropped voltage into heat dissipation. The switched-mode DC-DC converters are more and more used in some electronic devices such as DC-drive systems, electric traction, electric vehicles, machine tools, distributed power supply systems and embedded systems to extend battery life by minimizing power consumption (Rashid, 2001).

There are several topologies of DC-DC converters which can be classified into non-isolated and isolated topologies. The principle non-isolated structures of the DC-DC converters are the Buck, the Buck Boost, the Boost and the Cuk converters. The isolated topologies are used in applications where isolation is necessary between the input and the load. The isolation is insured by the use of an isolating transformer.

The DC-DC converters are designed to work in open-loop mode. However, these kinds of converters are nonlinear. This nonlinearity is due to the switch and the converter component characteristics.

For some applications, the DC-DC converters must provide a regulated output voltage with low ripple rate. In addition, the converter must be robust against load or input voltage variations and converter parametric uncertainties. Thus, for such case the regulation of the output voltage must be performed in a closed loop control mode. Proportional Integral and hysteretic control are the most used closed loop control solutions of DC-DC converters. This can be explained by the fact that these control techniques are not complicated and can be easily implemented on electronic circuit devises.

Nowadays, the control systems such as microcontrollers and programmable logic devises are sophisticated and allow the implementation of complex and time consuming control techniques.

The control theory provides several control solutions which can be classified into conventional and non-conventional controls. Many conventional controls, such as the PID control, were applied to DC-DC converters. The design of the linear controller is based on the linearized converter model around an equilibrium point near which the controller gives

good results. However, for some cases this control approach is not so efficient (Tse & Adams, 1992; Ahmed et al, 2003).
Sliding Mode Control (SMC) is a nonlinear control technique derived from variable structure control system theory and developed by Vladim UTKIN. Such control solution has several advantages such as simple implementation, robustness and good dynamical response. Moreover, such control complies with the nonlinear characteristic of the switch mode power supplies (Nguyen & Lee, 1996; Tan et al, 2005). Although, the drawback of SMC is the chattering phenomena. To overcome the chattering problem one solution consists into extending SMC to a Fuzzy Sliding Mode Control (FSMC), (Alouani, 1995).
Fuzzy Logic Control is a non-conventional and robust control law. It is suitable for nonlinear or complex systems characterized by parametric fluctuation or uncertainties (Kandel & Gideon, 1993; Passino, 1998). The advantage of the FSMC is that it is not directly related to a mathematical model of the controlled systems as the SMC.
This chapter aims to compare SMC and FSMC of DC-DC converters. The average models of Buck, Boost and Buck Boost converters are presented in section 2. Then in section 3, some classical sliding mode controls are presented and tested by simulations for the case of Buck and Buck Boost converters. In order to improve the DC-DC converters robustness against load and input voltage variations and to overcome the chattering problem, two approaches of FSMC are presented in section 4.

2. DC-DC converters modelling

The switching DC-DC converters are hybrid dynamical systems characterized by both continuous and discrete dynamic behaviour.
In the following, we present only a general modelling approach of DC-DC converters by application of the state space averaging technique of the Buck, Boost and Buck-Boost converters for the case of a continuous conduction mode.
Let us consider a switching converter which has two working topologies during a period *T*. When the switches are closed, the converter model is linear. The state-space equations of the circuit can be written and noted as follows (Middlebrook & Cuk, 1976):

$$\begin{cases} \dot{x} = A_1 x + B_1 u \\ y = C_1 x + E_1 u \end{cases} \tag{1}$$

When the switches are opened, the converter can be modelled by another linear state-space representation written and noted as follows:

$$\begin{cases} \dot{x} = A_2 x + B_2 u \\ y = C_2 x + E_2 u \end{cases} \tag{2}$$

From the equation (1) and (2) we can determine the averaged model given by equation (3) for an entire switching cycle *T*.

$$\begin{cases} \dot{\tilde{x}} = A(d)\tilde{x} + B(d)\tilde{u} \\ \tilde{y} = C(d)\tilde{x} + E(d)\tilde{u} \end{cases} \tag{3}$$

where the matrices $A(d)$, $B(d)$, $C(d)$ and $E(d)$ are defined as follows:

$$\begin{cases} A(d) = dA_1 + (1-d)A_2 \\ B(d) = dB_1 + (1-d)B_2 \\ C(d) = C_1 + (1-d)C_2 \\ E(d) = dE_1 + (1-d)E_2 \end{cases} \tag{4}$$

and $\tilde{x}$, $\tilde{y}$ and $\tilde{u}$ are respectively the average of x, y and u during the switching period T. Let us consider the Buck, Boost and Buck-Boost converters presented by Fig. 1, Fig. 2 and Fig. 3 respectively. The state space representation can be expressed for these converters as follows :

$$\begin{cases} \dot{\tilde{x}} = A(d)\tilde{x} + B(d)\tilde{u} \\ \tilde{v}_o = C(d)\tilde{x} \end{cases} \tag{5}$$

where

$\tilde{x} = \begin{pmatrix} \tilde{i}_L \\ \tilde{v}_o \end{pmatrix}$

$C(d) = (0 \quad 1)$

$\tilde{u} = V_{in}$

$\mathrm{d} = 1$ (Switch ON)

$\mathrm{d} = 0$ (Switch OFF)

However the matrix $A(d)$ and $B(d)$ depend on the kind of converter. Table 1 gives the expression of these matrixes for the considered converters.

Buck converter	Boost converter	Buck Boost converter
$A(d) = \begin{pmatrix} 0 & -\frac{1}{L} \\ \frac{1}{C} & -\frac{1}{RC} \end{pmatrix}$	$\mathrm{A(d)} = \begin{pmatrix} 0 & -\frac{1-\mathrm{d}}{\mathrm{L}} \\ \frac{1-\mathrm{d}}{\mathrm{C}} & -\frac{1}{\mathrm{RC}} \end{pmatrix}$	$A(d) = \begin{pmatrix} 0 & \frac{1-d}{L} \\ -\frac{1-d}{C} & -\frac{1}{RC} \end{pmatrix}$
$B(d) = \begin{pmatrix} \frac{d}{L} \\ 0 \end{pmatrix}$	$\mathrm{B(d)} = \begin{pmatrix} \frac{1}{\mathrm{L}} \\ 0 \end{pmatrix}$	$B(d) = \begin{pmatrix} \frac{d}{L} \\ 0 \end{pmatrix}$

Table 1. Matrix $A(d)$ and $B(d)$ expression for the Buck, Boost and Buck Boost converters

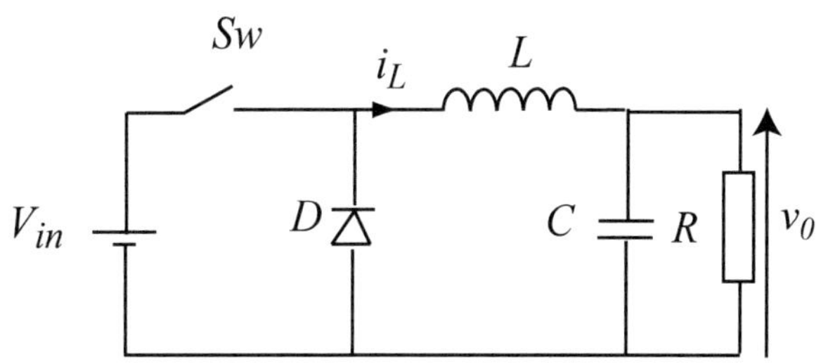

Fig. 1. Buck converter structure

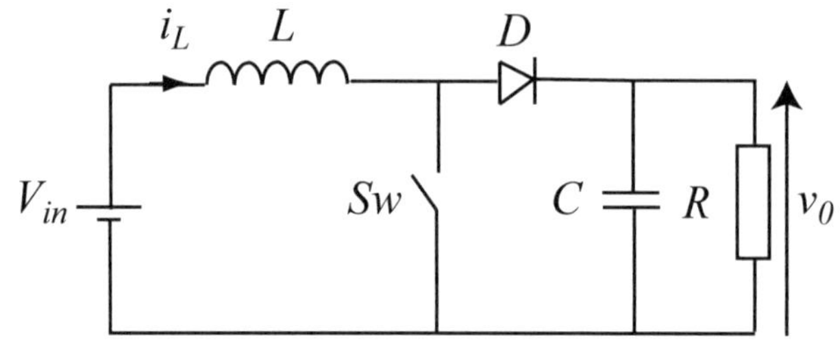

Fig. 2. Bosst converter structure

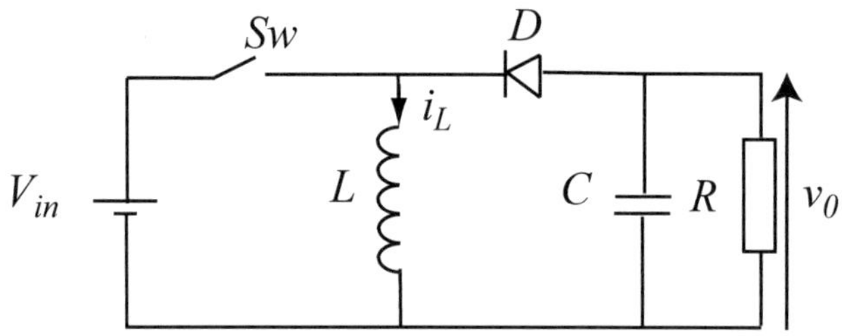

Fig. 3. Buck-Boost converter structure

The averaged modelling approach for the switching mode converter leads to an approximate non linear models. The linearization of this kind of models around the operating point allows the application of conventional control approach such as PID control and adaptive control. However, sliding mode control is considered to be the most adequate control solution because it complies with the nonlinear behaviour of the switching DC-DC converters and it is robust against all the modelling parametric uncertainties.

3. Sliding mode control for DC- DC converters

3.1 SMC general principle

SMC is a nonlinear control solution and a variable structure control (VSC) derived from the variable structure system theory. It was proposed by Vladim UTKIN in (Utkin, 1977).
SMC is known to be robust against modelling inaccuracies and system parameters fluctuations. It was successfully applied to electric motors, robot manipulators, power systems and power converters (Utkin, 1996). In this section, we will present the general principle of the SMC and the controller design principle.
Let us consider the nonlinear system represented by the following state equation:

$$\dot{x} = f(x,t) + g(x,t)u(t) \tag{6}$$

where x is n-dimensional column state vector, f and g are n dimensional continuous functions in x, u and t vector fields, u is the control input.
For the considered system the control input is composed by two components a discontinuous component u_n and a continuous one u_{eq} (Slotine & Li, 1991).

$$u = u_{eq} + u_n \tag{7}$$

The continuous component insures the motion of the system on the sliding surface whenever the system is on the surface. The equivalent control that maintains the sliding mode satisfies the condition

$$\dot{S} = 0 \tag{8}$$

Assuming that the matrix $\frac{\partial S}{\partial x} g(x,t)$ is non-singular, the equivalent control maybe calculated as:

$$u_{eq} = -\left(\frac{\partial S}{\partial x} g(x,t)\right)^{-1} \left(\frac{\partial S}{\partial t} + \frac{\partial S}{\partial x} f(x,t)\right) \tag{9}$$

The equivalent control is only effective when the state trajectory hits the sliding surface. The nonlinear control component brings the system states on to the sliding surface.
The nonlinear control component is discontinuous. It would be of the following general form (Slotine & Li, 1991; Bandyopadhyay & Janardhanan, 2005):

$$u_n = \begin{cases} u^+ \; with \; S > 0 \\ u^- \; with \; S < 0 \end{cases} \tag{10}$$

In the following SMC, will be applied to a buck and buck-boost converters.

3.2 SMC for Buck converter

3.2.1 Proposed SMC principle

For the Buck converter we consider the following sliding surface S:

$$S = ke + \dot{e} \tag{11}$$

where k is the sliding coefficient and e is the output voltage error defined as follows :

$$e = V_{ref} - v_0 \tag{12}$$

By considering the mathematical model of the Buck converter, the surface can be expressed by the following expression (Tan, et al, 2006; Ben Saad et al 2008):

$$S = -\frac{1}{C} i_L + \left(\frac{1}{RC} - k\right) v_o + KV_{ref} \tag{13}$$

and its derivative is given by :

$$\dot{S} = \left(\frac{1 - kRC}{RC^2}\right) i_L - \left(\frac{L - kRLC - R^2C}{R^2C^2L}\right) v_0 - \left(\frac{V_{in}}{LC}\right) u \tag{14}$$

The next step is to design the control input so that the state trajectories are driven and attracted toward the sliding surface and then remain sliding on it for all subsequent time.
The SMC signal u consists of two components a nonlinear component u_n and an equivalent component u_{eq}, (Ben Saad et al, 2008).

The equivalent control component constitutes a control input which, when exciting the system, produces the motion of the system on the sliding surface whenever the system is on the surface. The existence of the sliding mode implies that $\dot{S}=0$. So the equivalent control may be calculates as:

$$u_{eq} = \alpha_1 i_L - \alpha_2 v_o \tag{15}$$

where

$$\alpha_1 = \frac{L - kLRC}{RCV_{in}}$$

and

$$\alpha_2 = \frac{L - kRLC - R^2C}{R^2CV_{in}}$$

Let us consider the positive definite Lyapunov function V defined as follows:

$$V = \frac{1}{2}S^2 \tag{16}$$

The time derivative $\dot{V}$ of V must be negative definite $\dot{V}<0$ to insure the stability of the system and to make the surface S attractive. Such condition leads to the following inequality:

$$S\dot{S} = S\left(-\frac{V_{in}}{LC}u_n\right) < 0 \tag{17}$$

To satisfy such condition, the nonlinear control component can be defined as follows:

$$u_n = sign(S) \tag{18}$$

Fig. 4 presents the control diagram of the presented SMC.

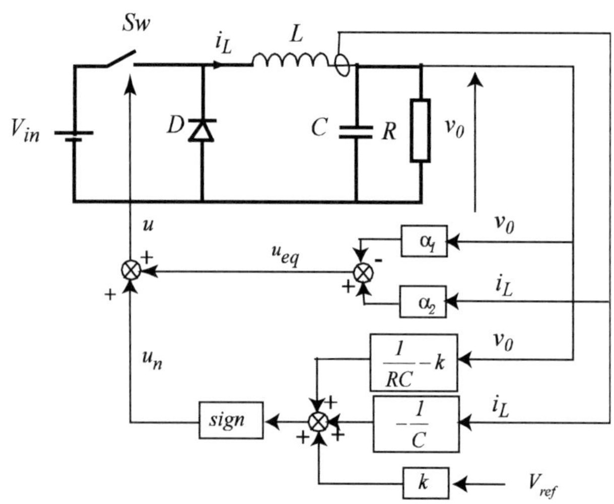

Fig. 4. SMC for Buck converter

3.2.2 Simulation and experimental results

The SMC is tested by simulation and experimentally using a dSAPCE control board. The test bench was built as shown in Fig. 10 and Fig. 11 around:

- a Buck converter,
- a computer equipped with a dSPACE DS1104 with its connector panel,
- a DC voltage power supply,
- two load resistances.

Fig. 5. Photo of the studied Buck converter

Fig. 6. Photo of the test bench

The dSPACE DS1104 controller board is a prototyping system. It is a real time hardware platform. It can be programmed with MATLAB/SIMULINK software through a real time interface allowing the generation of a real time code. Two ADC input channels of the DS1104, characterized by a 16 bits resolution, are used to acquire the Buck converter output voltage and the inductance current. The control board generates a digital PWM signal which is used to control the switch of the Buck converter.

The proposed SMC was applied to a Buck converter characterized by the parameters given in the table 2.

Parameters	Values
V_{in}	15 V
C	22 μF
L	3 mH
R	10 Ω
Switching frequency	10 kHz

Table 2. Studied buck converter parameters

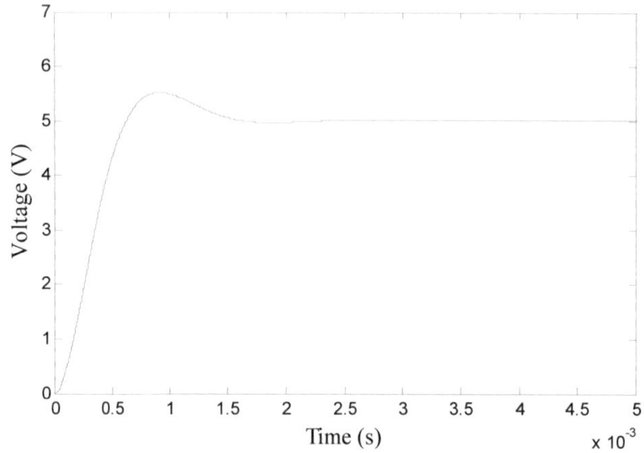

Fig. 7. Open loop responses of the buck converter by application of 16% PWM control signal

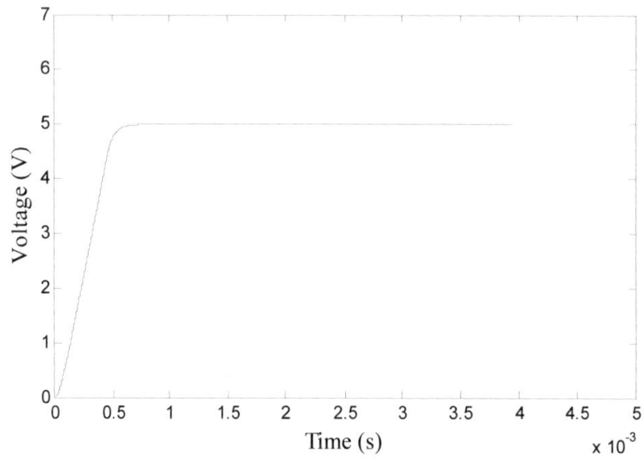

Fig. 8. Application of the SMC to the studied Buck converter ($Vref = 5V$)

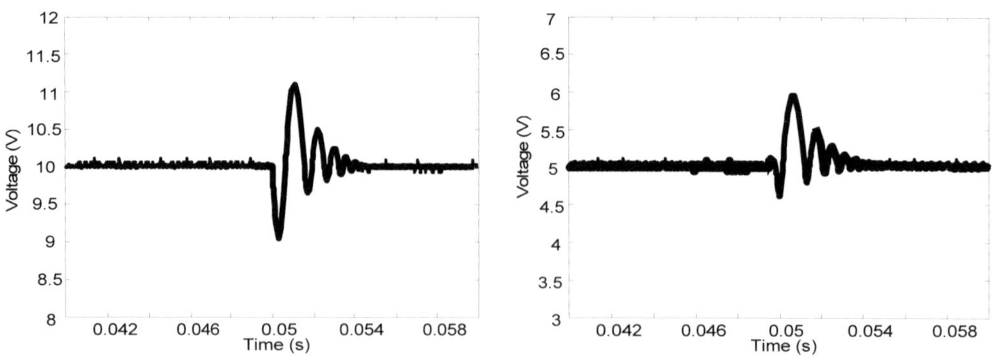

(a) Reference voltage 10 V (b) Reference voltage 5V

Fig. 9. Experimental test robustness of the SMC for the variation of the load from 10 Ω to 15 Ω

Fig. 7 presents the simulated output voltage by application of a PWM control signal of 16% duty cycle. The voltage response corresponds to a second order damped system response with an overshoot. Fig. 8 presents the obtained result by application of the proposed SMC to the studied controller for a 5 V voltage reference. We can see clearly that the observed voltage overshoot obtained on the open-loop response disappeared by application of the SMC.

The SMC is tested experimentally for the case of the load variation. Fig. 9 presents the obtained results for the case of the variation of the load resistance from 10 Ω to 15 Ω at 0.05s. It is clear that this perturbation is quickly rejected because the output voltage attends the reference voltage. The experimental test result for the case of the input voltage variation from 30 V to 20V, given in Fig. 10, shows the robustness of the applied SMC.

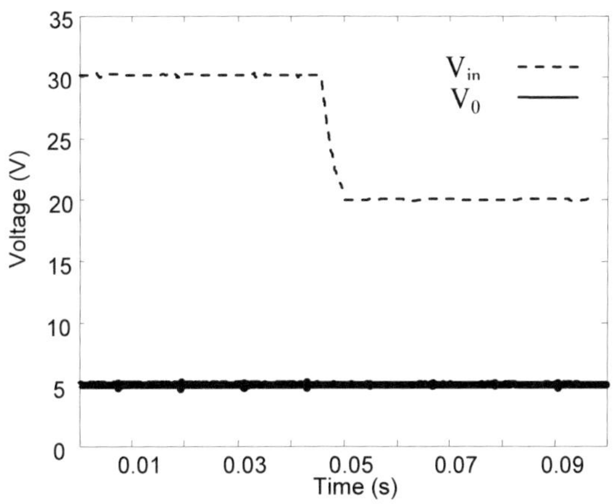

Fig. 10. Experimental output voltage evolution by application of the SMC for the variation of the input voltage from 30V to 20V

3.3 SMC for Buck-Boost converter

3.3.1 Proposed SMC principle

As for the Buck Converter, the Buck-Boost converter sliding surface and output voltage error are respectively defined by equations (8) and (9).

k can be chosen so that the outer voltage loop is enough to guarantee a good regulation of the output voltage with a near zero steady-state error and low overshoot.

Without any high frequency, when the system is on the sliding surface, we have $S = 0$ and $\dot{S} = 0$ (Hu et al, 2005; El Fadil et al, 2008).

As the control signal applied to the switch is pulse width modulated, we have only to determine the equivalent control component.

By considering the mathematical model of the DC-DC Boost converter, at the study state the variation of the surface can be expressed as:

$$\dot{S} = k\dot{e} = -k\dot{v}_0 = -k\left(\frac{1-u_{eq}}{C}i_L - \frac{v_0}{RC}\right) \qquad (19)$$

and then from equation (20) and by considering the condition $\dot{S} = 0$ we have:

$$k\dot{v}_0 + \frac{\dot{v}_0}{RC} = \frac{1-u_{eq}}{C}\dot{i}_L \qquad (20)$$

From the state representation (7) we can write the following relation:

$$\dot{v}_0(\frac{1}{RC} - k) = \frac{1-u_{eq}}{C}(\frac{u_{eq}-1}{L}v_0 + \frac{v_{in}}{L}) \qquad (21)$$

Then equivalent control component expression:

$$u_{eq} = 1 - \frac{v_{in} + \sqrt{v_{in}^{\ 2} + \frac{4kL}{R}(CRk - L)(v_{ref} - v_0)}}{2v_0} \qquad (22)$$

3.3.2 Simulation results

The proposed SMC was applied by simulation to the studied Buck-Boost converter characterized by the parameters given in Table 3. Fig. 11 presents the studied converter open-loop voltage and current responses. In Fig. 12 the output voltages evolution obtained by application of the SMC are presented for a reference voltages $V_{ref} = -20V$. So the application of the SMC allowed the elimination of the overshoot observed for the open-loop response. Fig. 13 presents the control signal. We can notice that it is strongly hatched. This is a consequence of the chattering phenomenon.

PARAMETERS	VALUES
V_{in}	20 V
C	22 μF
L	3 mH
R	10 Ω
SWITCHING FREQUENCY	10 kHz

Table 3. Studied Buck Boost converter parameters

To test the robustness of the SMC, we consider now the variations of the load resistance and the input voltage. Fig. 14 presents the evolution of the output voltage and the current in the load for the case of a sudden change of the load resistance from 30Ω to 20Ω. So by the application of the SMC, this perturbation was rejected in 10.10^{-3}s and the output voltage attends the reference voltage after. Fig. 15 illustrates the sudden variation of the input voltage from 15V to 10V at 0.05s. For such case we notice that the output voltage, presented by Fig. 16, attends after the rejection of the perturbation the desired value −20V and the converter work as boost one.

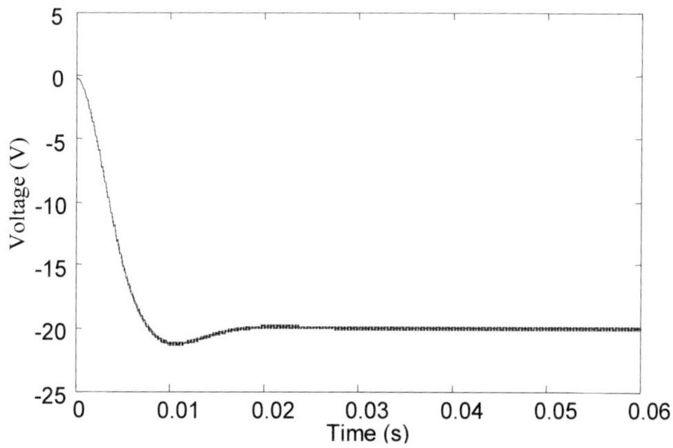

Fig. 11. Output voltage evolution of the Buck-Boost converter obtained by open-loop control.

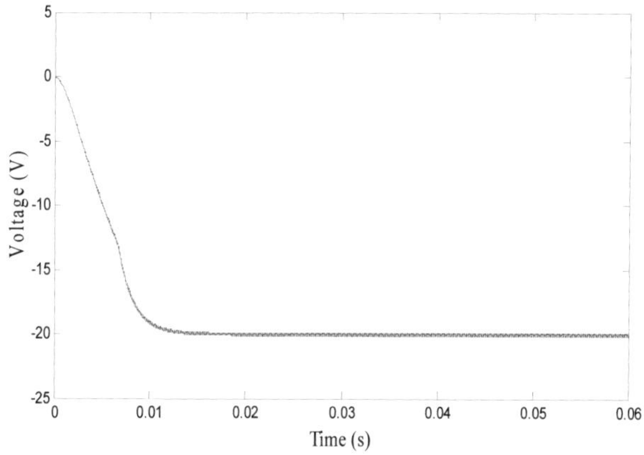

Fig. 12. Output vvoltage evolution obtained by application of the SMC

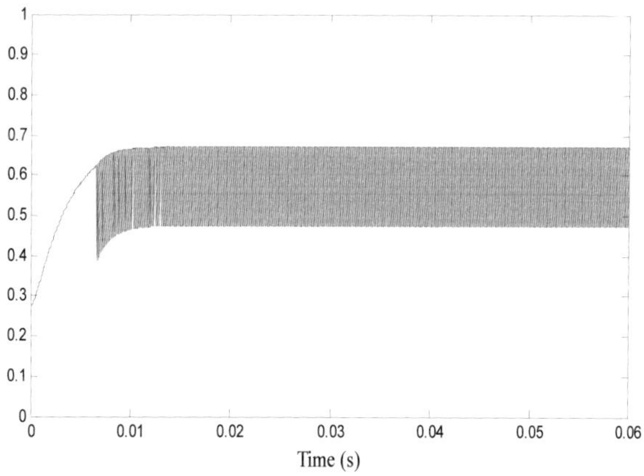

Fig. 13. Control signal evolution

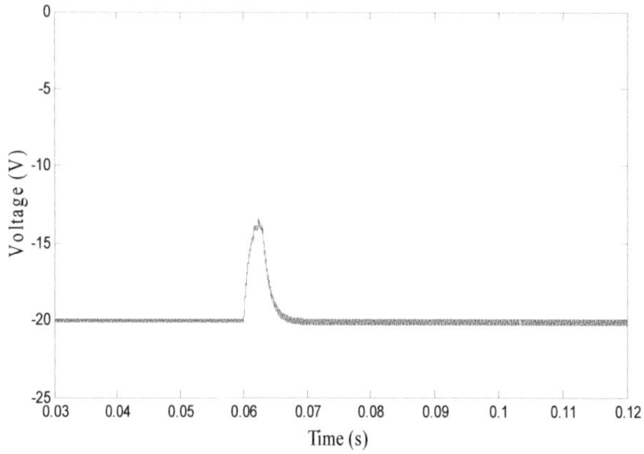

Fig. 14. Output voltage evolution by application of the SMC for the case of load variation from 30Ω to 20Ω

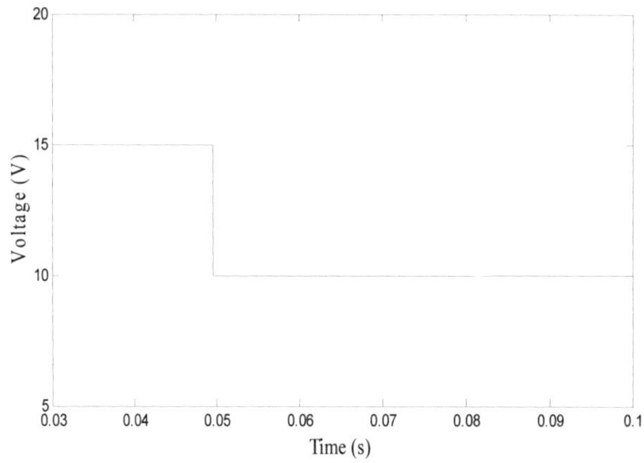

Fig. 15. Input voltage variation

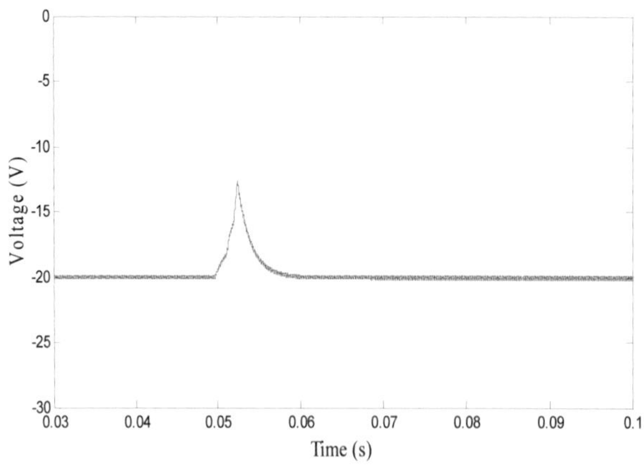

Fig. 16. Output voltage evolution by application of the SMC for the case of input voltage variation from 15 V to 10V

In order to overcome the problem of the chattering phenomena, it is possible to apply a high order SMC. However, the obtained analytical expression of the control component can be so complicated.
In the following, we will propose to apply a FSMC to the studied converters in order to improve the robustness of the SMC and to overcome the chattering problem.

4. FSMC for DC- DC converters

As SMC, Fuzzy Logic Control (FLC) is known to be robust. Moreover, it is considered to be an alternative to the chattering problem. FLC is an intelligent control complying with complex or uncertain systems. Some researchers show that FLC is a general form of variable structure control. Thus, some attempts have been made in order to integrate the SMC and FLC to a Fuzzy Sliding Mode Control (FSMC). However, the design of a fuzzy sliding mode controller for nonlinear system is a difficult problem. There have been quite a lot of researches on the combination of sliding mode control with fuzzy logic control techniques for improving the robustness and the performances of nonlinear systems with uncertainty (Qiao et al, 03).
We can distinguish two classes of control algorithms for FSMC. The first class is the fuzzy boundary layer SMC where the signum function is replaced by a fuzzy map so that the control input switch in a smooth manner to the equivalent control component. As consequence chattering is reduced. As an example of this kind of control a FSMC proposed by PALM is adopted.
The second class consists of the set of fuzzy control algorithms which approximate the input-output map of traditional sliding mode control (Alouani, 1995).

4.1 Fuzzy boundary layer SMC

For the first class of FSMC we present in the following the method proposed by PALM in (Palm, 1992).

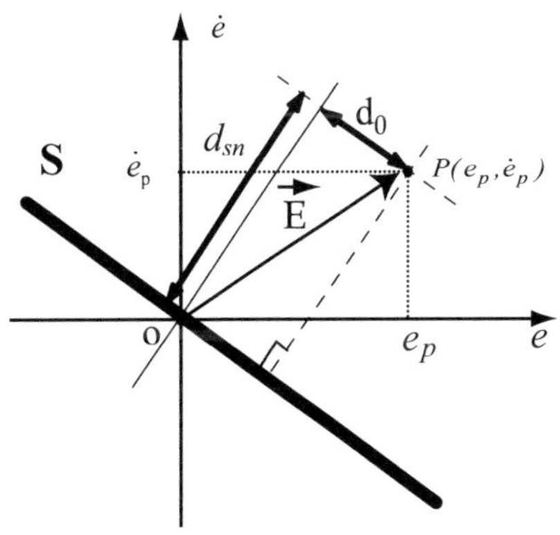

Fig. 17. Distances d_{sn} and d

Let us consider a second order system and as an example the sliding surface defined as follows (Sahbani et al, 2008):

$$S = EY^T \tag{23}$$

where $E=[e\ \dot{e}]$ and $Y=[k\ 1]$.
with k a constant gain and e the output system error.
The distance between the trajectory error and the sliding surface d_{sn} is defined as follows:

$$d_{sn}=\frac{\dot{e}p+kep}{\sqrt{1+k^2}} \tag{24}$$

d_{sn} is the normal distance between the point $P(e_p,\dot{e}_p)$ and the sliding surface. Such distance is illustrated graphically in Fig. 17 for an arbitrary point $P(e_p,\dot{e}_p)$.
Let $H(e_p,\dot{e}_p)$ be the intersection point of the switching line and its perpendicular passing through the point $P(e_p,\dot{e}_p)$.
d_o is defined as the distance between the point $H(e_p,\dot{e}_p)$ and the origin O.
The distance d_o is expressed as follows:

$$d_o=\sqrt{|E|^2-d_{sn}^2} \tag{25}$$

The presented FSMC has as inputs the two distances d_{sn} and d_o. The output signal is the control increment $\Delta U(k)$ which is used to update the control signal defined as follows:

$$U(k)=\Delta U(k)+U(k-1) \tag{26}$$

The control law is equivalent to an integral action allowing a steady state error.
The presented FSMC is a Mamdani fuzzy inference system composed by a fuzzification block, a rule base bloc and a defuzzification block.
Trapezoidal and triangular membership functions, denoted by N (Negative), Z (Zero) and P (Positive), are used for d_{sn}. The same shape of membership functions denoted by Z (Zero), PS (Positive small) and PB (Positive Big) are used for d_o. d_{sn} and d_o membership functions are presented respectively in Fig. 18 and Fig. 19 in the normalized domain $[-1\ \ 1]$ for d_{sn} and $[0\ \ 1]$ for d_o.

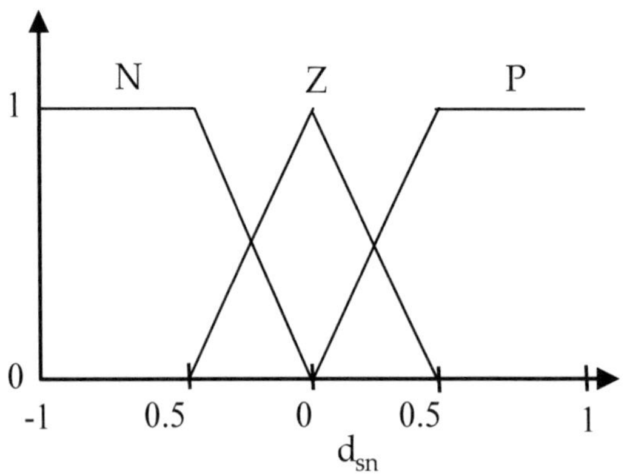

Fig. 18. d_{sn} membership functions

For the output signal of the proposed FSMC, fives triangular membership functions, denoted by NB (Negative Big), NM (Negative Middle), Z (Zero), PM (Positive Middle), PB (Positive Big) are used for the output signal Δd, Fig. 20. The rule base is given by table 4.

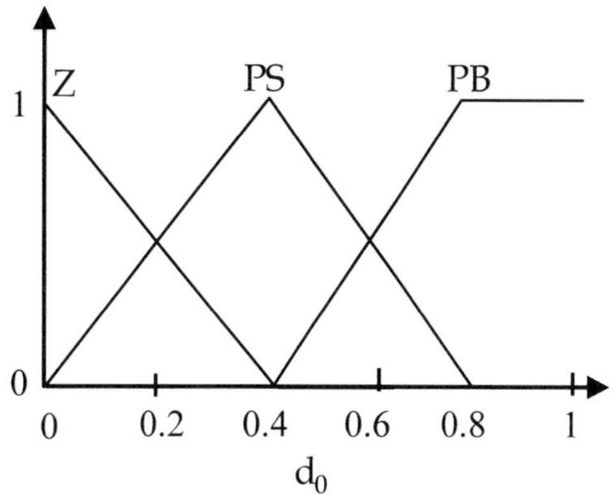

Fig. 19. d_o membership functions

d_o \ d_{sn}	N	Z	P
Z	PS	Z	NS
PS	PB	PS	NB
PB	PB	NS	NB

Table 4. Rule base of the proposed FSMC.

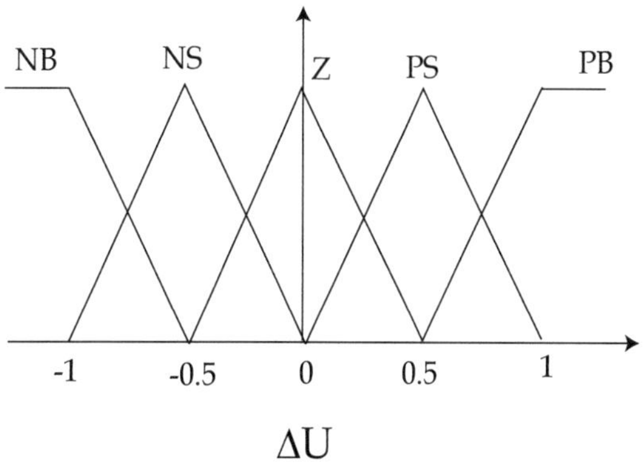

Fig. 20. ΔU membership functions

4.2 Fuzzy Lyapunov function SMC

The second class of FSMC uses the surface S and its variation $\dot{S}$ to define the changes on the control signal. The aim of this kind of FSMC is to insure the Lyapunov stability condition $S\dot{S}<0$.

Let us consider the sliding surface S. The proposed fuzzy sliding mode controller forces the derivative of the Lyapunov function to be negative definite. So, the rule base table is established to satisfy the inequality (17).

Intuitively, suppose that $S>0$ and $\dot{S}>0$, the duty cycle must increase. Also, if $S<0$ and $\dot{S}<0$ the duty cycle must decrease. Thus, the surface S and its variation $\dot{S}$ are the inputs of the proposed controller. The output signal is the control increment $\Delta U(k)$ which is used to update the control law. As for the Fuzzy boundary layer SMC the control signal is defined by equation (26). The proposed Fuzzy Sliding Mode Controller is a Sugeno fuzzy controller which is a special case of Mamdani fuzzy inference system. Only the antecedent part of the Sugeno controller has the "fuzzyness", the consequent part is a crisp function. In the Sugeno fuzzy controller, the output is obtained through weighted average of consequents.

As the proposed approach have to be implemented in practice, such choice can be motivated by the fact that Sugeno fuzzy controller is less time consuming than the Mamdani one

Trapezoidal and triangular membership functions, denoted by N (Negative), Z (Zero) and P (Positive), were used for both the surface and the surface change. They are respectively presented in Fig. 21 and Fig. 22 in the normalized domain $\left[-1 \;\; 1\right]$.

For the output signals, fives normalized singletons denoted by NB (Negative Big), NM (Negative Middle), Z (Zero), PM (Positive Middle), PB (Positive Big) are used for the output signal ΔU, Fig. 23.

The normalized control surface of the proposed FSMC, corresponding to the Rule Base given in table 5, is presented in Fig. 24. Such surface shows clearly the nonlinear characteristic of the proposed fuzzy control law.

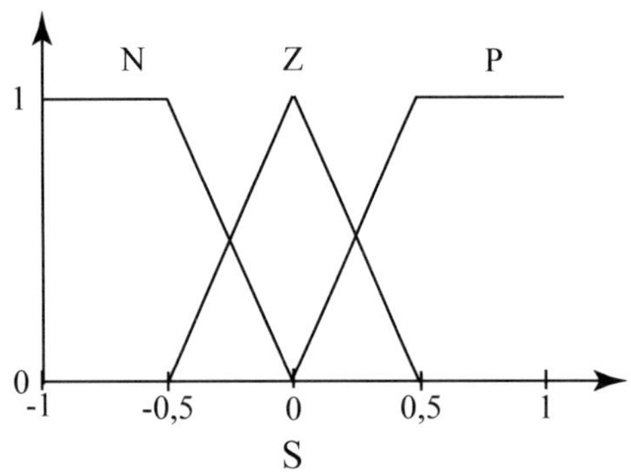

Fig. 21. Surface S membership functions

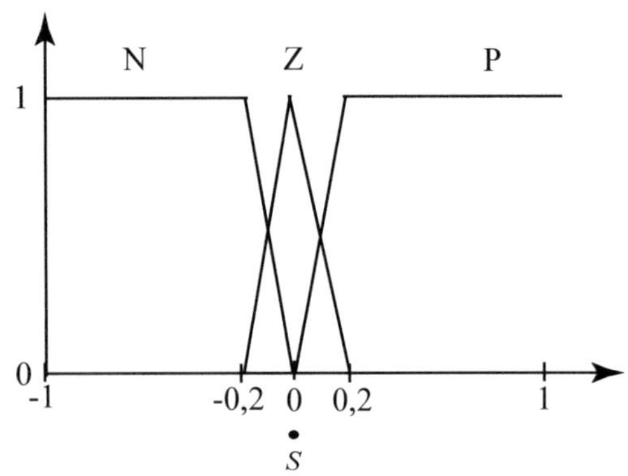

Fig. 22. Surface change $\dot{S}$ membership functions

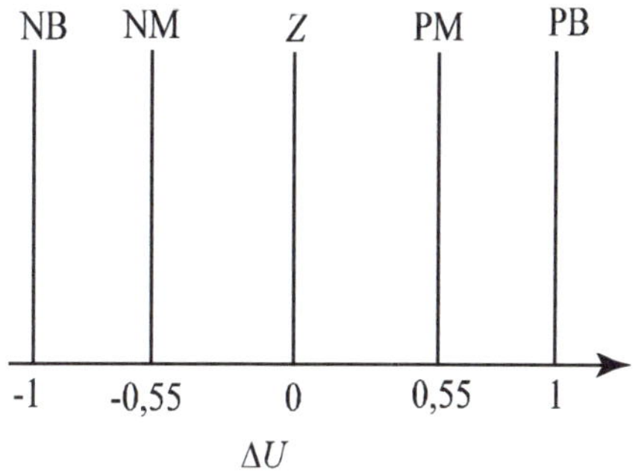

Fig. 23. Output singletons

$\dot{S}$ \ S	N	Z	P
P	Z	PM	PB
Z	NM	Z	PM
N	NB	NM	Z

Table 5. Rule base of the proposed FSMC

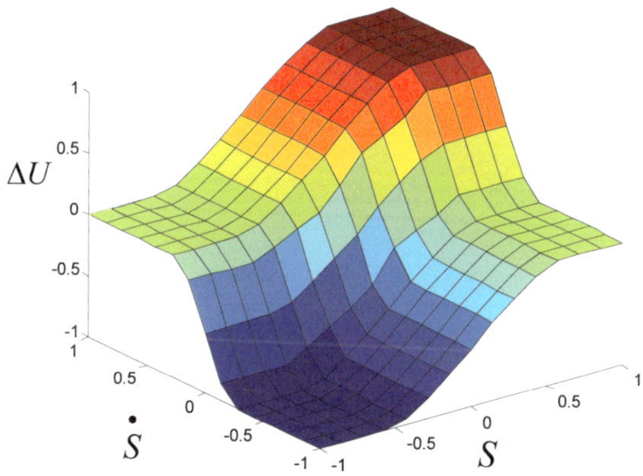

Fig. 24. FSMC control surface

In the following this second class of FSMC will be applied to the Buck and Buck Boost converters.

4.3 Application of the Fuzzy Lyapunov function SMC to Buck and Buck-Boost converters

The proposed Fuzzy Lyapunov function SMC is applied to the Buck and Buck-Boost converter to prove the efficiency of the proposed control law. The obtained results are compared to the classical SMC. As a fuzzy control, the main advantage of the FSMC is that it is not based on an analytical study.

4.3.1 Application of the Fuzzy Lyapunov function SMC to Buck

Fig. 25 presents the simulated output voltage and output current evolutions by application of the proposed FSMC for a reference voltage $V_{ref} = 5V$. The obtained result is similar to the one obtained by SMC.

As the SMC, the SMC is tested experimentally for the case of the load variation. Fig.26 presents the obtained results for the case of the variation of the load resistance from 10 Ω to 15 Ω at 0.05s. The perturbation is rejected and the output voltage attends the reference voltage. Moreover, the amplitudes of oscillations are smaller than those obtained by application of the SMC. As for the SMC, the experimental test result for the case of the input voltage variation from 30 V to 20V, given in Fig.27, shows the robustness of the applied FSMC for this kind of variation.

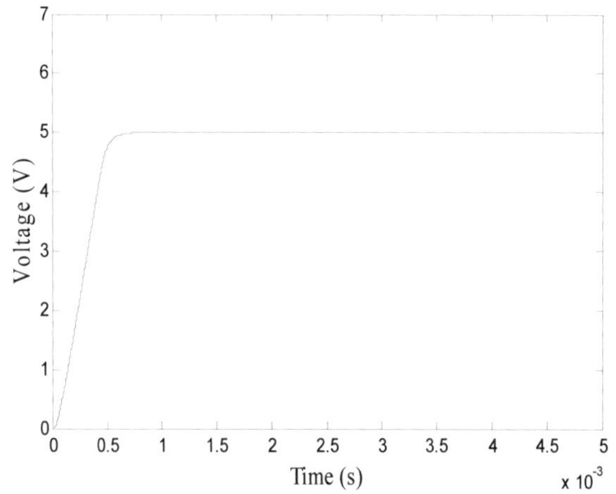

Fig. 25. Output voltage evolution by application of the SMC to the studied Buck converter ($Vref = 5V$)

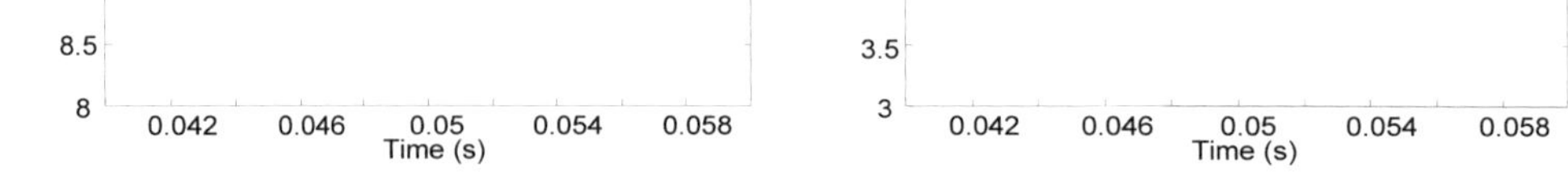

(a) Reference voltage 10 V

(b) Reference voltage 5 V

Fig. 26. Experimental test robustness of the FSMC for the variation of the load from 10 Ω to 15 Ω

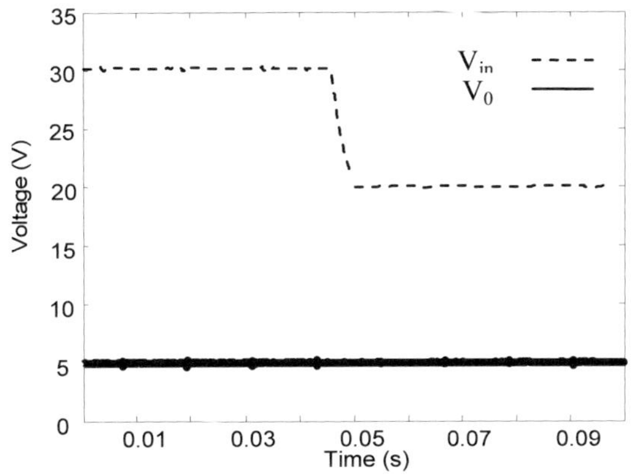

Fig. 27. Experimental output voltage evolution by application of the FSMC for the variation of the input voltage from 30V to 20V

4.3.2 Application of the Fuzzy Lyapunov function SMC to Buck-Boost converter

The proposed control is now applied to the studied Buck-Boost converter. Fig. 28 presents the simulated control signal obtained by application of the proposed FSMC. By comparison with the control signal obtained by application of the SMC and presented in Fig. 13, we notice that the control signal is smooth. So the chattering phenomenon obtained by application of the FSMC disappeared.

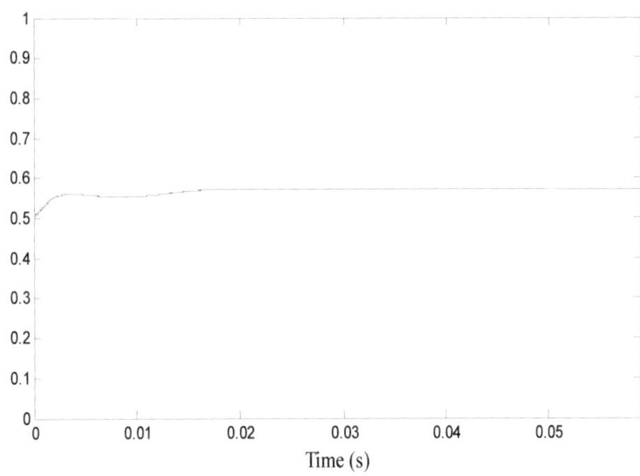

Fig. 28. Control signal by application of the FSMC

Fig. 29 presents the studied converter the output voltages evolution for $V_{ref} = -20V$. The obtained result is better than the one obtained by open-loop control. However, by comparing it with the output voltage presented by Fig. 12 we can notice a small oscillation. Fig. 30 presents the evolution of the output voltage and the current in the load for a change of the load resistance from 30Ω to 20Ω. By application of the FSMC, this perturbation was rejected and the output voltage attends the reference voltage after 30 10^{-3} s. For the case of a variation of the input voltage from 15V to 10V at 0.05s the output voltage, presented by Fig. 31, attends after the rejection of the perturbation the desired reference value −20V.

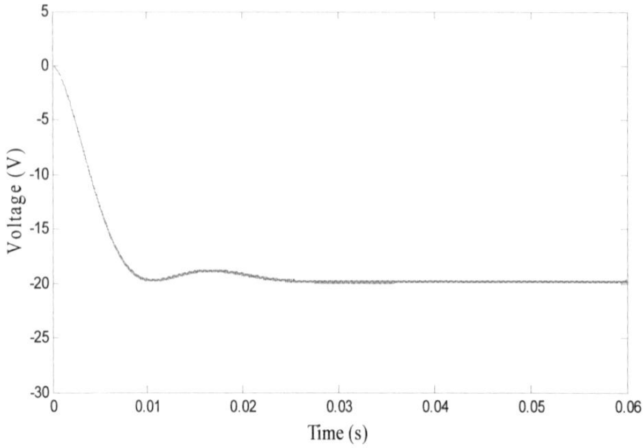

Fig. 29. Output voltage evolution obtained by application of the FSMC

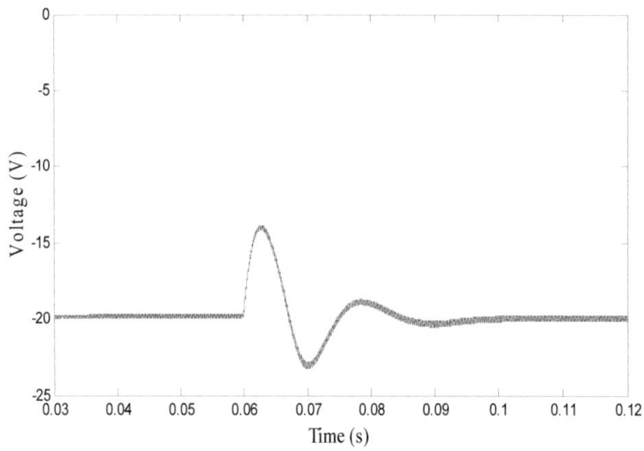

Fig. 30. Output voltage evolution by application of the SMC for the case of load variation from 30Ω to 20Ω

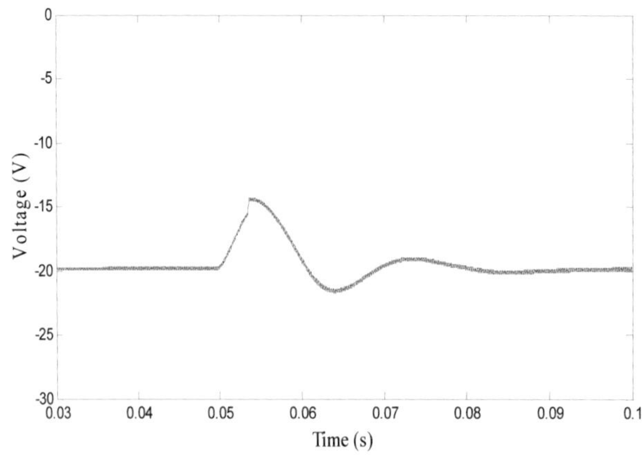

Fig. 31. Output voltage evolution by application of the SMC for the case of input voltage variation from 15 V to 10V

By comparing the robustness test results obtained by application of SMC with those obtained by FSMC we can notice that SMC allows a faster rejection of the perturbation than SMC for the case of the studied Buck-Boost converter.

5. Conclusion

In this chapter, Sliding Mode Control (SMC) and Fuzzy Sliding Mode Control (FSMC) for Buck, Boost and Buck-Boost converters are proposed, tested and compared. SMC is suitable for switched mode DC-DC converters. Moreover, such control approach leads to good results.

Two classical SMC are proposed respectively for Buck and Buck-Boost converters. The obtained simulation and practical results confirm the robustness of this control technique.

The extension of SMC into FSMC aims to improve the SMC robustness and to overcome the chattering problem. Two classes of FSMC are presented in this chapter. The first class of FSMC aimed to reduce chattering by changing the nonlinear component control by a fuzzy function. The second class of FSMC is based on a fuzzy control insuring the Lyapunov function stability. Then, a Fuzzy Lypunov based SMC is developed and applied to the Buck and Buck-Boost converters.

FSMC is not based on a rigorous analytical study as SMC. Thus, the same FSMC can be applied to Buck and Buck-Boost converters. In addition, FSMC allows the reduction of chattering for the case of the Buck-Boost converter thanks to the fuzzy control surface which allows a smooth and continuous control signal. However, the obtained results are nearly similar to those obtained by SMC.

6. References

Ahmed M.; Kuisma M.; Tolsa K. & Silventoinen P. (2003). Standard procedure for modelling the basic three converters (Buck, Boost, and Buck-boost) with PID algorithm applied, *Proceedings of the 13th International Symposium on Electrical Apparatus and Technologies (SIELA),* May, 2003, Plovdive

Alouani, A.T. (1995). Sliding mode control synthesis using fuzzy logic, *Proceedings of the American Control Conference on systems Man and cybernetics*, Vol. 2, Seattle, Washington, June 1995, pp. 1528-1532

Bandyopadhyay, B. & Janardhanan, B. (2005). *Discrete-time sliding mode control,* Springer, Berlin, 2005, ISBN-10 3-540-28140-1

Ben Saad, K.; Sahbani, A. & Benrejeb M. (2008). Design Procedure and Implementation of a Robust Fuzzy Sliding Mode Controller for Buck Converters, *International Review of Automatic Control,* Vol.1, No.3, pp. 303-310, September 2008, ISSN 1974-6059

El Fadil, H. ; Giri F., & Ouadi, H. (2008). Reducing Chattering Phenomenon in Sliding Mode Control of Buck-Boost Power Converters. *Proceedings of the IEEE International Symposium on Industrial Electronics*, Cambridge,2008

Hu Z.B.; Zhang B.; Du G.P.; Zhong L. & Deng W.H. (2005). Fast transient three-level converters with sliding-mode control, *Proceedings of the applied Power Electronics Conference and Exposition (APEC)*, Vol. 3, pp. 1436 - 1440, March, 2005

Kandel, A. & Gideon, L. (1993). *Fuzzy control systems*, CRC Press, New York, 1993

Middlebrook, R.D. & Cuk, S., (1976). A General Unified Approach to Modelling Switching Power Stages, *Proceedings of the IEEE Power Electronics Specialists Conference,* pp. 18-34, 1976

Ming-Chang, S. & Ching-Sham, L. (1993). Pneumatic servomotor drive a ball-screw with fuzzy sliding mode position control, *Proceedings of the IEEE Conference on Systems, Man and Cybernetics*, Vol.3, 1993, pp. 50-54, Le Touquet

Nguyen, V.M. & Lee, C.Q. (1996). Indirect implementations of sliding-mode control law in buck-type converters, *Proceedings of the IEEE Applied Power Electronics Conference and Exposition*, pp. 111-115, Vol. 1, March, 1996

Palm, R. (1992). Sliding mode fuzzy control, *Proceedings of the IEEE international conference on fuzzy systems*, San Diego CA pp. 519-526, 1992

Passino, M.K. (1998). *Fuzzy control*, Addison-Wesley, London

Qiao, F.; Zhu, Q.M.; Winfield, A. & Melhuish, C. (2003). Fuzzy sliding mode control for discrete nonlinear systems. *Transactions of China Automation Society*, Vol. 22, No. 2 June 2003

Sahbani, A.; Ben Saad, K. & Benrejeb, M. (2008). Chattering phenomenon suppression of buck boost dc-dc converter with Fuzzy Sliding Modes control, *International Journal of Electrical and Electronics Engineering*, 1:4, pp. 258-263, 2008

Sahbani, A., Ben Saad, K. & Benrejeb, M. (2008), Design procedure of a distance based Fuzzy Sliding Mode Control for Buck converter, *International Conference on Signals, Circuits & Systems, SCS* 08, Hammemet, 2008

Slotine, J.J.E & Li, W. (1991). *Applied nonlinear control*, Prentice Hall Englewood Cliffs, New Jersey

Tan, S.C.; Lai, Y.M.; Cheung, M.K.H. & Tse C.K. (2005). On the practical design of a sliding mode voltage controlled buck converters, *IEEE Transactions on Power Electronics*, Vol. 20, No. 2, pp. 425-437, March 2005, ISBN 0-13-040890-5

Tan, S.C.; Lai, Y.M. & Tse, C.K. (2006). An evaluation of the practicality of sliding mode controllers in DC-DC Converters and their general design issues, *Proceedings of the IEEE Power Electronics Specialists Conference*, pp.187-193, June 2006, Jeju

Tse, C.K. & Adams K.M. (1992). Quasi-linear analysis and control of DC-DC converters, *IEEE Transactions on Power Electronics*, Vol. 7 No. 2, pp. 315-323, April 1992

Rashid, M.H. (2001). *Power electronics handbook*, Academic Press, New York, ISBN 0849373360

Utkin, V.I. (1996). Sliding mode control design principles and applications to electric drives, *IEEE Transaction on Industrial Electronics*, Vol. 40, No. 1, pp. 1-14, 1996

Utkin, V.I. (1977). Variable structure systems with sliding modes, *IEEE Transactions on Automatic Control*, Vol. AC-22, No. 2, April, pp. 212-222, 1977

Wang, S.Y.; Hong, C.M.; Liu, C.C. & Yang, W.T. (1996), Design of a static reactive power compensator using fuzzy sliding mode control. *International Journal of control*, Vol. 2, No.63, 20 January 1996, pp. 393-413, ISSN 0020-7179

2

Investigation of Single-Phase Inverter and Single-Phase Series Active Power Filter with Sliding Mode Control

Mariya Petkova[1], Mihail Antchev[1] and Vanjo Gourgoulitsov[2]

[1]*Technical University - Sofia,*

[2]*College of Energetics and Electronics - Sofia*

Bulgaria

1. Introduction

The effective operation of the power converters of electrical energy is generally determined from the chosen operational algorithm of their control system. With the expansion of the application of the Digital Signal Processors in these control systems, gradually entering of novel operational principles such as space vector modulation, fuzzy logic, genetic algorithms, etc, is noticed. The method of sliding mode control is applicable in different power electronic converters - DC/DC converters, resonant converters (Sabanovic et al., 1986). The method' application is expanded in the quickly developing power electronic converters such as active power filters and compensators of the power factor (Cardenas et al., 1999; Hernandez et al, 1998; Lin et al., 2001; Mendalek et al., 2001).

In this chart, results of the study of a single-phase inverter and single-phase active power filter both with sliding mode control are discussed. A special feature is the use of control on only one output variable.

2. Single-phase inverter with sliding mode control

2.1 Schematic and operational principle

Different methods to generate sinusoidal voltage, which supplies different types of consumers, are known. Usually, a version of a square waveform voltage is generated in the inverter output and then using a filter the voltage first order harmonic is separated. Uni-polar or bi-polar pulse-width modulation, selective elimination of harmonics, several level modulation - multilevel inverters are applied to improve the harmonic spectrum of the voltage (Antchev, 2009; Mohan, 1994). Inverters with sinusoidal output voltage are applicable in the systems of reserve or uninterruptible electrical supply of critical consumers, as well as in the systems for distributed energy generation.

In this sub-chart an implementation of sliding mode control of a single-phase inverter using only one variable - the inverter output voltage passed to the load, is studied. As it is known, two single-phase inverter circuits - half-bridge and bridge, are mainly used in practice (see Fig.1). The inverters are supplied by a DC voltage source and a LC-filter is connected in their outputs. The output transformer is required at use of low DC voltage, and under certain conditions it may be missing in the circuit. The voltage passed to the load is monitored through a reverse bias using a voltage transducer. The use of a special measuring converter is necessitated by the need of correct and quick tracing of the changes in the waveform of the load voltage at the method used here. In the power electronic converters studied in the chart, measuring converter CV 3-1000 produced by LEM is applied.

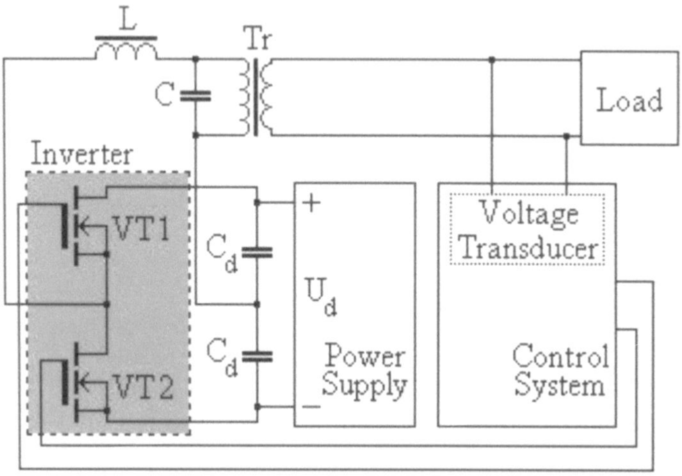

a)

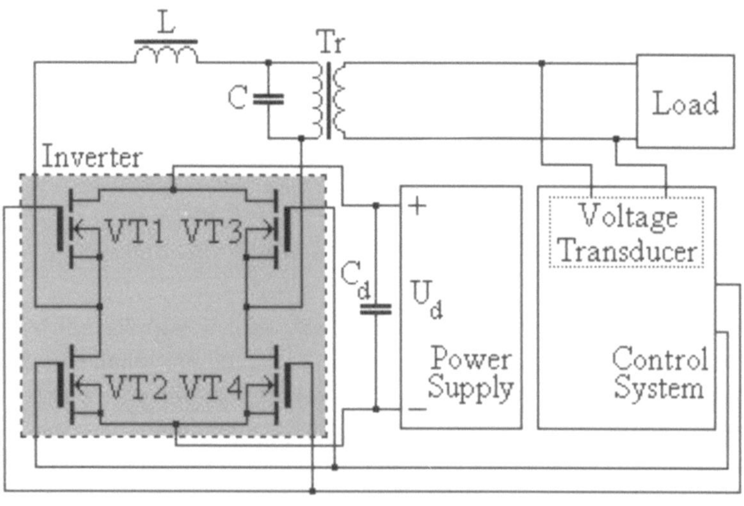

b)

Fig. 1. a) half-bridge and b) bridge circuits of an inverter

Fig.2 displays a block diagram of the control system of the proposed inverter. The control system consists of a generator of reference sinusoid $U_m \sin\Theta$ (it is not shown in the figure), a comparing device, a comparator with hysteresis and drivers. The control system compares the transitory values of the output voltage of the inverter to these values of the reference sinusoid and depending on the result (negative or positive difference) control signal is generated. The control signal is passed to the gate of the transistor VT1 or VT2 (for half-bridge circuit) or to the gates of the transistor pairs - VT1-VT4 or VT2-VT3 (for bridge circuit).

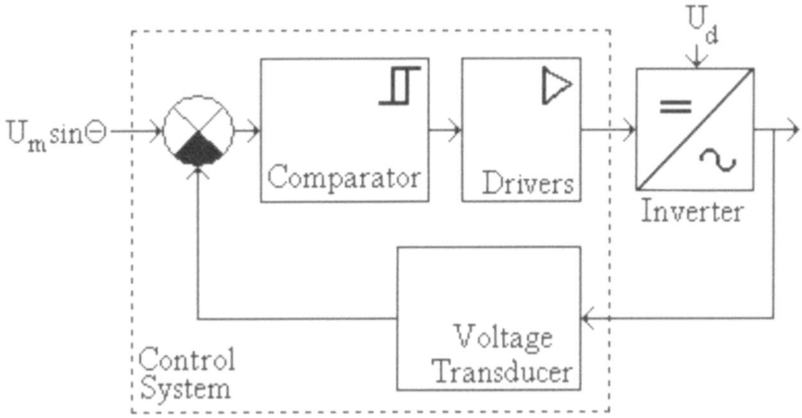

Fig. 2. Block diagram of the control system with hysteresis control

The process of sliding mode control is illustrated in Fig.3. Seven time moments are discussed - from t_0 to t_6. In the moment t_0 the transistor VT1 of the half-bridge schematic, transistors VT1 and VT4 of the bridge schematic, respectively, turns on. The voltage of the inverter output capacitor C increases fed by the DC voltage source. At the reach of the upper hysteresis borderline $U_m \sin\Theta + H$, where in H is the hysteresis size, at the moment t_1, VT1 turns off (or the pair VT1-VT4 turns off) and VT2 turns on (or the pair VT2-VT3).
The voltage of the capacitor C starts to decrease till the moment t_2, when it is equal to the lower hysteresis borderline $U_m \sin\Theta - H$. At this moment the control system switches over the transistors, etc. Therefore the moments t_0, t_2,, t_4 … and moments t_1, t_3, t_5… are identical.

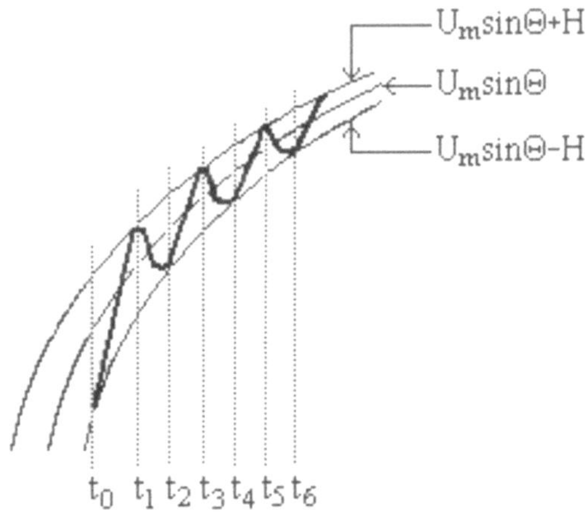

Fig. 3. Explanation of the sliding- mode control

2.2 Mathematical description

Fig.4 displays the circuit used to make the analysis of sliding mode control of the inverter. The power devices are assumed to be ideal and when they are switched over the supply voltage U_d with altering polarity is passed to the LC -filter.

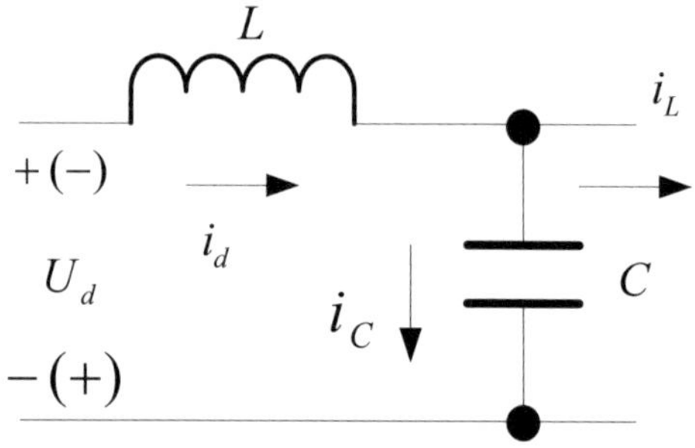

Fig. 4. Circuit used to make the analysis of sliding mode control of the inverter

The load current and the current of the output transformer, if it is connected in the schematic, is marked as i_L. From the operational principle, it is obvious that one output variable is monitored - the voltage of the capacitor u_C. Its transient value is changed through appling the voltage U_d with an altering sign. The task (the model) is:

$$u_{REF} = U_M.\sin \omega t \tag{1}$$

As a control variable, the production $u.U_d$ may be examined, where in:

$$\begin{aligned} u &= \operatorname{sgn}[(u_C - u_{REF}) - H] \\ u &= +1 \quad when(u_C - u_{REF}) < H \\ u &= -1 \quad when(u_C - u_{REF}) > H \end{aligned} \tag{2}$$

The following relationships are valid for the circuit shown in Fig.4:

$$\begin{aligned} U_d &= u_C + L\frac{di_d}{dt} \\ i_d &= i_C + i_L \\ i_C &= C\frac{du_C}{dt} \end{aligned} \tag{3}$$

Using (3) and after several transformations, it is found:

$$\dot{u}_C = \frac{du_C}{dt} = \frac{1}{L.C}\int U_d.dt - \frac{1}{LC}\int u_C dt - \frac{i_L}{C} \tag{4}$$

In conformity with the theory of sliding mode control, the following equations are written (Edwards & Spurgeon, 1998):

$$\begin{aligned} x_d &= u_{REF} \\ x &= u_C \\ \dot{x} &= \dot{u}_C \end{aligned} \tag{5}$$

The control variable u_{eq} corresponding to the so-called "equivalent control" may be found using the following equation (Utkin, 1977):

$$\dot{s} = \ddot{x} - \ddot{x}_d = 0 \tag{6}$$

Using (1) and (4) and taking in consideration (5) and (6), it is found:

$$u_{eq} = u.U_d = u_C + L.\frac{di_L}{dt} - L.C.\omega^2.U_M.\sin\omega.t \tag{7}$$

The value found may be considered as an average value when the switching is between the maximum U_{MAX} and minimum U_{MIN} values of the control variable (Utkin, 1977; Utkin, 1992). If they could change between $+\infty$ and $-\infty$, in theory, there is always the probability to achieve a mode of the sliding mode control in a certain range of a change of the output variable. In order to be such a mode, the following inequalities have to be fulfilled:

$$U_{MIN} < u_{eq} < U_{MAX} \tag{8}$$

for physically possible maximum and minimum values. In this case they are:

$$\begin{aligned} U_{MIN} &= -U_d \\ U_{MAX} &= +U_d \end{aligned} \tag{9}$$

Resolving (7) in respect to the variable, which is being monitored u_C, and substituting in (9), the boundary values of the existence of the sliding mode control could be found:

$$u_C = \pm U_d - L\frac{di_L}{dt} + LC.\omega^2.U_M.\sin\omega t \tag{10}$$

The equation (10) may be interpreted as follows: a special feature of the sliding mode control with one output variable - the capacitor voltage, is the influence of the load current changes upon the sliding mode, namely, at a sharp current change it is possible to break the sliding mode control within a certain interval. From this point of view, it is more suitable to operate with a small inductance value. As the load voltage has to alter regarding a sinusoid law, let (10) to be analyzed around the maximum values of the sinusoid waveform. It is found:

$$u_C = \pm U_d - L\left(\frac{di_L}{dt}\right)_{t=\pi/2.\omega} + L.C.\omega^2.U_M(\pm 1) \tag{11}$$

Where in (11) the positive sign is for the positive half period and the negative one - for the negative half period. After taking in consideration the practically used values of L *and* C (scores microhenrys and microfarads), the frequency of the supply source voltage $(f = 50 \quad or \quad 60Hz)$ and its maximum value U_M $(\approx 325 \quad or \quad 156V)$, it is obvious that the influence of the last term could be neglected. Thus the maximum values of the sinusoidal voltage of the load are mainly limited from the value of the supply voltage U_d and the speed of a change of the load current. So, from the point of view of the sliding mode control, it is good the value of U_d to be chosen bigger. Of course, the value is limited and has to be considered with the properties of the power switches implemented in the circuit.

2.3 Study through computer simulation

Study of the inverter operation is made using an appropriate model for a computer simulation. Software OrCad 10.5 is used to fulfill the computer simulation.

Fig.5 displays the schema of the computer simulation. The inverter operation is simulated with the following loads - active, active-inductive (with two values of the inductance - smaller and bigger ones) and with a considerably non-linear load (single-phase bridge uncontrollable rectifier with active-capacitive load). Only the load is changed during the simulations. The supply voltage of the inverter U_d is 250V, C= 120 μF and L = 10 μH.

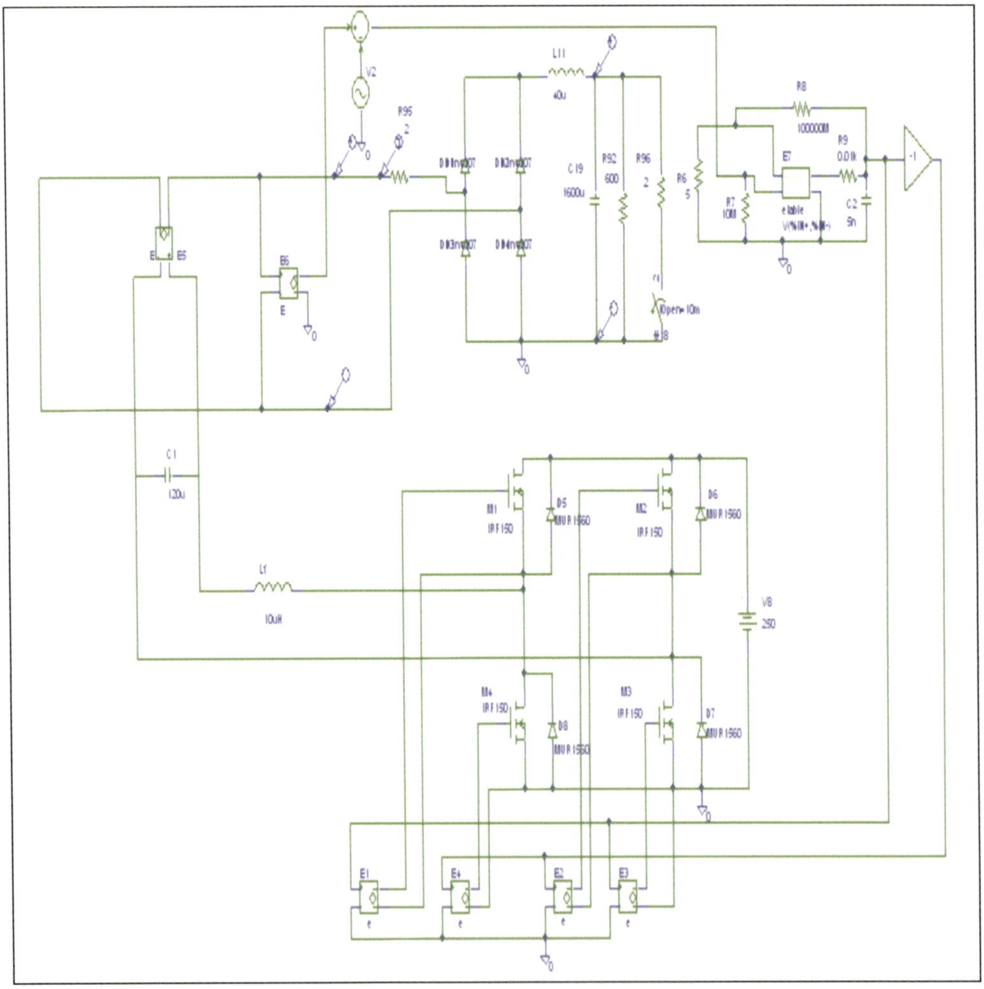

Fig. 5. Schematic for the computer simulation study of sliding mode control of the inverter

The simulation results are given in Fig.6, Fig.7, Fig.8 and Fig.9. The figures display the waveform of the voltage feeding the load, and the load current, which is displayed multiplied by 100 for the first three cases and by 40 for the last one.

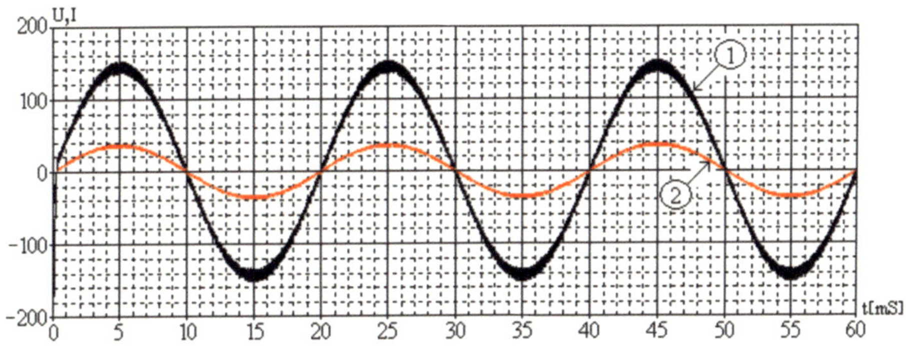

Fig. 6. Computer simulation results of the inverter operation with an active load equal to 500Ω using sliding mode control. Curve 1 – the voltage feeding the load, curve 2 – the load current

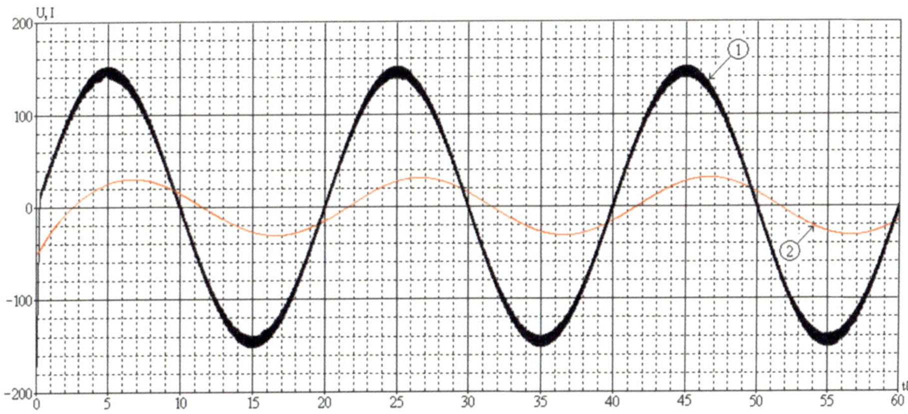

Fig. 7. Computer simulation results of the inverter operation with an active-inductive load equal to 400Ω/840μH using sliding mode control. Curve 1 – the voltage feeding the load, curve 2 – the load current

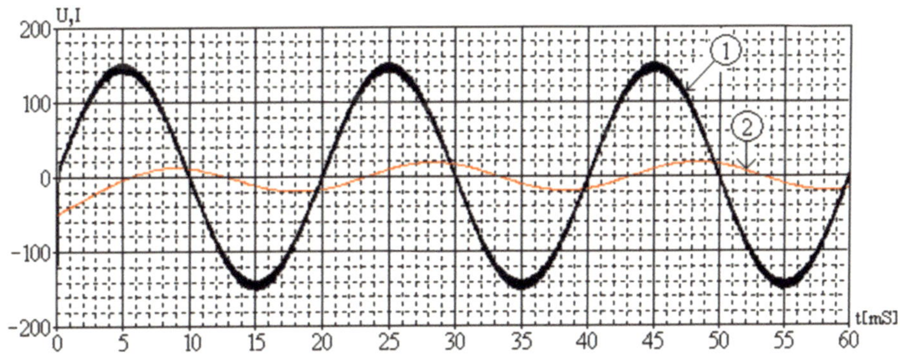

Fig. 8. Computer simulation results of the inverter operation with an active-inductive load equal to 400Ω/2H using sliding mode control. Curve 1 – the voltage feeding the load, curve 2 – the load current

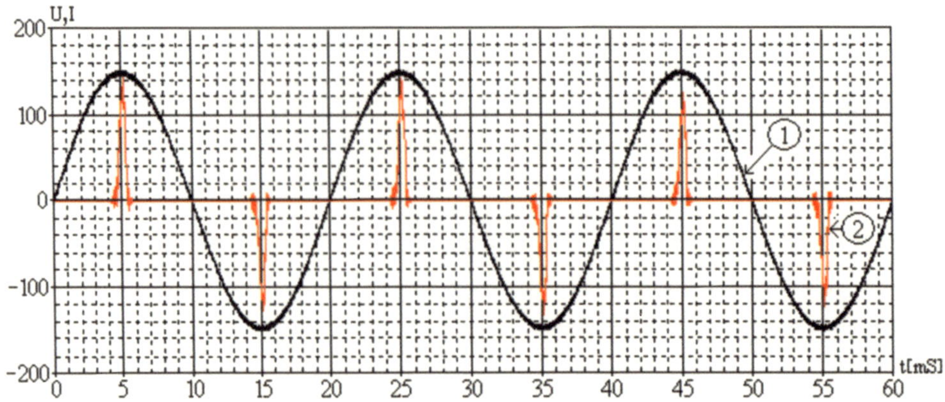

Fig. 9. Computer simulation results of the inverter operation with single-phase bridge uncontrolled rectifier load using sliding mode control. Curve 1 - the voltage feeding the load, curve 2 - the load current

The results support the probability using sliding mode control on one variable - the output voltage, in the inverter, to obtain a waveform close to sinusoidal one of the inverter output voltage feeding different types of load.

2.4 Experimental study

Based on the above-made study, single-phase inverter with output power of 600VA is materialized. The bridge schematic of the inverter is realized using IRFP450 transistors and transformless connection to the load. The value of the supply voltage of the inverter is 360V. Fig.10, Fig.11, Fig.12 and Fig.13 display the load voltage and load current waveforms for the load cases studied through the computer simulation.

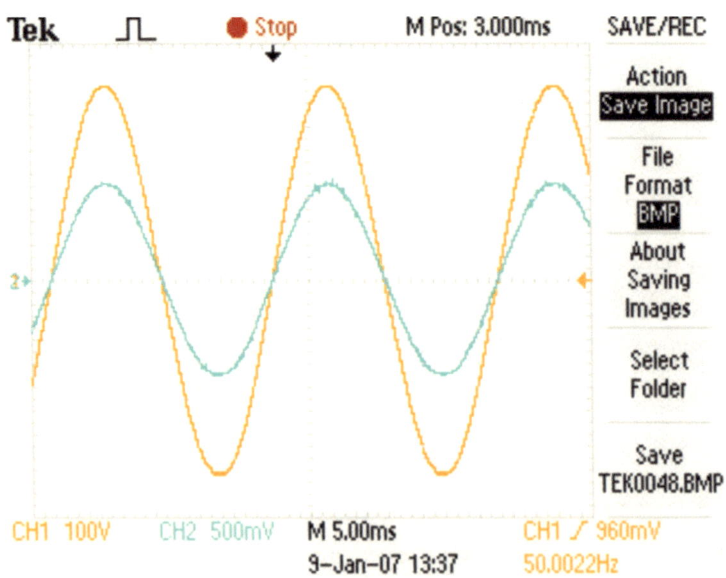

Fig. 10. The load voltage and load current in the case of active load

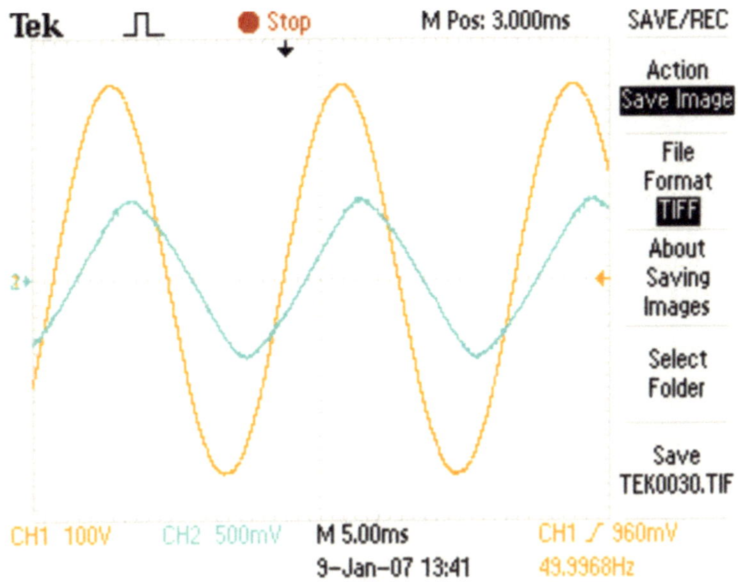

Fig. 11. The load voltage and load current in the case of active-inductive load with the smaller inductance

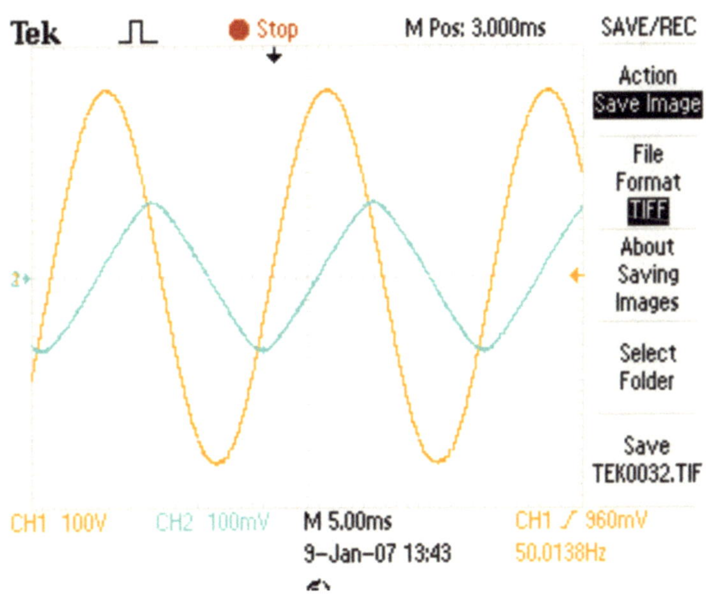

Fig. 12. The load voltage and load current in the case of active-inductive load with the bigger inductance

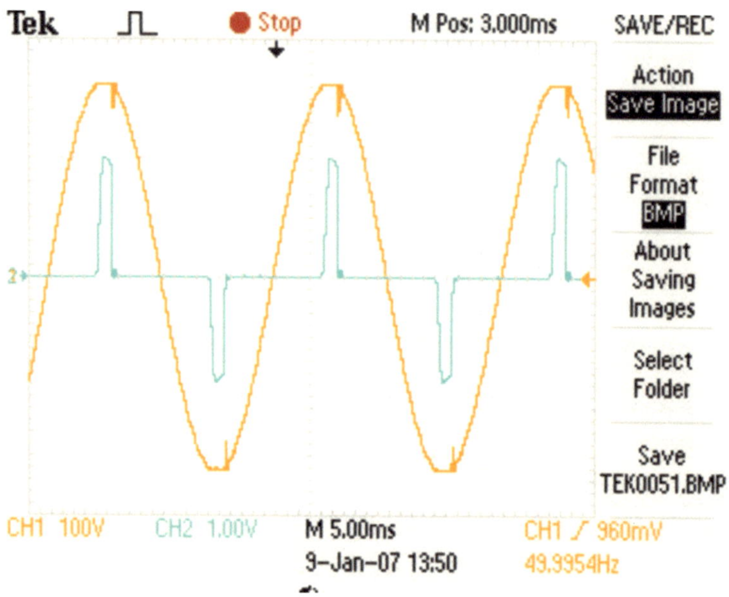

Fig. 13. The load voltage and load current in the case of single-phase bridge rectifier

All results show non-sinusoidal part of the output voltage less then 1.5% as well as high accuracy of the voltage value - (230V ± 2%).

3. Single-phase series active power filter with sliding mode control

3.1 Schematic and operational principle

Active power filters are effective means to improve the energy efficiency with respect to an AC energy source as well as to improve energy quality (Akagi, 2006). Series active power filters are used to eliminate disturbances in the waveform of a network source voltage in such a way that they compliment the voltage waveform to sinusoidal voltage regarding the load. Usually pulse-width modulation is used to control the filters, but also researches of sliding mode control of the filters on several variables are known (Cardenas et al, 1999; Hernandez et al, 1998). In this sub-chart sliding mode control of a single-phase series active power filter on one variable - the supply voltage of the load is studied (Antchev et al, 2007; Antchev et al, 2008).

Fig.14 shows the power schematic of the active power filter with the block diagram of its control system.

Synchronized to the source network and filtering its voltage, the first order harmonic of the source voltage is extracted. This harmonic is used as a reference signal U_{ref}. This signal is compared with a certain hysteresis to the transient value of the load voltage U_{real}. Depending on the sign of the comparison, the appropriate pair of diagonally connected transistors (VT1-VT4 or VT2-VT3) of the inverter is switched on.

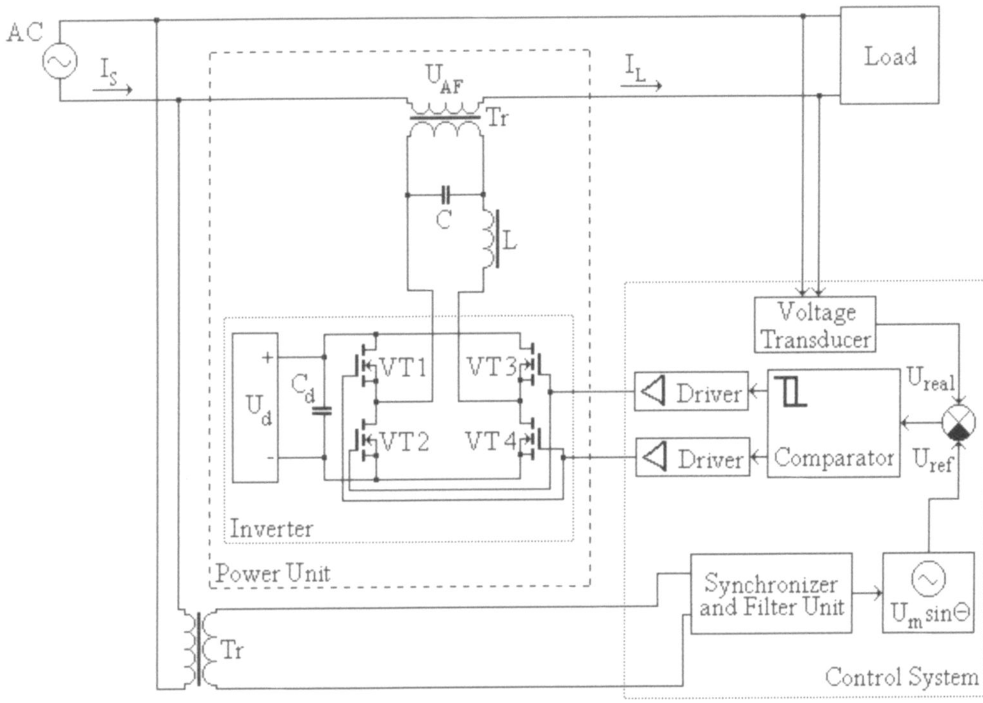

Fig. 14. Series active power filter with sliding mode control with hysteresis

3.2 Mathematical description

Fig.15 displays the circuit used to make the analysis of sliding mode control of the converter. The power switches are assumed to be ideal and in their switching the source voltage U_d with an altering polarity is passed to the LC-filter.

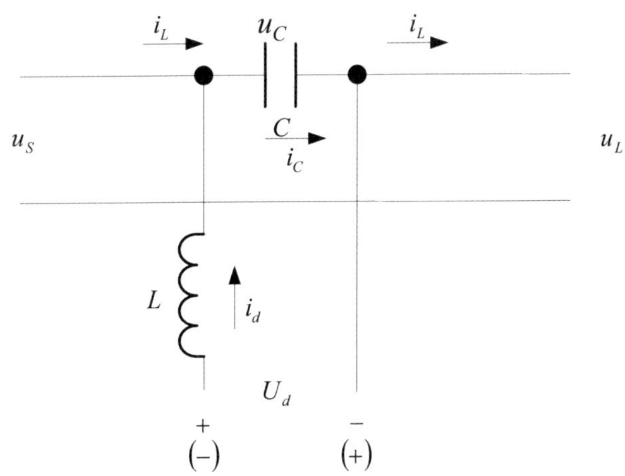

Fig. 15. Circuit used to make the analysis of sliding mode control of the series active power filter

The analysis is similar to those made for the single-phase inverter.
The load current is marked as i_L . From the operational principle, it is clear that one output variable - the load voltage u_L is monitored. Its transient value is changed through appling the voltage U_d with an altering sign. The task (the model) is:

$$u_{REF} = U_M.\sin \omega t \tag{12}$$

As a control variable, the production $u.U_d$ may be examined, where in:

$$\begin{aligned} &u = \operatorname{sgn}\left[\left(u_L - u_{REF}\right) - H\right] \\ &u = +1 \quad when\left(u_L - u_{REF}\right) < H \\ &u = -1 \quad when\left(u_L - u_{REF}\right) > H \end{aligned} \tag{13}$$

The following relationships are valid for the schematic shown in Fig.15:

$$\begin{aligned} &U_d = u_C + L\frac{di_d}{dt} \\ &i_C = i_L + i_d \\ &i_C = C\frac{du_C}{dt} \\ &u_S + u_C = u_L \end{aligned} \tag{14}$$

Using (14), it is found:

$$\dot{u}_L = \frac{du_L}{dt} = \dot{u}_S + \frac{1}{L.C}\int U_d.dt - \frac{1}{LC}\int u_C dt + \frac{i_L}{C} \tag{15}$$

In conformity with the theory of sliding mode control, the following equation is written (Edwards & Spurgeon, 1998):

$$\begin{aligned} &x_d = u_{REF} \\ &x = u_L \\ &\dot{x} = \dot{u}_L \end{aligned} \tag{16}$$

The control variable u_{eq} corresponding to the so-called "equivalent control" may be found using the following equation (Utkin, 1977, Utkin, 1992):

$$\dot{s} = \ddot{x} - \ddot{x}_d = 0 \tag{17}$$

Using (12) and (15) and taking in consideration (16) and (17), it is found:

$$u_{eq} = u.U_d = u_L - u_S - LC\ddot{u}_S - L\frac{di_L}{dt} - L.C.\omega^2.U_M.\sin \omega.t \tag{18}$$

The value found may be considered as an average value when the switching is between the maximum U_{MAX} and minimum U_{MIN} values of the control variable (Utkin, 1977; Utkin 1978). If they could change between $+\infty$ and $-\infty$, in theory, there is always the probability to achieve a mode of the sliding mode control in a certain range of a change of the output variable. In order to be such a mode, the following inequalities have to be fulfilled:

$$U_{MIN} < u_{eq} < U_{MAX} \tag{19}$$

for physically possible maximum and minimum values. In this case they are:

$$\begin{aligned} U_{MIN} &= -U_d \\ U_{MAX} &= +U_d \end{aligned} \tag{20}$$

Resolving (18) with respect to the variable, which is being monitored u_L, and substituting in (20), the boundary values of the existence of the sliding mode control could be found:

$$u_L = \pm U_d + u_S + L.\frac{di_L}{dt} + L.C.\ddot{u}_S + L.C.\omega^2.U_M.\sin\omega t \tag{21}$$

The equation (21) found could be interpreted in the following way: a special feature of the sliding mode control with one output variable - the load voltage, is the influence of the load current changes upon the sliding mode, namely, at a sharp current change it is possible to break the sliding mode control within a certain interval leading to distortion in the transient value of the voltage feeding the load. It is worthy to be mentioned that, for example, rectifiers with active-inductive load consume current with sharp changes in its transient value from the source. From this point of view, to reduce this influence it is more suitable to operate with a small inductance value. As the load voltage has to change regarding a sinusoid law, let (21) to be analyzed around the maximum values of the sinusoid waveform. It is found:

$$u_L = \pm U_d + \left(u_S\right)_{t=\pi/2.\omega} + L\left(\frac{di_L}{dt}\right)_{t=\pi/2.\omega} + L.C.\left(\ddot{u}_S\right)_{t=\pi/2.\omega} + L.C.\omega^2.U_M.(\pm 1) \tag{22}$$

Where in (22) the positive sign is for the positive half period and the negative one - for the negative half period. After taking in consideration the practically used values of L and C (scores microhenrys and microfarads), the frequency of the supply source voltage $(f = 50 \;\; or \;\; 60Hz)$ and its maximum value U_M $(\approx 311 \;\; or \;\; 156V)$, it is obvious that the influence of the last two terms could be neglected. Thus the maximum values of the sinusoidal voltage of the load is mainly limited from the value of the supply voltage U_d, the transient value of the load voltage and the speed of a change of the load current. So, from the point of view of the sliding mode control, it is good the value of U_d to be chosen bigger. Concerning the conclusion of the influence of the load current change made based on the equations (21) and (22), the following may be commented: let us assume the worst case - short circuit of the load terminals (for example during break down regime or commutation processes in the load). Then the speed of the current change will be of maximum value and it will be limited only by the impedance of the AC supply source in a case of transformless active power filter. When an output transformer is present, its inductance will be summed to this of the source and it will additionally decrease the speed. Taking in consideration the maximum value of the voltage of single-phase network at a low voltage, as well as the range of the inductance possible values, the speeds tentatively of $1\frac{A}{\mu S}$ order may be expected. The value of the filter inductance is within the range of 1 to 3mH. Therefore, the influence of the third term in equations (21) and (22) will be approximately 10 times lower then the influence of the supply voltage U_d.

3.3 Study through computer simulation

In this part, software PSIM is used to study single-phase active power filter. The operation of the single-phase active power filter is studied at a trapezoidal waveform of the voltage of the supply source. The computer simulation schematic is shown in Fig.16. The results from the simulation are shown in Fig.17. Total harmonic distortion of the source voltage is assumed to be 20%. The altitude of the trapezium is given equal to 300V. The values of the elements in the output of the single-phase uncontrolled rectifier are 1200 μF и 50 Ω .

At so chosen waveform of the AC source, the results put show good reaction of APF and also show its effective operation. So chosen trapezium form of the voltage is very close to the real cases of distortion of the source voltage. As it is seen from the results included, in this case of the source voltage waveform the system voltage supplying the load is obtained to be very closed to the ideal sinusoidal waveform without distortions around the maximum value of the sine wave and without presence of over voltages.

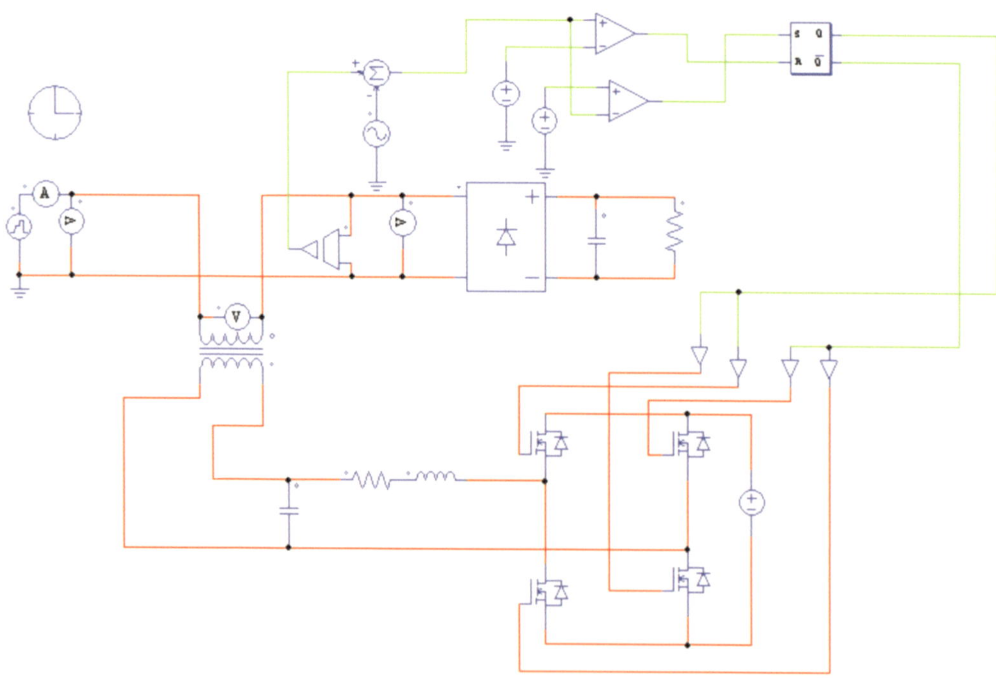

Fig. 16. Simulation schematic of operation of the single-phase APF with single-phase uncontrolled rectifier with active-capacitive load

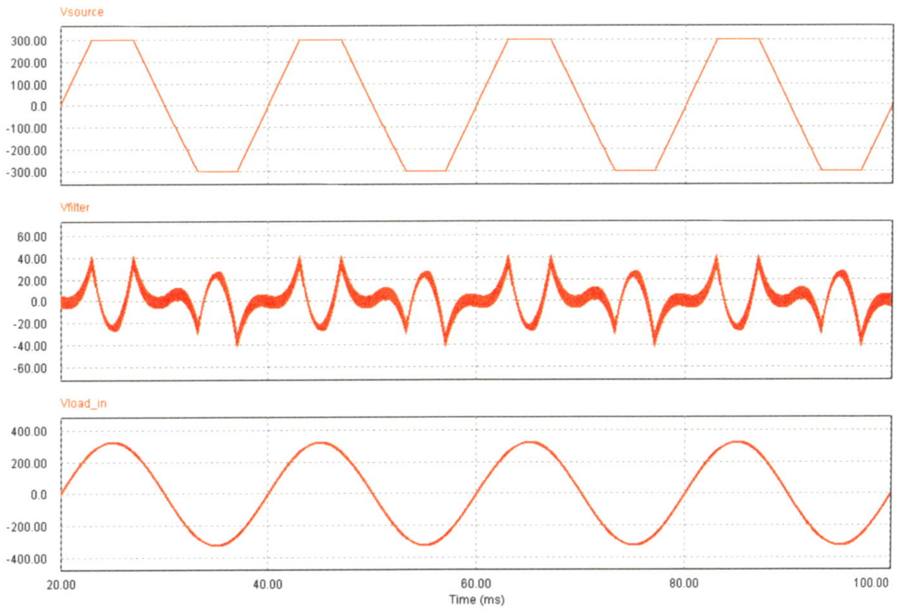

Fig. 17. Results of the simulation of the schematic shown in Fig.16. The upper waveform is the source voltage, the middle one - APF voltage, and the lower waveform - the voltage passed to the load

3.4 Experimental study

A precise stabilizer-filter for single-phase AC voltage for loads with power upto 3kVA is materialized. The device is realized using the block diagram shown in Fig.18. The source voltage U_{dc} for the active power filter is provided from a bi-directional converter connected to the network. Fig.19 shows the general appearance of the precise stabilizer-filter. Its basic blocks are marked.

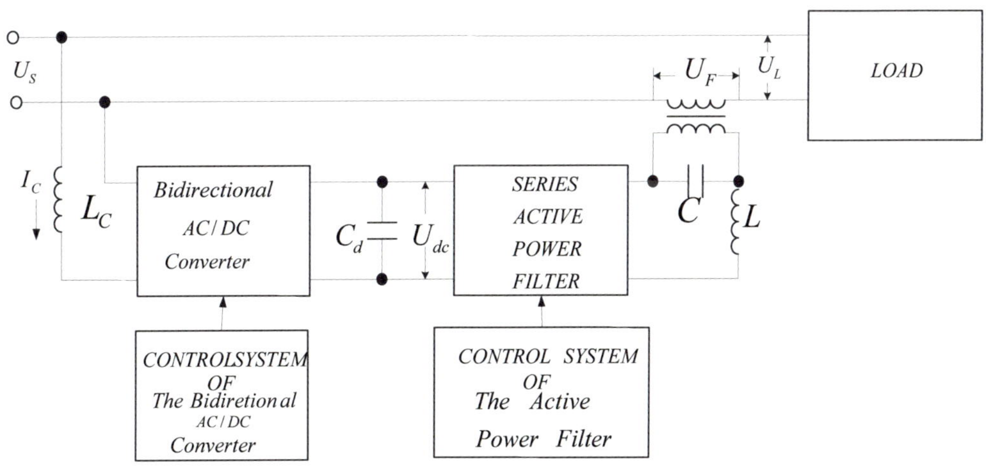

Fig. 18. Block diagram of a precise stabilizer-filter of AC voltage

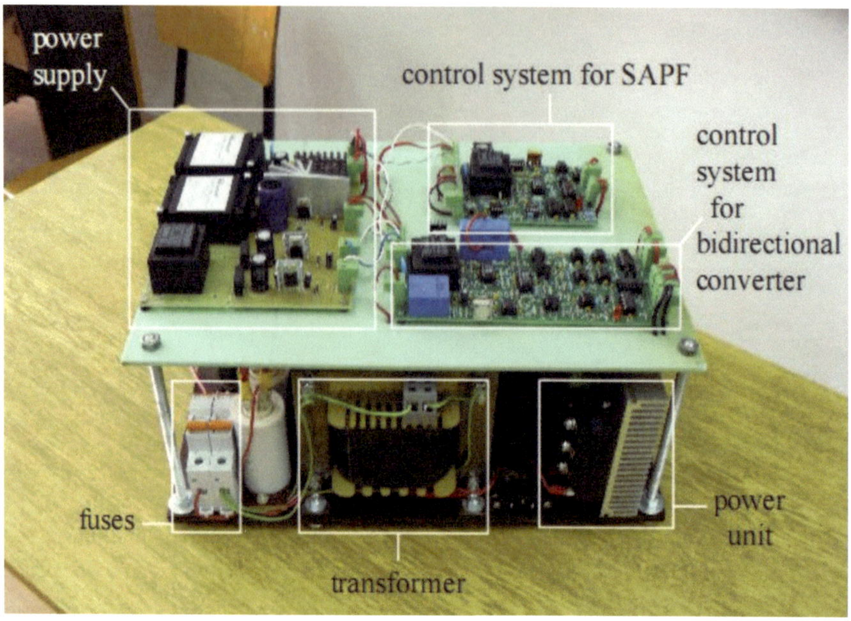

Fig. 19. Single-phase precise stabilizer-filter of AC voltage

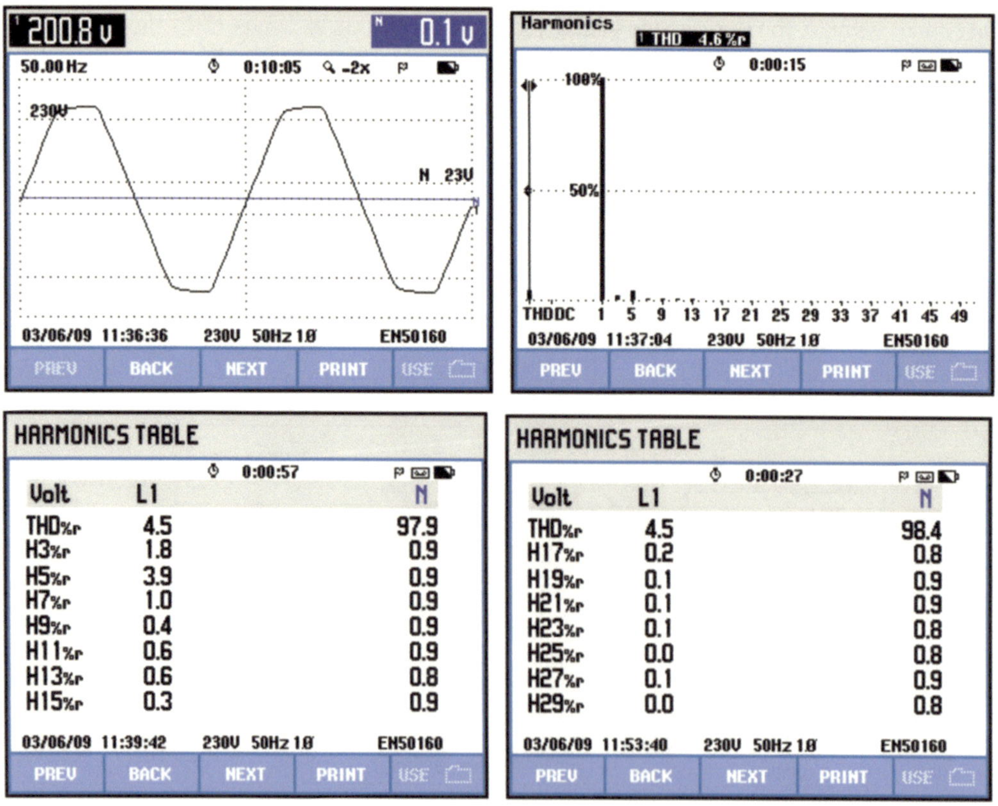

Fig. 20. Parameters of the load voltage when stabilizer - APF is switched off

Fig.20 displays the parameters of the load voltage - its value, harmonic spectrum, total harmonic distortion, when the stabilizer - APF is switched off. Fig.21 displays the same results when the stabilizer - APF is switched on. Fig.20 shows decreased effective value of the voltage with 13%, increased fifth harmonic and total harmonic distortion 4.5%. Fig.21 shows the stabilization of the effective value of the load voltage to (230V - 1.2%), decrease of the values of all harmonics, as well as a decrease in the total harmonic distortion to 1.6%.
Fig.22 displays results when the stabilizer - APF is switched off, the effective value of the source voltage is increased with approximately 10% and the total harmonic distortion of 3.2%. Fig.23 displays results when the stabilizer - APF is switched on. It is seen a stabilization of the voltage feeding the load to (230V +1.8%), decrease of the values of all harmonics, as well as a decrease in the total harmonic distortion to 1.8%.

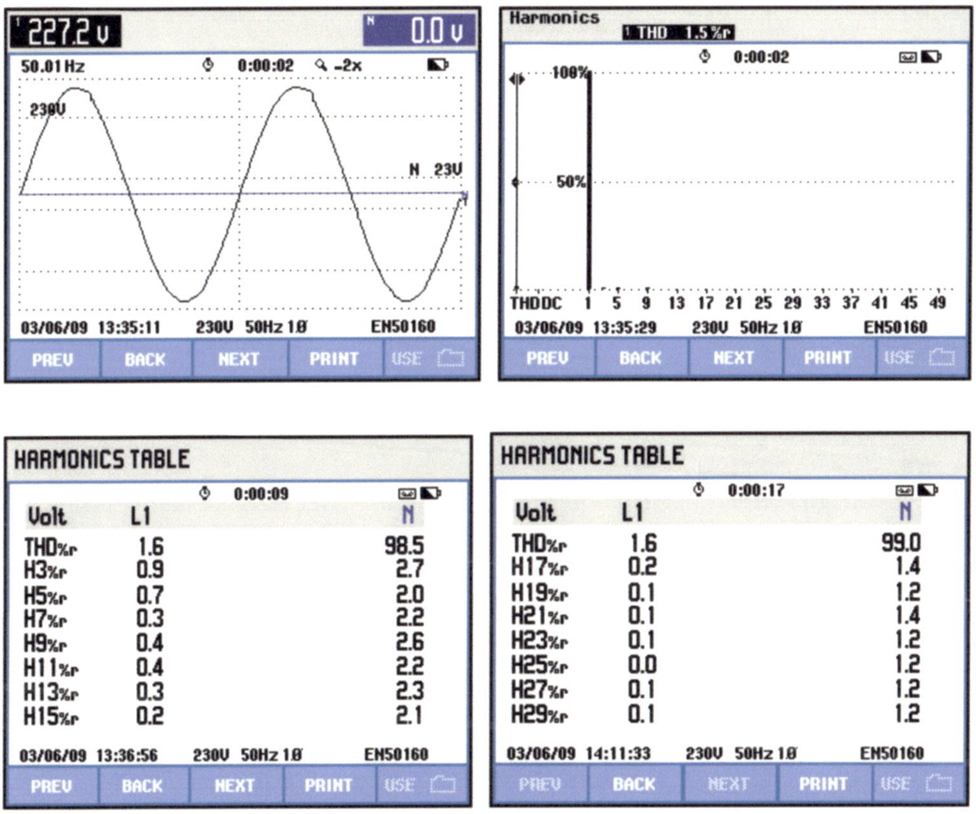

Fig. 21. Parameters of the load voltage when stabilizer - APF is switched on

Fig.24 shows transient processes at a sharp change of the source voltage. The reason that the sinusoidal waveform of the voltage is not seen is that the scale of the X-axis is 1s/div. The aim of this presentation is to be more clear that the value of the voltage feeding the load do not change significantly at a sharp change of the source voltage (both when its value decreases or increases) when the precise stabilizer-filter is switched on.

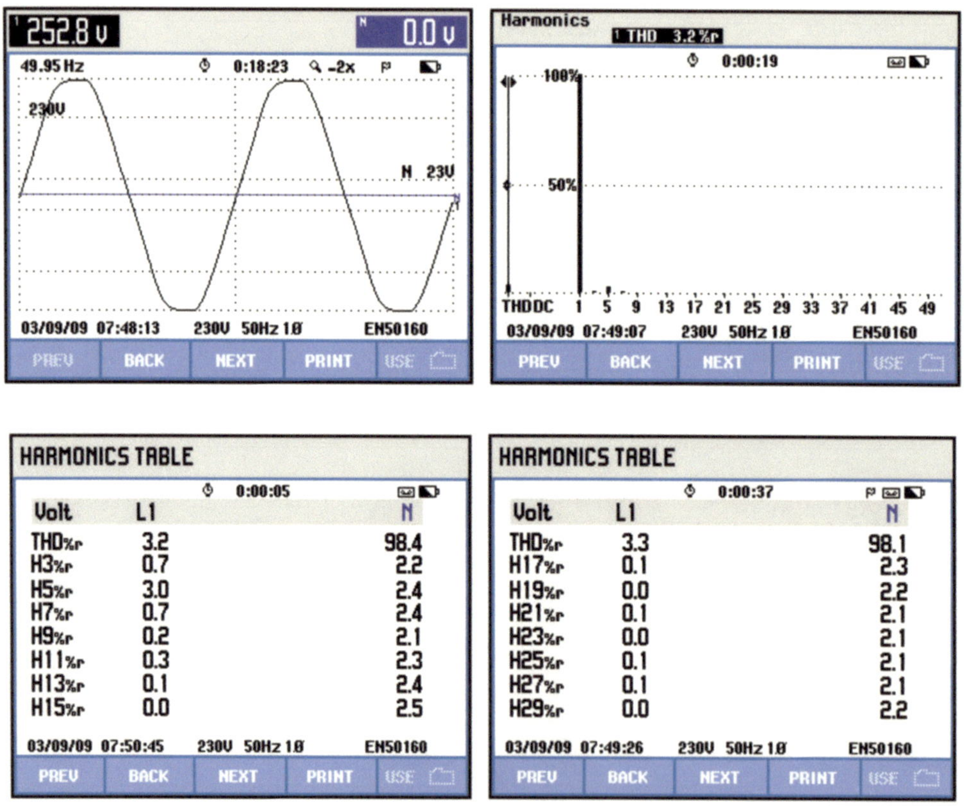

Fig. 22. Parameters of the load voltage when stabilizer - APF is switched off

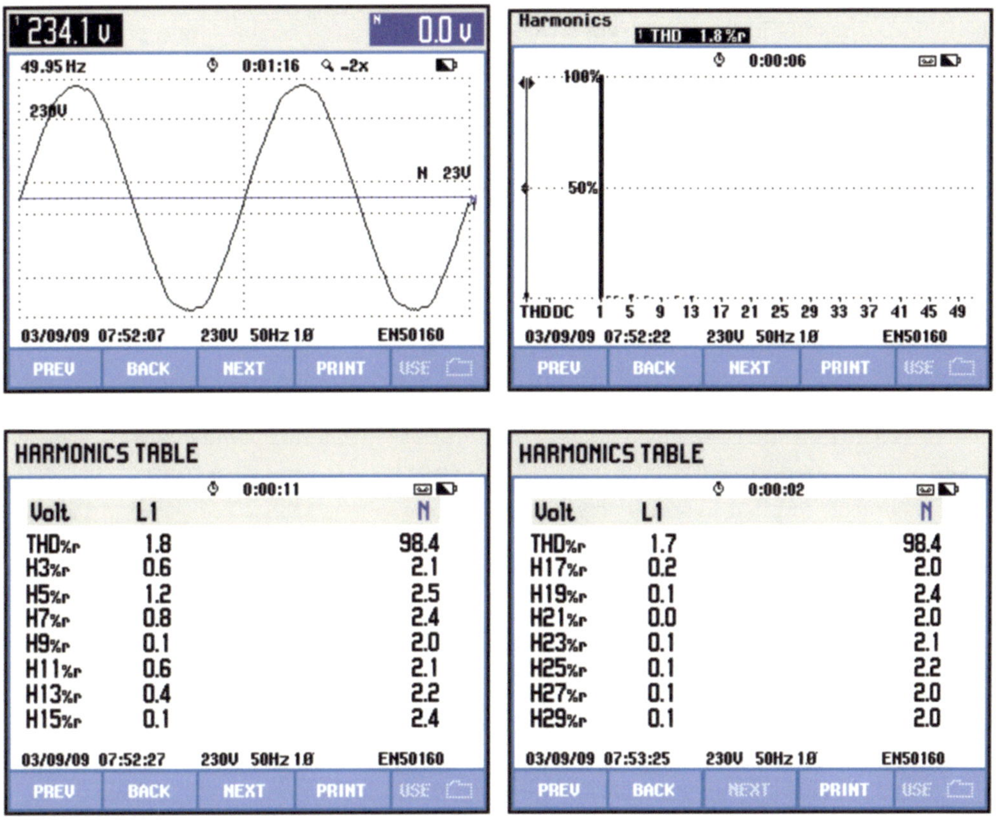

Fig. 23. Parameters of the load voltage when stabilizer – APF is switched on

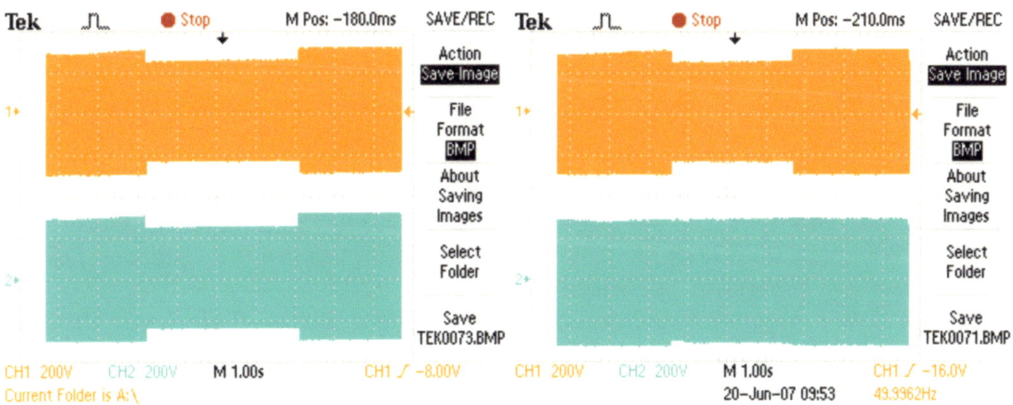

a b

Fig. 24. Experimental results at a sharp change of the value of the source voltage. a) without APF, b) with APF. The upper oscillograms present the source voltage, the lower ones – the load voltage.

4. Conclusion

The included results in the chart prove the effective operation of the single-phase inverter and single-phase active power filter studied with sliding mode control on one output variable - the voltage feeding the load.
The results found concerning the sliding mode control of inverters and series active power filters based on only one variable may be expanded and put into practice for three-phase inverters and three-phase series active power filters.

5. References

Akagi H. (2006). Modern active filters and traditional passive filters. *Bulletin of the Polish Academy of Sciences, Technical Sciences*, vol. 54, No 3, 2006.

Antchev, M.H., Petkova, M.P. & Gurgulitsov, V.T. (2007).Sliding mode control of a series active power filter, *IEEE conf. EUROCON 2007*, Proc., pp. 1344-1349, Warshaw, Poland.

Antchev, M.H., Gurgulitsov, V.T. & Petkova, ,M.P.(2008). Study of PWM and sliding mode controls implied to series active power filters, *conf. ELMAR 2008*, Zadar, Chroatia, 2008, pp.419 - 422.

Antchev, M. (2009). *Technologies for Electrical Power Conversion, Efficiency and Distribution, Methods and Processes*, IGI Global, USA.

Cardenas V.M., Nunez, C. & Vazquez, N. (1999). Analysis and Evaluation of Control Techniques for Active Power Filters: Sliding Mode Control and Proportional-Integral Control, *Proceedings of APEC'99*, vol.1, pp.649-654.

Edwards Ch. & Spurgeon, S. (1998). *Sliding mode control theory and applications*, Taylor and Francis

Hernandez, C., Varquez, N. & Cardenas, V. (1998). Sliding Mode Control for A Single Phase Active Power Filter, *Power Electronics Congress, 1998 CIEP 98*. VI IEEE International, pp.171-176.

Lin, B.-R., Tsay, S.-C. & Liao, M.-S. (2001). Integrated power factor compensator based on sliding mode controller, *Electric Power Applications, IEE Proceedings*, Vol. 148, No 3, May 2001.

Mendalek, N., Fnaiech, F. & Dessaint, L.A. (2001). Sliding Mode Control of 3-Phase 3-Wire Shunt Active Filter in the dq Frame, *Proceedings of Canadian Conf. Electrical and Computer Engineering*, vol.2. pp.765-769, 2001, Canada

Mohan R. (1994). *Power Electronics: Converters, Applications and Design,* John Wiley and Sons.

Sabanovic, A., Sabanovic, N. & Music, O. (1986) Sliding Mode Control Of DC-AC Converters, *IEEE 1986*, Energoinvest - Institute for Control and Computer Science Sarajevo, Yugoslavia, pp.560-566.

Utkin V.I. (1977). Variable structure system with sliding modes. *IEEE Trans. On A.C.*, Vol.AC-22, April 1977, pp. 212-222.

Utkin V.I.(1978). *Sliding Modes and Their Application in Variable Structure Systems*, Moskow, Mir.

Utkin V.I.(1992). *Sliding Modes in Control and Optimization*, Springer Ferlag, Berlin. ISBN 978-0387535166

3

Sliding Mode Control for Industrial Controllers

Khalifa Al-Hosani[1], Vadim Utkin[2] and Andrey Malinin[3]
[1,2]*The Ohio State University,*
[3]*IKOR*
USA

1. Introduction

This chapter presents sliding mode approach for controlling DC-DC power converters implementing proportional integral derivative (PID) controllers commonly used in industry. The core design idea implies enforcing sliding mode such that the output converter voltage contains proportional, integral and derivative components with the pre-selected coefficients. Traditionally, the method of pulse width modulation (PWM) is used to obtain a desired continuous output function with a discrete control command. In PWM, an external high frequency signal is used to modulate a low frequency desired function to be tracked. However, it seems unjustified to ignore the binary nature of the switching device (with ON/OFF as the only possible operation mode) in these power converters. Instead, sliding mode control utilizes the discrete nature of the power converters to full extent by using state feedback to set up directly the desired closed loop response in time domain. The most notable attribute in using sliding mode control is the low sensitivity to disturbances and parameter variations (Utkin, Guldner, & Shi, 2009), since uncertainty conditions are common for such control systems. An irritating problem when using sliding mode control however is the presence of finite amplitude and frequency oscillations called chattering (Utkin, Guldner, & Shi, 2009). In this chapter, the chattering suppression idea is based on utilizing harmonic cancellation in the so-called multiphase power converter structure. Moreover, the method is demonstrated in details for the design of two main types of DC-DC converter, namely the step-down buck and step-up boost converters.

Control of DC-DC step-down buck converters is a conventional problem discussed in many power electronics and control textbooks (Mohan, Undeland, & Robbins, 2003; Bose, 2006). However, the difficulty of the control problem presented in this chapter stems from the fact that the parameters of the buck converter such as the inductance and capacitance are unknown and the error output voltage is the only information available to the designer. The problem is approached by first designing a switching function to implement sliding mode with a desired output function. Chattering is then reduced through the use of multiphase power converter structure discussed later in the chapter. The proposed methodology is intended for different types of buck converters with apriory unknown parameters. Therefore the method of observer design is developed for estimation of state vector components and parameters simultaneously. The design is then confirmed by means of computer simulations.

The second type of DC-DC converter dealt with in this chapter is the step-up boost converter. It is generally desired that a sliding mode control be designed such that the output voltage tracks a reference input. The straightforward way of choosing the sliding surface is to use the output voltage error in what is called direct sliding mode control. This methodology leads to ideal tracking should sliding mode be enforced. However, as it will be shown, direct sliding mode control results in unstable zero dynamics confirming a non-minimum phase tracking nature of the formulated control problem. Thus, an indirect sliding mode control is proposed such that the inductor current tracks a reference current that is calculated to yield the desired value of the output voltage. Similar to the case of buck converters, chattering is reduced using multiphase structure in the sliding mode controlled boost converter. The results are also confirmed by means of computer simulations.

2. Modeling of single phase DC-DC buck converter

The buck converter is classified as a "chopper" circuit where the output voltage v_C is a scaled version of the source voltage E by a scalar smaller than unity. The ideal switch representation of a single-phase buck converter with resistive load is shown in Fig. 1. Simple applications of Kirchhoff's current and voltage laws for each resulting circuit topology from the two possible ideal switch's positions allow us to get the system of differential equations governing the dynamics of the buck converter as it is done in many control and power electronics textbooks. We first define a switch's position binary function u such that $u = 1$ when the ideal switch is positioned such that the end of the inductor is connected to the positive terminal of the input voltage source and $u = 0$ otherwise. With this, we get the following *unified* dynamical system:

$$\frac{di_L}{dt} = \frac{1}{L}\left(uE - v_C\right) \tag{1}$$

$$\frac{dv_C}{dt} = \frac{1}{C}\left(i_L - \frac{v_C}{R}\right) \tag{2}$$

Most often, the control objective is to regulate the output voltage v_C of the buck converter towards a desired average output voltage equilibrium value v_{sp}. In many applications, power converters are used as actuators for control system. Also, the dynamics of the power converters is much faster than that of the system to be controlled. Thus, it might be reasonable to assume that the desired output voltage v_{sp} is constant. By applying a discontinuous feedback control law $u \in \{0,1\}$, we command the position of the ideal switch in reference to an average value u_{avg}. Depending on the control algorithm, u_{avg} might take a constant value as in the case of Pulse Width Modulation PWM (referred to as duty ratio) or a time varying value as in the case of Sliding Mode Control SMC (referred to as equivalent control u_{eq}). In both cases (constant, or time varying), the average control u_{avg} takes values in the compact interval of the real line [0,1].

Generally, it is desired to relate the average value of state variables with the corresponding average value of the control input u_{avg}. This is essential in understanding the main *static* features of the buck converter. In steady state equilibrium, the time derivatives of the average current and voltage are set to zero and the average control input adopts a value given by u_{avg}. With this in mind, the following steady-state equilibrium average current and voltage are obtained:

$$\overline{v}_C = u_{avg} E \tag{3}$$

$$\overline{i}_L = \frac{\overline{v}_C}{R} = \frac{u_{avg} E}{R} \tag{4}$$

According to equation (3) and given the fact that the average control input u_{avg} is restricted to the interval [0,1], the output voltage v_C is a fraction of the input voltage E and the converter can't amplify it.

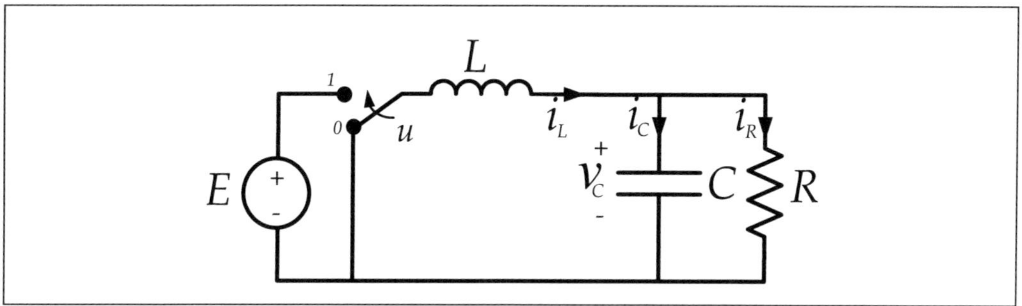

Fig. 1. Ideal switch representation of a single phase DC-DC Buck converter.

3. Sliding mode control of single phase DC-DC Buck converter

Control problems related to DC-DC converters are discussed in many textbooks. The control techniques used in these textbooks differ based on the problem formulation (e.g. available measured state variables as well as known and unknown parameters). PWM techniques are traditionally used to approach such problems. In PWM, low-power signal is amplified in average but in sliding mode, motion in some manifold with desired properties is enforced by discontinuous control. Moreover, sliding mode control provides a better solution over PWM due to the binary nature of sliding mode fitting the discrete nature of the available switches in modern power converters. In this section, the problem of regulating the output voltage v_c of a DC-DC buck converter towards a desired average output voltage v_{sp} is presented.

Consider the DC-DC buck converter shown in fig. 1. A control law u is to be designed such that the output voltage across the capacitor/resistive load v_C converges to a desired unknown constant reference voltage v_{sp} at a desired rate of convergence. The control u is to be designed under the following set of assumptions:

- The value of inductance L and Capacitance C are unknown, but their product $m = 1/LC$ is known.
- The load resistance R is unknown.
- The input voltage E is assumed to be constant floating in the range $[E_{\min}, E_{\max}]$.
- The only measurement available is that of the error voltage $e = v_C - v_{sp}$.
- The current flowing through the resistive load i_R is assumed to be constant.

The complexity of this problem stems from the fact that the voltage error $e = v_C - v_{sp}$ and constant m is the only piece of information available to the controller designer.

The dynamics of the DC-DC buck converter shown in fig. 1 are described by the *unified* dynamical system model in equations (1-2). For the case of constant load current i.e. $i_R \approx \text{constant}$, the model can be reduced to that given by equations (6-7) by introducing new variable i :

$$i = \frac{1}{C}(i_L - i_R) = \frac{1}{C}\left(i_L - \frac{v_C}{R}\right) \tag{5}$$

$$\frac{di}{dt} = \frac{1}{LC}(uE - v_C) = m(uE - v_C) \tag{6}$$

$$\frac{dv_C}{dt} = i \tag{7}$$

Note that the model described by equations (6-7) is only valid when the load resistance is constant and thus, the load current at steady state is constant. Alternatively, the load current i_R might be controlled through an independent controller such that it's always constant. For the case of changing load resistance, the model given by equations (6-7) becomes inaccurate and the converter must instead be modeled by equations (1-2) or equations (8-9) using the change of variable given by equation (5).

$$\frac{di}{dt} = m(uE - v_C) - \frac{1}{RC}i \tag{8}$$

$$\frac{dv_C}{dt} = i \tag{9}$$

A conventional method to approach this problem (v_C to track v_{sp}) is to design a PID controller with the voltage error $e = v_C - v_{sp}$ being the input to the controller. Here, a sliding mode approach to implement a PID controller is presented (Al-Hosani, Malinin, & Utkin, 2009). A block diagram of the controller is shown in Fig. 2. The dynamics of the controller is described by equations (10-11) where L_1 and L_2 are design constants properly chosen to provide stability (as will be shown later).

$$\frac{d\tilde{v}}{dt} = L_1(v_C - v_{sp}) \tag{10}$$

$$\frac{d\tilde{i}}{dt} = m\left[u - \tilde{v} + L_2(v_C - v_{sp})\right] \tag{11}$$

Sliding mode is to be enforced on the PID-like switching surface given by:

$$s = \frac{v_C - v_{sp}}{c\sqrt{m}} + \frac{\tilde{i}}{m} = 0 \tag{12}$$

where c is a constant parameter that is selected by the designer to provide desired system's characteristics when sliding mode is enforced. The control law based on this surface is given by:

$$u = \frac{1}{2}\left(1 - sign(s)\right) = \begin{cases} 0 & \text{if } s < 0 \\ 1 & \text{if } s < 0 \end{cases} \tag{13}$$

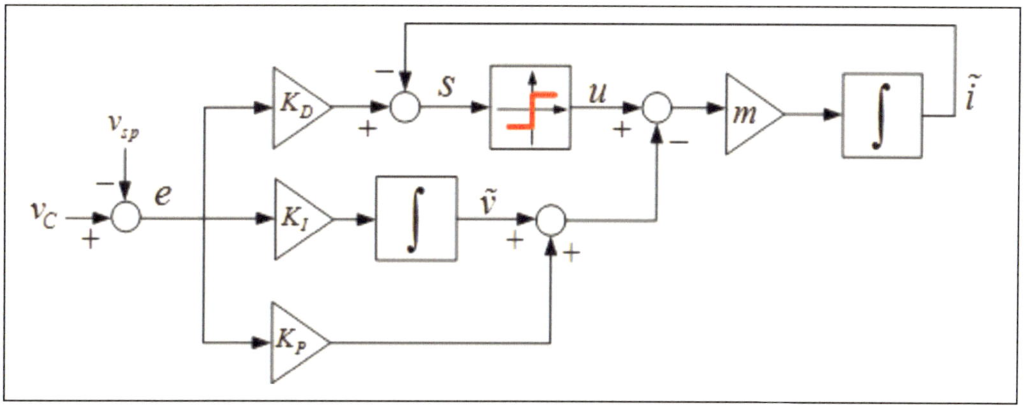

Fig. 2. Sliding Mode implementation of PID Controller.

For sliding mode to exist, the condition $s\dot{s} < 0$ must be always satisfied. Using the equivalent control method, the solution to equations $s = 0$ and $\dot{s} = 0$ must be substituted in equations (6-7) to get the motion equation after sliding mode is enforced along the switching plane $s = 0$ in the system's state space, thus

$$s = 0 \Rightarrow \tilde{i} = -\frac{\sqrt{m}}{c}\left(v_C - v_{sp}\right) \tag{14}$$

$$\dot{s} = 0 \Rightarrow u_{eq} = -L_2\left(v_C - v_{sp}\right) + L_1\int\left(v_C - v_{sp}\right)dt - \frac{1}{c\sqrt{m}}\dot{v}_C \tag{15}$$

As evident from equation (15), the equivalent control u_{eq} is indeed equivalent to a PID controller (Fig. 2) with respectively proportional, derivative, and integral gains $K_P = -L_2$, $K_I = L_1$, and $K_D = -1/c\sqrt{m}$. The equivalent control u_{eq} in (15) is then substituted into equations (6-7) and (10-11) resulting in the following 3rd order system equations:

$$\frac{d^2 v_C}{dt^2} = m\left[E\tilde{v} - \frac{E}{c\sqrt{m}}\frac{dv_C}{dt} - EL_2\left(v_C - v_{sp}\right) - v_C\right] \tag{16}$$

$$\frac{d\tilde{v}}{dt} = L_1\left(v_C - v_{sp}\right) \tag{17}$$

To analyze stability, we write the characteristic equation of the close loop system.

$$\lambda^3 + \frac{E\sqrt{m}}{c}\lambda^2 + m\left(EL_2 + 1\right)\lambda - mEL_1 = 0 \tag{18}$$

The constant parameters L_1, L_2, and c are chosen to provide stability of the system, i.e

$$c > 0, \quad EL_2 + 1 > 0, \quad L_1 < 0 \quad \frac{\sqrt{m}}{c}\left(EL_2 + 1\right) > -L_1 \tag{19}$$

Satisfying the above conditions will provide stability and the system will have the following equilibrium point:

$$v_c = v_{sp}, \quad \dot{v}_c = i = 0, \quad E\tilde{v} = v_{sp}. \tag{20}$$

Fig. 3 shows simulation of the system with parameters $C = 10\mu F$, $L = 2\mu H$, $c = 0.0045$, $L_1 = -10^4$, and $L_2 = 199.92$. The switching device is implemented using a hysteresis loop set such that the switching frequency is controlled to be about 100 KHz. As evident from fig. 3, the output voltage v_C converges (with finite amplitude oscillation or chattering) to the desired reference voltage v_{sp} at a rate set by the chosen controller's parameters.

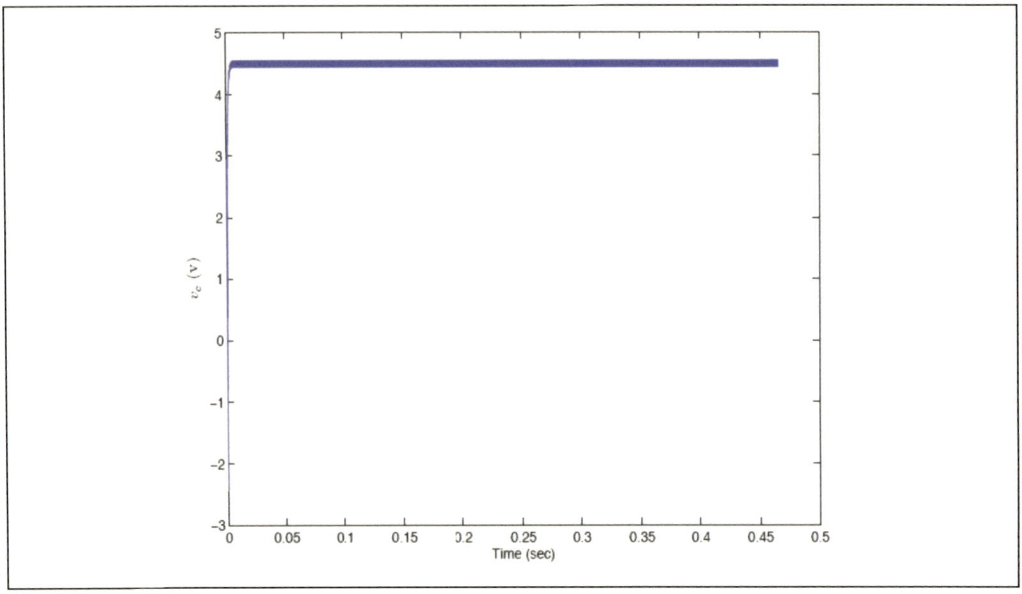

Fig. 3. Sliding mode PID control of single-phase buck converter.

4. Estimation in sliding mode PID controlled single phase DC-DC buck converter

The sliding mode PID controller described by equations (10-13) assumes only the knowledge of the parameter m and the voltage error measurement $e = v_C - v_{sp}$. However, the parameter m might be unknown or varying from one converter to another. Hence, the problem is to complement the sliding mode PID controller with a parameter estimator such that the controller's parameter can be selected automatically. For this, two types of observers are presented next (Al-Hosani, Malinin, & Utkin, 2009). The first type is a sliding mode observer, which assumes that both the reference voltage v_{sp}, and the voltage error $e = v_C - v_{sp}$ are known. On the other hand, the second type is an asymptotic observer that assumes that the voltage error $e = v_C - v_{sp}$ measurement is the only piece of information available.

A. *Sliding Mode Observer*

The converter's parameter m is assumed to be unknown in this case. Thus, we initially feed the sliding mode PID controller described by equations (10-13) with a guess value of m_0. In addition, both the voltage error $e = v_C - v_{sp}$ and the desired set point voltage v_{sp} are assumed to known. Defining the known function $G = Eu - v_C = Eu - e - v_{sp}$, equations (6-7) can be rewritten as:

$$\frac{di}{dt} = mG \tag{21}$$

$$\frac{dv_C}{dt} = i \tag{22}$$

The following sliding mode observer is now proposed:

$$\frac{d\hat{i}}{dt} = \hat{m}G \tag{23}$$

$$\frac{d\hat{m}}{dt} = -M \operatorname{sign}\left(G\bar{i}\right) \tag{24}$$

where $\hat{m}$ is the estimate of converter parameter m, and $\hat{i}$ is the estimate of variable i with mismatch $\bar{i} = \hat{i} - i$. Choosing the constant $M > m$ will enforce sliding mode along the sliding surface $\bar{i} = 0$, and the average value (easily obtained using a low pass filter) of $\hat{m}$ will tend to m. The only restriction in this observer design is that estimation should be fast enough (such that the real value of the estimated parameters is reached before $G = 0$). To overcome this problem, we introduce the following modification to equation (24):

$$\frac{d\hat{m}}{dt} = \begin{cases} -M \operatorname{sign}\left(G\bar{i}\right) & \text{if } |e| \geq \Delta \\ 0^* & \text{if } |e| < \Delta \end{cases} \quad \text{* in discrete time implementation} \tag{25}$$

$$\hat{m}_{k+1} = \hat{m}_k \text{ for } |e| < \Delta$$

The sliding mode observer described by equations (23) and (25) needs knowledge of the error derivative $i = de / dt$ which is not measured. A straightforward solution is to employ a first order filter with the voltage error $e = v_C - v_{sp}$ being the input and the output replacing the error derivative in equation (25). Several computer simulations are performed to confirm the operation of proposed observer. Fig. 4 shows simulation 1 result when using the following set of parameters: $L = 0.81\mu H$, $C = 0.6\mu F$, $v_{sp} = 8\text{V}$, $E = 12\text{V}$, $c = 1 \times 10^{-5}$, $L_1 = -1 \times 10^4$, $L_2 = 200$. For Simulation 2 shown in Fig. 5, the following set of parameters is used: $L = 0.5\mu H$, $C = 1.7\mu F$, $v_{sp} = 8\text{V}$, $E = 12\text{V}$, $c = 1 \times 10^{-5}$, $L_1 = -1 \times 10^4$, $L_2 = 200$. For Simulation 3 shown in Fig. 6, the following set of parameters is used: $L = 0.22\mu H$, $C = 0.76\mu F$, $v_{sp} = 8\text{V}$, $E = 12\text{V}$, $c = 1 \times 10^{-5}$, $L_1 = -1 \times 10^4$, $L_2 = 200$. As evident from these three different simulations, $\hat{m}_{eq}$ (average value of $\hat{m}$) converges to the real value of m with a desired rate of convergence set the observer parameters. The simulations also show that the estimation process was fast enough (estimation was complete before the average value of G reaches zero).

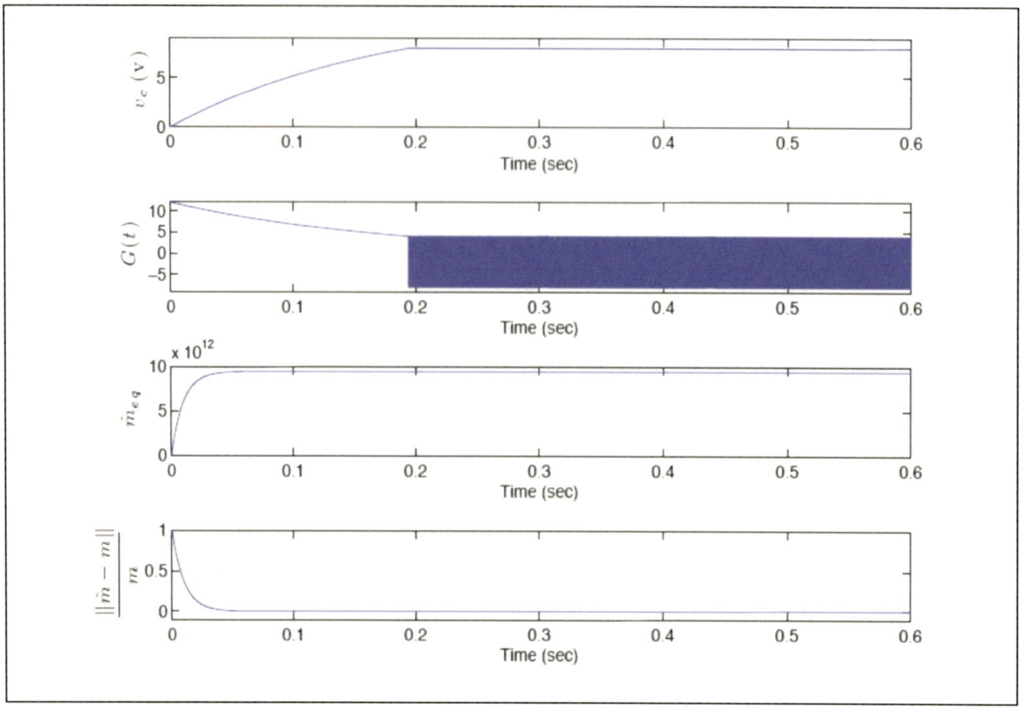

Fig. 4. Simulation 1 SM PID Control of DC-DC buck converter with SM observer.

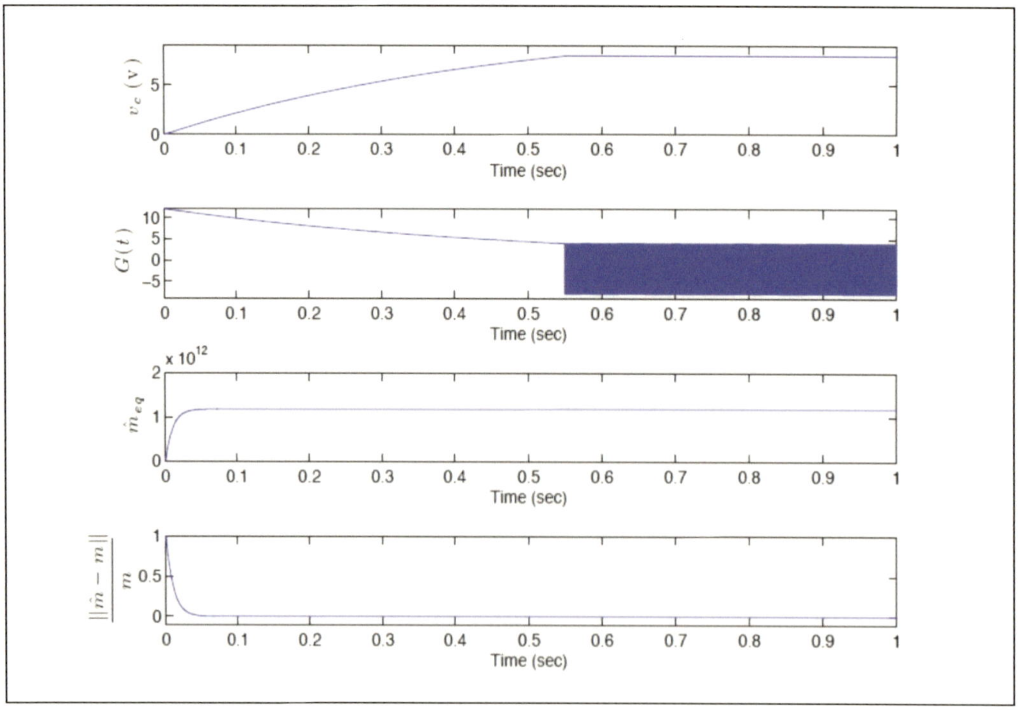

Fig. 5. Simulation 2 SM PID Control of DC-DC buck converter with SM observer.

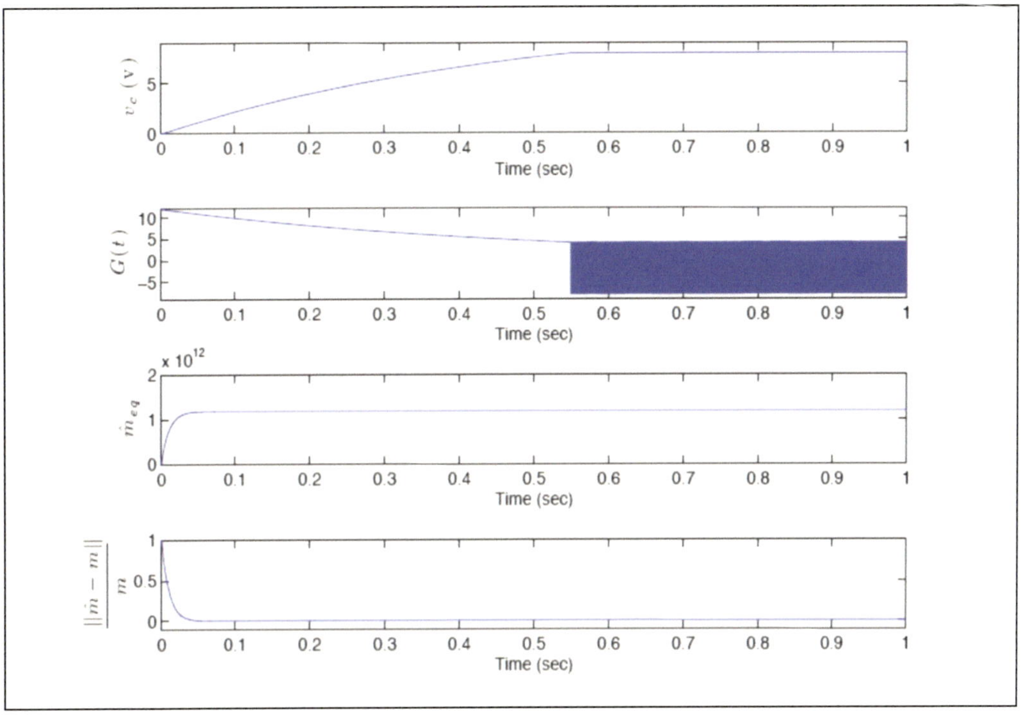

Fig. 6. Simulation 3 SM PID Control of DC-DC buck converter with SM observer.

B. *Asymptotic Observer*

In designing this observer, it is assumed that both the converter's parameter m and the reference set point voltage v_{sp} are unknown. Thus, an asymptotic observer with both parameters being simultaneously estimated is proposed. We first define $G = Eu - v_C = Eu - e - v_{sp} = \alpha(t) + \alpha_0$, where $\alpha(t) = Eu - e$ is a known function and α_0 is an unknown constant. With this, equation (21) can be rewritten as:

$$y = \frac{di}{dt} = m\alpha(t) + m\alpha_0 = m\alpha(t) + \beta \tag{26}$$

where $\beta = m\alpha_0$ and m are constants to be estimated. The following asymptotic observer is then proposed:

$$\hat{y} = \hat{m}\alpha(t) + \hat{\beta} \tag{27}$$

$$\frac{d\hat{m}}{dt} = -k\bar{y}\alpha(t) \tag{28}$$

$$\frac{d\hat{\beta}}{dt} = -k\bar{y} \tag{29}$$

where $\bar{y} = \hat{y} - y$, $\bar{\beta} = \hat{\beta} - \beta$, and $\bar{m} = \hat{m} - m$ are the estimation mismatches. The estimation convergence is proven by considering the Lyapunov candidate function:

$$V=\frac{1}{2}\left(\overline{m}^2+\overline{\beta}^2\right) \tag{30}$$

The time derivative of the Lyapunov function given by equation (30) is calculated to be:

$$\frac{dV}{dt}=-k\overline{y}^2\leq 0 \tag{31}$$

The estimation process is over when $\overline{y}=\overline{m}\alpha(t)+\overline{\beta}=0$. Since $\alpha(t)$ is a known time function, $\overline{m}$ and $\overline{\beta}$ has to be tending to zero ($\hat{m}$ and $\hat{\beta}$ are tending to constant values). Thus, values of $m=\hat{m}-\overline{m}$ and $\beta=\hat{\beta}-\overline{\beta}$ are found. To demonstrate the operation of the proposed observer, several computer simulations are presented in Figures 7-8 for the case of sliding mode control of DC-DC buck converter with an asymptotic observer as described by equations (27-29). In these simulations, the reference set point voltage v_{sp} is varied from 2V to 8V and the converter's parameter m is varied in the range $m_{\min}$ to $10m_{\min}$ where $m_{\min}=1.1765\times 10^{12}F^{-1}H^{-1}$. As the simulations demonstrate, estimates converge monotonously to the real values during the transient interval of the converter (before reaching $G=0$).

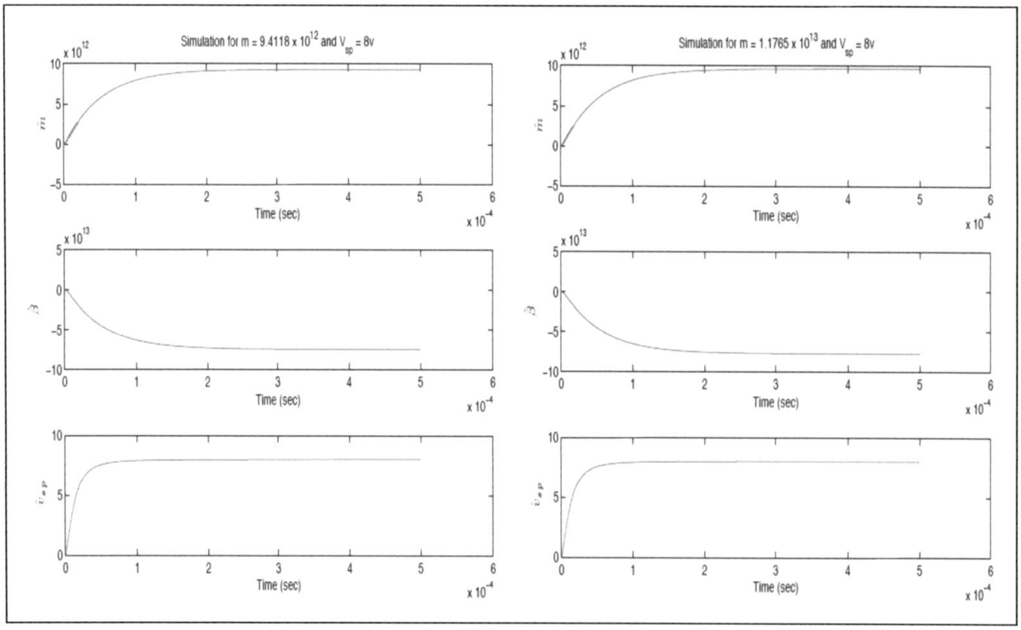

Fig. 7. Sliding Mode PID Control of buck converter with asymptotic observer.

5. Modeling and stability analysis of sliding mode controlled single phase DC-DC buck converter

In this section, a different type of DC-DC converter namely the boost converter is discussed. The boost converter is classified as an amplification circuit where the output voltage is a scaled version of the input voltage by a scalar greater than unity. The ideal switch representation of a single-phase boost converter with resistive load is shown in Fig. 9.

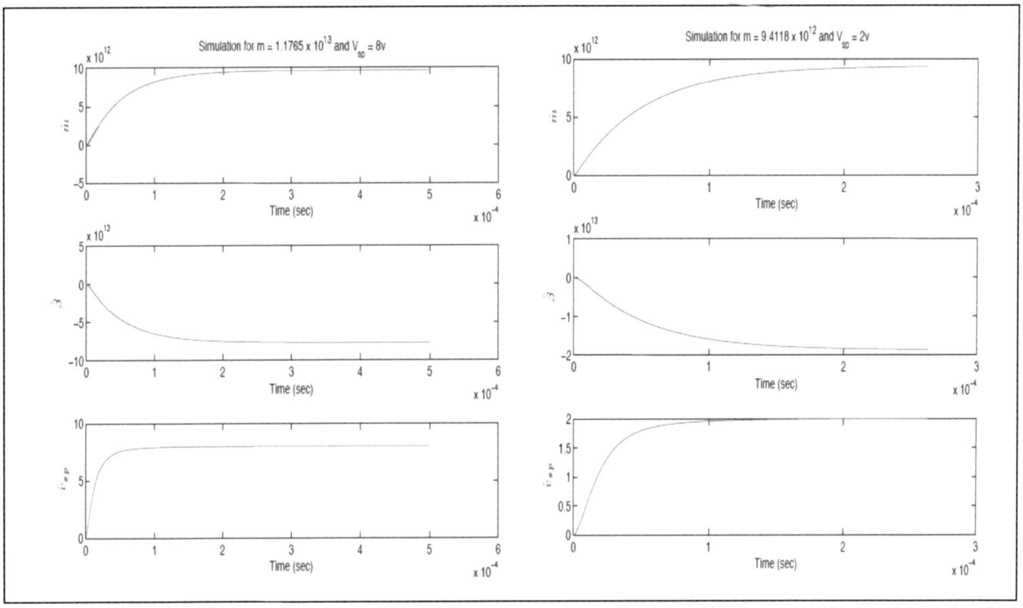

Fig. 8. Sliding Mode PID Control of buck converter with asymptotic observer.

Simple applications of Kirchhoff's current and voltage laws for each resulting circuit topology from the two possible ideal switch's positions allow us to get the system of differential equations governing the dynamics of the boost converter. Thus, we obtain the following *unified* dynamical system model:

$$\frac{dv_C}{dt} = \frac{1}{C}\left[(1-u)i_L - \frac{v_C}{R}\right] \tag{32}$$

$$\frac{di_L}{dt} = \frac{1}{L}\left[E - (1-u)v_C\right] \tag{33}$$

Most often, it is desired to regulate the output voltage v_C to a constant desired value v_{sp} or a to track a given reference signal. Next, we will distinguish between two types of control strategies when dealing with the DC-DC Boost converters. These are the direct and indirect control method. The control objective here is to drive the average of the output voltage v_C to a desired equilibrium value v_{sp}.

A. *Direct sliding Mode control of DC-DC Boost converter*

In the direct sliding mode control method, the output voltage v_C is used directly to synthesize a suitable sliding surface yielding the desired objective. For example, consider the following sliding surface and its associated control law:

$$s_{\text{direct}} = v_{sp} - v_C \tag{34}$$

$$u = \frac{1}{2}\left(1 - \text{sign}\left(s_{\text{direct}}\right)\right) \tag{35}$$

For sliding mode to exist, the condition $s_{\text{direct}}\dot{s}_{\text{direct}} < 0$ must always hold. Thus, we must have $i_L > v_C / R$ since

$$s_{\text{direct}} = 0 \Rightarrow v_C = v_{sp} \tag{36}$$

$$\dot{s}_{\text{direct}} = 0 \Rightarrow \left(1-u_{eq}\right)i_L = \frac{v_{sp}}{R} \Rightarrow \left(1-u_{eq}\right) = \frac{v_{sp}}{i_L R} \tag{37}$$

Thus, the motion in sliding mode is governed by the following first order differential equation:

$$\frac{di_L}{dt} = \frac{1}{L}\left(E - \frac{v_{sp}^2}{Ri_L}\right) \tag{38}$$

The system has an equilibrium point $i_L^{ss} = v_{sp}^2 / ER$. It is not difficult to see that the system exhibits an unstable equilibrium point (Sira-Ramírez, 2006) and we shall establish this via several approaches.
First, using an approximate linearization approach, the local stability around the equilibrium point of the zero dynamics $i_L^{ss} = v_{sp}^2 / ER$ will be investigated. Expanding the right-hand side of equation (38) into its Taylor series about the point $i_L = i_L^{ss}$, we obtain:

$$\frac{di_L}{dt} = \frac{1}{L}\left(E - \frac{v_{sp}^2}{Ri_L}\right)\Bigg|_{i_L = i_L^{ss}} + a_1\left(i_L = i_L^{ss}\right) + \mathrm{O}\left(\bar{i}\right) \tag{39}$$

where $a_1 = \frac{d}{di_L}\left(\frac{1}{L}\left(E - \frac{v_{sp}^2}{Ri_L}\right)\right)\Bigg|_{i_L = i_L^{ss}} = \frac{E^2 R}{v_{sp}^2}, \quad \bar{i} = i_L - i_L^{ss} \quad \lim\limits_{\bar{i}\to 0}\frac{\mathrm{O}\left(\bar{i}\right)}{\bar{i}} = 0$

$$\frac{d\bar{i}_L}{dt} = \frac{E^2 R}{v_{sp}^2}\bar{i}_L + H.O.T \tag{40}$$

If we restrict our attention to a sufficiently small neighborhood of the equilibrium point such that the higher order term $\mathrm{O}\left(\bar{i}\right)$ is negligible, then

$$\frac{d\bar{i}_L}{dt} = \frac{E^2 R}{v_{sp}^2}\bar{i}_L \tag{41}$$

Clearly, the equilibrium point $i_L^{ss} = v_{sp}^2 / ER$ is unstable in view of the fact the linearized zero dynamics exhibits a characteristic polynomial with a zero in the right-half part of the complex plane.
Another way of showing instability of the zero dynamics of a direct controlled DC-DC boost converter is to utilize Lyapunov stability theory. For example, consider the following positive definite Lyapunov candidate function:

$$V\left(i_L\right) = \frac{1}{2}\left(Ei_L - \frac{v_{sp}^2}{R}\right)^2 \tag{42}$$

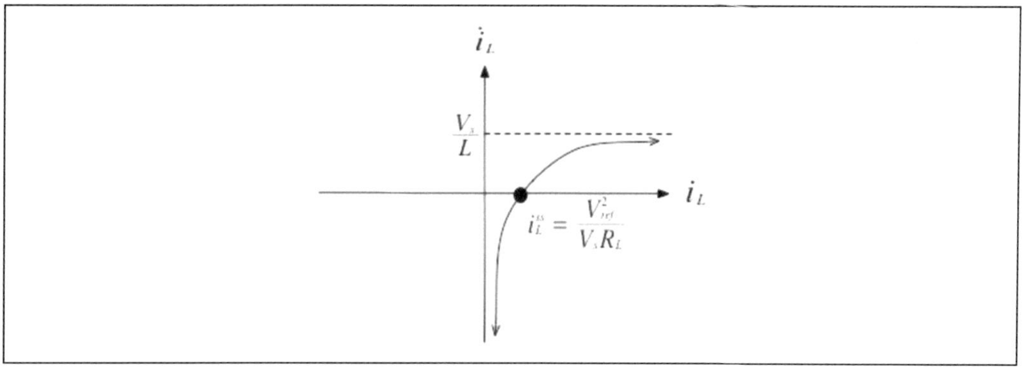

Fig. 9. Phase Portrait diagram for the case of Direct Sliding Mode control of boost converters

The time derivative of the above Lyapunov candidate function (taking into account that $i_L > 0$) is:

$$V(i_L) = \frac{E}{Li_L}\left(Ei_L - \frac{v_{sp}^2}{R}\right)^2 \geq 0 \tag{43}$$

Thus, the system exhibits an unstable zero dynamics. The phase portrait diagram shown in Fig. 9 also confirms this. The fact that the direct voltage controlled boost converters exhibits an unstable equilibrium point is addressed in the control system literature, by stating that the output voltage is a non-minimum phase output. On the contrary, as we will see next, the inductor current is said to be minimum phase output and thus, current controlled boost converter doesn't exhibit an unstable equilibrium point.

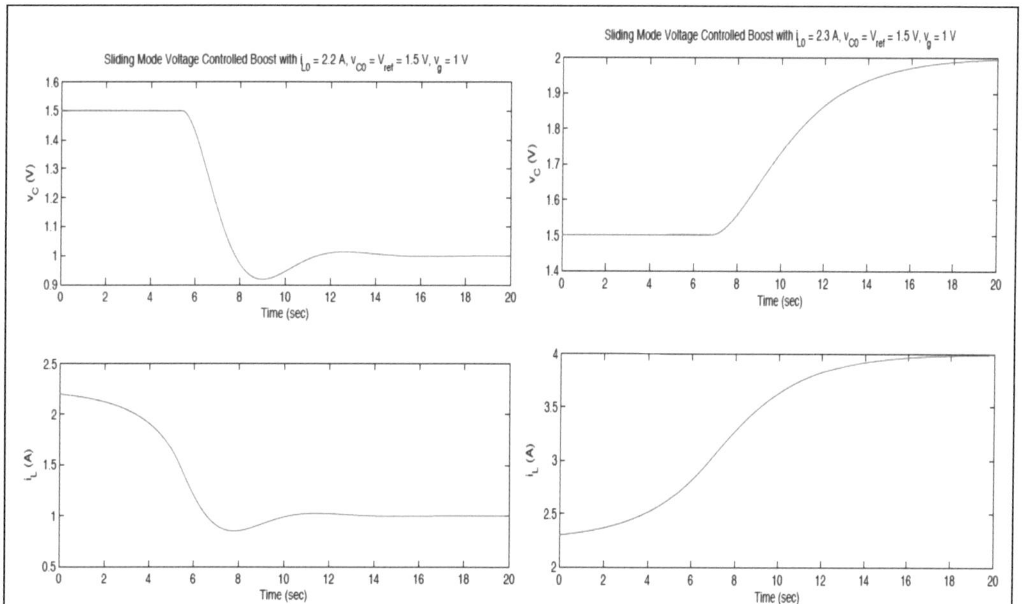

Fig. 10. Sliding Mode Voltage controlled (direct) boost converter

Simulation 1 (left) and 2 (right) in Fig. 10 demonstrates the instability of the zero dynamics for the case of direct voltage controlled DC-DC buck converters. For both of these simulations, the following converter parameters are used: $L = 1H$, $R = 1\Omega$, $C = 1F$, $E = 1\text{V}$, $v_{sp} = 1.5\text{V}$. Both simulations assume an initial voltage of $v_c(t=0) = 1.5\text{V}$. The initial inductor current value is $i_L(t=0) = 2.2\text{A} < i_L^{ss} = 2.25\text{A}$ for simulation 1 and $i_L(t=0) = 2.3\text{A} > i_L^{ss} = 2.25\text{A}$ for simulation 2.

B. *Indirect sliding Mode control of DC-DC boost converter*

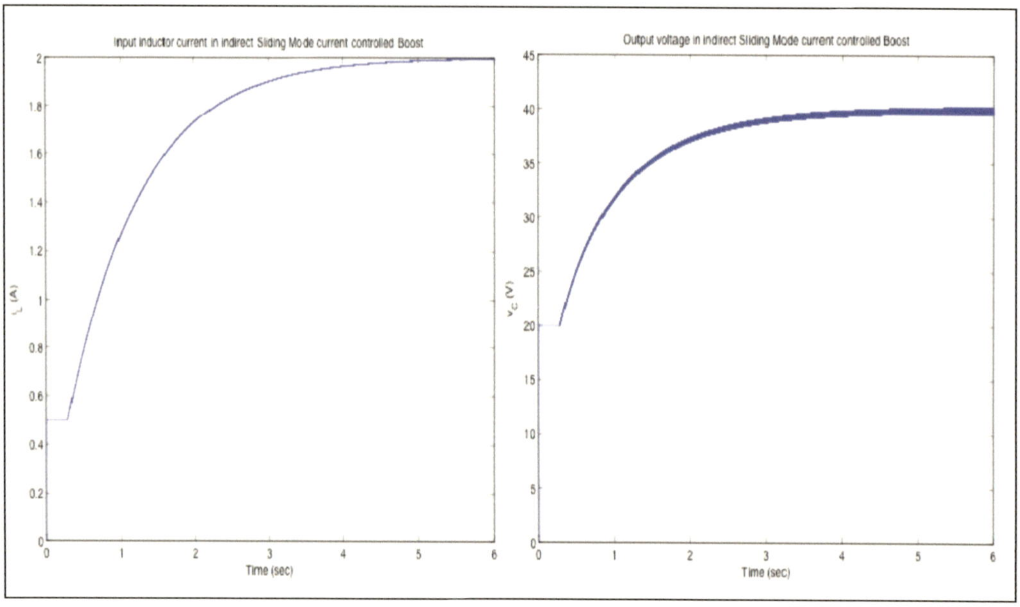

Fig. 11. Sliding Mode current controlled (indirect) boost converter

As demonstrated before, voltage-sliding mode controlled DC-DC boost converters exhibit unstable equilibrium points. The alternative is then to use a sliding surface based on the inductor current such that when it is set to zero, leads to a desired value of the input inductor current in correspondence with the desired output equilibrium voltage. Consider the following indirect current based sliding mode control:

$$s_{\text{indirect}} = i_L - i_{sp} \tag{44}$$

$$u = \frac{1}{2}\left(1 - \text{sign}\left(s_{\text{indirect}}\right)\right) \tag{45}$$

Where i_{sp} is a desired input inductor current calculated in accordance with desired output voltage v_{sp}. To calculate the needed i_{sp}, we compute the equilibrium point of the system under ideal sliding mode conditions, thus:

$$s_{\text{indirect}} = 0 \Rightarrow i_L = i_{sp},\ \dot{s}_{\text{direct}} = 0 \Rightarrow \left(1 - u_{eq}\right) = \frac{E}{v_C},\ \frac{dv_C}{dt} = 0 \Rightarrow \left(1 - u_{eq}\right) i_{sp} = \frac{v_C}{R} \Rightarrow i_{sp} = \frac{v_C^2}{ER} \tag{46}$$

For the desire an output voltage v_{sp}, the needed set point value for the inductor current is found as

$$i_{sp} = \frac{v_C^2}{ER} \tag{47}$$

The motion after sliding mode is enforced is governed by the following equation:

$$\frac{dv_C}{dt} = \frac{1}{C}\left[\frac{E}{v_C} i_{sp} - \frac{v_C}{R}\right] \tag{48}$$

It is evident that the unique equilibrium point of the zero dynamics is indeed an asymptotically stable one. To proof this, let's consider the following Laypunov candidate function:

$$V = \frac{1}{2}\left(v_C - v_{sp}\right)^2 \tag{49}$$

The time derivative of this Lyapunov candidate function is

$$\begin{aligned}
\dot{V} &= \left(v_C - v_{sp}\right)\frac{1}{C}\left[\frac{E}{v_C} i_{sp} - \frac{v_C}{R}\right] \\
&= \left(v_C - v_{sp}\right)\frac{1}{C}\left[\frac{v_{sp}^2}{v_C R} - \frac{v_C}{R}\right] \\
&= -\frac{1}{Cv_C}\left(v_C - v_{sp}\right)\left(v_C^2 - v_{sp}^2\right) \\
&= -\frac{1}{Cv_C}\left(v_C - v_{sp}\right)^2\left(v_C - v_{sp}\right)
\end{aligned} \tag{50}$$

The time derivative of the Lyapunov candidate function is negative around the equilibrium point v_{sp} given that $v_C > 0$ around the equilibrium. To demonstrate the efficiency of the indirect control method, figure 11 shows simulation result when using the following parameters: $L = 40mH$, $C = 4\mu F$, $R = 40\Omega$, $E = 20\text{V}$, $v_{sp} = 40\text{V}$.

6. Chattering reduction in multiphase DC-DC power converters

One of the most irritating problems encountered when implementing sliding mode control is chattering. Chattering refers to the presence of undesirable finite-amplitude and frequency oscillation when implementing sliding mode controller. These harmful oscillations usually lead to dangerous and disappointing results, e.g. wear of moving mechanical devices, low accuracy, instability, and disappearance of sliding mode. Chattering may be due to the discrete-time implementation of sliding mode control e.g. with digital controller. Another cause of chattering is the presence of unmodeled dynamics that might be excited by the high frequency switching in sliding mode.

Researchers have suggested different methods to overcome the problem of chattering. For example, chattering can be reduced by replacing the discontinuous control action with a continuous function that approximates the $\text{sign}(s(t))$ function in a boundary later layer of the manifold $s(t)=0$ (Slotine & Sastry, 1983; Slotine, 1984). Another solution (Bondarev, Bondarev, Kostyleva, & Utkin, 1985) is based on the use of an auxiliary observer loop rather than the main control loop to generate chattering-free ideal sliding mode. Others suggested the use of state dependent (Emelyanov, et al., 1970; Lee & Utkin, 2006) or equivalent control dependent (Lee & Utkin, 2006) gain based on the observation that chattering is proportional to the discontinuous control gain (Lee & Utkin, 2006). However, the methods mentioned above are disadvantageous or even not applicable when dealing with power electronics controlled by switches with "ON/OFF" as the only admissible operating states. For example, the boundary solution methods mentioned above replaces the discontinuous control action with a continuous approximation, but control discontinuities are inherent to these power electronics systems and when implementing such solutions techniques such as PWM has to be exploited to adopt the continuous control action. Moreover, commercially available power electronics nowadays can handle switching frequency in the range of hundreds of KHz. Hence, it seems unjustified to bypass the inherent discontinuities in the system by converting the continuous control law to a discontinuous one by means of PWM. Instead, the discontinuous control inputs should be used directly in control, and another method should be investigated to reduce chattering under these operating conditions.

The most straightforward way to reduce chattering in power electronics is to increase the switching frequency. As technology advances, switching devices is now manufactured with enhanced switching frequency (up to 100s KHz) and high power rating. However, power losses impose a new restriction. That is even though switching is possible with high switching frequency; it is limited by the maximum allowable heat losses (resulting from switching). Moreover, implementation of sliding mode in power converters results in frequency variations, which is unacceptable in many applications.

The problem we are dealing with here is better stated as follow. We would like first to control the switching frequency such that it is set to the maximum allowable value (specified by the heat loss requirement) resulting in the minimum possible chattering level. Chattering is then reduced under this fixed operating switching frequency. This is accomplished through the use of interleaving switching in multiphase power converters where harmonics at the output are cancelled (Lee, 2007; Lee, Uktin, & Malinin, 2009). In fact, several attempts to apply this idea can be found in the literature. For example, phase shift can be obtained using a transformer with primary and secondary windings in different phases. Others tried to use delays, filters, or set of triangular inputs with selected delays to provide the desired phase shift (Miwa, Wen, & Schecht, 1992; Xu, Wei, & Lee, 2003; Wu, Lee, & Schuellein, 2006). This section presents a method based on the nature of sliding mode where phase shift is provided without any additional dynamic elements. The section will first presents the theory behind this method. Then, the outlined method will be applied to reduce chattering in multiphase DC-DC buck and boost converters.

A. *Problem statement: Switching frequency control and chattering reduction in sliding mode power converters*

Consider the following system with scalar control :

$$\dot{x}=f(x,t)+b(x,t)u \quad x,f,b\in\mathbb{R}^n \tag{51}$$

Here, control is assumed to be designed as a continuous function of the state variables, i.e. $u_0(x)$. In electric motors with current as a control input, it is common to utilize the so-called "cascaded control". Power converters usually use PWM as principle operation mode to implement the desired control. One of the tools to implement this mode of operation is sliding mode control. A block diagram of possible sliding mode feedback control to implement PWM is shown in fig. 12. When sliding mode is enforced along the switching line $s = u_0(x) - u$, the output u tracks the desired reference control input $u_0(x)$. Sliding mode existence condition can be found as follow:

$$\begin{aligned} &s = u_0(x) - u, \quad \dot{u} = v = M\,\text{sign}(s), \quad M > 0 \\ &\Rightarrow \dot{s} = g(x) - M\,\text{sign}(s), \quad g(x) = \text{grad}^{\mathrm{T}}(u_0)(f + bu) \end{aligned} \tag{52}$$

Thus, for sliding mode to exist, we need to have $M > |g(x)|$.

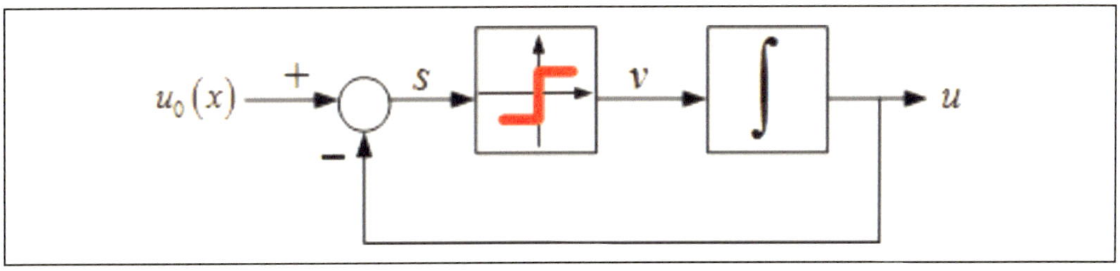

Fig. 12. Sliding mode control for simple power converter model

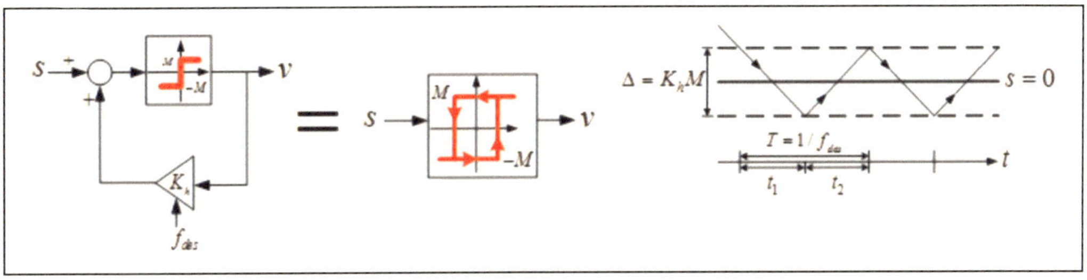

Fig. 13. Implementation of hysteresis loop with width $\Delta = K_h M$. Oscillations in the vicinity of the switching surface is shown in the right side of the figure. Frequency control is performed by changing the width of a hysteresis loop in switching devices (Nguyen & Lee, 1995; Cortes & Alvarez, 2002)

To maintain the switching frequency at a desired level f_{des}, control is implemented with a hysteresis loop (switching element with positive feedback as shown in fig. 13). Assuming that the switching frequency is high enough such that the state x can be considered constant within time intervals t_1 and t_2 in fig. 13, the switching frequency can be calculated as:

$$f = \frac{1}{t_1 + t_2}, \quad t_1 = \frac{\Delta}{M - g(x)}, \quad t_2 = \frac{\Delta}{M + g(x)} \tag{53}$$

Thus, the width of the hysteresis loop needed to result in a switching frequency f_{des} is:

$$\Delta = \frac{1}{f_{des}} \frac{M^2 - g^2(x)}{2M} \tag{54}$$

f_{des} is usually specified to be the maximum allowable switching frequency resulting in the minimum possible level of chattering. However, this chattering level may still not be acceptable. Thus, the next step in the design process is to reduce chattering under this operating switching frequency by means of harmonics cancellation, which will be discussed next.

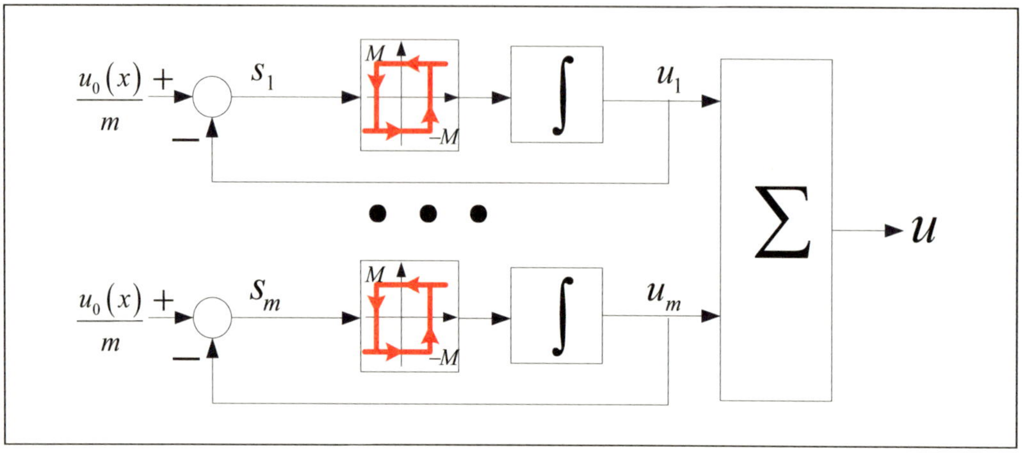

Fig. 14. m-phase power converter with evenly distributed reference input

Let's assume that the desired control $u_0(x)$ is implemented using m power converters, called "multiphase converter" (Fig. 14) with $s_i = u_0 / m - u_i$ where $i = 1, 2, \ldots, m$. The reference input in each power converter is u_0 / m. If each power converters operates correctly, the output u will track the desired control $u_0(x)$. The amplitude A and frequency f of chattering in each power converter are given by:

$$A = \frac{\Delta}{2}, \; f = \frac{M^2 - (g(x)/m)^2}{2M\Delta} \tag{55}$$

The amplitude of chattering in the output u depends on the amplitude and phase of chattering in each leg and, in the worst-case scenario, can be as high as m times that in each individual phase. For the system in Fig. 14, phases depend on initial condition and can't be controlled since phases in each channel are independent in this case. However, phases can be controlled if channels are interconnected (thus not independent) as we will be shown later in this section.

Now, we will demonstrate that by controlling the phases between channels (through proper interconnection), we can cancel harmonics at the output and thus reduce chattering. For now, let's assume that $m-$ phases power converter is designed such that the frequency of chattering in each channel is controlled such that it is the same in each phase $(f = 1/T)$. Furthermore, the phase shift between any two subsequent channels is assumed to be T/m. Since chattering is a periodic channel, it can be represented by its Fourier series with frequencies $\omega_k = k\omega$ where $\omega = 2\pi / T$ and $k = 1, 2, \ldots, \infty$. The phase difference between the first channel and ith channel is given by $\phi_i = 2\pi / (\omega m)$. The effect of the kth harmonic in the output signal is the sum of the individual kth harmonics from all channels and can be calculated as:

$$\sum_{i=0}^{m-1}\sin\left(\omega_k\left(t-\frac{2\pi}{\omega m}i\right)\right)=\sum_{i=0}^{m-1}\mathrm{Im}\left[\exp\left(j\omega_k\left(t-\frac{2\pi}{\omega m}i\right)\right)\right]$$
$$=\mathrm{Im}\left(Z\exp\left(j\omega_k t\right)\right) \tag{56}$$
$$\text{where}\quad Z=\sum_{i=0}^{m-1}\exp\left(-j\frac{2\pi k}{m}i\right)$$

Now consider the following equation:

$$Z\exp\left(-j\frac{2\pi k}{m}\right)=\sum_{i=0}^{m-1}\exp\left(-j\frac{2\pi k}{m}(i+1)\right)=\sum_{i'=1}^{m}\exp\left(-j\frac{2\pi k}{m}i\right)=Z \tag{57}$$

The solution to equation (57) is that either $\exp\left(-j2\pi k/m\right)=1$ or $Z=0$. Since we have $\exp\left(-j2\pi k/m\right)=1$ when k/m is integer, i.e. $k=m,2m,\ldots$, then we must have $Z=0$ for all other cases. This analysis means that all harmonics except for lm with $l=1,2,\ldots$ are suppressed in the output signal. Thus, chattering level can be reduced by increasing the number of phases (thus canceling more harmonics at the output) provided that a proper phases shift exists between any subsequent phases or channels.

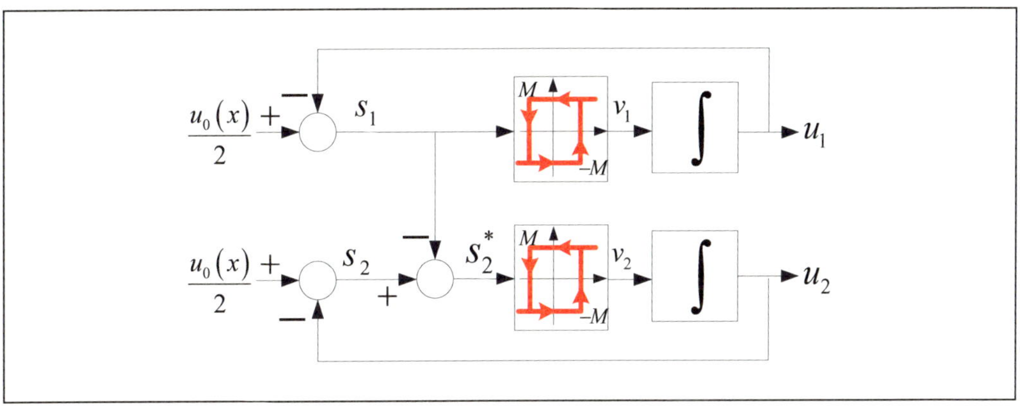

Fig. 15. Interconnection of channels in two-phase power converters to provide desired phase shift

The next step in design process is to provide a method of interconnecting the channels such that a desired phase shift is established between any subsequent channels. To do this, consider the interconnection of channels in the two-phase power converter model shown in Fig 15. The governing equations of this model are:

$$s_1=u_0/2-u_1,\quad s_2=u_0/2-u_2,\quad s_2^*=s_2-s_1 \tag{58}$$

The time derivative of s_1 and s_2^* is are given by:

$$\dot{s}_1=a-M\,\mathrm{sign}\left(s_1\right),\quad a=g\left(x\right)/m, m=2,\quad g\left(x\right)=\mathrm{grad}^T\left(u_0\right)\left(f+bu\right) \tag{59}$$

$$\dot{s}_2^*=M\,\mathrm{sign}\left(s_1\right)-M\,\mathrm{sign}\left(s_2^*\right) \tag{60}$$

Consider now the system behavior in the s_1, s_2^* plane as shown in Fig. 16 and 17. In Fig. 16, the width of hysteresis loops for the two sliding surfaces $s_1 = 0$ and $s_2^* = 0$ are both set to Δ. As can be seen from figure 16, the phase shift between the two switching commands v_1 and v_2 is always $T/4$ for any value of Δ, where T is the period of chattering oscillations $T = 2\Delta / m$. Also, starting from any initial conditions different from point 0 (for instance $0'$ in Fig. 16), the motion represented in Fig. 16 will appear in time less than $T/2$.

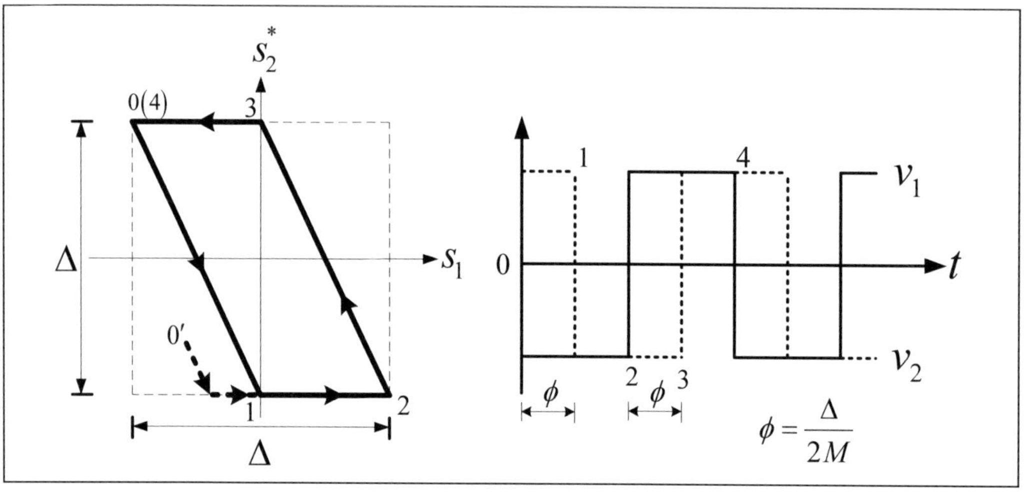

Fig. 16. System behavior in s plane with $\alpha = 1$ and $a > 0$

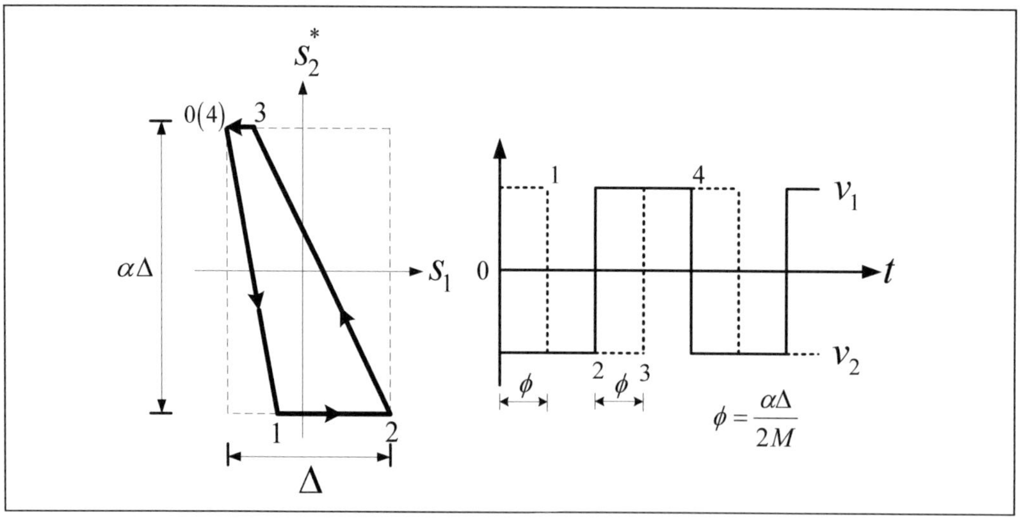

Fig. 17. System behavior in s plane with $\alpha > 1$ and $a > 0$

If the width of the hysteresis loop for the two sliding surface $s_1 = 0$ and $s_2^* = 0$ are set to Δ and $\alpha\Delta$ respectively (as in Fig. 17), the phase shift between the two switching commands v_1 and v_2 can be controlled by proper choice of α. The switching frequency f and phase shift ϕ are given by:

$$T = \frac{1}{f} = \frac{\Delta}{M-a} + \frac{\Delta}{M+a} = \frac{2M\Delta}{M^2 - a^2}, \tag{61}$$

$$\phi = \frac{\alpha\Delta}{2M}, \quad a = \frac{g(x)}{m}, \quad g(x) = g(x) = \text{grad}^T(u_0)(f + bu) \tag{62}$$

To preserve the switching cycle, the following condition must always be satisfied:

$$\frac{\alpha\Delta}{2M} \le \frac{\Delta}{M+|a|} \tag{63}$$

Thus, to provide a phase shift $\phi = T / m$, where m is the number of phases, we choose the parameter α as:

$$\alpha = \frac{4M^2}{m\left(M^2 - a^2\right)} \tag{64}$$

The function a is assumed to be bounded i.e. $|a| < a_{\max} < M$. With this, condition (63) can be rewritten as:

$$m > \frac{2\Delta}{M - a_{\max}} \quad \text{or} \quad |a| \le M\left(1 - \frac{2}{m}\right) m \ge 2 \tag{65}$$

It is important to make sure that the selected α doesn't lead to any violation of condition (63) or (65) which might lead to the destruction of the switching cycle. Thus, equation (63) is modified to reflect this restriction, i.e.

$$\alpha = \begin{cases} \dfrac{4M^2}{m\left(M^2 - a^2\right)} & \text{if} \quad |a| < M\left(1 - \dfrac{2}{m}\right) \\ \dfrac{2M}{M+|a|} & \text{if} \quad M\left(1 - \dfrac{2}{m}\right) \le |a| < M \end{cases} \tag{66}$$

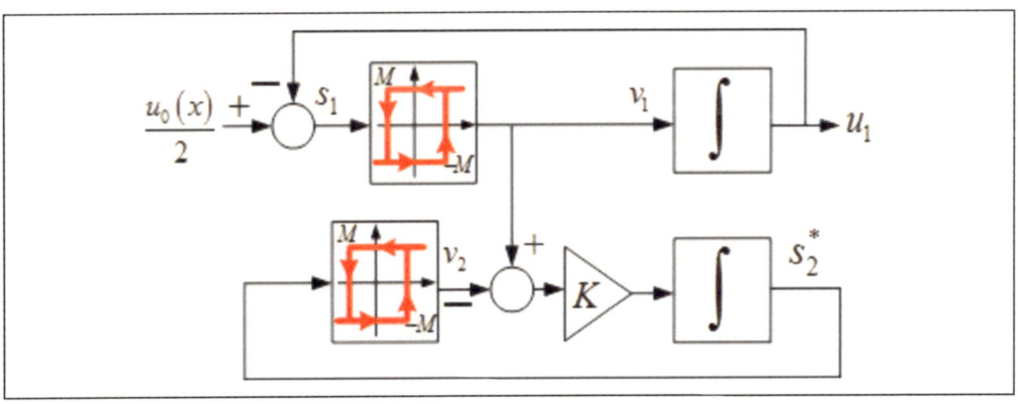

Fig. 18. Master-slave two-phase power converter model

Another approach in which a frequency control is applied for the first phase and open loop control is applied for all other phases is shown in Fig. 18. In this approach (called master-slave), the first channel (master) is connected to the next channel or phase (slave) through an additional first order system acting as a phase shifter. This additional phase shifter system acts such that the discontinuous control v_2 for the slave has a desired phase shift with respect to the discontinuous control v_1 for the master without changing the switching frequency. In this system, we have:

$$s_1 = \frac{1}{m}u_0 - u_1, \quad m = 2 \tag{67}$$

$$\dot{s}_1 = a - M\text{sign}(s_1), \; a = \frac{g(x)}{m}, \; g(x) = \text{grad}^T u_0(f + bu) \tag{68}$$

$$\dot{s}_2^* = KM\left(\text{sign}\left(s_1\right) - \text{sign}\left(s_2^*\right)\right), \tag{69}$$

The phase shift between v_1 and v_2 (based on s-plane analysis) is given by

$$\phi = \frac{\Delta}{2KM} \tag{70}$$

Thus, the value of gain K needed to provide a phase shift of T / m is:

$$K = \frac{m\left(M^2 - a^2\right)}{4M^2} \tag{71}$$

Please note that, $K = 1 / \alpha$ should be selected in compliances with equation (63) to preserve the switching cycle as discussed previously. To summarize, a typical procedures in designing multiphase power converters with harmonics cancellation based chattering reduction are:

1. Select the width of the hysteresis loop (or its corresponding feedback gain K_h) to maintain the switching frequency in the first phase at a desired level (usually chosen to be the maximum allowable value corresponding to the maximum heat power loss tolerated inn the system).
2. Determine the number of needed phases for a given range of function a variation.
3. Find the parameter α such that the phase shift between any two subsequent phases or channels is equal to $1 / m$ of the oscillation period of the first phase.

Next, we shall apply the outlined procedures to reduce chattering in sliding mode controlled multiphase DC-DC buck and boost converters.

7. Chattering reduction in multiphase DC-DC buck converters

Consider the multiphase DC-DC buck converter depicted in Fig. 19. The shown converter composed of $n = 4$ legs or channels that are at one end controlled by switches with switching commands $u_i \in \{0,1\}, i = 1,\ldots,n$, and all connected to a load at the other end. A $n-$dimensional control law $u = \left[u_1, u_2, \ldots, u_n\right]^T$ is to be designed such that the output voltage across the resistive load/capacitance converges to a desired unknown reference voltage v_{sp} under the following assumptions (similar to the single-phase buck converter discussed earlier in this chapter):

- Values of inductance L and capacitance C are unknown, but their product $m = 1/LC$ is known.
- Load resistance R and input resistance r are unknown.
- Input voltage E is assumed to be constant floating in the range $[E_{min}, E_{max}]$.
- The only measurement available is that of the voltage error $e = v_C - v_{sp}$.

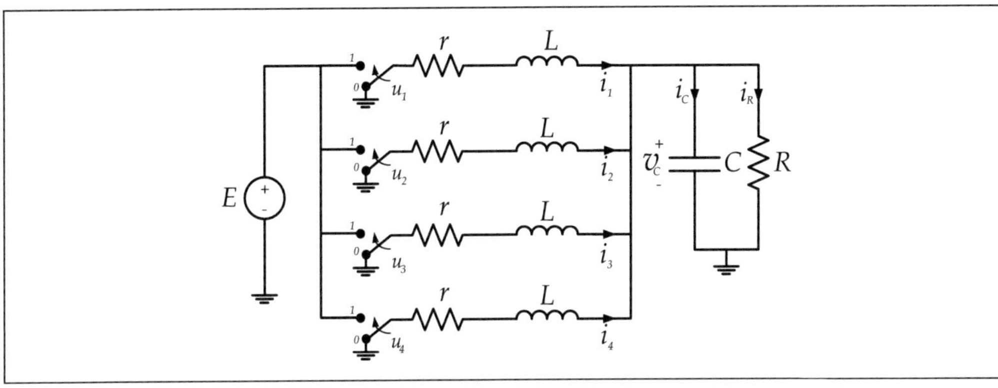

Fig. 19. 4-phase DC-DC buck converter

The dynamics of $n-$ phase DC-DC buck converter are governed by the following set of differential equations:

$$\dot{i}_k = \frac{1}{L}\left(Eu_k - ri_k - v_C\right), \quad k = 1,\dots,n \tag{72}$$

$$\dot{v}_C = \frac{1}{C}\left(\sum_{k=1}^{n} i_k - i_R\right) \tag{73}$$

A straightforward way to approach this problem is to design a PID sliding mode controlled as done before for the single-phase buck converter cases discussed earlier in this chapter. To reduce chattering, however, we exploit the additional degree of freedom offered by the multiphase buck converter in cancelling harmonics (thus reduce chattering) at the output. Based on the design procedures outlined in the previous section, the following controller is proposed:

$$\tilde{v} = L_1\left(v_C - v_{sp}\right) \tag{74}$$

$$\dot{\tilde{i}} = m\left[u_1 - \tilde{v} + L_2\left(v_C - v_{sp}\right)\right], \quad \text{m}=1/LC \tag{75}$$

$$s_1 = \frac{1}{c\sqrt{m}}\left(v_C - v_{sp}\right) + \frac{1}{m}\tilde{i} \tag{76}$$

$$\dot{s}_k = Kb\left[\text{sign}(s_{k-1}) - \text{sign}(s_k)\right], \quad k = 2,\dots,n, \quad b = \frac{1}{2} \tag{77}$$

$$u_k = \frac{1}{2}(1 - \text{sign}(s_k)), \quad k = 1,...,n \tag{78}$$

The time derivative of the sliding surface in the first phase is given by:

$$\dot{s}_1 = a - b\,\text{sign}(s_1) \quad a = \frac{1}{c\sqrt{m}C}\left(\sum_{k=1}^{n} i_k - i_R\right) + \frac{1}{2} + L_2\left(v_C - v_{sp}\right) - \tilde{v}, \quad b = \frac{1}{2} \tag{79}$$

To preserve the switching cycle, the gain $K = 1/\alpha$ is to be selected in accordance with equation (66) . Thus, we must have

$$K = \begin{cases} \frac{n}{4}\left(1 - \left(\frac{a}{b}\right)^2\right) & \text{if } \left|\frac{a}{b}\right| < \left(1 - \frac{2}{n}\right) \\ \frac{1}{2}\left(1 + \left|\frac{a}{b}\right|\right) & \text{if } \left(1 - \frac{2}{n}\right) \le \left|\frac{a}{b}\right| < 1 \end{cases} \tag{80}$$

As we can see from equation (80), the parameter a/b is needed to implement the controller. However, this parameter can be easily obtained by passing the signal $\text{sign}(s_1)$ as the input to a low pass filter. The output of the low pass filter is then a good estimate of the needed parameter. This is because when sliding mode is enforced along the switching surface $s_1 = 0$, we also have $\dot{s}_1 = 0$ leading to the conclusion that $(\text{sign}(s_1))_{eq} = a/b$. Of course, controller parameters c, L_1, and L_2 must be chosen to provide stability for the error dynamics similar to the single-phase case discussed earlier in this chapter. Using the above-proposed controller, the switching frequency is first controlled in the first phase. Then, a phase shift of T/n (where T is chattering period, and n is the number of phases) between any subsequent channels is provided by proper choice of gain K resulting in harmonics cancellation (and thus chattering reduction) at the output. Fig. 20 shows simulation result for a 4-phase DC-DC buck converter with sliding mode controller as in equations (74-78). The parameters used in this simulation are $L = 0.1\mu H$, $C = 8.5\mu F$, $R = 0.01\Omega$, $E = 12\text{V}$, $c = 9.2195 \times 10^{-4}$, $L_1 = -10^4$, and $L_2 = 199.92$. The reference voltage v_{sp} is set to be 2.9814V. As evident from the simulation result, the switching frequency is maintained at $f = 1/T = 100\text{KHz}$. Also, a desired phase shift of $T/4$ between any subsequent channels is provided leading to harmonics cancellation (and thus chattering reduction) at the output.

8. Chattering reduction in multiphase DC-DC boost converters

In this section, chattering reduction by means of harmonics cancellation in multiphase DC-DC boost converter is discussed (Al-Hosani & Utkin, 2009). Consider the multiphase ($n = 4$) DC-DC boost converter shown in Fig. 21. The shown converter is modeled by the following set of differential equations:

$$\dot{i}_k = \frac{1}{L}\left(E - (1 - u_k)v_C\right) \tag{81}$$

$$\dot{v}_C = \frac{1}{C}\left(\sum_{k=1}^{n}(1 - u_K)i_k - \frac{1}{R}v_C\right) \tag{82}$$

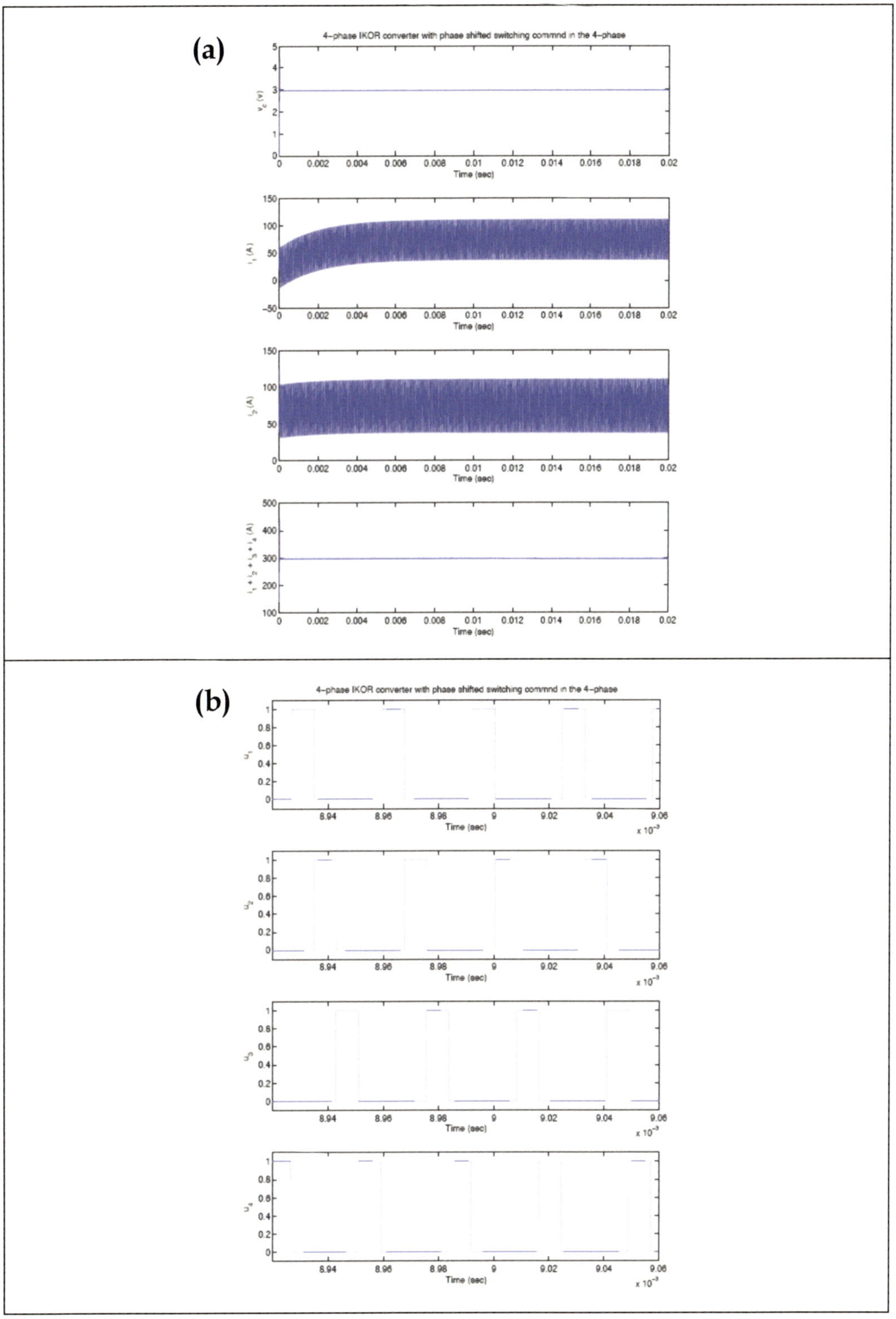
(a)
4-phase IKOR converter with phase shifted switching commnd in the 4-phase
v_c (v)
i_1 (A)
i_2 (A)
$i_1+i_2+i_3+i_4$ (A)
Time (sec)
(b)
4-phase IKOR converter with phase shifted switching commnd in the 4-phase
u_1
u_2
u_3
u_4
Time (sec)
x 10^{-3}

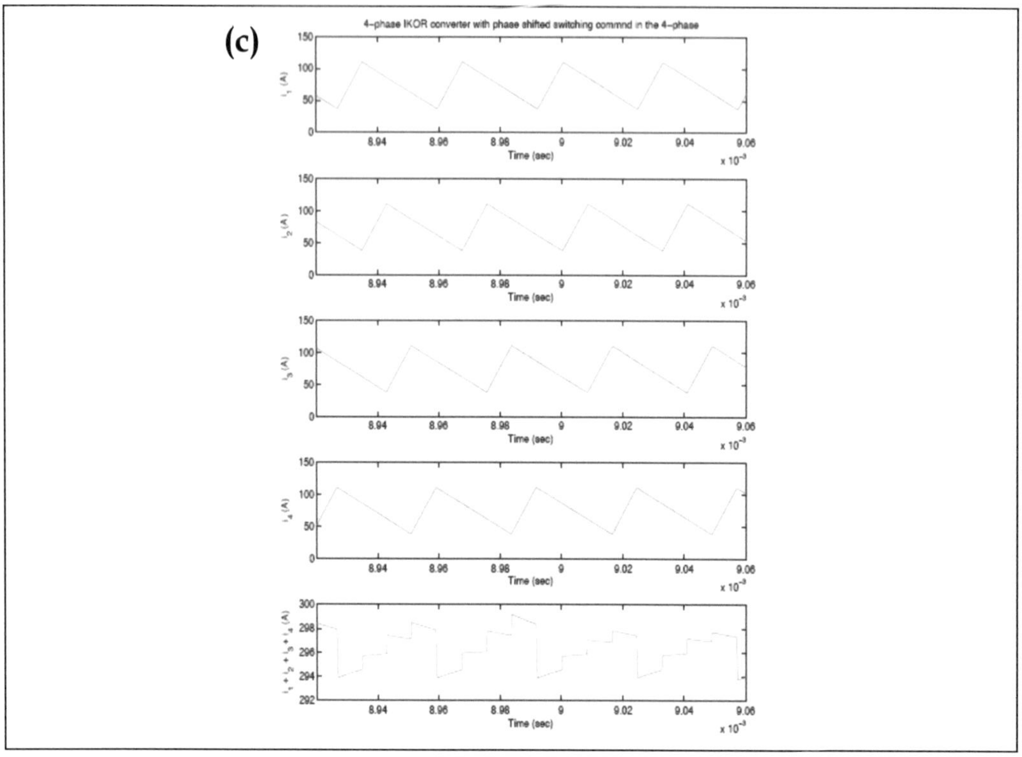

Fig. 20. Simulation of sliding mode controlled 4-phase DC-DC buck converter: (a): the top figure shows the output voltage across the resistive load/capacitor. Shown also are currents in the 1st and 2nd phases as well as current flowing through the load. (b): Switching Frequency is controlled in the first phase and a phase shift of a quarter period is provided between any two subsequent channels. (c): Zoomed in picture of the 4 currents as well as the output current going through the load. The amount of chattering is reduced at the output through harmonics cancellation provided by the phase-shifted currents.

where v_C is the voltage across the resistive load/capacitor, $i_k, k = 1,\ldots,n$ is the current flowing through kth leg, and E is the input voltage. A n-dimensional control law $u = [u_1, u_2, \ldots, u_n]^T$ is to be designed such that the output voltage v_C across the capacitor/resistive load converges to a desired known constant reference value v_{sp}. It is assumed that all currents $i_k, k = 1,\ldots,n$ and the output voltage v_C are measured. Also, the inductance L and capacitance C are assumed to be known.

The ultimate control's goal is to achieve a constant output voltage of v_{sp}. As discussed earlier in this chapter, direct control of the output voltage for boost converters results in a non-minimum phase system and therefore unstable controller. Instead, we control the output voltage indirectly by controlling the current flowing through the load to converge to a steady state value i_0 that results in a desired output voltage v_{sp}. By analyzing the steady-state behavior of the multiphase boost converter circuit, the steady state value of the sum of all individual currents in each phase is given by:

$$\sum_{k=1}^{n} i_k^{ss} = \frac{v_{sp}^2}{RE} = i_0 \tag{83}$$

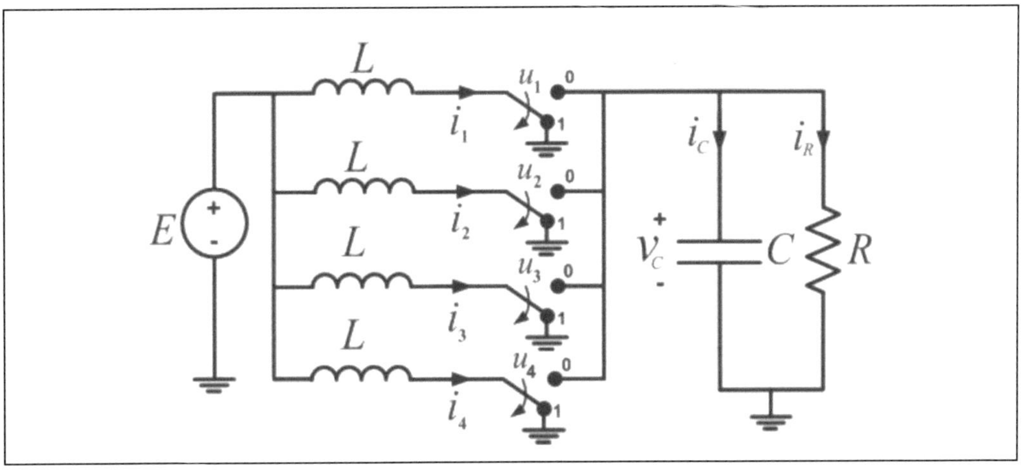

Fig. 21. 4-phase DC-DC boost converter

A sliding mode controller is designed such that each leg of the total n phases supplies an equal amount i_0 / n at steady state resulting in a total current of i_0 flowing through the resistive load at the output. To reduce chattering (through harmonics cancellation), the switching frequency $f = 1 / T$ is first controlled in the 1st phase. Then a phase shift of T / n is provided between any two subsequent phases. Assuming that the switching device is implemented with a hysteresis loop of width Δ for the first phase, and $\alpha\Delta$ for all other phases, we propose a controller with the following governing equations:

$$s_k = i_k - i_0 / n, \quad k = 1, \ldots, n \tag{84}$$

$$s_2^* = s_2 - s_1 \tag{85}$$

$$s_k^* = s_k - s_{k-1}^*, \quad k = 3, \ldots, n \tag{86}$$

The switching commands for each leg in the multiphase boost converter are given by:

$$u_1 = \frac{1}{2}(1 - \text{sign}(s_1)) \tag{87}$$

$$u_k = \frac{1}{2}\left(1 - \text{sign}\left(s_k^*\right)\right), \quad k = 2, \ldots, n \tag{88}$$

The time derivative of the switching surfaces are given by:

$$\dot{s}_1 = a - b\,\text{sign}\left(s_1\right), \quad a = E / L - v_C / 2L, \quad b = v_C / 2L \tag{89}$$

$$\dot{s}_2^* = b\,\text{sign}\left(s_1\right) - b\,\text{sign}\left(s_2^*\right) \tag{90}$$

$$\dot{s}_k^* = b\,\text{sign}\left(s_{k-1}^*\right) - b\,\text{sign}\left(s_k^*\right), \quad k = 3, \ldots, n \tag{91}$$

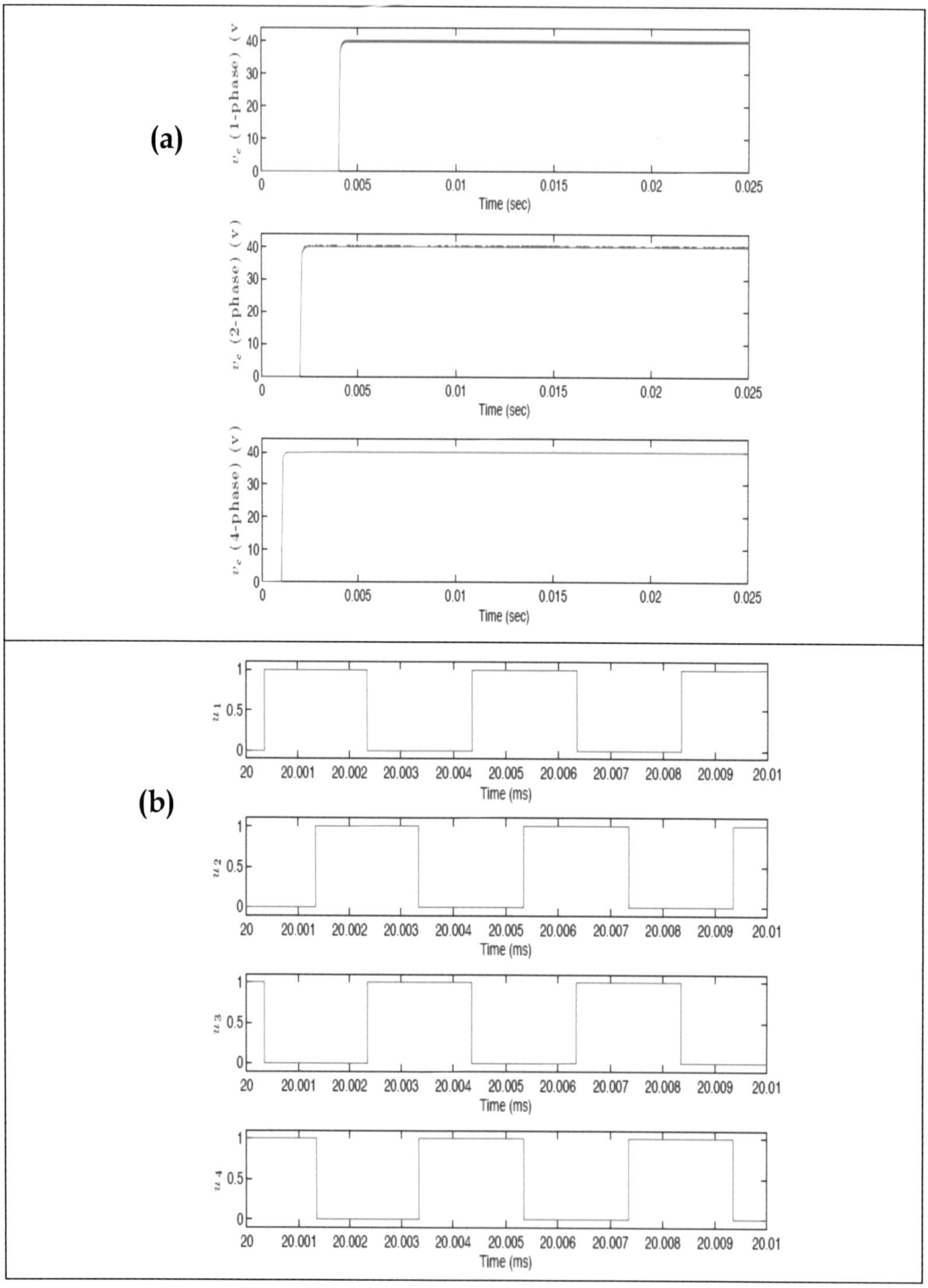

Fig. 22. Simulation of 1-phase, 2-phases and 4-phases Boost converter with $v_{sp} = 40\text{V}$. Figure (b) shows the switching command for the case of 4-phase boost converter. Clearly, a desired phase shift of one quarter chattering period is provided.

(a)

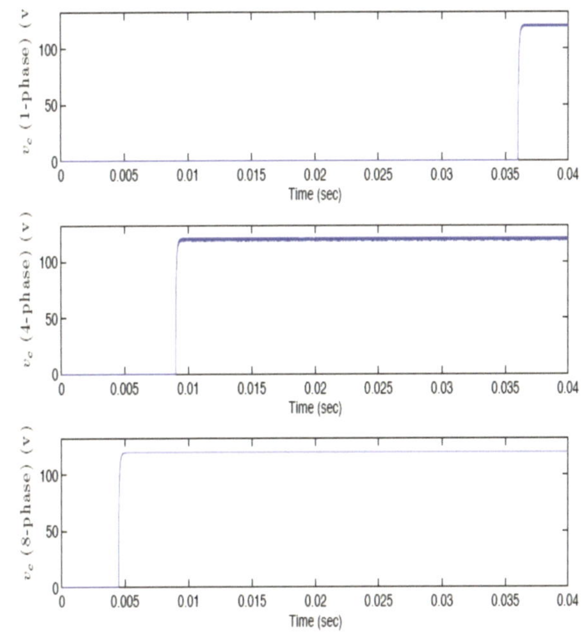

(b)

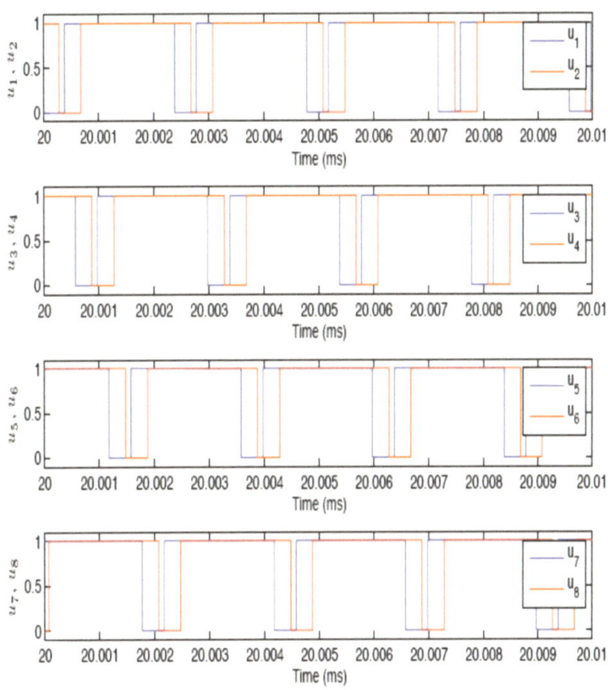

Fig. 23. Simulation of 1-phase, 4-phases and 8-phases Boost converter with $v_{sp} = 120\text{V}$. Figure (b) shows the switching command for the case of 8-phase boost converter. Clearly, a desired phase shift of one-eighth chattering period is provided.

The needed gain $K_h = 1/\alpha$ in figure 13 to implement a hysteresis loop of width $\alpha\Delta$ is calculated based on equation (66), i.e.,

$$\alpha = \frac{1}{K_h} = \begin{cases} \dfrac{4b^2}{n(b^2 - a^2)} & \text{if } \left|\dfrac{a}{b}\right| < \left(1 - \dfrac{2}{n}\right) \\ \dfrac{2b}{b + |a|} & \text{if } \left(1 - \dfrac{2}{n}\right) \le \left|\dfrac{a}{b}\right| < 1 \end{cases} \tag{92}$$

In figures 28-31, several simulations are conducted with converter's parameters:

$E = 20\text{V}$, $L = 40mH$, $C = 4\mu F$.

In the simulation in figure 28-29, the output voltage converges more rapidly to the desired set point voltage $v_{sp} = 40\text{V}$ for the case of 4-phases compared to the 2-phase and 1-phase cases. This is because of the fact that for a 4-phase power converter, only one fourth of the total current needed is tracked in each phase leg resulting in a faster convergence. It is also evident that a desired phase shift of $T/4$ is successfully provided with the switching frequency controlled to be $f = 1/T \approx 40\text{KHz}$. In simulations shown in figures 30-31, 4-phases is not enough to suppress chattering and thus eight phases is used to provide harmonics cancellation (for up to the seven harmonic) resulting in an acceptable level of chattering. The output voltage converges to the desired voltage $v_{sp} = 120\text{V}$ at a much faster rate than that for the 1-phase and 4-phases cases for the same reason mentioned earlier.

8. Conclusion

Sliding Mode Control is one of the most promising techniques in controlling power converters due to its simplicity and low sensitivity to disturbances and parameters' variations. In addition, the binary nature of sliding mode control makes it the perfect choice when dealing with modern power converters with "ON/OFF" as the only possible operation mode. In this paper, how the widely used PID controller can be easily implemented by enforcing sliding mode in the power converter. An obstacle in implementing sliding mode is the presence of finite amplitude and frequency oscillations called chattering. There are many factors causing chattering including imperfection in switching devices, the presence of unmodeled dynamics, effect of discrete time implementations, etc.

In this chapter, a method for chattering reduction based on the nature of sliding mode is presented. Following this method, frequency of chattering is first controlled to be equal to the maximum allowable value (corresponding to the maximum allowable heat loss) resulting in the minimum possible chattering level. Chattering is then reduced by providing a desired phase shift in a multiphase power converter structure that leads to harmonics elimination (and thus chattering reduction) at the output. The outlined theory is then applied in designing multiphase DC-DC buck and boost converters.

9. References

Al-Hosani, K., & Utkin, V. I. (2009). Multiphase power boost converters with sliding mode. *Multi-conference on Systems and Control* (pp. 1541-1544). Saint Petersburg, RUSSIA: IEEE.

Al-Hosani, K., Malinin, A. M., & Utkin, V. I. (2009). Sliding Mode PID Control and Estimation for Buck Converters. *International Conference on Electrical Drives and Power Electronics.* Dubrovnik, Croatia.

Al-Hosani, K., Malinin, A. M., & Utkin, V. I. (2009). Sliding mode PID control of buck converters. *European Control Conference.* Budapest, Hungary.

Bondarev, A. G., Bondarev, S. A., Kostyleva, N. E., & Utkin, V. I. (1985). Sliding Modes in Systems with Asymptotic Observer. *Automation Remote Control , 46*, 49-64.

Bose, B. K. (2006). *Power Electronics And Motor Drives: Advances and Trends* (1st Edition ed.). Academic Press.

Cortes, D., & Alvarez, J. (2002). Robust sliding mode control for the boost converter. *Power Electronics Congress, Technical Proceedings, Cooperative Education and Internship Program, VIII IEEE International* (pp. 208-212). Guadalajara, Mexico.

Emelyanov, S., Utkin, V. I., Taran, V., Kostyleva, N., Shubladze, A., Ezerov, V., et al. (1970). *Theory of Variable Structutre System.* Moscow: Nauka.

Lee, H. (2007). *PhD Thesis, Chattering Suppresion in Sliding Mode Control System.* Columbus, OH, USA: Ohio State University.

Lee, H., & Utkin, V. I. (2006). Chattering Analysis. In C. Edwards, E. C. Colet, & L. Fridman, *Advances in Variable Structure and Sliding Mode Control* (pp. 107-121). London.

Lee, H., & Utkin, V. I. (2006). The Chattering Analysis. *12th International Power Electronics and Motion Control Conference* (pp. 2014-2019). Portoroz, Slovenia: IEEE.

Lee, H., Uktin, V. I., & Malinin, A. (2009). Chattering reduction using multiphase sliding mode control . *International Journal of Control , 82* (9), 1720-1737.

Miwa, B., Wen, D., & Schecht, M. (1992). High Effciency Power Factor Correction Using Interleaving Techniques. *IEEE Applied Power Electronics Conference.* Boston, MA: IEEE.

Mohan, N., Undeland, T. M., & Robbins, W. P. (2003). *Power Electronics: Converters, Applications, and Designs* (3rd Edition ed.). John Wiley & Sons, Inc.

Nguyen, V., & Lee, C. (1995). Tracking control of buck converter using sliding-mode with adaptive hysteresis. *Power Electronics Specialists Conference, 26th Annual IEEE. 2*, pp. 1086 - 1093. Atlanta, GA: IEEE.

Sira-Ramírez, H. (2006). *Control Design Techniques in Power Electronics Devices.* Springer.

Slotine, J.-J. (1984). Sliding Controller Design for Nonlinear Systems. *International Journal of Control , 40* (2), 421-434.

Slotine, J.-J., & Sastry, S. S. (1983). Tracking Control of Nonlinear Systems using Sliding Surfaces, with Application to Robot Manipulator. *International Journal of Control , 38* (2), 465-492.

Utkin, V., Guldner, J., & Shi, J. (2009). *Sliding Mode Control in Electro-Mechanical Systems.* CRC Press, Taylor & Francis Group.

Wu, W., Lee, N.-C., & Schuellein, G. (2006). Multi-phase buck converter design with two-phase coupled inductors. *Applied Power Electronics Conference and Exposition, Twenty-First Annual IEEE* (p. 6). Seattle, WA: IEEE.

Xu, P., Wei, J., & Lee, F. (2003). Multiphase coupled-buck converter-a novel high efficient 12 V voltage regulator module. *IEEE Transactions on Power Electronics , 18* (1), 74-82.

4

The Synthetic Control of SMC and PI for Arc Welding/cutting Power Supply

Guo-Rong Zhu[1] and Yong Kang[2]
[1]*The Hong Kong Polytechnic University*
[2]*Huazhong University of Science and Technology*
China

1. Introduction

Nowadays, the manufacture of arc welding/cutting power source is mainly based on analog control in converters (Cho et al., 1996), in which component parameter flutters and performance varies with the changing of the environment and time. Owing to the fact that digital control technology is flexible, exact and reliable, it is the up-to-date method used in soft-switch arc welding/cutting power supply. The main circuit of high power arc welding/cutting power supply often uses Phase-Shift Full-Bridge (PS-FB) topology. As to PS-FB DC/DC converter circuit, there are generally three control methods: PID control, sliding mode control and fuzzy control. PID control is the most commonly used with simple algorithm, great steady-state performance and no steady-state error in the output, however, its dynamic performance is poor. Sliding mode control has excellent dynamic performance while it cannot guarantee no steady-state error in the output due to inertia that actual systems always have (He et al., 2004). Fuzzy control has good robustness, but its algorithm is complicated and its accuracy is low (Arulselvi et al., 2004). According to the basic characteristics of arc welding and cutting, from no-load to load, current is detected and it is expected that building current quickly with appropriate current overshoot to pilot arc easily and no steady-state error, PID control is more suitable for this case; from load to no-load, voltage is detected and it is expected that building voltage quickly with small voltage overshoot and it is not necessary high accuracy of voltage control, sliding mode control is more suitable for this case.

In this paper, the basic electrical characteristics and the needs of arc welding/cutting power supply, such as load current, short current and no-load voltage are analyzed. Considering to the grid voltage fluctuation, economical and personal safety, the arc welding/cutting power supply with synthetic control of Sliding Mode Control (SMC) and PI is researched and designed. Through demonstrating the external characteristic demands of welding/cutting power supply and analyzing the control algorithm, PI control is used on the current loop and SMC is introduced on the voltage loop. This method has not only effectively solved the voltage overshoot, but also realized a faster voltage resume to pilot arc again quickly. The control algorithm of phase shift full bridge indirect SMC based on the average state space model is deduced theoretically, and a direct phase shift PWM wave generation method is applied, which makes the control more practical and simpler. Some experiments on a 20 *kW* arc welding/cutting power source are conducted by digital control between the synthetic control of SMC and PI and the single PI control on TMS320LF2407. The results prove the effectiveness and robustness of the SMC and PI synthetic control.

2. The circuit topology and external characteristic

2.1 The circuit topology

In this soft-switch arc welding/cutting power supply, the Phase-Shift Full-Bridge ZVS (FB-ZVS-PWM) converter (Ruan et al., 2001) is employed. Although the volt-ampere characters and the ranges of voltage and current of arc welding machines and cutting machines are different, they both share the fundamental output characters of quickly slope voltage and invariable current; therefore, a machine with the multi-functions of arc welding and cutting can be developed. The secondary side of high frequency transformer can be shifted to output full-wave converter in arc welding, through which high current and low voltage can be obtained. It can be switched to output full-bridge converter in cutting, through which high voltage and low current can be obtained. Changing output converter mode means changing the voltage ration of the high frequency transformer, which can meet the two work situations only by shifting a switch (Zhu et al., 2007).

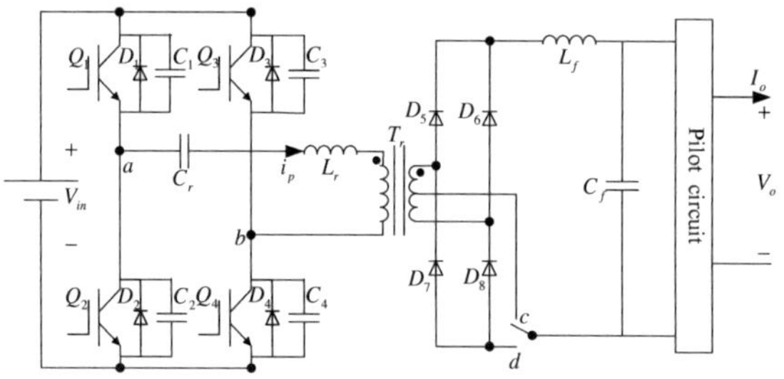

Fig. 1. The topology of the soft-switch arc welding/cutting power supply

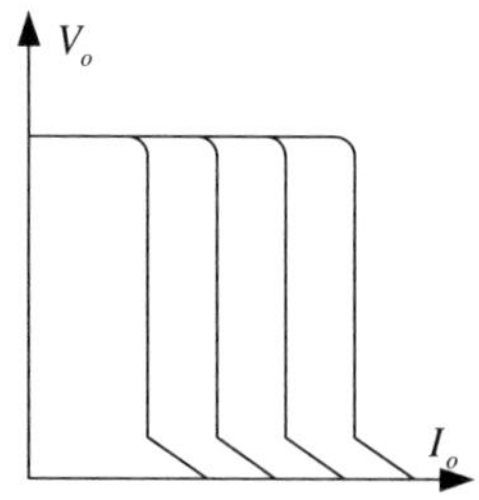

Fig. 2. Arc welding/cutting external characteristics curve

The topology of the soft-switch arc welding/cutting power supply is shown in Fig.1. In Fig.1, Q_1,D_1,C_1 and Q_2,D_2,C_2 are leading leg switches, Q_3,D_3,C_3 and Q_4,D_4,C_4 are lagging leg switches, and ZVS is realized by paralleling capacitors to the switches and resonance inductor. The topology shown in Fig.1 is mainly based on the following considerations: (1)The circuit is simple, which can realize the 4 switches ZVS without any more switches, which are advantage from the phase shift control method, and which changes the output voltage through changing phase shift angle.

(2)PWM (pulse width modulation) control strategy is adopted. Switching frequency is constant, so the designs of high frequency transformer and filtering links of input and output are easy.

In Fig.1, shifting the switch point to the c point means that the secondary side of high frequency transformer constructs output full-wave converter when working in arc welding. Shifting the switch point to the d point means that the secondary side of high frequency transformer constructs output full-bridge converter when working in cutting. A separate pilot arc circuit is in series to the main circuit in cutting, which can be started by the cutting gun.

The characteristic of phase-shift full-bridge soft switching power source is that the circuit structure is simple, compared with hard switching power source, only one resonant inductor is added which can make the four switches in the circuit work to realize ZVS.

2.2 The external characteristics of arc welding/cutting power supply

Arc welding/cutting power source has two working modes which alternate when arc welding/cutting power source is under work. One is constant-voltage control in the no-loaded mode while the other is working as constant-current source when loaded. Arc welding/cutting power source with good performance requires that the alteration between the two modes can be as fast as possible. As shown in Fig.2, if the external characteristic curve is steeper, the performance is better (Zheng et al., 2004). There, V_0 is the output voltage and I_0 is the output current.

3. Sliding mode control for arc welding/cutting power supply

Block diagram of converter system is shown in Fig.3. This paper selects two digital loop alternate control strategy (G.R.Zhu et al., 2007). Fig.3 shows that the digital system includes two control loops, one is current loop, and the other is voltage loop. Current loop samples from output current, and the sampling signal is processed by TMS320LF2407 DSP chip to get inverse feedback signal for the current digital regulator. The voltage loop samples from the output voltage, and the sampling signal is also processed by the DSP chip to get the inverse voltage feedback signal for the voltage digital regulator. The output voltage and current of the proposed converter are sensed by sensors and converted by A/D of DSP as feedback after being filtered by digital low pass filter. According to the basic characteristics of arc welding and cutting, from no-load to load, current is detected and it is expected that building current quickly with appropriate current overshoot to pilot arc easily and no steady-state error on work process, PID control is more suitable for this case; from load to no-load, voltage is detected and it is expected that building voltage quickly with small voltage overshoot, PID control has contradictory between the small overshoot and the fast response time, namely, overshoot will be increased due to fast response, which should be avoided in the voltage loop. Thus, a new control method is needed to solve voltage loop problem. Because of its good dynamic characteristic and small overshoot, sliding mode control can be applied in this field.

3.1 The fundamental principle of sliding mode Control

Sliding mode control is a control method in changing structure control system. Compared with normal control, it has a switching characteristic to change the structure of the system with time. Such characteristic can force system to make a fluctuation with small amplitude and

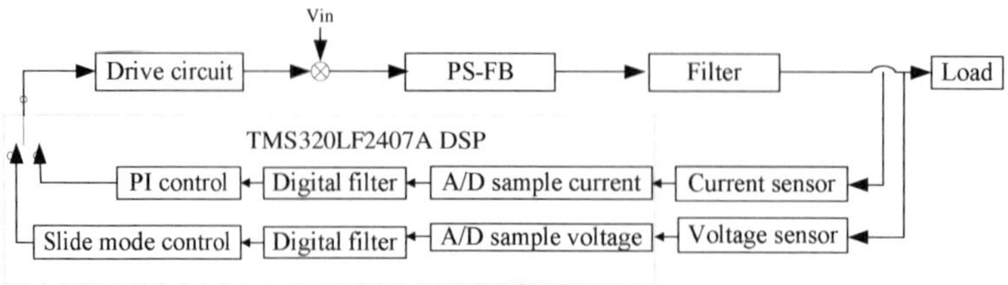

Fig. 3. Block diagram of converter system

high frequency along state track under determinate trait which can also be called as sliding mode movement. System under sliding mode has good robustness because the sliding mode can be designed and it has nothing to do with parameters and disturbances of the system. Theoretically speaking, sliding mode control has better robustness compared with normal continuous system, but it will result in fluctuation of the system due to the dis-continuousness of the switching characteristic. This is one of the main drawbacks of the sliding mode control and can't be avoiding as the switching frequency cannot be infinite. However, such effect can be ignored because high accuracy of voltage control is not required in arc welding/cutting power source.

3.2 The sliding mode digital control of phase shift full bridge

As the structure of phase-shift full-bridge main circuit is different when the switches are at different on-off state, it is suitable to use sliding mode control. Traditional sliding mode control is realized by hysteresis control, owing to the switching frequency is fixed in phase-shift full-bridge circuit, duty cycle is used for indirectly control instead of frequency directly using sliding mode control to control the switch. (Shiau et al., 1997)

When the switches Q_1 and Q_4 (or Q_2 and Q_3) in Fig.1 are switching on at the same time, its equivalent circuit is shown in Fig.4(a).

When the switches Q_1 and Q_4 (or Q_2 and Q_3) in Fig.1 are not switching on at the same time, its equivalent circuit is shown in Fig.4(b).

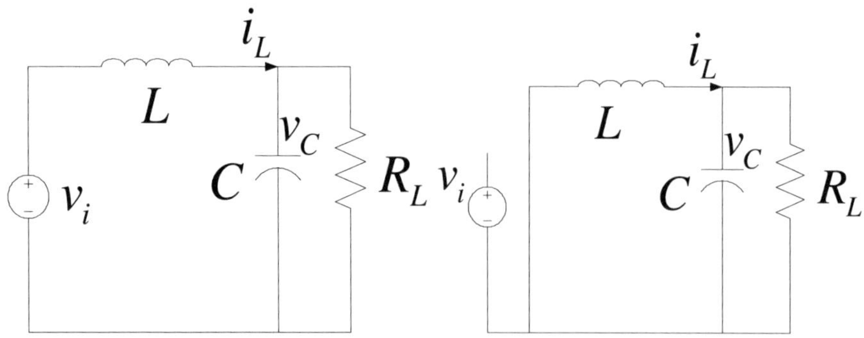

(a) Q1Q4(or Q2Q3) is switched on synchronously (b) Q1Q4(or Q2Q3) is not switched on no-synchronously

Fig. 4. The equivalent circuit of PS-FB-ZVS

Set inductance current and capacitance voltage as variables, by using state-space average method, the equation of phase-shift full-bridge is:

$$\begin{cases} \dot{i}_L = -\frac{1}{L}v_c + \frac{1}{L}dv_i \\ \dot{v}_c = \frac{1}{C}i_L - \frac{1}{R_L C}v_c \end{cases} \tag{1}$$

where d is duty cycle.
Choose voltage error as the state variable for the system:

$$e = v_c - v_{ref} \tag{2}$$

Then,

$$de/dt = \dot{v}_c = \frac{1}{C}i_L - \frac{1}{R_L C}v_c \tag{3}$$

Besides choose the switch function:

$$S = de/dt + ke \tag{4}$$

Take the Equation 2, Equation 3 into Equation 4, then can get:

$$S = \frac{1}{C}i_L - \frac{1}{R_L C}v_c + k(v_c - v_{ref}) \tag{5}$$

$$\dot{S} = \frac{1}{LC}dv_i + (\frac{k}{C} - \frac{1}{R_L C^2})i_L - (\frac{k}{R_L C} - \frac{1}{R_L^2 C^2} + \frac{1}{LC})v_c \tag{6}$$

Let

$$\dot{S} = 0 \tag{7}$$

then

$$d_{eq} = [(\frac{kL}{R_L} - \frac{L}{R_L^2 C} + 1)v_c - (kL - \frac{L}{R_L C})i_L]/v_i \tag{8}$$

Besides let $d = d_{eq} + d_n$čňto meet the requirements of sliding mode control $S \times \dot{S} < 0$, then $S \times \frac{1}{LC}d_n v_i < 0$, so:

$$d_n = a - bsgn(S) \tag{9}$$

where, sgn is the symbol function, a and b are selected by the implement systems. $R_L \to \infty$ when no-loadedčňwe can get:

$$d_{eq} = (v_c - kLi_L)/v_i \tag{10}$$

The block diagram of phase-shift full bridge SMC is shown in Fig.5.

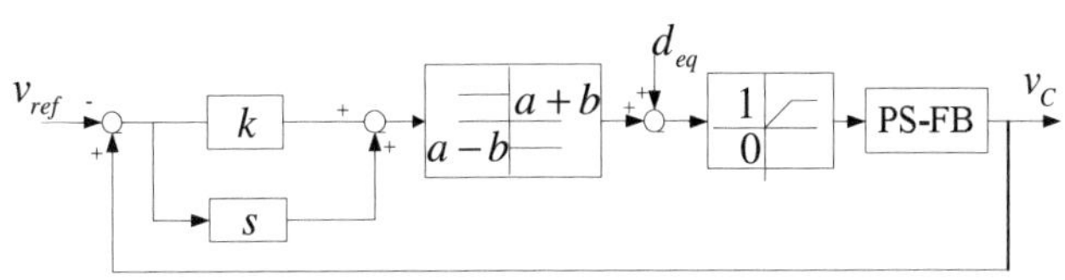

Fig. 5. Phase-shift full-bridge SMC chart

3.3 Digital control system structure and phase shift realization principle

Digital control system of the soft-switch arc welding/cutting power supply is shown in Fig.3. In this study, Digital Signal Processor (DSP) TMS320LF2407 provided by Texas Instruments is selected for implementation because of its function and simple architecture (TI et al., 2000). The features of this DSP are: A/D converter (10-bit), two event managers to generate PWM signals, 4 timer/counter (16-bit). The core of the hardware system is DSP, around which the circuits, which includes sampling circuit, protection circuit, DSP external circuit and drive circuit, are designed in detail. The output voltage and current of the proposed converter are sensed by sensors and converted by A/D of DSP as feedback after being filtered by digital low pass filter.

As to full bridge phase shift circuit, the most important problem is how to create phase shift pulse in the digital control system. A direct phase shift pulse method based on the DSP symmetric PWM waveform generation with full compare units is applied. The method is shown in Fig.6.

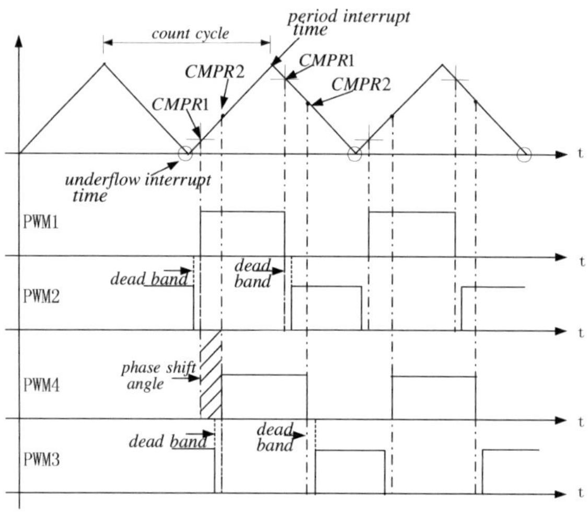

Fig. 6. Direct phase shift pulse methods with DSP full compare units

In Fig.6, the direct phase shift pulse method with DSP full compare units is that the two full compare units of the DSP Event Manager A (EVA) directly produce four PWM pulses. The fundamental theory of phase shift angle is that there is a periodic delay time from the leading leg drive to lagging leg drive. The two up/down switches drive pulses of the leading leg are produced by the full compare unit 1, and the two up/down switches drive pulses of the lagging leg are produced by the full compare unit 2. The up and down switching of each leg drive pulses are reverse and between them exists the dead band. If the given data of the leading leg register CMPR1 is fixed, the given data of phase shift angle register CMPR2 comes from full compare event, which can produce the lagging leg drive pulse. Therefore, this method can realize $0^0 - 180^0$ phase shift. The data of CMPR1 and CMPR2, which is the compare register of the two full compare units, varies in the underflow interrupt and period interrupt with the demand of the system regulator. The falling edge compare data is given in the underflow interrupt, rising edge compare data is given in the period interrupt, and the counter data is the pulse period.

In the program, the control register is set by symmetric PWM waveform generation with full compare units, Timer 1 must be put in the continuous up/down counting mode, and dead band can be set directly through Dead-Band Timer Control Register (DBTCR). In a word, the direct phase shift pulse method does not need more hardware to synthesize pulse (Kim et al., 2001), so it is very simple, flexible, convenient and reliable.

When fault comes out, such as over voltage or over current for the output, over current for the direct current bus, over voltage or under voltage for the input, overheating for the machine and etc, the peripheral hardware generates signal to lock-out the pulse amplifying circuit and the rectifier circuit, meanwhile generates PDPINTA signal to send to DSP within which PDPINTA interrupt is generated to lock-out pulse.

The phase shift PWM waves generated by the EVA module of the DSP and regulator are driven and amplified to control the power semiconductors IGBT of the high-frequency link converter. Moreover, the system can control the arc welding/cutting voltage and current by zero switching (Ben et al., 2005).

4. Experiment result

In this paper, a lab prototype of the 20W arc welding/cutting machine was built, and the specifications and designed components values are summarized in Table 1

Vin(input voltage)	DC 540V±20V
Po (output power)	20kVA
Unload voltage (arc welding)	70V
Output current (arc welding)	40A-500A (adjustable)
Unload voltage (cutting)	200V
Output current (cutting)	40A-100A (adjustable)
Switching frequency	20kHz
Controller	TMS320LF2407
Resonance inductor Lr	16uH
Leading leg parallel capacitor	8nF
Lagging leg parallel capacitor	4.7nF

Table 1. Specifications and compents used in experiment

Different switch functions are selected on different operation condition. $k = 1000$ when cutting, switch function $S = de/dt + 1000e$. It can be slided to sliding mode surface until stable output when the output voltage is 10-240V. So, $\frac{10}{330} \leq d_{eq} \leq \frac{240}{330}$, then $-\frac{1}{33} \leq d_n \leq \frac{9}{33}$, so $d_n = \frac{4}{33} - \frac{5}{33}sgn(S)$.

Similarly, $k = 600$ when arc welding, switch function $S = de/dt + 600e$. It can be slided to sliding mode surface until stable output when the output voltage is 10-80V, then, $d_n = \frac{5}{16} - \frac{6}{16}sgn(S)$.

Fig.7(a) shows the output voltage with SMC to control the no-load voltage from load to no-load mode when arc welding, while Fig.7(b) shows the wave with PI control is used from load to no-load mode when arc welding. In Fig.7(a) and Fig.7(b), we can see that sliding mode control can meet the requirement of fast voltage response and small voltage overshoot than PI control. Although PI control can also decrease the voltage overshoot by adjusting proportion factor, response time is affected, especially the regulation time increases from load to no-loaded mode.

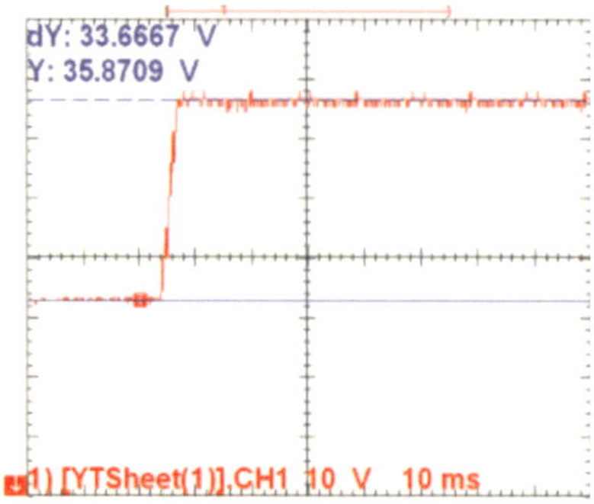

(a) SMC voltage from load to no-load on arc welding

(b) PI control voltage from load to no-load on arc welding

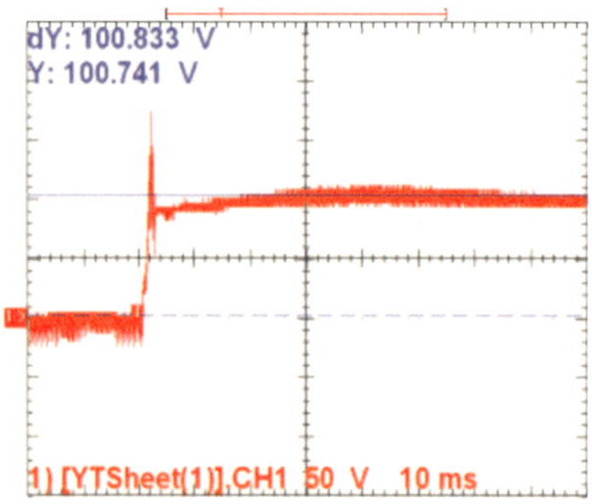

(c) Synthetic control current from no-load to load on cutting) is switched on

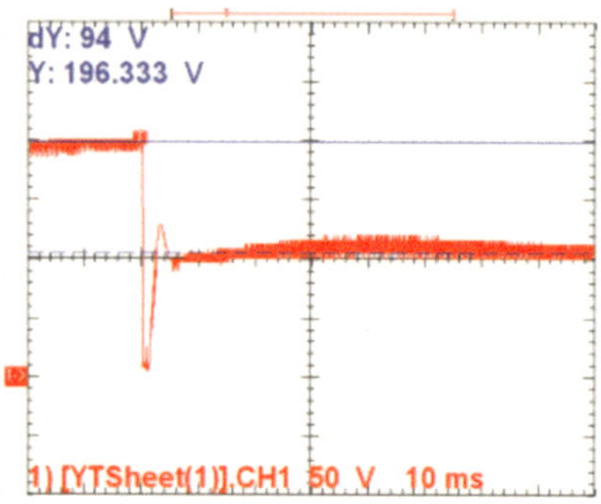

(d) Synthetic voltage from no-load to load on cutting

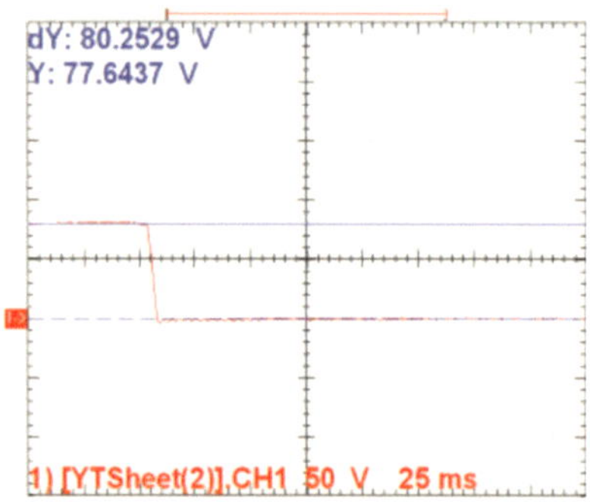

(e) Synthetic control current from load to no-load on cutting

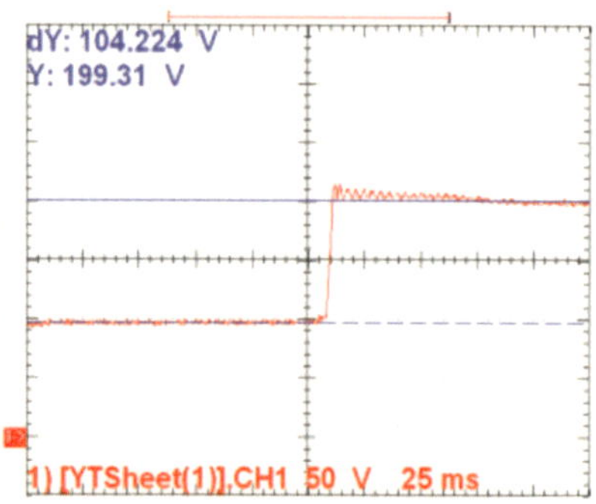

(f) Synthetic control voltage from load to no-load on cutting

Fig. 7. The experiment waves of output voltage and current when arc welding or cutting

Based on the synthetic control of SMC and PI when the machine is used to cut 25mm thick mild steel work piece, in which input voltage is 523V and output current is 100A, and no-loaded voltage is 200V, the waveforms of the output voltage and output current are shown in Fig.10. The output voltage and current waveforms in cutting process from no-load to load shown in Fig.7(c)-Fig.7(f). In Fig.7(c)(d), at first voltage loop is running when the current is zero, then current increases after pilot arc, when current is bigger than some giving value, current loop is running. There are some current overshoot with PI control in current loop, which is advantage to pilot success, and no steady-state error of current, which is advantage to steady arc welding/cutting process.In Fig.7(e)(f), the switch from load to no-load is quick and no any overshoot based on SMC control. From Fig.7(a)(f), since the scope of adjustable voltage is different, the k in sliding mode function is different, the speed of building voltage also is different, thus k is the adjustable parameter according to different system in SMC.

5. Conclusion

Applying PI control in current loop, some current overshoot can be favor for pilot success and there are not any current errors in arc welding/cutting process, which is advantage to working stability. Applying Sliding Mode Control in voltage loop for arc welding/cutting power supply can effectively decrease the overshoot of voltage loop without affecting the response time of current loop of arc welding/cutting power supply. Not any overshoot in SMC will decrease the stress of diode, which can decrease the cost of the power supply and reduce the threaten of personal safety. The Sliding Mode Control proposed in this paper has good dynamic performance and easily applied algorithm, and is very suitable for arc welding/cutting power supply which requires good dynamic performance for the control system. This paper provides a new idea to the control of arc welding/cutting power supply.

6. References

J. G. Cho, J. A. Sabate, G. C. Hua, "Zero-voltage and zero-current switching full bridge PWM converter for high-power applications," in *IEEE Transactions on Power Electronics*,vol. 11, no. 4, pp. 622-628, 1996.

Y. He and F. L. Luo, "Study of Sliding Mode Control for DC-DC Converter," in *International Conference on Power System Technology*, pp. 1969-1174, 2004.

S. Arulselvi, A. Biju and G. Uma, "Design and Simulation of Fuzzy Logic Controller for a Constant Frequency Quasi-Resonant DC-DC Converter," in *IEEE ICISIP*, PP. 472-476, 2004.

X. B. Ruan, Y. G. Yan,"The Soft Switching Technology of PWM DCčŕDC Full Bridge Converters," in *Beijing: Science Press*, 2001.

G. R. Zhu, Z. Liu, X. Li, S. X. Duan and Y. Kang, "The multi-functional arc welding cutting inverter based on PS-FB-ZVZCS," in *IEEE Conference in Industrial Electronics and Applications"*, pp.1912-1916, 2007.

Y. T. Zheng, S .S. Huang, "Arc Welding Power Source," in *Beijing: China Machine Press*, 2004.

G. R. Zhu, Z. Liu, X. Li, S. X. Duan, Y. Kang and J.Chen, "The Digital Multi-functional Arc Welding/Cutting Inverter Based on ZVS," in *International Power Engineering Conference*, IPEC2007.

L. G. Shiau, J. L. Lin, "Direct and indirect SMC control schemes for DC-DC switching converters," in *Proceedings of the 36th SICE Annual Conference*, pp. 1289-1294, 1997.

T.I.TI, "TMS320F/C240 DSP Controllers Reference Guide,"' *Texas Instruments Incorporated*, vol. 6, pp. 2-95, 2000.

E. S. Kim, T. J. Kim, Y. B. Byun, T. G. Koo,Y. H. Kim, "High power full bridge DC/DC converter using digital-to-phase-shift PWM circuit," in *Power Electronics Specialists Conference*, vol. 1, pp. 17-21, 2001.

H. Q. Ben, S. Y. Yang, S. B. Yuan, J. B. Zhao, "Application research on ZVZCS full-bridge converter technique," in *China Welding Transaction*, vol. 14,no. 2, pp. 113-116, 2005.

5

Sliding Mode Control of Fuel Cell, Supercapacitors and Batteries Hybrid Sources for Vehicle Applications

M. Y. Ayad[1], M. Becherif[2], A. Aboubou[3] and A. Henni[4]
[1]*Industrial Hybrid Vehicle Applications,*
[2]*FEMTO-FC-Labs, UTBM University,*
[3]*LMSE Laboratory, Biskra University,*
[4]*UTBM University,*
[1, 2, 4]*France*
[3]*Algeria*

1. Introduction

Fuel Cells (FC) produce electrical energy from an electrochemical reaction between a hydrogen-rich fuel gas and an oxidant (air or oxygen) (Kishinevsky & Zelingher, 2003) (Larminie & Dicks, 2000). They are high-current, and low-voltage sources. Their use in embedded systems becomes more interesting when using storage energy elements, like batteries, with high specific energy, and supercapacitors (SC), with high specific power. In embedded systems, the permanent source which can either be FC's or batteries must produce the limited permanent energy to ensure the system autonomy (Pischinger et al., 2006) (Moore et al., 2006) (Corrêa et al., 2003). In the transient phase, the storage devices produce the lacking power (to compensate for deficit in power required) in acceleration function, and absorbs excess power in braking function. FC's, and due to its auxiliaries, have a large time constant (several seconds) to respond to an increase or decrease in power output. The SCs are sized for the peak load requirements and are used for short duration load levelling events such as fuel starting, acceleration and braking (Rufer et al, 2004) (Thounthong et al., 2007). These short durations, events are experienced thousands of times throughout the life of the hybrid source, require relatively little energy but substantial power (Granovskii et al.,2006) (Benziger et al., 2006).
Three operating modes are defined in order to manage energy exchanges between the different power sources. In the first mode, the main source supplies energy to the storage device. In the second mode, the primary and secondary sources are required to supply energy to the load. In the third, the load supplies energy to the storage device.
We present in this work two hybrids DC power sources using SC as auxiliary storage device, a Proton Exchange Membrane-FC (PEMFC) as main energy source (Ayad et al., 2010) (Becherif et al., 2010) (Thounthong et al., 2007) (Rufer et al., 2004). The difference between the two structures is that the second contains a battery DC link. A single phase DC machine is connected to the DC bus and used as load. The general structures of the studied systems are presented and a dynamic model of the overall system is given in a state space

model. The control of the whole system is based on nonlinear sliding mode control for the DC-DC SCs converter and a linear regulation for the FC converter. Finally, simulation results using Matlab are given.

2. State of the art and potential application

2.1 Fuel cells

A. Principle

The developments leading to an operational FC can be traced back to the early 1800's with Sir William Grove recognized as the discoverer in 1839.

A FC is an energy conversion device that converts the chemical energy of a fuel directly into electricity. Energy is released whenever a fuel (hydrogen) reacts chemically with the oxygen of air. The reaction occurs electrochemically and the energy is released as a combination of low-voltage DC electrical energy and heat.

Types of FCs differ principally by the type of electrolyte they utilize (Fig. 1). The type of electrolyte, which is a substance that conducts ions, determines the operating temperature, which varies widely between types.

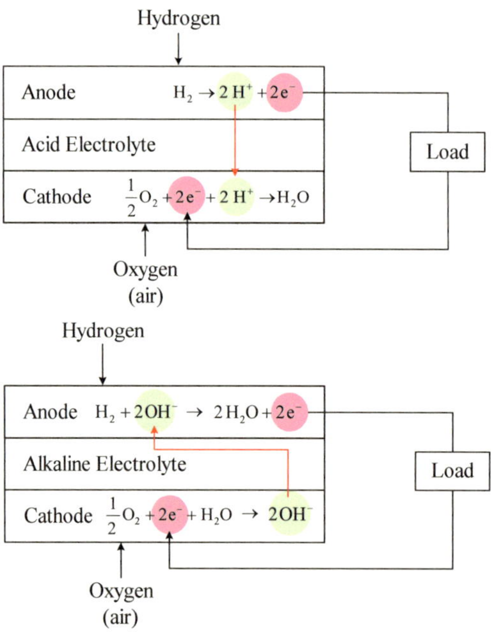

Fig. 1. Principle of acid (top) and alkaline (bottom) electrolytes fuel cells

Proton Exchange Membrane (or "solid polymer") Fuel Cells (PEMFCs) are presently the most promising type of FCs for automotive use and have been used in the majority of prototypes built to date.

The structure of a cell is represented in Fig. 2. The gases flowing along the x direction come from channels designed in the bipolar plates (thickness 1-10 mm). Vapour water is added to the gases to humidify the membrane. The diffusion layers (100-500 μm) ensure a good distribution of the gases to the reaction layers (5-50 μm). These layers constitute the electrodes of the cell made of platinum particles, which play the role of catalyst, deposited within a carbon support on the membrane.

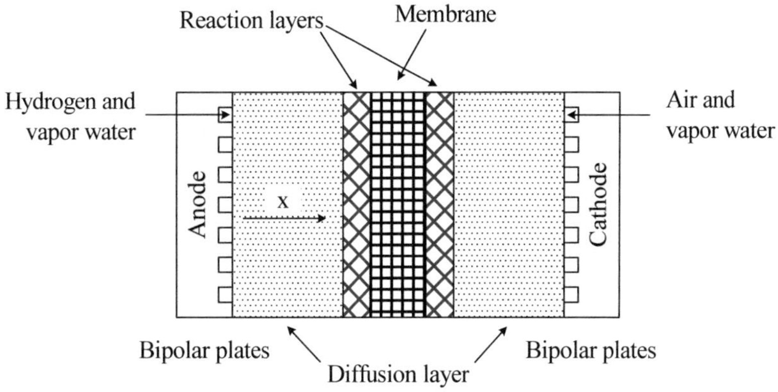

Fig. 2. Different layers of an elementary cell

Hydrogen oxidation and oxygen reduction:

$$\begin{aligned} & H_2 \rightarrow 2H^+ + 2e^- && \text{anode} \\ & 2H^+ + 2e^- + \frac{1}{2}O_2 \rightarrow H_2O && \text{cathode} \end{aligned} \tag{1}$$

The two electrodes are separated by the membrane (20-200 μm) which carries protons from the anode to the cathode and is impermeable to electrons. This flow of protons drags water molecules as a gradient of humidity leads to the diffusion of water according to the local humidity of the membrane. Water molecules can then go in both directions inside the membrane according to the side where the gases are humidified and to the current density which is directly linked to the proton flow through the membrane and to the water produced on the cathode side.

Electrons which appear on the anode side cannot cross the membrane and are used in the external circuit before returning to the cathode. Proton flow is directly linked to the current density:

$$J_{H^+} = \frac{i}{F} \tag{2}$$

where F is the Faraday's constant.

The value of the output voltage of the cell is given by Gibb's free energy ΔG and is:

$$V_{rev} = -\frac{\Delta G}{2.F} \tag{3}$$

This theoretical value is never reached, even at no load condition. For the rated current (around 0.5 A.cm^{-2}), the voltage of an elementary cell is about 0.6-0.7 V.

As the gases are supplied in excess to ensure a good operating of the cell, the non-consumed gases have to leave the FC, carrying with them the produced water.

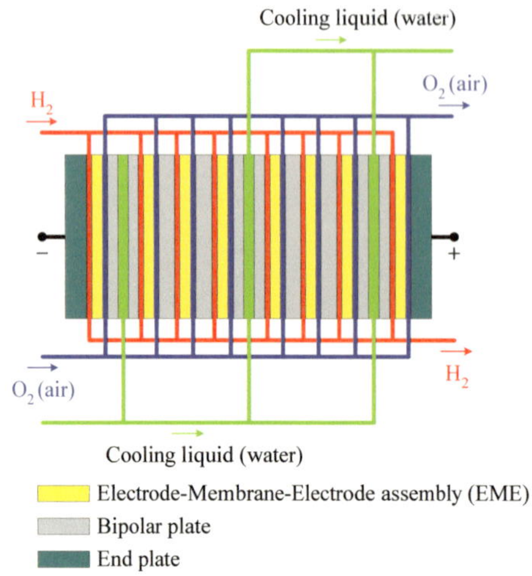

Fig. 3. External and internal connections of a PEMFC stack

Generally, a water circuit is used to impose the operating temperature of the FC (around 60-70 °C). At start up, the FC is warmed and later cooled as at the rated current nearly the same amount of energy is produced under heat form than under electrical form.

B. Modeling Fuel Cell

The output voltage of a single cell V_{FC} can be defined as the result of the following static and nonlinear expression (Larminie & Dicks, 2000):

$$V_{FC} = E - V_{act} - V_{ohm} - V_{concent} \tag{4}$$

where E is the thermodynamic potential of the cell and it represents its reversible voltage, V_{act} is the voltage drop due to the activation of the anode and of the cathode, V_{ohm} is the ohmic voltage drop, a measure of the ohmic voltage drop associated with the conduction of the protons through the solid electrolyte and electrons through the internal electronic resistances, and $V_{concent}$ represents the voltage drop resulting from the concentration or mass transportation of the reacting gases.

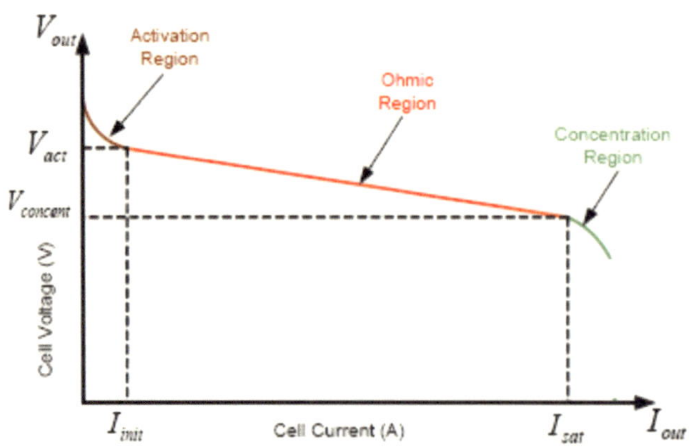

Fig. 4. A typical polarization curve for a PEMFC

In (4), the first term represents the FC open circuit voltage, while the three last terms represent reductions in this voltage to supply the useful voltage of the cell V_{FC}, for a certain operating condition. Each one of the terms can be calculated by the following equations,

$$\begin{aligned} V_{act} &= A\log\left(\frac{i_{FC}-i_n}{i_0}\right) \\ V_{ohm} &= R_m\left(i_{FC}-i_n\right) \\ V_{concent} &= b\log\left(1-\frac{i_{FC}-i_n}{i_{lim}}\right) \end{aligned} \tag{5}$$

Hence, i_{FC} is the delivered current, i_0 is the exchange current, A is the slope of the Tafel line, i_{Lim} is the limiting current, B is the constant in the mass transfer, i_n is the internal current and R_m is the membrane and contact resistances.

2.2 Electric double-layer supercapacitors

A. Principle

The basic principle of electric double-layer capacitors lies in capacitive properties of the interface between a solid electronic conductor and a liquid ionic conductor. These properties discovered by Helmholtz in 1853 lead to the possibility to store energy at solid/liquid interface. This effect is called electric double-layer, and its thickness is limited to some nanometers (Belhachemi et al., 2000).
Energy storage is of electrostatic origin, and not of electrochemical origin as in the case of accumulators. So, supercapacitors are therefore capacities, for most of marketed devices. This gives them a potentially high specific power, which is typically only one order of magnitude lower than that of classical electrolytic capacitors.

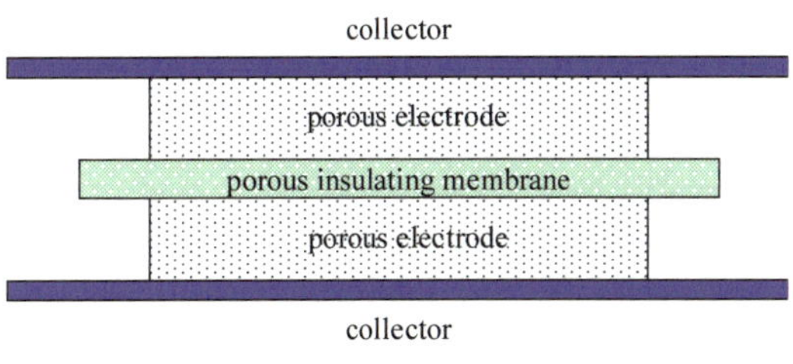

Fig. 5. Principle of assembly of the supercapacitors

In SCs, the dielectric function is performed by the electric double-layer, which is constituted of solvent molecules. They are different from the classical electrolytic capacitors mainly because they have a high surface capacitance (10-30 $\mu F.cm^{-2}$) and a low rated voltage limited by solvent decomposition (2.5 V for organic solvent). Therefore, to take advantage of electric double-layer potentialities, it is necessary to increase the contact surface area between electrode and electrolyte, without increasing the total volume of the whole.
The most widespread technology is based on activated carbons to obtain porous electrodes with high specific surface areas (1000-3000 $m^2.g^{-1}$). This allows obtaining several hundred of farads by using an elementary cell.
SCs are then constituted, as schematically presented below in Fig. 5, of:

- two porous carbon electrodes impregnated with electrolyte,
- a porous insulating membrane, ensuring electronic insulation and ionic conduction between electrodes,
- metallic collectors, usually in aluminium.

B. Modeling and sizing of suparcpacitors

Many applications require that capacitors be connected together, in series and/or parallel combinations, to form a "bank" with a specific voltage and capacitance rating. The most critical parameter for all capacitors is voltage rating. So they must be protected from over voltage conditions. The realities of manufacturing result in minor variations from cell to cell. Variations in capacitance and leakage current, both on initial manufacture and over the life of the product, affect the voltage distribution. Capacitance variations affect the voltage distribution during cycling, and voltage distribution during sustained operation at a fixed voltage is influenced by leakage current variations. For this reason, an active voltage balancing circuit is employed to regulate the cell voltage.

It is common to choose a specific voltage and thus calculating the required capacitance. In analyzing any application, one first needs to determine the following system variables affecting the choice of SC,

- the maximum voltage, V_{SCMAX}
- the working (nominal) voltage, V_{SCNOM}
- the minimum allowable voltage, V_{SCMIN}
- the current requirement, I_{SC}, or the power requirement, P_{SC}
- the time of discharge, t_d
- the time constant
- the capacitance per cell, C_{SCcell}
- the cell voltage, V_{SCcell}
- the number of cell needs, n

To predict the behavior of SC voltage and current during transient state, physics-based dynamic models (a very complex charge/discharge characteristic having multiple time constants) are needed to account for the time constant due to the double-layer effects in SC. The reduced order model for a SC cell is represented in Fig. 6. It is comprised of four ideal circuit elements: a capacitor C_{SCcell}, a series resistor R_S called the equivalent series resistance (ESR), a parallel resistor R_P and a series stray inductor L of ~nH. The parallel resistor R_P models the leakage current found in all capacitors.

This leakage current varies starting from a few milliamps in a big SC under a constant current as shown in Fig. 7.

A constant discharging current is particularly useful when determining the parameters of the SC.

Nevertheless, Fig. 7 should not be used to consider sizing SCs for constant power applications, such as common power profile used in hybrid source.

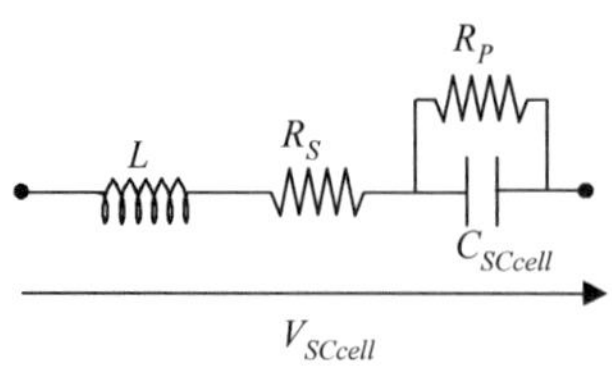

Fig. 6. Simple model of a supercapacitor cell

To estimate the minimum capacitance C_{SCMIN}, one can write an energy equation without losses (R_{ESR} neglected) as,

$$\frac{1}{2}C_{SCMIN}\left(V_{SCNOM}^2 - V_{SCMIN}^2\right) = P_{SC}t \tag{6}$$

with

$$P_{SC}(t) = V_{SC}(t)i_{SC}(t) \tag{7}$$

Then,

$$C_{SCMIN} = \frac{2P_{SC}t_d}{V_{SCNOM}^2 - V_{SCMIN}^2} \tag{8}$$

From (6) and (7), the instantaneous capacitor voltage and current are described as,

$$\begin{cases} V_{SC}(t) = V_{SCNOM}\sqrt{1-\left(1-\left(\dfrac{V_{SCNOM}}{V_{SCMIN}}\right)^2\right)\dfrac{t}{t_d}} \\ i_{SC}(t) = \dfrac{P_{SC}}{\sqrt{1-\left(1-\left(\dfrac{V_{SCNOM}}{V_{SCMIN}}\right)^2\right)\dfrac{t}{t_d}}} \end{cases} \tag{9}$$

Since the power being delivered is constant, the minimum voltage and maximum current can be determined based on the current conducting capabilities of the SC. (6) and (7) can then be rewritten as,

$$\begin{cases} V_{SCMIN} = \sqrt{V_{SCNOM}^2 - \dfrac{2P_{SC}t_d}{C_{MIN}}} \\ I_{SCMIN} = \dfrac{P_{SC}}{\sqrt{V_{SCNOM}^2 - \dfrac{2P_{SC}t_d}{C_{MIN}}}} \end{cases} \tag{10}$$

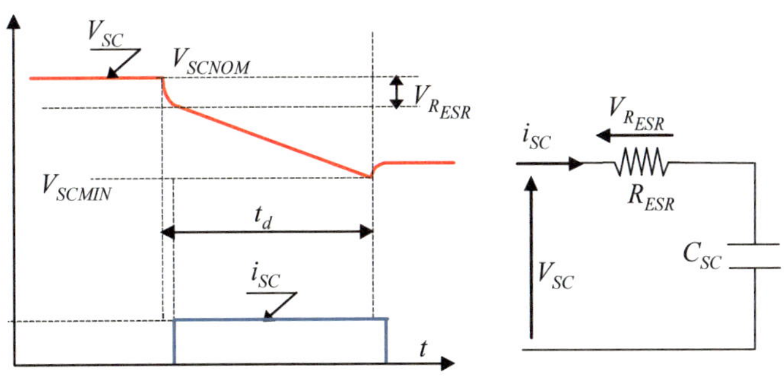

Fig. 7. Discharge profile for a SC under constant current.

The variables V_{SCMAX} and C_{SC} are indeed related by the number of cells n. The assumption is that the capacitors will never be charged above the combined maximum voltage rating of all the cells. Thus, we can introduce this relationship with the following equations,

$$\begin{cases} V_{SCMAX} = nV_{SCcell} \\ C_{SC} = \dfrac{C_{SCcell}}{n} \end{cases} \tag{11}$$

Generally, V_{SCMIN} is chosen as $V_{SCMAX}/2$, from (6), resulting in 75% of the energy being utilized from the full-of-charge (SOC[1] = 100%). In applications where high currents are drawn, the effect of the R_{ESR} has to be taken into account. The energy dissipated W_{loss} in the R_{ESR}, as well as in the cabling, and connectors could result in an under-sizing of the number of capacitors required. For this reason, knowing SC current from (6), one can theoretically calculate these losses as,

$$W_{loss} = \int_0^{t_d} i_C^2(\tau) R_{ESR} d\tau = P_{SC} R_{ESR} C_{MIN} \ln\left(\frac{V_{SCNOM}}{V_{SCMIN}}\right) \tag{12}$$

To calculate the required capacitance C_{SC}, one can rewrite (6) as,

$$\frac{1}{2} C_{SCMIN}\left(V_{SCMAX}^2 - V_{SCMIN}^2\right) = P_{SC} t + W_{loss} \tag{13}$$

From (6) and (13), one obtains

$$\begin{cases} C_{SC} = C_{SCMIN}(1+X) \\ X = \dfrac{W_{loss}}{P_{SC} t} \end{cases} \tag{14}$$

where X is the energy ratio.
From the equations above, an iterative method is needed in order to get the desired optimum value.
The differential capacitance can be represented by two capacitors: a constant capacitor C_0 and a linear voltage dependent capacitor kV_0. k is a constant corresponding to the slope voltage. The SC is then modelled by:

$$\begin{cases} \dfrac{dV_0}{dt} = \dfrac{1}{C_0 + kV_0} i_{SC} \\ V_{SC} = R_{RSR} i_{SC} + V_0 \end{cases} \tag{15}$$

Where $C_0 + kV_0 > 0$

C. State of the art and potential application

Developed at the end of the seventies for signal applications (for memory back-up for example), SCs had at that time a capacitance of some farads and a specific energy of about 0.5 $Wh.kg^{-1}$.

---

[1] State Of Charge

High power SCs appear during the nineties and bring high power applications components with capacitance of thousand of farads and specific energy and power of several $Wh.kg^{-1}$ and $kW.kg^{-1}$.
In the energy-power plan, electric double layers SCs are situated between accumulators and traditional capacitors.
Then these components can carry out two main functions:

- the function "source of energy", where SCs replace electrochemical accumulators, the main interest being an increase in reliability,
- the function "source of power", for which SCs come in complement with accumulators (or any other source limited in power), for a decrease in volume and weight of the whole system.

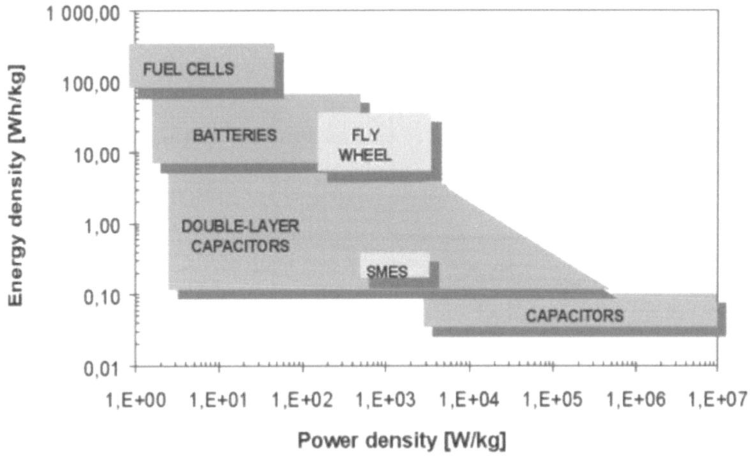

Fig. 8. Comparison between capacitors, supercapacitors, batteries and Fuel cell

2.3 State of the art of battery in electric vehicles

An electric vehicle (EV) is a vehicle that runs on electricity, unlike the conventional vehicles on road today which are major consumers of fossil fuels like gasoline. This electricity can be either produced outside the vehicle and stored in a battery or produced on board with the help of FC's.
The development of EV's started as early as 1830's when the first electric carriage was invented by Robert Andersen of Scotland, which appears to be appalling, as it even precedes the invention of the internal combustion engine (ICE) based on gasoline or diesel which is prevalent today. The development of EV's was discontinued as they were not very convenient and efficient to use as they were very heavy and took a long time to recharge.
This led to the development of gasoline based vehicles as the one pound of gasoline gave equal energy as a hundred pounds of batteries and it was relatively much easier to refuel and use gazoline. However, we today face a rapid depletion of fossil fuel and a major concern over the noxious green house gases their combustion releases into the atmosphere causing long term global crisis like climatic changes and global warming. These concerns are shifting the focus back to development of automotive vehicles which use alternative fuels for operations. The development of such vehicles has become imperative not only for the scientists but also for the governments around the globe as can be substantiated by the Kyoto Protocol which has a total of 183 countries ratifying it (As on January 2009).

A. Batteries technologies

A battery is a device which converts chemical energy directly into electricity. It is an electrochemical galvanic cell or a combination of such cells which is capable of storing chemical energy. The first battery was invented by Alessandro Volta in the form of a voltaic pile in the 1800's. Batteries can be classified as primary batteries, which once used, cannot be recharged again, and secondary batteries, which can be subjected to repeated use as they are capable of recharging by providing external electric current. Secondary batteries are more desirable for the use in vehicles, and in particular traction batteries are most commonly used by EV manufacturers. Traction batteries include Lead Acid type, Nickel and Cadmium, Lithium ion/polymer , Sodium and Nickel Chloride, Nickel and Zinc.

	Lead Acid	Ni - Cd	Ni - MH	Li – Ion	Li - polymer	Na - $NiCl_2$	Objectives
Specific Energy (Wh/Kg)	35 – 40	55	70 – 90	125	155	80	200
Specific Power (W/Kg)	80	120	200	260	315	145	400
Energy Density (Wh/m^3)	25 – 35	90	90	200	165	130	300
Cycle Life (No. of charging cycles)	300	1000	600	+ 600	+ 600	600	1000

Table 1. Comparison between different baterries technologies.

The battery for electrical vehicles should ideally provide a high autonomy (i.e. the distance covered by the vehicle for one complete discharge of the battery starting from its potential) to the vehicle and have a high specific energy, specific power and energy density (i.e. light weight, compact and capable of storing and supplying high amounts of energy and power respectively). These batteries should also have a long life cycle (i.e. they should be able to discharge to as near as it can be to being empty and recharge to full potential as many number of times as possible) without showing any significant deterioration in the performance and should recharge in minimum possible time. They should be able to operate over a considerable range of temperature and should be safe to handle, recyclable with low costs. Some of the commonly used batteries and their properties are summarized in the Table 1.

B. Principle

A battery consists of one or more voltaic cell, each voltaic cell consists of two half-cells which are connected in series by a conductive electrolyte containing anions (negatively charged ions) and cations (positively charged ions). Each half-cell includes the electrolyte and an electrode (anode or cathode). The electrode to which the anions migrate is called the anode and the electrode to which cations migrate is called the cathode. The electrolyte connecting these electrodes can be either a liquid or a solid allowing the mobility of ions.

In the redox reaction that powers the battery, reduction (addition of electrons) occurs to cations at the cathode, while oxidation (removal of electrons) occurs to anions at the anode. Many cells use two half-cells with different electrolytes. In that case each half-cell is enclosed in a container, and a separator that is porous to ions but not the bulk of the electrolytes prevents mixing. The figure 10 shows the structure of the structure of Lithium-Ion battery using a separator to differentiate between compartments of the same cell utilizing two respectively different electrolytes

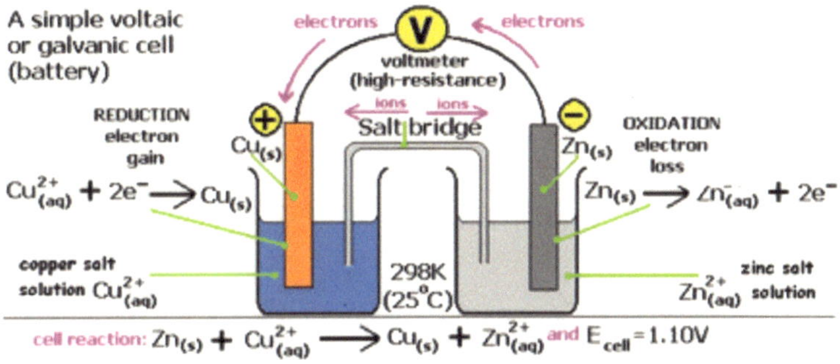

Fig. 9. Showing the apparatus and reactions for a simple galvanic Electrochemical Cell

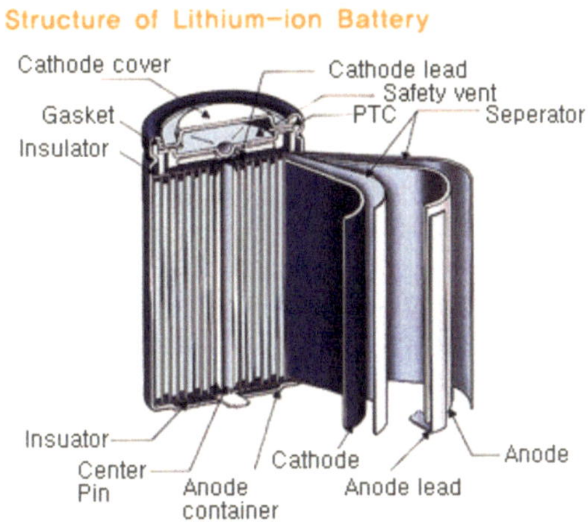

Fig. 10. Structure of Lithium-Ion Battery

Each half cell has an electromotive force (or emf), determined by its ability to drive electric current from the interior to the exterior of the cell. The net emf of the battery is the difference between the emfs of its half-cells. Thus, if the electrodes have emfs E_1 and E_2, then the net emf is $E_{cell} = E_2 - E_1$. Therefore, the net emf is the difference between the reduction potentials of the half-cell reactions.

The electrical driving force or ΔV_{Bat} across the terminals of a battery is known as the terminal voltage and is measured in volts. The terminal voltage of a battery that is neither charging nor discharging is called the open circuit voltage and equals the emf of the battery.

An ideal battery has negligible internal resistance, so it would maintain a constant terminal voltage until exhausted, then dropping to zero. If such a battery maintained 1.5 volts and stored a charge of one Coulomb then on complete discharge it would perform 1.5 Joule of work.

$$\text{Work done by battery (W)} = -\ \text{Charge X Potential Difference} \tag{16}$$

$$\text{Charge} = \frac{\text{Coulomb}}{\text{Mole} \quad \text{Electrons}} \text{Moles} \quad \text{Electrons} \tag{17}$$

$$W = -nFEcell \tag{18}$$

Where $\boldsymbol{n}$ is the number of moles of electrons taking part in redox, $\boldsymbol{F = 96485}$ coulomb/mole is the Faraday's constant i.e. the charge carried by one mole of electrons.
The open circuit voltage, $\boldsymbol{E_{cell}}$ can be assumed to be equal to the maximum voltage that can be maintained across the battery terminals. This leads us to equating this work done to the Gibb's free energy of the system (which is the maximum work that can be done by the system)

$$\Delta G = W\max = -nFEcell \tag{19}$$

C. Model of battery

Non Idealities in Batteries: Electrochemical batteries are of great importance in many electrical systems because the chemical energy stored inside them can be converted into electrical energy and delivered to electrical systems, whenever and wherever energy is needed. A battery cell is characterized by the open-circuit potential ($\boldsymbol{V_{OC}}$), i.e. the initial potential of a fully charged cell under no-load conditions, and the cut-off potential ($\boldsymbol{V_{cut}}$) at which the cell is considered discharged. The electrical current obtained from a cell results from electrochemical reactions occurring at the electrode-electrolyte interface. There are two important effects which make battery performance more sensitive to the discharge profile:

- Rate Capacity Effect: At zero current, the concentration of active species in the cell is uniform at the electrode-electrolyte interface. As the current density increases the concentration deviates from the concentration exhibited at zero current and state of charge as well as voltage decrease (Rao et al., 2005)
- Recovery Effect: If the cell is allowed to relax intermittently while discharging, the voltage gets replenished due to the diffusion of active species thereby giving it more life (Rao et al., 2005)

D. Equivalent electrical circuit of battery

Many electrical equivalent circuits of battery are found in literature. (Chen at al., 2006) presents an overview of some much utilized circuits to model the steady and transient behavior of a battery. The Thevenin's circuit is one of the most basic circuits used to study the transient behavior of battery is shown in figure 11.

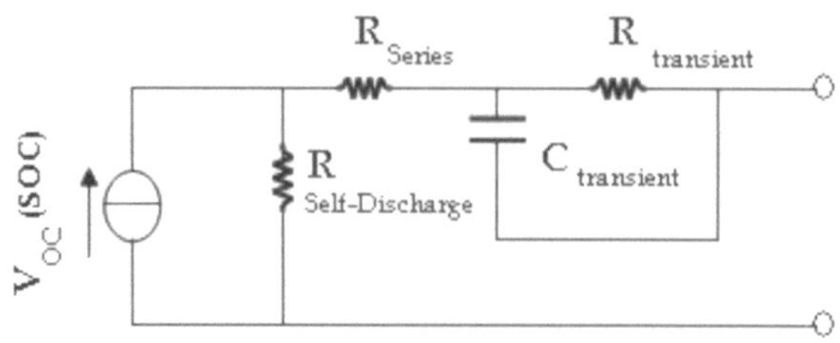

Fig. 11. Thevenin's model

It uses a series resistor (R_{series}) and an RC parallel network ($R_{transient}$ and $C_{transient}$) to predict the response of the battery to transient load events at a particular state of charge by assuming a constant open circuit voltage [V_{oc}(SOC)] is maintained. This assumption unfortunately does not help us analyze the steady-state as well as runtime variations in the battery voltage. The improvements in this model are done by adding more components in this circuit to predict the steady-state and runtime response. For example, (Salameh at al., 1992) uses a variable capacitor instead of V_{oc} (SOC) to represent nonlinear open circuit voltage and SOC, which complicates the capacitor parameter.

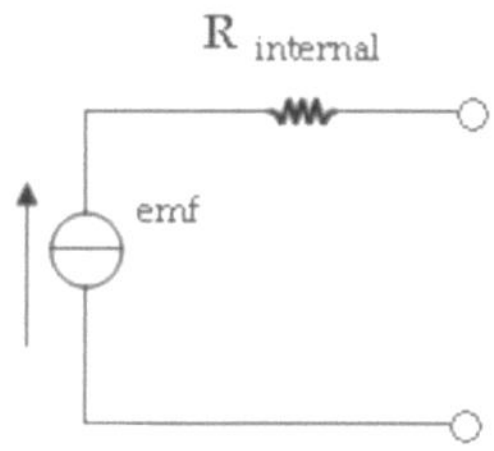

Fig. 12. Circuit showing battery emf and internal resistance $R_{internal}$

However, in our study we are mainly concerned with the recharging of this battery which occurs while breaking. The SC coupled with the battery accumulates high amount of charge when breaks are applied and this charge is then utilized to recharge the battery. Therefore, the design of the battery is kept to a simple linear model which takes into account the internal resistance (**R**$_{internal}$) of the battery and assumes the emf to be constant throughout the process (Figure. 12).

3. Control of the hybrid sources based on FC, SCs and batteries

3.1 Structures of the hybrid power sources

As shown in Fig. 13, the first hybrid source comprises a DC link supplied by a PEMFC and an irreversible DC-DC converter which maintains the DC voltage V_{DL} to its reference value, and a supercapacitive storage device, which is connected to the DC link through a current reversible DC-DC converter allowing recovering or supplying energy through SC.

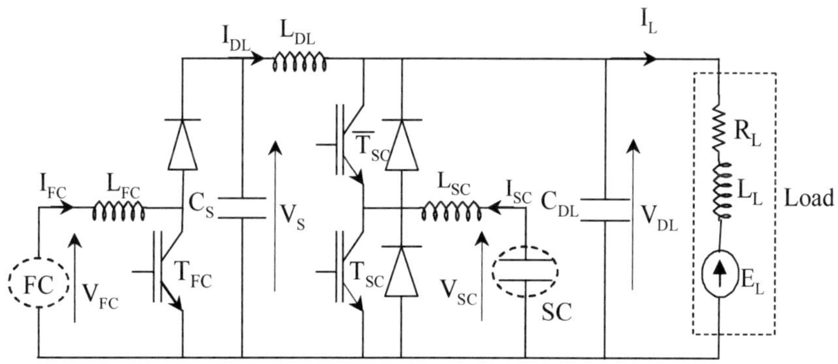

Fig. 13. Structure of the first hybrid source

The second system, shown in Fig. 14, comprises of a DC link directly supplied by batteries, a PEMFC connected to the DC link by means of boost converter, and a supercapacitive storage device connected to the DC link through a reversible current DC-DC converter. The role of FC and the batteries is to supply mean power to the load, whereas the storage device is used as a power source: it manages load power peaks during acceleration and braking.

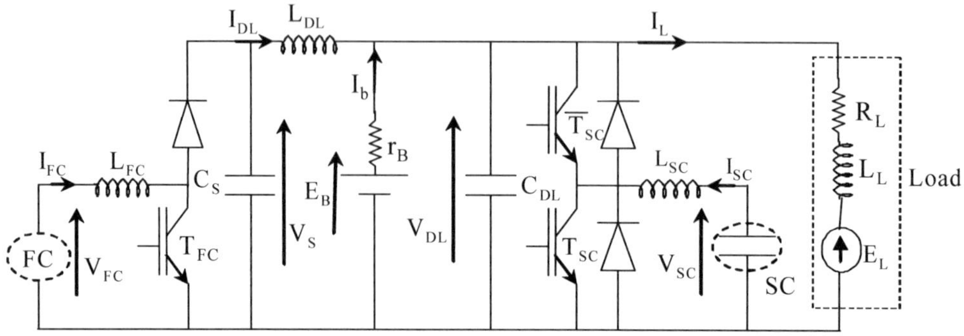

Fig. 14. Structure of the second hybrid source

3.2 Problem formulation

Both structures are supplying energy to the DC bus where a DC machine is connected. This machine plays the role of the load acting as a motor or as a generator when breaking.
The main purpose of the study is to present a control technique for the two hybrid source with two approaches. Two control strategies, based on sliding mode control have been considered, the first using a voltage controller and the second using a current controller. The second aim is to maintain a constant mean energy delivered by the FC, without a significant power peak, and the transient power is supplied by the SCs. A third purpose consists in recovering energy throw the charge of the SC.
After system modeling, equilibrium points are calculated in order to ensure the desired behavior of the system. When steady state is reached, the load has to be supplied only by the FC source. So the controller has to maintain the DC bus voltage to a constant value and the SCs current has to be cancelled. During transient, the power delivered by the DC source has to be as constant as possible (without a significant power peak), and the transient power has to be delivered through the SCs. The SCs in turn, recover their energy during regenerative braking when the load provides current.
At equilibrium, the SC has to be charged and then the current has to be equal to zero.

3.3 Sliding mode control of the hybrid sources

Due to the weak request on the FC, a classical PI controller has been adapted for the boost converter. However, because of the fast response in the transient power and the possibility of working with a constant or variable frequency, a sliding mode control (Ayad et al., 2007). has been chosen for the DC-DC bidirectional SC converter. The bidirectional property allows the management of charge- discharge cycles of the SC tank.
The current supplied by the FC is limited to a range [I_{MIN}, I_{MAX}]. Within this interval, the FC boost converter ensures current regulation (with respect to reference). Outside this interval, i.e. when the desired current is above I_{MAX} or below I_{MIN}, the boost converter saturates and the surge current is then provided or absorbed by the storage device. Hence the DC link current is kept equal to its reference level. Thus, three modes can be defined to optimize the functioning of the hybrid source:

- ***The normal mode***, for which load current is within the interval [I_{MIN}, I_{MAX}]. In this mode, the FC boost converter ensures the regulation of the DC link current, and the control of the bidirectional SC converter leads to the charge or the discharge of SC up to a reference voltage level V_{SCREF},
- ***The discharge mode***, for which load power is greater than I_{MAX}. The current reference of the boost is then saturated to I_{MAX}, and the FC DC-DC converter ensures the regulation of the DC link voltage by supplying the lacking current, through SC discharge,
- ***The recovery mode***, for which load power is lower than I_{MIN}. The power reference of the FC boost converter is then saturated to I_{MIN}, and the FC DC-DC converter ensures the regulation of the DC link current by absorbing the excess current, through SC charge.

A. DC-DC boost FC converter control principle

Fig. 15 presents the synoptic control of the first hybrid FC boost. The FC current reference is generated by means of a PI voltage loop control on a DC link voltage and its reference:

$$I_{FC}^{*} = k_{PF1}\left(V_{DL}^{*} - V_{DL}\right) + k_{IF1}\int_{0}^{t}\left(V_{DL}^{*} - V_{DL}\right)dt \tag{20}$$

With, k_{PF1} and k_{IF1} are the proportional and integral gains.

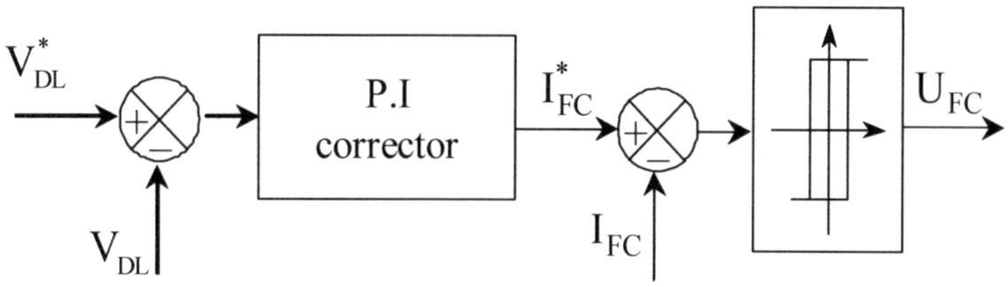

Fig. 15. Control of the FC converter

The second hybrid source FC current reference I_{FC}^{*} is generated by means of a PI current loop control on a DC link current and load current and the switching device is controller by a hysteresis comparator:

$$I_{FC}^{*} = k_{PF2}\left(I_{L} - I_{DL}\right) + k_{IF2}\int_{0}^{t}\left(I_{L} - I_{DL}\right)dt \tag{21}$$

With, k_{PF2} and k_{IF2} are the proportional and integral gains.

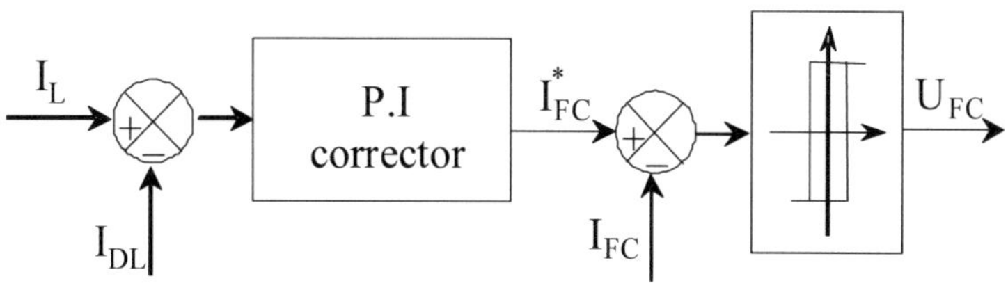

Fig. 16. Control of the FC converter

The switching device is controlled by a hysteresis comparator.

B. DC-DC Supercapacitors converter control principle

To ensure proper functioning for the three modes, we have used a sliding mode control strategy for the DC-DC converter. Here, we define a sliding surface S, for the first hybrid source, as a function of the DC link voltage V_{DL}, its reference V_{DL}^*, the SCs voltage V_{SC}, its reference V_{SC}^*, and the SCs current I_{SC}:

$$S_1 = k_{11}\left(V_{DL} - V_{DL}^*\right) + k_{21} \cdot \left(I_{SC} - I_1\right) \tag{22}$$

with

$$I = k_{ps1}\left(V_{SC} - V_{SC}^*\right) + k_{is1}\int_0^t \left(V_{SC} - V_{SC}^*\right)dt \tag{23}$$

With, k_{ps1} and k_{is1} are the proportional and integral gains.
The FC PI controller ensures that V_{DL} tracks V_{DL}^*. The SC PI controller ensures that V_{SC} tracks its reference V_{SC}^*.
k_{11}, k_{21} are the coefficients of proportionality, which ensure that the sliding surface equal zero by tracking the SC currents to its reference I when the FC controller can't ensures that V_{DL} tracks V_{DL}^*.
In steady state condition, the FC converter ensures that the first term of the sliding surface is null, and the integral term of equation (23) implies $V_{SC} = V_{SC}^*$. Then, imposing $S_1 = 0$ leads to $I_{SC} = 0$, as far as the boost converter output current I_{DL} is not limited. So that, the storage element supplies energy only during power transient and I_{DL} limitation.
For the second hybrid source, we define a sliding surface S_2 as a function of the DC link current I_{DL}, The load current I_L, the SC voltage V_{SC}, its reference V_{SC}^*, and the SC current I_{SC}:

$$S_2 = k_{12}\left(I_{DL} - I_L\right) + k_{22}\left(I_{SC} - I_2\right) \tag{24}$$

with

$$I_2 = k_{ps2}\left(V_{SC} - V_{SC}^*\right) + k_{is2}\int_0^t \left(V_{SC} - V_{SC}^*\right)dt \tag{25}$$

With, k_{ps2} and k_{is2} are the proportional and integral gains.
The FC PI controller ensures that I_{DL} tracks I_L. The SC PI controller ensures that V_{SC} tracks its reference V_{SC}^*.
k_{12}, k_{22} are the coefficients of proportionality, which ensure that the sliding surface equal zero by tracking the SC currents to its reference I when the FC controller can't ensures that I_{DL} tracks I_L.
In the case of a variable frequency control, a hysteresis comparator is used with the sliding surface S as input. In the case of a constant frequency control, the general system equation can be written as:

$$\dot{X}_i = A_i X_i + B_i U_i + C_i + \xi_i \tag{26}$$

with i=1,2

With for the first system:

$$X_1 = \begin{bmatrix} V_{DL} & I_{SC} & V_{SC} & I \end{bmatrix}^T \tag{27}$$

and

$$A_1 = \begin{bmatrix} 0 & 1/C_{DL} & 0 & 0 \\ -1/L_{SC} & -r_{SC}/L_{SC} & 1/L_{SC} & 0 \\ 0 & -1/C_{SC} & 0 & 0 \\ 0 & -k_{ps1}/C_{SC} & k_{is1} & 0 \end{bmatrix}, \ B_1 = \begin{bmatrix} \dfrac{-I_{SC}}{C_{DL}} & \dfrac{V_{DL}}{L_{SC}} & 0 & 0 \end{bmatrix}^T,$$

$$U_1 = U_{SC}, \ \xi_1 = \begin{bmatrix} 0 & 0 & 0 & 0 \end{bmatrix}^T \text{ and } C_1 = \begin{bmatrix} \dfrac{(I_{DL} - I_L)}{C_{DL}} & 0 & 0 & -k_{is} V_{SC}^* \end{bmatrix}^T$$

If we note:

$$G_1 = \begin{bmatrix} k_{11} & k_{21} & 0 & -k_{21} \end{bmatrix} \tag{28}$$

the sliding surface is then given by:

$$S_1 = G_1 \cdot (X_1 - X_{ref1}) \tag{29}$$

with

$$X_{ref1} = \begin{bmatrix} V_{DL}^* & 0 & 0 & 0 \end{bmatrix}^T$$

With for the second system:

$$X_2 = \begin{bmatrix} V_{DL} & I_{SC} & V_{SC} & I \end{bmatrix}^T$$

and

$$A_2 = \begin{bmatrix} -1/(r_B.C_{DL}) & 1/C_{DL} & 0 & 0 \\ -1/L_{SC} & -r_{SC}/L_{SC} & 1/L_{SC} & 0 \\ 0 & -1/C_{SC} & 0 & 0 \\ 0 & -k_{ps2}/C_{SC} & k_{is2} & 0 \end{bmatrix}, B_2 = \begin{bmatrix} \dfrac{-I_{SC}}{C_{DL}} & \dfrac{V_{DL}}{L_{SC}} & 0 & 0 \end{bmatrix}^T,$$

$$\xi_2 = \begin{bmatrix} \dfrac{(I_{DL} - I_L)}{C_{DL}} & 0 & 0 & 0 \end{bmatrix}^T, \ C_2 = \begin{bmatrix} \dfrac{E_B}{(r_B C_{DL})} & 0 & 0 & -k_{is} V_{SC}^* \end{bmatrix}^T, U_2 = U_{SC}$$

If we note:

$$G_2 = \begin{bmatrix} k_{12} & k_{22} & 0 & -k_{22} \end{bmatrix} \tag{30}$$

the sliding surface is then given by:

$$S_2 = C_{DL} \xi_2 + G_2 X_2 \tag{31}$$

In order to set the system dynamic, we define the reaching law:

$$\dot{S}_i = -\lambda_i S_i - K_i \operatorname{sign}(S_i) \tag{32}$$

with i=1,2

$$K_i = 0 \qquad \text{if} \left\| S_i \right\| < \varepsilon_i \,. \tag{33}$$

and

$$K_i = n_i \lambda_i \varepsilon_i \quad \text{if} \left\| S_i \right\| > \varepsilon_i \,. \tag{34}$$

The linear term $-\lambda_i S_i(X)$ imposes the dynamic to remain inside the error bandwidth ε_i. The choice of a high value of λ_i ($\le f_C/2$) ensures a small static error when $\left\| S_i \right\| < \varepsilon_i$. The nonlinear term $-K_i.\text{sign}(S_i)$ permits to reject perturbation effects (uncertainty of the model, variations of the working conditions). This term allows compensating high values of error $\left\| S_i \right\| > \varepsilon_i$ due to the above mentioned perturbations. The choice of a small value of ε_i leads to high current undulation (chattering effect) but the static error remains small. A high value of ε obliges to reduce the value of λ_i to ensure the stability of the system and leads to higher static error. Once the parameters (λ_i, K_i, ε_i) of the reaching law are determined, it is possible to calculate the continuous equivalent control, which allows to maintain the state trajectory on the sliding surface. We use the equations (28), (27) and (29), we find for the first system:

$$U_{SC1} = (G_1 B_1)^{-1} \{ -G_1 A_1 X_1 - G_1 C_1 + G_1 X_{ref1} - \lambda_1 G_1 X_1 + \lambda_1 G_1 X_{ref1} - K_1 \text{sign}(S_1) \} \tag{35}$$

Equations (26), (28) and (30) are used, we find for the second system:

$$U_{SC2} = (G_2 B_2)^{-1} \{ -G_2 A_2 X_2 - G_2 C_2 - \lambda_2 G_2 X_2 - K_2 \text{sign}(S_2) - C_{DL} [\dot{\xi}_2 + \lambda_2 \xi_2] \} \tag{36}$$

The control laws (35) and (36) contain the attractive and the equivalent controls. These equations (35) and (36) give for both hybrid sources the equation:

$$A_{eqi} = A_i - B_i (G_i B_i)^{-1} G_i A_i - B_i (G_i B_i)^{-1} \lambda_i G_i \tag{37}$$

The equation (27) allows finding poles of the systems during the sliding motion as a function of λ_i, k_{1i} and k_{2i}. The parameters k_{isi} and k_{psi} are then determined by solving S_i=0, equation justified by the fact that the sliding surface dynamic is hugely much greater than SC voltage variation.

C. Stability

Consider the following Lyapunov function:

$$V_i = \frac{1}{2} S_i^2 \tag{38}$$

With *S* is the sliding surface, *i=1,2*.
The derivative of the Lyapunov function along the trajectory of (15) is:

$$\dot{V}_i = S_i \dot{S}_i = -\lambda_i S_i^2 - K_i S_i \text{sign}(S_i) \le 0 \tag{39}$$

With $\lambda_i, K_i > 0$

Hence, the origin, with the sliding surface giving by (22) and (24), is globally asymptotically stable since the Lyapunov function (38) is radially unbounded and its derivative is strictly negative when $S_i \ne 0$ and $V_i = 0 \Leftrightarrow S_i = 0$.

3.4 Simulation results of the hybrid sources control

The whole system has been implemented in the Matlab-Simulink Software with the following parameters associated to the hybrid sources:

- FC parameters: P_{FC} = 130 W.
- DC link parameters: V_{DL}= 24 V.
- SC parameters: C_{SC} = 3500/6 F, $V_{SC}^{*} = 15V$.

The results presented in this section have been carried out by connecting the hybrid source to a "R_L, L_L and E_L" load representing a single phase DC machine.

Figures 17 and 18 present the behavior of currents I_{DL}, I_L, I_{SC}, and the DC link voltage V_{DL} for transient responses obtained while moving from the normal mode to the discharge mode, using sliding mode control. The test is performed by sharply changing the e.m.f load voltage E_L in the interval of $t \in [1.5s, 5s]$. The load current I_L changes from 16.8A to 24A. The current load I_L = 16.8A corresponds to a normal mode and the current load I_L = 24A to a discharge mode.

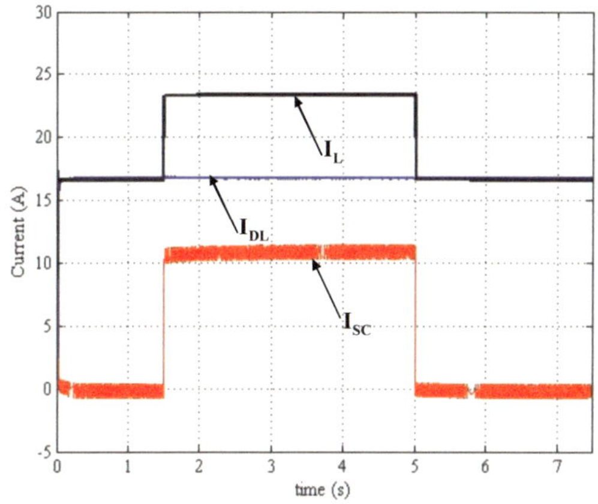

Fig. 17. FC, SCs and load currents

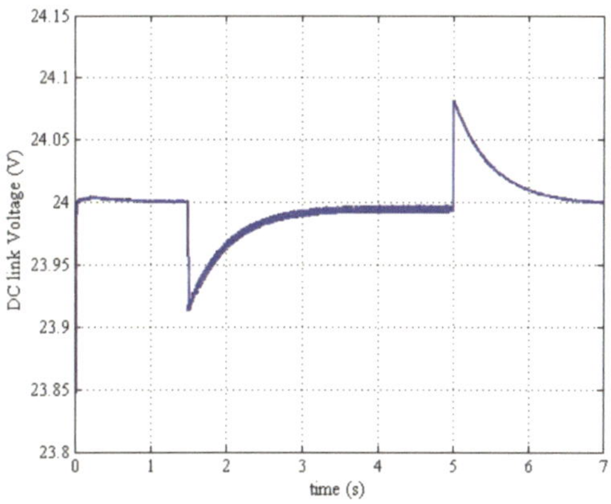

Fig. 18. DC link voltage

At the starting of the system, only FC provides the mean power to the load. The storage device current reference is equal to zero, when we are in normal mode. In the transient state, the load current IL becomes lower than the DC link current IDL. The DC link voltage reference is set at 24V. The DC link voltage tracks the reference well during the first second, after which, a very small overshoot is observed when the load current becomes negative. Then, the storage device current reference becomes negative because the controller compensates the negative load current value by the difference between the SC voltage and its reference. This is the recovering mode. After the load variation (t > 5s), the current in the DC link becomes equal to the load current. The SC current I_{SC} becomes null.

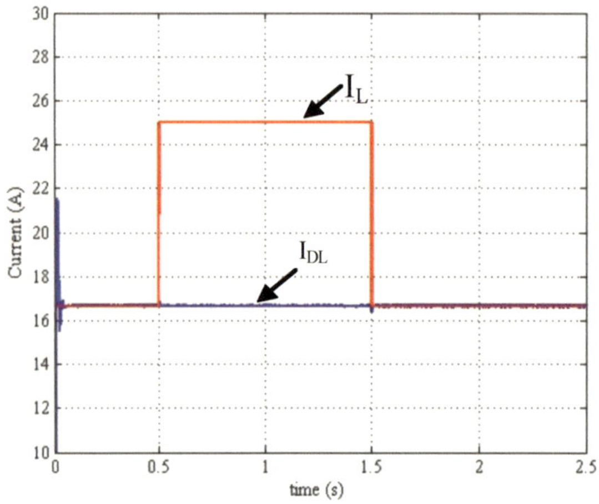

Fig. 19. Load and DC link currents

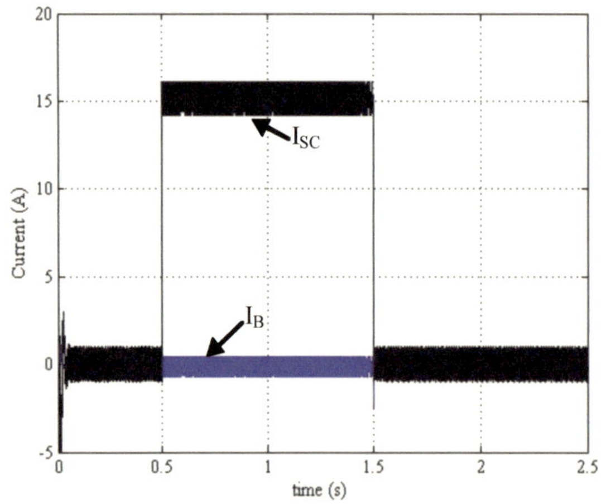

Fig. 20. SC and batteries currents

Figures 19, 20 and 21 present the behavior of currents I_{DL}, I_L, I_{SC}, I_B and the DC link voltage V_{DL} for transient responses obtained for a transition from the normal mode to the discharge mode applying using sliding mode control. The test is performed by changing sharply the e.m.f load voltage E_L in the interval of $t \in [0.5s, 1.5s]$. The load current I_L changes from 16.8A to 25A. The current load I_L = 16.8A corresponds to a normal mode and the current load I_L = 25A to a discharge mode.

At the starting of the system, only the FC provides the mean power to the load. The storage device current reference is equal to zero, we are in normal mode. In the transient state, the load current I_L became greater then the DC link current I_{DL}. The storage device current reference became positive thanks to control function which compensates this positive value by the difference between the SC voltage and its reference. We are in discharging mode. After the load variation *(t > 1.5s)*, the current in the DC link became equal to the load current. The SC current I_{SC} became null. We have a small variation in the batteries currents.

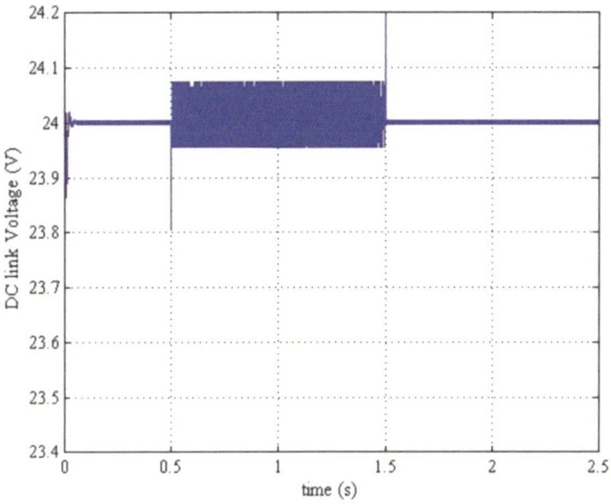

Fig. 21. DC link voltage

3. Conclusion

In this paper, the modeling and the control principles of two DC hybrid source systems have been presented. These systems are composed of a fuel cell source, SuperCapacitor source and with or without batteries on DC link. The state space models are given for both structures. These sources use the fuel cell as mean power source and supercapacitors as auxiliary transient power sources.

For the two hybrid structures, Sliding Mode Control principles have been applied in order to obtain a robustness control strategy. The sliding surface is generated as a function of multiple variables: DC link voltage and current, supercapacitors current and voltage, Load current.

Global asymptotic stability proofs are given and encouraging simulation results has been obtained.

Many benefits can be expected from the proposed structures such as supplying and absorbing the power peaks by using supercapactors which also allows recovering energy.

4. References

Kishinevsky, Y. & Zelingher, S. (2003). Coming clean with fuel cells, IEEE Power & Energy Magazine, vol. 1, issue: 6, Nov.-Dec. 2003, pp. 20-25.

Larminie, J. & Dicks, A. (2000). Fuel cell systems explained, Wiley, 2000.

Pischinger, S.; Schönfelder, C. & Ogrzewalla, J. (2006). Analysis of dynamic requirements for fuel cell systems for vehicle applications, J. Power Sources, vol. 154, no. 2, pp. 420-427, March 2006.

F. Belhachemi, S. Rael and B. Davat "A Physical based model of power elctric double layer supercapacitors", IAS 2000, 35th IEEE Industry Applications Conference, Rome, 8-12 October

Moore, R. M.; Hauer, K. H.; Ramaswamy, S. & Cunningham, J. M. (2006). Energy utilization and efficiency analysis for hydrogen fuel cell vehicles, J. Power Sources, 2006.

Corbo, P.; Corcione, F. E.; Migliardini, F. & Veneri, O. (2006). Experimental assessment of energy-management strategies in fuel-cell propulsion systems, J. Power Sources, 2006.

Rufer, A.; Hotellier, D. & Barrade, P. (2004). A Supercapacitor-Based Energy-Storage Substation for Voltage - Compensation in Weak Transportation Networks," IEEE Trans. Power Delivery, vol. 19, no. 2, April 2004, pp. 629-636.

Thounthong, P.; Raël, S. & Davat, B. (2007). A new control strategy of fuel cell and supercapacitors association for distributed generation system, IEEE Trans. Ind. Electron, Volume 54, Issue 6, Dec. 2007 Page(s): 3225 - 3233

Corrêa, J. M.; Farret, F. A.; Gomes, J. R. & Simões, M. G. (2003). Simulation of fuel-cell stacks using a computer-controlled power rectifier with the purposes of actual high-power injection applications, IEEE Trans. Ind. App., vol. 39, no. 4, pp. 1136-1142, July/Aug. 2003.

Benziger, J. B.; Satterfield, M. B.; Hogarth, W. H. J.; Nehlsen, J. P. & Kevrekidis; I. G. (2006). The power performance curve for engineering analysis of fuel cells, J. Power Sources, 2006.

Granovskii, M.; Dincer, I. & Rosen, M. A. (2006). Environmental and economic aspects of hydrogen production and utilization in fuel cell vehicles, J. Power Sources, vol. 157, pp. 411-421, June 19, 2006

Ayad, M. Y.; Pierfederici, S.; Raël, S. & Davat, B. (2007). Voltage Regulated Hybrid DC Source using supercapacitors, Energy Conversion and Management, Volume 48, Issue 7, July 2007, Pages 2196-2202.

Rao, V.; Singhal, G.; Kumar, A. & Navet, N. (2005). Model for Embedded Systems Battery, Proceedings of the 18th International Conference on VLSI Design held jointly with 4th International Conference on Embedded Systems Design (IEEE-VLSID'05), 2005.

Chen, M.; Gabriel, A.; Rincon-Mora. (2006). Accurate Electrical Battery Model Capable of Predicting Runtime and *I–V* Performance. . *IEEE Trans. Energy Convers*, Vol. 21, No.2, pp.504-511 June 2006.

Salameh, Z.M.; Casacca, M.A. & Lynch, W.A. (1992). A mathematical model for lead-acid batteries, *IEEE Trans. Energy Convers.*, vol. 7, no. 1, pp. 93–98, Mar. 1992.

Jonathan J. Awerbuch and Charles R. Sullivan, "Control of Ultracapacitor-Battery Hybrid Power Source for Vehicular Applications", Atlanta, Georgia, USA 17-18 November 2008

Ayad, M. Y.; Becherif, M.; Henni, A.; Wack, M. and Aboubou A. (2010). "Vehicle Hybridization with Fuel Cell, Supercapacitors and batteries by Sliding Mode Control", Proceeding IEEE-ICREGA'10 – March 8th-10th Dubai

Becherif, M.; Ayad, M. Y.; Henni, A.; Wack, M. and Aboubou A. (2010). "*Control of Fuel Cell, Batteries and Solar Hybrid Power Source*", Proceeding IEEE-ICREGA'10-- March 8th-10th Dubai

M. Chen, A. Gabriel and Rincon-Mora, "Accurate Electrical Battery Model Capable of Predicting Runtime and *I–V* Performance", *IEEE Trans. Energy Convers*, Vol. 21, No.2, pp.504-511 June 2006.

Z.M. Salameh, M.A. Casacca and W.A. Lynch, "A mathematical model for lead-acid batteries", *IEEE Trans. Energy Convers.*, vol. 7, no. 1, pp. 93–98, Mar. 1992.

6

Sensorless First- and Second-Order Sliding-Mode Control of a Wind Turbine-Driven Doubly-Fed Induction Generator

Ana Susperregui, Gerardo Tapia and M. Itsaso Martinez
University of the Basque Country (UPV/EHU)
Spain

1. Introduction

The doubly-fed induction generator (DFIG) is a wound-rotor electric machine on which about 75% of the wind turbines installed nowadays are based. As sketched in Fig. 1, when generating power, its stator is directly connected to the grid, while a back-to-back double-bridge converter —comprising both the rotor- (RSC) and grid-side (GSC) converters— interfaces its rotor with the grid, hence allowing the flow of slip power both from the grid to the rotor —at subsynchronous speeds— and vice-versa —at supersynchronous speeds— within a certain speed range.

Given that only the slip power has to be managed by the bidirectional rotor converter, it is sufficient to size it so that it typically supports between 25% and 30% of the DFIG rated power (Ekanayake et al., 2003; Peña et al., 1996). This is more than probably the main reason for the success of the DFIG in the field of variable-speed wind generation systems.

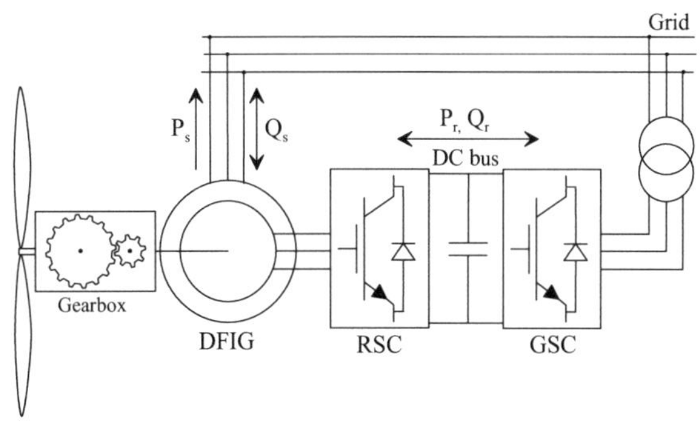

Fig. 1. Structure of a DFIG-based wind turbine

Standard field oriented control (FOC) schemes devised to command wind turbine-driven DFIGs comprise proportional-integral (PI)-controlled cascaded current and power loops, which require the use of an incremental encoder (Tapia et al., 2003). Although stator-side active and reactive powers can be independently governed by adopting those control schemes, the system transient performance degrades as the actual values of the DFIG resistances and inductances deviate from those based on which the control system tuning was carried out during commissioning (Xu & Cartwright, 2006). In addition, the optimum power curve tracking achievable using PI-based control schemes shows a considerable room for improvement. Even if feedforward decoupling control terms are traditionally incorporated to enhance the closed-loop DFIG dynamic response, they are extremely dependent on DFIG parameters (Tapia et al., 2006; Xu & Cartwright, 2006).

In this framework, alternative high dynamic performance power control schemes for DFIGs are being proposed, among of which a strong research line focuses on the so-called direct power control (DPC) (Xu & Cartwright, 2006; Zhi & Xu, 2007). Several others explore the alternative of applying sliding-mode control (SMC), both standard —first-order— (Beltran et al., 2008; Susperregui et al., 2010), and higher-order (Beltran, Ahmed-Ali & Benbouzid, 2009; Beltran, Benbouzid & Ahmed-Ali, 2009; Ben Elghali et al., 2008).

Moreover, since, as already mentioned, the back-to-back rotor converter is sized to manage a slip power up to 25% or 30% of the wind generator rated power, DFIGs are kept connected to the grid provided that their rotational speed remains within a certain range. Accordingly, connection of DFIGs to the grid is only accomplished if the wind is strong enough to extract energy from it profitably. In particular, the four-pole 660-kW DFIG considered in this chapter is not connected to the grid until its rotational speed exceeds the threshold value of 1270 rpm.

Yet, connecting the DFIG stator to the grid is not straightforward. In fact, although wind-turbine-driven DFIGs are asynchronous machines, owing to the double-bridge rotor converter managing the slip power, they behave as real synchronous generators. Accordingly, prior to connecting the stator of a DFIG to the grid, the voltage induced at its stator terminals must necessarily be synchronized to that of the grid.

However, even though control of wind turbine-driven DFIGs is a topic extensively covered in the literature, not many contributions outline or describe in some detail possible strategies for smooth connection of DFIGs to the grid. So far, the synchronization problem has been approached from different viewpoints, hence giving rise to alternative methods, as open-loop stator voltage control (Peña et al., 2008), closed-loop regulation of rotor current (Peresada et al., 2004; Tapia et al., 2009), and phase-locked loop (PLL) (Abo-Khalil et al., 2006; Blaabjerg et al., 2006) or even direct torque control (DTC) of the voltage induced at the open stator (Arnaltes & Rodríguez, 2002).

Considering those precedents, together with the robustness and tracking ability naturally conferred by SMC, both a first-order and a higher-order sensorless SMC algorithms, conceived to command the RSC feeding the rotor of a DFIG, are described and evaluated in this chapter. Those two algorithms are not only aimed at governing active and reactive power exchange between the DFIG stator and the grid, but also at ensuring the synchronization required for smooth connection of the DFIG stator to the grid.

The chapter is organized as follows. Given that the DFIG exhibits different dynamics depending on whether its stator is connected to the grid or not (Tapia et al., 2009), the mathematical model corresponding to each of those two operating conditions is first briefly presented. Conditions to reach synchronization are also provided. After selection of the switching functions associated, respectively, to the power control and synchronization objectives, a global first-order sliding-mode control (1-SMC) algorithm, based on Utkin's research work on various other types of electric machines (Utkin et al., 1999; Utkin, 1993; Yan et al., 2000), is described in detail. Stability analyses are also provided for both the power control and synchronization operation regimes. An overall second-order sliding-mode control (2-SMC) algorithm, alternative to the previous one, is next presented. Special attention is paid to the derivation of effective tuning equations for all its gains and constants. The practical issue related to bumpless transition between the controllers in charge of synchronization and power control, at the instant of connecting the DFIG stator to the grid, is then tackled. Adaptation of the model reference adaptive system (MRAS) observer put forward in (Peña et al., 2008), so that it remains valid for sensorless control during synchronization, is also dealt with. Sensorless versions of the two SMC algorithms proposed are evaluated via real-time hardware-in-the-loop (HIL) emulation over a virtual 660-kW DFIG prototype. The chapter finishes with a conclusion section, devoted to analyze the results arising from the HIL emulation tests carried out.

2. Review of DFIG model and grid synchronization

Focused on a 660-kW DFIG, the main objective of the two alternative versions of the control system presented along this chapter consists in succeeding in the achievement of the maximum active power the machine is able to generate at each rotational speed; i.e., to track the DFIG optimum power curve. As a secondary goal, but still essential from the point of view of the electricity supply quality, the reactive power the machine generates or absorbs from the grid is also managed.

Before raising the modelling of the machine, and, to get ride of misunderstandings due to the diverse nomenclature used to identify the reference frames taking part in FOC, Fig. 2 presents the terminology that is going to be adopted hereafter. It can be observed that the stator direct and quadrature axes are represented as sD and sQ, respectively, and that the rotor reference frame, which forms the θ_r turning angle with respect to sD axis, is denominated $r\alpha$-$r\beta$. Since the machine is going to be rotor-side controlled, the magnitudes will be referred to a frame, labeled as x-y, whose direct axis is aligned with the stator flux, $\vec{\psi}_s$ —and, therefore, with the stator magnetizing current, $\vec{i}_{ms}$. The latter reference frame is turned ρ_s with respect to the sD-sQ plane.

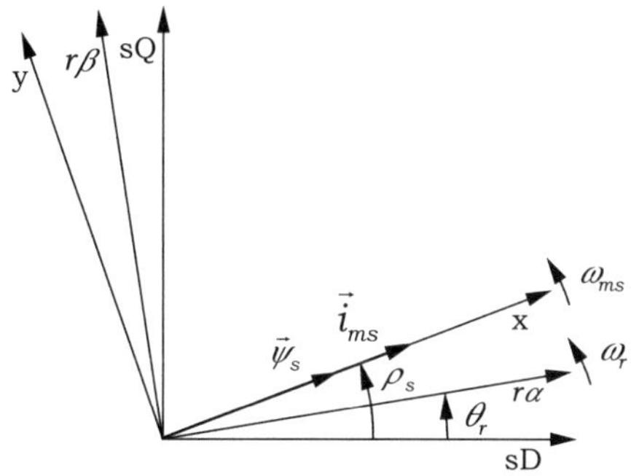

Fig. 2. Stator-flux-oriented reference frame

When connected to the grid, the rotor-side-voltage generator model regarding the stator-flux-oriented reference frame —x-y— may be expressed as (Tapia et al., 2009; Vas, 1998)

$$v_{rx} = R_r i_{rx} + L_r' \frac{di_{rx}}{dt} + \frac{L_m}{L_s}\frac{d|\vec{\psi}_s|}{dt} - \omega_{sl} L_r' i_{ry} \quad (1)$$

$$v_{ry} = R_r i_{ry} + L_r' \frac{di_{ry}}{dt} + \omega_{sl}\frac{L_m}{L_s}|\vec{\psi}_s| + \omega_{sl} L_r' i_{rx}, \quad (2)$$

where v_{rx} and v_{ry} are the components of the rotor voltage, i_{rx} and i_{ry} represent the rotor currents, and $\omega_{sl} = \omega_{ms} - \omega_r$ stands for the slip frequency between the $r\alpha$ and x axes. R_r, L_s

and L_m denote the rotor resistance, and the stator and magnetizing inductances, respectively. Finally, $L_r' = \sigma L_r$ symbolizes the transient inductance of the rotor, where $\sigma = 1 - L_m^2/(L_s L_r)$ is the total leakage factor.

Taking into account that power generation is not profitable at low speeds —less than 1270 rpm in this particular case—, the generator will not be connected to the grid until this threshold value is exceeded. Therefore, a new "grid-non-connected" state appears where the machine dynamic behaviour differs from that in which its stator is connected to the grid, and, consequently, the model changes. Moreover, the transition between the disconnected and connected states is not trivial, since the grid voltage and that induced at the open stator of the DFIG may present magnitude and/or phase differences. At this point, aiming at removing the risk of short circuit, it can be taken advantage of a properly controlled "grid-non-connected" state, turning it into a synchronization stage.

Let a new x'-y' reference frame be defined when the stator is disconnected from the grid, where, as shown in Fig. 3, its y' quadrature axis and the grid voltage space-phasor are collinear. Moreover, assuming steady-state regime, and, if rotor current is stable, it can be demonstrated (Tapia et al., 2009) that the stator flux and voltage space-vectors are collinear to x and y axes, respectively; i.e., $\vec{\psi}_s \perp \vec{v}_s$.

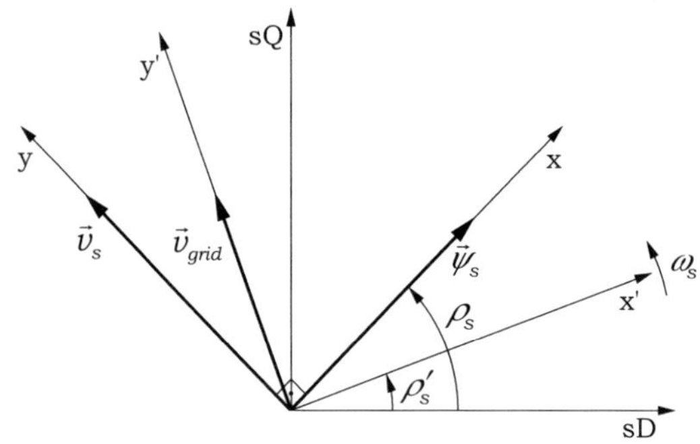

Fig. 3. New reference frame for synchronization

Bearing in mind that stator current is null when disconnected from the grid, the new open-stator model, expressed according to the x'-y' reference frame, arises (Tapia et al., 2009):

$$v_{rx'} = R_r i_{rx'} + L_r \frac{di_{rx'}}{dt} - \omega_{sl} L_r i_{ry'} \rightarrow \frac{di_{rx'}}{dt} = \frac{v_{rx'}}{L_r} - \frac{R_r}{L_r} i_{rx'} + \omega_{sl} i_{ry'} \tag{3}$$

$$v_{ry'} = R_r i_{ry'} + L_r \frac{di_{ry'}}{dt} + \omega_{sl} L_r i_{rx'} \rightarrow \frac{di_{ry'}}{dt} = \frac{v_{ry'}}{L_r} - \frac{R_r}{L_r} i_{ry'} - \omega_{sl} i_{rx'}. \tag{4}$$

As evidenced in Fig. 3, synchronization may be achieved if x-y and x'-y' reference frames are aligned. However, for a complete match-up, the grid and stator voltage space-vectors must be not only collinear but also identical in magnitude. The two conditions are satisfied if the following rotor current values are achieved (Tapia et al., 2009):

$$i_{rx'\,ref} = \frac{\left|\vec{v}_{grid}\right|}{\omega_s L_m};\ i_{ry'\,ref} = 0, \tag{5}$$

and, consequently, synchronization is ensured; i.e.:

$$i_{rx} = i_{rx'} = \frac{\left|\vec{v}_{grid}\right|}{\omega_s L_m};\ i_{ry} = i_{ry'} = 0;\ \rho_s = \rho_s'. \tag{6}$$

Furthermore, if the current values presented in (6) are substituted into the stator-side reactive and active power expressions given next (Vas, 1998):

$$Q_s = \frac{3}{2}\frac{|\vec{v}_s|}{L_s}(|\vec{\psi}_s| - L_m i_{rx});\ P_s = -\frac{3}{2}\frac{L_m}{L_s}|\vec{v}_s| i_{ry}, \tag{7}$$

it follows that, at the instant of the connection, zero power-exchange is achieved.
For the sake of a proper performance of the whole system, each state must be commanded with its own controller. Moreover, the transition between the two states must carefully be followed up, depending on the control strategy being applied, in order to achieve a bumpless connection. This aspect will be thoroughly described in a later section.

3. Sensorless sliding-mode control arrangement for the DFIG

Aiming to track the optimum power curve of the DFIG, sliding-mode control theory has been adopted, which provides the system with superior tracking ability and high robustness despite uncertainties or parameter variations. The basis of SMC is the judicious election of a switching variable, which usually depends on a linear combination of the error of the variable to be commanded and its subsequent time derivatives. Here, the proposed switching variables for optimum power control —stator connected to the grid— are

$$s_{Q_s} = e_{Q_s} + c_Q \int e_{Q_s} dt \tag{8}$$

$$s_{P_s} = e_{P_s} + c_P \int e_{P_s} dt, \tag{9}$$

where $e_{Q_s} = Q_{s\ ref} - Q_s$ and $e_{P_s} = P_{s\ ref} - P_s$ represent de errors in reactive and active powers, respectively, and the integral terms, weighted by c_Q and c_P positive constants, are added for steady-state response improvement (Utkin et al., 1999).
Examining the set-points proposed in (5), it can be derived that rotor current regulation must be carried out if synchronization is required. Therefore, when the stator is disconnected from the grid, and, similarly to the previous case, the following switching functions are suggested:

$$s_{i_{rx'}} = e_{i_{rx'}} + c_{x'} \int e_{i_{rx'}} dt \tag{10}$$

$$s_{i_{ry'}} = e_{i_{ry'}} + c_{y'} \int e_{i_{ry'}} dt, \tag{11}$$

where $e_{i_{rx'}} = i_{rx'\,ref} - i_{rx'}$ and $e_{i_{ry'}} = i_{ry'\,ref} - i_{ry'}$ represent the errors in $i_{rx'}$ and $i_{ry'}$, respectively, and $c_{rx'}$ and $c_{ry'}$ are positive constants.
The switching variable defines the relative degree of a system, and, as a result, the order of the applicable SMC (Levant, 1993). As the system is of first-order relative degree in both states, connected and disconnected from the grid, it may be commanded applying 1-SMC or 2-SMC (Bartolini et al., 1999). The design of the two controllers is detailed in subsequent sections.

3.1 First-order sliding-mode control

In this section, a 1-SMC scheme is proposed. Due to the different dynamic behaviours presented by the DFIG when disconnected or connected to the grid, a different DFIG model is considered to conceive the control of each of those two cases, and a first-order sliding-mode controller is accordingly synthesized for each of them.

In particular, the 1-SMC applied is that based on V. I. Utkin's research work (Utkin et al., 1999; Utkin, 1993; Yan et al., 2000), which sets out the following: most of the electrical systems must modulate the control signals in order to command the transistors' gates of their converters; so, why not directly generate those gating signals thus eluding the use of pulse-width modulation (PWM) or space-vector modulation (SVM) techniques? (Yan et al., 2008) This theory fits perfectly the present case, in which controllers for the RSC of the back-to-back configuration are designed for the two possible connection states of the DFIG.

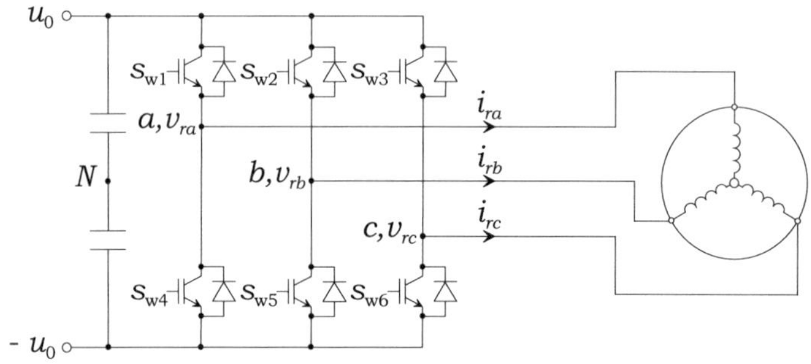

Fig. 4. Rotor-side converter scheme

Depending on whether or not the DFIG stator is connected to the grid, its model and controllers do vary, but the RSC to be commanded, displayed in Fig. 4, remains obviously the same. Analyzing this scheme (Utkin et al., 1999), it is possible to find a link between:

- the signals generated by controllers based on a synchronous frame, v_{rx} and v_{ry}, and those between the midpoints of the converter legs and the DC link, v_{raN}, v_{rbN} and v_{rcN}:

$$\overbrace{\begin{bmatrix} v_{rx} \\ v_{ry} \end{bmatrix}}^{\mathbf{V}_{xy}} = \overbrace{\begin{bmatrix} \cos\rho & \cos(\rho - \frac{2\pi}{3}) & \cos(\rho + \frac{2\pi}{3}) \\ -\sin\rho & -\sin(\rho - \frac{2\pi}{3}) & -\sin(\rho + \frac{2\pi}{3}) \end{bmatrix}}^{\mathbf{D}} \overbrace{\begin{bmatrix} v_{raN} \\ v_{rbN} \\ v_{rcN} \end{bmatrix}}^{\mathbf{V}_{abc}} . \tag{12}$$

If the opposite relation is needed, the inverse of $\mathbf{D}$ matrix must exist. But, as it is not square, Moore-Penrose pseudo-inverse concept (Utkin et al., 1999) may be used to calculate its inverse, $\mathbf{D}^+ = \mathbf{D}^T(\mathbf{D}\mathbf{D}^T)^{-1}$, resulting the previous matrix expression in:

$$\begin{bmatrix} v_{raN} \\ v_{rbN} \\ v_{rcN} \end{bmatrix} = \overbrace{\begin{bmatrix} \cos\rho & -\sin\rho \\ \cos(\rho - \frac{2\pi}{3}) & -\sin(\rho - \frac{2\pi}{3}) \\ \cos(\rho + \frac{2\pi}{3}) & -\sin(\rho + \frac{2\pi}{3}) \end{bmatrix}}^{\mathbf{D}^+} \begin{bmatrix} v_{rx} \\ v_{ry} \end{bmatrix} , \tag{13}$$

where $\rho = \rho_s - \theta_r$.

- the voltages v_{raN}, v_{rbN} and v_{rcN}, and the transistors' gating signals, s_{w1}, s_{w2}, s_{w3}, s_{w4}, s_{w5} and s_{w6}:

$$\begin{aligned} s_{w1} &= 0.5(1 + v_{raN}/u_0) & s_{w4} &= 1 - s_{w1} \\ s_{w2} &= 0.5(1 + v_{rbN}/u_0) & s_{w5} &= 1 - s_{w2} . \\ s_{w3} &= 0.5(1 + v_{rcN}/u_0) & s_{w6} &= 1 - s_{w3} \end{aligned} \tag{14}$$

The following sections describe the design of the control scheme for the cases mentioned above: DFIG connected to and disconnected from the grid.

3.1.1 DFIG connected to the grid —Optimum power generation

Once synchronization is completed and the DFIG is connected to the grid, it is going to be commanded applying the following multivariable control law, in order to achieve optimum power generation:

$$\mathbf{V}_{abc} = -u_0 \mathrm{sgn}(\mathbf{S}), \tag{15}$$

where $\mathbf{S} = \begin{bmatrix} s_1 \; s_2 \; s_3 \end{bmatrix}^T$ contains the switching variable expressions represented in *a-b-c* three-phase reference frame. Note that, as the system to be controlled presents negative gain, that of the control law must also be negative if stability is pursued.

Aiming to ease the design of the controllers and, subsequently, to demonstrate the stability of the closed-loop system, the model can be transferred to subspace $\mathbf{S}_{QP} = \begin{bmatrix} s_{Q_s} \; s_{P_s} \end{bmatrix}^T$, if the time derivatives of (8) and (9) are taken, and making use of (1)-(2):

$$\overbrace{\begin{bmatrix} \dot{s}_{Q_s} \\ \dot{s}_{P_s} \end{bmatrix}}^{\dot{\mathbf{S}}_{QP}} = \overbrace{\begin{bmatrix} F_1 \\ F_2 \end{bmatrix}}^{\mathbf{F}_{QP}} + a \overbrace{\begin{bmatrix} v_{rx} \\ v_{ry} \end{bmatrix}}^{\mathbf{V}_{xy}} \tag{16}$$

where $F_1 = f(\dot{Q}_{s\ ref}, |\vec{v}_s|, |\vec{\psi}_s|, Q_{s\ ref}, i_{rx}, w_{sl}, i_{ry})$, $F_2 = f(\dot{P}_{s\ ref}, |\vec{v}_s|, |\vec{\psi}_s|, P_{s\ ref}, i_{rx}, w_{sl}, i_{ry})$, and $a = \frac{3}{2}\frac{L_m}{L_s L_r'}|\vec{v}_s|$.

It is possible to relate the new model in (16) to the voltage signals between the midpoints of the converter legs and the DC link, if $\mathbf{D}$ transformation matrix in (12) is applied

$$\dot{\mathbf{S}}_{QP} = \mathbf{F}_{QP} + \overbrace{a\mathbf{D}}^{\mathbf{D}_a} \mathbf{V}_{abc}. \tag{17}$$

It can be noticed that control signals are transformed from *a-b-c* to the stator-flux-oriented reference frame by means of $\mathbf{D}_a$ matrix. Now, it seems logical to derive the $\mathbf{S}$ in (15) by arranging (8) and (9) in matrix format, $\mathbf{S}_{QP} = \begin{bmatrix} s_{Q_s} \; s_{P_s} \end{bmatrix}^T$, and then transforming $\mathbf{S}_{QP}$ by means of $\mathbf{D}_a^+$:

$$\mathbf{S} = \mathbf{D}_a^+ \mathbf{S}_{QP}. \tag{18}$$

This allows obtaining the three-phase control signals as:

$$\mathbf{V}_{abc} = u_0 \begin{bmatrix} \mathrm{sgn}(s_{P_s} \sin\rho - s_{Q_s} \cos\rho_s) \\ \mathrm{sgn}(s_{P_s} \sin(\rho - \frac{2\pi}{3}) - s_{Q_s} \cos(\rho - \frac{2\pi}{3})) \\ \mathrm{sgn}(s_{P_s} \sin(\rho + \frac{2\pi}{3}) - s_{Q_s} \cos(\rho + \frac{2\pi}{3})) \end{bmatrix}, \tag{19}$$

where $1/a$ constant should appear multiplying the terms inside every sgn function. However, as its value is always positive, it does not affect the final result, and this is the reason why it has been removed from (19). To conclude, the transistor gating signals are achieved just by replacing (19) in (14).

Due to the discontinuous nature of the generated command signals —which are in fact the transistors' gating signals—, a bumpless transition between synchronization and optimum generation states takes place spontaneously, without requiring the use of further control techniques, as that proposed in (Tapia et al., 2009).

3.1.1.1 Stability proof

In order to confirm that the designed control signals assure the zero-convergence of the switching variables, the following positive-definite Lyapunov function candidate is proposed:

$$V = \frac{1}{2}\mathbf{S}_{QP}^T\mathbf{S}_{QP}, \tag{20}$$

and, as it is well-known, its time derivative must be negative-definite:

$$\dot{V} = \frac{1}{2}\left(\dot{\mathbf{S}}_{QP}^T\mathbf{S}_{QP} + \mathbf{S}_{QP}^T\dot{\mathbf{S}}_{QP}\right) = \mathbf{S}_{QP}^T\dot{\mathbf{S}}_{QP} < 0. \tag{21}$$

Considering (17) and (18), the Lyapunov function time derivative can be rewritten as:

$$\dot{V} = \mathbf{S}^T\mathbf{F} - \frac{4}{9}a^2u_0\begin{bmatrix} s_1 \\ s_2 \\ s_3 \end{bmatrix}^T\begin{bmatrix} \text{sgn}(s_1) - 0.5\text{sgn}(s_2) - 0.5\text{sgn}(s_3) \\ \text{sgn}(s_2) - 0.5\text{sgn}(s_3) - 0.5\text{sgn}(s_1) \\ \text{sgn}(s_3) - 0.5\text{sgn}(s_1) - 0.5\text{sgn}(s_2) \end{bmatrix}, \tag{22}$$

where $\mathbf{F} = \mathbf{D}_a^T\mathbf{F}_{QP} = [F_1^* \; F_2^* \; F_3^*]^T$.
Taking into account that the elements of $\mathbf{V}_{abc}$ will never coincide in sign at every moment, nor will $\mathbf{S}$ components, as it can be inferred from (15). Therefore, $\text{sgn}(s_l) \neq \text{sgn}(s_m) = \text{sgn}(s_n)$, where $l \neq m \neq n$, for $l, m, n \in \{1, 2, 3\}$. Let $l = 1$, $m = 2$ and $n = 3$; moreover, suppose that $\text{sgn}(s_1) = +1 \neq \text{sgn}(s_2) = \text{sgn}(s_3)$, then (22) could be transformed into:

$$\dot{V} = \underbrace{s_1F_1^* + s_2F_2^* + s_3F_3^*}_{p} - \underbrace{\frac{4}{9}a^2u_0\left(2|s_1| + |s_2| + |s_3|\right)}_{q}. \tag{23}$$

If $\dot{V} < 0$ must be guaranteed, it can be stated that $|q| > |p|$. Furthermore, if the most restrictive case is considered, the following condition must be derived:

$$\frac{4}{9}a^2u_0\left(2|s_1| + |s_2| + |s_3|\right) > |s_1||F_1^*| + |s_2||F_2^*| + |s_3||F_3^*|. \tag{24}$$

Comparing each accompanying term of $|s_1|$, $|s_2|$ and $|s_3|$, u_0 can be fixed by guaranteing that

$$u_0 > \frac{9}{4a^2}\max\left(\frac{|F_1^*|}{2}, |F_2^*|, |F_3^*|\right) \tag{25}$$

is satisfied. Nevertheless, bearing in mind the remaining signs combinations between switching functions s_1, s_2 and s_3, and taking into account the most demanding case, the above proposed condition turns out to be:

$$u_0 > \frac{9}{4a^2}\max\left(|F_1^*|, |F_2^*|, |F_3^*|\right). \tag{26}$$

Provided that the controller supplies the convenient voltage, derived from (26), the system is robust even in the presence of disturbances, guarantying thus the asymptotic convergence of s_{Q_s} and s_{P_s} to zero. (26) presents a very conservative condition, but, in practice, a lower value of u_0 is usually enough to assure the stability of the whole system.

3.1.2 DFIG disconnected from the grid —Synchronization stage

When rotor speed-threshold is achieved, the control system activates the synchronization stage. As mentioned before, in order to avoid short circuit, the goal is to match the stator and grid voltages in magnitude and phase by requesting the reference values presented in (5). Let the same multivariable control law structure exposed in (15) be employed, considering, of course, the new subspace where it must be applied. Combining (3) and (4) with the time derivatives of (10) and (11), the model can be transferred to the above-mentioned new subspace $\mathbf{S}_{x'y'} = \begin{bmatrix} s_{i_{rx'}} & s_{i_{ry'}} \end{bmatrix}^T$:

$$\overbrace{\begin{bmatrix} \dot{s}_{i_{rx'}} \\ \dot{s}_{i_{ry'}} \end{bmatrix}}^{\dot{\mathbf{S}}_{x'y'}} = \overbrace{\begin{bmatrix} M_1 \\ M_2 \end{bmatrix}}^{\mathbf{M}_{x'y'}} + b \overbrace{\begin{bmatrix} v_{rx'} \\ v_{ry'} \end{bmatrix}}^{\mathbf{V}_{x'y'}}, \tag{27}$$

where $M_1 = f(\dot{i}_{rx'\,ref}, i_{rx'\,ref}, i_{rx'}, w_{sl}, i_{ry'})$, $M_2 = f(\dot{i}_{ry'\,ref}, i_{ry'\,ref}, i_{ry'}, w_{sl}, i_{rx'})$, and $b = -1/L_r$.

Following a similar procedure to that presented in 3.1.1, the switching variables referred to *a-b-c* reference frame will be obtained as:

$$\mathbf{S} = \mathbf{D}_b^+ \mathbf{S}_{x'y'}, \tag{28}$$

where $\mathbf{D}_b^+$ is the Moore-Penrose pseudo-inverse of $\mathbf{D}_b = b\mathbf{D}$. Substituting (28) in the proposed control law (15), the three-phase command signals to be generated turn out to be:

$$\mathbf{V}_{abc} = u_0 \begin{bmatrix} \operatorname{sgn}(s_{i_{rx'}} \cos\rho' - s_{i_{ry'}} \sin\rho') \\ \operatorname{sgn}(s_{i_{rx'}} \cos(\rho' - \frac{2\pi}{3}) - s_{i_{ry'}} \sin(\rho' - \frac{2\pi}{3})) \\ \operatorname{sgn}(s_{i_{rx'}} \cos(\rho' + \frac{2\pi}{3}) - s_{i_{ry'}} \sin(\rho' + \frac{2\pi}{3})) \end{bmatrix}, \tag{29}$$

where $\rho' = \rho_s' - \theta_r$. The gating signals should easily be achieved by replacing (29) in (14).

3.1.2.1 Stability proof

Analogous to the case in 3.1.1.1, the asymptotic zero-convergence of switching functions is assured if the following condition is accomplished:

$$u_0 > \frac{9}{4b^2} \max\left(|M_1^*|, |M_2^*|, |M_3^*|\right), \tag{30}$$

where $\mathbf{M} = \mathbf{D}_b^T \mathbf{M}_{x'y'} = [M_1^* \; M_2^* \; M_3^*]^T$, and the positive-definite Lyapunov function candidate is selected as:

$$V = \frac{1}{2} \mathbf{S}_{x'y'}^T \mathbf{S}_{x'y'}. \tag{31}$$

3.2 Higher-order sliding-mode controller

The proposed structure based on 1-SMC leads to a variable switching frequency of the RSC transistors (Susperregui et al., 2010), which may inject broadband harmonics into the grid, complicating the design of the back-to-back converter itself, as well as that of the grid-side AC filter (Zhi & Xu, 2007). As an alternative to the 1-SMC, higher-order sliding-mode control (HOSMC) could be adopted. In particular, and owing to the relative order the system presents, a 2-SMC realization, known as the *super-twisting algorithm* (STA), may be employed (Bartolini et al., 1999; Levant, 1993). The control signal comprises two terms; one guaranteing that switching surface $s = 0$ is reached in finite time, and another related to the integral of the switching variable sign. Namely,

$$u = -\lambda |s|^{\rho} \text{sgn}(s) - w \int \text{sgn}(s) dt, \tag{32}$$

where $\rho = 0.5$ assures a real second-order sliding-mode. This technique gives rise to a continuous control signal, which not only alleviates or completely removes the "chatter" from the system, but must also be modulated. To this effect, SVM may be applied, therefore obtaining a fixed switching frequency which results in elimination of the above-mentioned drawback.
As previously remarked, two controllers must be designed in order to command the performance of the DFIG when connected and disconnected from the grid.

3.2.1 DFIG connected to the grid —Optimum power generation

Considering the time derivatives of (8) and (9) together with expressions (3), (4) and (7), it turns out that

$$\begin{aligned} \dot{s}_{Q_s} &= \dot{Q}_{s\,ref} - \frac{3}{2}\frac{1}{L_s}|\vec{v}_s| \left[c_Q |\vec{\psi}_s| + \left(\frac{R_r}{L_r'} - c_Q \right) L_m i_{rx} - \omega_{sl} L_m i_{ry} \right] + c_Q Q_{s\,ref} + \\ &+ \frac{3}{2}\frac{L_m}{L_s L_r'} |\vec{v}_s| v_{rx} \end{aligned} \tag{33}$$

$$\begin{aligned} \dot{s}_{P_s} &= \dot{P}_{s\,ref} + \frac{3}{2}\frac{L_m}{L_s}|\vec{v}_s| \left[\left(c_P - \frac{R_r}{L_r'} \right) i_{ry} - \omega_{sl} i_{rx} - \omega_{sl} \frac{L_m}{L_s L_r'} |\vec{\psi}_s| \right] + c_P P_{s\,ref} + \\ &+ \frac{3}{2}\frac{L_m}{L_s L_r'} |\vec{v}_s| v_{ry}, \end{aligned} \tag{34}$$

where $\frac{d|\vec{\psi}_s|}{dt}$ has been neglected due to the fact that the DFIG is grid connected. Aiming to track the optimum power curve, the voltage to be applied to the rotor may be derived according to control law

$$v_{rx} = v_{rx_{ST}} + v_{rx_{eq}}; \quad v_{ry} = v_{ry_{ST}} + v_{ry_{eq}}, \tag{35}$$

where the terms with subscript *'ST'* are computed, through application of the STA, as:

$$v_{rx_{ST}} = \frac{2}{3}\frac{L_s L_r'}{|\vec{v}_s| L_m} \left[-\lambda_Q |s_{Q_s}|^{0.5} \text{sgn}(s_{Q_s}) - w_Q \int \text{sgn}(s_{Q_s}) dt \right] \tag{36}$$

$$v_{ry_{ST}} = \frac{2}{3}\frac{L_s L_r'}{|\vec{v}_s| L_m} \left[-\lambda_P |s_{P_s}|^{0.5} \text{sgn}(s_{P_s}) - w_P \int \text{sgn}(s_{P_s}) dt \right] \tag{37}$$

with λ_Q, w_Q, λ_P and w_P being positive parameters to be tuned. The gain premultiplying the algorithms —the inverse of that affecting control signal in (33) and (34)— is exclusively applied for assisting in the process of tuning the foregoing parameters. The addends with subscript *'eq'* in (35), which correspond to equivalent control terms, are derived by letting $\dot{s}_{P_s} = \dot{s}_{Q_s} = 0$ (Utkin et al., 1999). As a result,

$$v_{rx_{eq}} = -\frac{2}{3}\frac{L_s L_r'}{|\vec{v}_s| L_m} \left[\dot{Q}_{s\,ref} + c_Q (Q_{s\,ref} - Q_s) \right] + R_r i_{rx} - L_r' \omega_{sl} i_{ry} \tag{38}$$

$$v_{ry_{eq}} = -\frac{2}{3}\frac{L_s L_r'}{|\vec{v}_s| L_m} \left[\dot{P}_{s\,ref} + c_P (P_{s\,ref} - P_s) \right] + R_r i_{ry} + \frac{L_m}{L_s} \omega_{sl} |\vec{\psi}_s| + L_r' \omega_{sl} i_{rx}. \tag{39}$$

It should be noted that sliding regime in manifold $s_{P_s} = s_{Q_s} = \dot{s}_{P_s} = \dot{s}_{Q_s} = 0$ can also be attained by applying only the control terms in (35) corresponding to the STA. Accordingly, the equivalent control terms in (38) and (39) are not strictly necessary, but, once included, the more accurately they are computed, the lower is the control effort let to be done by the STA. Equivalent control terms are hence incorporated not only to improve the system transient response (Rashed et al., 2005), but also to ease selection of constants c_P and c_Q, as well as tuning of the STA $\lambda_{Q,P}$ and $w_{Q,P}$ gains.
In effect, substituting control law (35) into (33) and (34) produces:

$$\dot{s}_{Q_s} = -\lambda_Q \left|s_{Q_s}\right|^{0.5} \operatorname{sgn}(s_{Q_s}) - w_Q \int \operatorname{sgn}(s_{Q_s})\, dt \tag{40}$$

$$\dot{s}_{P_s} = -\lambda_P \left|s_{P_s}\right|^{0.5} \operatorname{sgn}(s_{P_s}) - w_P \int \operatorname{sgn}(s_{P_s})\, dt. \tag{41}$$

Now, given that $\operatorname{sgn}(s) = s/\left|s\right|$, taking the time derivatives of (40) and (41) leads to:

$$\ddot{s}_{Q_s} = -0.5\lambda_Q \left|s_{Q_s}\right|^{-0.5} \dot{s}_{Q_s} - w_Q \left|s_{Q_s}\right|^{-1} s_{Q_s} \tag{42}$$

$$\ddot{s}_{P_s} = -0.5\lambda_P \left|s_{P_s}\right|^{-0.5} \dot{s}_{P_s} - w_P \left|s_{P_s}\right|^{-1} s_{P_s}. \tag{43}$$

Let us assume that, thanks to the first addend in the STA, the reaching phase is satisfactorily completed and the sliding regime is entered. From that moment on, $\left|s_{Q_s,P_s}\right| \leq \delta_{Q,P}$, with $\delta_{Q,P}$ close to zero. Considering the most unfavorable case, in which $\left|s_{Q_s,P_s}\right| = \delta_{Q,P}$, and using the definition of s_{Q_s,P_s} in (8) and (9), the following expressions can respectively be worked out from (42) and (43):

$$\ddot{e}_Q + \overbrace{\left(0.5\delta_Q^{-0.5}\lambda_Q + c_Q\right)}^{a_2(c_Q,\,\lambda_Q)} \dot{e}_Q + \overbrace{\left(0.5\delta_Q^{-0.5}\lambda_Q c_Q + \delta_Q^{-1} w_Q\right)}^{a_1(c_Q,\,\lambda_Q,\,w_Q)} e_Q + \overbrace{\delta_Q^{-1} w_Q c_Q}^{a_0(c_Q,\,w_Q)} \int e_Q dt = 0 \tag{44}$$

$$\ddot{e}_P + \overbrace{\left(0.5\delta_P^{-0.5}\lambda_P + c_P\right)}^{b_2(c_P,\,\lambda_P)} \dot{e}_P + \overbrace{\left(0.5\delta_P^{-0.5}\lambda_P c_P + \delta_P^{-1} w_P\right)}^{b_1(c_P,\,\lambda_P,\,w_P)} e_P + \overbrace{\delta_P^{-1} w_P c_P}^{b_0(c_P,\,w_P)} \int e_P dt = 0. \tag{45}$$

Taking the time derivatives of (44) and (45), the following differential equations reflecting the e_Q and e_P error dynamics while in sliding regime are obtained:

$$\dddot{e}_Q + a_2\ddot{e}_Q + a_1\dot{e}_Q + a_0 e_Q = 0 \tag{46}$$

$$\dddot{e}_P + b_2\ddot{e}_P + b_1\dot{e}_P + b_0 e_P = 0. \tag{47}$$

Hence, once $\delta_{Q,P}$ is fixed, adequate selection of $c_{Q,P}$, $\lambda_{Q,P}$ and $w_{Q,P}$ allows attaining certain target error dynamics established through the third-order characteristic equation given next:

$$\left(p^2 + 2\xi\omega_n p + \omega_n^2\right)(p + \alpha\xi\omega_n) = p^3 + \underbrace{(2+\alpha)\,\xi\omega_n}_{d_2} p^2 + \underbrace{\left(1 + 2\alpha\xi^2\right)\omega_n^2}_{d_1} p + \underbrace{\alpha\xi\omega_n^3}_{d_0} = 0 \tag{48}$$

which, provided that α is selected high enough —typically $\alpha \geq 10$—, gives rise to a pair of dominant poles with respect to a third one placed at $p = -\alpha\xi\omega_n$. As a result, it can be considered that target error dynamics are entirely defined via ξ damping coefficient and ω_n natural frequency. Those designer-defined error dynamics would theoretically be

achieved just by tuning $c_{Q,P}$, $\lambda_{Q,P}$ and $w_{Q,P}$ so that $a_2 = d_{2Q}(\xi_Q,\omega_{nQ})$, $a_1 = d_{1Q}(\xi_Q,\omega_{nQ})$, $a_0 = d_{0Q}(\xi_Q,\omega_{nQ})$, $b_2 = d_{2P}(\xi_P,\omega_{nP})$, $b_1 = d_{1P}(\xi_P,\omega_{nP})$ and $b_0 = d_{0P}(\xi_P,\omega_{nP})$ are simultaneously fulfilled. Note that both ξ and ω_n could in general take different values if different dynamic behaviours for reactive and active power errors were required.
Considering the expressions for a_2, a_1, a_0, b_2, b_1 and b_0 provided in (44) and (45), as well as those for d_2, d_1 and d_0 reflected in (48), the latter conditions lead to the following tuning equations:

$$c_Q^3 - d_{2Q}c_Q^2 + d_{1Q}c_Q - d_{0Q} = 0;\; c_P^3 - d_{2P}c_P^2 + d_{1P}c_P - d_{0P} = 0 \tag{49}$$

$$\lambda_Q = 2\left(d_{2Q} - c_Q\right)\delta_Q^{0.5};\; \lambda_P = 2\left(d_{2P} - c_P\right)\delta_P^{0.5} \tag{50}$$

$$w_Q = \left[d_{1Q} - c_Q(d_{2Q} - c_Q)\right]\delta_Q;\; w_P = \left[d_{1P} - c_P(d_{2P} - c_P)\right]\delta_P. \tag{51}$$

It is important to note that the coefficients in (49) coincide with those of target characteristic equation (48), except for the signs of the squared and independent terms, which are negative. It therefore turns out that the three possible values for $c_{Q,P}$ are equal to the roots —poles— of target characteristic equation (48), although their real parts have opposite signs. Since the real parts of the desired poles must necessarily be negative to ensure stability, the latter implies that the real parts of the three possible values for $c_{Q,P}$ will always be positive. As a result, given that expressions in (49) are third-order equations, it is guaranteed that at least one of the three solutions for $c_{Q,P}$ will be both real and positive, as required.
Specifically, depending on the value chosen for $\xi_{Q,P}$, one of the following three cases arises:

1. If $0 < \xi_{Q,P} < 1$, only one of the three solutions for $c_{Q,P}$ is both real and positive, $c_{Q,P} = \alpha\xi_{Q,P}\omega_{nQ,P}$.
2. If $\xi_{Q,P} = 1$, two different acceptable solutions for $c_{Q,P}$ are obtained, $c_{1Q,P} = \omega_{nQ,P}$ and $c_{2Q,P} = \alpha\omega_{nQ,P}$.
3. If $\xi_{Q,P} > 1$, the three solutions for $c_{Q,P}$ are real and positive, $c_{1Q,P} = \omega_{nQ,P}\left(\xi_{Q,P} - \sqrt{\xi_{Q,P}^2 - 1}\right)$, $c_{2Q,P} = \omega_{nQ,P}\left(\xi_{Q,P} + \sqrt{\xi_{Q,P}^2 - 1}\right)$ and $c_{3Q,P} = \alpha\xi_{Q,P}\omega_{nQ,P}$.

For cases 2 and 3, two or three possible sets of values for $c_{Q,P}$, $\lambda_{Q,P}$ and $w_{Q,P}$ are respectively obtained. The set of parameters leading to the best performance may, for example, be identified through simulation.

3.2.2 DFIG disconnected from the grid —Synchronization stage

The design and tuning process of the current controllers, which synchronizes the voltage induced at the open stator to that of the grid, is analogous to that presented in the preceding section 3.2.1. Let the control law with respect to x'-y' reference frame be

$$v_{rx'} = v_{rx'_{ST}} + v_{rx'_{eq}};\; v_{ry'} = v_{ry'_{ST}} + v_{ry'_{eq}}. \tag{52}$$

Taking the time derivatives of (10) and (11), the expressions given next arise if (3) and (4) are considered:

$$\dot{s}_{i_{rx'}} = \dot{i}_{rx'\,ref} + c_{rx'}i_{rx'\,ref} + \left(\frac{R_r}{L_r} - c_{rx'}\right)i_{rx'} - \omega_{sl}i_{ry'} - \frac{1}{L_r}v_{rx'} \tag{53}$$

$$\dot{s}_{i_{ry'}} = \dot{i}_{ry'\,ref} + c_{ry'}i_{ry'\,ref} + \left(\frac{R_r}{L_r} - c_{ry'}\right)i_{ry'} + \omega_{sl}i_{rx'} - \frac{1}{L_r}v_{ry'}. \tag{54}$$

The terms corresponding to the STA may be obtained as

$$v_{rx'_{ST}} = L_r \left[\lambda_{x'} |s_{i_{rx'}}|^{0.5} \mathrm{sgn}(s_{i_{rx'}}) + w_{x'} \int \mathrm{sgn}(s_{i_{rx'}}) dt \right] \tag{55}$$

$$v_{ry'_{ST}} = L_r \left[\lambda_{y'} |s_{i_{ry'}}|^{0.5} \mathrm{sgn}(s_{i_{ry'}}) + w_{y'} \int \mathrm{sgn}(s_{i_{ry'}}) dt \right], \tag{56}$$

where $\lambda_{x'}$, $w_{x'}$, $\lambda_{y'}$ and $w_{y'}$ are positive parameters to be tuned. As far as equivalent control terms are concerned, they are derived by zeroing (53) and (54), thus yielding

$$v_{rx'_{eq}} = L_r \left[\dot{i}_{rx'\,ref} + \frac{R_r}{L_r} i_{rx'} - \omega_{sl} i_{ry'} + c_{rx'} (i_{rx'\,ref} - i_{rx'}) \right] \tag{57}$$

$$v_{ry'_{eq}} = L_r \left[\dot{i}_{ry'\,ref} + \frac{R_r}{L_r} i_{ry'} + \omega_{sl} i_{rx'} + c_{ry'} (i_{ry'\,ref} - i_{ry'}) \right]. \tag{58}$$

Again, note that all the control terms are premultiplied by a $-L_r$ gain in this case, which is the inverse of that affecting control signals in (53) and (54). As mentioned before, its only purpose is to facilitate the tuning of the parameters involved in the commanding algorithm.
Substitution of control law (52) into (53) and (54) leads to

$$\dot{s}_{i_{rx'}} = -\lambda_{x'} |s_{i_{rx'}}|^{0.5} \mathrm{sgn}(s_{i_{rx'}}) - w_{x'} \int \mathrm{sgn}(s_{i_{rx'}}) dt \tag{59}$$

$$\dot{s}_{i_{ry'}} = -\lambda_{y'} |s_{i_{ry'}}|^{0.5} \mathrm{sgn}(s_{i_{ry'}}) - w_{y'} \int \mathrm{sgn}(s_{i_{ry'}}) dt, \tag{60}$$

expressions which turn out to be identical to those presented in (40) and (41) provided that $'i_{rx'}'$ and $'x''$ subscripts are respectively replaced by $'Q_s'$ and $'Q'$, and, likewise, $'i_{ry'}'$ and $'y''$ subscripts are interchanged with $'P_s'$ and $'P'$. Therefore, the same reasoning detailed in section 3.2.1 can be followed in order to achieve the tuning equations of $\lambda_{x'}$, $w_{x'}$, $c_{x'}$, $\lambda_{y'}$, $w_{y'}$ and $c_{y'}$ parameters. As a result,

$$c_{x'}^3 - d_{2x'} c_{x'}^2 + d_{1x'} c_{x'} - d_{0x'} = 0; \; c_{y'}^3 - d_{2y'} c_{y'}^2 + d_{1y'} c_{y'} - d_{0y'} = 0 \tag{61}$$

$$\lambda_{x'} = 2 \left(d_{2x'} - c_{x'} \right) \delta_{x'}^{0.5}; \; \lambda_{y'} = 2 \left(d_{2y'} - c_{y'} \right) \delta_{y'}^{0.5} \tag{62}$$

$$w_{x'} = \left[d_{1x'} - c_{x'} (d_{2x'} - c_{x'}) \right] \delta_{x'}; \; w_{y'} = \left[d_{1y'} - c_{y'} (d_{2y'} - c_{y'}) \right] \delta_{y'} \tag{63}$$

3.2.3 Bumpless connection

Considering that the DFIG presents different dynamics when disconnected or connected to the grid, two STA-based controllers have been designed for generating a continuous command signal. So far, the performance for each state has only been considered, but undesirable phenomena may appear if the switch between the two controllers is not properly carried out. If a direct transition is accomplished, a discontinuity arises in the command signal at the instant of connection, due to the magnitude mismatch between the rotor voltages generated by the two controllers. This effect produces high stator current values, leading the machine to an excessive power exchange with the grid. Aiming to avoid this "bump", it is possible to apply the same value of the control signal previous to and just after the transition —$k-1$ and k instants respectively—; i.e.,

$$v_{rx} = v_{rx'} \tag{64}$$

$$v_{ry} = v_{ry'}. \tag{65}$$

However, this way the "bump" is only delayed one sample time and it actually takes effect in the next sampling instant —$k+1$.
A bumpless transition may take place if the above proposed solution is slightly modified (Åström & Hägglund, 1995). Setting the focus on the rotor voltage components when the DFIG is connected to the grid, appropriate combination of (36), (38) and (37), (39) produces

$$v_{rx} = -\frac{2}{3}\frac{L_s L_r'}{|\vec{v}_s| L_m}\left[\lambda_Q |s_{Q_s}|^{0.5}\text{sgn}(s_{Q_s}) + w_Q \int \text{sgn}(s_{Q_s})dt + \dot{Q}_{s\,ref} + c_Q(Q_{s\,ref} - Q_s)\right] + \\ + R_r i_{rx} - L_r' \omega_{sl} i_{ry} \quad (66)$$

$$v_{ry} = -\frac{2}{3}\frac{L_s L_r'}{|\vec{v}_s| L_m}\left[\lambda_P |s_{P_s}|^{0.5}\text{sgn}(s_{P_s}) + w_P \int \text{sgn}(s_{P_s})dt + \dot{P}_{s\,ref} + c_P(P_{s\,ref} - P_s)\right] + \\ + R_r i_{ry} + \frac{L_m}{L_s}\omega_{sl}|\vec{\psi}_s| + L_r' \omega_{sl} i_{rx}. \quad (67)$$

Two integral terms, $I_{sgn(s_{Q_s})} = \int \text{sgn}(s_{Q_s})dt$ and $I_{sgn(s_{P_s})} = \int \text{sgn}(s_{P_s})dt$, can be observed. Their initial values, which are set to zero when connection occurs, are the source of the mentioned "bump". Aiming at lessening or even eliminating this effect, it can be taken advantage of (64) and (65) to calculate those initial values at connection time. Substituting (66) and (67) into (64) and (65), respectively, leads to

$$I_{\text{sgn}(s_{Q_s})_0} = -\frac{3}{2}\frac{|\vec{v}_s| L_m}{L_s L_r' w_Q}\left(v_{rx'} - R_r i_{rx} + L_r' \omega_{sl} i_{ry}\right) - \\ - \frac{\lambda_Q |s_{Q_s}|^{0.5}\text{sgn}(s_{Q_s}) + \dot{Q}_{s\,ref} + c_Q(Q_{s\,ref} - Q_s)}{w_Q} \quad (68)$$

$$I_{\text{sgn}(s_{P_s})_0} = -\frac{3}{2}\frac{|\vec{v}_s| L_m}{L_s L_r' w_P}\left(v_{ry'} - R_r i_{ry} - \frac{L_m}{L_s}\omega_{sl}|\vec{\psi}_s| - L_r' \omega_{sl} i_{rx}\right) - \\ - \frac{\lambda_P |s_{P_s}|^{0.5}\text{sgn}(s_{P_s}) + \dot{P}_{s\,ref} + c_P(P_{s\,ref} - P_s)}{w_P}. \quad (69)$$

3.3 Sensorless scheme —Adaptation for synchronization

Both the 1-SMC and 2-SMC designs are combined with the MRAS observer proposed by (Peña et al., 2008) in order to build two alternative sensorless control schemes. As a result, the controller is provided with the estimated rotor electrical speed and position, thus avoiding both the use of mechanical components —encoders— and the initial rotor positioning required for the synchronization of the stator and grid voltages (Tapia et al., 2009). Morover, observers may be used for “chattering” phenomenon alleviation (Utkin et al., 1999; Utkin, 1993).
It is worth pointing out that the MRAS observer must be adapted for the case of being disconnected from the grid. On the one hand, since stator currents are null in this state, calculation of stator flux in the stationary s_D-s_Q frame must be slightly modified

$$\psi_{sQ} = \int (v_{sQ} - R_s i_{sQ})dt \rightarrow \psi_{sQ} = \int v_{sQ} dt \quad (70)$$

$$\psi_{sD} = \int (v_{sD} - R_s i_{sD})dt \rightarrow \psi_{sD} = \int v_{sD} dt. \quad (71)$$

On the other hand, the stator voltages are those induced by the rotor currents for synchronization, and present a considerable noise. For a proper control, the affected signals are filtered out by means of a 500-Hz bandwidth second-order Butterworth filter, which in turn produces a phase lag of $\gamma = 8.1297°$ at 50Hz. This lag must be compensated in order to estimate the components of the actual stator voltage phasor. Considering Fig. 5, it can be stated that

$$|\vec{v}_s| = |\vec{v}_{s\ filtered}| = \sqrt{v^2_{sD\ filtered} + v^2_{sQ\ filtered}} \tag{72}$$

$$\beta = \arctan \frac{v_{sQ\ filtered}}{v_{sD\ filtered}}. \tag{73}$$

Furthermore, given that the γ phase lag is known deriving the argument of $\vec{v}_s$ from Fig. 5 as $\arg(\vec{v}_s) = \beta + \gamma$, the components of the actual stator voltage space-phasor given next arise:

$$v_{sD} = |\vec{v}_s| \cos(\beta + \gamma) \tag{74}$$

$$v_{sQ} = |\vec{v}_s| \sin(\beta + \gamma). \tag{75}$$

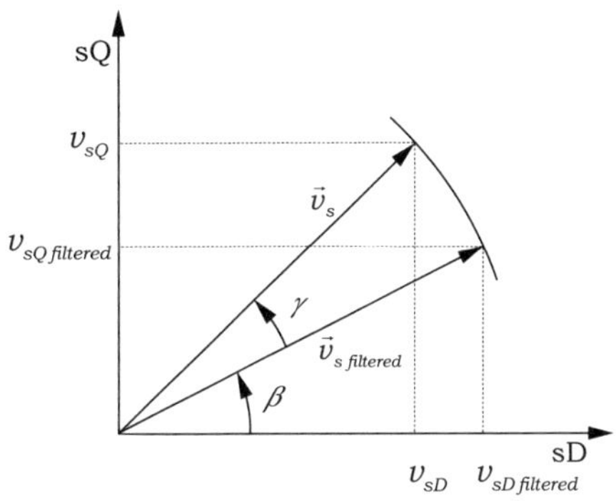

Fig. 5. Actual and filtered stator voltage space-phasors

4. Hardware-in-the-loop results

The presented sensorless 1-SMC and 2-SMC algorithms are evaluated, through real-time HIL emulation, over a full-detail virtual DFIG prototype running on eMEGAsim OP4500 F11-13 simulator by OPAL-RT. The electric parameters of the 660-kW DFIG under consideration are collected in Table 1.

Aiming at showing some of the most illustrative results of the two alternative control algorithms put forward, the test whose main events are reflected in Table 2 is conducted. It should be pointed out that what causes the DFIG control system to generate the order of connection to the grid, taking place at second 0.474, is the DFIG rotational speed exceeding the already mentioned threshold of 1270 rpm.

The 1-SMC algorithm itself is implemented on a Virtex-II Pro series FPGA by Xilinx, which allows reaching the 40-kHz sampling rate required to avoid causing excessive chatter. Direct measurement of the grid voltage allows accurately computing angle ρ'_s, and, as a result, identifying the exact position of the x'-y' reference frame. On the other hand, the ρ_s angle, which provides the location of the stator-flux-oriented x-y reference frame, is derived from the direct (ψ_{sD}) and quadrature (ψ_{sQ}) stationary-frame components of the stator flux. These are in turn estimated by integration of the stator voltage minus the resistive drop. A digital bandpass filter is used as a modified integrator to avoid drift (Peña et al., 2008). Regarding

PARAMETER	VALUE
Rated r.m.s. stator voltage	398/690 V
Rated peak rotor voltage	380 V
Rated peak rotor current	400 A
Stator resistance per phase, R_s	6.7 mΩ
Stator inductance per phase, L_s	7.5 mH
Magnetizing inductance, L_m	19.4 mH
Rotor resistance per phase, R_r	39.9 mΩ
Rotor inductance per phase, L_r	52 mH
General turns ratio, n	0.3806
Number of pole pairs, P	2

Table 1. DFIG electric parameters

EVENT	TIME INSTANT (S)
Order of connection to the grid; start of synchronization process; initial convergence of the observer	0.474
End of synchronization process; connection to the grid at zero power	1.474
Start of power generation	1.974
Sudden increase of wind speed	7
Sudden decrease of wind speed	12

Table 2. Main events of the designed test

switching variables, in this particular case, c_Q and c_P are set to 10, while $c_{x'}$ and $c_{y'}$ are made equal to 0. In addition, the integral terms in s_{P_s} and s_{Q_s} are discretized by applying Tustin's trapezoidal method (Kuo, 1992).

Fig. 6(a) displays a general portrait of the synchronization stage. In addition, details at both its beginning and its end are reflected in Figs. 6(b) and 6(c), respectively. The former evidences the rapid dynamic response of v_{sA}, v_{sB} and v_{sC} voltages, induced at the terminals of the DFIG open stator, when synchronizing with $v_{grid\,A}$, $v_{grid\,B}$ and $v_{grid\,C}$ grid voltages. As expected, no active and reactive powers are exchanged between the DFIG stator and the grid during synchronization, as corroborated by Figs. 7(a) and 7(b).

The DFIG is then connected to the grid at zero power, until power generation according to the optimum power curve is launched 0.5 s later. Fig. 7(a) shows the excellent performance of the stator-side active power, P_s, when, as a result of the two sudden wind speed changes occurring at seconds 7 and 12, a great part of the optimum power curve is tracked both up and downwards. The also superior tracking of the target stator-side reactive power, $Q_{s\,ref}$, is evidenced in Fig. 7(b). The instantaneous reference value for Q_s is fixed so that the DFIG operates with a 0.95 leading —capacitive— power factor all through the test. Chatter in P_s and Q_s represents only ±3% of the rated power.

Given that, as indicated above, control signals are updated at a 40-kHz sample rate, gating signals s_{wk}; $k = 1, 2 \ldots 6$, are able to toggle every 25 μs, if required. This results in a maximum switching frequency of 20 kHz with 50% duty cycle. However, switching frequency of the RSC insulated gate bipolar transistors (IGBTs) is variable, as dictated by switching functions s_1, s_2 and s_3. This is clearly observable in Fig. 8(a), where the frequency spectrum corresponding to

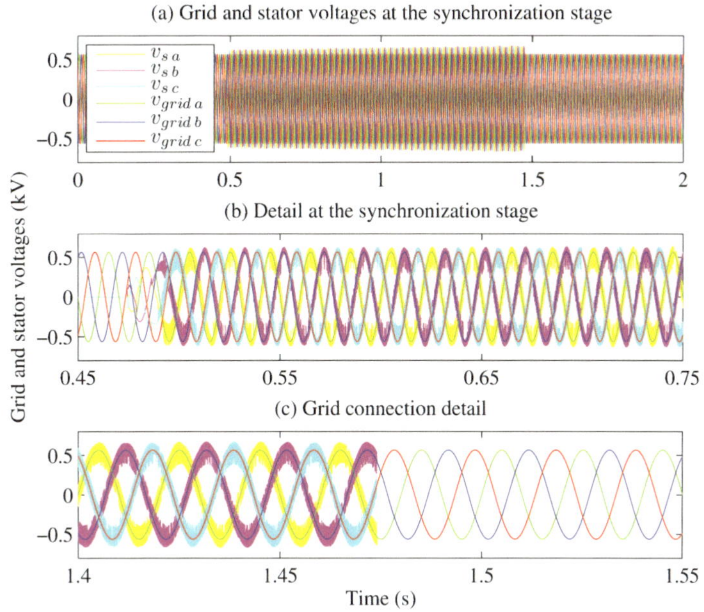

Fig. 6. Stator and grid voltages of the 1-SMC controller-driven DFIG at the synchronization stage

gating signal s_{w1} is displayed. The three-phase rotor current resulting from the gating signals applied to the RSC IGBTs is that displayed in Fig. 8(b). As expected, it turns out to be variable in magnitude, frequency and phase.

Fig. 9 reflects the performance of the digital MRAS observer, which operates at a 1-kHz sample rate. In this particular case, it is incorporated not only for sensorless control, but also as a supporting tool for chatter attenuation (Utkin et al., 1999; Utkin, 1993). The observer is launched once the order of connection to the grid is automatically generated. As evidenced in Fig. 9(a), the estimated rotor mechanical speed converges rapidly to its actual value at the earlier part of the synchronization stage, and, from that point onwards, keeps track of it satisfactorily in spite of the transition from the disconnected state to the connected one taking place at second 1.474. Moreover, a detail illustrating the fast convergence of the estimated rotor electrical position to its actual value is displayed in Fig. 9(b).

As far as the sensorless 2-SMC algorithm is concerned, it is programmed in C language on a DSP-based board. Control signals v_{rx} and v_{ry} are demodulated to derive the $v_{r\alpha}$ and $v_{r\beta}$ voltage components, expressed in the rotor natural reference frame, which are then supplied as inputs to the SVM algorithm generating the gating signals of RSC IGBTs. Angles ρ_s and ρ_s', required both to estimate equivalent control terms and to demodulate v_{rx} and v_{ry} control signals, are derived in the same manner as for the 1-SMC algorithm. Both the 2-SMC and SVM algorithms operate at a 5-kHz sample rate, while the MRAS observer runs, as in the preceding case, at 1 kHz. The integral terms included in both the switching functions and the STA algorithm itself are digitally implemented based on Tustin's trapezoidal method. Yet, aiming to elude the risk of causing derivative "ringing" (Åström & Hägglund, 1995), Euler's rectangular method is applied to discretize the time derivatives of $P_{s\,ref}$ and $Q_{s\,ref}$ appearing in equivalent control terms.

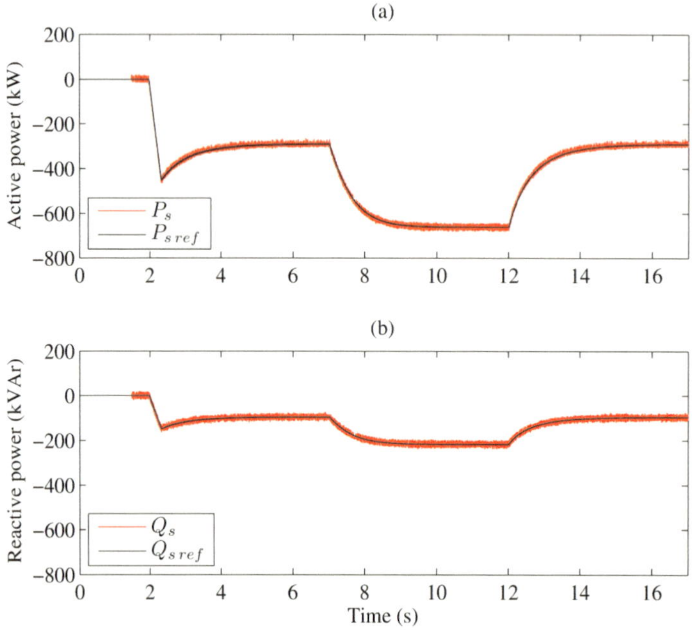

Fig. 7. Active and reactive powers of the DFIG commanded by the 1-SMC controller

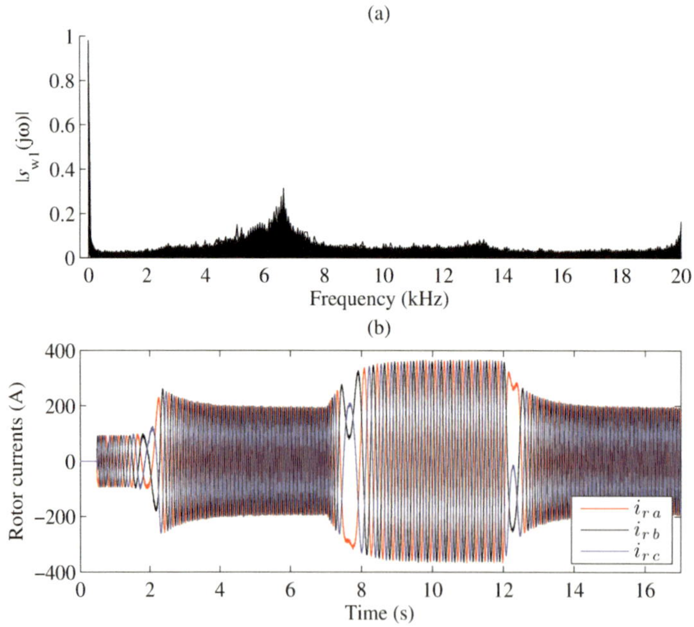

Fig. 8. Frequency spectrum of s_{w1} gating signal, and resulting three-phase rotor current

Selecting a $\delta_{x',y'} = 0.01$ A, control parameters for synchronization are adjusted seeking to reach closed-loop rotor current error dynamics exhibiting a unit damping coefficient and a $\omega_{nx',y'} = 55.2381$ rad/s natural frequency while in sliding regime. As a result, if different

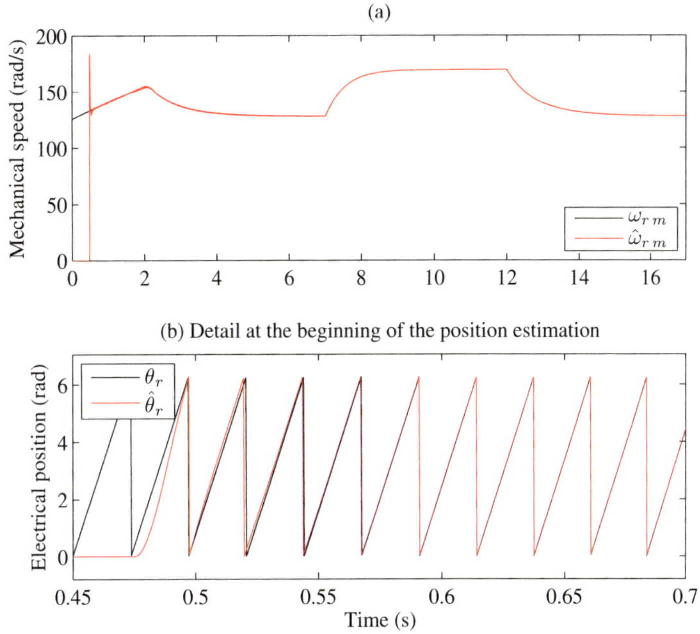

Fig. 9. Actual and estimated mechanical speed and electrical position of the DFIG rotor

from zero, errors in rotor current $i_{rx'}$ and $i_{ry'}$ components would vanish, according to the 2% criterion (Ogata, 2001), in 105 ms, showing no overshoots. Similarly, forcing $\delta_{P,Q}$ to be equal to 0.1 kW, and specifying $\xi_{P,Q} = 1$ and $\omega_{nP,Q} = 82.8571$ rad/s, respectively, as target damping coefficient and natural frequency for closed-loop power error dynamics, possible errors arising in active and reactive powers would theoretically decay to zero in 70 ms, with no overshoots. If α is made equal to 10, the values resulting for the 2-SMC algorithm parameters are those collected in Table 3.

PARAMETER	VALUE
$c_{x'}$, $c_{y'}$	55.2381
$\lambda_{x'}$, $\lambda_{y'}$	121.5238
$w_{x'}$, $w_{y'}$	305.1247
c_P, c_Q	82.8571
λ_P, λ_Q	$1.8229 \cdot 10^4$
w_P, w_Q	$6.8653 \cdot 10^6$

Table 3. Values for the parameters belonging to the 2-SMC algorithm

The most significant results corresponding to the applied sensorless SVM-based 2-SMC are illustrated by Figs. 10, 11 and 12. Given that those figures are very similar to their corresponding 1-SMC counterparts —Figs. 6, 7 and 8, respectively—, only the main differences between them will accordingly be commented on.

Comparison of Figs. 6 and 7 with Figs. 10 and 11, respectively, reveals that the resulting chatter is somewhat lower for the 2-SMC case. In addition, synchronization of the voltage induced in the DFIG open stator to that of the grid is achieved faster when applying the 1-SMC algorithm,

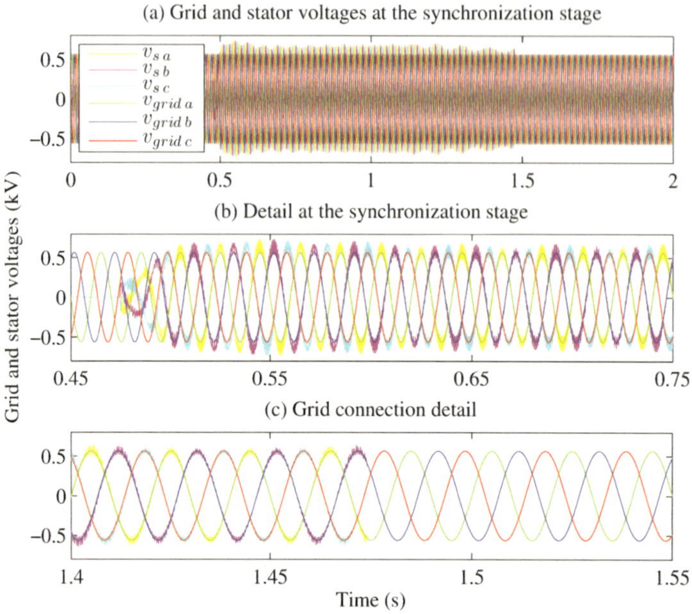

Fig. 10. Stator and grid voltages of the 2-SMC controller-driven DFIG at the synchronization stage

as evidenced by Figs. 6 and 10. Furthermore, even though the 2-SMC algorithm is provided with bumpless transfer from the disconnected state to the connected one, Figs. 7 and 11 prove that power exchange with the grid at the instant of connection is considerably lower for the case of the 1-SMC.

On the other hand, since the SVM-based 2-SMC leads to a 5-kHz constant switching frequency of the RSC IGBTs, in Fig. 12(a) the frequency spectrum of Fig. 8(a) has been replaced with the $v_{r\alpha}$ and $v_{r\beta}$ voltage components supplied as inputs to the SVM algorithm. The smoothness of $v_{r\alpha}$ and $v_{r\beta}$ in Fig. 12(a) indicates that, like in classical PI controller-based FOC schemes, chatter in P_s and Q_s observable in Fig. 11 is just attributable to SVM, not to the 2-SMC algorithm itself.

To conclude, as it turns out that the MRAS observer performance is extremely similar to that resulting in the case of the sensorless 1-SMC, it is not included here to avoid reiteration.

5. Conclusion

Real-time HIL emulation results obtained by running sensorless versions of the 1-SMC and 2-SMC arrangements presented in this chapter reveal that excellent tracking of a predefined rotor speed-dependent optimum power curve is reached in both cases. In addition, prior to connecting the DFIG stator to the grid, they are also capable of achieving satisfactory synchronization of the voltage induced at the open stator terminals to that of the grid.

In any case, it may be of interest to contrast both SMC algorithms, so as to identify the strengths and weaknesses associated to each of them. This section will hence focus on that comparison.

As far as the complexity of the algorithm itself is concerned, the 1-SMC version turns out to be considerably simpler than the 2-SMC one. Given that the control signals generated

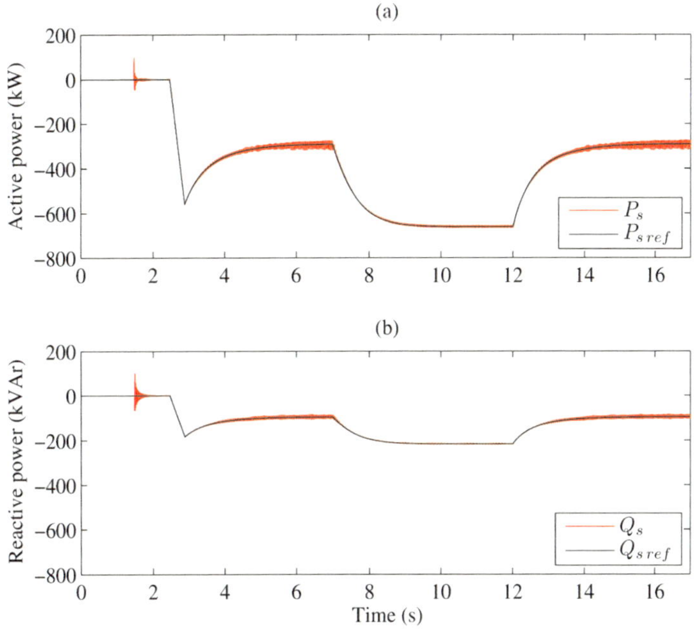

Fig. 11. Active and reactive powers of the DFIG commanded by the 2-SMC controller

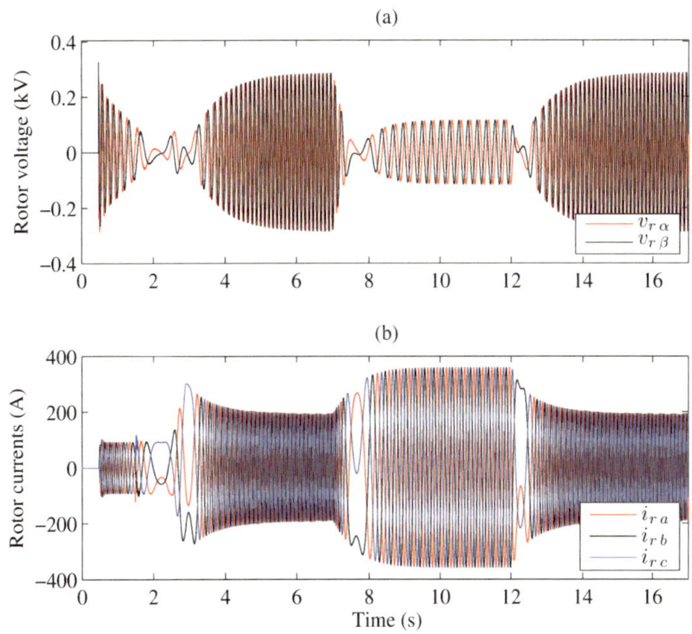

Fig. 12. Rotor voltage components fed into the SVM algorithm, and resulting three-phase rotor current

by the 1-SMC correspond to the gating signals of the RSC IGBTs, no additional modulation techniques —such as pulse-width modulation (PWM) or SVM— are required. In contrast, the

control signals produced by the proposed 2-SMC correspond to continuous voltage direct and quadrature components to be applied to the rotor by means of the RSC, which implies the use of intermediate SVM modulation. Furthermore, providing an additional procedure for bumpless transition between the algorithms devoted to synchronization and power control is indispensable for the case of the 2-SMC, but it is not required for the 1-SMC scheme.
Regarding parameter tuning, only the c constants included in the four switching functions considered need to be tuned for the case of the 1-SMC. It therefore turns out that satisfactory parameter adjustment is easily achieved by mere trial and error. However, in addition to those c constants, the λ and w gains present in the STAs must also be tuned for the 2-SMC variant. Even though, as stated in (Bartolini et al., 1999), it is actually the most common practice, trial and error tuning is not particularly effective in this latter case, as it may become highly time-consuming. Therefore, it is believed that there exists a strong need for development of alternative methods for STA-based 2-SMC tuning.
Concerning the switching frequency of the RSC IGBTs, it is fixed at 5 kHz in the case of the 2-SMC. On the contrary, it turns out to be variable, within the range from 0 to 20 kHz, for the 1-SMC algorithm. This feature complicates the design of both the back-to-back converter feeding the DFIG rotor and the grid-side AC filter, since broadband harmonics may be injected into the grid. As a result of the 25-μs sample time selected for the 1-SMC scheme, which leads to the aforementioned maximum switching frequency of 20 kHz, chatter observable in stator-side active and reactive powers is somewhat lower than $\pm 3\%$ of the DFIG 660-kW rated power. Even a lower level of chatter arises from application of the SVM-based 2-SMC algorithm put forward. Furthermore, that chatter, or at least great part of it, is caused by the SVM, not by the 2-SMC algorithm itself.
Apart from the superior optimum power curve tracking achieved with both alternative SMC designs, the dynamic performance resulting from realization of the proposed 1-SMC scheme is noticeably better than that to which application of its 2-SMC counterpart leads. In effect, focusing on the state in which the DFIG stator is disconnected from the grid, HIL emulation results demonstrate that synchronization is reached faster by employing the 1-SMC algorithm. On the other hand, the power exchange between the DFIG and the grid taking place at the initial instants after connection is significantly lower when adopting the 1-SMC algorithm put forward, hence evidencing that its dynamic performance is also better for the stage during which power control is dealt with. The excellent dynamic performance reachable by means of its application supports the 1-SMC approach as a potential candidate for DFIG control under grid faults, where rapidity of response becomes crucial.
The main conclusions drawn from the comparison conducted in this section are summarized in Table 4.

	1-SMC ALGORITHM	2-SMC ALGORITHM
ALGORITHM COMPLEXITY	Relatively simple	More complex
PWM/SVM	Not required	Required
BUMPLESS PROCEDURE	Not required	Required
PARAMETER TUNING	Straightforward	Complex
SWITCHING FREQUENCY	Variable from 0 to 20 kHz	Fixed at 5 kHz
CHATTER LEVEL	$\pm 3\%$ of the rated power	Lower
DYNAMIC PERFORMANCE	Excellent	Very good

Table 4. Comparison between the two SMC algorithms put forward

6. References

Abo-Khalil, A. G., Lee, D.-C. & Lee, S.-H. (2006). Grid connection of doubly fed induction generators in wind energy conversion system, *Proceedings of the CES/IEEE 5th International Power Electronics and Motion Control Conference (IPEMC 2006)*, Shanghai, China, vol. 3, pp. 1–5.

Arnaltes, S. & Rodríguez, J. L. (2002). Grid synchronisation of doubly fed induction generators using direct torque control, *Proceedings of the IEEE 28th Annual Conference of the Industrial Electronics Society (IECON 2002)*, Seville, Spain, pp. 3338–3343.

Åström, K. J. & Hägglund, T. (1995). *PID Controllers: Theory, Design and Tuning*, Instrument Society America, USA.

Bartolini, G., Ferrara, A., Levant, A. & Usai, E. (1999). On second order sliding mode controllers, *in* K. Young & Ü. Özgüner (eds), *Variable Structure Systems, Sliding Mode and Nonlinear Control*, Springer Verlag, London, UK, pp. 329–350.

Beltran, B., Ahmed-Ali, T. & Benbouzid, M. E. H. (2008). Sliding mode power control of variable-speed wind energy conversion systems, *IEEE Transactions on Energy Conversion* 23(2): 551–558.

Beltran, B., Ahmed-Ali, T. & Benbouzid, M. E. H. (2009). High-order sliding-mode control of variable-speed wind turbines, *IEEE Transactions on Industrial Electronics* 56(9): 3314–3321.

Beltran, B., Benbouzid, M. E. H. & Ahmed-Ali, T. (2009). High-order sliding mode control of a DFIG-based wind turbine for power maximization and grid fault tolerance, *Proceedings of the IEEE International Electric Machines and Drives Conference (IEMDC 2009)*, Miami, USA, pp. 183–189.

Ben Elghali, S. E., Benbouzid, M. E. H., Ahmed-Ali, T., Charpentier, J. F. & Mekri, F. (2008). High-order sliding mode control of DFIG-based marine current turbine, *Proceedings of the IEEE 34th Annual Conference of the Industrial Electronics Society (IECON 2008)*, Orlando, USA, pp. 1228–1233.

Blaabjerg, F., Teodorescu, R., Liserre, M. & Timbus, A. V. (2006). Overview of control and grid synchronization for distributed power generation systems, *IEEE Transactions on Industrial Electronics* 53(5): 1398–1409.

Ekanayake, J. B., Holdsworth, L., Wu, X. & Jenkins, N. (2003). Dynamic modeling of doubly fed induction generator wind turbines, *IEEE Transactions on Power Systems* 18(2): 803–809.

Kuo, B. C. (1992). *Digital Control Systems*, Oxford University Press, New York, USA.

Levant, A. (1993). Sliding order and sliding accuracy in sliding mode control, *International Journal of Control* 58(6): 1247–1263.

Ogata, K. (2001). *Modern Control Engineering*, Prentice Hall, Englewood Cliffs (New Jersey), USA.

Peña, R., Cárdenas, R., Proboste, J., Asher, G. & Clare, J. (2008). Sensorless control of doubly-fed induction generators using a rotor-current-based MRAS observer, *IEEE Transactions on Industrial Electronics* 55(1): 330–339.

Peña, R., Clare, J. C. & Asher, G. M. (1996). Doubly fed induction generator using back-to-back PWM converters and its application to variable-speed wind-energy generation, *IEE Proceedings - Electric Power Applications* 143(3): 231–241.

Peresada, S., Tilli, A. & Tonielli, A. (2004). Power control of a doubly fed induction machine via output feedback, *Control Engineering Practice* 12(1): 41–57.

Rashed, M., Dunnigan, M. W., MacConell, P. F. A., Stronach, A. F. & Williams, B. W. (2005). Sensorless second-order sliding-mode speed control of a voltage-fed induction-motor drive using nonlinear state feedback, *IEE Proceedings-Electric Power Applications* 152(5): 1127–1136.

Susperregui, A., Tapia, G., Zubia, I. & Ostolaza, J. X. (2010). Sliding-mode control of doubly-fed generator for optimum power curve tracking, *IET Electronics Letters* 46(2): 126–127.

Tapia, A., Tapia, G., Ostolaza, J. X. & Sáenz, J. R. (2003). Modeling and control of a wind turbine driven doubly fed induction generator, *IEEE Transactions on Energy Conversion* 12(2): 194–204.

Tapia, G., Santamaría, G., Telleria, M. & Susperregui, A. (2009). Methodology for smooth connection of doubly fed induction generators to the grid, *IEEE Transactions on Energy Conversion* 24(4): 959–971.

Tapia, G., Tapia, A. & Ostolaza, J. X. (2006). Two alternative modeling approaches for the evaluation of wind farm active and reactive power performances, *IEEE Transactions on Energy Conversion* 21(4): 901–920.

Utkin, V., Guldner, J. & Shi, J. (1999). *Siliding Mode Control in Electromechanical Systems*, Taylor & Francis, London, UK.

Utkin, V. I. (1993). Sliding mode control design principles and applications to electric drives, *IEEE Transactions on Industrial Electronics* 40(1): 23–36.

Vas, P. (1998). *Sensorless Vector and Direct Torque Control of AC Machines*, Oxford University Press, New York, USA.

Xu, L. & Cartwright, P. (2006). Direct active and reactive power control of DFIG for wind energy generation, *IEEE Transactions on Energy Conversion* 21(3): 750–758.

Yan, W., Hu, J., Utkin, V. & Xu, L. (2008). Sliding mode pulsewidth modulation, *IEEE Transactions on Power Electronics* 23(2): 619–626.

Yan, Z., Jin, C. & Utkin, V. I. (2000). Sensorless sliding-mode control of induction motors, *IEEE Transactions on Industrial Electronics* 47(6): 1286–1297.

Zhi, D. & Xu, L. (2007). Direct power control of DFIG with constant switching frequency and improved transient performance, *IEEE Transactions on Energy Conversion* 22(1): 110–118.

Part 2

Sliding Mode Control of Electric Drives

Sliding Mode Control Design for Induction Motors: An Input-Output Approach

John Cortés-Romero[1], Alberto Luviano-Juárez[2] and Hebertt Sira-Ramírez[3]
[1]*Universidad Nacional de Colombia. Facultad de Ingeniería, Departamento de Ingeniería Eléctrica y Electrónica. Carrera 30 No. 45-03 Bogotá*
[1,2,3]*Cinvestav IPN, Av. IPN No. 2508, Departamento de Ingeniería Eléctrica, Sección de Mecatrónica*
[1]*Colombia*
[2,3]*México*

1. Introduction

Three-phase induction motors have been widely used in a variety of industrial applications. Induction motors have been able to incrementally improve energy efficiency to satisfy the requirements of reliability and efficiency, Melfi et al. (2009). There are well known advantages of using induction motors over permanent magnet DC motors for position control tasks; thus, efforts aimed at improving or simplifying feedback controller design are well justified.
There exists a variety of control strategies that depend on *difficult to measure* motor parameters while their closed loop behavior is found to be sensitive to their variations. Even adaptive schemes tend to be sensitive to speed-estimation errors, yielding to a poor performance in the flux and torque estimation, especially during low-speed operation, Harnefors & Hinkkanen (2008).
Generally speaking, the designed feedback control strategies have to exhibit a certain robustness level in order to guarantee an acceptable performance. It is possible to (on-line or off-line) obtain estimates of the motor parameters, Hasan & Husain (2009); Toliyat et al. (2003), but some of them can be subject to variation when the system is undergoing actual operation. Frequent misbehavior is due to external and internal disturbances, such as generated heat, that significantly affect some of the system parameter values. An alternative to overcome this situation is to use robust feedback control techniques which take into account these variations as unknown disturbance inputs that need to be rejected. In this context, sliding mode techniques are a good alternative due to their disturbance rejection capability (see for instance, Utkin et al. (1999)).
In this chapter, we consider a two stage control scheme, the first one is devoted to the control of the rotor shaft position. This analog control is performed by means of the stator current inputs, in a configuration of an observer based control. The mathematical model of the rotor dynamics is a simplified model including additive, completely unknown, lumping nonlinearities and external disturbances whose effect is to be determined in an on-line fashion by means of linear observers. The gathered knowledge will be used in the appropriate canceling of the assumed perturbations themselves while reducing the underlying control problem to a simple linear feedback control task. The control scheme thus requires a rather reduced set of parameters to be implemented.

The observation scheme for the modeled perturbation is based on an extension of the Generalized Proportional Integral (GPI) controller, Fliess, Marquez, Delaleau & Sira-Ramírez (2002) to their dual counterpart: the GPI observer which corresponds to a class of extended Luenberger-like observers, Luviano-Juárez et al. (2010). Such observers were introduced in, Sira-Ramirez, Feliu-Batlle, Beltran-Carbajal & Blanco-Ortega (2008) in the context of Sigma-Delta modulation observer tasks for the detection of obstacles in flexible robotics. Under reasonable assumptions, the observation technique consists in viewing the measured output of the plant as generated by an equivalent perturbed pure integration dynamics with an additive perturbation input lumping, in a single function, all the nonlinearities of the output dynamics. The linear GPI observer, is set to approximately estimate the states of the pure integration system as well as the evolution of the, state dependent, perturbation input. This observer allows one to approximately estimate, on the basis of the measured output, the states of the nonlinear system, as well as to closely estimate the unknown perturbation input. The proposed observation scheme allows one to solve, rather accurately, the disturbance estimation problem.

Here, these observers are used in connection with a robust controller design application within the context of high gain observation. This approach is prone to overshot effects and may be deemed sensitive to saturation input constraints, specially when used in a high gain oriented design scheme via the choice of large eigenvalues. Such a limitation is, in general, an important weakness in many practical situations. However, since our control scheme is based on a linear observer design that can undergo temporary saturations and smooth "clutchings" into the feedback loop, its effectiveness can be enhanced without affecting the controller structure and the overall performance. We show that the observer-based control, overcomes these adverse situations while enhancing the performance of the classical GPI based control scheme.

The linear part of the controller design is based on the Generalized Proportional Integral output feedback controller scheme established in terms of Module Theory.

In the second design stage, the designed current signals of the first stage are deemed as reference trajectories, and a discontinuous feedback control law for the input voltages is sought which tracks the reference trajectories. Since the electrical subsystem is faster than the mechanical, we propose a sliding mode control approach based on a class of filtered sliding surfaces which consist in regarding the traditional surface with the addition of a low pass filter, without affecting the relative degree condition of the sliding surface. The "chattering effect" related to the sliding mode application is eased by means of a first order low-pass filter as proposed in, Utkin et al. (1999).

GPI control has been established as an efficient linear control technique (See Fliess et al., Fliess, Marquez, Delaleau & Sira-Ramírez (2002)); it has been shown, in, Sira-Ramírez & Silva-Ortigoza (2006), to be intimately related to classical compensator networks design. The main limitation of this approach lies in the assumption that the available output signal coincides with the system's flat output (See Fliess *et al.*Fliess et al. (1995), and also Sira-Ramírez and Agrawal, Sira-Ramírez & Agrawal (2004)) and, hence, the underlying system is, both, controllable and, also, observable from this special output. Nevertheless, this limitation is lifted for the case of the induction motor system.

The controller design is carried out with the philosophy of the classical field oriented controller scheme and implemented through a flux simulator, or reconstructor (see Chiasson, Chiasson (2005)). The methodology is tested and illustrated in an actual laboratory implementation of the induction motor plant in a position trajectory tracking task.

The rest of the chapter is presented as follows: Section 2 describes each of the methodologies to use along the chapter such as the sliding mode control method, the Generalized Proportional Integral control and the disturbance observer. The modeling of the motor and the problem formulation are given in Section 3, and the proposed methodologies are joined to solve the problem in Section 4. The results of the approach are obtained in an experimental framework, as depicted in Section 5. Finally some concluding remarks are given.

2. Some preliminary aspects

2.1 Sliding mode control using a proportional integral surface: Introductory example

Consider the following first order system:

$$\dot{y} = u + \xi(t) \tag{1}$$

where y is the output of the system, $\xi(t)$ can be interpreted as a disturbance input (which may be state dependent) and $u \in \{-W, W\}$ is a switched class input. We propose here to take as a sliding surface coordinate function the following expression in Laplace domain s:

$$\begin{aligned} \sigma &= -\frac{s+z}{s}e \\ e &= y - y^* \end{aligned} \tag{2}$$

with $z > 0$.
The switched control is defined as

$$\begin{aligned} u &= W\text{sign}(\sigma), \\ W &> 0 \end{aligned} \tag{3}$$

We propose the following Lyapunov candidate function:

$$V = \frac{1}{2}\sigma^2 \tag{4}$$

whose time derivative is $\dot{V} = \sigma\dot{\sigma}$. From (2)

$$\dot{\sigma} = -\dot{e} - ze \tag{5}$$

We have

$$\begin{aligned} \sigma\dot{\sigma} &= -\sigma\dot{e} - ze\sigma \\ &= -\sigma\dot{y} + \sigma\dot{y}^* - ze\sigma \\ &= -W|\sigma| - \sigma\xi(t) + \sigma\dot{y}^* - ze\sigma \end{aligned}$$

since the term $-\sigma\xi(t) + \sigma\dot{y}^* - ze\sigma$ does not depend on the input, by setting W in such a way that we can ensure that $\dot{V} < 0$, the sliding condition for σ is achieved.
The classical interpretation of the output feedback controller suggests, immediately, the following discontinuous feedback control scheme:

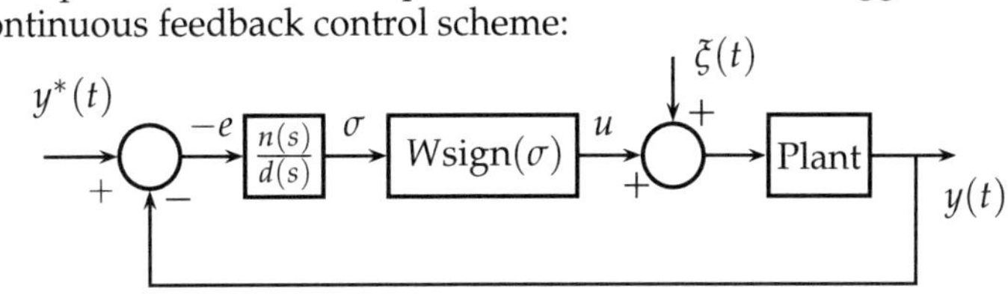

Fig. 1. GPI control scheme.

where $n(s) = s + z$ regulates the dynamic behavior of the tracking error and $d(s) = s$ acts as a "filter" of the sliding surface.
The equivalent control is obtained from the invariance conditions:

$$\sigma = \dot{\sigma} = 0$$

i.e,

$$u_{eq} = \dot{y}^* - ze \tag{6}$$

in other words, the proposed sliding surface has, in the equivalent control sense, the same behavior of the traditional proportional sliding surface of the form $\sigma_1 = ze$. However, the closed loop behavior of the system with the smooth sliding surface, presents some advantages as shown in, Slotine & Li (1991). Since this class of controls induce a "chattering effect", to reduce this phenomenon, we insert in the control law output a first order low-pass filter, which, in some cases, needs and auxiliary control loop (as shown in the integral sliding mode control design, Utkin et al. (1999)). In our case, the architecture of the control system based on two control loops and disturbance observers will act as the auxiliary control input.

2.2 Generalized Proportional Integral Control

GPI control, or Control based on Integral Reconstructors, Fliess & Sira-Ramírez (2004), is a recent development in the literature on automatic control. Its main line of development rests within the finite dimensional linear systems case, with some extensions to linear delayed differential systems and to nonlinear systems (see Fliess *et al.*, Fliess, Marquez, Delaleau & Sira-Ramírez (2002), Fliess *et al.*, Fliess, Marquez & Mounier (2002) and Hernández and Sira-Ramírez, Hernández & Sira-Ramírez (2003)).
The main idea of this control approach is the use of structural reconstruction of the state vector. This means that states of the system are obtained modulo the effect of unknown initial conditions as well as constant, ramp, parabolic, or, in general, polynomial, additive external perturbation inputs. The reconstructed states are computed solely on the basis of inputs and outputs. These state reconstructions may be used in a linear state feedback controller design, provided the feedback controller is complemented with a sufficient number of iterated output, or input, integral error compensation which structurally match the effects of the neglected perturbation inputs and initial states.
To clarify the idea behind GPI control, consider the following elementary example,

$$\begin{aligned} \ddot{y} &= u + \xi \\ y(0) &= y_0 \\ \dot{y}(0) &= \dot{y}_0 \end{aligned} \tag{7}$$

with ξ being an unknown constant disturbance input. The control problem consists in obtaining an output feedback control law, u, that forces y to track a desired reference trajectory, given by $y^*(t)$, in spite of the presence of the unknown disturbance signal and the unknown value of $\dot{y}(0)$.
Let $e_y \triangleq y - y^*(t)$ be the reference trajectory tracking error and let u^* be a feed-forward input nominally given by $\ddot{y}^*(t) = u^*(t)$. The input error is defined as $e_u \triangleq u - u^*(t) = u - \ddot{y}^*(t)$.
Integrating equation (7) we have,

$$\dot{y} = \int_0^t u(\tau)d\tau + \dot{y}(0) + \xi t \tag{8}$$

The integral reconstructor of $\dot{y}$ is defined to be:

$$\hat{\dot{y}} = \int_0^t u(\tau)d\tau \tag{9}$$

The relation between the structural estimate of $\dot{y}$ of the velocity and the actual value of the velocity state is given by,

$$\hat{\dot{y}} = \dot{y} - \dot{y}(0) - \xi t \tag{10}$$

The presence of an unstable ramp error between the integral reconstructor of the velocity and the actual velocity value, prompts us to use a complementary double integral compensating control action on the basis of the position tracking error. We have the following result:

Proposition 1. *Given the perturbed dynamical system, described in (7), the following dynamical feedback control law*

$$\begin{aligned} u &= \ddot{y}^* - k_3(\hat{\dot{y}} - \dot{y}^*) - k_2 e_y(t) - k_1 \int_0^t e_y(\tau)d\tau \\ &\quad -k_0 \int_0^t \int_0^\tau e_y(\sigma)d\sigma d\tau \end{aligned} \tag{11}$$

with $\hat{\dot{y}}$ defined by (9), forces the output y to asymptotically exponentially track the desired reference trajectory, $y^*(t)$.

Proof. Substituting equation (11) into equation (7), yields the following closed loop tracking error dynamics:

$$\begin{aligned} &\ddot{e}_y + k_3(\hat{\dot{y}} - \dot{y}^*) + k_2 e_y + k_1 \int_0^t e_y(\tau)d\tau \\ &+ k_0 \int_0^t \int_0^\tau e_y(\sigma)d\sigma d\tau = 0 \end{aligned} \tag{12}$$

Using (10) one obtains,

$$\begin{aligned} &\ddot{e}_y + k_3\dot{e}_y + k_2 e_y + k_1 \int_0^t e_y(\tau)d\tau \\ &+ k_0 \int_0^t \int_0^\tau e_y(\sigma)d\sigma d\tau = k_3(\dot{y}(0) + \xi t) \end{aligned} \tag{13}$$

Taking two time derivatives in (13) the introduced disturbance due to the integral reconstructor is annihilated as follows:

$$e_y^{(4)} + k_3 e_y^{(3)} + k_2 \ddot{e}_y + k_1 \dot{e}_y + k_0 e_y = 0 \tag{14}$$

The justification of this last step is readily obtained by defining the following state variables along with their initial conditions,

$$\begin{aligned}
\rho_1 &= \int_0^t e_y(\tau)d\tau - (k_3/k_1)y(0), \\
\rho_1(0) &= -(k_3/k_1)y(0) \\
\rho_2 &= \int_0^t \int_0^\tau e_y(\lambda)d\lambda d\tau - (k_3/k_0)\xi t, \\
\rho_2(0) &= 0 \\
\rho_3 &= \dot{\rho}_2 = \int_0^t e_y(\tau)d\tau - (k_3/k_0)\xi, \\
\rho_3(0) &= -(k_3/k_0)\xi
\end{aligned}$$

The closed loop system reads then as follows,

$$\frac{d}{dt}\chi = A\chi$$

with $\chi = (e_y, \dot{e}_y, \rho_1, \rho_2, \rho_3)^T$ and

$$A = \begin{bmatrix} 0 & 1 & 0 & 0 \\ -k_2 & -k_3 & -k_1 & -k_0 \\ 1 & 0 & 0 & 0 \\ 0 & 0 & 0 & 1 \\ 1 & 0 & 0 & 0 \end{bmatrix}$$

The characteristic polynomial associated with the matrix A is readily found to be given by

$$P_A(s) = s^4 + k_3 s^3 + k_2 s^2 + k_1 s + k_0 \tag{15}$$

Finally, by choosing k_3, k_2, k_1, k_0 such that the polynomial (15) has all its roots located on the left half of the complex plane, $\mathbb{C}$, the tracking error, e_y, decreases exponentially asymptotically to zero as a function of time. □

Remark 2. *Notice that the GPI controller (11) can also be written as a classic compensation network (expressed in the frequency domain). From* (11) *and* (9),

$$u(t) = u^*(t) - k_3 \int_0^t e_u(\tau)d\tau - k_2 e_y(t) - k_1 \int_0^t e_y(\tau)d\tau - k_0 \int_0^t \int_0^\tau e_y(\sigma)d\sigma d\tau \tag{16}$$

Using the fact that, $e_u = u - u^*(t)$, *and applying the Laplace transform to the last expression, we have,*

$$e_u(s) = -k_3 \frac{e_u(s)}{s} - k_2 e_y(s) - k_1 \frac{e_y(s)}{s} - k_0 \frac{e_y(s)}{s^2} \tag{17}$$

Re-ordering the last equation we have:

$$e_u(s) = -\left[\frac{k_2 s^2 + k_1 s + k_0}{s(s+k_3)}\right] e_y(s) \tag{18}$$

In other words

$$u(s) = s^2 y^*(s) - \left[\frac{k_2 s^2 + k_1 s + k_0}{s(s+k_3)} \right] e_y(s) \tag{19}$$

2.3 Generalized proportional integral observers

Consider the following n-th order scalar nonlinear differential equation,

$$y^{(n)} = \phi(t, y, \dot{y}, \ddot{y}, \cdots, y^{(n-1)}) + ku \tag{20}$$

where ϕ is a smooth nonlinear scalar function, $k \in \mathbb{R}$ and u is a control input. We state the following definitions and assumptions:

Definition 3. *Define the following time function:*

$$\varphi : t \mapsto \phi(t, y(t), \dot{y}(t), \ddot{y}(t), \cdots, y^{(n-1)}(t)) \tag{21}$$

i.e., denote by $\varphi(t)$, the value of ϕ for a certain solution $y(t)$ of (20) for a fixed set of finite initial conditions. In other words; $\varphi(t) = \phi(t, y(t), \dot{y}(t), \ddot{y}(t), \cdots, y^{(n-1)}(t))$, where $y(t)$ is a smooth bounded solution of Eq. (20) from a certain set of finite initial conditions.

Assumptions 4.

- *We assume that a unique, smooth, bounded solution, $y(t)$, exists for the nonlinear differential equation, (20), for every given set of finite initial conditions.*
- *The values of the function, $\varphi(t)$, are unknown, except for the fact that they are known to be uniformly, absolutely, bounded for every smooth bounded function, $y(t)$, which is a solution of Eq. (20).*
- *For any positive integer p, we can find a small positive, real number, δ_p, such that $\varphi^{(p)}(t)$ is uniformly absolutely bounded, i.e.,*

$$\sup_{t \geq 0} |\varphi^{(p)}(t)| < \delta_p, \ \forall p \in \mathbb{Z}^+ < \infty \tag{22}$$

- *The following system*

$$y^{(n)} = \varphi(t) + ku \tag{23}$$

  with u as a known system input, and $\varphi(t)$ unknown but bounded with negligible high order derivatives after some integer order p is assumed to capture, from a signal processing viewpoint, all the essential features of the nonlinear system (20).

2.3.1 A GPI observer approach to state estimation of unknown dynamics

We formulate the state estimation problem for the system (20) via GPI observers as follows: *Under the above assumptions, given the noise-free measurement of $y(t), u(t)$, it is desired to estimate the natural state variables (or: phase variables) of the system (20), given by $y(t), \dot{y}(t)$, $\ddot{y}(t), ..., y^{(n-1)}(t)$, via the use of the natural equivalence of system (20) with the simplified uncertain system given by (23).*

The solution to the simultaneous state and perturbation estimation problem can be achieved via the use of an extended version of the traditional linear Luenberger observer, that we address here as GPI observer, as follows.

Proposition 5. *Luviano-Juárez et al. (2010) Under the assumptions given above. For a system of the form (23), the following observer*

$$
\begin{aligned}
\dot{\hat{y}}_1 &= \lambda_{p+n-1}(y - \hat{y}_1) + \hat{y}_2 \\
\dot{\hat{y}}_2 &= \lambda_{p+n-2}(y - \hat{y}_1) + \hat{y}_3 \\
&\vdots \\
\dot{\hat{y}}_n &= \lambda_p\,(y - \hat{y}_1) + ku + \rho_1 \\
\dot{\rho}_1 &= \lambda_{p-1}\,(y - \hat{y}_1) + \rho_2 \\
\dot{\rho}_2 &= \lambda_{p-2}\,(y - \hat{y}_1) + \rho_3 \\
&\vdots \\
\dot{\rho}_{p-2} &= \lambda_2\,(y - \hat{y}_1) + \rho_{p-1} \\
\dot{\rho}_{p-1} &= \lambda_1\,(y - \hat{y}_1) + \rho_p \\
\dot{\rho}_p &= \lambda_0\,(y - \hat{y}_1) \\
\hat{y}_i &= \hat{y}^{(i-1)}
\end{aligned}
\tag{24}
$$

asymptotically exponentially reconstructs, via the observer variables: $\hat{y}_1,\ \hat{y}_2,\ \cdots,\ \hat{y}_n$*, the phase variables* $y,\ \dot{y},\ \cdots,\ y^{(n-1)},\ y^{(n)}$*, while the observer variables* ρ_1*,* ρ_2*,..., respectively, reconstruct in an asymptotically exponentially fashion, the perturbation input* $\varphi(t)$ *and its time derivatives* $\dot{\varphi}(t)$*,... modulo a small error, uniformly bounding the reconstruction error* $\varepsilon = y - \hat{y}(t) = y - \hat{y}_1$*, and its first* $n-1$ *- th order time derivatives provided the design parameters,* $\lambda_0, \cdots, \lambda_{p+n-1}$ *are chosen so that the roots of the associated polynomial in the complex variable s:*

$$P(s) = s^{p+n} + \lambda_{p+n-1}s^{p+n-1} + \lambda_{p+n-2}s^{p+n-2} + \ldots + \lambda_1 s + \lambda_0 \tag{25}$$

are all located deep in the left half of the complex plane.

Proof. Define, as suggested in the Proposition, the estimation error as follows:

$$\varepsilon(t) \triangleq y(t) - \hat{y}_1(t) \tag{26}$$

taking $p+n$ time derivatives in last equation, and using the reconstruction error dynamics for ε, derivable from the observer equations, leads to the following perturbed reconstruction error dynamics:

$$\varepsilon^{(p+n)} + \lambda_{p+n-1}\varepsilon^{(p+n-1)} + \lambda_{p+n-2}\varepsilon^{(p+n-2)} + \cdots + \lambda_1\dot{\varepsilon} + \lambda_0\varepsilon = \varphi^{(p)}(t) \tag{27}$$

which is a perturbed $n+p$ - th order linear time invariant system, whose perturbation input is given by $\varphi^{(p)}(t)$. Given that the characteristic polynomial $P(s)$, corresponding to the unperturbed output reconstruction error system, has its roots in the left half of the complex plane, then the Bounded Input Bounded Output (BIBO) stability condition is assured, Kailath (1979) since, uniformly in t, $\left|\varphi^{(p)}(t)\right| < \gamma_p$. Thus, the output reconstruction error, ε, and its first $n+p-1$ time derivatives are ultimately constrained to a disk in the reconstruction error phase space of arbitrary small radius which is further decreased as the roots of the dominating characteristic polynomial are chosen farther and farther into the left half of the complex plane. □

3. Problem formulation

Consider the following dynamic model describing the two-phase equivalent model of a three-phase motor controlled by the phase voltages u_{Sa} and u_{Sb} with state variables given by: θ, describing the rotor angular position, ω being the rotor angular velocity, ψ_{Ra} and ψ_{Rb}, representing the unmeasured rotor fluxes, while i_{Sa} and i_{Sb} are taken to be the stator currents.

$$\begin{aligned}
\frac{d\theta}{dt} &= \omega \\
\frac{d\omega}{dt} &= \mu(i_{Sb}\psi_{Ra} - i_{Sa}\psi_{Rb}) - \frac{\tau_L}{J} \\
\frac{d\psi_{Ra}}{dt} &= -\eta\psi_{Ra} - n_p\omega\psi_{Rb} + \eta M i_{Sa} \\
\frac{d\psi_{Rb}}{dt} &= -\eta\psi_{Rb} + n_p\omega\psi_{Ra} + \eta M i_{Sb} \\
\frac{di_{Sa}}{dt} &= \eta\beta\psi_{Ra} + \beta n_p\omega\psi_{Rb} - \gamma i_{Sa} + \frac{u_{Sa}}{\sigma L_S} \\
\frac{di_{Sb}}{dt} &= \eta\beta\psi_{Rb} - \beta n_p\omega\psi_{Ra} - \gamma i_{Sb} + \frac{u_{Sb}}{\sigma L_S}
\end{aligned} \tag{28}$$

with

$$\eta := \frac{R_R}{L_R},\ \beta := \frac{M}{\sigma L_R L_S},\ \mu := \frac{n_p M}{J L_R},$$
$$\gamma := \frac{M^2 R_R}{\sigma L_R^2 L_S} + \frac{R_S}{\sigma L_S}, \sigma := 1 - \frac{M^2}{L_R L_S}$$

R_R and R_S are, respectively, the rotor and stator resistances, L_R and L_S represent, respectively, the rotor and stator inductances, M is mutual inductance constant, J is the moment of inertia and n_p is the number of pole pairs. The signal τ_L is the unknown load torque perturbation input. We adopt the complex notation like in Sira-Ramirez, Beltran-Carbajal & Blanco-Ortega (2008). Define the following complex variables:

$$\begin{aligned}
\psi_R &= \psi_{Ra} + j\psi_{Rb} = |\psi_R|\, e^{j\theta_\psi} \\
u_S &= u_{Sa} + ju_{Sb} = |u_S|\, e^{j\theta_u} \\
i_S &= i_{Sa} + ji_{Sb} = |i_S|\, e^{j\theta_i}
\end{aligned}$$

The induction motor dynamics is rewritten as

$$\begin{aligned}
\frac{d^2\theta}{dt^2} &= \mu \mathcal{I}m(\overline{\psi}_R i_S) - \tau_L(t) \\
\frac{d|\psi_R|^2}{dt} &= -2\eta\,|\psi_R|^2 + 2\eta M \mathcal{R}e\left(\overline{\psi}_R i_S\right) \\
\frac{d\theta_\psi}{dt} &= n_p\omega + \frac{R_r M}{L_r|\psi_R|^2}\mathcal{I}m(\overline{\psi}_R i_S) \\
\frac{di_S}{dt} &= \beta(\eta + n_p\omega)\psi_R - \gamma i_S + \frac{1}{\sigma L_s}u
\end{aligned}$$

where $\overline{\psi}_R$ denotes the complex conjugate of the complex rotor flux ψ_R.
We have, thus, established explicit, separate, dynamics for the squared rotor flux magnitude and for the rotor flux phase angle. This representation, clearly exhibits a *decoupling property* of the model which allows one to, independently, control the square of the flux magnitude and the angular position by means of the stator currents acting as auxiliary control input variables. This representation also establishes that the complex flux phase angle is largely determined by the manner in which the angular position is controlled by the stator currents.
The problem formulation is as follows: Given the induction motor dynamics, given a desired constant reference level for the rotor flux magnitude $|\psi_R^*| > 0$, and given a smooth reference trajectory $\theta^*(t)$ for the angular position of the motor shaft, the control problem consists in finding a feedback control law for the phase voltages u_{Sa} and u_{Sb} in such a way that θ is forced to track the given reference trajectory, θ^*, while the rotor flux magnitude stabilizes around the desired value, $|\psi_R^*|$. Such objectives are to be achieved in spite of the presence of unknown but bounded perturbation inputs represented by 1) the load torque, $\tau_L(t)$, in the rotor shaft dynamics and 2) the effects of motor nonlinearities acting on the current dynamics through possibly unknown parameters.

4. Control strategy

The GPI observer-controller design considerations will be based on the following simplified, linear, models lumping the external load disturbances and the system nonlinearities in the form of components of an unknown perturbation input vector, as follows:

$$\frac{d^2\theta}{dt^2} = \mu \mathcal{I}m(\overline{\psi}_R i_S) + \xi_1(t) \tag{29}$$

$$\frac{d\,|\psi_R|^2}{dt} = -2\eta\,|\psi_R|^2 + 2\eta M \mathcal{R}e\left(\overline{\psi}_R i_S\right)$$

$$\frac{d\theta_\psi}{dt} = n_p\omega + \frac{R_r M}{L_r|\psi_R|^2}\mathcal{I}m(\overline{\psi}_R i_S)$$

$$\frac{di_S}{dt} = \frac{1}{\sigma L_s}u_S + \xi(t) \tag{30}$$

where $\xi_1(t) = -\tau_L(t)$, $\xi(t) = \xi_2(t) + j\xi_3(t)$ are considered as disturbance inputs, with $\xi_1(t)$ representing the unknown load perturbation input, and $\xi(t)$ represents nonlinear and linear additive dissipation terms, depending on the stator currents i_{Sa}, i_{Sb} and the angular velocity. The currents i_{Sa} and i_{Sb} can be directly measured; on the other hand, rotor fluxes must be estimated. For the flux estimation, we used a real time simulation of the rotor flux equation dynamics. Parameters η, n_p, M need to be known; on the other hand, the lumped parameter μ must be estimated. Nevertheless, in our control scheme, such a task is not entirely necessary due to the remarkable robustness of the scheme and a reasonable guess can be used in the controller expression for such parameters. The disturbance functions $\xi_1(t)$, $\xi(t)$ can be envisioned to contain the rest of the system dynamics, including some un-modeled dynamics (which can be of a rather complex nonlinear character). In these terms, we also lump disturbances of additive nature such as frictions and the effects generated by parameter variations during the system operation and even the effects of inaccurate parameter estimations. These perturbation inputs, however, do not contain any control terms.
For the correct tracking of angular position, it is necessary to provide additional control loops for other variables. As it is customary, the flux modulus has to be regulated to a certain value

in order to assure thc efficient operation of the induction machine avoiding possible saturation effects.
The proposed control scheme consists in a two stage feedback controller design. The first stage controls the angular position of the motor shaft to track the reference signal $\theta^*(t)$ by means of the stator currents taken as auxiliary control inputs. As a collateral objective it is desired to have the flux magnitude converging towards a given constant value $|\psi_R^*|$[1]. For this stage, the control strategy is implemented by means of a GPI based observer controller, Cortés-Romero et al. (2009). As a result of the first stage a set of desirable current trajectories is synthesized. The obtained currents are thus taken as output references for the second multi-variable stage. The second stage designs a discontinuous feedback controller to force the actual currents to track the obtained current references in the first stage. In the second stage the stator voltages are the control inputs. The following section deals with the flux reconstructor.

4.1 Flux reconstruction

Note that the complex rotor flux ψ_R satisfies the following dynamics:

$$\frac{d\psi_R}{dt} = -(\eta + jn_p\omega)\psi_R + \eta M i_S$$

A simple reconstruction dynamics, with self stable reconstruction error dynamics, is given by

$$\frac{d\widehat{\psi}_R}{dt} = -(\eta + jn_p\omega)\widehat{\psi}_R + \eta M i_S$$

The complex reconstruction error $e = \psi_R - \widehat{\psi}_R$ satisfies then the linear dynamics:

$$\frac{de}{dt} = -(\eta + jn_p\omega)e$$

whose unique eigenvalue has a strictly negative real part (and a time varying complex part). Thus the complex error, e, satisfies $e \to 0$ in an exponentially asymptotic manner. Thus, henceforth, when we use ψ in the expressions it is implicitly assumed that it is obtained from the proposed reconstructor undergoing the exponential convergence process $\widehat{\psi} \to \psi$.

4.2 Outer loop controller design stage

For this first design stage we consider the following dynamics:

$$\frac{d^2\theta}{dt^2} = \mu \mathcal{I}m(\overline{\psi}_R i_S) + \xi_1(t) \tag{31}$$

$$\begin{aligned} \frac{d\,|\psi_R|^2}{dt} &= -2\eta\,|\psi_R|^2 + 2\eta M \mathcal{R}e\,(\overline{\psi}_R i_S) \\ \frac{d\theta_\psi}{dt} &= n_p\omega + \frac{R_r M}{L_r|\psi_R|^2}\mathcal{I}m(\overline{\psi}_R i_S) \end{aligned} \tag{32}$$

with the complex stator current i_S acting as auxiliary control input.
We propose the following complex controller:

[1] Inaccurate parameters may cause minimal variations in the flux regulation, however, the angular position remains unaffected due to the robustness of the controller.

$$i_S = \frac{\psi_R}{|\psi_R|^2}\left[\frac{|\psi_R^*|^2}{M} + jv\right]$$

with $|\psi_R^*|$ being the desired flux magnitude reference value and v is an auxiliary control input. In closed loop, the squared modulus of the rotor flux satisfies

$$\frac{d\,|\psi_R|^2}{dt} = -2\eta\left[|\psi_R|^2 - |\psi_R^*|^2\right]$$

then, $|\psi_R| \to |\psi_R^*| =$ constant, in an exponential asymptotic manner.
On the other hand, the angular position dynamics satisfies, in closed loop, the perturbed dynamics.

$$\frac{d^2\theta}{dt^2} = \mu v + \xi_1(t)$$

with $\xi_1(t)$ assumed to be time-varying, unknown but bounded signal, directly related to the load torque, and v being a control input yet to be specified. The specification of the auxiliary control input v is made on the basis of a GPI controller. With an abuse in the notation, we use the following GPI observer based controller:

$$v = \frac{1}{\mu}\left[\ddot{\theta}^* - \frac{k_{1\theta}s + k_{0\theta}}{s + k_{2\theta}}(\theta - \theta^*) - \hat{\xi}_1\right]$$

where $\hat{\xi}_1$ is the on-line estimate of the unknown signal $\xi_1(t)$. For the estimation of the disturbance function $\xi_1(t)$, we assume that $\xi_1(t)$ is bounded with bounded low order time derivatives and negligible higher order time derivatives. Such signals may be locally approximated in a *self-updated* manner, thanks to the *internal model principle*, by a generic representative of a *family of time polynomial signals*, of fixed finite, relatively low degree, and free coefficients. Thus, modeling $\xi_1(t)$ by means of, say, a 5th degree family of polynomials, the following GPI observer, containing a suitable internal model of the perturbation input, is proposed:

$$\begin{aligned}
\frac{d\hat{\theta}}{dt} &= \lambda_7(\theta - \hat{\theta}) + \hat{\dot{\theta}} \\
\frac{d\hat{\dot{\theta}}}{dt} &= \lambda_6(\theta - \hat{\theta}) + \mu v + \rho_{1\theta} \\
\dot{\rho}_{1\theta} &= \lambda_5(\theta - \hat{\theta}) + \rho_{2\theta} \\
\dot{\rho}_{2\theta} &= \lambda_4(\theta - \hat{\theta}) + \rho_{3\theta} \\
\dot{\rho}_{3\theta} &= \lambda_3(\theta - \hat{\theta}) + \rho_{4\theta} \\
\dot{\rho}_{4\theta} &= \lambda_2(\theta - \hat{\theta}) + \rho_{5\theta} \\
\dot{\rho}_{5\theta} &= \lambda_1(\theta - \hat{\theta}) + \rho_{6\theta} \\
\dot{\rho}_{6\theta} &= \lambda_0(\theta - \hat{\theta}) \\
\hat{\xi}_1 &= \rho_{1\theta}
\end{aligned} \qquad (33)$$

The estimation error is defined as $e_\theta := \theta - \hat{\theta}$ satisfies the following injected dynamics:

$$\begin{aligned}
&e_\theta^{(8)} + \lambda_7 e_\theta^{(7)} + \lambda_6 e_\theta^{(6)} + \lambda_5 e_\theta^{(5)} + \lambda_4 e_\theta^{(4)} \\
&+ \lambda_3 e_\theta^{(3)} + \lambda_2 \ddot{e}_\theta + \lambda_1 \dot{e}_1 + \lambda_0 e_\theta = \xi_1^{(6)}(t)
\end{aligned}$$

By an appropriate choice of the coefficients, $\lambda_i; i = 0, 1, \ldots, 7$ the characteristic polynomial in the complex variable s

$$\begin{aligned} p_{\hat{\xi}_1}(s) =& s^8 + \lambda_7 s^7 + \lambda_6 s^6 + \lambda_5 s^5 + \lambda_4 s^4 \\ &+ \lambda_3 s^3 + \lambda_2 s^2 + \lambda_1 s + \lambda_0 \end{aligned}$$

can be made into a Hurwitz polynomial. The estimation error is assured to be ultimately bounded by a small disk around the origin in the estimating error state space which can be further reduced by adjusting the observer gains to produce eigenvalues sufficiently far at the left half of the complex plane. Under these circumstances, $\hat{\theta} \to \theta$ and $\hat{\dot{\theta}} \to \dot{\theta}$ modulo an arbitrarily small error and, subsequently, it is clear that $\rho_{1\theta} \to \xi_1$ with the same convergence rate (See Sira-Ramírez et al., Feliu-Battle, Sira-Ramirez, Feliu-Batlle, Beltran-Carbajal & Blanco-Ortega (2008)).
The closed loop characteristic polynomial for the angular position tracking error response is just:

$$p_\theta(s) = s^3 + k_{2\theta}s^2 + k_{1\theta}s + k_{0\theta}$$

while the characteristic polynomial governing the exponential convergence of the squared norm of the flux towards the desired constant value, is given by

$$p_{|\psi_R|^2}(s) = s + 2\eta$$

4.3 Inner loop controller design stage

Consider now the simplified perturbed stator currents dynamics (30):

$$\frac{di_S}{dt} = \frac{1}{\sigma L_S} u_{Sav} + \xi(t) \tag{34}$$

We regard this simplified dynamics as an average perturbed representation of the stator current dynamics which is to be regulated by means of a switched input voltages strategy, similar in nature to those arising from the variable structure, sliding mode controller, approach (the reader is referred to, Castillo-Toledo et al. (2008); Utkin et al. (1999) and references therein).
Denote the three phase currents, and three phase voltages, respectively, by i_1, i_2, i_3, and u_1, u_2, u_3. Define the stator current vector as an $\mathbb{R}^2$ vector with components, i_{Sa}, i_{Sb}, while the stator voltage vector, is also defined as a vector with components: u_{Sa}, u_{Sb}, with the perturbation vector also defined as: $\xi = [\xi_2 \ \xi_3]^T$. These quantities are transformed from the phase current vector, i, the phase voltage vector, u, and the phase sliding surface, σ, as follows:

$$i = \begin{bmatrix} i_1 \\ i_2 \\ i_3 \end{bmatrix} = \sqrt{\frac{3}{2}} \begin{bmatrix} 2/3 & 0 \\ -1/3 & 1/\sqrt{3} \\ -1/3 & -1/\sqrt{3} \end{bmatrix} \begin{bmatrix} i_{Sa} \\ i_{Sb} \end{bmatrix} = P \begin{bmatrix} i_{Sa} \\ i_{Sb} \end{bmatrix}$$

$$u = \begin{bmatrix} u_1 \\ u_2 \\ u_3 \end{bmatrix} = \sqrt{\frac{3}{2}} \begin{bmatrix} 2/3 & 0 \\ -1/3 & 1/\sqrt{3} \\ -1/3 & -1/\sqrt{3} \end{bmatrix} \begin{bmatrix} u_{Sa} \\ u_{Sb} \end{bmatrix} = P \begin{bmatrix} u_{Sa} \\ u_{Sb} \end{bmatrix}$$

$$\overline{\xi} = \sqrt{\frac{3}{2}} \begin{bmatrix} 2/3 & 0 \\ -1/3 & 1/\sqrt{3} \\ -1/3 & -1/\sqrt{3} \end{bmatrix} \begin{bmatrix} \xi_2 \\ \xi_3 \end{bmatrix} = P \begin{bmatrix} \xi_2 \\ \xi_3 \end{bmatrix}$$

Let us define the following vector sliding surface in operational calculus terms:

$$\overline{\sigma} = \begin{bmatrix}\sigma_1 & \sigma_2 & \sigma_3\end{bmatrix}^T = -\left(\frac{s+z}{s}\right) e_i \tag{35}$$

$$e_i = i - i^*(t)$$

and the switched control input is given by

$$u = W\text{sign}(\overline{\sigma}) = W \begin{bmatrix}\text{sign}(\sigma_1)\\ \text{sign}(\sigma_2)\\ \text{sign}(\sigma_3)\end{bmatrix} \tag{36}$$

Proceeding as in the introductory example, let us consider the following Lyapunov candidate function

$$V = \frac{1}{2}\overline{\sigma}^T\overline{\sigma} \tag{37}$$

We have:

$$\dot{V} = \overline{\sigma}^T\dot{\overline{\sigma}} = -\overline{\sigma}^T\dot{e}_i - z\overline{\sigma}^T e_i \tag{38}$$

$$\dot{e}_i = \frac{W}{\sigma L_s}\text{sign}(\overline{\sigma}) + \overline{\xi} - \frac{di^*}{dt} \tag{39}$$

We have:

$$\dot{V} = -\overline{\sigma}^T\dot{e}_i - z\overline{\sigma}^T e_i = \tag{40}$$

$$= -\frac{W}{\sigma L_s}\left(|\sigma_1| + |\sigma_2| + |\sigma_3|\right) - \overline{\sigma}^T\overline{\xi} + \overline{\sigma}^T\frac{di^*}{dt} - z\overline{\sigma}^T e_i \tag{41}$$

Thus, for a large enough voltages amplitude W, the sliding condition $\overline{\sigma}^T\dot{\overline{\sigma}} < 0$ is satisfied and the vector of phase sliding coordinates converges towards $\sigma_1 = 0$, $\sigma_2 = 0$, $\sigma_3 = 0$, in, a finite time under the switching control u.
The closed loop behavior under sliding mode condition can be obtained using the equivalent control method. Using the invariance conditions $\overline{\sigma} = \dot{\overline{\sigma}} = 0$, we have:

$$\dot{e}_i + ze_i = 0 \tag{42}$$

Therefore, the tracking error converges to zero asymptotically. A schematic diagram of the control methodology is given in figure 2.

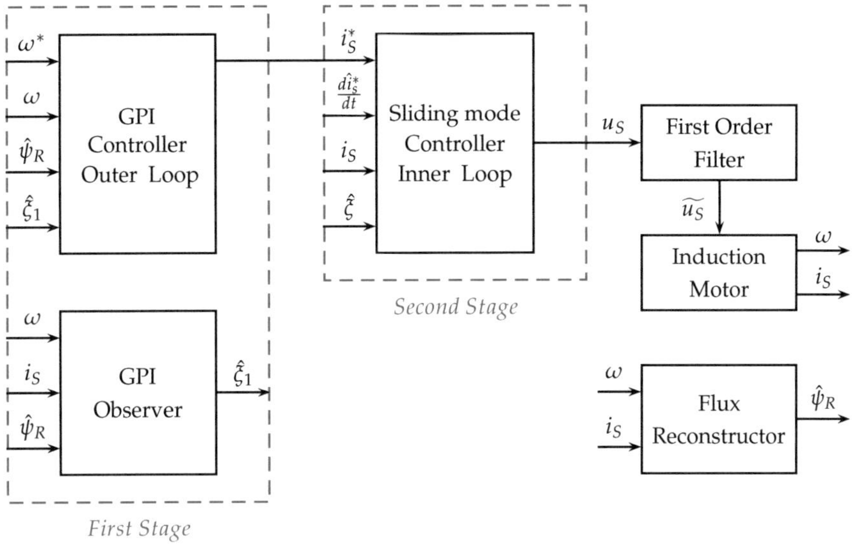

Fig. 2. Control schematics.

5. Experimental results

We illustrate the proposed control approach by some experiments on an actual induction motor test bed. The experimental induction motor prototype includes the following parameters: $J = 4.5 \times 10^{-4}$ $[Kg\ m^2]$, $n_p = 1$, $M = 0.2768$ [H], $L_R = 0.2919$ $[H]$, $L_S = 0.2919$ [H], $R_S = 5.12$ $[\Omega]$, $R_R = 2.23$ $[\Omega]$. The flux absolute desired value was 0.5872 [Wb].
The sliding mode surface parameter was $z = 350$. The output of the sliding control was filtered by means of a first order low pass filter of Bessel type with cut frequency of 750 rad/s. The angle measurement was obtained using an incremental encoder with 10000 PPR. The desired closed loop tracking error was set in terms of the characteristic polynomial $P_\theta(s) = (s^2 + 2\zeta\omega_n + \omega_n^2)(s + p)$, with $\zeta = 1$, $\omega_n = 330$, $p = 320$, and the observer injection error characteristic polynomial was $P_{\tilde{\xi}} = (s^2 + 2\zeta_1\omega_{n1} + \omega_{n1}^2)^4$, with $\zeta_1 = 2$, $\omega_{n1} = 27$.
The controller was devised in a MATLAB - xPC Target environment using a sampling period of .1 [ms]. The communication between the plant and the controller was performed by two data acquisition devises. The analog data acquisition was performed by a National Instruments PCI-6025E data acquisition card, and the digital outputs as well as the encoder reading for the position sensor were performed in a National Instruments PCI-6602 data acquisition card. The voltage and current signals are conditioned for adquisition system by means of low pass filters with cut frequency of 1 [kHz]. The interconnection of the modules can be appreciated in a block diagram form as depicted in figure 3.
The output reference trajectory to be tracked, was set to be a biased sinusoidal wave of the form:

$$\theta^* = \begin{cases} 0 & 0 \le t < 2 \\ 1 + \sin(t - \pi/2) & 2 \le t \le 10 \end{cases} \tag{43}$$

Figure 4 shows an accurate position tracking with respect to the desired trajectory. As we can see in figure 5, the control loops indirectly regulate the flux magnitude whose error is under 5×10^{-3} [Wb]. The sliding mode control induces a slight high frequency wave envelope. However, as depicted in figure 6, the average current tracks perfectly the desired reference currents in both phases (a and b). Notice that the control voltages (figure 7) are continuous, but they remain affected by the discontinuities despite the filtering effect.

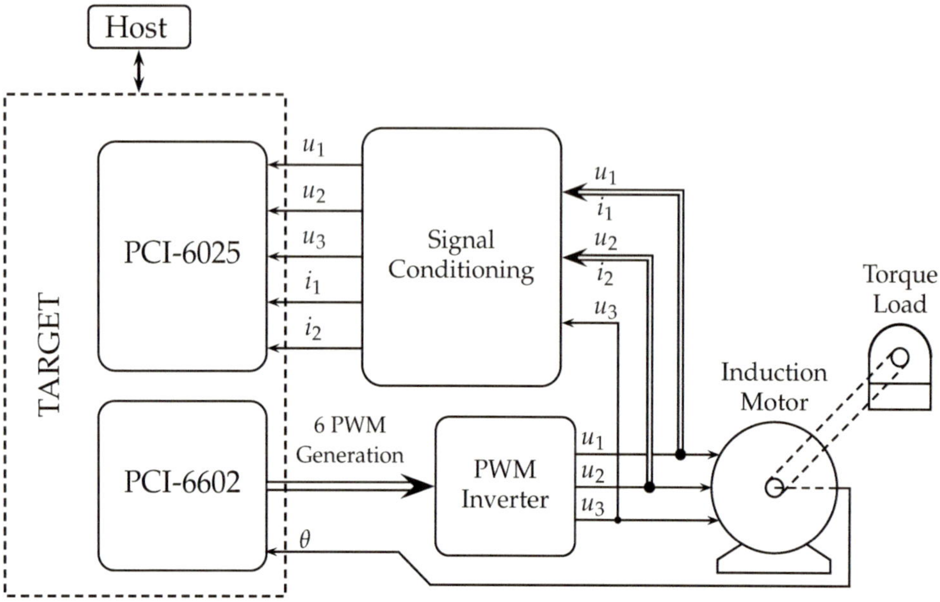

Fig. 3. Block diagram of the control system

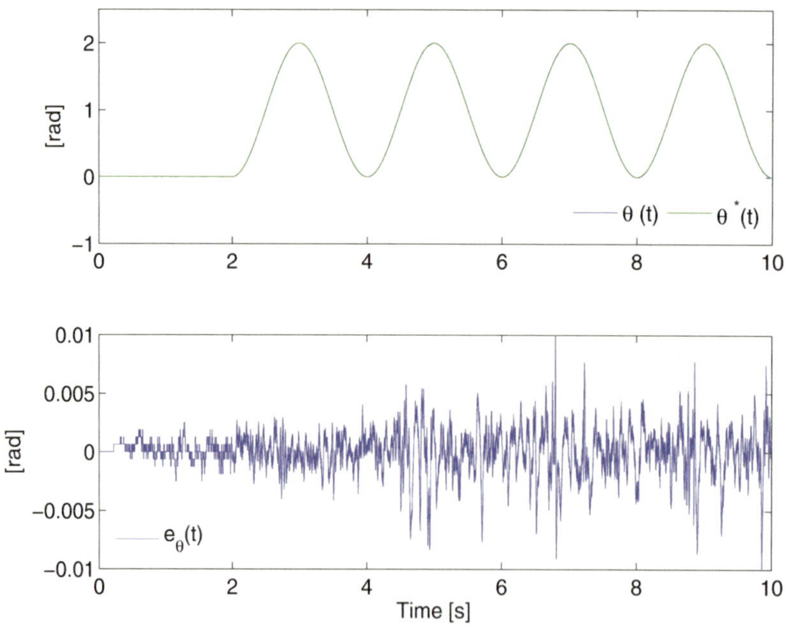

Fig. 4. Position trajectory tracking.

Finally, to illustrate the robustness of the strategy, we applied a load torque with a voltage controlled brake. The applied voltage was implemented by a variable resistor array, where its value was randomly adjusted by a manual tuning. The disturbance estimation, as well as the load torque observation and the auxiliary input v can be seen in figure 8.

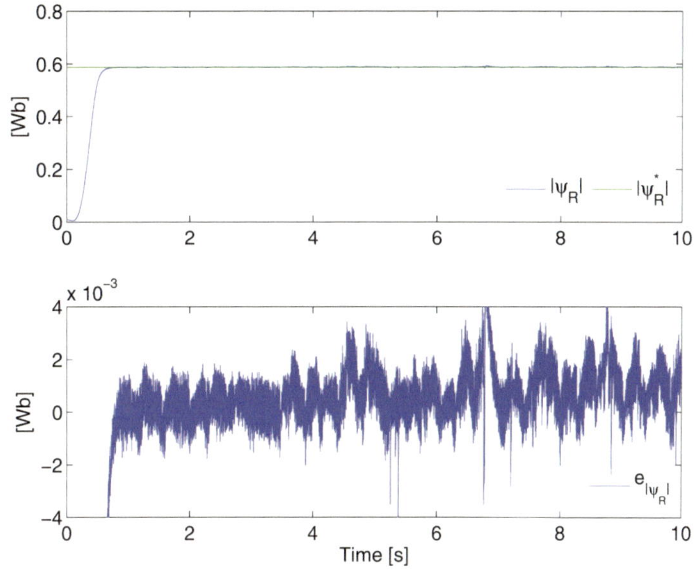

Fig. 5. Flux magnitude regulation.

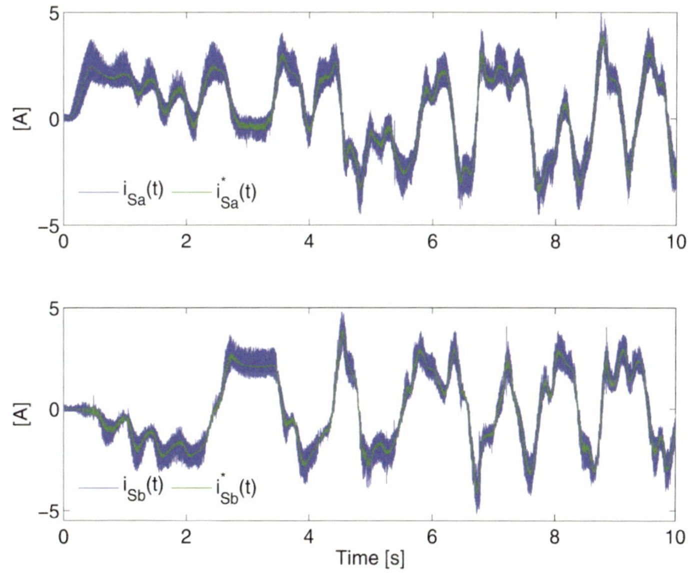

Fig. 6. Current tracking.

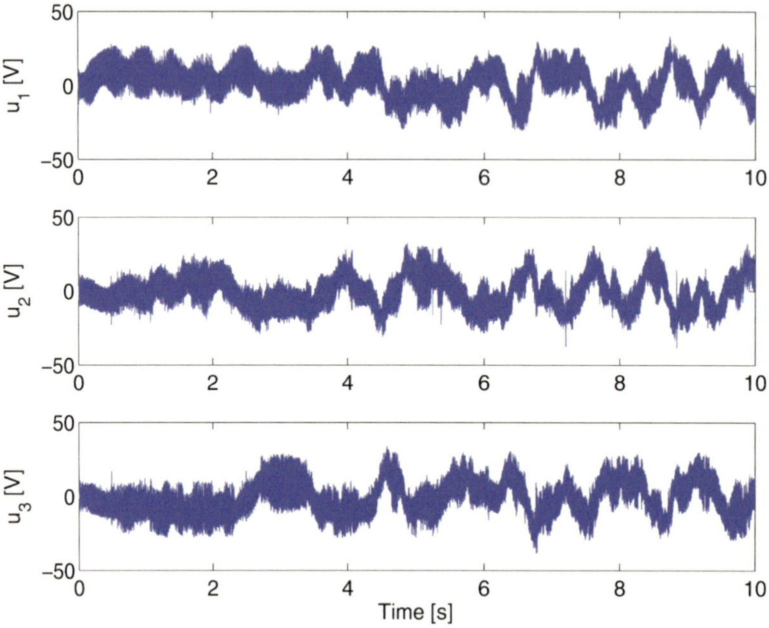

Fig. 7. Stator voltages.

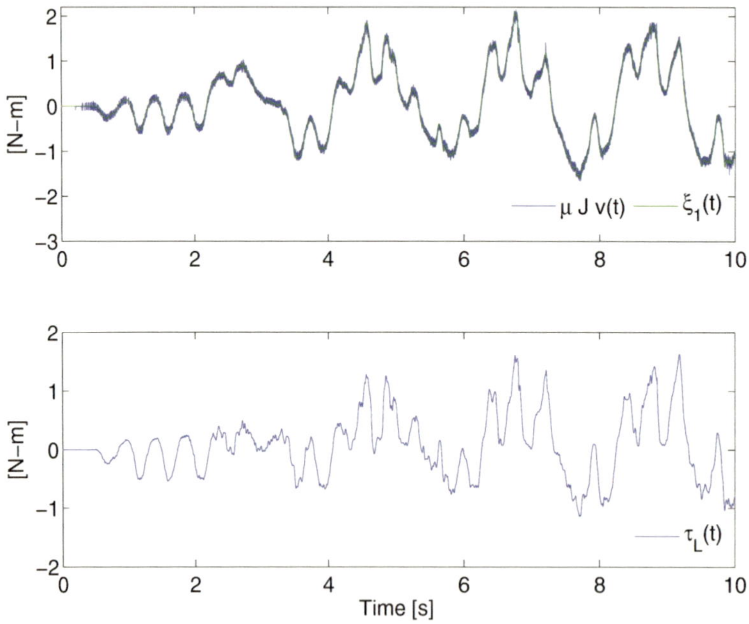

Fig. 8. Load torque, disturbance estimation and auxiliary control input v.

6. Concluding remarks

In this work, a combination of two control loops, one discontinous sliding mode control and another based on the combination of GPI control and GPI disturbance observer was proven to be quite suitable for robust position control and tracking tasks in an induction motor system. An experimental test was carried out where the plant is subject to unforseen external disturbances and un-modeled nonlinear state dependent perturbations. Here, we used the disturbance estimates to carry out the disturbance rejection and for canceling the effects of un-modeled disturbance inputs in the motor, in the case of the mechanical subsystem.
Since the strategy regulates the flux, as a collateral task, since the current variables are well regulated, then the experimental flux variable showed accurate results.
The behavior of the proposed scheme is based upon the correct setting of the characteristic polynomial of the observer which guarantees the correct cancelation of disturbance terms by means of its estimation process.

7. References

Castillo-Toledo, B., Gennaro, S. D., Loukianov, A. & Rivera, J. (2008). Hybrid control of induction motors via sampled closed representations, *IEEE Transactions on Industrial Electronics* 55(10): 3758–3771.

Chiasson, J. (2005). *Modeling and High-Performance Control of Electric Machines*, New York: Wiley.

Cortés-Romero, J., Luviano-Juárez, A. & Sira-Ramírez, H. (2009). Robust GPI controller for trajectory tracking for induction motors, *Proceedings of the 2009 IEEE International Conference on Mechatronics*, Málaga, Spain.

Fliess, M., Levine, J., Martin, P. & Rouchon, P. (1995). Flatness and defect of non-linear systems: introductory theory and examples, *International Journal of Control* 61(6): 1327–1361.

Fliess, M., Marquez, R., Delaleau, E. & Sira-Ramírez, H. (2002). Correcteurs proportionnels integraux généralisés, *ESAIM: Control, Optimisation and Calculus of Variations* 7: 23–41. **URL:** *http://dx.doi.org/doi/10.1051/cocv:2002002*

Fliess, M., Marquez, R. & Mounier, H. (2002). An extension of predictive control, pid regulators and smith predictors to some linear delay systems, *International Journal of Control* 75: 728Ű743.

Fliess, M. & Sira-Ramírez, H. (2004). Reconstructeurs d'etat, *C.R. Acad. Sci. Paris* 338(1): 91–96. t. 332, Série I.

Harnefors, L. & Hinkkanen, M. (2008). Complete stability of reduced-order and full-order observers for sensorless IM drives, *IEEE Transactions on Industrial Electronics* 55(3): 1319 –1329.

Hasan, S. & Husain, I. (2009). A Luenberger-Sliding Mode Observer for Online Parameter Estimation and Adaptation in High-Performance Induction Motor Drives, *IEEE Transactions on Industry Applications* 45(2): 772–781.

Hernández, V. M. & Sira-Ramírez, H. (2003). Generalized p.i. control for global robust position regulation of rigid robot manipulators, *2003 American Control Conference*, Denver Colorado, USA.

Kailath, T. (1979). *Linear Systems (Prentice-Hall Information and System Science Series)*, Prentice Hall.
URL: *http://www.amazon.ca/exec/obidos/redirect?tag=citeulike09-20&path=ASIN/0135369614*

Luviano-Juárez, A., Cortés-Romero, J. & Sira-Ramírez, H. (2010). Synchronization of chaotic oscilators by means of proportional integral observers, *International Journal of Bifurcation and Chaos* 20(5): 1509Ű–1517.

Melfi, M., Evon, S. & McElveen, R. (2009). Induction versus permanent magnet motors, *IEEE Industry Applications Magazine* 15(6): 28 –35.

Sira-Ramírez, H. & Agrawal, S. (2004). *Differentially flat systems*, Marcel Dekker Inc.

Sira-Ramirez, H., Beltran-Carbajal, F. & Blanco-Ortega, A. (2008). A robust generalized proportional integral controller for the induction motor, *International Workshop on Variable Structure Systems VSS '08*. pp. 89–94.

Sira-Ramirez, H., Feliu-Batlle, V., Beltran-Carbajal, F. & Blanco-Ortega, A. (2008). Sigma-delta modulation sliding mode observers for linear systems subject to locally unstable inputs, *Proceedings of the 16th Mediterranean Conference on Control and Automation*, pp. 344–349.

Sira-Ramírez, H. & Silva-Ortigoza, R. (2006). *Control design techniques in power electronics devices*, Power systems, Springer-Verlag, London.

Slotine, J. & Li, W. (1991). *Applied Nonlinear Control*, Prentice Hall.

Toliyat, H., Levi, E. & Raina, M. (2003). A review of RFO induction motor parameter estimation techniques, *IEEE Transactions on Energy Conversion* 18(2): 271–283.

Utkin, V., Guldner, J. & Shi, J. (1999). *Sliding Mode Control Electromechanical System*, CRC Press.

8

Cascade Sliding Mode Control of a Field Oriented Induction Motors with Varying Parameters

Abdellatif REAMA, Fateh MEHAZZEM and Arben CELA
ESIEE Paris- Paris Est University
France

1. Introduction

The induction machine is widely used in industry, because of its mechanical robustness, low maintenance requirement, and relatively low cost. However, from control point of view, control of the induction machines is one of the most challenging topics. Its control is complex because the dynamic of the induction machine is nonlinear, multivariable, and highly coupled. Furthermore, there are various parameter uncertainties and disturbances in the system. The rotor resistances, for example, can vary up to 100% because of rotor heating during operation. In the last few years, many versions of a nonlinear state feedback control schemes, such as, input-output feedback linearization ((Marino et al., 1993)), passivity-based control ((Ortega & Espinoza, 1993; Ortega et al., 1996)) and Backstepping (Kanellakopoulos et al. (1991); Krstic et al. (1995)) have been applied to the IM drive. Adaptive versions of most of those nonlinear control schemes are also available for the effective compensation of the parameter uncertainties and disturbances in the induction motor systems (Ebrahim & Murphy (2006); Marino et al. (1993); Ortega et al. (1993); Rashed et al. (2006)). A fundamental problem in the design of feedback controllers is that of stabilizing and achieving a specified transient performance in the presence of external disturbances and plant parameter variations.
Since the publication of the survey paper by (Utkin (1977)), significant interest on Sliding mode control has been generated in the control research community worldwide. This interest is increased in the last two decades due to the possibility to implement this control in industrial applications with the advances of the power electronics technology and the availability of cheap and fast computation.
One of the most intriguing aspects of sliding mode is the discontinuous nature of the control action whose primary function of each of the feedback channels is to switch between two distinctively different system structures (or components) such that a new type of system motion, called sliding mode, exists in a manifold. This peculiar system characteristic is claimed to result in superior system performance which includes insensitivity to parameter variations, and complete rejection of disturbances (Young et al. (1999)).
In this paper, a nonlinear adaptive Sliding mode speed and rotor flux control scheme combined with field orientation for the induction-motor drive has been developed. Some sliding surfaces are chosen for which an appropriate logic commutation associated to these surfaces is determined. One important characteristic of the proposed controller is its cascade structure; witch gives a high performance using simple sliding surfaces. Furthermore, in order to reduce chattering phenomenon, smooth control functions with appropriate threshold have

been chosen. The rotor flux is estimated using the rotor-circuit model and, thus, is insensitive to the stator resistance. A stable model reference adaptive system (MRAS) rotor-resistance and load torque estimators have been designed using the measured stator current and rotor speed, and voltage command.

2. Indirect field-oriented control of the IM

Assuming linear magnetic circuits and balanced three phase windings, the fifth-order nonlinear model of IM (Krauss (1995)), expressed in the stator frame is :

$$\begin{aligned}
\frac{d\omega}{dt} &= \frac{3n_pL_m}{2JL_2}(\lambda_{2a}i_{1b} - \lambda_{2b}i_{1a}) - \frac{T_L}{J} \\
\frac{d\lambda_{2a}}{dt} &= -\frac{R_2}{L_2}\lambda_{2a} - n_p\omega\lambda_{2b} + \frac{R_2}{L_2}L_m i_{1a} \\
\frac{d\lambda_{2b}}{dt} &= -\frac{R_2}{L_2}\lambda_{2b} + n_p\omega\lambda_{2a} + \frac{R_2}{L_2}L_m i_{1b} \\
\frac{di_{1a}}{dt} &= \frac{L_mR_2}{\sigma L_1L_2^2}\lambda_{2a} + \frac{n_pL_m}{\sigma L_1L_2}\omega\lambda_{2b} - \frac{L_m^2R_2 + L_2^2R_1}{\sigma L_1L_2^2}i_{1a} + \frac{1}{\sigma L_1}u_{1a} \\
\frac{di_{1b}}{dt} &= \frac{L_mR_2}{\sigma L_1L_2^2}\lambda_{2b} - \frac{n_pL_m}{\sigma L_1L_2}\omega\lambda_{2a} - \frac{L_m^2R_2 + L_2^2R_1}{\sigma L_1L_2^2}i_{1b} + \frac{1}{\sigma L_1}u_{1b}
\end{aligned} \tag{1}$$

We can see that the model described by (1) is highly coupled, multivariable and nonlinear system. It is very difficult to control the IM directly based on this model. State transformation to simplify the system representation is required. A well-known method to this end is the transformation of the field orientation principle. It involves basically a change of the representations of the state vector $(i_{1a}, i_{1b}, \lambda_{2a}, \lambda_{2b})$ in the fixed stator frame (a, b) into a new state vector in a frame (d, q) which rotates along with the flux vector $(\lambda_{2a}, \lambda_{2b})$. Mathematically, the field oriented transformation can be described as:

$$\begin{aligned}
i_{1d} &= \frac{\lambda_{2a}i_{1a} + \lambda_{2b}i_{1b}}{\sqrt{\lambda_{2a}^2 + \lambda_{2b}^2}}, \quad i_{1q} = \frac{\lambda_{2a}i_{1b} - \lambda_{2b}i_{1a}}{\sqrt{\lambda_{2a}^2 + \lambda_{2b}^2}} \\
\lambda_{2d} &= \sqrt{\lambda_{2a}^2 + \lambda_{2b}^2}, \quad \lambda_{2q} = 0, \quad \rho = \arctan\left(\frac{\lambda_{2b}}{\lambda_{2a}}\right)
\end{aligned} \tag{2}$$

Where much simplification is gained by the fact that $\lambda_{2q} = 0$. Using this transformation and the notations in the Nomenclature, the state equations (1) can be rewritten in the new state variables as:

$$\begin{aligned}
\frac{dw}{dt} &= \mu\lambda_{2d}i_{1q} - \frac{T_L}{J} \\
\frac{di_{1q}}{dt} &= -\eta_1 i_{1q} - \beta n_p\omega\lambda_{2d} - n_p\omega i_{1d} - R_2(\eta_2 i_{1q} + \alpha L_m\frac{i_{1q}i_{1d}}{\lambda_{2d}}) + \frac{1}{\sigma L_1}u_{1q} \\
\frac{\lambda_{2d}}{dt} &= -\alpha R_2\lambda_{2d} + \alpha L_m R_2 i_{1d}
\end{aligned}$$

$$\frac{di_{1d}}{dt} = -\eta_1 i_{1d} + n_p \omega i_{1q} + R_2(-\eta_2 i_{1d} + \alpha\beta\lambda_{2d} + \alpha L_m \frac{i_{1q}^2}{\lambda_{2d}}) + \frac{1}{\sigma L_1} u_{1d} \quad (3)$$

$$\frac{d\rho}{dt} = n_p \omega + \alpha L_m R_2 \frac{i_{1q}}{\lambda_{2d}}$$

The decoupling control method with compensation is to choose inverter output voltages such that:

$$u_{1d}^* = \left(K_p + K_i \frac{1}{s}\right)(i_{1d}^* - i_{1d}) - \rho L_1 \sigma i_{1q}^* + L_1 \sigma \frac{di_{1q}^*}{dt}$$

$$u_{1q}^* = \left(K_p + K_i \frac{1}{s}\right)\left(i_{1q}^* - i_{1q}\right) + \rho L_1 \sigma i_{1d}^* + \rho \frac{L_m}{L_2} \hat{\lambda}_{2d} \quad (4)$$

Where K_p, K_i are PI controller gains.
For that, we need the estimation of the rotor flux as given by

$$\dot{\hat{\lambda}}_{2d} = -\alpha \hat{R}_2 \hat{\lambda}_{2d} + \alpha L_m \hat{R}_2 i_{1d} \quad (5)$$

Fig.1 shows the implemented diagram of an induction motor indirect field-oriented control (IFOC).

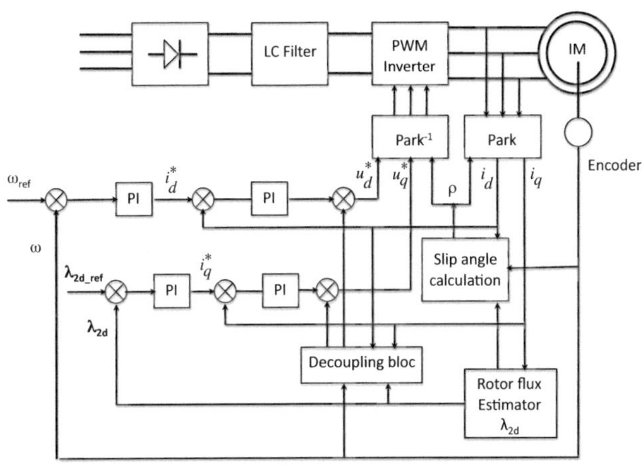

Fig. 1. bloc diagram of IFOC for the IM

3. Cascade sliding mode control

From the above IFOC structure, the sliding mode control mechanisms for the rotor angular speed regulation and the flux generation can be better applied to replace the traditional nonlinear feedback PI control of the field oriented control technique for better performance. Basically, the sliding mode control objectives consist mainly of the following steps:

3.1 Design of the switching surfaces

We choose the sliding surface to obtain a sliding mode regime which guarantees the convergence of the state x to its desired value x_d according to the relation (6) (Slotine & Li (1991):

$$S(x) = \left(\frac{d}{dt} + \lambda\right)^{r-1} e \tag{6}$$

Where $e = x_d - x$: tracking error
λ: positive coefficient
r: relative degree
Such in IFOC structure, we have four PI controllers; we will define four sliding surfaces (Mahmoudi et al. (1999)):

$$\begin{aligned} S_1(\omega) &= \omega_{ref} - \omega \\ S_2(\lambda_{2d}) &= \lambda_{2dref} - \lambda_{2d} \\ S_3(i_{1q}) &= i_{1q}^* - i_{1q} \\ S_4(i_{1d}) &= i_{1d}^* - i_{1d} \end{aligned} \tag{7}$$

3.2 Control calculation

Two parts have to be distinguished in the control design procedure. The first one concerns the attractivity of the state trajectory to the sliding surface and the second represents the dynamic response of the representative point in sliding mode. This latter is very important in terms of application of nonlinear control techniques, because it eliminates the uncertain effect of the model and external perturbation. For that, the structure of a sliding mode controller includes two terms:

$$u_c = u_{eq} + u_n \tag{8}$$

Where
-u_{eq} is called equivalent control which is used when the system state is in the sliding mode. It is calculated from $\dot{S}(x) = 0$.
-u_n is given to guarantee the attractivity of the variable to be controlled towards the commutation surface. This latter is achieved by the condition (Slotine & Li (1991); Utkin (1993)).

$$S(x).\dot{S}(x) < 0 \tag{9}$$

A simple form of the control action using sliding mode theory is a relay function; witch has a discontinuous form given by:

$$u_n = -k.sgn(S(x)) \tag{10}$$

k is a constant and is chosen positive to satisfy attractivity and stability conditions.

3.2.1 For the rotor flux regulation

$$\dot{S}_2(\lambda_{2d}) = 0 \Rightarrow i_{1deq} = \frac{\lambda_{2d} + T_r\dot{\lambda}_{2dref}}{L_m} \tag{11}$$

$$\dot{S}_2(\lambda_{2d})\,S_2(\lambda_{2d}) < 0 \Rightarrow i_{1dn} = K_\phi.sign\left(S_2(\lambda_{2d})\right) \tag{12}$$

Thus the controller is

$$i_{1dc} = i_{1deq} + i_{1dn} \tag{13}$$

3.2.2 For the direct current regulation

$$\begin{aligned} &\dot{S}_4(i_{1d}) = 0 \Rightarrow \\ &\quad u_{1deq} = \sigma L_1\dot{i}_{1dref} + R_{sm}i_{1d} - \sigma L_1\omega_s i_{1q} - \frac{L_m}{L_2T_r}\lambda_{2d} \end{aligned} \tag{14}$$

$$\dot{S}_4(i_{1d})\,S_4(i_{1d}) < 0 \Rightarrow u_{1dn} = K_d.sign\left(S_4(i_{1d})\right) \tag{15}$$

The controller is given by

$$u_{1dc} = u_{1deq} + u_{1dn} \tag{16}$$

3.2.3 For the speed regulation

$$\dot{S}_1(\omega) = 0 \Rightarrow i_{1qeq} = \frac{J\dot{\omega}_{ref} + F\omega + T_L}{\frac{pL_m}{L_2}\lambda_{2d}} \tag{17}$$

$$\dot{S}_1(\omega)\,S_1(\omega) < 0 \Rightarrow i_{1qn} = K_\omega.sign\left(S_1(\omega)\right) \tag{18}$$

The controller is given by

$$i_{1qc} = i_{1qeq} + i_{1qn} \tag{19}$$

3.2.4 For the quadrature current regulation

$$\begin{aligned} &\dot{S}_3(i_{1q}) = 0 \Rightarrow \\ &\quad u_{1qeq} = \sigma L_1\dot{i}_{1qref} + R_{sm}i_{1q} + \sigma L_1\omega_s i_{1d} + \frac{L_m}{L_2}\omega\lambda_{2d} \end{aligned} \tag{20}$$

$$\dot{S}_3(i_{1q})\,S_3(i_{1q}) < 0 \Rightarrow u_{1qn} = K_q.sign\left(S_3(i_{1q})\right) \tag{21}$$

The controller is given by

$$u_{1qc} = u_{1qeq} + u_{1qn} \tag{22}$$

To satisfy stability condition of the system, all of the following gains $(K_d, K_q, K_\phi, K_\omega)$ should be chosen positive.

4. Load torque estimator

Since the load torque is not exactly known, its estimation is introduced. The mechanical equation gives

$$\hat{T}_L = \frac{pL_m}{L_2}\lambda_{2d}i_{1q} - J\frac{d\omega}{dt} - F\omega \tag{23}$$

5. Mras rotor resistance identification

As the estimated rotor flux is sensitive to rotor-resistance variation, a stable rotor-resistance MRAS estimator can be developed. We can rewrite the dynamic model of an induction motor given before by equations (11) as a compact form given by (Leonhard (1984); Pavlov & Zaremba (2001)) :

$$\frac{dw}{dt} = \frac{3}{2}\frac{n_p L_m}{JL_2} i_1^T M \lambda_2 - \frac{T_L}{J} \tag{24}$$

$$\frac{d\lambda_2}{dt} = \left(-\frac{R_2}{L_2}I + n_p w M\right)\lambda_2 + \frac{R_2}{L_2}Ji_1 \tag{25}$$

$$\frac{di_1}{dt} = -\frac{J}{\sigma L_1 L_2}\left(-\frac{R_2}{L_2}I + n_p w M\right)\lambda_2 - \frac{1}{\sigma L_1}\left(R_1 + \frac{J^2 R_2}{L_2^2}\right)i_1 + \frac{1}{\sigma L_1}u_1 \tag{26}$$

Where

$$I = \begin{pmatrix} 1 & 0 \\ 0 & 1 \end{pmatrix}, \qquad M = \begin{pmatrix} 0 & -1 \\ 1 & 0 \end{pmatrix}$$

In order to design a rotor resistance identifier equations (25) (26) are transformed to eliminate the unobservable rotor flux. At this point an assumption is made that the rotor speed changes significantly slower relative to the rotor flux. Thus it may be treated as a constant parameter. First differentiating (26) and eliminating λ_2 gives

$$\frac{d^2 i_1}{dt^2} = (\alpha_1 I + w\beta_1 M)\frac{di_1}{dt} + (\alpha_2 I + w\beta_2 M)\,i_1 + (\alpha_3 I + w\beta_3 M)\,u_1 + \alpha_4\frac{du_1}{dt} \tag{27}$$

where

$$\alpha_1 = -\frac{1}{\sigma L_1}\left(R_1 + \frac{L_m^2 R_2}{L_2^2}\right) - \frac{R_2}{L_2}, \quad \beta_1 = n_p,$$
$$\alpha_2 = -\frac{R_2 R_1}{\sigma L_2 L_1}, \quad \beta_2 = \frac{n_p R_1}{\sigma L_1},$$
$$\alpha_3 = \frac{R_2}{\sigma L_1 L_2}, \quad \beta_3 = -\frac{n_p}{\sigma L_1}, \quad \alpha_4 = \frac{1}{\sigma L_1}.$$

Adding $cdi_1/dt, c > 0$, to both sides of (27) and formally dividing by (d/dt + c) transforms (27) to

$$\frac{di_1}{dt} = a + \omega b + \epsilon \tag{28}$$

where $\epsilon \rightarrow 0$ exponentially and the functions a and b are linear combinations of the filtered stator current and stator voltage command:

$$a = (c+\alpha_1)\, i_{11} + \alpha_2 i_{10} + \alpha_3 u_{10} + \alpha_4 u_{11}$$
$$b = M\left(\beta_1 i_{11} + \beta_2 i_{10} + \beta_3 u_{10}\right)$$

where

$$i_{10} = \frac{1}{s+c} i_1, \quad i_{11} = \frac{s}{s+c} i_1$$
$$u_{10} = \frac{1}{s+c} u_1, \quad u_{11} = \frac{s}{s+c} u_1 \qquad (29)$$

Here s denotes d/dt.
To obtain a reference model for the rotor resistance identification the part of the right-hand side of (28) containing R_2 is separated:

$$\frac{di_1}{dt} = f_1 + R_2 f_2 + \omega M f_3 + \epsilon \qquad (30)$$

where

$$f_1 = (c+\rho_1)\, i_{11} + \rho_2 u_{11}$$
$$f_2 = \gamma_1 i_{11} + \gamma_2 i_{10} + \gamma_3 u_{10} \qquad (31)$$
$$f_3 = \beta_1 i_{11} + \beta_2 i_{10} + \beta_3 u_{10}$$

Coefficients in (31) are calculated according to the following formulae:

$$\rho_1 = -\frac{R_1}{\sigma L_1}, \quad \rho_2 = \frac{1}{\sigma L_1},$$
$$\beta_1 = n_p, \quad \beta_2 = \frac{n_p R_1}{\sigma L_1}, \quad \beta_3 = -\frac{n_p}{\sigma L_1},$$
$$\gamma_1 = -\frac{1}{L_2}\left(1 + \frac{L_m^2}{\sigma L_1 L_2}\right), \quad \gamma_2 = \frac{\rho_1}{L_2}, \quad \gamma_3 = \frac{\rho_2}{L_2}.$$

Since ω is available for measurement the design of an R_2 identifier is straightforward. It is based on the MRAS identification approach (Sastry & Bodson (1989)) with(30) being a reference model.
Within this approach an identifier consists of a tuning model depending on an estimate of the unknown parameter and a mechanism to adjust the estimate. This adjustment is performed to make the output of the tuning model asymptotically match the output of the reference model. In our case the tuning model is given by

$$\frac{d\hat{i}_1}{dt} = -L\left(\hat{i}_1 - i_1\right) + f_1 + \omega M f_3 + \hat{R}_2 f_2 \qquad (32)$$

where $L > 0$ is a constant and $\hat{R}_2$ is the estimate of R_2. The dynamics of the error $e = \hat{i}_1 - i_1$ is the following

$$\frac{de}{dt} = -Le + \left(\hat{R}_2 - R_2\right) f_2 - \epsilon \qquad (33)$$

The adjustment equation for $\hat{R}_2$ is

$$\frac{d\hat{R}_2}{dt} = -\gamma \left(\hat{i}_1 - i_1\right)^T f_2, \quad \gamma > 0. \tag{34}$$

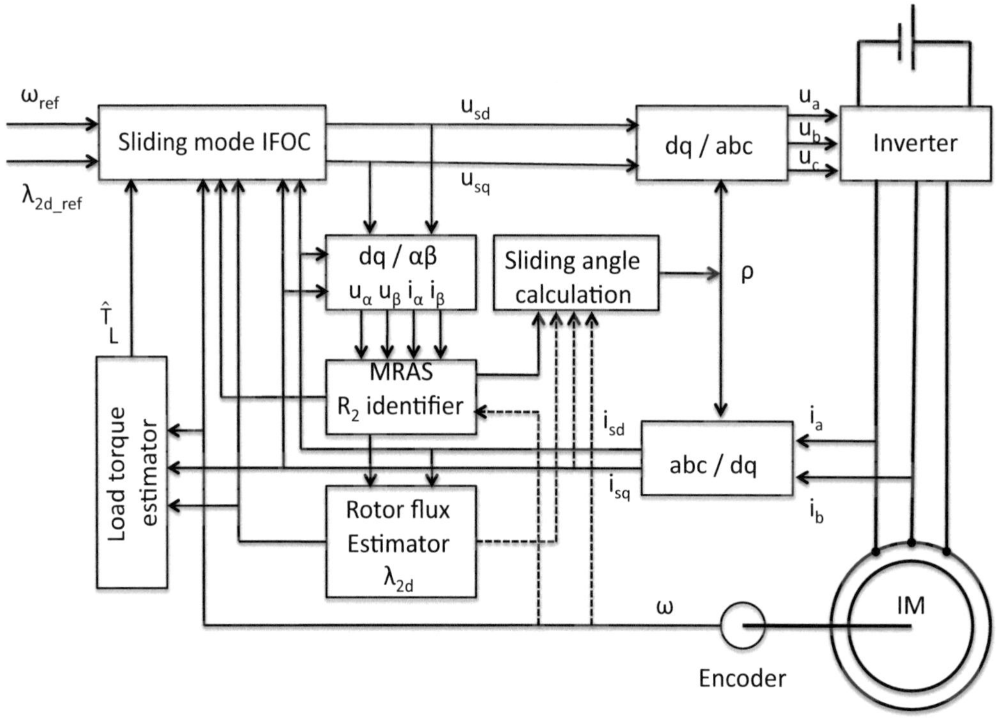

Fig. 2. Overall bloc diagram of the control scheme for IM

6. Simulations results

The overall configuration of the control system for IM is shown in Fig.2. The effectiveness of the proposed controller combined with the rotor resistance and load torque estimations has been verified by simulations in Matlab/Simulink. The parameters of the induction motor used are given in Appendix. The simulation results have been obtained under a constant load torque of 10 Nm.

Parameters of the Sliding mode controllers are: $K_d = 500, K_q = 500, K_\phi = 400$ and $K_\omega = 300$.

Parameters of the MRAS identifier are: $\gamma = 0.2, L = 100$ and $c = 0.01$.

Results obtained are shown in Fig.3. The reference speed is set to 200 rad/s until t=4s, when it is reversed to -200rad/s to allow drive to operate in the generating mode. The reference flux is set to 0.4wb. The load torque is changed from 0 to 10 Nm at t=0.6s. The rotor resistance and the load torque estimators are activated. It can be seen that the estimated load torque converges very quickly to the actual value. In addition, the estimated values of R_2 follow its actual value very closely.

In order to show the convergence capability of the MRAS rotor resistance estimator, at t=1.2 s, the interne value of R_2 in the MAS model has been disturbed and varied intentionally 30%

of its initial value and held constant until t=2.5s, at the same moment, the rotor resistance estimator is disconnected from the control structure.
It can be noted that the error in the estimated value of R_2 produces a steady state error in the speed and the rotor flux control and also generates an error in the estimated value of the load torque.
At t=2.5s, the initial value of R_2 in MAS model has restored and the rotor resistance estimator has been reconnected and the rotor resistance is seen to converge to the actual value and also the other system variables.
At t=4s, the reference speed is reversed to -200 rad/s to allow the drive to operate in the generating mode.
The results show a stable operation of the drive in the various operating modes.

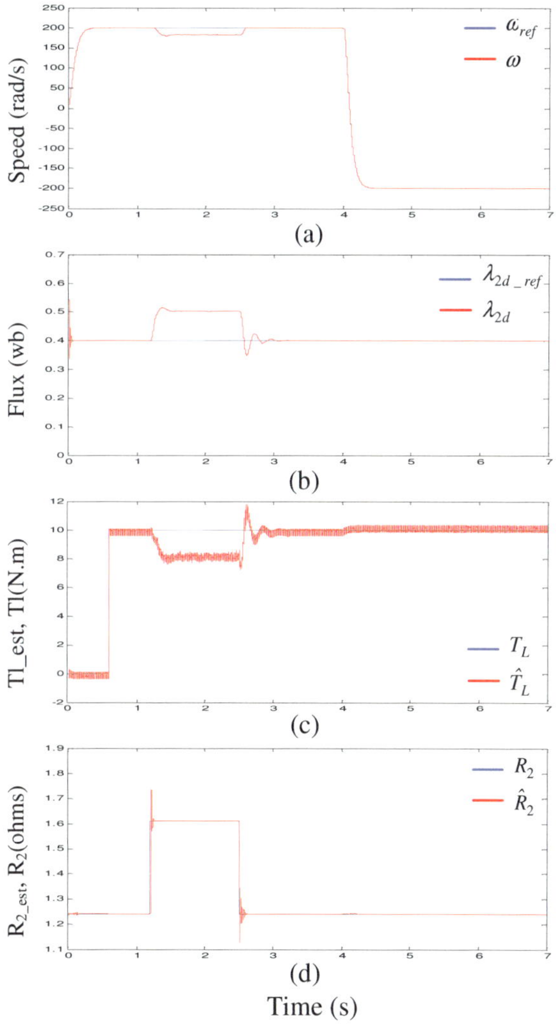

Fig. 3. Tracking Performance and parameters estimates : a)reference and actual motor speeds, b)reference and actual rotor flux, c)T_L and $\hat{T}_L$(N.m), d)R_2 and $\hat{R}_2(\Omega)$.

Furthermore, simulation results have been performed to show the capability of the load torque estimator to track the rapid load torque changes and also to show the performance of flux and speed control. The results are shown in Fig. 4. These results have been obtained using the same parameters used for the results in Fig. 3. The rotor resistance and load torque estimators are activated at t=0.2s. The load torque has been reversed from 10 Nm to -10 Nm at t=2s. It can be seen from Fig. 4 that the estimated load torque converges rapidly to its actual value and the rotor resistance estimator is stable. In addition, the results in Fig. 4 show an excellent control of rotor flux and speed.

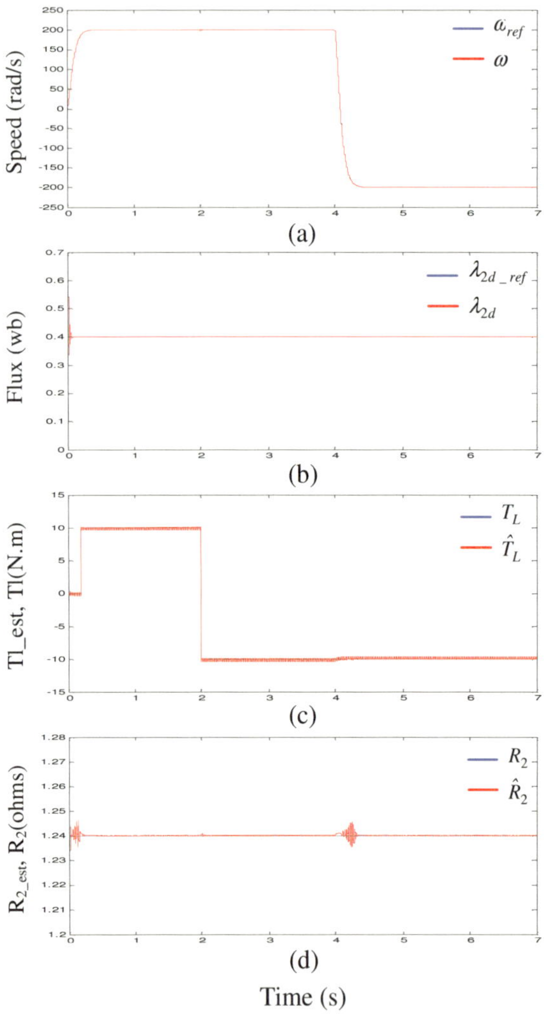

Fig. 4. Tracking Performance and parameters estimates : a)reference and actual motor speeds, b)reference and actual rotor flux, c)T_L and $\hat{T}_L$(N.m), d)R_2 and $\hat{R}_2(\Omega)$.

Thus, the simulation results confirm the robustness of the proposed scheme with respect to the variation of the rotor resistance and load torque.

7. Conclusions and future works

In this paper, a novel scheme for speed and flux control of induction motor using online estimations of the rotor resistance and load torque have been described. The nonlinear controller presented provides voltage inputs on the basis of rotor speed and stator currents measurements and guarantees rapid tracking of smooth speed and rotor flux references for unknown parameters (rotor resistance and load torque) and non-measurable state variables (rotor flux). In simulation results, we have shown that the proposed nonlinear adaptive control algorithm achieved very good tracking performance within a wide range of the operation of the IM. The proposed method also presented a very interesting robustness properties with respect to the extreme variation of the rotor resistance and reversal of the load torque. The other interesting feature of the proposed method is that it is simple and easy to implement in real time.

From a practical point of view, in order to reduce the chattering phenomenon due to the discontinuous part of the controller, the sign(.) functions have been replaced by the saturation functions $\frac{(.)}{(.)+0.01}$ (Slotine & Li (1991).

It would be meaningful in the future work to implement in real time the proposed algorithm in order to verify its robustness with respect to the discretization effects, parameter uncertainties and modelling inaccuracies.

Induction motor data		
Stator resistance	1.34	Ω
Rotor resistance	1.24	Ω
Mutual inductance	0.17	H
Rotor inductance	0.18	H
Stator inductance	0.18	H
Number of pole pairs	2	H
Motor load inertia	0.0153	Kgm^2

Nomenclature	
R_1, R_2	rotor, stator resistance
i_1, i_2	rotor, stator current
λ_1, λ_2	rotor, stator flux linkage
λ_{2d}	amlpitude of rotor flux linkage
u_1, u_2	rotor, stator voltage input
ω	rotor angular speed
ω_s	stator angular frequency
ρ	rotor flux angle
n_p	number of pole pairs
L_1, L_2	rotor, stator inductance
T_r	rotor time constant
L_m	mutual inductance
J, T_L	inercia, load torque
F	coefficient of friction
$(.)_d, (.)_q$	in (d,q) frame
$(.)_a, (.)_b$	in (a,b) frame
$(\hat{.})$	estimate of (.)
$(.^*)$	reference of (.)

$$\sigma = 1 - \frac{L_m^2}{L_1 L_2}, \alpha = \frac{1}{L_2}, \beta = \frac{L_m}{\sigma L_1 L_2}, \eta_1 = \frac{R_1}{\sigma L_1}$$
$$\eta_2 = \frac{L_m^2}{\sigma L_1 L_2^2}, \mu = \frac{3 n_p L_m}{2 J L_2}, R_{sm} = R_1 + \frac{L_m^2}{L_2^2} R_2$$

8. References

Ebrahim, A. & Murphy, G. (2006). Adaptive backstepping control of an induction motor under time-varying load torque and rotor resistance uncertainty, *Proceedings of the 38th Southeastern Symposium on System Theory Tennessee Technological University Cookeville* .

Kanellakopoulos, I., Kokotovic, P. V. & Morse, A. S. (1991). Systematic design of adaptive controllers for feedback linearizable systems, *IEEE Trans. Automatic Control* 36: 1241–1253.

Krauss, P. C. (1995). Analysis of electric machinery, *IEEE Press* 7(3): 212–222.

Krstic, M., Kannellakopoulos, I. & Kokotovic, P. (1995). Nonlinear and adaptive control design, *Wiley and Sons Inc., New York* pp. 1241–1253.

Leonhard, W. (1984). Control of electric drives, *Springer Verlag* .

Mahmoudi, M., Madani, N., Benkhoris, M. & Boudjema, F. (1999). Cascade sliding mode control of field oriented induction machine drive, *The European Physical Journal* pp. 217–225.

Marino, R., Peresada, S. & Valigi, P. (1993). Adaptive input-output linearizing control of induction motors, *IEEE Transactions on Automatic Control* 38(2): 208–221.

Ortega, R., Canudas, C. & Seleme, S. (1993). Nonlinear control of induction motors: Torque traking with unknown disturbance, *IEEE Transaction On Automatic Control* 38: 1675–1680.

Ortega, R. & Espinoza, G. (1993). Torque regulation of induction motor, *Automatica* 29: 621–633.

Ortega, R., Nicklasson, P. & Perez, G. E. (1996). On speed control of induction motor, *Automatica* 32(3): 455–460.

Pavlov, A. & Zaremba, A. (2001). Real-time rotor and stator resistances estimation of an induction motor, *Proceedings of NOLCOS-01, St-Petersbourg* .

Rashed, M., MacConnell, P. & Stronach, A. (2006). Nonlinear adaptive state-feedback speed control of a voltage-fed induction motor with varying parameters, *IEEE TRANSACTIONS ON INDUSTRY APPLICATIONS* 42(3): 1241–1253.

Sastry, S. & Bodson, M. (1989). Adaptive control: stability, c onvergence, and robustness, *Prentice Hall. New Jersey* .

Slotine, J. J. & Li, W. (1991). Applied nonlinear control, *Prentice Hall. New York* .

Utkin, V. (1993). Sliding mode control design principles and applications to electric drives, *IEEE Transactions On Industrial Electronics* 40: 26–36.

Utkin, V. I. (1977). Variable structure systems with sliding modes, *IEEE Transaction on Automatic Control* AC-22: 212–222.

Young, K. D., Utkin, V. I. & Ozguner, U. (1999). A control engineer's guide to sliding mode control, *IEEE Transaction on Control Systems Technology* 7(3): 212–222.

9

Sliding Mode Control of DC Drives

Dr. B. M. Patre[1], V. M. Panchade[2] and Ravindrakumar M. Nagarale[3]
[1]*Department of Instrumentation Engineering*
S.G.G.S. Institute of Engineering and Technology, Nanded
[2]*Department of Electrical Engineering*
G.H. Raisoni Institute of Engineering &Technology Wagholi, Pune
[3]*Department of Instrumentation Engineering*
M.B.E.Society's College of Engineering, Ambajogai
India

1. Introduction

Variable structure control (VSC) with sliding mode control (SMC) was first proposed and elaborated in early 1950 in USSR by Emelyanov and several co researchers. VSC has developed into a general design method for wide spectrum of system types including nonlinear system, MIMO systems, discrete time models, large scale and infinite dimensional systems (Carlo et al., 1988; Hung et al., 1993; Utkin, 1993). The most distinguished feature of VSC is its ability to result in very robust control systems; in many cases invariant control systems results. Loosely speaking, the term "invariant" means that the system is completely insensitive to parametric uncertainty and external disturbances.
In this chapter the unified approach to the design of the control system (speed, Torque, position, and current control) for DC machines will be presented. This chapter consists of parts: dc motor modelling, sliding mode controller of dc motor i.e. speed control, torque, position control, and current control. As will be shown in each section, sliding mode control techniques are used flexibly to achieve the desired control performance. All the design procedures will be carried out in the physical coordinates to make explanations as clear as possible. Drives are used for many dynamic plants in modern industrial applications.
The simulation result depicts that the integral square error (ISE) performance index for reduced order model of the system with observer state is better than reduced order with measured state.
Simulation results will be presented to show their agreement with theoretical predications. Implementation of sliding mode control implies high frequency switching. It does not cause any difficulties when electric drives are controlled since the "on –off" operation mode is the only admissible one for power converters.

2. Dynamic modelling of DC machine

Fig. 1 shows the model of DC motor with constant excitation is given by following state equations (Sabanovic et al., 1993; Krause, 2004).

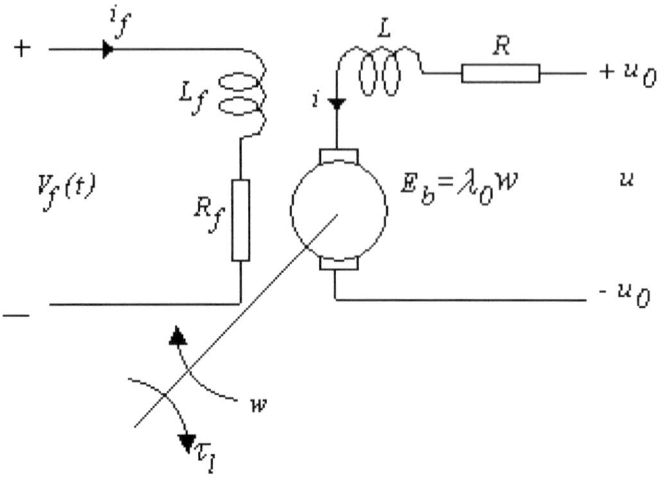

Fig. 1. Model of DC motor with constant excitation.

$$L\frac{di}{dt} = u - Ri - \lambda_0 w$$
$$j\frac{dw}{dt} = k_t i - \tau_l \tag{1}$$

where

i armature current
w shaft speed
R armature resistance
λ_0 back emf constant
τ_l load torque
u terminal voltage
j inertia of the motor rotor and load
L armature inductance
k_t torque constant.

Its motion is governed by second order equations (1) with respect to armature current i and shaft speed w with voltage u and load torque k_t. A low power-rating device can use continuous control. High power rating system needs discontinuous control. Continuously controlled voltage is difficult to generate while providing large current.

3. Sliding mode control design

DC motors have been dominating the field of adjustable speed drives for a long time because of excellent operational properties and control characteristics. In this section different sliding mode control strategies are formulated for different objectives e.g. current control, speed control, torque control and position control.

3.1 Current control

Let i^* be reference current providing by outer control loop and i be measured current. Fig. 2 illustrates Cascaded control structure of DC motors.

Consider a current control problem, by defining switching function

$$s = i^* - i \tag{2}$$

Design a discontinuous control as

$$u = u_0 sign(s) \tag{3}$$

Where u_0 denotes the supplied armature voltage.

$$s\dot{s} = s(\frac{di^*}{dt} + \frac{R}{L}i + \frac{\lambda_0}{L_0}w) - \frac{1}{L}u_0|s| \tag{4}$$

Choice of control u_0 as

$$u_0 > \left| L\frac{di^*}{dt} + Ri + \lambda_0 w \right| \quad \text{Makes (4)}$$

$s\dot{s} < 0$ Which means that sliding can happen in $s = 0$.

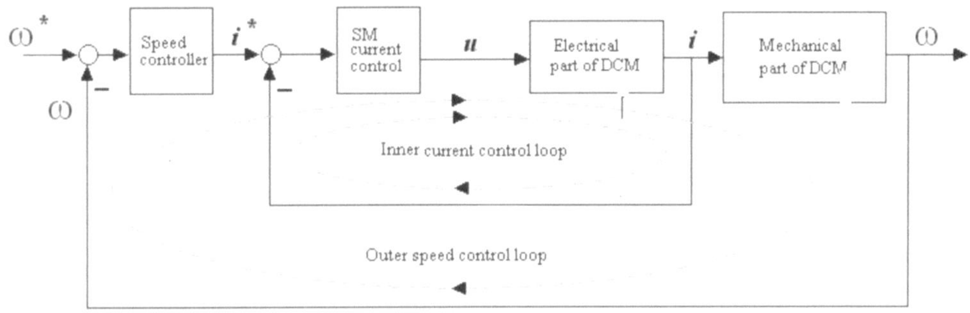

Fig. 2. Cascaded control structure of DC motors.

3.2 Speed control

w^* be the reference shaft speed, then the second order motion equation with respect to the error $e = w^* - w$ is of form. State variable $x_1 = e$ & $x_2 = \dot{e}$

$$\begin{aligned} \dot{x}_1 &= x_2 \\ \dot{x}_2 &= -a_1 x_1 - a_2 x_2 + f(t) - bu \end{aligned} \tag{5}$$

Where

$$a_1 = \frac{k_t \lambda_0}{jL}, a_2 = \frac{R}{L} \ \& \ b = \frac{k_t}{jL}$$

$$f(t) = \ddot{w} + a_2 \dot{w}^* + a_1 w^* + \frac{R\tau_l}{jL} + \frac{\dot{\tau}_l}{j}$$

are constant values.

The sliding surface and discontinuous control are designed as

$$s = c(w^* - w) + \frac{d}{dt}(w^* - w) \tag{6}$$

$$u = u_0 sign(s)$$

This design makes the speed tracking error e converges to zero exponentially after sliding mode occurs in $s = 0$, where c is a positive constant determining the convergence rate for implementation of control (6), angle of acceleration $x_2 = \dot{e}$ is needed. The system motion is independent of parameters a_1, a_2, b and disturbances in g(t).
Combining (1) & (6) produces

$$\begin{aligned} \dot{s} &= c\dot{w}^* + \ddot{w} - \frac{c}{j}(k_t - \tau_l) + \frac{1}{j}\tau_l + \frac{k_t}{jL}(Ri + k_t w) - \frac{k_t}{jL}u \\ &= g(t) - \frac{k_t}{jL}u \end{aligned} \tag{7}$$

$$g(t) = c\dot{w}^* + \ddot{w} - \frac{c}{j}(k_t - \tau_l) + \frac{1}{j}\tau_l + \frac{k_t}{jL}(Ri + k_t w) - \frac{k_t}{jL}u \tag{8}$$

$$if\ u_0 > \frac{jL}{k_t}|g(t)|,\ s\dot{s} < 0 \tag{9}$$

Then sliding mode will happen (Utkin,1993).
The mechanical motion of a dc motor is normally much slower then electromagnetic dynamics. It means that $L << j$ in (1).
a. Reduced -order Speed control with measured speed
Following reduced order control methods proposed below will solve chattering problem without measuring of current and acceleration (x_2).
Speed tracking error is $w_e = w^* - w$. The dc motor model (1) in terms of w_e:

$$\begin{aligned} L\frac{di}{dt} &= u - Ri - \lambda_0(w^* - w_e) \\ j\frac{dw_e}{dt} &= -k_t i + \tau_l + j\dot{w}^* \end{aligned} \tag{10}$$

Let L be equal to zero due to $L << j$. Then (10) becomes with $L = 0$

$$i = -\frac{\lambda_0}{R}(w^* - w_e) - \frac{k_t}{R}u + \tau_l + j\dot{w}^* \tag{11}$$

Substituting (11) into (10) results in

$$j\frac{dw_e}{dt} = \frac{k_t\lambda_0}{R}(w^* - w_e) - \frac{k_t}{R}u + \tau_l + j\dot{w}^* \tag{12}$$

Equation (12) is a reduced order (first order) model of dc motor. The discontinuous control is designed as

$$u = u_0 sign(w_e) \tag{13}$$

and the existence condition for the sliding mode $w_e = 0$ will be

$$u_0 > \left| \lambda_0 (w^* - w_e) + \frac{\tau_l R}{k_t} + \frac{jR\dot{w}^*}{k_t} \right| \tag{14}$$

The principle advantage of the reduced order based method is that the angle acceleration $(x_2 = \dot{e})$ is not needed for designing sliding mode control (Carlo et al.,1988).

b. Reduced -order Speed control with observer speed

The unmodelled dynamics (1) may excite non-admissible chattering.

Let us design an a asymptotic observer to estimate w_e (Utkin,1993).

$$j\frac{d\hat{w}_e}{dt} = \frac{k_t \lambda_0}{R}(w^* - \hat{w}_e) - \frac{k_t}{R}u + \hat{\tau}_l + j\dot{w}^* - l_1(\hat{w} - w_e)$$

$$j\frac{dw_e}{dt} = \frac{k_t \lambda_0}{R}(w^* - w_e) - \frac{k_t}{R}u + \tau_l + j\dot{w}^* \tag{15}$$

$$\frac{d\hat{\tau}_l}{dt} = -l_2(w^* - w_e)$$

where

$\hat{w}_e$ Estimated error

$\overline{w}_e = \hat{w}_e - w_e$ Speed tracking error

l_1, l_2 Observer gain

The discontinuous control designed using estimate state $\hat{w}_e$ (Utkin,1993) will be

$$u = u_0 sign(\hat{w}_e) \tag{16}$$

The sliding mode will happen if

$$u_0 > \left| \lambda_0 (w^* - \hat{w}_e) + \frac{\hat{\tau}_l R}{k_t} + \frac{jR\dot{w}^*}{k_t} + \frac{l_1 R}{k_t}(w^* - w_e) \right| \tag{17}$$

And $\hat{w} = 0 \,\&\, \hat{\tau} = 0$.

Under the control scheme, chattering is eliminated, but robustness provided by the sliding mode control is preserved within accuracy of $\frac{L}{j} << 1$. The observer gains l_1, l_2 should be chosen to yields mismatch dynamics slower than the electrical dynamics of the dc motors to prevent chattering. Since the estimated $\hat{w}$ is close to w, the real speed w tracks the desired speed w^*. Fig. 3 shows the control structure based on reduced order model and observed state. Chattering can be eliminated by using reduce observer states. The sliding mode occurs in the observer loop, which does not contain unmodelled dynamics.

3.3 Position control

To consider the position control issue, it is necessary to augment the motor equations (1) with

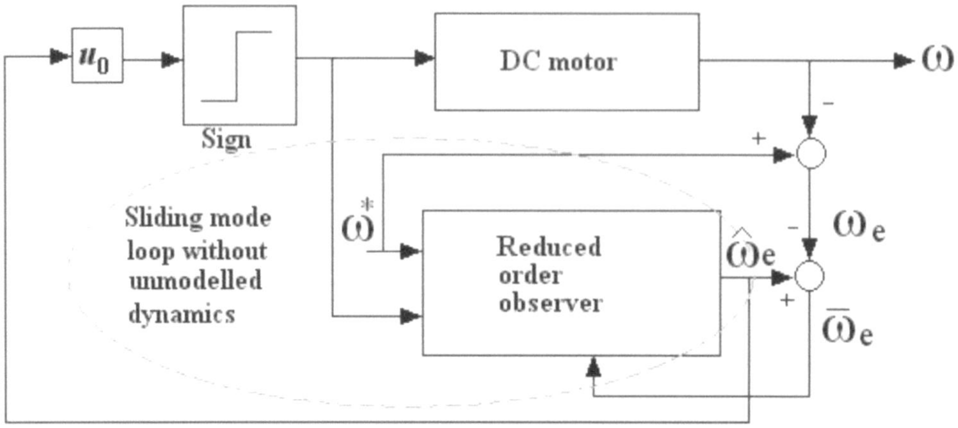

Fig. 3. Speed control based on reduced order model and observed state.

$$\frac{d\theta}{dt} = w \tag{18}$$

where θ denotes the rotor position.
The switching function s for the position control is selected as

$$\dot{s} = (\ddot{\theta}^* - \ddot{\theta}) + c_1(\dot{\theta}^* - \dot{\theta}) + c_2(\theta^* - \theta) \tag{19}$$

and the discontinuous control is

$$u = u_0 sign(s) \tag{20}$$

Combining (1), (18), (19)

$$\dot{s} = h(t) - \frac{k_t}{jL}u \tag{21}$$

where

$$h(t) = \ddot{w}^* + c_1\dot{w}^* + c_2 w^* - \frac{c_1}{j}(k_t i - \tau_l) - c_2 w + \frac{1}{j}\dot{\tau}_l + \frac{k_t}{jL}(Ri + k_t w) \tag{22}$$

Choice of u_0 as

$$u_0 > \frac{jL}{k_t}|h(t)| \tag{23}$$

Makes $s\dot{s} < 0$ which means that sliding mode can happen $s = 0$ with properly chosen c_1, c_2. We can make velocity tracking error $e = w^* - w$ converges to zero.

3.4 Torque control

The torque control problem by defining switching function

$$s = \tau^* - \tau \tag{24}$$

As the error between the reference torque τ^* and the real torque τ developed by the motor. Design a discontinuous control as

$$u = u_0 sign(s) \tag{25}$$

Where u_0 is high enough to enforce the sliding mode in $s = 0$, which implies that the real torque τ tracks the reference torque τ^*.

$$\begin{aligned} \dot{s} &= \dot{\tau}^* - k_t i \\ &= \dot{\tau}^* + \frac{k_t R i}{L} + \frac{k_t \lambda_0 w}{L} - \frac{k_t}{L} u \\ &= f(t) - \frac{k_t}{L} u \end{aligned} \tag{26}$$

Where

$$f(t) = \dot{\tau}^* + \frac{k_t R i}{L} + \frac{k_t \lambda_0 w}{L}$$

depending on the reference signal for

$$\begin{aligned} u_0 &> \frac{L}{k_t} |f(t)| \\ s\dot{s} &= s f(t) - \frac{k_t}{L} u_0 |s| < 0 \end{aligned} \tag{27}$$

So sliding mode can be enforced in $s = 0$.

4. Simulation results

4.1 Simulation results of current control

Examine inequality (2 & 4), if reference current is constant, the link voltage u_0 needed to enforce sliding mode should be higher than the voltage drop at the armature resistance plus back emf, otherwise the reference current i^* cannot be followed.

Figure 4 depicts a simulation result of the proposed current controller. The sliding mode controller has been already employed in the inner current loop thus, if we were to use another sliding mode controller for speed control, the output of speed controller i^* would be discontinuous, implying an infinite $\frac{di^*}{dt}$ and therefore destroying in equality (4) for any implement able u_0

4.1 Simulation results of speed control

To show the performance of the system the simulation result for the speed control of dc machine is depicted. Rated parameters of the dc motor used to verify the design principle are 5hp, 240V, R=0.5Ω, L=1mH, j=0.001kgm2, $k_t = 0.008$NmA-1 , $\lambda_0 = 0.001$v rad s^{-1} and $\tau_l = Bw$ where B=0.01 Nm rads-1.

Figure 5 depicts the response of sliding mode reduced order speed control with measured speed. It reveals that reduced order speed control with measured speed produces larger overshoot & oscillations.

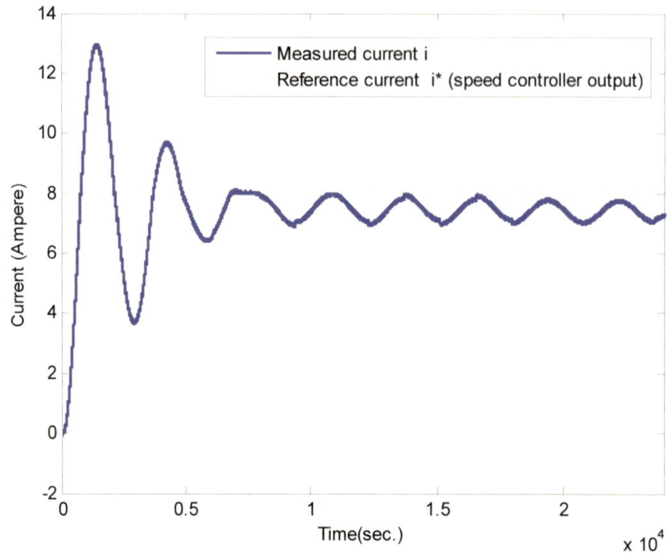

Fig. 4. Cascade current control of dc motor.

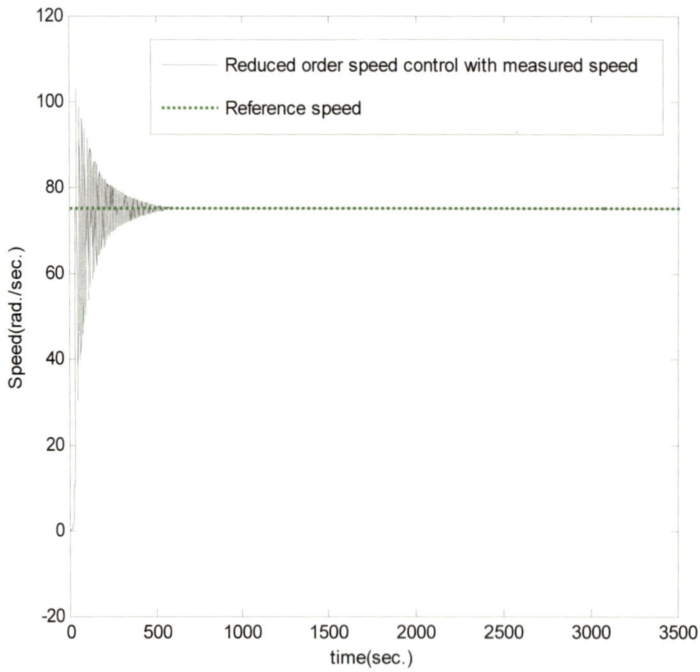

Fig. 5. Response of sliding mode reduced order speed control with measured speed.

Figure 6 depicts the response of sliding mode reduced order speed control with observer speed. It reveals that reduced order speed control with observer speed produces smaller overshoot & less oscillation.

Fig. 7(a), 7(b), 7(c) & 7(d) depicts the simulation result of response of sliding mode speed control, variation of error, squared error and integral square error (ISE) with reference speed of 75 radian per sec respectively. Fig. 8 reveals that variation of the controller for reduced order speed control with observed speed at load condition. Fig. 9 reveals that variation of the controller for reduced order speed control with measured speed at load condition. Fig. 10 depicts the response of sliding surface in sliding mode control. Fig. 11 depicts the robustness (insensitivity) to parameters (+10%) variation. Fig. 12 depicts the robustness (insensitivity) to parameters (-10%) variation. The high frequency chatter is due to neglecting the fast dynamics i.e. dynamics of the electric of the electric part. In order to reduce the weighting of the large initial error & to Penalise small error occurring later in response move heavily, the following performance index is proposed.The integral square error (ISE) is given by (Ogata,1995).

$$\text{ISE} = \int_{0}^{T} e^{2}(t)dt \tag{28}$$

The minimum value of ISE is obtained as gain tends to infinity.

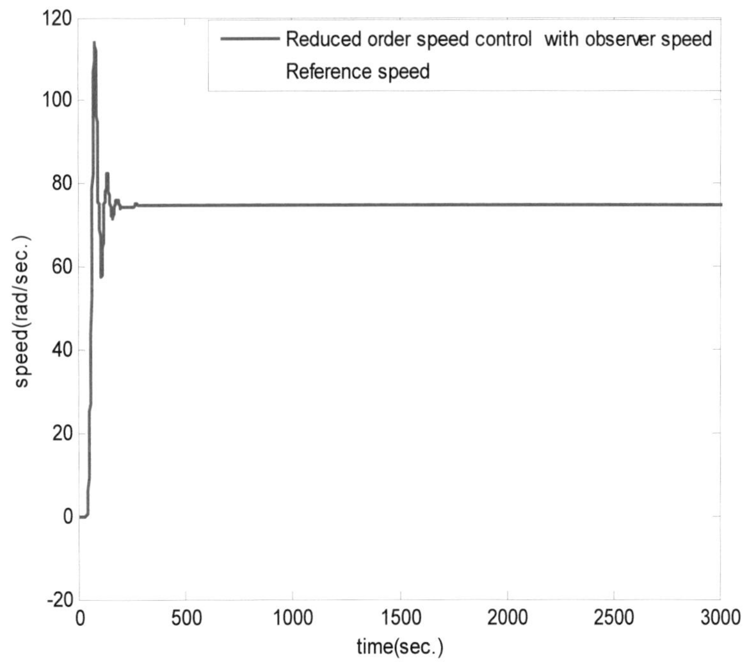

Fig. 6. Response of sliding mode reduced order speed control with observer speed.

5. Conclusions

The SMC approach to speed control of dc machines is discussed. Both theoretical and implementation result speed control based on reduced order with measured speed and reduced order with observer speed, using simulation are conducted. Besides, reduced order observer deals with the chattering problem, en-counted often in sliding mode. Control area Selection of the control variable (angular position, speed, torque) leaves basic control structures unchanged. Inspection of Fig. 7(a), 7(b), 7(c) & 7(d) reveals that reduced order speed control with observer speed produces smaller overshoot & oscillation than the reduced order speed with measure speed (Panchade et al.,2007). The system is proven to be robust to the parameters variations, order reduction, fast response, and robustness to disturbances.

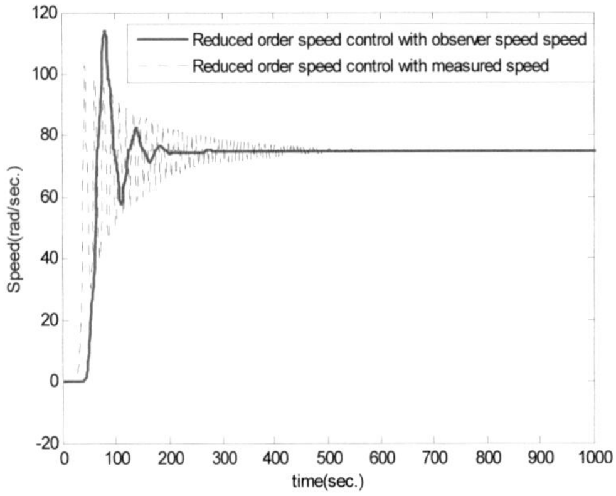

(a)

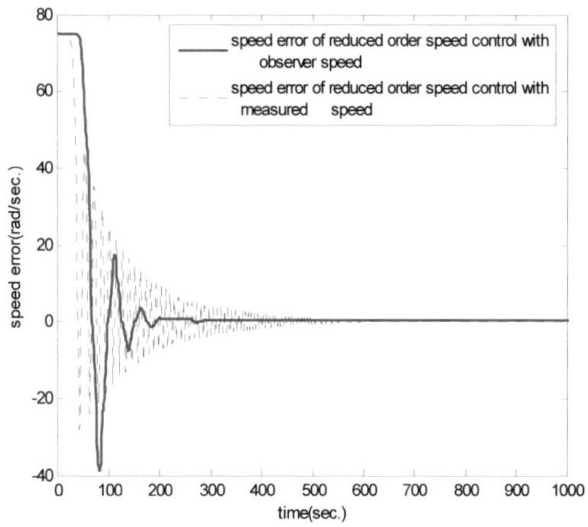

(b)

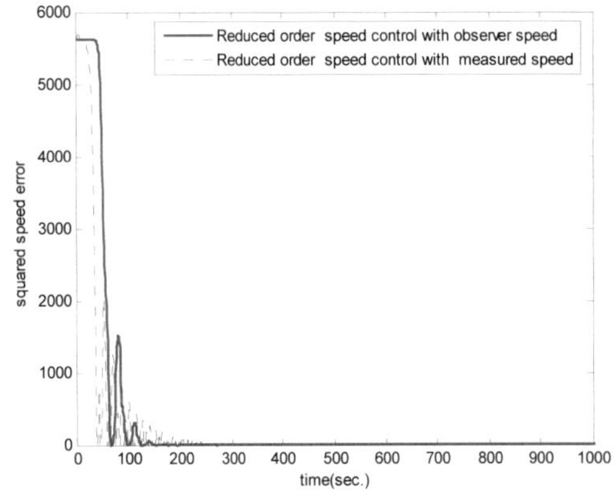

(c)

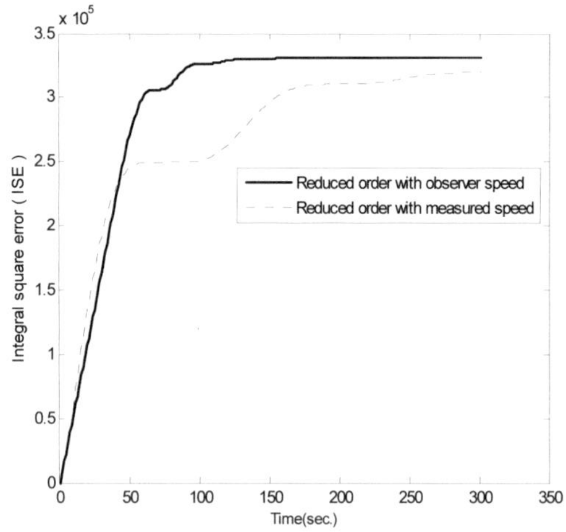

(d)

Fig. 7. (a) Response of the sliding mode speed control with reference speed 75 radian per sec. (b) The variation error with reference speed 75 radian per sec. (c) The variation of squared error with reference speed 75 radian per sec (d) Integral square error (ISE)

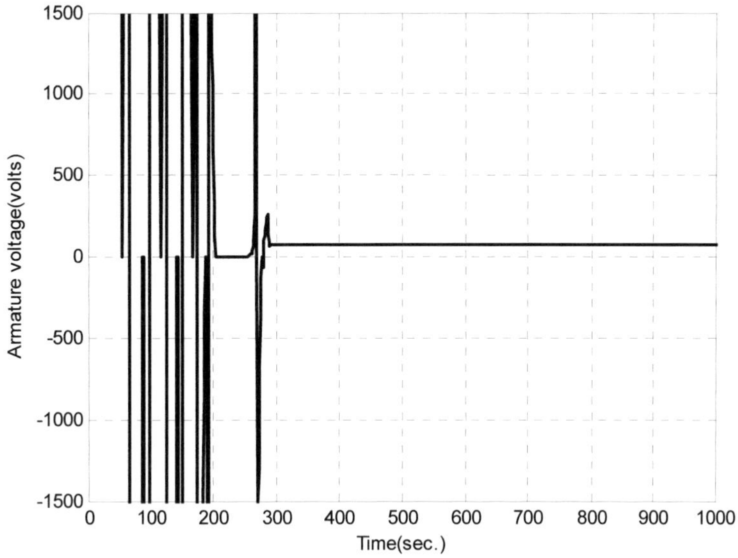

Fig. 8. Variation of the controller for reduced order speed control with observed speed at load condition

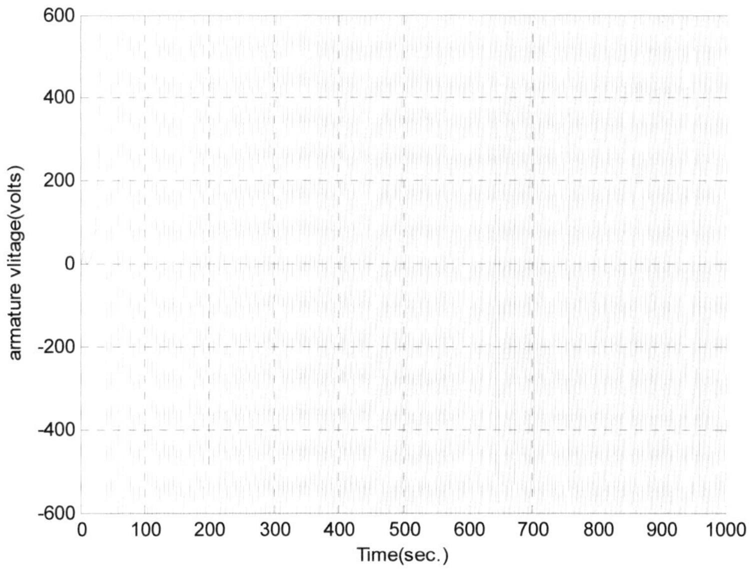

Fig. 9. Variation of the controller for reduced order speed control with measured speed at load condition

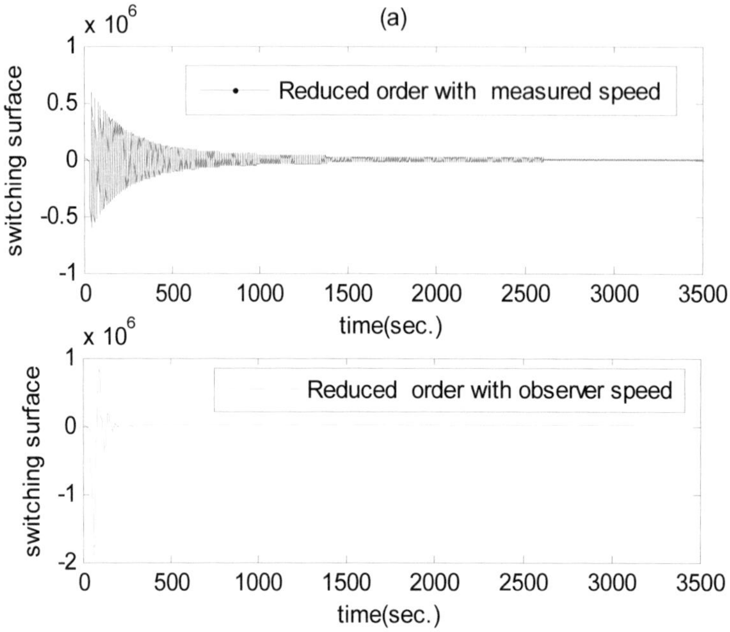

Fig. 10. Response of sliding surface in sliding mode speed control.

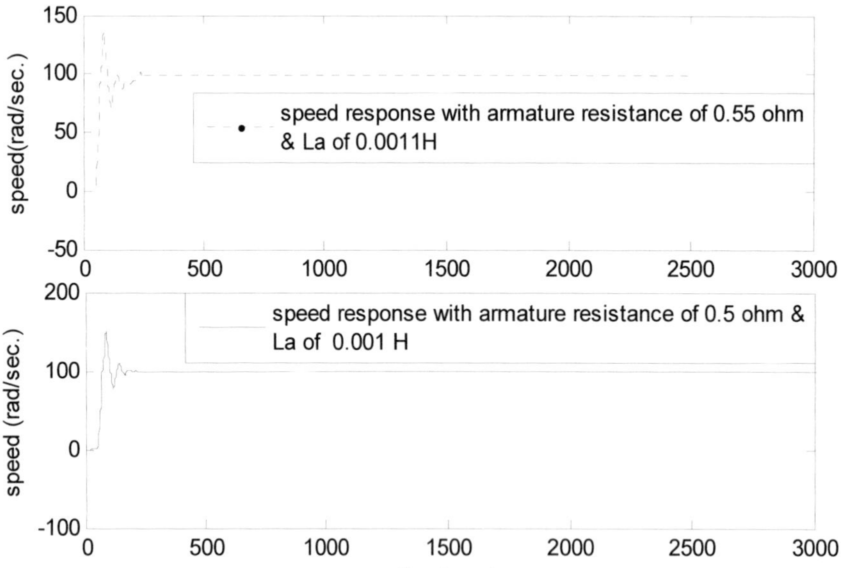

Fig. 11. Robust (Insensitivity) to parameters (+10%) variation.

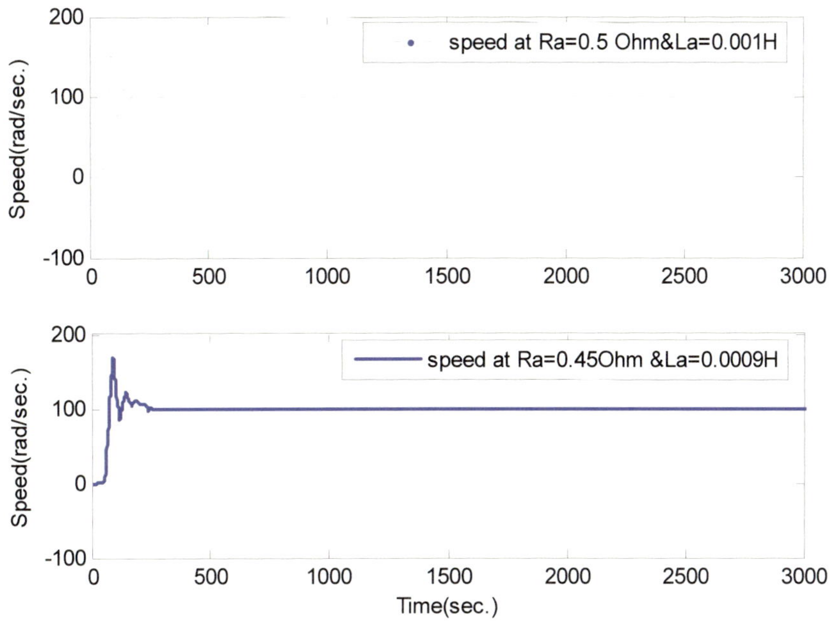

Fig. 12. Robust (insensitivity) to the parameters (-10%) variation

6. Acknowledgements

We are extremely grateful to Shri Sunil Raisoni, Chairmen of Raisoni Group of Institutions and Ajit Tatiya, Director, Raisoni group of Institutions (Pune campus) for their holehearted support.

We would also like to express my deep sense of gratitude to Dr. R. D. Kharadkar, Principal of G. H. Raisoni Institute of Engineering & Technology Pune and Prof. B. I. Khadakbhavi Principal College of Engineering Ambajogai for their support, kind co-operation and encouragement for pursuing this work.

7. References

John Y. Hung, W. Gao, J.C. Hung. Variable Structure Control: A survey, *IEEE Trans. on Industrial Electronics*, Vol.40, no.1, pp.1- 22, February 1993.

De Carlo, Zak S. H. Mathews G.P. Variable structure control of nonlinear multivariable systems: A Tutorial, *Proceedings of IEEE*, Vol.76, No.3, pp. 212-232, March 1988.

V.I. Utkin.Sliding mode control design principles and applications to Electric drives, *IEEE Trans on Industrial Electronics*, Vol.40, no.1, pp. 23-36, February 1993.

A. Sabanovic, KenzoWada, Faruk Ilalovic, Milan Vujovic. Sliding modes in Electrical machines control systems *Proceedings of IEEE*, vol., No., pp.73-78, 1993.

N. Sabanovic, A. Sabanovic,k. Jezernik O.M.Vujovic.Current control in three phase switching converters and AC electrical machines, in *Proceedings of the International IECON'94*, pp. 581-586.

P.C. Krause, O.Wasynczuk, Scottd D. Sudhoff. *Analysis of Electric Machinery and Drive systems*, IEEE Press John Wiley & Sons Inc. Publication, 2nd edition 2004, pp. 67-96.

V. Utkin , Jurgen Guldner, J. Shi, *Sliding mode control in Electromechanical Systems*, Taylor and Francis London, 1999.

Katsuhiko Ogata. *Modern Control Engineering*, Prentice Hall of India Pvt. Ltd. New Delhi, 2nd edition, 1995, pp. 295-303.

V.M.Panchade, L.M.Waghmare,B.M.Patre,P.P.Bhogle. Sliding Mode Control of DC Drives , *Proceedings of 2007 IEEE International Conference on Mechatronics and Automation,* ISBN 1-4244-0828-8 ,pp.1576-1580, August 5 - 8, 2007, Harbin, China.

10

Sliding Mode Position Controller for a Linear Switched Reluctance Actuator

António Espírito Santo,
Maria do Rosário Calado and Carlos Manuel Cabrita
University of Beira Interior – Electromechanical Engineering Department
Portugal

1. Introduction

More than never, the automation of industrial processes has high technical requirements. Today, most of the fabrication processes must operate with an efficient and precise control of different parameters like: velocity, position and torque. Simultaneously, the controller must be immune enough to the outside world perturbations typically found in industry. At the same time, the kind of movement required by today processes is beyond the simple rotational configuration. Linear actuators are making their appearance in the industry, being already a reality and a truly available option that designers and engineers can consider. The traditional conversion method used to transform rotational motion into linear displacement is no more acceptable. In the old days, linear displacement was obtained from a rotational motor shaft, after mechanical conversion by a specific mechanism containing pulleys, worm gears and belts. The presence of these components diminishes the robustness and reliability of the industrial processes.

The ac induction motor has good robustness and low fabrication cost. Over the past decades it has replaced, with great success, the conventional brushed DC motor in servo-type applications. Although this change allowed process reliability improvement, for instance, problems related with the motor brushes are eliminated, the introduction of an electronic power drive increases systems complexity, raising other problems.

The switched reluctance machine (SRM) can be classified as a current-controlled stepping motor of the variable-reluctance type. This technology is one of the most recent options in the field of variable speed actuators. Consumer products, aerospace, and automobile industries are today taking advantage from SRM drives characteristics.

Advances in power electronics and the use of microelectronics and microprocessors allowed the development of different control strategies, such as nonlinear, adaptive, variable structure, and fuzzy, contributing to the popularity that SRM drives actually enjoy.

The SRM is a rugged and reliable actuator, that can be produced at a low cost, presents a simple and robust structure and can operate in a wide speed range, in all four-quadrant, without a considerable reduction of efficiency. These characteristics make it an attractive alternative to permanent-magnet brushless and induction motor drives (Corda, J. et al. (1993)). Its main constructive characteristic is the absence of winding on the rotor of the machine, giving it a potential advantage over conventional machines. Furthermore, the SRM combines different advantages, such as high power density, low-cost maintenance, the ability to operate even with a phase shorted or opened, and the possibility of direct drive without mechanical transmissions. The power converter design can be very simplified because of the unidirectional stator current. Although most of the control strategies applied to the SRM require the perfect knowledge of rotor position, sensorless operation techniques can also be adopted to control it (Gan, C. et al. (2003)), (Espírito Santo et al. (2005)), (Espírito Santo et al. (2008)).

Nevertheless, the control complexity of these actuators, the high torque ripple presented in output and the acoustic noise produced under normal operation, until recent years, lead to some lack of interest in adopting SRM as a driving technology.

The output power density is increased when the SRM is operated with high saturation levels. Because of this, the SRM flux linkage, inductance, back EMF and phase torque, are highly nonlinear parameters. Furthermore, the SRM cannot operate directly from an AC or DC power supply, requiring a specific electronic controller.

This chapter describes the development of a sliding mode position controller for a linear switched reluctance actuator (LSRA). The main goal of the presented chapter is to control the position of the primary of the actuator. The actuator test bench used is shown in Fig. 1.

The remaining of this chapter describes the energy conversion process of a LSRA. Based on a finite elements analysis, both traction and attraction maps are derived. This information is useful to understand the device working principle and its performance for different excitation and working conditions. The proposed control strategy, based on the sliding mode control technique, and the electronics to drive the LSRA are described. The control strategy was implemented with the TMS320F2812 eZdsp Start Kit, taking advantage of the microprocessor internal peripherals and resources. Finally, the validation of the proposed control strategy is discussed through the presentation of experimental results.

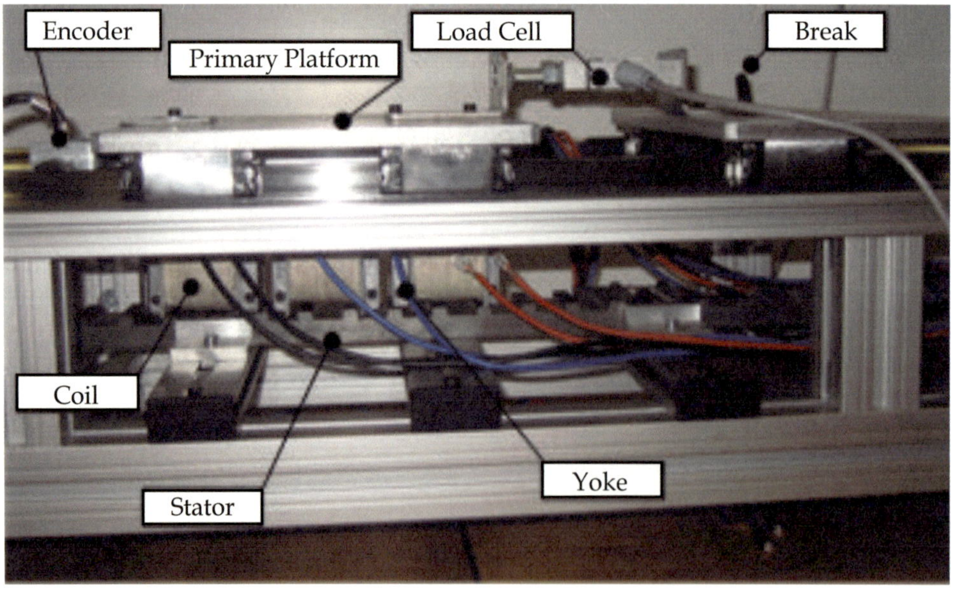

Fig. 1. Actuator experimental test bench.

2. The switched reluctance actuator

2.1 Introduction to the switched reluctance working principle

The SRM working principle takes advantage from the fact that an electromagnetic system always tries to adopt the geometrical configuration corresponding to the minimum reluctance of the magnetic circuit. In the beginning of the IXX century, scientists from all over the world performed experiments in this field, trying to use the magnetic reluctance principle to produce a continuous movement. The first reference to a SRM appears in a scientific paper published in 1969 by Nasar (Nasar, S. A. (1969)). The first SRM was commercially available in the UK in 1983, and was marketed by TASC Drivers.

In 1838, W.H. Taylor registered two patents of an electromagnetic motor, one in the U.S and the other in the UK (Taylor, W.H. (1840)). Another pioneer in this field was the Scottish engineer Robert Davidson that, in 1838, developed an actuator based in the switched reluctance principle. The actuator constructed by him was used to power an electric locomotive in the Edimburgo-Glasgow railway. Two patents registered in the U.S. by B.D. Bedford and R.G. Hoft, in 1971 and 1972, describe some of the functionalities that can be found in actual SRMs. For instance, the regulation electronics is synchronized with the rotor position. With this information, the motor phases are sequentially energized. Another important step was performed by L. E. Unnewehr and W. H. Koch, from Ford Motor Company Scientific Research Staff Company; they developed an axial SRM controlled by thyristors (Unnewehr, L. E. et al. (1974)). In Europe, others researchers registered several patents, e.g., J. V. Byrne and P. J. Lawrence (Byrne, J.V. (1979)), (Byrne , J.V. et al. (1976)).

A SRM can be used to produce a rotational movement, a linear displacement, or a more complex combination of these movements. The machine primary can be inside of the structure, or outside, and it can be stationary or allowed to move. Typically, the magnetic circuit is energized by independent electric circuits (phases). In turn, the electromagnetic flux from each phase may, or may not, share the same magnetic circuit. Several works related with the SRM can be found in the scientific literature, mainly with respect to the rotational configuration. However, while some of them are related with the linear configuration, papers describing the usage of switched reluctance technology in the production of compound movements are almost non-existent.

Although the SRM is normally used as an actuator, it can also be used as a generator. The phases of a SRM working as motor are energized when the inductance of the phase is increasing, which is to say, when the teeth of the rotor are approaching the poles of the stator. Phases are powered off at the vicinity of the aligned position. The described procedure is inverted when the SRM is operated as a generator. Inductance evolution taking as reference rotor position can be observed in Fig. 2, where aligned and unaligned characteristic positions are identified. It should be observed that the inductance changes not only with rotor position, but also with current. Due to the magnetic saturation effects, at the same position, the inductance value decreases when current value increases.

With the commercial appearance of the SRM, becomes clear the problems related with the acoustic produced noise. The study presented in (Krishnan R. et al. (1998)) allowed to conclude that the acoustic noise has its sources in the vibrations developed by the radial forces acting in the stator. The acoustic noise is amplified when the vibrating frequency matches the mechanical resonant frequency. Acoustic noise can also result from inaccurate mechanical construction or produced by the action of the actuator electronic drive (Chi-Yao Wu et al. (1995)).

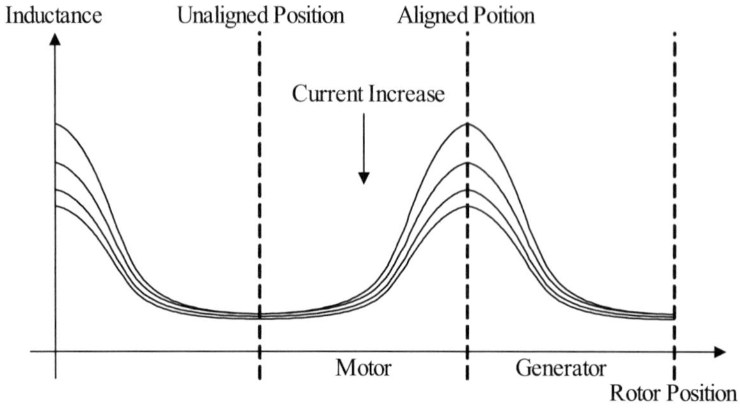

Fig. 2. Inductance variation, taking as reference the rotor position.

A typical geometrical configuration of a SRM can be observed in Fig. 3. This configuration receives the 6/4 designation, because it has six poles in the stator and four teeth in the rotor. Each phase comprises a pair of coils (A_1, A_2), (B_2, B_2) and (C_1, C_2), so there are three phases (A, B, C) in the actuator. Each pair of coils are supported by poles geometrically opposed and are electrically connected in order that the magnetic flux created is additive.

The magnetic force vectors F_1 and F_2 can be decomposed in two orthogonal components. The axial components F_{a1} and F_{a2} are always equal in magnitude, with opposite directions, and cancel each other out all the time. The transversal components F_{t1} and F_{t2} produce a mechanical torque than changes with current and angular position. If the rotor is withdrawn from the aligned position, in whatever direction (Fig. 3a)), a resistant torque will be created, tending to put the phase at the aligned position again.

At the aligned position (Fig. 3b)) the produced magnetic force doesn't have transversal components and the resulting torque is zero. At this position, the magnetic reluctance has its minimum value; this fulfils the necessary conditions to observe the saturation of the magnetic circuit.

The magnetization curves of a typical SRM are represented in Fig. 4 (Miller, T. J. E, (1993)). The magnetization curves for the intermediate positions are placed between the curves corresponding to those of the aligned and unaligned positions.

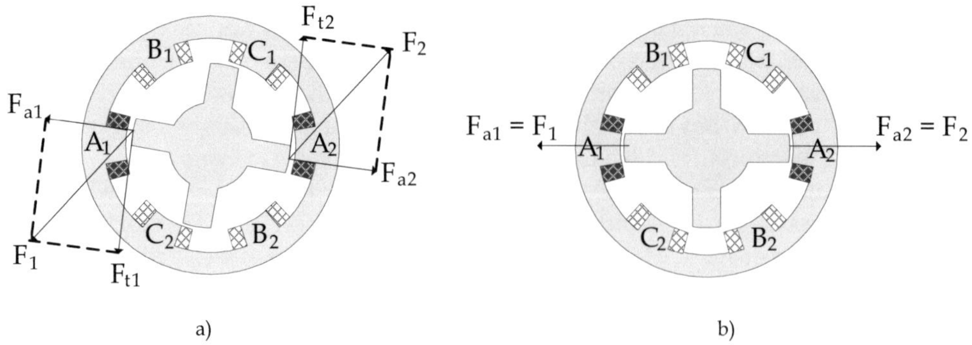

Fig. 3. SRM with phase A energized: a) unaligned position, b) aligned position.

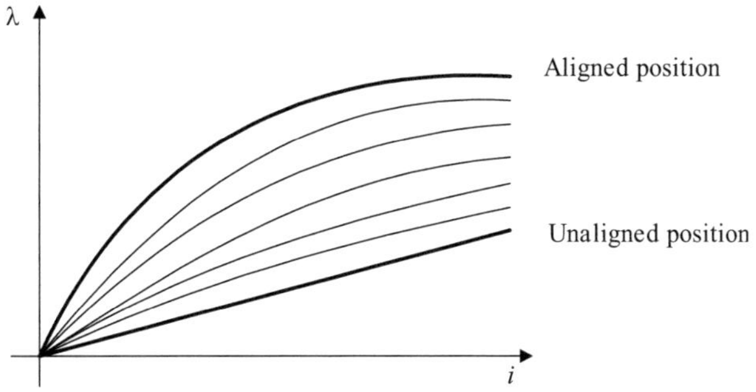

Fig. 4. Typical SRM magnetization curves.

At the aligned and unaligned positions the phase can not produce torque and it is unavailable to start the movement. When a phase is aligned, the others two will be in good position to start the movement in both clockwise and anti-clockwise directions. The previously described situation helps to realize that the SRM can only start by itself, in both directions, if the minimum number of phases is three.

To understand how the switched reluctance actuator works we first need to observe the energy conversion process. The electrical and the mechanical systems are interconnected through the magnetic coupling field. The way how energy flows from the power source to the mechanical load is explained next.

The magnetic reluctance is a measure of the opposition to the magnetic flux crossing a magnetic circuit. If one of the magnetic circuit parts is allowed to move, then, system will try to reconfigure itself to a geometrical shape corresponding to the minimal magnetic reluctance. Fig. 5 illustrates the different stages associated with the energy conversion procedure, when a very fast movement from position x to position $x + dx$ occurs.

Because the movement is fast, it is assumed that the linkage flux does not change. The magnetic field energy W_{fe} at the beginning of the movement is given by the area established in Fig. 5a) as

$$W_{fe}(a') = \text{area}\{o,a',c',o\}. \tag{1}$$

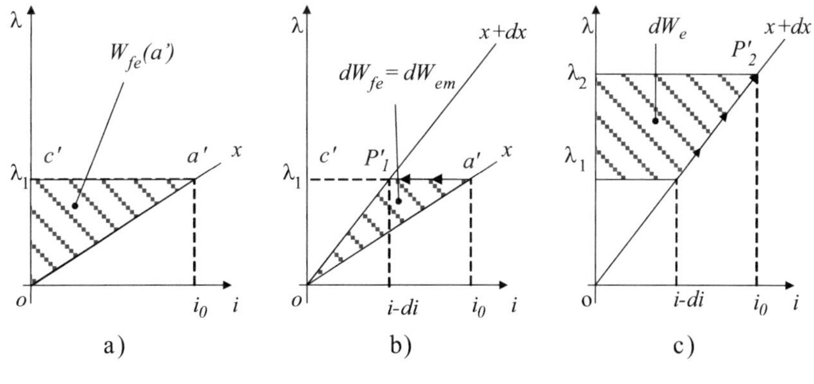

Fig. 5. Quasi-instantaneous movement from position x to position $x + dx$. a) initial system condition, b) actuator movement, c) power source restoring energy.

When actuator fast displaces from position x to position $x + dx$, represented in Fig. 5b), some of the energy stored in the magnetic coupling field is converted into mechanical energy. This amount of energy is equivalent to the area given by

$$dW_{fe} = \text{area}\{o, a', P_1', o\}. \tag{2}$$

It can be observed that the energy stored at the coupling field decreases. Simultaneously, current also decreases. If voltage from the supply source is constant, the current value will return to its initial value i_0, through the path shown in Fig. 5c), restoring the energy taken from the magnetic coupling field during the fast movement.
Because linkage flux is constant throughout the movement, there is no induced voltage and, therefore, the magnetic coupling field does not receive energy from the voltage source.
An electromagnetic device can convert electrical energy into mechanical energy, or vice-versa. The energy responsible for actuator motion, taken from the magnetic coupling field, is expressed by

$$dW_{em} = -dW_{fe} \Rightarrow f_{em} = -\left.\frac{dW_{fe}}{dx}\right|_{\lambda=\text{constant}}. \tag{3}$$

The variation of the energy coupling field is equal, but with opposite signal, to the mechanical energy used to move the actuator. The mechanical force f_{em} can be represented by

$$W_{fe} = \frac{1}{2}\frac{\lambda^2}{L} \Rightarrow f_{em} = \frac{\lambda^2}{2L^2}\frac{dL}{dx} = \frac{i^2 dL}{2dx}. \tag{4}$$

Because i^2 is always positive, the force applied to the actuator part allowed to move, in the direction x, is also positive as long as the inductance L is increasing in the x direction. So, the mechanical force acts in the same direction, which also increases the magnetic circuit inductance.
The mechanical force can also be calculated by changing the magnetic circuit reluctance with position. If the linkage flux λ is constant, then the flux φ in the magnetic circuit is also constant and, therefore, the mechanical force is given by

$$W_{fe} = \frac{\Re\varphi^2}{2} \Rightarrow f_{em} = -\frac{\varphi^2}{2}\frac{d\Re}{dx}. \tag{5}$$

The mechanical force f_{em} acts in the direction that puts the actuator in a geometric configuration that corresponds to the path with lower reluctance. The SRM base their work on this basic principle.
We can also define the co-energy as

$$W_{fe}'(i,x) = i\lambda - W_{fe}, \tag{6}$$

depending on current i and position x,

$$dW_{fe}'(i,x) = \frac{\partial W_{fe}'(i,x)}{\partial i}di + \frac{\partial W_{fe}'(i,x)}{\partial x}dx. \tag{7}$$

As i and x are independent variables, the linkage flux λ, inductance L and the mechanical force f_{em} can be obtained from the device co-energy map, applying

$$\begin{cases} \lambda = \dfrac{\partial W_{fe}'(i,x)}{\partial i} \Rightarrow L = \dfrac{\lambda}{i} \\ f_{em} = \dfrac{\partial W_{fe}'(i,x)}{\partial x} \end{cases} . \tag{8}$$

Thus, as can be seen, if a change in the linkage flux occurs, the system energy will also change. This variation can be promoted by means of a variation in excitation, a mechanical displacement, or both. The coupling field can be understood as an energy reservoir, which receives energy from the input system, in this case the electrical system, and delivers it to the output system, in this case the mechanical system.

The instantaneous torque produced by the actuator is not constant but, instead, it changes according to the current pulses supplied. This problem can be avoided, either by making a proper inspection of the current that flows through the phases, either by increasing the number of phases of the machine.

The operating principle used by the SRM can also be used to build linear actuators. The movement, as in the rotational version, is achieved by the tendency that the system has to reduce the reluctance of its magnetic circuit. In the rotational version, the normal attraction force between the stator and the rotor is counterbalanced by the normal force of the attraction developed by the phase windings which are placed in the diametrically opposite positions, thus contributing to the reduction of the electrodynamic efforts. The linear topology also experiences this situation and because it develops a very high attraction force, designers must have special attention to prevent possible mechanical problems.

The LSRA can have a longitudinal or a transverse configuration (Corda , J. et al. (1993)). In both, the force developed between the primary and the secondary can be vectorially decomposed into attraction and traction force, being the latter one responsible for the displacement. While in the longitudinal configuration the magnetic flux path has a direction parallel to the axis of motion, in the transversal configuration, the magnetic flux has an orientation perpendicular to the axis of movement.

The performance of the two previous configurations can be diminished by the influence of the force of attraction between the primary and secondary. As a consequence, the mechanical robustness of the actuator must be increased. Simultaneously, as already stated, these configurations are more problematic in what concerns the acoustic noise. A symmetric version, with a dual primary, avoids the problems caused by the attraction force. The attraction force developed through a phase, and applied on one side of the secondary, is counterbalanced by the attraction force in the opposite direction, and also applied in the secondary.

A tubular configuration can also introduce significant improvements that minimize some of the problems identified in previous paragraphs. The resultant of the radial forces developed in the tubular actuator will be null. It is therefore possible to use smaller airgaps, because there are no mechanical deformations and thereby maximize the performance of the actuator. In general, the use of ferromagnetic material is maximized. In addition, the construction of the actuator becomes much simpler. The coils can be self-supported and the entire assembly of the structure is greatly simplified. In low-speed applications, eddy currents can be ignored, since the magnetic flux changes occur more slowly and, therefore, the construction of the magnetic circuit with ferromagnetic laminated material is not mandatory.

2.2 LSRA characterization through finite element analysis

Some industrial processes can take advantage from actuators with the ability for doing linear displacements with precision. The switched reluctance driving technology is a valid solution justified by the qualities previously enumerated. The problem to solve will be the development of a new design, not only able to perform linear movements but, simultaneously, that allows the accurate control of its position.

An operational schematic of a LSRA based on the concepts previously introduced can be observed in Fig. 6. This actuator is classified as belonging to the longitudinal class, because the magnetic flux has the same direction as the movement.

The force F developed by each phase can be decomposed in the traction force F_t and the attraction force F_a. One of the tasks performed during design procedure is the increase of the traction force and, at the same time, the reduction of the attraction force. While the traction force contributes to the displacement, the attraction force does not produce any useful work and has adverse effects in the mechanical structure, as for example, the changing in the airgap length. Attraction force effects can be minimized through the change of the geometrical configuration of the pole head.

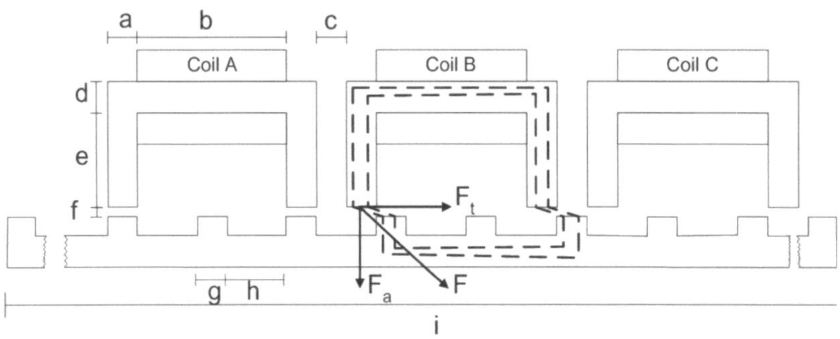

Fig. 6. LSRA physical dimensions.

The physical dimensions of the actuator are listed in Table 1.

Yoke pole width (a)	10
Coil length (b)	50
Space between phases (c)	10
Yoke thickness (d)	10
Yoke pole depth (e)	30
Airgap length (f)	0.66
Stator pole width (g)	10
Stator slot width (h)	20
LSRA length (i)	2000
LSRA stack width	50

Table 1. LSRA physical dimensions [mm]

Finite elements tools allowed to study the complexity of this technology (Ohdachi, Y., (1997)), (Brisset, S. et al. (1998)). But there is still missing a standard method to assist the design of this class of actuators, although several proposals have been published until now (Krishnan, R. et al. (1998)), (Anwar , M. N. et al. (2001)).
A Finite Elements Model (FEM) of a single-phase LSRA was constructed using FLUX2D. FEM construction starts with geometric model definition, where each specific region is defined through points and lines. The FEM constructed for the analysis of a single-phase actuator can be observed in Fig. 7.

a)

b)

Fig. 7. Finite Elements Model of a single-phase actuator: a) global view and b) polar region detail.

One of the used finite elements software FLUX2D features is the translation displacement possibility, allowing the longitudinal displacement of one, or a group, of regions, without the need of redefining the geometry of the model and respective finite element mesh. Two regions (in magenta) are defined between the set of regions that must be displaced in the longitudinal direction throughout the static simulations. These regions are the actuator primary (in blue), the phase coil that carries current in the positive direction (inner region) and negative direction (outer region) (both in red), and respective surrounding air (in white). Translation function demands also the definition of two narrow airgap regions (in yellow), located in both side of the translations regions, and used for the displacement of the regions defined between them.
All regions use triangular elements in the finite elements mesh creation. The exceptions are the two displacement regions that use quadrangular elements. Mesh creation must also

observe that only a single layer of elements can be established in the translations air-gap regions.
Primary and secondary regions are associated to materials with ST-37 steel magnetic characteristics. All other regions inherit the vacuum magnetic characteristics. Inside the coil regions, a current source is defined with a positive value for the region that carries the current in the positive direction and an identical negative value for the region that carries the current in the negative direction. Dirichlet conditions are formulated in the model boundary, imposing a null flux across it.
Several simulations were performed for a set of primary positions and currents. Obtained phase traction force F_t and attraction force F_a maps are presented in Fig. 8.
As can be observed, when actuator poles are aligned (position $x = 0$ mm) with the secondary teeth the traction force is zero. This same situation occurs also at non-aligned positions (position $x = 15$ mm and $x = -15$ mm). From the analysis of the attraction force map is possible to conclude that maximum attraction force is obtained at the aligned position (position $x = 0$ mm).

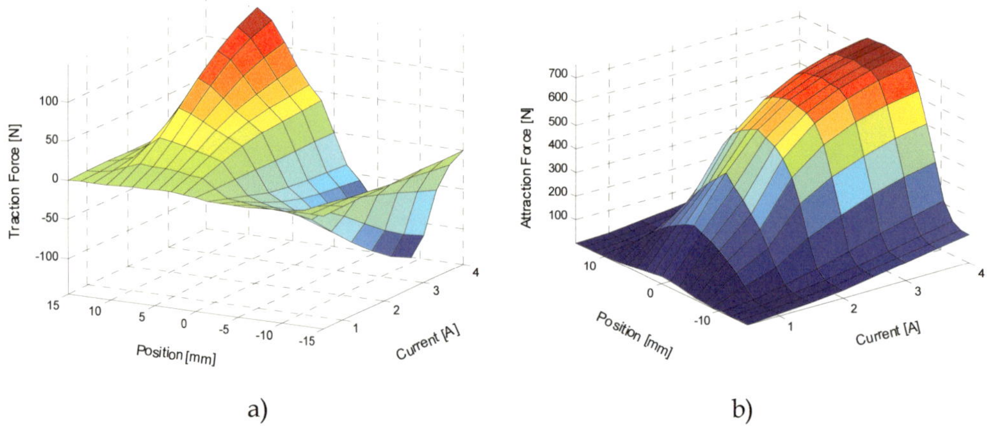

Fig. 8. LSRA force maps as a function of position and current: a) traction force and b) attraction force.

2.3 LSRA step-by-step operation control

The configuration of the LSRA with independent phases allows to increase the robustness of the actuator and improves its performance, because more than one phase can operate simultaneously, without any kind of disturbance among them. The number of phases of the actuator can also be easily changed.
The sequential energization of the actuator phases is a simple control methodology, which can be implemented without requiring large computational resources. Typically, turn-on and turn-off positions are established. As demonstrated by the FEM simulations, the traction force depends on position and current. For each phase, the attraction force developed at the aligned and unaligned positions is always zero. As a consequence, for these positions, the phase can't contribute to displace the actuator. Between these two positions, the traction force direction depends of the relative positions between phase poles and the nearest stator teeth.

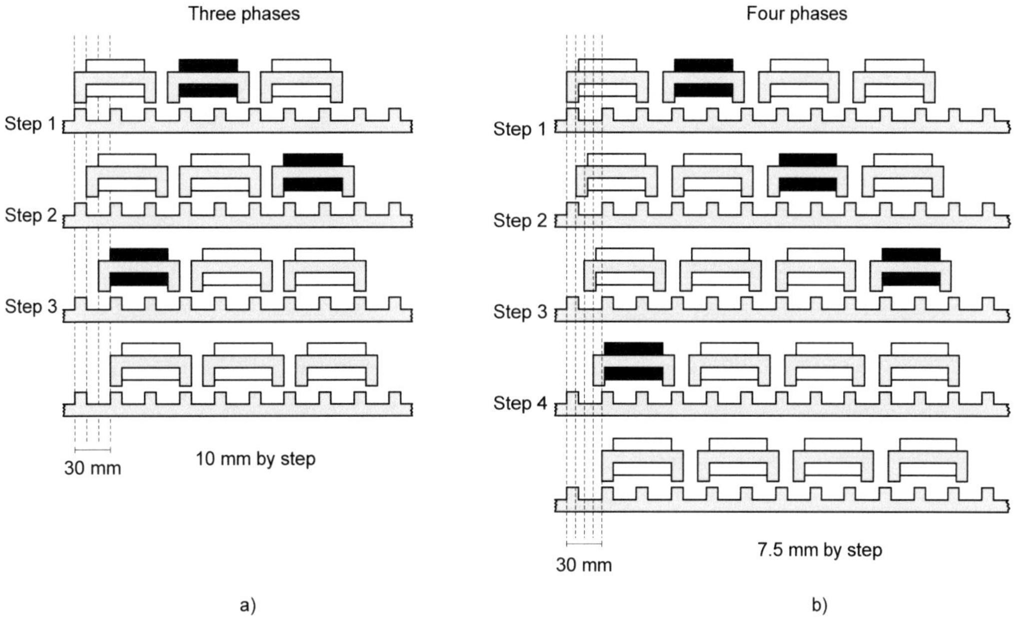

Fig. 9. LSRA step-by-step motion: a) three phases configuration; b) four phases configuration.

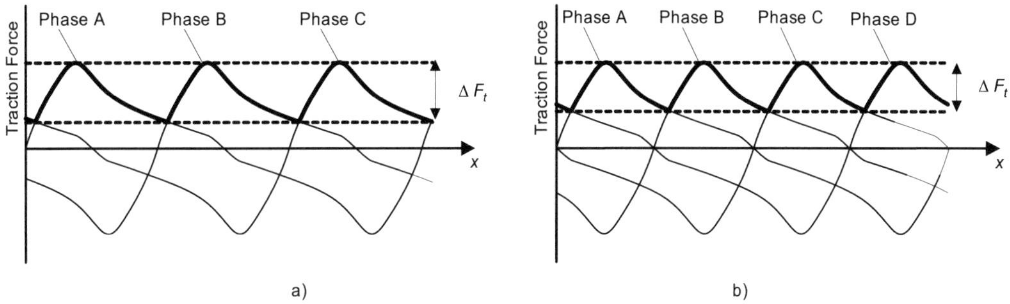

Fig. 10. LSRA step-by-step motion: a) three phases configuration; b) four phases configuration.

Considering phase A at the aligned position as the displacement reference (x = 0 mm), the motion can be started in both directions, depending on the energization of the others two phases. If phase B is energized, the actuator will be moved to the right. The movement to the left can be achieved through the energization of phase C.

Operation of the actuator in three phase configuration is represented in Fig. 9a). With this configuration the actuator can perform displacements with 10 mm of resolution. In Fig. 10a) is illustrated a schematic representation of the total traction force developed by all the three phases, considering that they are energized with a constant value of current. It can be observed that the traction force isn't constant and changes with position. In the example, the actuator performs a displacement of 30 mm to the right. During this movement all phases are energized in B-C-A order.

Actuator performance can be improved if the number of phases is increased. A 30 mm displacement is represented in Fig 9b), for an actuator with four phases energized in

B-C-D-A order. It can be observed that the resolution of the displacement was increased to 7.5 mm. At the same time, the total traction force ripple was also decreased, see Fig. 10b). The major drawback observed is that the size and the height of the actuator increase.

3. Sliding mode position controller

The sliding mode technique was used by others to control the behaviour of SRM. Through it, problems as the acoustic noise, binary flicker, or velocity control were solved with success or, at least, machine performance was improved.

A direct torque control algorithm for a SRM using sliding mode control is reported in (Sahoo, S.K. et al. (2005)). A control strategy for an energy recovery chopper in a capacitor-dump is proposed in (Bolognani, S. et al. (1991)). A flux-linkage controller, using the sliding mode technique, with integral compensation, is proposed in (Wanfeng Shang et al. (2009)) for torque ripple minimization. A sliding mode controller is used in (Pan, J. et al. (2005)) to control the position of a two-dimensional (2D) switched reluctance planar motor.

An approximate sliding mode input power controller and another feedforward sliding mode speed controller are combined with space voltage vector modulation in (Tzu-Shien et al. (1997)) to implement a robust speed control. Another robust speed controlled drive system using sliding mode control strategy is presented in (John, G. et al. (1993)).

A sliding-mode observer is proposed in (Zhan, Y.J. et al. (1999)) for indirect position sensing. A sliding mode binary observer is also used in (Yang, I.-W. et al. (2000)) to estimate the rotor speed and position. Current and voltage are used by a sliding mode observer described in (Islam, M.S. et al. (2000)) to estimate the position and the speed to minimize the torque-ripple. The conduction angles are controlled in (McCann, R.A. et al. (2001)) by a sliding mode observer that estimates the rotor position and the velocity.

A sliding mode controller is used in (Xulong Zhang et al. (2010)) in order to reduce cost, simplify the system structure and increase the reliability of a switched reluctance generator.

The LSRA electromagnetic model to n = {1,2,3} phases is described by (9), where R_n is the phase coil resistance, v_n the supplied voltage and i_n the phase current. Mathematical expression (10) describes actuator mechanical behaviour, where a is the device acceleration and M the mass. Actuator non-linearity is taken into account because both inductance $L(i,x)$ and traction force $F(i,x)$ change not only with position x, but also with current i.

$$0 = R_n i_n(t) + \frac{dL(i_n,x)i_n(t)}{dt} - v_n = \frac{di_n}{dt}\left[L(i_n,x) + i_n(t)\frac{\delta L(i_n,x)}{\delta i}\right] + \frac{dx}{dt}\left[\frac{\delta L(i_n,x)}{\delta x}\right] R_n i_n(t) - v_n \quad (9)$$

$$a = \frac{F(x,t)}{M} \quad (10)$$

The state space is defined as:

$$\begin{cases} \dfrac{dx}{dt} = y \\ \dfrac{dy}{dt} = \dfrac{F}{M} \\ \dfrac{di_n}{dt} = \dfrac{y\beta_n + R_n i_n - v_n}{\alpha_n} \end{cases} \quad (11)$$

where α_n and β_n stands for:

$$\begin{cases} \alpha_n = L(i_n, x) + i_n \dfrac{\delta L(i_n, x)}{\delta i} \\ \beta_n = i_n \dfrac{\delta L(i_n, x)}{\delta x} \end{cases} \tag{12}$$

The control strategy previously presented allows to reach discrete positions. Intermediate positions aren't reachable. The concept of variable structure is introduce next, and applied to the LSRA to improve the displacement capacities the actuator. The Variable Structure System (VSS) as defined below

$$\dot{X} = A(X,t) + B(X,t)u\,, \tag{13}$$

belongs to a particular case of automatic control systems. Intentional commutation is introduced between two different control actions, in one or more channels of control inputs. The possible type of motion of a VSS in the state space is manifested by the appearance of the sliding mode regime. This kind of control was successively applied to the rotational switching reluctance motors as described previously. To achieve sliding motion regime, a switching surface is defined by $s(X,t) = 0$, the control structure $u(X,t)$ stated in (14) changes from one structure to another. When $s(X,t) > 0$ the variable structure control is changed in order to decrease $s(X,t)$. The same kind of action is taken when $s(X,t) < 0$. The control main goal is to keep the system state space sliding in the surface $s(X,t) = 0$. During this sliding motion, the system behaves like a reduced-order system, being insensitive to disturbances and parameters changes. The control structure can be expressed by:

$$u(X,t) = \begin{cases} u^{+}(X,t) & \Leftarrow \quad s(X) > 0 \\ u^{-}(X,t) & \Leftarrow \quad s(X) < 0 \end{cases}. \tag{14}$$

The previously explained concept is used to develop the LSRA position controller. At a specific moment, it is assumed that traction force can be developed in both directions with proper choice of actuator phase. After turning off, an actuator phase still has the ability to produce traction force. This situation occurs because current phase do not goes down instantaneously, but diminishes by the free wheel diodes path, with a time constant that depends on phase inductance and resistance. This behaviour is responsible for the introduction of a delay in controller response, contributing to increase the oscillations (chattering) around the sliding surface.

The movement of the actuator can be expressed as a VSS with two possible control actions. Considering that the primary of the actuator has the M mass. A control action produces a traction force in the left direction F_l, while the other control action produces a traction force in the right direction F_r. Under this control action the actuator will have the horizontal displacement x, at velocity y, given by

$$\begin{cases} \dot{x} = y \\ \dot{y} = \dfrac{u}{M} \end{cases}. \tag{15}$$

The control law $u(t)$ is established as

$$u(t) = \begin{cases} F_r & \Leftarrow \quad s(e,\dot{e}) > 0 \\ F_l & \Leftarrow \quad s(e,\dot{e}) < 0 \end{cases}, \tag{16}$$

The VSS model has the following formulation

$$\begin{cases} \dot{x} = y \\ \dot{y} = k\dfrac{F_\gamma}{M} + (1-k)\dfrac{F_l}{M} \end{cases}, \tag{17}$$

with $k \in \{0,1\}$.
The system will be defined by

$$\dot{x} = f(x) + g(x)k\,, \tag{18}$$

with

$$f(x) = \begin{bmatrix} y \\ F_l/M \end{bmatrix} \quad and \quad g(x) = \begin{bmatrix} 0 \\ (F_r - F_l)/M \end{bmatrix}. \tag{19}$$

The commutation function $s(e,\dot{e})$ depends on the position error e and the derivative of the position error $\dot{e}$, and is defined by

$$s(e,\dot{e}) = me + \dot{e}\,, \tag{20}$$

where m is a positive constant, experimentally obtained.
The controller selects from the lookup Table 2 which phase will provide the desired control action, in order to maintain $s(e,\dot{e}) = 0$. The position reference (x = 0) is taken as the aligned position for phase A.

Traction Force	**[0,10[**	**[10,20[**	**[20,30[**
Left direction (F_l)	Phase A	Phase B	Phase C
Right direction (F_r)	Phase B	Phase C	Phase A

Table 2. Relative actuator position [mm]

4. Regulation and command electronic driver

For the actuator could be able to perform the predefined task with the required performance, a power converter must be designed for driving it and to apply the proposed control strategy. The half-bridge configuration has the required versatility to permit the usually adopted operation modes in applications with SRM: single-pulse, soft-chopping and hard-chopping. An example of a functional structure of the controller for one phase can be seen in Fig. 11.
The developed power converter has three main blocks (Fig. 12), which are (1) the Distribution Unit, responsible to make the interface with the microprocessor using proper electronics and to manage the operation of the other two blocks, (2) the Regulation Units receive orders concerning the states that must be imposed to the power switches; making use of the reference

signal available from the microprocessor, have the ability to control the shape of the current that flows in each coil of the actuator phases, (3) the Power Units, which, receiving orders from the state of the power switches, can properly feed the phase coils.

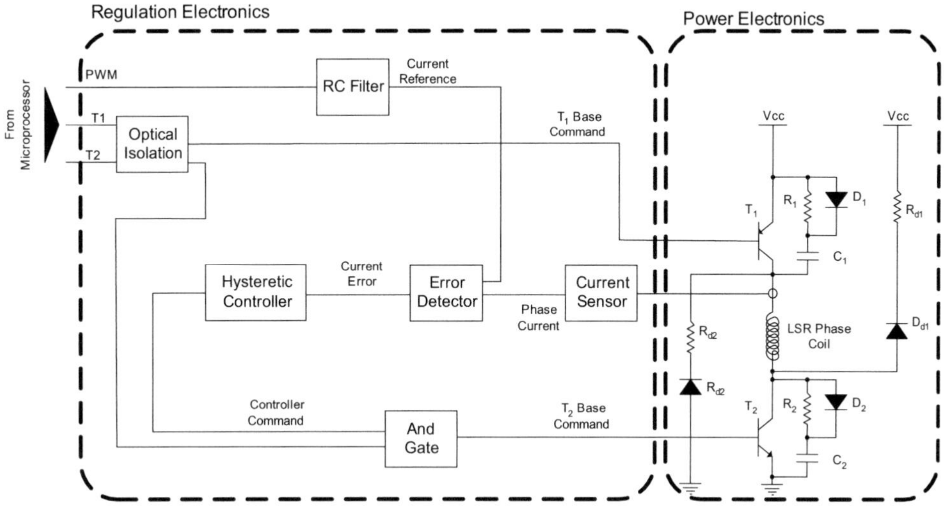

Fig. 11. Power converter general topology for one phase.

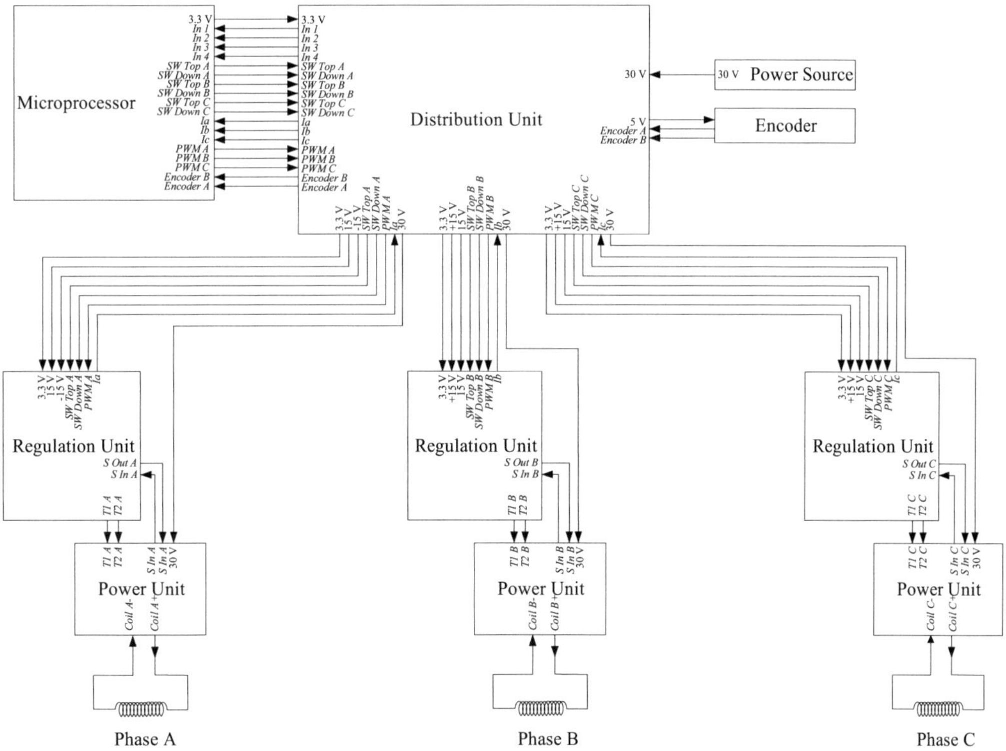

Fig. 12. Global visualization of the actuator driver.

All the information is centralized in the Distribution Unit, which shares that information with the processor, and for each phase is possible to know the required PWM (Pulse-Width Modulation) signal used to generate the controller current reference, the state of the switch T_1, the state of the switch T_2, and the signal corresponding to the current in the coil of the actuator phase. Beyond these knowledge, the Distribution Unit receives from the encoder two lines with pulse sequences (Encoder A and Encoder B), which are sent to the microprocessor, allowing the determination of both the position and the direction of the displacement.

The analog signal, reference of the current controller, is generated by a DAC (Digital-to-Analog Converter), by using a PWM. The signal, after being filtered by a low-pass filter is returned as a signal depending on the band width of the PWM. The DAC resolution is stated by the length of the counter used to generate the PWM signal.

The resolution R_{bit}, in number of bits, can be calculated as

$$R_{bit} = Log_2(R) = Log_2\left(\frac{N_{\max}}{C}\right) = \frac{\ln\left(\frac{N_{\max}}{C}\right)}{\ln(2)}, \tag{21}$$

with $R = N_{\max} / C$, N_{max} the maximum number of counts and C the minimum change of duty-cycle. As an example, with a maximum count of 512 and 2 the minimum change of the duty-cycle, the obtained resolution is of 256 different levels for the analog output. The corresponding resolution in bits is

$$\mathrm{R_{bit}} = \log_2(256) = 8\,\mathrm{bits} \tag{22}$$

The required frequency of the PWM is established by the rate of change stated by the DAC, as each change in the duty-cycle corresponds to a sample of the DAC. The frequency of the counter f_c depends on the frequency f_{PWM}, stated for the PWM signal, and on its resolution R_{bit}. So, it can be written that

$$f_c = 2^{R_{bit}} f_{PWM} \tag{23}$$

Thus, in order to obtain an 8 bits equivalent DAC, able to generate signals with 8 kHz frequency samples, the PWM signal must have a similar frequency. Using the equation (23), the counter must receive a clock input with a frequency of 2.048 MHz. The cut-off frequency of the low-pass filter must be defined quite below the frequency of the PWM signal, so that the noise generated by the signal commutation can be canceled. Nevertheless, it can't be so small that restricts the rate of change of the DAC output by imposing a high time constant.

The acquisition of the current flowing in the phase coil is achieved by using a Hall effect sensor with dynamic characteristic that enables it to follow the variations in the signal to be acquired.

The comparator is used to compare the voltages applied to its two input ports and return a value depending on the sign of the difference between those values (Franco, S. (2001)), (Williams, J. (1990)). This device can be faced as an analog to digital converter of one single bit.

When the amplifier feedback signal is positive, it can be said that it is in regenerative operation mode. In this situation, the circuit acts in order to amplify the effect of any disturbance. The amplifier output can then adopt one of two states: high or low levels. The obtained hysteresis characteristic can be used in implementing the on-off type control.

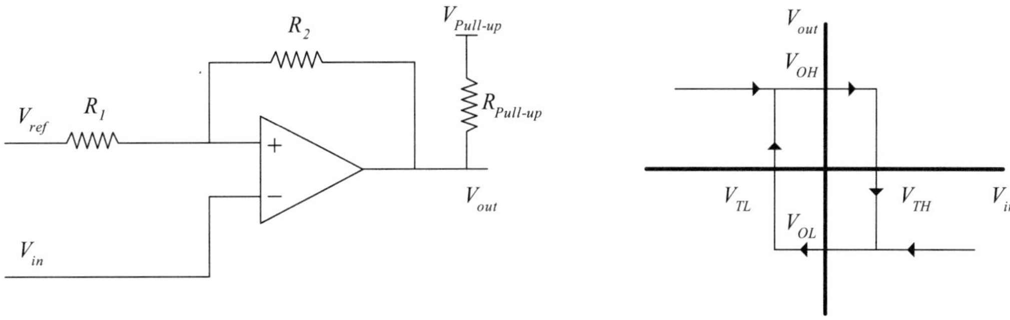

Fig. 13. Scheme of hysteresis control circuit in inverting assembly

In the electrical scheme of Fig 13, the signal V_{in} is applied to the inverting input and the resistance R_2 is much higher than the resistance R_1. In the special case of having infinite resistance R_2 there will be no hysteresis, and the circuit switch on with the reference of voltage V_{ref}.

$$
\begin{aligned}
&V_{TL} = \left(R_2 V_{\mathrm{Ref}} + R_1 V_{OL}\right) / \left(R_1 + R_2\right) \\
&V_{TH} = \left(R_2 V_{\mathrm{Ref}} + R_1 V_{OH}\right) / \left(R_1 + R_2\right) \\
&\text{Hysteresis gap} = V_{TH} - V_{TL} = R_1\left(V_{OH} - V_{OL}\right)\left(R_1 + R_2\right)
\end{aligned}
\tag{24}
$$

All the electronic chain used to control the current flowing in the phase coil is represented in Fig. 14, being the way how it is integrated in the power converter operation described as follows.

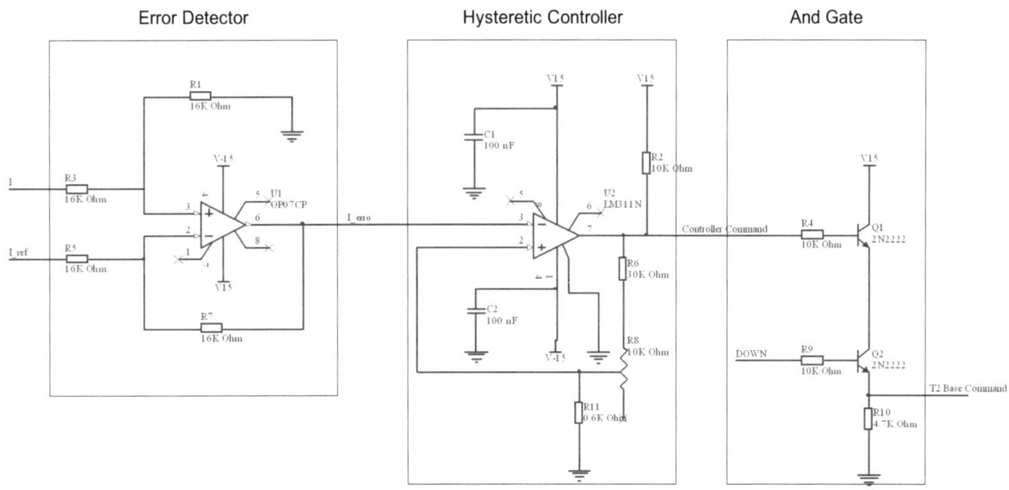

Fig. 14. Circuit used to control the current flowing in the actuator phase coil.

Once the current *I* flowing in the actuator phase coil is acquired and conditioned, the value of the reference current *I_ref* is subtracted; the resulting value from that operation, *I_error*, is sent to the histeresis controller, with the histeresis range set by the potentiometer R_8. The resistance R_2 is used to make the pull-up of the control output signal. The output signal is available to the following electronic structure, which operates as an AND logical gate. If the current flowing through the base-emitter diode is enough high to saturate it, the collector voltage can be less than 1 volt, which is considered as a logic TTL zero. The result from that logical operation is the basis to establish the control of the power switches of the half-bridge power converter T_2. Thus, the switch is only activated if there are simultaneous orders from the microprocessor and from the histeresis controller.
With the current controller it can be established a current profile. In Fig. 15 is possible to observe how voltage at the phase coil is switched for different current profiles.

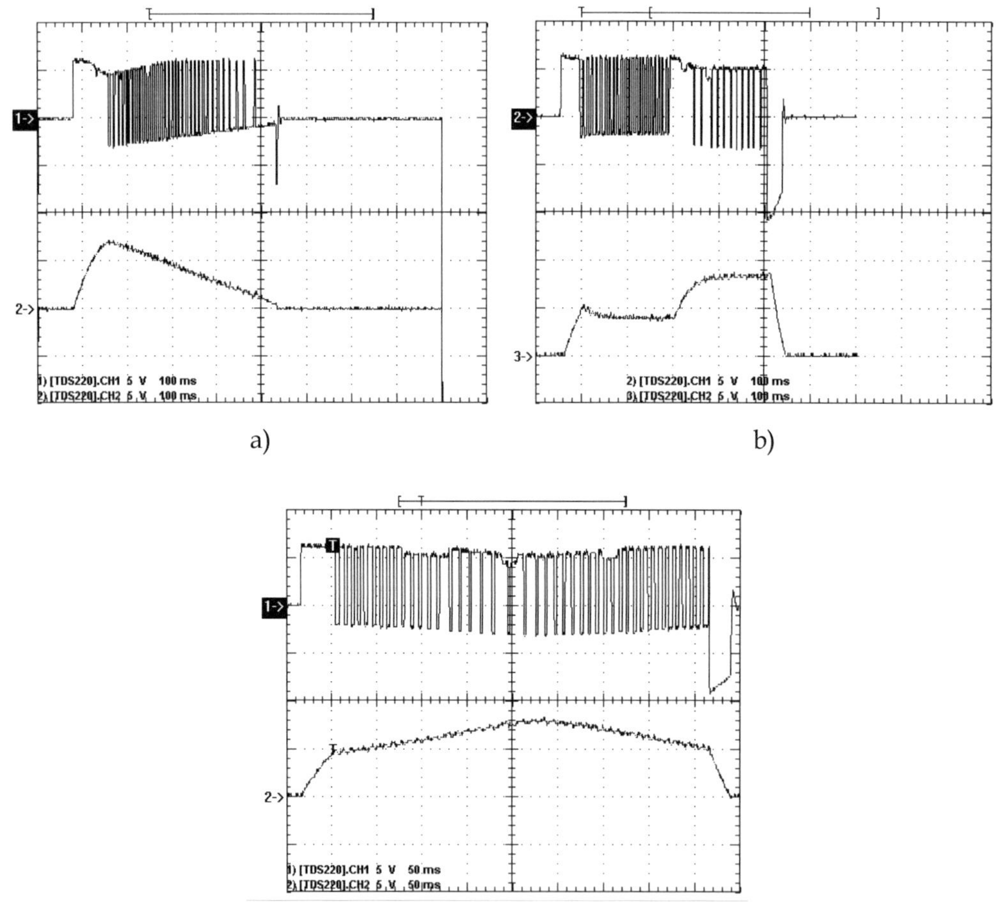

a) b)

c)

Fig. 15. Actuator hardware current controller operation: voltage (ch. 1) e current (ch. 2): a) current decreases linearly, b) two different current stages, c) increasing and decreasing the current linearly.

5. Development of the controller firmware

To verify the applicability of the here proposed control methodology, an experimental setup was constructed based on the TMS320F2812 eZdsp Start Kit. Event Manager EVA is used to generate the PWM signals from where the Current Reference signals are obtained. Each current phase signal is acquired by the on-chip ADC and saved in a buffer memory. At this moment current phase information is not used by the controller.
Actuator position is feedback to the TMS320F2812 QEP Unit by the incremental encoder. From the Quadrature Encoder Pulse (QEP) unit data, actuator velocity and position are derived. The sliding mode controller establishes the switching strategy, used to turn-on and turn-off the LSRA phases. Using Microcontroller GPIO, each phase signal lines T_1 and T_2 are properly switched. Data lines shared between the eZdsp and the LSRA regulation electronics are represented in Fig. 16a).

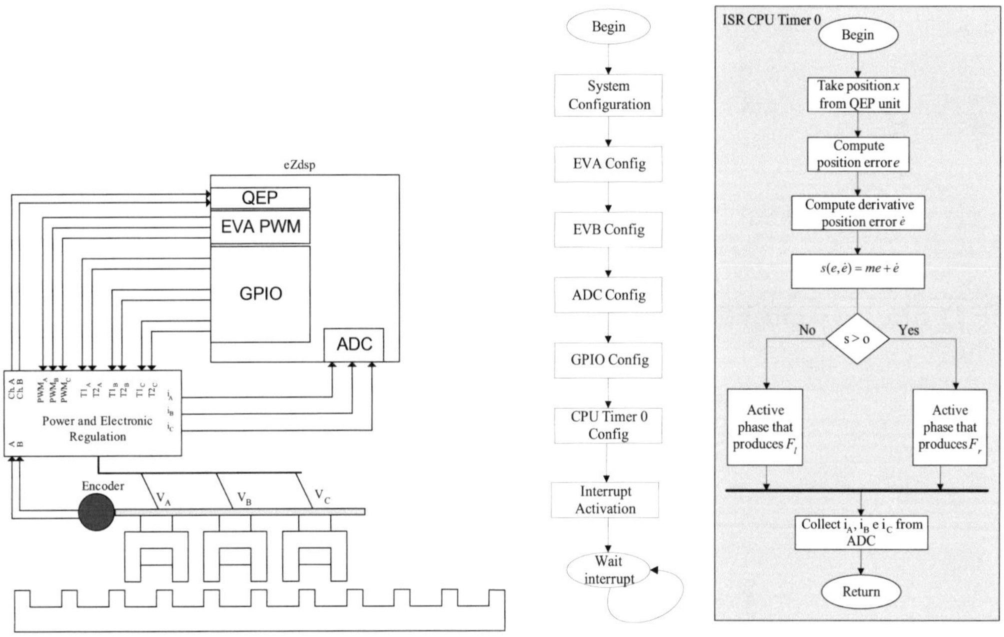

a) b)

Fig. 16. a) eZdsp interface with LSRA electronic regulation and b) TMS320F2812 code flow.

Developed code to the TMS320F2812 implements the control methodology previously described. The most important software blocks are represented at Fig. 16b). Software begins with the configuration of each peripheral and before enter in a wait state activates the CPU Timer 0 interrupt. This interrupt possess an ISR that based in QEP information determines the actuator position and, based on it, the corresponding position error and derivative position error. For each CPU Timer 0 interrupt a service routine is executed. The information needed to realize the control procedure is obtained. Based on it, the control action is derived and applied to the proper phase, as specified in the lookup table. Status system information (position and phase current) is saved in memory. After the operation action, the functionalities of Code Composer Studio are used to collect the results from the microprocessor memory and saved it on file for posterior analysis.

6. Results and discussion

The results returned by the application of the sliding mode control methodology to the LSRA are presented next. The primary of the actuator always start from the initial position ($x = 0$) with the poles of the phase A aligned with the stator teeth. The information on the displacement that the primary of the actuator must perform is provided to the sliding mode controller. Position evolution of primary of the actuator is presented in Fig. 17 for different required final positions.
For one of the previous displacements (25 mm) the phase portrait is presented in Fig. 18.

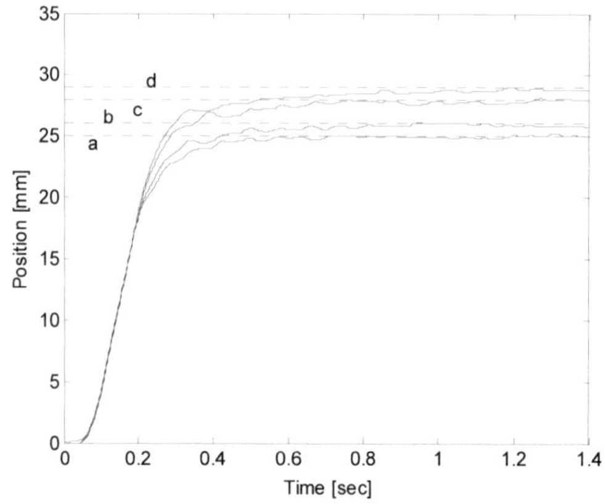

Fig. 17. Actuator position for small displacements: a) 25 [mm], d) 26 [mm], c) 28 [mm], d) 29 [mm].

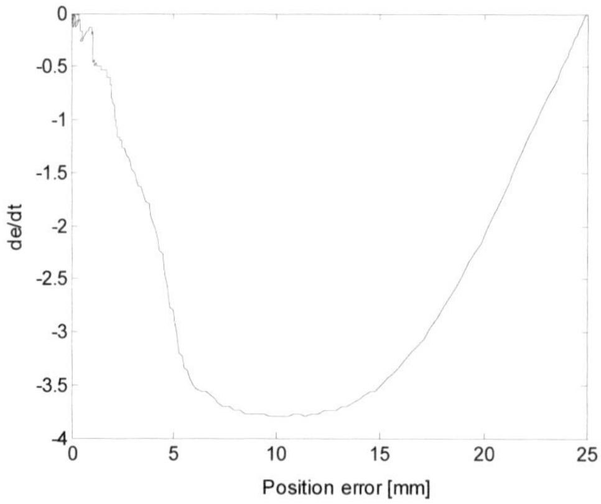

Fig. 18. Phase portrait for a 25 [mm] displacement.

7. Conclusion

The finite elements analysis allowed to understand the working principle of a three phase Linear Switched Reluctance Actuator developed for robotics applications. Using this tool, traction and attraction map were obtained. This information allows to characterize the actuator behaviour. Based on the obtained results a prototype was constructed with correspondent power and regulation electronics. The establishment of proper position control for a high performance device was achieved through a proposed strategy based on sliding mode control. That task was performed by implementing the developed methodology on a TMS320F2812 eZdsp Start Kit, taking advantage from their built-in peripherals. Experimental results allowed to conclude that actuator can realize displacements with 1 [mm] resolution.

8. References

Anwar , M. N.; Husain I.; Radun A. V., (2001). "A Comprehensive Design Methodology for Switched Reluctance Machines," *IEEE Transactions on Industry Applications*, Vol. 37, No. 6, pp. 1684-1692, November/December 2001.

Bolognani, S.; Ognibeni, E.; Zigliotto, M.; (1991). "Sliding mode control of the energy recovery chopper in a C-dump switched reluctance motor drive," in *Proceedings of Applied Power Electronics Conference and Exposition*, 1991. APEC '91. Conference Proceedings, 1991., Sixth Annual , vol., no., pp.188-194, 10-15 Mar.

Brisset, S.; Brochet, P., (1998). "Optimization of Switched Reluctance Motors using Deterministic Methods with Static And Dynamic Finite Element Simulations," *IEEE Transactions on Magnetics*, Vol. 34, No. 5, pp. 2853-2856, September.

Byrne , J.V.; Lacy, J.C., (1976). United States Patent No. 3,956, 678, May 11.

Byrne, J.V., (1979). German Patent No. DT 20 30789 B2, 23 June.

Chi-Yao Wu, Pollock, C., (1995). "Analysis and Reduction of Vibration and Acoustic Noise in the Switched Reluctance Drive, (1995)." *IEEE Transactions on Industry Applications*, Vol. 31, No. 1, pp. 91-98, January/ February.

Corda , J.; Skopljak , E., (1993) "Linear Switched Reluctance Actuator," in *Proceedings of the Sixth International Conference on Electrical Machines and Drives*, pp.535-539, September.

Corda, J. ; Skopljak, E, (1993). "Linear switched reluctance actuator," in *Proceedings of the Sixth International Conference on Electrical Machines and Drives.*

Espírito Santo, A.; Calado M. R. A. ; Cabrita, C. M. P., (2005). "Variable Reluctance Linear Actuator Dynamics Analysis Based on Co-energy Maps for Control Optimization," in *Proceedings of Linear Drives for Industry Application*.

Espírito Santo, A.; Calado M. R. A. ; Cabrita, C. M. P., (2008). "Analysis and Position Control of a Linear Switched Reluctance Actuator Based on Sliding Mode Control," in *Proceedings of the 13th International Conference on Power Electronics and Motion Control*, September.

Franco, S., (2001). *Design With Operational Amplifiers and Analog Integrated Circuits*, McGraw-Hill.

Gan, C. ; Cheung, N. C. ; Qiu, L., (2003). "Position Control of Linear Switched Reluctance," *IEEE Transactions on Industry Applications*, vol. 39, no. 5, September/October.

Hung, J. Y. ; Gao, W.; Hung, C, (1993). "Variable Structure Control: A Survey," *IEEE Transactions on Industrial Electronics*, vol. 40, no. 1, February.

Islam, M.S.; Husain, J., (2000). "Torque-ripple minimization with indirect position and speed sensing for switched reluctance motors," *IEEE Transactions on Industrial Electronics*, vol.47, no.5, pp.1126-1133, Oct.

John, G.; Eastham, A.R., (1993). "Robust speed control of a switched reluctance drive," in *Proceedings of Electrical and Computer Engineering*, 1993. Canadian Conference on , vol., no., pp.317-320 vol.1, 14-17 Sep.

Krishnan R.; Vijayraghavan, P., (1998). "State of the Art: Acoustic Noise in Switched Reluctance Motor Drives," in *Proceedings of the 24th Annual Conference of the Industrial Electronics Society* - IECON98, pp. 929-934.

Krishnan, R., Arumugam, R., Lindsay J. F., (1998). "Design Procedure for Switched-Reluctance Motors," *IEEE Transactions on Industry Applications*. Vol. 24, No. 3, pp. 456-461, May/June.

McCann, R.A.; Islam, M.S.; Husain, I., (2001). "Application of a sliding-mode observer for position and speed estimation in switched reluctance motor drives," *IEEE Transactions on Industry Applications*, vol.37, no.1, pp.51-58, Jan/Feb.

Miller, T. J. E, (1993). *Switched Reluctance Motors and their Control*, Magna Physics Publishing Oxford Science Publications.

Nasar, S. A., (1969). "DC Structures Reluctance Motor," *IEE Proceedings* Vol. 116, No. 1048-9.

Ohdachi, Y., (1997) "Optimum Design of Switched Reluctance Motors Using Dynamic Finite Element Analysis," *IEEE Transactions on Magnetics*, Vol. 33, No. 2, pp. 2033-2036, March.

Pan, J.; Cheung, N.C.; Jinming Yang, (2005). "High-precision position control of a novel planar switched reluctance motor," *IEEE Transactions on Industrial Electronics*, vol.52, no.6, pp. 1644- 1652, Dec.

Sahoo, S.K.; Panda, S.K.; Xu, J.X., (2005). "Direct Torque Controller for Switched Reluctance Motor Drive using Sliding Mode Control," in *Proceedings of Power Electronics and Drives Systems*, 2005. PEDS 2005. International Conference on , vol.2, no., pp. 1129-1134, 28-01 Nov.

Taylor, W.H., (1840). "Obtaining Motive Power," Patent No.8255, England, 2nd May 1840.

Tzu-Shien Chuang; Pollock, C., (1997). "Robust speed control of a switched reluctance vector drive using variable structure approach," *IEEE Transactions on Industrial Electronics*, vol.44, no.6, pp.800-808, Dec.

Unnewehr, L. E.; W. H. Koch, (1974)."An Axial Air-Gap Reluctance Motor For Variable Speed Applications," *IEEE Transactions on Power Apparatus and Systems*, Volume: PAS-93, Issue: 1, pp. pp. 367-376, January.

Wanfeng Shang; Shengdun Zhao; Yajing Shen; Ziming Qi, (2009). "A Sliding Mode Flux-Linkage Controller With Integral Compensation for Switched Reluctance Motor," *IEEE Transactions on Magnetics*, vol.45, no.9, pp.3322-3328, Sept.

Williams, J. (1990). "High-speed Comparator Techniques," *Application Note AN-13*, Linear Technology.

Xulong Zhang; Guojun Tan; Songyan Kuai; Qihu Wang, (2010). "Position Sensorless Control of Switched Reluctance Generator for Wind Energy Conversion," in *Proceedings of Power and Energy Engineering Conference* (APPEEC), 2010 Asia-Pacific , vol., no., pp.1-5, 28-31 March.

Yang, I.-W.; Kim, Y.-S., (2000). "Rotor speed and position sensorless control of a switched reluctance motor using the binary observer," in *IEE Proceedings Electric Power Applications*, vol.147, no.3, pp.220-226, May.

Zhan, Y.J.; Chan, C.C.; Chau, K.T., (1999). "A novel sliding-mode observer for indirect position sensing of switched reluctance motor drives," *IEEE Transactions on Industrial Electronics*, vol.46, no.2, pp.390-397, Apr.

11

Application of Sliding Mode Control to Friction Compensation of a Mini Voice Coil Motor

Shir-Kuan Lin[1], Ti-Chung Lee[2] and Ching-Lung Tsai[1]
[1]*Department of Electrical Engineering, National Chiao Tung University*
[2]*Department of Electrical Engineering, Minghsin University*
Taiwan

1. Introduction

This chapter deals with the position control of a mini voice coil motor (VCM) mounted on a compact camera module (CCM) of a mobile phone. Mini VCMs are increasingly popular nowadays in 3C electronic gadgets such as mobile phones, digital cameras, web cams, etc. (Yu et al., 2005). The common requirements of these gadgets are miniaturization and high performance. Miniaturized VCM faces the challenge of accuracy position control. Sliding mode control will be adopted to compensate for the nonlinear friction in the actuator of the VCM. Experimental results in this chapter will show that good position control performance is achieved by sliding mode control.

Fig. 1 shows such a typical VCM with the size of 8.5×8.5×4.6 mm^3 and the stroke of 0.35mm. In Fig. 1(b), the congeries of the magnet (a), the yoke (d), and the lens holder (e) forms the actuator, while the guide pins (b), the coils (c), and the CMOS sensor cover (f) are stationary parts. The current through the coils generates force to move the actuator along the guide pins, which induces nonlinear friction. It is known that friction is the cause of stick slip oscillations during the motion when a usual PI controller is applied to the VCM.

The work (Bona & Indri, 2005) presents a comprehensive survey of different kinds of friction compensation schemes, and indicates that types A and B solutions are suitable for cost-sensitive applications because of its limited calculation burden. Several other methods in the literature are a nonlinear proportional controller with bang-bang force in specified region to compensate for the stick slip friction (Southward et al., 1991) , a look-up table position controller with higher gain for smaller position error and lower gain for larger position error to eliminate stick slip oscillations (Hsu et al., 2007), and an anti-windup PI controller, incorporated with the disturbance observer, to control a VCM (Lin et al., 2008).

To overcome the load variation due to tilt attitude of the CCM and the nonlinear friction force of the VCM, a dedicated sliding mode controller will be designed for the position control. High accuracy repeatability under 10 μm, fast settling time, and free of stick slip oscillations are the control goals. The challenge of the sliding mode controller design for the VCM is to select the control gains such that the error state variable in the sliding surface $s = 0$ will approach zero as time approaches infinite. The final value approach is used to make sure that the error state variable is bounded and can be made as small as possible by increasing control gains. In practical implementation, if the allowable steady-state error is given, the control gains can be easily calculated out.

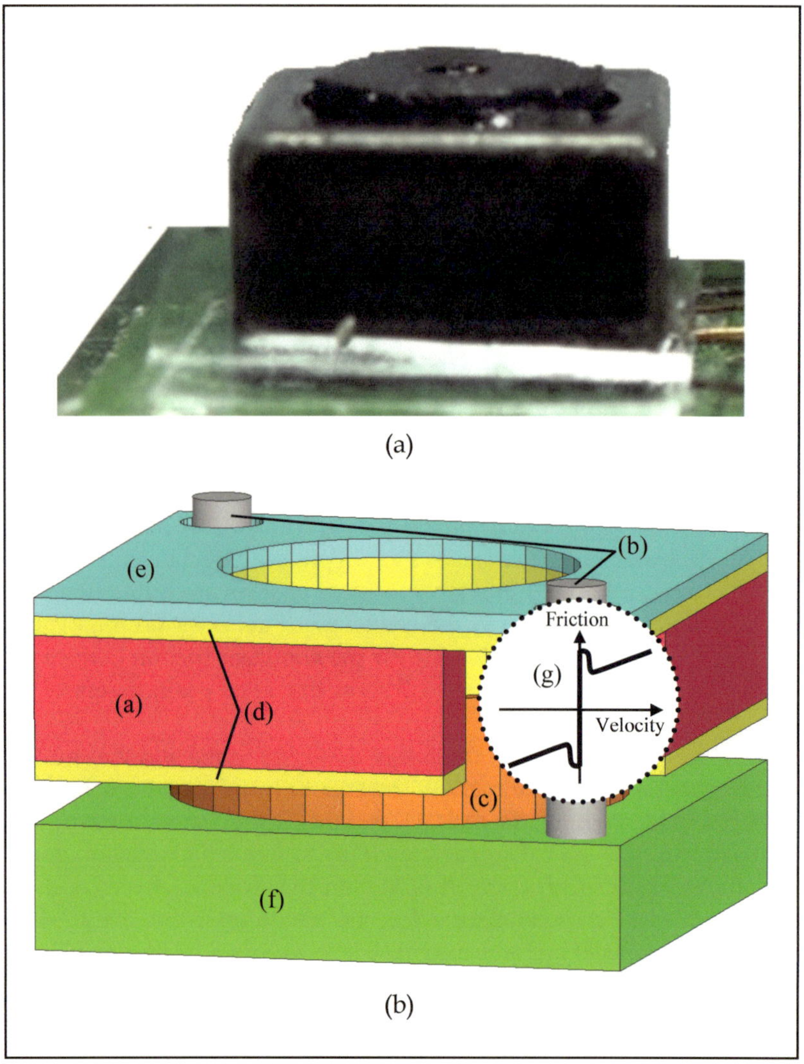

Fig. 1. Mini voice-coil motor with 8.5×8.5×4.6 mm³: (a) photo; (b) illustration of the components

2. Mathematical model of the VCM

Let the position of the actuator be d, and the current of the coils be i. The mathematical model of the VCM in Fig. 1 can be described by the dynamical and the electrical equations as follows:

$$m\ddot{d} + B\dot{d} = K_C i - f_D \tag{1}$$

$$L\dot{i} + Ri = u - K_b\dot{d} \tag{2}$$

where m and f_D are, respectively, the mass and the friction of the actuator, B is the viscous coefficient, L and R are, respectively, the inductance and the resistance of the coils, K_C is the magnetic force constant, K_b is the back-emf constant, and u is the input voltage.
Assume that the desired position is d^*. To transform Eqs. (1) and (2) into the form of state equations, we introduce the state variables of

$$x_1 = d - d^*, \; x_2 = \dot{d}, \; x_3 = i \tag{3}$$

and the parameters of

$$\alpha_1 = \frac{-B}{m}, \alpha_2 = \frac{K_C}{m}, \alpha_3 = \frac{-1}{m}, \alpha_4 = \frac{-K_b}{L}, \alpha_5 = \frac{-R}{L}, \alpha_6 = \frac{1}{L} \tag{4}$$

The VCM model of Eqs. (1) and (2) can then be rewritten in the form of

$$\begin{cases} \dot{x}_1 = x_2 \\ \dot{x}_2 = \alpha_1 x_2 + \alpha_2 x_3 + \alpha_3 F_D \\ \dot{x}_3 = \alpha_4 x_2 + \alpha_5 x_3 + \alpha_6 u \end{cases} \tag{5}$$

In the VCM system, x_1 is the output as well. The system turns out to be an output regulation problem with a mismatched condition, since F_D and u are in the different equations. The usual compensation and cancellation method cannot be used to eliminate F_D.
It is known (Canudas de Wit et al., 1995) that the friction F_D consists of the stiffness and the damping parts in the form of

$$F_D = \sigma_0 z + \sigma_1 \dot{z} \tag{6}$$

where σ_0 and σ_1 are the stiffness and the damping coefficients of Stribeck effect, and z is average deflection of the bristles. Let F_C be the coulomb friction force, F_S be the static friction force and v_S be the Stribeck velocity. Define the function $g(x_2)$ as

$$g(x_2) = \frac{1}{\sigma_0}\left(F_C + (F_S - F_C)e^{-\left(x_2/v_S\right)^2} \right) \tag{7}$$

Thus, the dynamic equation of z is

$$\dot{z} = x_2 - \frac{|x_2|}{g(x_2)} z \tag{8}$$

3. Sliding mode control law

The key technique of sliding mode control is to find a sliding surface in which any value of the state x_1 will move toward zero, i.e., zero position error. And then a control law is designed to drive any state variables outside the sliding surface to drop on the surface and to adhere to the surface. In such a way, the sliding mode position controller regulates the position of the actuator d to the desired one d^*.

It will be shown later that any x_1 in the sliding surface $S = 0$ will eventually approach zero, where

$$S = x_2 - \beta_1 x_1 - \beta_2 x_3 \quad (9)$$

with proper constants β_1 and β_2. The surface $S = 0$ is then the desired sliding surface of the VCM model (5).

The next mission is to design a switching input u in Eq. (5) that drives the state variables of the system to the sliding surface $S = 0$. We define $V(s) = S^2/2$, which is greater than 0 for $S \neq 0$. According to Lyapunov's stability theorem, if we can find a controller $u(x_1, x_2, x_3)$ such that $\dot{V}(0) = 0$ and $\dot{V}(s) = S\dot{S} < 0, \forall S \neq 0$, then $S = 0$ is an asymptotically stable equilibrium. Taking derivative of Eq. (9) and substituting Eq. (5) into it, we obtain

$$\dot{S} = (\alpha_1 - \alpha_4\beta_2 - \beta_1)x_2 + (\alpha_2 - \alpha_5\beta_2)x_3 - \alpha_6\beta_2 u + \alpha_3 F_D \quad (10)$$

The first two terms on the right-hand side of Eq. (10) can be easily eliminated by directly inserting them in $u(x_1, x_2, x_3)$, since x_2 and x_3 are available states and can be used as feedback signals. However, the value of F_D is not available, so that to eliminate it needs a sufficiently large constant value. This motivates us to select

$$u = \frac{1}{\alpha_6\beta_2}\left[(\alpha_1 - \alpha_4\beta_2 - \beta_1)x_2 + (\alpha_2 - \alpha_5\beta_2)x_3 + c_1^+ \operatorname{sgn}(S) + c_2^+ S\right] \quad (11)$$

where sgn(S) is the sign of S, and c_1^+ and c_2^+ are nonnegative constants. Substituting Eq. (11) into Eq. (10) yields

$$\dot{S} = -c_2^+ S - c_1^+ \operatorname{sgn}(S) + \alpha_3 F_D \quad (12)$$

It is apparent that c_1^+ sgn(S) can be used to cancel the divergent part of $\alpha_3 F_D$, while c_2^+ provides a freedom to adjust the convergent speed. Finally,

$$\begin{aligned} \dot{V} = S\dot{S} &= -c_2^+ S^2 - c_1^+ S \operatorname{sgn}(S) + \alpha_3 F_D S \\ &\leq -c_1^+ |S| + |\alpha_3 F_D S| = -\left(c_1^+ - |\alpha_3 F_D|\right) |S| \\ &\leq -\left(c_1^+ - |\alpha_3| F_D^{\max}\right) |S| \end{aligned} \quad (13)$$

where $F_D^{\max}$ is the static friction and $F_D^{\max} \geq |F_D|$. Choose

$$c_1^+ > |\alpha_3| F_D^{\max} \quad (14)$$

to obtain $\dot{V} < 0$ for $S \neq 0$. Consequently, $u(x_1, x_2, x_3)$ in Eq. (11) with c_1^+ satisfying Eq. (14) is a controller for the asymptotically stability of $S = 0$. The approaching speed can be assigned by $c_2^+ > 0$. Moreover, we have the following theorem.

Theorem 1. Consider the VCM model of Eq. (5). Suppose that the upper bound of $F_D^{\max} \geq |F_D|$ is known. The controller $u(x_1, x_2, x_3)$ in Eq. (11) with S defined in Eq. (9), $c_1^+ > |\alpha_3| F_D^{\max}$, and $c_2^+ \geq 0$ makes the steady state value $x_1(\infty)$ of the system converge to a bounded region of

$$\left|x_1(\infty)\right| \le \frac{\left|\alpha_3\right| F_D^{\max}}{\lambda^2} \tag{15}$$

where $\lambda > 0$ is a constant, if β_1 and β_2 in (9) are

$$\beta_1 = \frac{-\lambda^2}{2\lambda + \alpha_1}, \quad \beta_2 = \frac{-\alpha_2}{2\lambda + \alpha_1} \tag{16}$$

Proof. We just need to prove that any states in the sliding surface S=0 will eventually converge to the region of Eq. (15).
It follows from Eq. (9) that in the sliding surface $S = 0$,

$$x_3 = \frac{x_2}{\beta_2} - \frac{\beta_1}{\beta_2} x_1 \tag{17}$$

Substituting Eq. (17) into the VCM model of Eq. (5), we reduce the state equation to a second-order differential equation:

$$\ddot{x}_1 - \left(\alpha_1 + \frac{\alpha_2}{\beta_2}\right)\dot{x}_1 + \frac{\alpha_2\beta_1}{\beta_2} x_1 = \alpha_3 F_D \tag{18}$$

We take Laplace transform of above equation to obtain

$$X_1(s) = \frac{s x_1(0) + \dot{x}_1(0) - \left(\alpha_1 + \frac{\alpha_2}{\beta_2}\right) x_1(0) + L\left[cF_D(t)\right]}{s^2 - \left(\alpha_1 + \frac{\alpha_2}{\beta_2}\right)s + \frac{\alpha_2\beta_1}{\beta_2}} \tag{19}$$

Substituting Eq. (16) into the characteristic equation of (19) yields

$$s^2 + 2\lambda s + \lambda^2 = 0 \tag{20}$$

which has double roots of $-\lambda < 0$. The time-domain solution to Eq. (19) is then (Golnaraghi & Kuo, 2009)

$$x_1(t) = k_1 e^{-\lambda t} + k_2 t e^{-\lambda t} + \int_0^{\infty} \alpha_3 F_D(t-\tau)\tau e^{-\lambda\tau} d\tau \tag{21}$$

where k_1 and k_2 are some constants. The final value of of x_1 as $t \to \infty$ is then

$$\begin{aligned} \left|x_1(\infty)\right| &= \lim_{t\to\infty}\left|\int_0^{\infty} \alpha_3 F_D(t-\tau)\tau e^{-\lambda\tau} d\tau\right| \\ &\le \left|\alpha_3\right| F_D^{\max} \int_0^{\infty} \tau e^{-\lambda\tau} d\tau = \left|\alpha_3\right| F_D^{\max}\left(-\frac{\tau e^{-\lambda\tau}}{\lambda} - \frac{e^{-\lambda\tau}}{\lambda^2}\right)\Bigg|_0^{\infty} \\ &= \frac{\left|\alpha_3\right| F_D^{\max}}{\lambda^2} \end{aligned} \tag{22}$$

This completes the proof. ■

Theoretically, the bounded region Eq. (15) of the steady state value $x_1(\infty)$ can be made as small as possible by increasing λ. In a practical problem, the bound of $F_D^{\max}$ and the value of α_3 are known *a priori*, so λ can be calculated out from Eq. (15) for a given bound of $x_1(\infty)$. However, the larger λ is, the larger is the absolute value of β_1, and then the larger is those of S in Eq. (9) and the controller u in Eq. (11). To limit the controller u, a control gain switching strategy is implemented. A threshold value $x_{th} > 0$ is defined first. As the sliding mode control starts up, a low-value λ is used until $|x_1| < x_{th}$. Thereafter a high-value λ is used to reduce the convergent bounded region. It can be expected that $|x_2|$ is small after $|x_1| < x_{th}$, since x_2 is the time derivative of x_1. This imples that the absolute values of $\beta_1 x_1$ and $\beta_1 x_2$ are small after $|x_1| < x_{th}$., and so are S and u. The overall sliding mode control law incorporated with the control gain switcihing strategy is illustrated in Fig. 2. There are two controllers in Fig. 2. One with low gains is outputted to the VCM for $|x_1| \geq x_{th}$, while the output to the VCM for $|x_1| < x_{th}$ is the other with high gains.

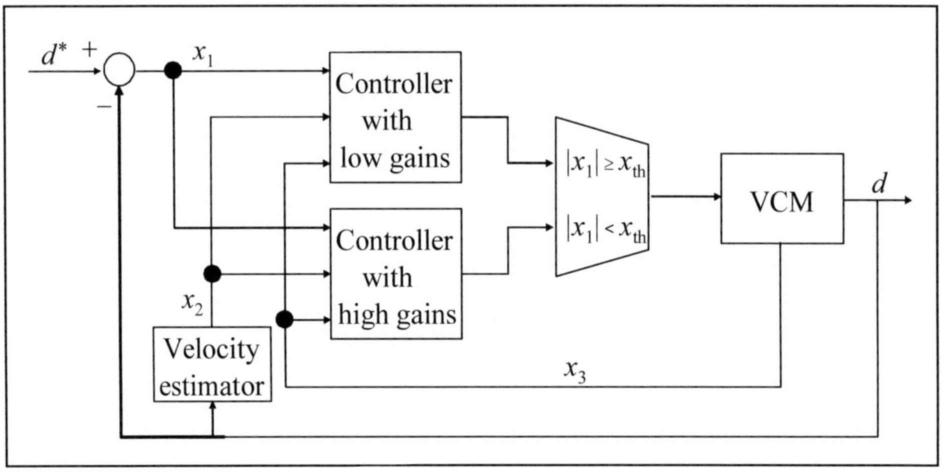

Fig. 2. Block diagram of the overall control law.

It should be remarked that the undesired chattering of the sliding mode control can be alleviated by replacing sgn(s) in Eq. (11) with the following saturation function of

$$\text{sat}(S) = \begin{cases} 1, & \text{for } S > k \\ \dfrac{S}{k}, & \text{for } |S| \leq k \\ -1, & \text{for } S < -k \end{cases} \tag{23}$$

where $k > 0$ represents the thickness of the boundary layer.

4. Simulations

Consider a real VCM which will be used in the experiments. The parameters of the VCM are $\alpha_1 = -24$, $\alpha_2 = 800$, $\alpha_3 = -1000$, $\alpha_4 = -2666.7$, $\alpha_5 = 66666.7$, $\alpha_6 = 3333.3$, $F_D^{\max} = 0.011$. Assume that the design goal is to make the steady state error smaller than 0.4 μm, which in turn asks $\lambda = 5244.044$ by Eq. (15). Substituting the value of λ into Eq. (16), we obtain $\beta_1 = -2628.036$

and β_2 = -0.076. Of course, these are high control gains. The low control gains are assigned as λ = 1172.6, β_1 = -592.36, and β_2 = -0.3446, while x_{th} = 1.5 um is chosen. We let $c_2^+ = 0$ to observe the effectiveness of c_1^+. First, choose $c_1^+ = 70 > |\alpha_3| F_D^{\max} = 11$.

In the computer simulation, the VCM is modelled by Eq. (5) with the friction model of Eqs. (6)-(8). The Simulation result for the proposed controller in Theorem 1 is shown in Fig. 3. There is a steady state error $x_1(\infty) = d(\infty) - d^*$ of 0.0973 μm, which is less than the design goal of 0.4μm. This shows that the controller proposed in Theorem 1 can have the response satisfying an assigned steady state error by choosing λ from Eq. (15). On the other hand, the value of c_1^+ has the ability to drive the system states to the sliding surface $S = 0$, which can also be seen from Fig. 3, too. Actually, $S(t) = 0$ for $t > 0.009$ s. It is remarkable that the usual stick slip phenomena of the friction does not appear in the response of the proposed sliding mode controller.

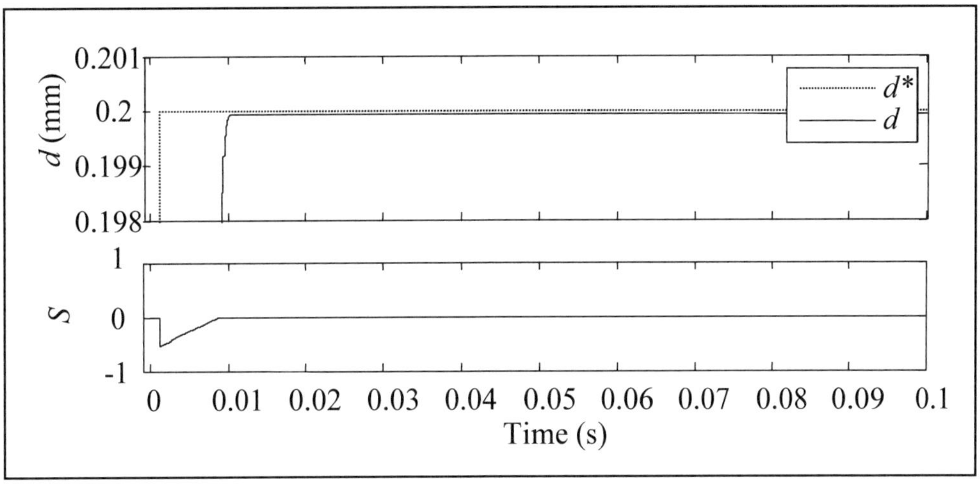

Fig. 3. Simulation response of the proposed sliding mode controller.

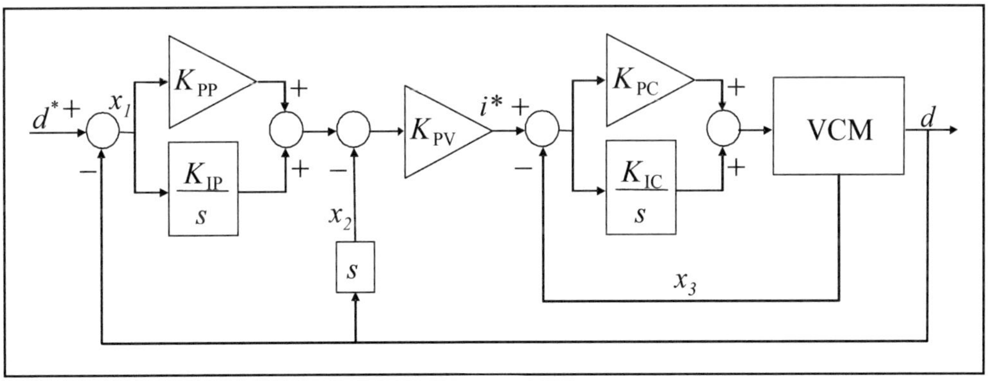

Fig. 4. Block diagram of a classic PI control law.

To show that stick slip phenomena of the friction appear in a usual controller, a classic PI control law is also simulated. The block diagram of the PI controller is shown in Fig. 4. There are two PI control loops. The inner one is the current control loop:

$$U(s) = \left(K_{\mathrm{PC}} + \frac{K_{\mathrm{IC}}}{s} \right) \left(I^*(s) - X_3(s) \right) \tag{24}$$

where $I^*(s)$ is the Laplace transform of $i^*(t)$, which is the output of the position control loop:

$$I^*(s) = K_{\mathrm{PV}} \left[\left(K_{\mathrm{PP}} + \frac{K_{\mathrm{IP}}}{s} \right) X_1(s) - X_2(s) \right] \tag{25}$$

Fig. 5(a) shows the simulation result of the classic PI controller for $0 \text{ s} \le t \le 1.5 \text{ s}$, while its transient part before $t < 0.02$ s is shown in Fig. 5(b). It is apparent from Fig. 5(a) that there are stick slip oscillations in the steady-state of the classic PI controller. For the purpose of comparison, Fig. 5(c) and 5(d) show the counterparts of the simulation result of the sliding mode controller. It can be seen from Fig. 5(c) that the proposed sliding mode controller does not induce any stick slip oscillations. Besides the ability to compensate for the nonlinear friction force of the VCM, the proposed controller also has faster transient response, which can be obtained by comparing Fig. 5(d) with 5(b).

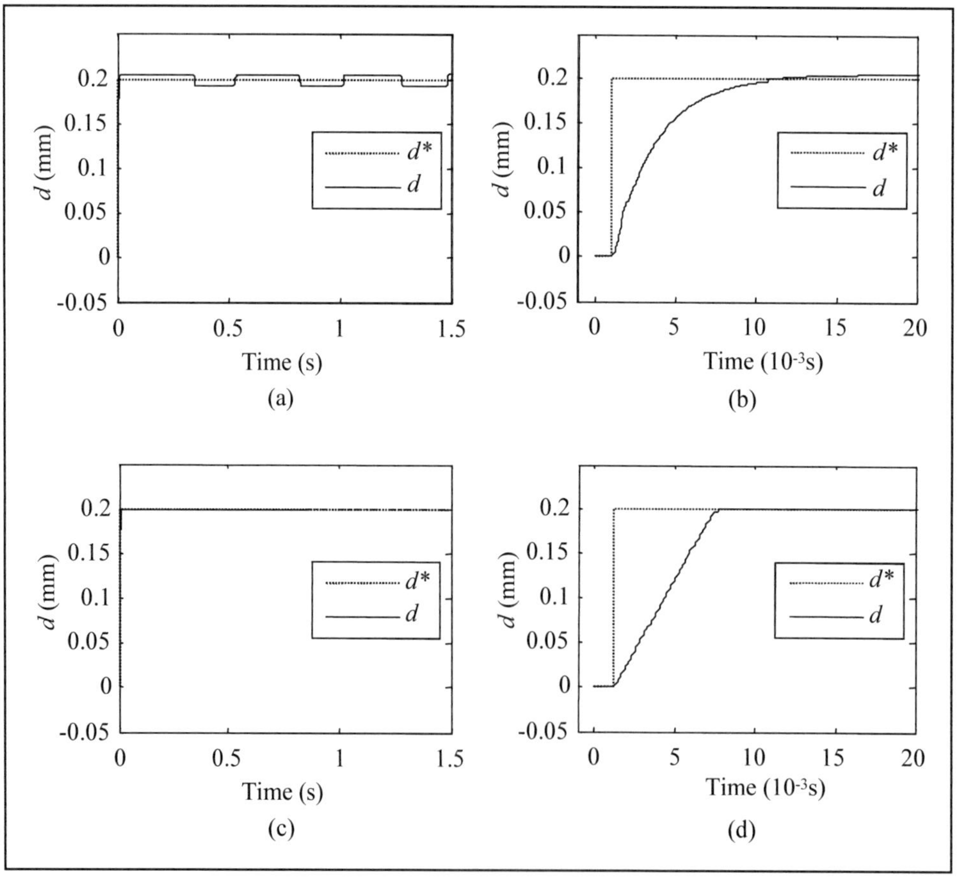

Fig. 5. Classic PI controller: (a) steady response with stick slip and (b) transient response; sliding mode controller: (c) steady response without stick slip and (d) transient response.

Furthermore, a simulation is performed to observe how c_2^+ in Eq. (12) affects the approaching speed to the sliding surface $S = 0$. We increase the value of c_2^+ from 0 to 200 and 800, while retain all other control gains. The simulation responses of S for these three values of c_2^+ are shown in Fig. 6. As was expected from Eq. (12), the settling time for S decreases with the increase of c_2^+. The response for $c_2^+ = 800$ is almost the same as that of a first order homogenous differential equation, since the other terms on the right side of Eq. (12) are much smaller than $c_2^+ S$ in absolute value. Actually, the value of c_1^+ has similar effect on the approaching speed to the sliding surface.

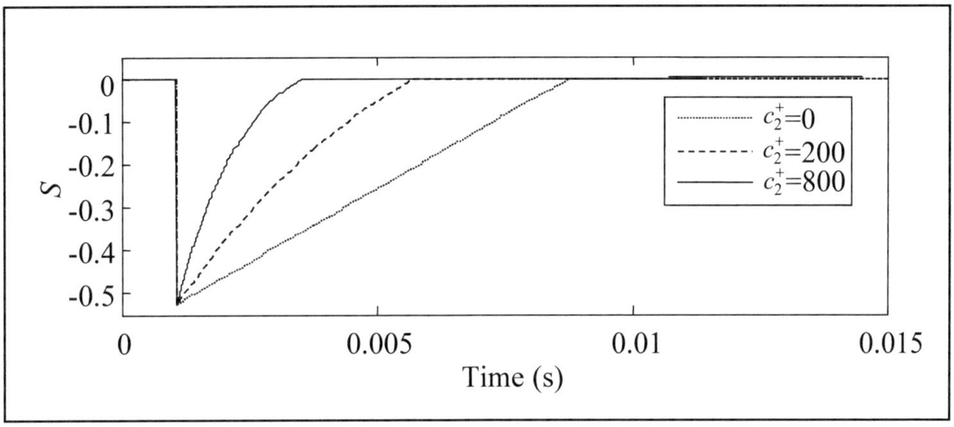

Fig. 6. Simulation responses of the sliding mode controller with various c_2^+.

We fix $c_2^+ = 0$ and change c_1^+ from 70 to 140 and 280. The simulation responses in Fig. 7 reveal that the settling time for S also decreases with the increase of c_1^+. However, the responses in Fig. 7 are not so smooth as those in Fig. 6. This indicates that c_2^+ is still a useful parameter to adjust the convergent speed.

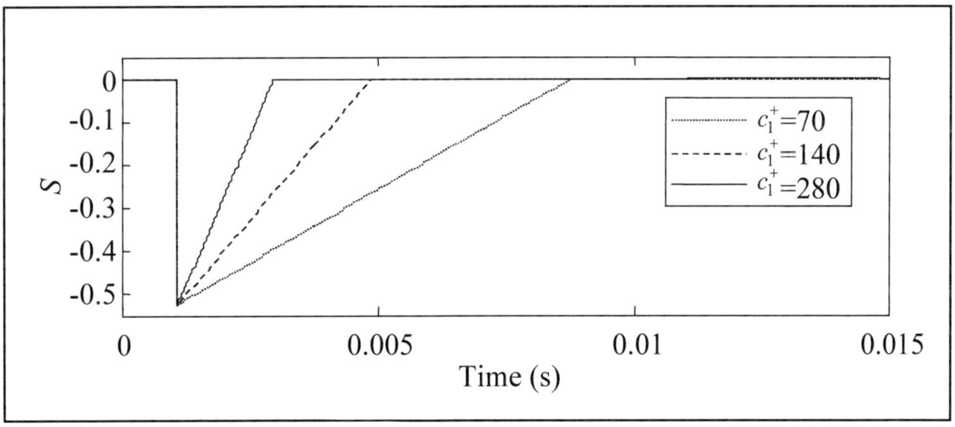

Fig. 7. Simulation responses of the sliding mode controller with various c_1^+.

Finally, we are interested in the chattering phenomenon of sliding mode control. The chattering parasitizes in the response of the sliding function S after the states reach the sliding surface $S = 0$. The states driven by the controller of Eq. (11) will go out of and back to the sliding surface, since the term associated with sgn(S) changes the control effort with the direction of S. The simulation response in Fig. 8 demonstrates the existence of the chattering phenomenon in S and u for the controller with sgn(S). If we replace sgn(S) in Eq. (12) with sat(S) in Eq. (23), the chattering can be alleviated as shown by the response for the controller with sat(S) in Fig. 8.

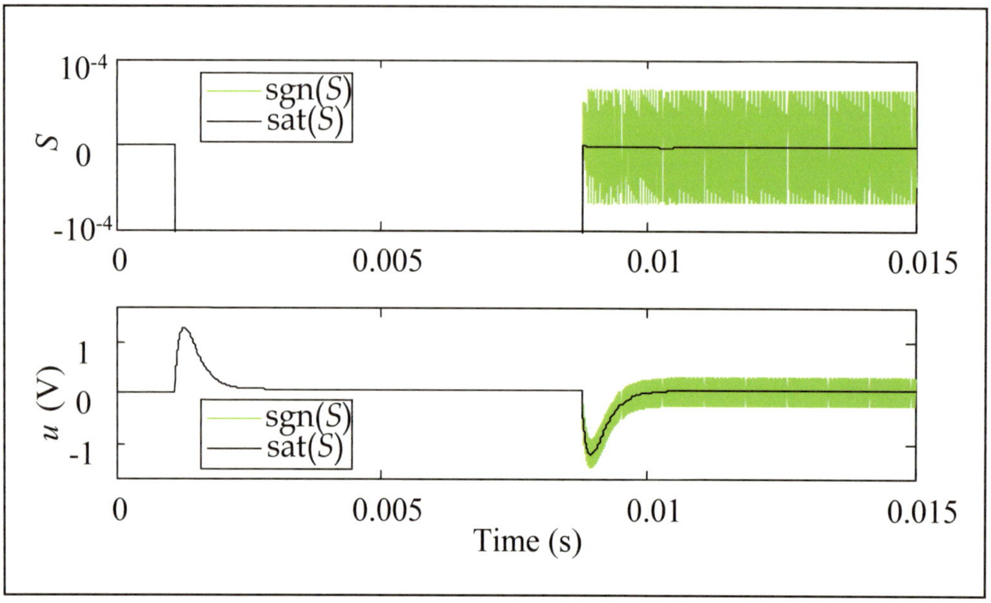

Fig. 8. Simulation responses of the sliding mode controller with sgn(S) and sat(S).

5. Experiments

The experimental setup is shown in Fig 9. The experimental system consists of a PC motherboard with X86 CPU and a FPGA board. There is a parallel ATA cable as a data communication interface. The FPGA board takes charge of generating the driving voltage of the VCM and measuring the coil current i and the actuator position d of the VCM. The PWM (Pulse Width Modulation) algorithm in the FPGA will generate the PWM signals to drive the full bridge (Chen et al., 2003) and then output the command voltage to the VCM. The full bridge plays the role of a power converter. Two ADC (analog-to-digital converter) circuits are used to sense the coil current and the actuator position, respectively. The sensed signals are filtered by the IIR filter algorithm in the FPGA. The PC motherboard reads the filtered current and position signals through the parallel ATA interface. The controller algorithm is programmed and executed in the PC motherboard. The current and position signals are the feedback signals of the controller, and are used to calculate out the controller output u. The output voltage u is then sent back, via the ATA interface, to the PWM modular of the FPGA, which transfers the voltage command to PWM signals and drives the VCM through the full bridge.

The tuning method introduced in (Ellis, 2004) is used to tune the control gains of the classic PI controller Eqs. (24)-(25). Fig. 10(a) and 10(b) shows the experimental result of the PI controller. It can be seen from Fig. 10(a) that there are stick slip oscillations after 0.15 s. The transient response in Fig. 10(b) is similar to the one of the simulation result in Fig. 5(b).

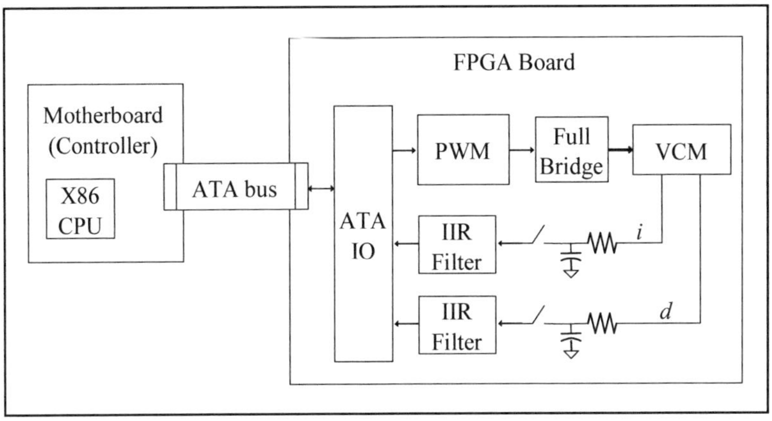

Fig. 9. Experiment system of the VCM Controller.

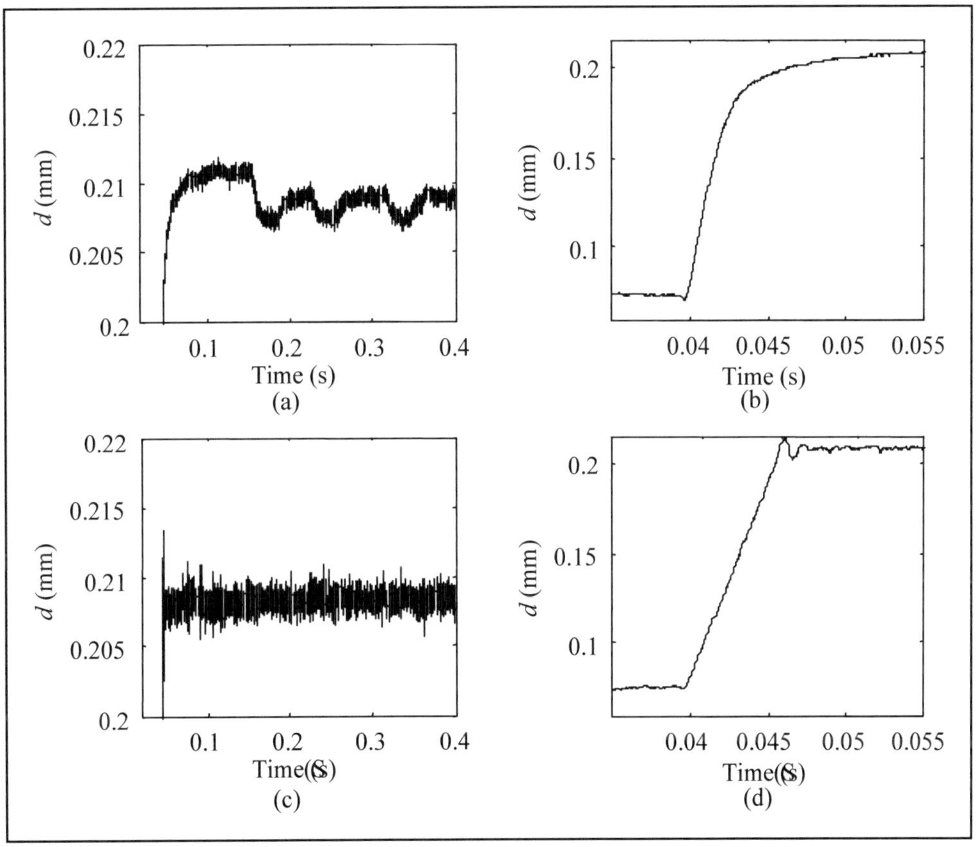

Fig. 10. Classic PI controller: (a) steady response with stick slip and (b) transient response; sliding mode controller: (c) steady response without stick slip and (d) transient response.

For the same desired position command, the experimental result of the proposed sliding mode controller has no stick slip oscillations as shown in Fig. 10(c). However, the sliding mode controller induces small overshoot in the transient response (see Fig. 10(d)), although it has a faster response. Such a fast response is able to support the advanced AF (auto focus) algorithm capable of 60 frame-rate. It should be remarked that the response in Fig. 10(c) is noisier than that in Fig. 10(a). This is caused by the chattering of sliding mode control with sgn(S).

The difference between the responses of the sliding mode controller with sgn(S) and with sat(S) can be obtained by the experimental results shown in Fig. 11. In this experiment, the VCM is first hold in the position of d = 0.07 mm, and then is driven to the position of d = 0.22 mm. The chattering phenomenon dominates in the measured feedback coil current i in Fig. 11(b) for the controller with sgn(S). Fig. 11(d) shows that the controller with sat(S) diminishes the chattering amplitude in the coil current i. On the other hand, comparing Fig. 11(c) with Fig. 11(a) reveals that the transient response from d = 0.07 mm to d = 0.22 mm for the controller with sat(S) is smoother than that with sgn(S).

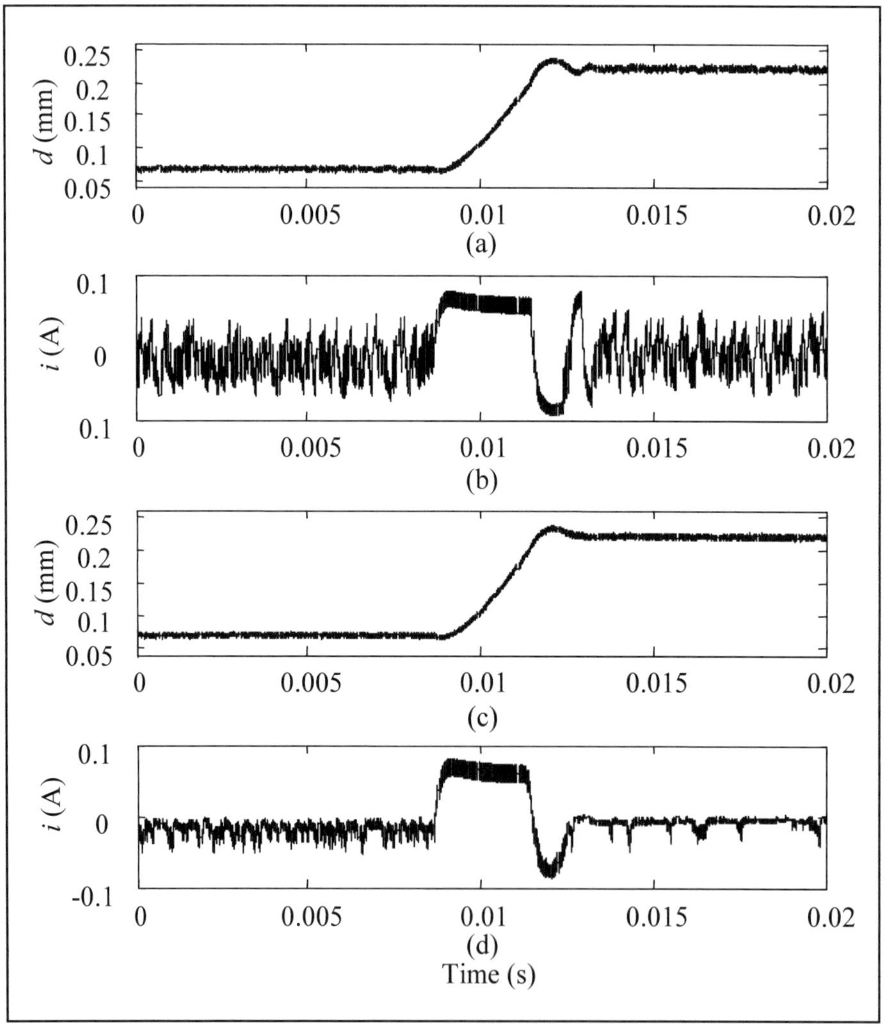

Fig. 11. Chattering phenomena: (a) and (b) responses of the controller with sgn(S); (c) and (d) responses of the controller with sat(S).

6. Repeatability tests

Repeatability is a critical specification for the VCM. In a camera, an AF (Auto Focus) algorithm detects the sharpness of images in multiple positions over the full optical stroke, and then asks the actuator to the position with the sharpest image. Poor repeatability would degrade the AF performance because the actuator would go to a wrong position different from the one with sharpest image. Thus, repeatability tests are inevitable for the VCM to be mounted in a compact camera module.

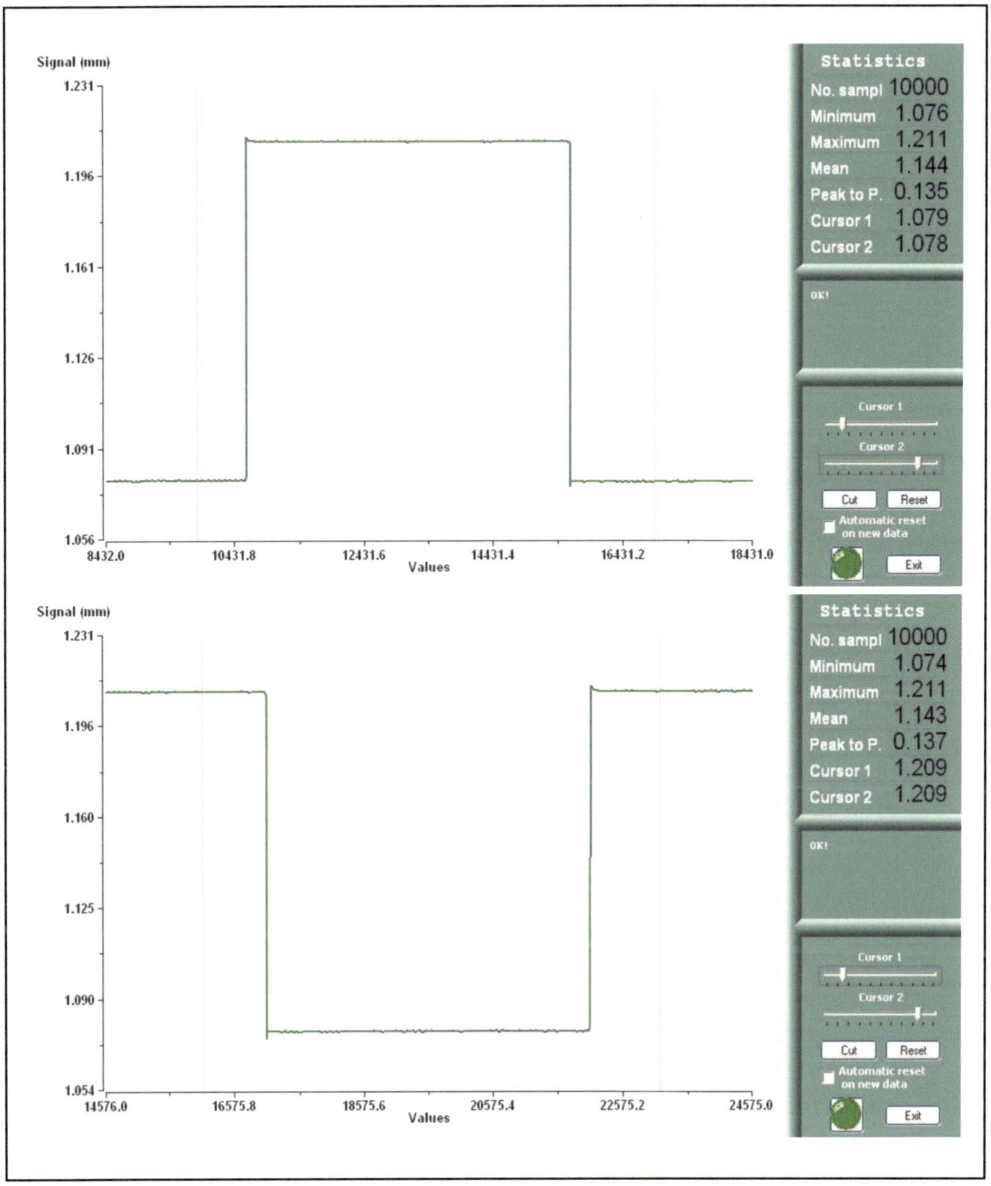

Fig. 12. LDM records for the vertical movements.

We utilize a laser displacement meter (LDM) to measure the physical position of the actuator in repeatability experiments. The controller used in the repeatability tests is only the proposed sliding mode law with sat(S). The movement of the actuator is tested both in the vertical and horizontal directions. The LDM records for the vertical movements are shown Fig. 12, while those for horizontal ones are in Fig. 13.

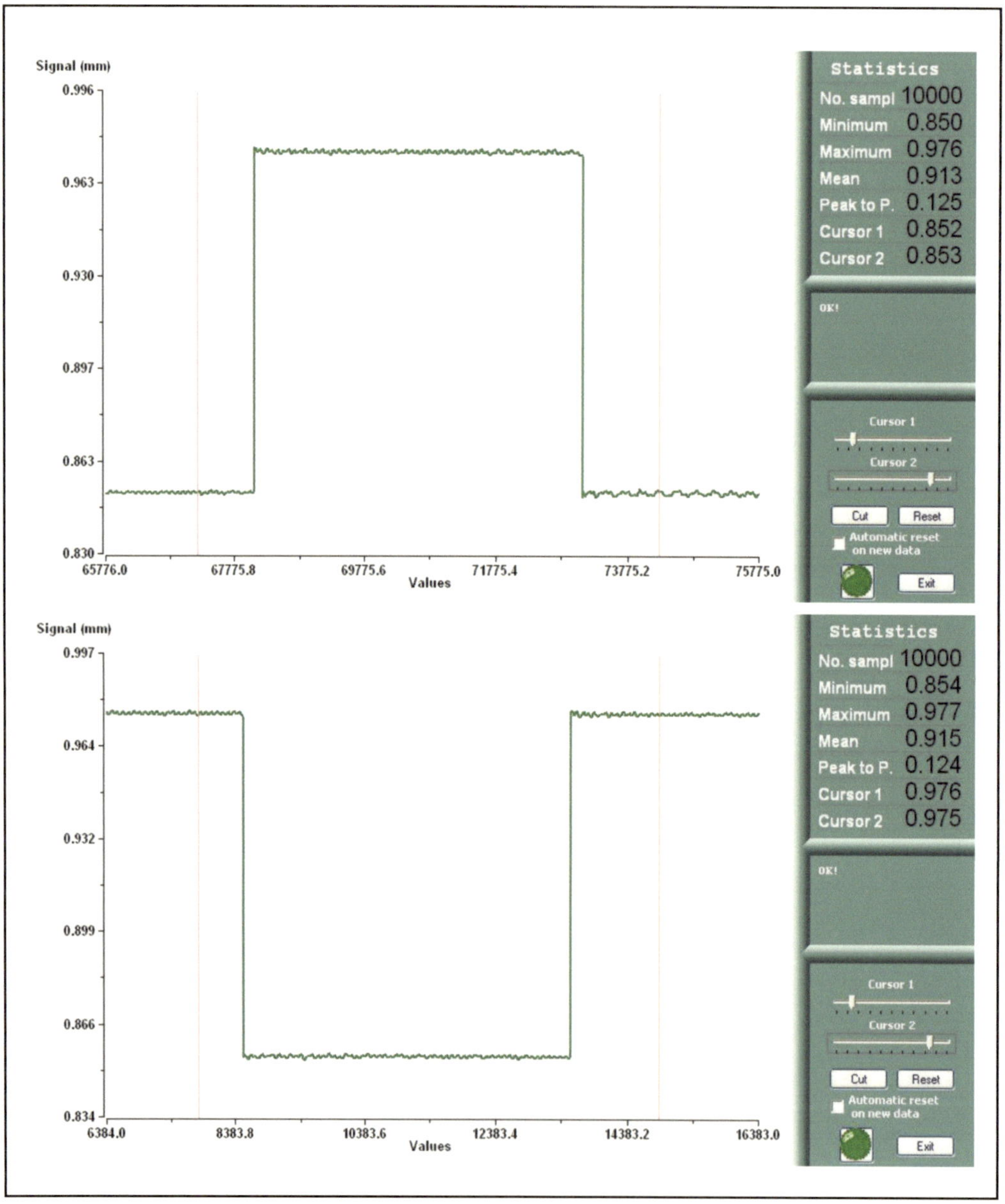

Fig. 13. LDM records for the horizontal movements.

In each record, the actuator leaves its original position and then comes back. Cursor 1 of the LDM picks the position value at an arbitrary time before leaving, while Cursor 2 picks another one after coming back. The difference of these two values is the repeatability. The records both in Fig. 12 and in Fig. 13 have the repeatability less than or equal to 1 μm. Comparing Fig. 13 with Fig. 12, we also find that the repeatability in horizontal direction is slightly poorer. The cause is that unbalanced friction force is more largely applied on the two guide pins in this horizontal orientation. However, the nominal requirement of the repeatability for a mini camera is about 10 μm. It is of no doubt that the proposed sliding mode controller with sat(S) is suitable to drive the VCM for the AF system of a camera.

7. Conclusions

In this chapter, a sliding mode controller is proposed to compensate for the nonlinear friction force of the mini VCM mounted on a compact camera module. It is known that the stick slip friction phenomenon would result in significant image-shaking and is not allowed in the camera application. In the proposed control system, stick slip oscillations disappear and the steady state error can be designed in arbitrarily small by the pole placement in the dynamical equation of the sliding surface. The experimental results show that the transient response is less than 10 ms, no stick slip oscillations occur in the steady state response, and the repeatability performance is also excellent. Consequently, the proposed control scheme works well and is reliable.

8. Acknowledgment

This work was partially supported by the National Science Council, Taiwan under grant no. NSC 97-2221-E-009-086-MY3 and by JD Components Co., Ltd. under grant no.99C201

9. References

Bona, B. & Indri, M. (2005). Friction compensation in robotics: an overview. *Proceedings of the 44th IEEE Conference on Decision and Control and the European Control Conference 2005*, pp. 12-15, Spain, December 2005.

Canudas de Wit C.; Olsson H.; Åström K. & Lischinsky P. (1995). A new model for control of systems with friction. *IEEE Trans. on Automatic Control*, Vol. 40, No. 3, pp. 419–425.

Chen J.; Prodić A.; Erickson R. W. & Maksimović D. (2003). Predictive digital current programmed control. *IEEE Trans. on Power Electronics*, Vol. 18, No. 1, pp. 411-419.

Ellis G. H. (2004). *Control System Design Guide*, 3rd Edition, Academic Press, ISBN: 978-0-12-237461-6, San Diego.

Golnaraghi, F. & Kuo, B. C. (2009). *Automatic Control Systems*, 9th Edition, John Wiley & Sons, ISBN: 978-0470-04896-2, New York.

Hsu J. D.; Tsai C. L. & Tzou Y. Y. (2007). Design and implementation of a voice-coil motor servo control ic for auto-focus mobile camera applications. *Proceedings of 2007 IEEE 38th Annual Power Electronics Specialists Conference*, pp. 1357-1362, ISBN: 978-1-4244-0654-8, Florida, USA, June 2007.

Lin S. K.; Wang C. M. & Wang S. J. (2008). Design and implementation of anti-handshaking position control for a voice coil motor. *Journal of Applied Physics*, Vol. 103, No. 7, Art. No. 07F128, 2008.

Southward S., Radcliffe C., & MacCluer C. (1991). Robust nonlinear stick-slip friction compensation. *ASME Journal of Dynamic Systems, Measurement and Control*, Vol. 113, No. 4, pp. 639–645.

Yu H. C.; Lee T. Y.; Wang S. J.; Lai M. L.; Ju J. J.; Huang D. R. & Lin S. K. (2005). Design of a voice coil motor used in the focusing system of a digital video camera. *IEEE Transactions on Magnetics*, Vol. 41, No. 10, pp. 3979-3981.

Part 3

Sliding Mode Control of Robotic Systems

12

Sliding Mode Control for Visual Servoing of Mobile Robots using a Generic Camera

Héctor M. Becerra and Carlos Sagüés
Instituto de Investigación en Ingeniería de Aragón, Universidad de Zaragoza
Spain

1. Introduction

At present, vision sensors represent a very good option for the control of robots since they provide at a low cost a lot of information from the environment. The feasibility of using a vision system as the only source of feedback information has been shown by many approaches (Chaumette & Hutchinson (2006), Chaumette & Hutchinson (2007)). Particularly, incorporating machine vision for the control of mobile robots can improve their navigation capabilities (DeSouza & Kak (2002)). The approach of closing the control loop through a vision system is called visual servoing (VS). The schemes in this control approach can be classified according to the nature of the feedback information. Image data can be used directly in the control loop (image-based schemes IBVS), for instance (Abdelkader et al. (2005), Benhimane & Malis (2006)), or can be used to compute an estimate of pose parameters (position-based schemes: PBVS) as in (Das et al. (2001), Fontanelli et al. (2009)). Hybrid schemes combining these both approaches have been performed as well (Malis et al. (1999), Fang et al. (2005)).
Most of the current efforts of the research on visual servoing focus on applications of monocular vision. This chapter presents an IBVS approach to drive a wheeled mobile robot equipped with a monocular camera onboard to a desired pose (position and orientation). The desired pose is specified by a target image previously acquired, i.e., the teach-by-showing strategy. In this context, a good way to relate the current and the target view is through a geometric constraint: epipolar geometry or the homography model. A geometric constraint is imposed on images in which there exist correspondences between features (Hartley & Zisserman (2000)). The information provided by a geometric constraint can be used directly as measurement for output feedback control. Comparing this two-view geometric constraints, the epipolar geometry is a more general approach because it is not constrained to planar objects or planar scenes. Currently, there also exist approaches that use three views (Becerra et al. (2010)).
This chapter focuses on exploiting the epipolar geometry (EG) in an IBVS approach. This constraint has been applied in some works (Basri et al. (1998), Rives (2000), Mariottini et al. (2007), López-Nicolás et al. (2008)). These works deal with the teach-by-showing problem, in which the target pose must be reached using only image data provided from the current and target images. In (Basri et al. (1998)) and (Rives (2000)) are reported visual servoing schemes based on epipolar geometry for manipulators. In the field of mobile robots, an epipolar-based VS approach that takes into account the nonholonomic nature of the robots is introduced in

(Mariottini et al. (2007)). The resultant motion in this approach steers the robot away from the target while the lateral error is corrected, and after that, the robot moves backward to the target position with a different control. This maneuvers are carried out in order to avoid a singularity problem that is induced by the epipolar geometry. The problem arises when the interaction matrix relating the robot velocities and the rate of change of the epipoles becomes singular. The motion strategy has been improved by driving the robot directly toward the target in the approach presented in (López-Nicolás et al. (2008)), however, in this work, one of the control inputs is not computed when the singularity occurs.
Although some of the previous works claim to achieve good robustness against camera parameters uncertainty, there is not a theoretical support of it. This chapter presents a sliding mode (SM) control law that drives the robot moving always toward the target and deals with the singularity problem. Thanks to the SM control, the robot is able to pass through the singularity caused by the epipolar geometry using bounded control inputs. Moreover, the visual control can be performed even when the initial robot pose is just on the singular point. Additionally, the SM control provides the required robustness to the closed loop control in this type of application. It is particularly important in the case of conventional perspective cameras because the presence of camera calibration uncertainty. This has been tackled through SM control in (Kim et al. (2006)) and (Becerra & Sagues (2008)). In this chapter, the last work is extended to calibrated omnidirectional images given by a generic camera. This type of camera is considered as an imaging system that approximately obeys the central projection model (Geyer & Daniilidis (2000)). The use of a generic camera provides the important advantage of keeping the target in the field of view. Wide field of view cameras have been applied for the control of mobile robots, for instance in (Abdelkader et al. (2005)) and (Mariottini & Prattichizzo (2008)). Although the scheme described herein is a calibrated approach, the benefits of SM control are present in the treatment of the singularity and the robustness against image noise and the uncertainty in a control parameter (the distance between the current and target locations).
The chapter is organized as follows. Section 2 introduces the mobile robot model, summarizes the model of generic cameras and describes the way to estimate the epipolar geometry of this type of cameras. Section 3 details the design procedure for the sliding mode control law. Section 4 presents an stability analysis. Section 5 shows the performance of the closed-loop control system via realistic simulations and finally, Section 6 provides the conclusions.

2. Mathematical modeling

2.1 Robot kinematics

Many wheeled mobile platforms can be represented as differential-drive robots, whose kinematic model is expressed as the affine system $\dot{\mathbf{z}} = f(t,\mathbf{z}) + B(t,\mathbf{z})\mathbf{u}$. The differential kinematics of the robot to be controlled, in accordance with the frame defined in Fig. 1(a), is as follows

$$\begin{bmatrix} \dot{x} \\ \dot{y} \\ \dot{\phi} \end{bmatrix} = \begin{bmatrix} -\sin\phi & 0 \\ \cos\phi & 0 \\ 0 & 1 \end{bmatrix} \begin{bmatrix} v \\ \omega \end{bmatrix}, \tag{1}$$

where, $\mathbf{z} = (x, y, \phi)^T$ represents the state of the robot, $x(t)$ and $y(t)$ are the robot position in the plane and $\phi(t)$ is the orientation. Additionally, $v(t)$ and $\omega(t)$ are the translational and rotational input velocities. The affine model (1) has the particularity that $f(t,\mathbf{z}) = \mathbf{0}$. Hence, this is a driftless system (i.e. no motion takes place under zero input, or in control

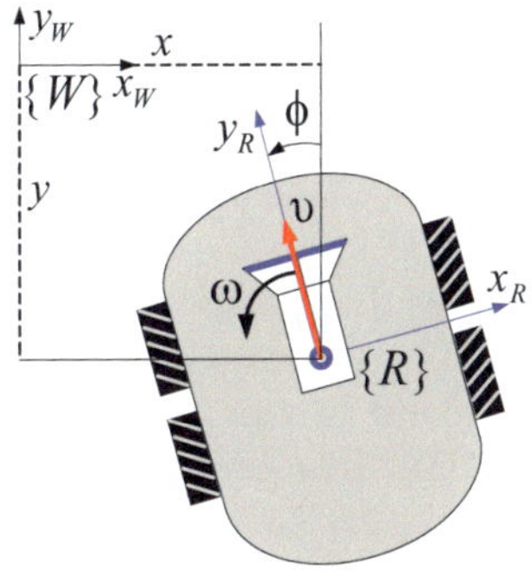

Fig. 1. Upper view of a mobile robot with a camera onboard and important variables of the system, where $\{W\}$ referes to the world frame and $\{R\}$ to the robot frame.

theory concepts, any state is an equilibrium point under zero input). Furthermore, the corresponding linear approximation in any point $\mathbf{z}(t) \in \Re^3$ does not have the property of controlability. However, it fulfills the Lie Algebra rank condition (Isidori (1995)), in such a way that controlability can be demonstrated from a nonlinear point of view.

2.2 Generic camera model

The constrained field of view of conventional cameras can be enhanced using wide field of view imaging systems such as full view omnidirectional cameras, which capture images as the one in Fig. 2(a). This can be achieved using some optic arrangements that combine mirrors and lens, i.e., catadioptric imaging systems (Fig. 2(b)). These systems use hyperboloidal, paraboloidal or ellipsoidal mirrors and have been well studied in the field of computer vision (Baker & Nayar (1999)). According to this theory, all of them satisfy the fixed view point constraint. In practice, with a careful construction of the system, it is realistic to assume a central configuration and many robotic applications have proven its effectiveness (Abdelkader et al. (2005), Benhimane & Malis (2006), Mariottini & Prattichizzo (2008), Guerrero et al. (2008)).

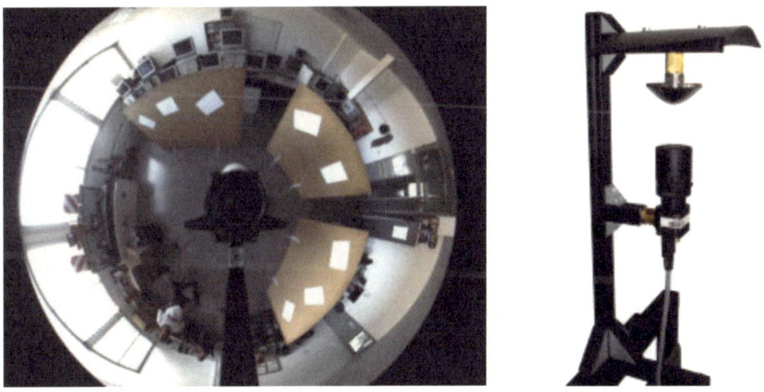

(a) Omnidirectional image (b) Catadioptric imaging system

Fig. 2. Example of an omnidirectional image and the system to capture this type of images.

It is known that the imaging process performed by conventional and catadioptric cameras can be modeled by a unique representation (Geyer & Daniilidis (2000)). Such unified projection model works properly for imaging systems having a single center of projection (central cameras). Although fisheye cameras do not accomplish such property, some recent experimental results have shown that the unified projection model is able to represent their image formation process with the required accuracy for robotic applications (Courbon et al. (2007)).

The unified projection model describes the image formation as a composition of two central projections (Geyer & Daniilidis (2000)). The first is a central projection of a 3D point onto a virtual unitary sphere and the second is a perspective projection onto the image plane. According to (Barreto & Araujo (2005)), this generic model can be parameterized by (ξ,λ) which are parameters describing the type of imaging system and by the matrix $\mathbf{K}$ containing the intrinsic parameters

$$\mathbf{K}_c = \begin{bmatrix} \alpha_x & s & x_0 \\ 0 & \alpha_y & y_0 \\ 0 & 0 & 1 \end{bmatrix}, \tag{2}$$

where α_x and α_y represent the focal length of the camera in terms of pixel dimensions in the x and y directions respectively, s is the skew parameter and (x_0, y_0) are the coordinates of the principal point.

The parameter ξ encodes the nonlinearities of the image formation in the range $0 \leq \xi \leq 1$ for the cases of catadioptric cameras and $\xi > 1$ for fisheye cameras. The parameter λ can be seen as a factor for the focal length and it is already included in its estimated value. Thus, the parameter ξ and the camera parameters can be obtained through a calibration process using an algorithm for central catadioptric cameras like the one in (Mei & Rives (2007)).

The mapping of a point $\mathbb{X}$ in the 3D world with coordinates $\mathbf{X} = \begin{bmatrix} X, Y, Z \end{bmatrix}^T$ in the camera frame F_c resulting in the image point $\mathbf{x}_{ic}$ with homogeneous coordinates $\mathbf{x}_{ic}^h$ can be divided into three steps (refer to Fig. 3):

1. The world point is projected onto the unit sphere on a point $\mathbb{X}_c$ with coordinates $\mathbf{X}_c$ in F_c, which are computed as $\mathbf{X}_c = \mathbf{X} / \|\mathbf{X}\|$.
2. The point coordinates $\mathbf{X}_c$ are then changed to a new reference frame O_c centered in $O = \begin{bmatrix} 0, 0, -\xi \end{bmatrix}^T$ and perspectively projected onto the normalized image plane $Z = 1 - \xi$:

$$\begin{aligned} \mathbf{x}^h &= \begin{bmatrix} \mathbf{x}^T, 1 \end{bmatrix}^T = \begin{bmatrix} x, y, 1 \end{bmatrix}^T \\ &= \begin{bmatrix} \frac{X}{Z+\xi\|\mathbf{X}\|}, \frac{Y}{Z+\xi\|\mathbf{X}\|}, 1 \end{bmatrix}^T. \end{aligned}$$

3. The image coordinates expressed in pixels are obtained after a collineation $\mathbf{K}$ of the 2D projected point $\mathbf{x}_{ic}^h = \mathbf{K}\mathbf{x}^h$.

The matrix $\mathbf{K}$ can be written as $\mathbf{K} = \mathbf{K}_c\mathbf{M}$, where $\mathbf{K}_c$ has been given in (2) and $\mathbf{M}$ is the following diagonal matrix

$$\mathbf{M} = \begin{bmatrix} \lambda - \xi & 0 & 0 \\ 0 & \xi - \lambda & 0 \\ 0 & 0 & 1 \end{bmatrix}. \tag{3}$$

Notice that, setting $\xi = 0$, the general projection model becomes the well known perspective projection model. Images also depend on the extrinsic parameters $\mathbf{C} = (x, y, \phi)$, i.e. the camera pose in the plane relative to a global reference frame. Then an image is denoted by $I(\mathbf{K}, \mathbf{C})$.

2.3 Epipolar geometry

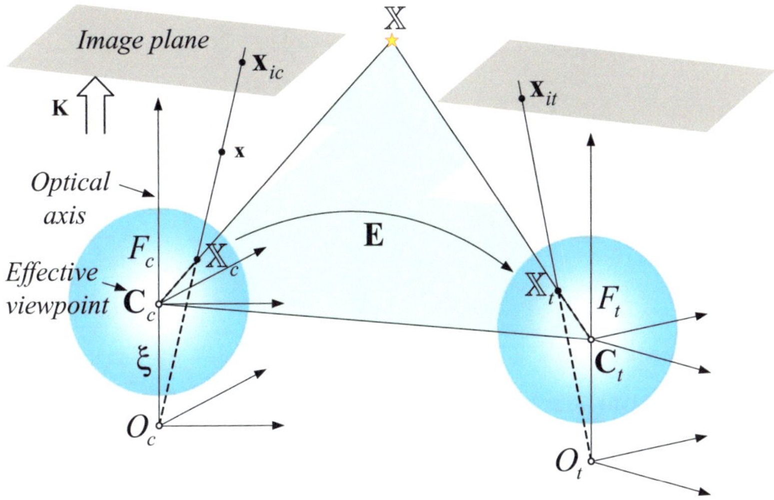

Fig. 3. Generic model of the image formation and epipolar geometry between generic central cameras.

Regarding to Fig. 3, let $\mathbb{X}$ be a 3D point and let $\mathbf{X}_c$ and $\mathbf{X}_t$ be the coordinates of that point projected onto the unit spheres of the current F_c and target frame F_t. The epipolar plane contains the effective viewpoints of the imaging systems $\mathbf{C}_c$ and $\mathbf{C}_t$, the 3D point $\mathbb{X}$ and the points $\mathbb{X}_c$ and $\mathbb{X}_t$. The coplanarity of these points leads to the well known epipolar constraint

$$\mathbf{X}_c^T \, \mathbf{E} \, \mathbf{X}_t = 0, \tag{4}$$

being $\mathbf{E}$ the essential matrix relating the pair of normalized virtual cameras. Normalized means that the effect of the known calibration matrix has been removed and virtually, the cameras can be represented as perspective. As typical, from this geometry it is possible to compute the epipoles as the points lying on the base line and intersecting the corresponding virtual image plane. Figure 4(a) shows the epipolar geometry for a pair of catadioptric systems and Fig. 4(b) depicts the projection of the epipoles in the produced omnidirectional images. The virtual representation of these imaging systems as perspective cameras is shown in Fig. 4(c) considering the planar motion constraint. A global reference frame centered in the origin $\mathbf{C}_t = (0,0,0)$ of the target viewpoint is defined, as well as important parameters. Then, the current camera location with respect to this reference is $\mathbf{C}_c = (x, z, \phi)$. Assuming the described framework in Fig. 1, where the camera location coincides with the robot location, the epipoles can be written as a function of the robot state as follows:

$$\begin{aligned} e_{cx} &= \alpha_x \frac{x\cos\phi + y\sin\phi}{y\cos\phi - x\sin\phi}, \\ e_{tx} &= \alpha_x \frac{x}{y}. \end{aligned} \tag{5}$$

Cartesian coordinates x and y can be expressed as a function of the polar coordinates d and ψ as

$$x = -d\sin\psi, \; y = d\cos\psi, \tag{6}$$

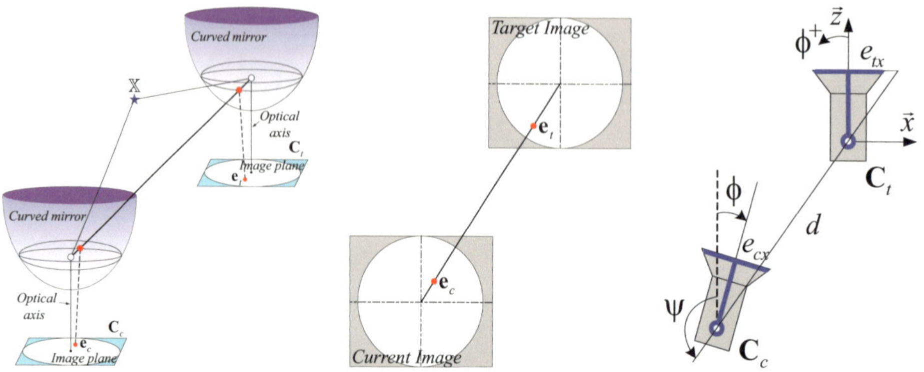

(a) 3D epipolar geometry (EG). (b) EG in omnidirectional images. (c) Planar EG framework.

Fig. 4. Generic cameras can be treated as virtual perspective cameras, in which the epipolar geometry is estimated as typical when the points on the unitary sphere are used.

with $\psi = -\arctan(e_{tx}/\alpha_x)$, $\phi - \psi = \arctan(e_{cx}/\alpha_x)$ and $d^2 = x^2 + y^2$. For the case of normalized cameras $\alpha_x = 1$ in (5) and in the subsequent equations.

3. Sliding mode control law

The goal of this work is to steer a mobile robot to a target pose by using the feedback information provided by the x-coordinate of the epipoles for any type of central camera. The visual servoing problem is transformed in a tracking problem for a nonlinear system, where the references for the epipoles are defined. A robust tracking control law under image noise and uncertainty of parameters is designed on the basis of SM control theory. We propose to perform a smooth motion toward the target position by tracking sinusoidal references to drive the epipoles to zero.

The main concern of the proposal is to deal with the singularity problem that arises because the decoupling matrix of the system becomes singular in a point of the state trajectory. This causes the translational velocity to grow unbounded when the system evolves near to that point. This problem is considered by (Mariottini et al. (2007)), where reaching the singular value is avoided during the servoing by using a particular motion strategy. In (López-Nicolás et al. (2008)), one of the control inputs is not computed when the singularity happens. Our strategy is able to pass through the singularity by switching to a bounded SM control law, instead of avoiding to reach to it. Furthermore, this approach can be also used when the initial robot pose is just on the singularity.

Let us define the outputs of the system using the x-coordinates of the epipoles for the current image $I_c(\mathbf{K}, \mathbf{C}_2(t))$ and the target one $I_t(\mathbf{K}, \mathbf{0})$. Then, the two-dimensional output of the system is

$$\mathbf{y} = h(\mathbf{x}) = \left[e_{cx}, e_{tx} \right]^T. \tag{7}$$

It can be seen from (5) that if both epipoles are zero implies $x = 0$, $\phi = 0$ and y (depth error) may be different to zero. From a control theory point of view this means that, when the epipoles reach to zero the so-called *zero dynamics* is achieved in the robot system. Zero dynamics is described by a subset of the state space that makes the output to be identically zero (Sastry (1999)). Thus, in the particular case of the robot system (1) with output vector (7), this set is given as follows

$$\begin{aligned} Z^* &= \left\{ \left[x, y, \phi \right]^T \mid e_{cx} \equiv 0,\ e_{tx} \equiv 0 \right\} \\ &= \left\{ \left[0, y, 0 \right]^T,\ y \in \mathbb{R} \right\}. \end{aligned} \tag{8}$$

Zero dynamics in this control system makes necessary a second step in which the remaining depth error must be corrected. We address the depth correction by using a constant translational velocity and the stop condition is given by thresholding the image error between corresponding points of the current and the target views. The *image error* is defined as the mean squared error between the p corresponding image points of the current image and points of the target image, i.e.,

$$\epsilon = \frac{1}{p} \sum_{j=1}^{p} \left\| \mathbf{x}_{t,j} - \mathbf{x}_{c,j} \right\|. \tag{9}$$

In order to design the appropriate control law the following tracking error system (**r**-system) is obtained by using the change of variables $r_c = e_{cx} - e_{cx}^d$, $r_t = e_{tx} - e_{tx}^d$ and the polar coordinates (6).

$$\begin{bmatrix} \dot{r}_c \\ \dot{r}_t \end{bmatrix} = \begin{bmatrix} -\frac{\alpha_x \sin(\phi-\psi)}{d\cos^2(\phi-\psi)} & \frac{\alpha_x}{\cos^2(\phi-\psi)} \\ -\frac{\alpha_x \sin(\phi-\psi)}{d\cos^2(\psi)} & 0 \end{bmatrix} \begin{bmatrix} v \\ \omega \end{bmatrix} - \begin{bmatrix} \dot{e}_{cx}^d \\ \dot{e}_{tx}^d \end{bmatrix}. \tag{10}$$

The system (10) has the form $\dot{\mathbf{r}} = \mathbf{M}(\phi, \psi)\mathbf{u} - \dot{\mathbf{e}}^d$, where $\mathbf{M}(\phi, \psi)$ corresponds to the decoupling matrix and $\dot{\mathbf{e}}^d$ represents a known disturbance. It is evident that the decoupling matrix loses rank if $\phi - \psi = n\pi$ with $n \in \mathbb{Z}$. For all the rest of the state space this matrix is invertible, with inverse matrix

$$\mathbf{M}^{-1}(\phi, \psi) = \frac{1}{\alpha_x} \begin{bmatrix} 0 & -\frac{d\cos^2(\psi)}{\sin(\phi-\psi)} \\ \cos^2(\phi - \psi) & -\cos^2(\psi) \end{bmatrix}. \tag{11}$$

We faced the tracking problem as an stabilization problem of the error system (10).

3.1 Decoupling-based control law

Firstly, in order to design a SMC law, we have to define suitable sliding surfaces. The simplest way to do it for the **r**-system (10) is to use directly the errors as sliding surfaces, in such a way that if there exist switching feedback gains that make the states to evolve in $\mathbf{s} = 0$, then the tracking problem is solved. Thus, the sliding surfaces are the following

$$\mathbf{s} = \begin{bmatrix} s_c \\ s_t \end{bmatrix} = \begin{bmatrix} r_c \\ r_t \end{bmatrix} = \begin{bmatrix} e_{cx} - e_{cx}^d \\ e_{tx} - e_{tx}^d \end{bmatrix}. \tag{12}$$

Next, the *equivalent control method* is used to find switching feedback gains to drive the state trajectory to $\mathbf{s} = 0$ and to maintain it there. The equivalent control method consists in working out the value of inputs from the equation $\dot{\mathbf{s}} = 0$. The so-called equivalent control is then

$$\mathbf{u}_{eq} = \mathbf{M}^{-1}(\phi, \psi)\dot{\mathbf{e}}^d. \tag{13}$$

A decoupling-based SMC law that ensures global stabilization of the $\mathbf{r}$-system has the form $\mathbf{u}_{sm} = \mathbf{u}_{eq} + \mathbf{u}_{disc}$, where $\mathbf{u}_{disc}$ is a two-dimensional vector containing switching feedback gains. We propose the simplest form of these gains as follows

$$\mathbf{u}_{disc} = \mathbf{M}^{-1}(\phi, \psi) \begin{bmatrix} -k_c^{sm} sign\,(s_c) \\ -k_t^{sm} sign(s_t) \end{bmatrix}, \tag{14}$$

where $k_c^{sm} > 0$ and $k_t^{sm} > 0$ are control gains. Although $\mathbf{u}_{sm}$ can achieve global stabilization of the $\mathbf{r}$-system, it needs high gains and, consequently, the state trajectory may not reach the sliding surfaces in a smooth way. This could cause a non-smooth behavior in the robot state that is not valid in real situations. We add a pole placement term in the control law to alleviate this problem

$$\mathbf{u}_{pp} = \mathbf{M}^{-1}(\phi, \psi) \begin{bmatrix} -k_c & 0 \\ 0 & -k_t \end{bmatrix} \begin{bmatrix} s_c \\ s_t \end{bmatrix}, \tag{15}$$

where $k_c > 0$ and $k_t > 0$ are control gains. Finally, the complete SMC law that achieves robust global stabilization of the system (10) is as follows

$$\mathbf{u}_{db} = \begin{bmatrix} \upsilon_{db} \\ \omega_{db} \end{bmatrix} = \mathbf{u}_{eq} + \mathbf{u}_{disc} + \mathbf{u}_{pp}. \tag{16}$$

3.2 Bounded control law

The control law (16) utilizes the decoupling matrix and it presents the singularity problem for the condition $e_{cx} = 0$ ($\phi - \psi = n\pi$ with $n \in \mathbb{Z}$), which means that the camera axis of the robot at its current pose is aligned with the baseline. We can note from (11) that the singularity only affects the translational velocity computation. In order to pass through this singularity we propose to commute to a direct sliding mode controller when $\phi - \psi$ is near to $n\pi$. This kind of controller has been studied for output tracking through singularities (Hirschorn (2002)). The direct sliding mode controller is as follows

$$\mathbf{u}_b = \begin{bmatrix} \upsilon_b \\ \omega_b \end{bmatrix} = \begin{bmatrix} -M sign\,(s_t\, b\,(\phi, \psi)) \\ -N sign(s_c) \end{bmatrix}, \tag{17}$$

where M and N are suitable gains and $b\,(\phi, \psi)$ is a function that describes the change in sign of the translational velocity when the state trajectory crosses the singularity. We can find out this function from (10) as follows

$$\begin{aligned} \dot{r}_c &= b_1\,(\phi, \psi)\,\upsilon + b_2\,(\phi, \psi)\,\omega - \dot{e}_{cx}^d, \\ \dot{r}_t &= b_3\,(\phi, \psi)\,\upsilon - \dot{e}_{tx}^d, \end{aligned} \tag{18}$$

where $b_1 = -\frac{\alpha_x \sin(\phi - \psi)}{d \cos^2(\phi - \psi)}$, $b_2 = \frac{\alpha_x}{\cos^2(\phi - \psi)}$, $b_3 = -\frac{\alpha_x \sin(\phi - \psi)}{d \cos^2(\psi)}$. According to that, b_2 is always positive, and $sign(b_1) = sign(b_3) = sign(-\sin(\phi - \psi))$. Hence,

$$b\,(\phi, \psi) = -\sin(\phi - \psi). \tag{19}$$

The control law (17) with $b\,(\phi, \psi)$ as in (19) locally stabilizes the system (10) and is always bounded.

3.3 Desired references of the epipoles

As main requirement, the references to track must provide a smooth zeroing of the epipoles starting from their initial values. Figure 5(a) shows two configurations of robot locations for cases in which $sign\,(e_{23}) \neq sign\,(e_{32})$. From these conditions, the epipoles are naturally reduced to zero as the robot moves smoothly toward the target. Because of the nonholomic motion constraint, any direct path reaching the target implies $sign\,(e_{23}) \neq sign\,(e_{32})$. Therefore, locations starting $sign\,(e_{23}) = sign\,(e_{32})$ need to be controlled to the situation of $sign\,(e_{23}) \neq sign\,(e_{32})$. This allows getting an adequate orientation from the very beginning (Fig. 5(b)) in order to be able to align the robot with the target at the end of the first step. It is worth emphasizing that this initial rotation is autonomously carried out through the control inputs given by the described controllers. Thus, we define the following desired trajectories, which are always opposite in sign each other

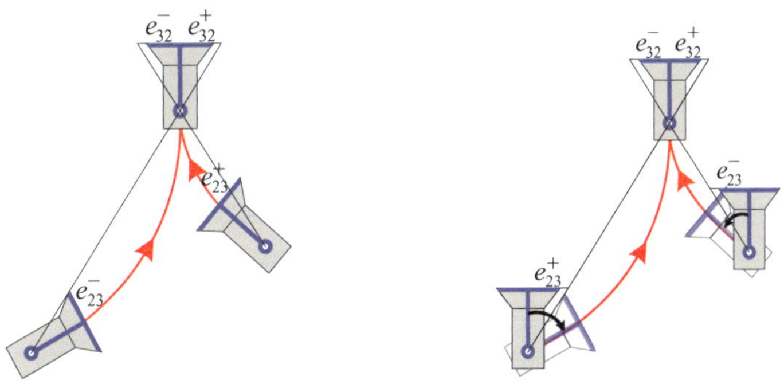

(a) Condition where $sign\,(e_{tx}) \neq sign\,(e_{cx})$ (b) Condition where $sign\,(e_{tx}) = sign\,(e_{cx})$

Fig. 5. Motion strategy for different initial locations. For the cases in (a) a direct motion toward the target is carried out and for those in (b), the robot rotates initially to reach the same condition as in (a).

$$
\begin{aligned}
e_{cx}^{d}\,(t) &= S\frac{e_{cx}(0)}{2}\left(1+\cos\left(\frac{\pi}{\tau}t\right)\right), && 0 \leq t \leq \tau \\
e_{cx}^{d}\,(t) &= 0, && \tau < t < \infty \\
e_{tx}^{d}\,(t) &= \frac{e_{tx}(0)}{2}\left(1+\cos\left(\frac{\pi}{\tau}t\right)\right), && 0 \leq t \leq \tau \\
e_{tx}^{d}\,(t) &= 0, && \tau < t < \infty
\end{aligned} \tag{20}
$$

where $S = -sign\,(e_{cx}(0)e_{tx}(0))$ and T is the time to reach epipoles to zero. We highlight that since our scheme is an IBVS approach, the desired trajectories in the image space play an important role in the resulting Cartesian path. In fact, by only changing the trajectory for the target epipole (related to the translational velocity) is possible to run our approach for car-like robots. Thus, the references may also be used to constraint the target in the field of view. However, as mentioned previously, this problem does not appear when using omnidirectional vision.

4. Stability analysis

In this section, the stability of both proposed sliding mode control laws is analyzed. Given that the relationship between zeroing the epipoles and the robot state has been established through the zero dynamics, we focus on demonstrating the convergence property of the tracking control law as stated in the following proposition.

Proposition 1. *A commuted control law that combines the decoupling-based control (16) by switching to the bounded control (17) whenever $|\phi - \psi| < n\pi + T_h$, where T_h is a suitable threshold value and $n \in \mathbb{Z}$, achieves global stabilization of the system (10). Moreover, global stabilization is achieved even with uncertainty in parameters.*

Proof. Stabilization of the system (10) is proved by showing that the sliding surfaces can be reached in a finite time (existence conditions of sliding modes). Let be the natural Lyapunov function for a sliding mode controller

$$V = V_1 + V_2, \;\; V_1 = \frac{1}{2}s_c^2, \;\; V_2 = \frac{1}{2}s_t^2, \tag{21}$$

which accomplish $V(s_c = 0, s_t = 0) = 0$ and $V > 0$ for all $s_c \neq 0, s_t \neq 0$.

$$\dot{V} = \dot{V}_1 + \dot{V}_2 = s_c\dot{s}_c + s_t\dot{s}_t. \tag{22}$$

We analyze each term of (22) for the decoupling-based controller (16)

$$\dot{V}_1 = s_c\left(-\frac{\alpha_x}{\alpha_{x_e}}\left(k_c^{sm}sign\left(s_c\right) + k_c s_c\right) + A\right) = -\left(\frac{\alpha_x}{\alpha_{x_e}}\left(k_c^{sm}\left|s_c\right| + k_c s_c^2\right) - s_c A\right),$$
$$\dot{V}_2 = s_t\left(-\frac{\alpha_x d_e}{\alpha_{x_e} d}\left(k_t^{sm}sign(s_t) + k_t\xi_t\right) + B\right) = -\left(\frac{\alpha_x d_e}{\alpha_{x_e} d}\left(k_t^{sm}\left|s_t\right| + k_t s_t^2\right) - s_t B\right),$$

where $A = \frac{\alpha_x}{\alpha_{xe}}\left(\frac{d_e}{d} - 1\right)\left(\dot{e}_{tx}^d - k_t^{sm}sign(s_t) - k_t s_t\right)\frac{\cos^2(\psi)}{\cos^2(\phi-\psi)} + \left(\frac{\alpha_x}{\alpha_{xe}} - 1\right)\dot{e}_{cx}^d$, $B = \left(\frac{\alpha_x d_e}{\alpha_{xe} d} - 1\right)\dot{e}_{tx}^d$, and α_{x_e}, d_e represent estimated values for the corresponding system parameter. We can see that

$$\dot{V}_1 \leq -\left(\frac{\alpha_x}{\alpha_{x_e}}\left(k_c^{sm} + k_c\left|s_c\right|\right) - \left|A\right|\right)\left|s_c\right|,$$
$$\dot{V}_2 \leq -\left(\frac{\alpha_x d_e}{\alpha_{x_e} d}\left(k_t^{sm} + k_t\left|s_t\right|\right) - \left|B\right|\right)\left|s_t\right|.$$

$\dot{V}_1$ and $\dot{V}_2$ are negative definite iff the following inequalities are guaranteed for all $s_c \neq 0$, $s_t \neq 0$

$$k_c^{sm} + k_c\left|s_c\right| > \frac{\alpha_{x_e}}{\alpha_x}\left|A\right|, \tag{23}$$
$$k_t^{sm} + k_t\left|s_t\right| > \frac{\alpha_{x_e} d}{\alpha_x d_e}\left|B\right|.$$

Therefore, $\dot{V} < 0$ iff both inequalities (23) are fulfilled. Global convergence to the sliding surfaces can be achieved regardless of uncertainty in parameters.

Now, let us develop the existence conditions of sliding modes for the bounded controller (17). The same Lyapunov function (21) is used. For each term of (22) after some basic operations we have

$$\dot{V}_1 = -N\frac{\alpha_x}{\cos^2(\phi-\psi)}|s_c| - s_c\dot{e}^d_{cx} - s_c C,$$
$$\dot{V}_2 = -M\frac{\alpha_x|b(\phi,\psi)|}{d\cos^2(\psi)}|s_t| - s_t\dot{e}^d_{tx},$$

where $C = M\frac{\alpha_x|b(\phi,\psi)|}{d\cos^2(\phi-\psi)}sign(s_t)$ and $b(\phi,\psi)$ is given in (19). We can see that

$$\dot{V}_1 \leq -\left(N\frac{\alpha_x}{\cos^2(\phi-\psi)} - \left|\dot{e}^d_{cx}\right| - |C|\right)|s_c|,$$
$$\dot{V}_2 \leq -\left(M\frac{\alpha_x|b(\phi,\psi)|}{d\cos^2(\psi)} - \left|\dot{e}^d_{tx}\right|\right)|s_t|.$$

$\dot{V}_1$ and $\dot{V}_2$ are negative definite iff the following inequalities are assured for all $s_c \neq 0$, $s_t \neq 0$

$$N > \frac{\cos^2(\phi-\psi)}{\alpha_x}\left(|C| + \left|\dot{e}^d_{cx}\right|\right), \tag{24}$$
$$M > \frac{d\cos^2(\psi)}{\alpha_x|b(\phi,\psi)|}\left|\dot{e}^d_{tx}\right|.$$

Therefore, $\dot{V} < 0$ iff both inequalities (24) are fulfilled. The bounded controller does not need any information of system parameters and thus, its robustness is implicit.
According to the existence conditions of sliding modes, the bounded controller (17) is able to locally stabilize the system (10); its region of attraction is bigger as long as the control gains M and N are higher. Nevertheless, this controller can not achieve the smooth behavior demanded for real situations and it is only used to cross the singularity. Due to the control strategy commutes between two switching control laws and each one acts inside of its region of attraction, respectively, the commutation between the control laws does not affect the stability of the control system. The decoupling-based controller ensures entering to the region of attraction of the bounded one.
Once sliding surfaces are reached for any case of SMC law, the system's behavior is independent of uncertainties and disturbances. It is clear that uncertainties in the system (10) fulfill the matching condition and then, robustness of the control system is accomplished.
∎

5. Performance evaluation

The evaluation of the approach has been carried out through realistic simulations. These simulations have been performed in Matlab with a sampling time of 100 ms. The results show that the main objective of driving the robot to a desired pose ((0,0,0^o) in all the cases) is attained in spite of passing through the singularity that occurs in the first step for some initial poses, and moreover, the task is accomplished even when the robot starts exactly in a singular pose. The good performance of the approach with noise in the images is also reported.

Regarding to the parameters of the control law, the initial distance between the current and target locations (d_e) is fixed to 10 m in all the cases. The threshold to switch to the bounded control law (T_h) is set to 0.03 rad. Related to the control gains, they are set to $k_c = 2$, $k_t = 1$, $k_c^{sm} = 0.2$, $k_t^{sm} = 0.2$, $M = 0.1$ and $N = 0.06$. Synthetic images of size 640×480 pixels are used to estimate the epipoles at each instant time. These images are obtained by using adequate camera parameters in the generic model of Section 2.2. We present results with hypercatadioptric, paracatadioptric and also fisheye cameras, which can be approximately represented with the same model (Courbon et al. (2007)).

The simulations are carried out for four different initial locations: (-5,-9,-50°), (-4,-14,0^o), (8,-16,10^o) and (2.5,-12,11.77^o) and consequently, the fixed value of d_e represents a significative uncertainty in the control parameter. In spite of that, the good behavior of the approach can be seen in the image space through the pictures in Fig. 6. This figure shows the motion of the point features for the different types of synthetic images used. We can notice that the points of the images at the end of the motion (marker "×") are practically the same as the ones in the target images (marker "O").

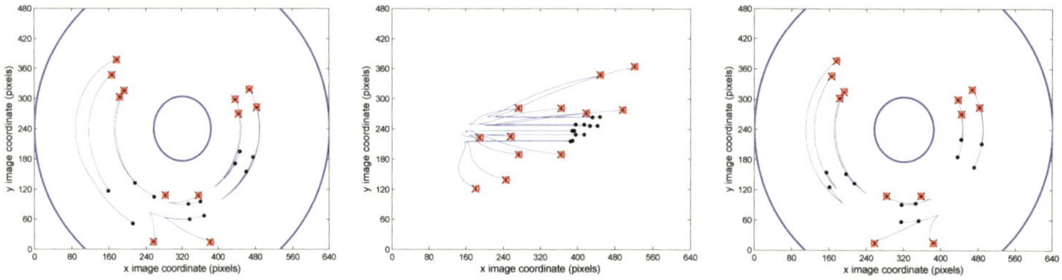

(a) (-4,-14,0^o) Paracatad. (b) (8,-16,10^o) Fisheye. (c) (2.5,-12,11.77^o) Hypercatad.

Fig. 6. Behavior of the approach in the image space for different omnidirectional images.

Figure 7(a) shows the resultant paths and the evolution of the epipoles for each one of the initial locations. The case (-5,-9,-50°) corresponds to an initial location from where the robot can exert a direct navigation to the target and has been tested using a hypercatadioptric camera. In the cases (-4,-14,0^o) and (8,-16,10^o), the robot starts with $sign\,(e_{cx}) = sign\,(e_{tx})$ and by driving the epipoles to the desired trajectories, e_{cx} changes its sign during the first seconds (Fig. 7(b)). It causes a rotation of the robot, and then, it begins a direct motion toward the target. These cases are tested using paracatadiotric and fisheye cameras respectively.

The initial location (2.5,-12,11.77^o), tested with hypercatadioptric images, corresponds to a special case where the state trajectory just starts on the singularity $e_{cx} = 0$. The line from the initial position to the target shows that the camera axis is aligned with the baseline for this location. When the robot starts just on the singularity, we assign a suitable amplitude to the desired trajectory for the current epipole. Given that $|\phi - \psi|$ is less than the threshold, the bounded controller takes the system out of the singularity and then, the epipoles evolve as shown in Fig. 7(b). In all the cases both epipoles reach to zero in $\tau = 60$ s, which is fixed through the desired trajectories.

From the graphics of the epipoles, it can be seen that the state trajectory crosses the singularity $e_{cx} = 0$ for the initial locations (-4,-14,0^o) and (8,-16,10^o). The behavior of the robot state is presented in Fig. 8(a) for the former case. This is obtained using the bounded input velocities of Fig. 8(b). It is worth noting that the control inputs are maintained bounded even when the epipoles are close to zero after 45 s, which ensures entire correction of orientation and lateral position. It takes approximately 3 s more to correct the remaining depth error using a constant

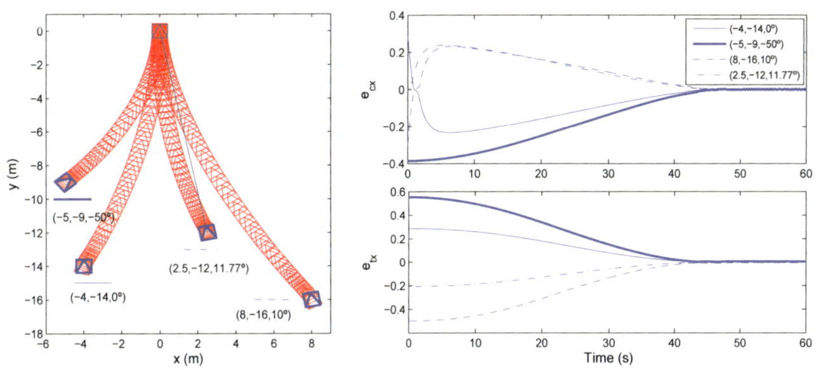

(a) Paths on the $x - y$ plane. (b) Current and target epipoles.

Fig. 7. Behavior of the approach in the Cartesian space and evolution of epipoles for different initial locations.

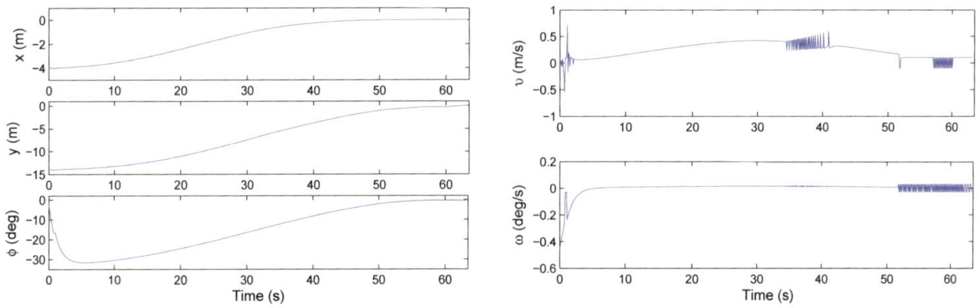

(a) Evolution of the robot state. (b) Input velocities.

Fig. 8. Evolution of the position and orientation of the robot and the velocities given by the sliding mode control law for a case where the singularity is crossed, (-4,-14,0^o) of Fig. 7.

translational velocity $\upsilon = 0.1$ m/s in this case, but this time may be different for each initial location with the same velocity. The stop condition is given by thresholding the mean squared error (9) between the corresponding image points of the current image and the points of the target image.

Finally, Fig. 9(a) shows the performance of the approach under image noise for the initial location (5,-13,15^o). An image noise with standard deviation of 0.5 pixels has been added and the time to reach the alignment with the target is set to $\tau = 80$ s. During the remaining 9 s, depth correction is carried out by using a constant translational velocity and then, each one of the state variables reaches zero (Fig. 9(b)). It is clear the presence of the noise in the motion of the image points in Fig. 9(c). It can be seen in Fig. 10(a) that the estimated epipoles are more affected by the noise as the robot approaches to the target and eventually it turns out to be unstable (problem of short baseline). However, after 80 s only the sign of e_{cx} is used to compute the rotational velocity that keeps the robot aligned to the target (Fig. 10(b)).

According to these results when testing the performance of the proposed visual control scheme, the use of SM control provides good benefits in order to solve the singularity problem and robustness against image noise. Additionally, it is worth noting that the target location is always reached with an accuracy in the order of centimeters for position and negligible

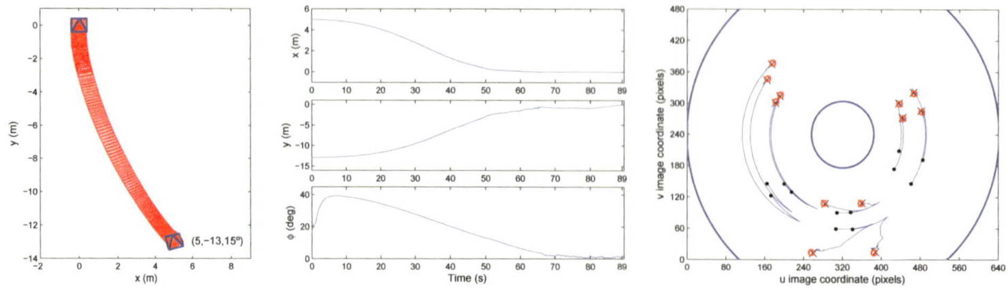

(a) Path on the plane. (b) Evolution of the robot state. (c) Motion of the image points.

Fig. 9. Robustness under image noise using a hypercatadioptric imaging system.

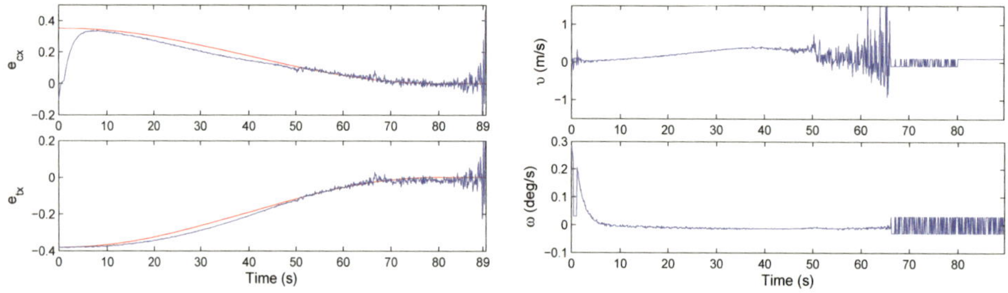

(a) Evolution of the epipoles. (b) Input velocities.

Fig. 10. Performance of the reference tracking and the velocities given by the sliding mode control law for the servoing task of Fig. 9.

orientation error. This is achieved in spite of the uncertainty in the distance between the current and the target locations (d). As mentioned before, it is enough to fix this value in the controller thanks to the robustness of the control law. We claim that the second phase regarding to depth correction may be carried out exploiting also the information provided by the epipolar geometry. This could be a way to avoid the switching to a totally different approach for depth correction.

6. Conclusions

In this chapter, a robust control law to perform image-based visual servoing for differential-drive mobile robots has been presented. The visual control utilizes the usual teach-by-showing strategy, in which the desired location is specified by a target image previously acquired. The mobile robot is driven toward the target by comparing a set of visual features in the current view of the onboard camera and those on the target image. The visual features are gathered through the epipolar geometry and exploited in a sliding mode control law, which provides good robustness against image noise and uncertainty in camera parameters.

The major contribution of this work is the validity of the approach for generic imaging systems. This extends the applicability of the proposed control scheme given that a generic camera allows a major maneuverability of the robot than a conventional camera because its wide field of view. Additionally, the use of sliding mode control allows to solve the problem of passing through a singularity induced by the epipoles, maintaining bounded inputs. Furthermore, the visual control accomplishes its goal even when the robot starts on the singularity. The good performance of the approach has been evaluated through realistic simulations using virtual images.

7. Acknowledgment

This work has been supported by project MICINN DPI 2009-08126 and grants of Banco Santander-Universidad de Zaragoza and Conacyt-Mexico.

8. References

Abdelkader, H. H., Mezouar, Y., Andreff, N. & Martinet, P. (2005). Image-based control of mobile robot with central catadioptric cameras, *IEEE International Conference on Robotics and Automation*, pp. 3522–3527.

Baker, S. & Nayar, S. K. (1999). A theory of single-viewpoint catadioptric image formation, *International Journal of Computer Vision* 35(2): 175–196.

Barreto, J. P. & Araujo, H. (2005). Geometric properties of central catadioptric line images and their application in calibration, *IEEE Transactions on Pattern Analysis and Machine Intelligence* 27(8): 1327–1333.

Basri, R., Rivlin, E. & Shimshoni, I. (1998). Visual homing: Surfing on the epipoles, *IEEE Conference on Computer Vision*, pp. 863–869.

Becerra, H. M., López-Nicolás, G. & Sagues, C. (2010). Omnidirectional visual control of mobile robots based on the 1D trifocal tensor, *Robotics and Autonomous Systems* 58(6): 796–808.

Becerra, H. M. & Sagues, C. (2008). A sliding mode control law for epipolar visual servoing of differential-drive robots, *IEEE/RSJ International Conference on Intelligent Robots and Systems*, pp. 3058–3063.

Benhimane, S. & Malis, E. (2006). A new approach to vision-based robot control with omni-directional cameras, *IEEE International Conference on Robotics and Automation*, pp. 526–531.

Chaumette, F. & Hutchinson, S. (2006). Visual servo control Part I: Basic approaches, *IEEE Robotics and Automation Magazine* 13(4): 82–90.

Chaumette, F. & Hutchinson, S. (2007). Visual servo control Part II: Advance approaches, *IEEE Robotics and Automation Magazine* 14(1): 109–118.

Courbon, J., Mezouar, Y., Eck, L. & Martinet, P. (2007). A generic fisheye camera model for robotics applications, *IEEE/RSJ International Conference on Intelligent Robots and Systems*, pp. 1683–1688.

Das, A. K., Fierro, R., Kumar, V., Southall, B., Spletzer, J. & Taylor, C. J. (2001). Real-time vision-based control of a nonholonomic mobile robot, *IEEE/RSJ International Conference on Intelligent Robots and Systems*, pp. 1714–1718.

DeSouza, G. & Kak, A. (2002). Vision for mobile robot navigation: A survey, *IEEE Transactions on Pattern Analysis and Machine Intelligence* 24(2): 237–267.

Fang, Y., Dixon, W. E., Dawson, D. M. & Chawda, P. (2005). Homography-based visual servo regulation of mobile robots, *IEEE Transactions on Systems, Man, and Cybernetics, Part B* 35(5): 1041–1050.

Fontanelli, D., Danesi, A., Belo, F. A. W., Salaris, P. & Bicchi, A. (2009). Visual servoing in the large, *The International Journal of Robotics Research* 28(6): 802–814.

Geyer, C. & Daniilidis, K. (2000). An unifying theory for central panoramic systems and practical implications, *European Conference on Computer Vision*, pp. 445–461.

Guerrero, J., Murillo, A. & Sagüés, C. (2008). Localization and matching using the planar trifocal tensor with bearing-only data, *IEEE Transactions on Robotics* 24(2): 494–501.

Hartley, R. & Zisserman, A. (2000). *Multiple View Geometry in Computer Vision*, Cambridge University Press, Cambridge.

Hirschorn, R. M. (2002). Output tracking through singularities, *IEEE Conference on Decision and Control*, pp. 3843–3848.

Isidori, A. (1995). *Nonlinear Control Systems*, Springer, Great Britain.

Kim, J. K., Kim, D. W., Choi, S. J. & Won, S. C. (2006). Image-based visual servoing using sliding mode control, *SICE-ICASE International Joint Conference*, pp. 4996–5001.

López-Nicolás, G., Sagüés, C., Guerrero, J., Kragic, D. & Jensfelt, P. (2008). Switching visual control based on epipoles for mobile robots, *Robotics and Autonomous Systems* 56(7): 592–603.

Malis, E., Chaumette, F. & Boudet, S. (1999). 2 1/2 D visual servoing, *IEEE Transactions on Robotics and Automation* 15(2): 234–246.

Mariottini, G. L., Oriolo, G. & Prattichizzo, D. (2007). Image-based visual servoing for nonholonomic mobile robots using epipolar geometry, *IEEE Transactions on Robotics* 23(1): 87–100.

Mariottini, G. L. & Prattichizzo, D. (2008). Image-based visual servoing with central catadioptric cameras, *The International Journal of Robotics Research* 27(1): 41–56.

Mei, C. & Rives, P. (2007). Single view point omnidirectional camera calibration from planar grids, *IEEE International Conference on Robotics and Automation*, pp. 3945–3950.

Rives, P. (2000). Visual servoing based on epipolar geometry, *IEEE/RSJ International Conference on Intelligent Robots and Systems*, pp. 602–607.

Sastry, S. (1999). *Nonlinear Systems: Analysis, Stability and Control*, Springer, New York.

13

Super-Twisting Sliding Mode in Motion Control Systems

Jorge Rivera[1], Luis Garcia[2], Christian Mora[3], Juan J. Raygoza[4] and Susana Ortega[5]

[1,2,3,4]*Centro Universitrio de Ciencias Exactas e Ingenierías, Universidad de Guadalajara*
[5]*Centro de Investigación y Estudios Avanzados del I.P.N. Unidad Guadalajara*
México

1. Introduction

Nowadays, the major advancements in the control of motion systems are due to the automatic control theory. Motion control systems are characterized by complex nonlinear dynamics and can be found in the robotic, automotive and electromechanical area, among others. In such systems it is always wanted to impose a desired behavior in order to cope with the control objectives that can go from velocity and position tracking to torque and current tracking among other variables. Motion control systems become vulnerable when the output tracking signals present small oscillations of finite frequency known as chattering. The chattering problem is harmful because it leads to low control accuracy; high wear of moving mechanical parts and high heat losses in power circuits. The chattering phenomenon can be caused by the deliberate use of classical sliding mode control technique. This control technique is characterized by a discontinuous control action with an ideal infinite frequency. When fast dynamics are neglected in the mathematical model such phenomenon can appear. Another situation responsible for chattering is due to implementation issues of the sliding mode control signal in digital devices operating with a finite sampling frequency, where the switching frequency of the control signal cannot be fully implemented. Despite of the disadvantage presented by the sliding mode control, this is a popular technique among control engineer practitioners due to the fact that introduces robustness to unknown bounded perturbations that belong to the control sub-space; moreover, the residual dynamic under the sliding regime, i.e., the sliding mode dynamic, can easily be stabilized with a proper choice of the sliding surface. A proof of their good performance in motion control systems can be found in the book by Utkin et al. (1999). A solution to this problem is the high order sliding mode (HOSM) technique, Levant (2005). This control technique maintains the same sliding mode properties (in this sense, first-order sliding mode) with the advantage of eliminating the chattering problem due to the continuous-time nature of the control action. The actual disadvantage of this control technique is that the stability proofs are based on geometrical methods since the Lyapunov function proposing results in a difficult task, Levant (1993). The work presented in Moreno & Osorio (2008) proposes quadratic like Lyapunov functions for a special case of second-order sliding mode controller, the super-twisting sliding mode controller (STSMC), making possible to obtain an explicit relation for the controller design parameters.

In this chapter, two motion control problems will be addressed. First, a position trajectory tracking controller for an under-actuated robotic system known as the Pendubot will be designed. Second, a rotor velocity and magnetic rotor flux modulus tracking controller will be designed for an induction motor.
The Pendubot (see Spong & Vidyasagar (1989)) is an under-actuated robotic system, characterized by having less actuators than links. In general, this can be a natural design due to physical limitations or an intentional one for reducing the robot cost. The control of such robots is more difficult than fully actuated ones. The Pendubot is a two link planar robot with a dc motor actuating in the first link, with the first one balancing the second link. The purpose of the Pendubot is research and education inside the control theory of nonlinear systems. Common control problems for the Pendubot are swing-up, stabilization and trajectory tracking. In this work, a super-twisting sliding mode controller for the Pendubot is designed for trajectory tracking, where the proper choice of the sliding function can easily stabilize the residual sliding mode dynamic. A novel Lyapunov function is used for a rigorous stability analysis of the controller here designed. Numeric simulations verify the good performance of the closed-loop system.
In the other hand, induction motors are widely used in industrial applications due to its simple mechanical construction, low service requirements and lower cost with respect to DC motors that are also widely used in the industrial field. Therefore, induction motors constitute a classical test bench in the automatic control theory framework due to the fact that represents a coupled MIMO nonlinear system, resulting in a challenging control problem. It is worth mentioning that there are several works that are based on a mathematical induction motor model that does not consider power core losses, implying that the induction motor presents a low efficiency performance. In order to achieve a high efficiency in power consumption one must take into consideration at least the power core losses in addition to copper losses; then, to design a control law under conditions obtained when minimizing the power core and copper losses. With respect to loss model based controllers, there is a main approach for modeling the core, as a parallel resistance. In this case, the resistance is fixed in parallel with the magnetization inductance, increasing the four electrical equations to six in the (α, β) stationary reference frame, Levi et al. (1995). In this work, one is compelled to design a robust controller-observer scheme, based on the super-twisting technique. A novel Lyapunov function is used for a rigorous stability analysis. In order to yield to a better performance of induction motors, the power core and copper losses are minimized. Simulations are presented in order to demonstrate the good performance of the proposed control strategy.
The remaining structure of this chapter is as follows. First, the sliding mode control will be revisited. Then, the Pendubot is introduced to develop the super-twisting controller design. In the following part, the induction motor model with core loss is presented, and the super-twisting controller is designed in an effort of minimizing the power losses. Finally some comments conclude this chapter.

2. Sliding mode control

The sliding mode control is a well documented control technique, and their fundamentals can be founded in the following references, Utkin (1993), Utkin et al. (1999), among others. Therefore in this section, the main features of this control technique are revisited in order to introduce the super-twisting algorithm.
The first order or classical sliding mode control is a two-step design procedure consisting of a sliding surface ($S = 0$) design with relative degree one w.r.t. the control (the control

appears explicitly in $\dot{S}$), and a discontinuous control action that ensures a sliding regime or a sliding mode. When the states of the system are confined in the sliding mode, i.e., the states of the system have reached the surface, the convergence happens in a finite-time fashion, moreover, the matched bounded perturbations are rejected. From this time instant the motion of the system is known as the sliding mode dynamic and it is insensitive to matched bounded perturbations. This dynamic is commonly characterized by a reduced set of equations. At the initial design stage, one must predict the sliding mode dynamic structure and then to design the sliding surface in order to stabilize it. It is worth mentioning that the sliding mode dynamic (commonly containing the output) is commonly asymptotically stabilized. This fact is sometimes confusing since one can expect to observe the finite-time convergence at the output of the system, but as mentioned above the finite-time convergence occurs at the designed surface. The main disadvantage of the classical sliding mode is the chattering phenomenon, that is characterized by small oscillations at the output of the system that can result harmful to motion control systems. The chattering can be developed due to neglected fast dynamics and to digital implementation issues.

In order to overcome the chattering phenomenon, the high-order sliding mode concept was introduced by Levant (1993). Let us consider a smooth dynamic system with an output function S of class C^{r-1} closed by a constant or dynamic discontinuous feedback as in Levant & Alelishvili (2007). Then, the calculated time derivatives $S, \dot{S}, \ldots, S^{r-1}$, are continuous functions of the system state, where the set $S = \dot{S} = \ldots = S^{r-1} = 0$ is non-empty and consists locally of Filippov trajectories. The motion on the set above mentioned is said to exist in r-sliding mode or r_{th} order sliding mode. The r_{th} derivative S^r is considered to be discontinuous or non-existent. Therefore the high-order sliding mode removes the relative-degree restriction and can practically eliminate the chattering problem.

There are several algorithms to realize HOSM. In particular, the 2_{nd} order sliding mode controllers are used to zero the outputs with relative degree two or to avoid chattering while zeroing outputs with relative degree one. Among 2_{nd} order algorithms one can find the sub-optimal controller, the terminal sliding mode controllers, the twisting controller and the super-twisting controller. In particular, the twisting algorithm forces the sliding variable S of relative degree two in to the 2-sliding set, requiring knowledge of $\dot{S}$. The super-twisting algorithm does not require $\dot{S}$, but the sliding variable has relative degree one. Therefore, the super-twisting algorithm is nowadays preferable over the classical siding mode, since it eliminates the chattering phenomenon.

The actual disadvantage of HOSM is that the stability proofs are based on geometrical methods, since the Lyapunov function proposal results in a difficult task, Levant (1993). The work presented in Moreno & Osorio (2008) proposes quadratic like Lyapunov functions for the super-twisting controller, making possible to obtain an explicit relation for the controller design parameters. In the following lines this analysis will be revisited.

Let us consider the following SISO nonlinear scalar system

$$\dot{\sigma} = f(t, \sigma) + u \tag{1}$$

where $f(t,\sigma)$ is an unknown bounded perturbation term and globally bounded by $|f(t,\sigma)| \leq \delta|\sigma|^{1/2}$ for some constant $\delta > 0$. The super-twisting sliding mode controller for perturbation and chattering elimination is given by

$$\begin{aligned} u &= -k_1\sqrt{|\sigma|}sign(\sigma) + v \\ \dot{v} &= -k_2 sign(\sigma). \end{aligned} \tag{2}$$

System (1) closed by control (2) results in

$$\begin{aligned}\dot{\sigma} &= -k_1\sqrt{|\sigma|}sign(\sigma) + v + f(t,\sigma)\\ \dot{v} &= -k_2 sign(\sigma).\end{aligned} \tag{3}$$

Proposing the following candidate Lyapunov function:

$$\begin{aligned}\mathcal{V} &= 2k_2|\sigma| + \frac{1}{2}v^2 + \frac{1}{2}(k_1|\sigma|^{1/2}sign(\sigma) - v)^2\\ &= \xi^T P \xi\end{aligned}$$

where $\xi^T = (|\sigma|^{1/2}sign(\sigma)\ v)$ and

$$P = \frac{1}{2}\begin{pmatrix} 4k_2 + k_1^2 & -k_1 \\ -k_1 & 2 \end{pmatrix},$$

Its time derivative along the solution of (3) results as follows:

$$\dot{\mathcal{V}} = -\frac{1}{|\sigma^{1/2}|}\xi^T Q \xi + \frac{f(t,\sigma)}{|\sigma^{1/2}|}q_1^T \xi$$

where

$$Q = \frac{k_1}{2}\begin{pmatrix} 2k_2 + k_1^2 & -k_1 \\ -k_1 & 1 \end{pmatrix},$$

$$q_1^T = \left(2k_2 + \tfrac{1}{2}k_1^2 \ \ -\tfrac{1}{2}k_1\right).$$

Applying the bounds for the perturbations as given in Moreno & Osorio (2008), the expression for the derivative of the Lyapunov function is reduced to

$$\dot{\mathcal{V}} = -\frac{k_1}{2|\sigma^{1/2}|}\xi^T \tilde{Q} \xi$$

where

$$\tilde{Q} = \begin{pmatrix} 2k_2 + k_1^2 - (\frac{4k_2}{k_1} + k_1)\delta & -k_1 + 2\delta \\ -k_1 + 2\delta & 1 \end{pmatrix}.$$

In this case, if the controller gains satisfy the following relations

$$k_1 > 2\delta, \quad k_2 > k_1\frac{5\delta k_1 + 4\delta^2}{2(k_1 - 2\delta)},$$

then, $\tilde{Q} > 0$, implying that the derivative of the Lyapunov function is negative definite.

3. STSMC for an under-actuated robotic system

In this section a super-twisting sliding mode controller for the Pendubot is designed. The Pendubot is schematically shown in Figure 1.

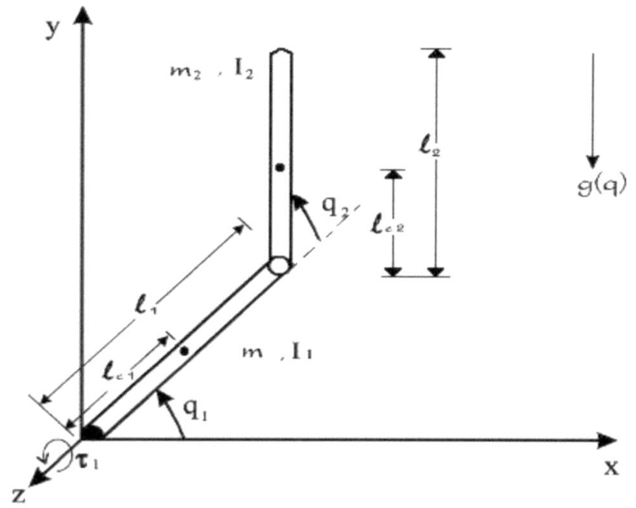

Fig. 1. Schematic diagram of the Pendubot.

3.1 Mathematical model of the Pendubot

The equation of motion for the Pendubot can be described by the following general Euler-Lagrange equation Spong & Vidyasagar (1989):

$$D(q)\ddot{q} + C(q,\dot{q}) + G(q) + F(\dot{q}) = \tau \tag{4}$$

where $q = [q_1, q_2]^T \in \Re^n$ is the vector of joint variables (generalized coordinates), $q_1 \in \Re^m$ represents the actuated joints, and $q_2 \in \Re^{(n-m)}$ represents the unactuated ones. $D(q)$ is the $n \times n$ inertia matrix, $C(q,\dot{q})$ is the vector of Coriolis and centripetal torques, $G(q)$ contains the gravitational terms, $F(\dot{q})$ is the vector of viscous frictional terms, and τ is the vector of input torques. For the Pendubot system, the dynamic model (4) is particularized as

$$\begin{bmatrix} D_{11} & D_{12} \\ D_{12} & D_{22} \end{bmatrix} \begin{bmatrix} \ddot{q}_1 \\ \ddot{q}_2 \end{bmatrix} + \begin{bmatrix} C_1 \\ C_2 \end{bmatrix} + \begin{bmatrix} G_1 \\ G_2 \end{bmatrix} + \begin{bmatrix} F_1 \\ F_2 \end{bmatrix} = \begin{bmatrix} \tau_1 \\ 0 \end{bmatrix}$$

where $D_{11}(q_2) = m_1 l_{cl}^2 + m_2(l_1^2 + l_{c2}^2 + 2l_1 l_{c2}\cos q_2) + I_1 + I_2$, $D_{12}(q_2) = m_2(l_{c2}^2 + l_1 l_{c2}\cos q_2) + I_2$, $D_{22} = m_2 l_{c2}^2 + I_2$, $C_1(q_2,\dot{q}_1,\dot{q}_2) = -2m_2 l_1 l_{c2}\dot{q}_1\dot{q}_2 \sin q_2 - m_2 l_1 l_{c2}\dot{q}_2^2 \sin q_2$, $C_2(q_2,\dot{q}_1) = m_2 l_1 l_{c2}\dot{q}_1^2 \sin q_2$, $G_1(q_1,q_2) = m_1 g l_{c1}\cos q_1 + m_2 g l_1 \cos q_1 + m_2 g l_{c2}\cos(q_1+q_2)$, $G_2(q_1,q_2) = m_2 g l_{c2}\cos\,(q_1+q_2)$, $F_1(\dot{q}_1) = \mu_1\dot{q}_1$, $F_2(\dot{q}_2) = \mu_2\dot{q}_2$, with m_1 and m_2 as the mass of the first and second link of the Pendubot respectively, l_1 is the length of the first link , l_{c1} and l_{c2} are the distance to the center of mass of link one and two respectively, g is the acceleration of gravity, I_1 and I_2 are the moment of inertia of the first and second link respectively about its centroids, and $\mu_1$and μ_2 are the viscous drag coefficients. The nominal values of the parameters are taken as follows: $m_1 = 0.8293$, $m_2 = 0.3402$, $l_1 = 0.2032$, $l_{c1} = 0.1551$, $l_{c2} = 0.1635125$, $g = 9.81$, $I_1 = 0.00595035$, $I_2 = 0.00043001254$, $\mu_1 = 0.00545$, $\mu_2 = 0.00047$. Choosing $x = \begin{pmatrix} x_1 & x_2 & x_3 & x_4 \end{pmatrix}^T = \begin{pmatrix} q_1 & q_2 & \dot{q}_1 & \dot{q}_2 \end{pmatrix}^T$ as the state vector, $u = \tau_1$ as the input, and x_2 as the output, the description of the system can be given in state space form as:

$$\dot{x}(t) = f(x) + g(x)u(t) \tag{5}$$
$$e(x,w) = x_2 - w_2$$
$$\dot{w} = s(w) \tag{6}$$

where $e(x,w)$ is output tracking error, $w = (w_1, w_2)^T$, and w_2 as the reference signal generated by the known exosystem (6),

$$f(x) = \begin{pmatrix} f_1(x_3) \\ f_2(x_4) \\ f_3(x) \\ f_4(x_1,x_2,x_3) \end{pmatrix} = \begin{pmatrix} x_3 \\ x_4 \\ b_3(x_2)p_1(x) \\ b_4(x_2)p_2(x) \end{pmatrix},$$

$$g(x) = \begin{pmatrix} b_1 \\ b_2 \\ b_3(x_2) \\ b_4(x_2) \end{pmatrix} = \begin{pmatrix} 0 \\ 0 \\ \frac{D_{22}}{D_{11}(x_2)D_{22}-D_{12}^2(x_2)} \\ \frac{-D_{12}(x_2)}{D_{11}(x_2)D_{22}-D_{12}^2(x_2)} \end{pmatrix},$$

$$s(w) = \begin{pmatrix} \alpha w_2 \\ -\alpha w_1 \end{pmatrix},$$

$$p_1(x) = \frac{D_{12}(x_2)}{D_{22}}\left(C_2(x_2,x_3) + G_2(x_1,x_2) + F_2(x_4)\right) - C_1(x_2,x_3,x_4) - G_1(x_1,x_2) - F_1(x_3),$$

$$p_2(x) = \frac{D_{11}(x_2)}{D_{12}}\left(C_2(x_2,x_3) + G_2(x_1,x_2) + F_2(x_4)\right) - C_1(x_2,x_3,x_4) - G_1(x_1,x_2) - F_1(x_3).$$

3.2 Control design

Now, the steady-state zero output manifold $\pi(w) = (\pi_1(w), \pi_2(w), \pi_3(w), \pi_4(w))^T$ is introduced. Making use of its respective regulator equations:

$$\frac{\partial \pi_1(w)}{\partial w} s(w) = \pi_3(w) \tag{7}$$
$$\frac{\partial \pi_2(w)}{\partial w} s(w) = \pi_4(w) \tag{8}$$
$$\frac{\partial \pi_3(w)}{\partial w} s(w) = b_3(\pi_2(w))p_1(\pi(w)) + b_3(\pi_2(w))c(w) \tag{9}$$
$$\frac{\partial \pi_4(w)}{\partial w} s(w) = b_4(\pi_2(w))p_2(\pi(w)) + b_4(\pi_2(w))c(w) \tag{10}$$
$$0 = \pi_2(w) - w_2 \tag{11}$$
$$\pi/2 = \pi_1(w) + \pi_2(w) \tag{12}$$

with $c(w)$ as the steady-state value for $u(t)$ that will be defined in the following lines. From equation (11) one directly obtains $\pi_2(w) = w_2$, then, replacing $\pi_2(w)$ in equation (8) yields to $\pi_4(w) = -\alpha w_1$. For calculating $\pi_1(w)$ and $\pi_3(w)$, the solution of equations (7) and (9) are needed, but in general this is a difficult task, that it is commonly solved proposing an approximated solution as in Ramos et al. (1997) and Rivera et al. (2008). Thus, one proposes the following approximated solution for $\pi_1(w)$

$$\pi_1(w) = a_0 + a_1 w_1 + a_2 w_2 + a_3 w_1^2 + a_4 w_1 w_2 + a_5 w_2^2 + a_6 w_1^3 + a_7 w_1^2 w_2 + a_8 w_1 w_2^2 + a_9 w_2^3 + O^4(\|w\|_1) \quad (13)$$

replacing (13) in (7) and chosing $\alpha = 0.3$ yields the approximated solution for $\pi_3(w)$

$$\pi_3(w) = 0.3a_1 w_2 - 0.3a_2 w_1 + 0.6a_3 w_1 w_2 + 0.3a_4 w_2^2 - 0.3a_4 w_1^2 - 0.6a_5 w_2 w_1 + 0.9a_6 w_1^2 w_2 + 0.6a_7 w_1 w_2^2 - 0.3a_7 w_1^3 + 0.3a_8 w_2^3 - 0.6a_8 w_1^2 w_2 - 0.9a_9 w_2^2 w1 + O^4(\|w\|_1). \quad (14)$$

Calculating from (10) $c(w) = -p_2(\pi(w)) - \alpha^2 w_2 / b_4(\pi_2(w))$, and using it along with (14) in equation (9) and performing a series Taylor expansion of the right hand side of this equation around the equilibrium point $(\pi/2, 0, 0, 0)^T$, then, one can find the values a_i $(i = 0, \ldots, 9)$ if the coefficients of the same monomials appearing in both side of such equation are equalized. In this case, the coefficients results as follows: $a_0 = 1.570757$, $a_1 = -0.00025944$, $a_2 = -1.001871$, $a_3 = 0.0$, $a_4 = 0.0$, $a_5 = 0.0$, $a_6 = 0.0$, $a_7 = 0.001926$, $a_8 = 0.0$, $a_9 = -0.00001588$. It is worth mentioning that there is a natural steady-state constraint (12) for the Pendubot (see Figure 1), i.e., the sum of the two angles, q_1 and q_2 equals $\pi/2$. Using such constraint one can easily calculate $\pi_{1a}(w) = \pi/2 - \pi_2(w)$, and replacing $\pi_{1a}(w)$ in equation (7) yields to $\pi_{3a}(w) = \alpha w_1$, where the sub-index a refers to an alternative manifold. This result was simulated yielding to the same results when using the approximate manifold, which is to be expected if the motion of the pendubot is forced only along the geometric constraints.

Then, the variable $z = x - \pi(w) = (z^1, z^2)^T$ is introduced, where

$$\begin{aligned} z^1 &= (z_1, z_2, z_3)^T = (x_1 - \pi_1, x_2 - \pi_2, x_3 - \pi_3)^T \\ z^2 &= z_4 = x_4 - \pi_4. \end{aligned} \quad (15)$$

Then, system (5) is represented in the new variables (15) as

$$\begin{aligned} \dot{z}_1 &= z_3 + \pi_3 - \frac{\partial \pi_1}{\partial w} s(w) \\ \dot{z}_2 &= z_4 + \pi_4 - \frac{\partial \pi_2}{\partial w} s(w) \\ \dot{z}_3 &= b_3(z_2 + \pi_2) p_1(z + \pi) + b_3(z_2 + \pi_2) u - \frac{\partial \pi_3}{\partial w} s(w) \\ \dot{z}_4 &= b_4(z_2 + \pi_2) p_2(z + \pi) + b_4(z_2 + \pi_2) u - \frac{\partial \pi_4}{\partial w} s(w) \\ e(z, w) &= z_2 + \pi_2 - w_2 \\ \dot{w} &= s(w). \end{aligned} \quad (16)$$

We now define the sliding manifold:

$$\sigma = z_4 + \Sigma_1 (z_1, z_2, z_3)^T, \; \Sigma_1 = (k_1, k_2, k_3) \quad (17)$$

and by taking its derivative along the solution of system (16) results in

$$\dot{\sigma} = \phi(w, z) + \gamma(w, z) u \quad (18)$$

where

$$\phi(w,z) = b_4(z_2+\pi_2)p_2(z+\pi) - \frac{\partial \pi_4}{\partial w}s(w) + k_1(z_3+\pi_3 - \frac{\partial \pi_1}{\partial w}s(w))$$
$$+ k_2(z_4+\pi_4 - \frac{\partial \pi_2}{\partial w}s(w)) + k_3(b_3(z_2+\pi_2)p_1(z+\pi) - \frac{\partial \pi_3}{\partial w}s(w)),$$
$$\gamma(w,z) = b_4(z_2)+\pi_2 + k_3 b_3(z_2+\pi_2),$$

moreover, one can assume that $\phi(w,z)$ is an unknown perturbation term bounded by $|\phi(w,z)| \leq \delta_\phi$ with $\delta_\phi > 0$. At this point, one can propose the super-twisting controller as follows:

$$u = (-\rho_1\sqrt{|\sigma|}sign(\sigma) + v)/\gamma(w,z)$$
$$\dot{v} = -\rho_2 sign(\sigma), \tag{19}$$

and the system (18) closed-loop by control (19) results in

$$\dot{\sigma} = -\rho_1\sqrt{|\sigma|}sign(\sigma) + v + \phi(w,z)$$
$$\dot{v} = -\rho_2 sign(\sigma), \tag{20}$$

where the controller gains ρ_1 and ρ_2 are determined in a similar fashion to the procedure outlined in the previous section.

When the sliding mode occurs, i.e., $\sigma = 0$ one can easily determine from (17) that

$$z_4 = -k_1 z_1 - k_2 z_2 - k_3 z_3$$

moreover the order of system (16) reduces in one, obtaining the sliding mode dynamic, i.e.,

$$\dot{z}^1 = \phi_{sm}(w,z) = f_1(w,z) + g_1(w,z)u_{eq}|_{z_4=-k_1z_1-k_2z_2-k_3z_3} \tag{21}$$
$$e(z,w) = z_2+\pi_2 - w_2$$
$$\dot{w} = s(w).$$

with

$$f_1 = \begin{pmatrix} z_3+\pi_3 - \frac{\partial \pi_1}{\partial w}s(w) \\ z_4+\pi_4 - \frac{\partial \pi_2}{\partial w}s(w) \\ b_3(z_2+\pi_2)p_1(z+\pi) - \frac{\partial \pi_3}{\partial w}s(w) \end{pmatrix}, \quad g_1(w,z) = \begin{pmatrix} 0 \\ 0 \\ b_3(z_2+\pi_2) \end{pmatrix},$$

and u_{eq} as the equivalent control calculated from $\dot{\sigma} = 0$ as

$$u_{eq} = -\frac{\phi(w,z)}{b_4(w,z) + k_3 b_3(w,z)}.$$

The sliding function parameters k_1, k_2 and k_3 should stabilize the sliding mode dynamic (21). For a proper choice of such constant parameters one can linearize the sliding mode dynamic

$$\dot{z}^1 = A_{sm}(\kappa)z^1$$

where $A_{sm}(\kappa) = \partial\phi_{sm}/\partial z^1 \,|_{z^1=0}$, with $\kappa = (k_1,k_2,k_3)$. In order to choose the design parameters, a polynomial with desired poles is proposed, $p_d(s) = (s-\lambda_1)(s-\lambda_2)(s-\lambda_3)$, such that, the coefficients of the characteristic equation that results from the matrix $A_{sm}(\kappa)$ are equalized with the ones related with $p_d(s)$, i. e., $det(sI - A_{sm}(\kappa)) = p_d(s)$, in such manner one can find explicit relations for κ. In this case $lim_{t\to\infty}z = 0$, accomplishing with the control objective.

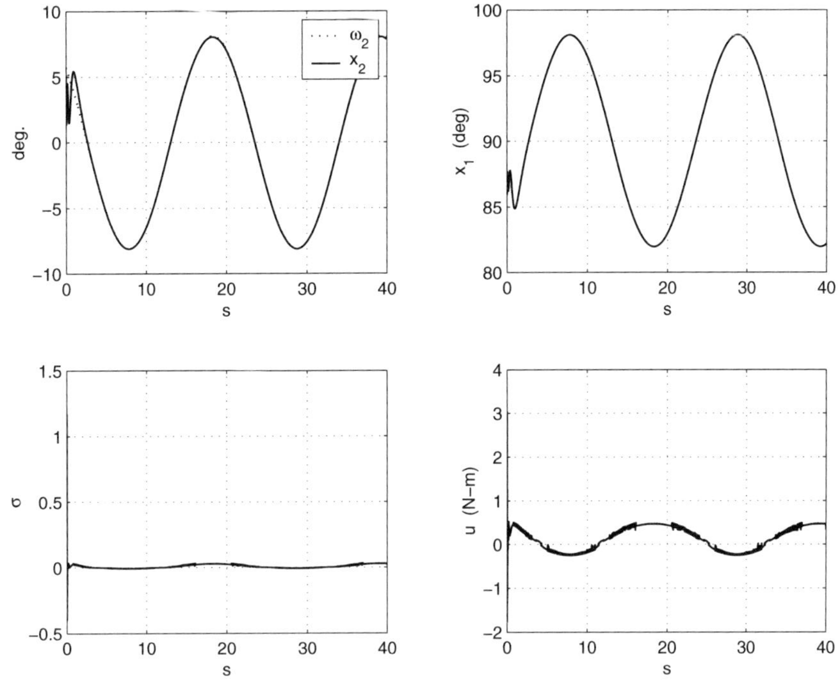

Fig. 2. a) Output tracking of the angle of the second link. b) Angle of the first link. c) Sliding surface. d) Control signal.

3.3 Simulations

In order to show the performance of the control methodology here proposed, simulations are carried out. The initial condition for the Pendubot is chosen as follows: $x_1(0) = 1.5$, $x_2(0) = 0.09$. Moreover, plant parameter variations are considered from time $t = 0$, due to possible measurement errors, therefore, the mass of the second link is considered as $m_2 = 0.5$, the moment of inertias of the first and second link are assumed to be $I_1 = 0.007$ and $I_2 = 0.0006$ respectively and the frictions of the first and second link are $\mu_1 = 0.01$ and $\mu_2 = 0.001$ respectively. The results are given in Fig. 2, where the robust performance of the super-twisting controller is put in evidence.

4. STSMC for induction motors with core loss

4.1 Induction motor model with core loss

In this section a super-twisting sliding mode controller for the induction motor is designed for copper and core loss minimization. Now we show the nonlinear affine representation for the induction motor with core loss in the stationary (α, β) reference frame taken from Rivera Dominguez et al. (2010):

$$\frac{d\omega}{dt} = \eta_0(\psi_\alpha i_\beta - \psi_\beta i_\alpha) - \frac{Tl}{J}$$

$$\frac{d\psi_\alpha}{dt} = -\eta_4\psi_\alpha - Np\omega\psi_\beta + \eta_4 L_m i_{\alpha,Lm}$$

$$
\begin{aligned}
\frac{d\psi_\beta}{dt} &= -\eta_4\psi_\beta + Np\omega\psi_\alpha + \eta_4 L_m i_{\beta,Lm} \\
\frac{di_{\alpha,Lm}}{dt} &= -(\eta_1+\eta_2)i_{\alpha,Lm} + \frac{\eta_1}{L_m}\psi_\alpha + \eta_2 i_\alpha \\
\frac{di_{\beta,Lm}}{dt} &= -(\eta_1+\eta_2)i_{\beta,Lm} + \frac{\eta_1}{L_m}\psi_\beta + \eta_2 i_\beta \\
\frac{di_\alpha}{dt} &= -(R_s\eta_3+\eta_5)i_\alpha - \eta_1\eta_3\psi_\alpha \\
&\quad + (\eta_5+\eta_1\eta_3 L_m)i_{\alpha,Lm} + \eta_3 v_\alpha \\
\frac{di_\beta}{dt} &= -(R_s\eta_3+\eta_5)i_\beta - \eta_1\eta_3\psi_\beta \\
&\quad + (\eta_5+\eta_1\eta_3 L_m)i_{\beta,Lm} + \eta_3 v_\beta
\end{aligned} \tag{22}
$$

where

$$
\eta_0 = \frac{3L_m Np}{2J(L_r-L_m)}, \quad \eta_1 = \frac{R_c}{L_r-L_m},
$$
$$
\eta_2 = \frac{R_c}{L_m}, \quad \eta_3 = \frac{1}{L_s-L_m},
$$
$$
\eta_4 = \frac{R_r}{L_r-L_m}, \quad \eta_5 = \frac{R_c}{L_s-L_m}.
$$

with ω as the rotor velocity, v_α, v_β are the stator voltages, i_α, i_β are the stator currents, $i_{\alpha,Lm}, i_{\beta,Lm}$ are the magnetization currents and ψ_α, ψ_β are the rotor fluxes, with N_p as the number of pole pairs, R_s, R_r and R_c as the stator, rotor and core resistances respectively, Ll_s, Ll_r and L_m as the stator leakage, rotor leakage and magnetizing inductances respectively.

4.2 Transformation to the (d,q) rotating frame

Now, the induction motor model (22) will be transformed to the well known (d,q) reference frame by means of the following change of coordinates

$$
\begin{bmatrix} i_d \\ i_q \end{bmatrix} = e^{-J\theta_\psi}\begin{bmatrix} i_\alpha \\ i_\beta \end{bmatrix}, \quad \begin{bmatrix} \psi_d \\ \psi_q \end{bmatrix} = e^{-J\theta_\psi}\begin{bmatrix} \psi_\alpha \\ \psi_\beta \end{bmatrix}
$$
$$
\begin{bmatrix} i_{dL_m} \\ i_{qL_m} \end{bmatrix} = e^{-J\theta_\psi}\begin{bmatrix} i_{\alpha L_m} \\ i_{\beta L_m} \end{bmatrix}, \quad \begin{bmatrix} v_\alpha \\ v_\beta \end{bmatrix} = e^{J\theta_\psi}\begin{bmatrix} v_d \\ v_q \end{bmatrix}
$$

where

$$
e^{-J\theta_\psi} = \begin{bmatrix} \cos\theta_\psi & \sin\theta_\psi \\ -\sin\theta_\psi & \cos\theta_\psi \end{bmatrix}
$$

with

$$
\theta_\psi = \arctan\left(\frac{\psi_\beta}{\psi_\alpha}\right).
$$

The field oriented or (d, q) model of the induction motor with core loss is now shown

$$
\begin{aligned}
\dot{\theta}_{\psi} &= n_p\omega + \frac{\eta_4 L_m i_{qL_m}}{\psi_d} \\
\frac{d\omega}{dt} &= \eta_0 i_q \psi_d - \frac{Tl}{J} \\
\frac{d\psi_d}{dt} &= -\eta_4\psi_d + \eta_4 L_m i_{dL_m} \\
\frac{di_{dL_m}}{dt} &= -\left(\eta_1+\eta_2\right) i_{dL_m} + \frac{\eta_1\psi_d}{L_m} + \eta_2 i_d + i_{qL_m}\dot{\theta} \\
\frac{di_{qL_m}}{dt} &= -\left(\eta_1+\eta_2\right) i_{qL_m} + \eta_2 i_q + i_{dL_m}\dot{\theta} \\
\frac{di_d}{dt} &= -\left(R_s\eta_3+\eta_5\right) i_d - \eta_1\eta_3\psi_d + \left(\eta_5+\eta_1\eta_3 L_m\right) i_{dL_m} + \eta_3 v_d + i_q\dot{\theta} \\
\frac{di_q}{dt} &= -\left(R_s\eta_3+\eta_5\right) i_q + \left(\eta_5+\eta_1\eta_3 L_m\right) i_{qL_m} + \eta_3 v_q - i_d\dot{\theta}
\end{aligned} \tag{23}
$$

The *control problem* is to force the rotor angular velocity ω and the square of the rotor flux modulus $\psi_m = \psi_\alpha^2 + \psi_\beta^2$ to track some desired references ω_r and $\psi_{m,r}$, ensuring at the same time load torque rejection. The control problem will be solved in a subsequent subsection by means of a super-twisting sliding mode controller.

4.2.1 Optimal rotor flux calculation

The copper and core losses are obtained by the corresponding resistances and currents. Therefore, the power lost in copper and core are expressed as follows:

$$P_L = \frac{3}{2}R_s\left(i_{d,s}^2 + i_{q,s}^2\right) + \frac{3}{2}R_r\left(i_{d,r}^2 + i_{q,r}^2\right) + \frac{3}{2}R_c\left(i_{d,Rc}^2 + i_{q,Rc}^2\right)$$

where $i_{d,r}$ and $i_{q,r}$ are the currents flowing through the rotor, $i_{d,Rc}$ and $i_{q,Rc}$ are the currents flowing through the resistance that represents the core. Since P_L is a positive-definite function can be considered as a cost function and then to be minimized with any desired variables, in this case the most suitable is the rotor flux, i. e.,

$$\frac{\partial P_L}{\partial \psi_d} = 0.$$

The resulting rotor flux component is given of the following form

$$\psi_{d,o} = \left(\frac{R_r L_m}{R_r + R_c} + \frac{R_c Lr}{R_r + R_c}\right) i_{dL_m} - \frac{R_c\left(Lr - L_m\right) i_d}{R_r + Rc}$$

4.3 Control design

In order to solve the posed control problem using the super-twisting sliding mode approach, we first derive the expression of the tracking error dynamics $z_1 = \omega - \omega_r$, $z_2 = \psi_m - \psi_{d,o}$ which are the output which we want to force to zero. The error tracking dynamic for the rotor velocity results as

$$\dot{z_1} = \eta_0\psi_d i_q - \frac{Tl}{J} - \dot{\omega}_r. \tag{24}$$

Proposing a desired dynamic for z_1 of the following form

$$\dot{z_1} = \eta_0 \psi_d i_q - \frac{Tl}{J} - \dot{\omega}_r = k_1 z_1$$

one can calculate i_q as a reference signal, i. e., i_{qr}

$$i_{qr} = \frac{\left(k_1 z_1 + \frac{Tl}{J} + \dot{\omega}_r\right)}{\eta_0 \psi_d} \tag{25}$$

in order to force the current component i_q to track its reference current, one defines the following trackng error

$$\xi_2 = i_q - i_{qr} \tag{26}$$

and tanking the derivative of this error

$$\dot{\xi}_2 = \phi_q + \eta_3 v_q. \tag{27}$$

where

$$\phi_q = -\left(R_s \eta_3 + \eta_5\right) i_q + \left(\eta_5 + \eta_1 \eta_3 L_m\right) i_{qL_m} - i_d \dot{\theta_\psi} - \dot{i_q r}$$

is considered to be a bounded unknown perturbation term, i.e., $|\phi_q| \leq \delta_q$ with $\delta_q > 0$. The control law is proposed of the following form:

$$\begin{aligned} v_q &= (-\rho_{q,1}\sqrt{|\xi_2|}sign(\xi_2) + \nu_q)/\eta_3 \\ \dot{\nu}_q &= -\rho_{q,2} sign(\xi_2), \end{aligned} \tag{28}$$

and the system (27) closed-loop by control (28) results in

$$\begin{aligned} \dot{\xi}_2 &= -\rho_{q,1}\sqrt{|\xi_2|}sign(\xi_2) + \nu_q + \phi_q \\ \dot{\nu}_q &= -\rho_{q,2} sign(\xi_2), \end{aligned} \tag{29}$$

where the controller gains $\rho_{q,1}$ and $\rho_{q,2}$ are determined in a similar fashion to procedure outlined in the previous section. Now, from (26) one can write i_q as follows

$$i_q = \xi_2 + i_{qr}$$

and when substituting it along with (25) in (24) yields to

$$\dot{z_1} = k_1 z_1 + \eta_0 \psi_d \xi_2.$$

Finally, collecting the equations

$$\begin{aligned} \dot{z_1} &= k_1 z_1 + \eta_0 \psi_d \xi_2 \\ \dot{\xi}_2 &= -\rho_{q,1}\sqrt{|\xi_2|}sign(\xi_2) + \nu_q + \phi_q \\ \dot{\nu}_q &= -\rho_{q,2} sign(\xi_2). \end{aligned}$$

When the sliding mode occurs, i.e., $\xi_2 = 0$, the sliding mode dynamic results as:

$$\dot{z_1} = k_1 z_1$$

and that with a proper choice of k_1, one can lead to $z_1 = 0$.
Let us consider the second output z_2, where its dynamic results as follows:

$$\dot{z_2} = -\eta_4\psi_d + \eta_4 L_m i_{dL_m} - \dot{\psi_{dr}}, \tag{30}$$

note that the relative degree for z_2 is three, therefore in order to cope with the relative degree of z_1, one proposes the following desired dynamic for z_2

$$\dot{z_2} = -\eta_4\psi_d + \eta_4 L_m i_{dL_m} - \dot{\psi_{dr}} = k_2 z_2 + z_3$$

where the new variable z_3 is calculated as:

$$z_3 = -\eta_4\psi_d + \eta_4 L_m i_{dL_m} - \dot{\psi_{dr}} - k_2 z_2.$$

Taking the derivative of z_3 and assigning a desired dynamic

$$\begin{aligned}\dot{z_3} = {} & i_d L_m \left(-\eta_4{}^2 L_m - \eta_4\eta_1 L_m - \eta_4\eta_2 L_m - k_2\eta_4 L_m\right) + \psi_d\left(\eta_4{}^2 + \eta_4\eta_1 + k_2\eta_4\right) \\ & + \eta_4\eta_2 L_m i_d + \eta_4 L_m i_{qLm}\dot{\theta}_\psi - \ddot{\psi_{dr}} + k_2\dot{\psi_{dr}} = k_3 z_3 \end{aligned} \tag{31}$$

then, one can calculate i_d as a reference current, i. e., i_{dr}

$$\begin{aligned} i_{dr} = {} & \frac{k_3 z_3 - i_d L_m\left(-\eta_4{}^2 L_m - \eta_4\eta_1 L_m - \eta_4\eta_2 L_m - k_2\eta_4 L_m\right) - \psi_d\left(\eta_4{}^2 + \eta_4\eta_1 + k_2\eta_4\right)}{\eta_4\eta_2 L_m} \\ & + \frac{-\eta_4 L_m i_{qLm}\dot{\theta}_\psi + \ddot{\psi_{dr}} - k_2\dot{\psi_{dr}}}{\eta_4\eta_2 L_m}. \end{aligned} \tag{32}$$

Defining the tracking error for the current d component

$$\xi_1 = i_d - i_{dr} \tag{33}$$

and by taking its derivative, i. e.,

$$\dot{\xi_1} = \phi_d + \eta_3 v_d \tag{34}$$

where

$$\phi_d = -\left(R_s\eta_3 + \eta_5\right) i_d - \eta_1\eta_3\psi_d + \left(\eta_5 + \eta_1\eta_3 L_m\right) i_{dL_m} + i_q\dot{\theta}_\psi - \dot{i_d}r$$

is considered to be a bounded unknown perturbation term, i.e., $|\phi_d| \leq \delta_d$ with $\delta_d > 0$. The control law is proposed of the following form:

$$\begin{aligned} v_d &= (-\rho_{d,1}\sqrt{|\xi_1|}sign(\xi_1) + \nu_d)/\eta_3 \\ \dot{\nu}_d &= -\rho_{d,2} sign(\xi_1), \end{aligned} \tag{35}$$

and the system (34) closed-loop by control (35) results in

$$\begin{aligned} \dot{\xi_1} &= -\rho_{d,1}\sqrt{|\xi_1|}sign(\xi_1) + \nu_d + \phi_d \\ \dot{\nu}_d &= -\rho_{d,2} sign(\xi_1), \end{aligned} \tag{36}$$

where the controller gains $\rho_{d,1}$ and $\rho_{d,2}$ are determined in a similar fashion to procedure outlined in the previous section. Now, from (33) one can write

$$i_d = \xi_1 + i_{dr}$$

and replacing it in (31) along with (32) yields to

$$\dot{z_3} = k_3 z_3 + \eta_4 \eta_2 L_m \xi_1.$$

Finally, collecting the equations

$$\begin{aligned}
\dot{z_2} &= k_2 z_2 + z_3 \\
\dot{z_3} &= k_3 z_3 + \eta_4 \eta_2 L_m \xi_1 \\
\dot{\xi_1} &= -\rho_{d,1}\sqrt{|\xi_1|}sign(\xi_1) + \nu_d + \phi_d \\
\dot{\nu}_d &= -\rho_{d,2} sign(\xi_1).
\end{aligned}$$

When the sliding mode occurs, i.e., $\xi_1 = 0$, the closed-loop channel reduces its order:

$$\dot{z_2} = k_2 z_2 + z_3$$

$$\dot{z_3} = k_3 z_3$$

one can see that the determination of k_2, k_3, is easily achieved in order to lead to $z_2 = 0$.

4.4 Observer design

The first problem with the control strategy here developed is that the measurements of the rotor fluxes and magnetization currents are not possible. This problem is solved using an sliding mode observer. The second problem concerns the estimation of the load torque, where a classical Luemberger observer is designed.

The proposed sliding mode observer for rotor fluxes and magnetization currents is proposed based on (22) as follows:

$$\begin{aligned}
\frac{d\hat{\psi}_\alpha}{dt} &= -\eta_4 \hat{\psi}_\alpha - Np\omega\hat{\psi}_\beta + \eta_4 L_m \hat{i}_{\alpha,Lm} + \rho_\alpha \nu_\alpha \\
\frac{d\hat{\psi}_\beta}{dt} &= -\eta_4 \hat{\psi}_\beta + Np\omega\hat{\psi}_\alpha + \eta_4 L_m \hat{i}_{\beta,Lm} + \rho_\beta \nu_\beta \\
\frac{d\hat{i}_{\alpha,Lm}}{dt} &= -(\eta_1 + \eta_2)\hat{i}_{\alpha,Lm} + \frac{\eta_1}{L_m}\hat{\psi}_\alpha + \eta_2 i_\alpha + \lambda_\alpha \nu_\alpha \\
\frac{d\hat{i}_{\beta,Lm}}{dt} &= -(\eta_1 + \eta_2)\hat{i}_{\beta,Lm} + \frac{\eta_1}{L_m}\hat{\psi}_\beta + \eta_2 i_\beta + \lambda_\beta \nu_\beta \\
\frac{d\hat{i}_\alpha}{dt} &= -(R_s\eta_3 + \eta_5)\hat{i}_\alpha - \eta_1\eta_3\hat{\psi}_\alpha \\
&\quad + (\eta_5 + \eta_1\eta_3 L_m)\hat{i}_{\alpha,Lm} + \eta_3 v_\alpha + \nu_\alpha \\
\frac{d\hat{i}_\beta}{dt} &= -(R_s\eta_3 + \eta_5)\hat{i}_\beta - \eta_1\eta_3\hat{\psi}_\beta \\
&\quad + (\eta_5 + \eta_1\eta_3 L_m)\hat{i}_{\beta,Lm} + \eta_3 v_\beta + \nu_\beta
\end{aligned}$$

where ρ_α, ρ_β, λ_α and λ_β are the observer design parameters, and ν_α and ν_β are the injected inputs to the observer that will be defined in the following lines. Now one defines the estimation errors, $\tilde{\psi}_\alpha = \psi_\alpha - \hat{\psi}_\alpha$, $\tilde{\psi}_\beta = \psi_\beta - \hat{\psi}_\beta$, $\tilde{i}_{\alpha,Lm} = i_{\alpha,Lm} - \hat{i}_{\alpha,Lm}$, $\tilde{i}_{\beta,Lm} = i_{\beta,Lm} - \hat{i}_{\beta,Lm}$, $\tilde{i}_\alpha = i_\alpha - \hat{i}_\alpha$ and $\tilde{i}_\beta = i_\beta - \hat{i}_\beta$, whose dynamics can be expressed as:

$$\begin{aligned}
\frac{d\tilde{\psi}_\alpha}{dt} &= -\eta_4\tilde{\psi}_\alpha - Np\omega\tilde{\psi}_\beta + \eta_4 L_m\tilde{i}_{\alpha,Lm} - \rho_\alpha\nu_\alpha \\
\frac{d\tilde{\psi}_\beta}{dt} &= -\eta_4\tilde{\psi}_\beta + Np\omega\tilde{\psi}_\alpha + \eta_4 L_m\tilde{i}_{\beta,Lm} - \rho_\beta\nu_\beta \\
\frac{d\tilde{i}_{\alpha,Lm}}{dt} &= -(\eta_1+\eta_2)\tilde{i}_{\alpha,Lm} + \frac{\eta_1}{L_m}\tilde{\psi}_\alpha - \lambda_\alpha\nu_\alpha \\
\frac{d\tilde{i}_{\beta,Lm}}{dt} &= -(\eta_1+\eta_2)\tilde{i}_{\beta,Lm} + \frac{\eta_1}{L_m}\tilde{\psi}_\beta - \lambda_\beta\nu_\beta \\
\frac{d\tilde{i}_\alpha}{dt} &= -(R_s\eta_3+\eta_5)\tilde{i}_\alpha - \eta_1\eta_3\tilde{\psi}_\alpha \\
&+ (\eta_5+\eta_1\eta_3 L_m)\tilde{i}_{\alpha,Lm} - \nu_\alpha \\
\frac{d\tilde{i}_\beta}{dt} &= -(R_s\eta_3+\eta_5)\tilde{i}_\beta - \eta_1\eta_3\tilde{\psi}_\beta \\
&+ (\eta_5+\eta_1\eta_3 L_m)\tilde{i}_{\beta,Lm} - \nu_\beta.
\end{aligned} \tag{37}$$

Since the stator currents are measurable variables, one can choose the observer injection as $\nu_\alpha = l_\alpha sign(\tilde{i}_\alpha)$ and $\nu_\beta = l_\beta sign(\tilde{i}_\beta)$. From the derivative of the following Lyapunov candidate function $\mathcal{V}_o = \frac{1}{2}(\tilde{i}_\alpha^2 + \tilde{i}_\beta^2)$ along the trajectories of (37), one can easily determine the following bounds, $l_\alpha > |\eta_1\eta_3\tilde{\psi}_\alpha - (\eta_5+\eta_1\eta_3 L_m)\tilde{i}_{\alpha,Lm}|$ and $l_\beta > |\eta_1\eta_3\tilde{\psi}_\beta - (\eta_5+\eta_1\eta_3 L_m)\tilde{i}_{\beta,Lm}|$ that guarantees the convergence of $\tilde{i}_\alpha$ and $\tilde{i}_\beta$ towards zero in finite time. When the sliding mode occurs, i. e., $\tilde{i}_\alpha = \tilde{i}_\beta = 0$ one can calculate the equivalent control for the injected signals from $\dot{\tilde{i}}_\alpha = 0$ and $\dot{\tilde{i}}_\beta = 0$ as $\nu_{\alpha,eq} = -\eta_1\eta_3\tilde{\psi}_\alpha + (\eta_5+\eta_1\eta_3 L_m)\tilde{i}_{\alpha,Lm}$, $\nu_{\beta,eq} = -\eta_1\eta_3\tilde{\psi}_\beta + (\eta_5 + \eta_1\eta_3 L_m)\tilde{i}_{\beta,Lm}$, then, the sliding mode dynamic can be obtained by replacing the calculated equivalent controls, resulting in a linear time-variant dynamic system, $\dot{\epsilon} = A_o(\omega)\epsilon$, where $\epsilon = \left(\tilde{\psi}_\alpha\ \tilde{\psi}_\beta\ \tilde{i}_{\alpha,Lm}\ \tilde{i}_{\beta,Lm}\right)^T$,

$$A_o = \begin{pmatrix} A_{o,11} & A_{o,12} \\ A_{o,21} & A_{o,22} \end{pmatrix}$$

with

$$\begin{aligned}
A_{o,11} &= \begin{pmatrix} \rho_\alpha\eta_1\eta_3 - \eta_4 & -Np\omega \\ Np\omega & \rho_\beta\eta_1\eta_3 - \eta_4 \end{pmatrix}, \\
A_{o,12} &= \begin{pmatrix} \eta_4 L_m - \rho_\alpha\gamma & 0 \\ 0 & \eta_4 L_m - \rho_\beta\gamma \end{pmatrix}, \\
A_{o,21} &= \begin{pmatrix} \frac{\eta_1}{L_m} + \lambda_\alpha\eta_1\eta_3 & 0 \\ 0 & \frac{\eta_1}{L_m} + \lambda_\beta\eta_1\eta_3 \end{pmatrix}, \\
A_{o,22} &= \begin{pmatrix} -\eta_1 - \eta_2 - \lambda_\alpha\gamma & 0 \\ 0 & -\eta_1 - \eta_2 - \lambda_\beta\gamma \end{pmatrix}, \\
\gamma &= \eta_5 + \eta_1\eta_3 L_m.
\end{aligned}$$

In order to choose the design parameters, a polynomial with desired poles is proposed, $p_d(s) = (s - p_1)(s - p_2)(s - p_3)(s - p_4)$, such that, the coefficients of the characteristic equation that results from the matrix A_o are equalized with the ones related with $p_d(s)$, i. e., $det(sI - A_o) = p_d(s)$, moreover, one can assume that the rotor velocity is constant, therefore the design parameters are easily determined. This will guarantee that $lim_{t\to\infty}\epsilon(t) = 0$.
For the load torque estimation we consider that it is slowly varying, so one can assume it is constant, i. e., $\dot{Tl} = 0$. This fact can be valid since the electric dynamic of the motor is faster than the mechanical one. Therefore, one proposes the following observer based on rotor velocity and stator current measurements

$$\frac{d\hat{\omega}}{dt} = \eta_0(\hat{\psi}_\alpha i_\beta - \hat{\psi}_\beta i_\alpha) - \frac{\hat{Tl}}{J} + l_1(\omega - \hat{\omega})$$
$$\frac{d\hat{Tl}}{dt} = l_2(\omega - \hat{\omega}).$$

Defining the estimation errors as $e_\omega = \omega - \hat{\omega}$ and $e_{Tl} = Tl - \hat{Tl}$ one can determine the estimation error dynamic

$$\begin{pmatrix} \dot{e}_\omega \\ \dot{e}_{Tl} \end{pmatrix} = \begin{pmatrix} -l_1 & -\frac{1}{J} \\ -l_2 & 0 \end{pmatrix} \begin{pmatrix} e_\omega \\ e_{Tl} \end{pmatrix} + \eta_0 \begin{pmatrix} \tilde{\psi}_\alpha i_\beta - \tilde{\psi}_\beta i_\alpha \\ 0 \end{pmatrix}. \quad (38)$$

When the estimation errors for the rotor fluxes in (37) are zero, equation (38) reduces to

$$\begin{pmatrix} \dot{e}_\omega \\ \dot{e}_{Tl} \end{pmatrix} = \begin{pmatrix} -l_1 & -\frac{1}{J} \\ -l_2 & 0 \end{pmatrix} \begin{pmatrix} e_\omega \\ e_{Tl} \end{pmatrix} \quad (39)$$

where l_1 and l_2 can easily be determined in order to yield to $lim_{t\to\infty}e_\omega(t) = 0$ and $lim_{t\to\infty}e_{Tl}(t) = 0$.

4.5 Simulations

In this section we verify the performance of the proposed control scheme by means of numeric simulations.
We consider an induction motor with the following nominal parameters: $R_r = 10.1\ \Omega$, $R_s = 14\ \Omega$, $R_c = 1\ k\Omega$, $L_s = 400 \times 10^{-3}$ H, $L_r = 412.8 \times 10^{-3}$ H, $L_m = 377 \times 10^{-3}$ H, $J = 0.01$ Kg m^2.
Hence, $\eta_1 = 27,932.96\ \Omega/H$, $\eta_2 = 2,652.51\ \Omega/H$, $\eta_3 = 43.47\ H^{-1}$, $\eta_4 = 282.12\ \Omega/H$ and $\eta_5 = 43,478.26\ \Omega/H$.
A load torque Tl of 5 Nm, with decrements of 1 Nm and 2 Nm at 8 s and 12 s respectively, has been considered in simulations. The reference velocity signal increases from 0 to 188.5 rad/s in the first 5 s and then remains constant, while the rotor flux modulus reference signal is directly taken from the calculated optimal flux.
A good tracking performance by the proposed controller can be appreciated in Figures 3 and 4. In Fig. 5 the power lost in copper and core is shown in the case of using the optimal flux modulus and the predicted open-loop steady state values in the cases of considering or not the core, this is a common practice when dealing with the control of the rotor flux in induction motors. From this figure one can observe a low power lost in copper and core when using the optimal flux, also one can note in Fig. 4 that the less is the load torque the lower is the flux level and as a consequence the power lost is reduced.

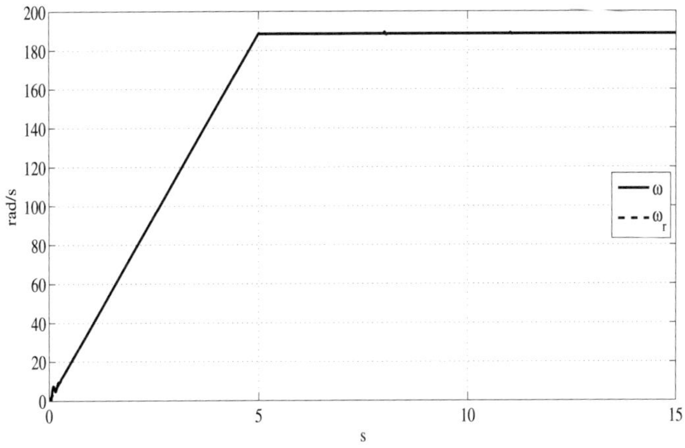

Fig. 3. Closed-loop velocity tracking of the proposed controller.

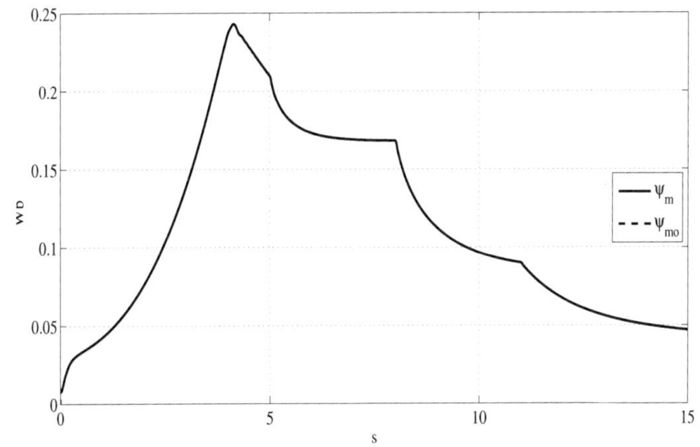

Fig. 4. Closed-loop optimal flux tracking of the proposed controller.

5. Conclusions

In this chapter the super-twisting algorithm and its application to motion control systems is shown. An under-actuated robotic system known as the Pendubot was closed-loop with a super-twisting controller. The procedure can easily be generalized to such type of motion systems. For that, one must consider the following generic steps: find the steady state for all states, then, based on the dynamic of the steady state errors one proposes an sliding function that linearly stabilizes the sliding mode dynamic. For the induction motor motion control, the (d, q) reference frame allows to decouple the control problem simplifying the control design. In each channel, a cascade strategy of defining first the output tracking error and then a desired current that shapes the dynamic of such output. Therefore, the sliding surface is simply chosen as a deviation of the current and its desired current. This strategy can be applied to all type of electric motors. In both scenarios, the super-twisting algorithm facilitates the motion control design and eliminates the chattering phenomenon at the outputs.

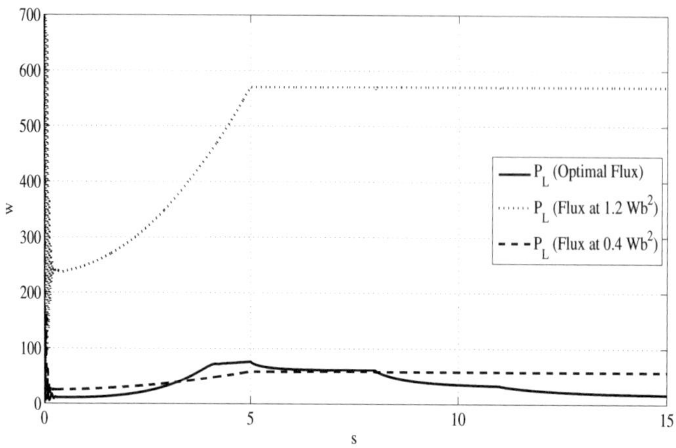

Fig. 5. Comparison of the power lost in copper and core using the optimal flux modulus, and the steady-state open-loop values for the flux modulus predicted by the classical fifth-order model and the seventh-order model here presented.

6. References

Levant, A. (1993). Sliding order and sliding accuracy in sliding mode control, *Int. J. Control* 58(6): 1247–1263.

Levant, A. (2005). Quasi-continuous high-order sliding-mode controllers, *IEEE Transactions on Automatic Control* 50(11): 1812–1816.

Levant, A. & Alelishvili, L. (2007). Integral high-order sliding modes, *IEEE Transactions on Automatic Control* 52(7): 1278–1282.

Levi, E., Boglietti, A. & Lazzar, M. (1995). Performance deterioration in indirect vector controlled induction motor drives due to iron losses, *Proc. Power Electronics Specialists Conf.*

Moreno, J. A. & Osorio, M. A. (2008). A lyapunov approach to second-order sliding mode controllers and observers, *Proceedings of the 47th IEEE Conference on Decision and Control*.

Ramos, L. E., Castillo-Toledo, B. & Alvarez, J. (1997). Nonlinear regulation of an underactuated system, *International Conference on Robotics and Automation*.

Rivera Dominguez, J., Mora-Soto, C., Ortega, S., Raygoza, J. J. & De La Mora, A. (2010). Super-twisting control of induction motors with core loss, *Variable Structure Systems (VSS), 2010 11th International Workshop on*, pp. 428 –433.

Rivera, J., Loukianov, A. & Castillo-Toledo, B. (2008). Discontinuous output regulation of the pendubot, *Proceedings of the 17th world congress The international federation of automatic control*.

Spong, M. W. & Vidyasagar, M. (1989). *Robot Dynamics and Control*, John Wiley and Sons, Inc., New York.

Utkin, V., Guldner, J. & Shi, . (1999). *Sliding mode control in electromechanical systems*, CRC Press.

Utkin, V. I. (1993). Sliding mode control design principles and applications to electric drives, *IEEE Trans. Ind. Electron.* 40(1): 23Ð36.

14

Non-Adaptive Sliding Mode Controllers in Terms of Inertial Quasi-Velocities

Przemyslaw Herman and Krzysztof Kozlowski
Poznan University of Technology
Poland

1. Introduction

Trajectory control problem arises if the manipulator is required to follow a desired trajectory. In the robotic literature mainly two approaches are used: computer torque (inverse dynamic control) and sliding mode control Sciavicco & Siciliano (1996); Slotine & Li (1991). The system under inverse dynamics controller is linear and decoupled with respect to the newly obtained input. In robotics literature very popular is the sliding mode method described by Slotine & Li (1987; 1991). The approach differs from the previous one because even if the parameters are exactly known, the manipulator equations of motion are not linearized by the control law. The sliding mode control strategies are used in the manipulator joint space as well as in its operational space Sciavicco & Siciliano (1996); Slotine & Li (1987; 1991). From the practical point of view to track the position of the end-effector of the manipulator is more convenient than the joint position tracking because the task is realized directly. The motion control problem in the manipulator joint space and the operational space is investigated also in newer references Kelly & Moreno (2005); Moreno & Kelly (2003); Moreno et al. (2003); Moreno-Valenzuela & Kelly (2006). Sometimes also a friction model is taken into account, e.g. Moreno et al. (2003); Moreno-Valenzuela & Kelly (2006). One of known applications of the sliding mode approach allows one to control a shape Mochiyama et al. (1999). In order to design various versions of control laws strict Lyapunov functions for a class of global regulators for robot manipulators are introduced Santibanez & Kelly (1997); Spong (1992) or in terms of the IQV also in Herman (2009b).

Classical description leads to obtaining second-order nonlinear differential equations of motion. The equations involve both generalized position vector and velocity vector which represent a joint space of the manipulator. However, for control purposes first-order equations of motion with diagonal mass matrix seem more convenient than the second-order equations. It is possible to consider the dynamics of mechanical systems using quasi-velocities and differential geometry Kwatny & Blankenship (2000). The obtained first-order equations of motion are the Poincaré's form of the Lagrange's equations. One of useful solutions which leads to the diagonal or the unit inertia matrix is introducing so called inertial quasi-velocities (IQV). There exist several methods which enable such decomposition (e.g.Hurtado (2004); Jain & Rodriguez (1995); Junkins & Schaub (1997); Loduha & Ravani (1995); Sovinsky et al. (2005)). The method presented in Hurtado (2004) is associated with the Cholesky decomposition Sovinsky et al. (2005). In the method described by Jain & Rodriguez (1995) the normalized quasi-velocities (NQV) and the unnormalized quasi-velocities (UQV) were introduced. The next method Junkins & Schaub (1997) is based on the eigenvalues and eigenvectors calculation of the inertia matrix. The Loduha and Ravani offer the generalized velocity components (GVC) which can be related to the modified Kane's equations given e.g. in Kane & Levinson (1983). Finally, also the normalized generalized velocity components (NGVC) are considered in references Herman (2005b; 2006). The NGVC are a useful form of the GVC.

The key idea of the paper is a survey of selected non-adaptive sliding mode controllers expressed in terms of the inertial quasi-velocities (IQV). The IQV mean that the quasi-velocities contain the kinematic and dynamic parameters of a rigid manipulator as well as its geometrical dimensions. In spite that there exist several IQV, only some of them are considered here, namely: the GVC described in Loduha & Ravani (1995), the NQV given in Jain & Rodriguez (1995), and the NGVC presented in Herman (2005b; 2006). It is because these kind of IQV very well explain the idea of non-adaptive sliding mode control in terms of the quasi-velocities. The second aim is to point at some advantages which offers the sliding mode control scheme in using the IQV. It is also shown which benefits are observable if the system under the proposed control law is considered. One of advantages arises from the fact that the IQV are decoupled in the kinetic energy sense and they lead to decoupling of the inertia matrix of the manipulator. Consequently, the inertia which takes into account also dynamical coupling can be determined. Moreover, some disadvantages of the IQV control approach are indicated. The third objective is to show that the sliding mode controllers are realized both in the manipulator joint space and the operational space. Additionally, it is possible to take into consideration disturbances (here represented by a viscous damping function) which, in prospect, it allows one to extend the results for use of various friction models.

The paper is organized as follows. Section 2 gives diagonalized equations of motion in terms of the IQV. In Section 3 the sliding mode controllers in the joint space of a manipulator as well as in its operational space are presented. Simulation results comparing performance between the new control schemes and the classical controllers for two models of rigid serial manipulator, namely 3 D.O.F. spatial DDArm robot and Yasukawa-like robot are contained in Section 4. The last section offers conclusions and future research.

2. Dynamics in terms of inertial quasi-velocities

2.1 Notation

$\theta, \dot{\theta}, \ddot{\theta} \in R^{\mathcal{N}}$ - vectors of generalized positions, velocities, and accelerations, respectively,
$\mathcal{N}$ - number of degrees of freedom,
$M(\theta) \in R^{\mathcal{N}\times\mathcal{N}}$ - system inertia matrix,
$C(\theta,\dot{\theta})\dot{\theta} \in R^{\mathcal{N}}$ - vector of Coriolis and centrifugal forces in classical equations of motion,
$G(\theta) \in R^{\mathcal{N}}$ - vector of gravitational forces in classical equations of motion,
$f(\dot{\theta}) = F\dot{\theta} \in R^{\mathcal{N}}$ - vector of forces due to friction (viscous damping) which depends on the joint velocity vector $\dot{\theta}$ where $F = diag\{F_1, \ldots, F_{\mathcal{N}}\}$ is a positive definite diagonal matrix containing the damping coefficients for all joints,
$Q \in R^{\mathcal{N}}$ - vector of generalized forces,
$N \in R^{\mathcal{N}\times\mathcal{N}}$ - diagonal system inertia matrix in terms of the GVC,
$u, \dot{u} \in R^{\mathcal{N}}$ - vector of generalized velocity components and its time derivative, respectively,
$\Upsilon = \Upsilon(\theta) \in R^{\mathcal{N}\times\mathcal{N}}$ - upper triangular transformation matrix between the velocity vector $\dot{\theta}$ and the generalized velocity components vector u Loduha & Ravani (1995),
$\dot{\Upsilon}(\theta) \in R^{\mathcal{N}\times\mathcal{N}}$ - time derivative of the matrix $\Upsilon(\theta)$,
$C_u(\theta, u)u \in R^{\mathcal{N}}$ - vector of Coriolis and centrifugal forces in terms of the GVC,
$G_u(\theta) \in R^{\mathcal{N}}$ - vector of gravitational forces in terms of the GVC,

$f_u(\theta, \dot{\theta}) \in R^{\mathcal{N}}$ - vector of friction damping forces in terms of the GVC,
$\pi \in R^{\mathcal{N}}$ - vector of quasi-forces in terms of the GVC,
$\vartheta, \dot{\vartheta} \in R^{\mathcal{N}}$ - vector of quasi-velocities, i.e. the NGVC and its time derivative, respectively,
$\Phi = \Phi(\theta) \in R^{\mathcal{N} \times \mathcal{N}}$ - upper triangular velocity transformation matrix in terms of the NGVC,
$\dot{\Phi}(\theta) \in R^{\mathcal{N} \times \mathcal{N}}$ - time derivative of the matrix $\Phi(\theta)$,
$C_\vartheta(\theta, \vartheta)\vartheta \in R^{\mathcal{N}}$ - vector of Coriolis and centrifugal forces in terms of the NGVC,
$G_\vartheta(\theta) \in R^{\mathcal{N}}$ - vector of gravitational forces in terms of the NGVC,
$f_\vartheta(\theta, \dot{\theta}) \in R^{\mathcal{N}}$ - vector of friction damping forces in terms of the NGVC,
$\varpi \in R^{\mathcal{N}}$ - vector of quasi-forces in terms of NGVC,
$\nu \in R^{\mathcal{N}}$ vector of normalized quasi-velocities,
$C_\nu(\theta, \nu)\nu \in R^{\mathcal{N}}$ vector of Coriolis and centrifugal forces in the NQV,
$G_\nu(\theta) \in R^{\mathcal{N}}$ vector of gravitational forces in the NQV,
$m = m(\theta) \in R^{\mathcal{N} \times \mathcal{N}}$ spatial operator (matrix) - "square root" of the inertia matrix $M(\theta)$,
$\dot{m}(\theta) \in R^{\mathcal{N} \times \mathcal{N}}$ time derivative of the matrix $m(\theta)$,
$\epsilon \in R^{\mathcal{N}}$ vector of normalized quasi-forces (in terms of the NQV),
$D \in R^{\mathcal{N} \times \mathcal{N}}$ articulated inertia about joint axes matrix Jain & Rodriguez (1995),
$(.)^T$ - transpose operation.

2.2 Equations of motion

Recall that the classical equations of motion for a manipulator can be written in the following form Sciavicco & Siciliano (1996); Slotine & Li (1987; 1991):

$$M(\theta)\ddot{\theta} + C(\theta, \dot{\theta})\dot{\theta} + G(\theta) = Q. \tag{1}$$

In terms of the IQV the equations of motion depend on the used decomposition of the inertia matrix $M(\theta)$. The first of here considered decomposition methods is based on the generalized velocity components (GVC) Loduha & Ravani (1995). In this method $M(\theta) = \Upsilon^{-T} N \Upsilon^{-1}$. The equations were proposed by Loduha & Ravani (1995).

$$N\dot{u} + C_u(\theta, u)u + G_u(\theta) = \pi, \tag{2}$$
$$\dot{\theta} = \Upsilon(\theta)u, \tag{3}$$

where matrices and vectors are given as follows:

$$N = \Upsilon^T M(\theta)\Upsilon, \quad \dot{u} = \Upsilon^{-1}\ddot{\theta} + \dot{\Upsilon}^{-1}\dot{\theta}, \tag{4}$$
$$C_u(\theta, u) = \Upsilon^T[M(\theta)\dot{\Upsilon} + C(\theta, \dot{\theta})\Upsilon], \tag{5}$$
$$G_u(\theta) = \Upsilon^T G(\theta), \tag{6}$$
$$\pi = \Upsilon^T Q. \tag{7}$$

Equations (2) and (3) provide a closed set of first-order differential equations for manipulator in terms of GVC.
In the second considered method assuming the inertia matrix decomposition method given in Jain & Rodriguez (1995), which leads to the NQV and with $M(\theta) = mm^T$, we obtain the following equations of motion:

$$\dot{\nu} + C_\nu(\theta, \nu)\nu + G_\nu(\theta) = \epsilon, \tag{8}$$
$$\nu = m^T(\theta)\dot{\theta}, \tag{9}$$

where

$$\dot{v} = \dot{m}^T(\theta)\dot{\theta} + m^T(\theta)\ddot{\theta}, \quad (10)$$
$$C_v(\theta, v) = [m^{-1}(\theta)C_v(\theta, \dot{\theta}) - \dot{m}^T(\theta)](m^{-1}(\theta))^T, \quad (11)$$
$$G_v(\theta) = m^{-1}(\theta)G(\theta), \quad (12)$$
$$\epsilon = m^{-1}(\theta)Q. \quad (13)$$

As results from Jain & Rodriguez (1995) we have also the relationship

$$v^T C_v(\theta, v)v = 0. \quad (14)$$

However, we can prove this property. The time derivative of the mass matrix is $\dot{M} = \dot{m}m^T + m\dot{m}^T$. Using (9), (11), and taking into account the above assumption one can calculate:

$$v^T C_v(\theta, v)v = v^T[m^{-1}C_v(\theta, \dot{\theta}) - \dot{m}^T](m^T)^{-1}v = \dot{\theta}^T C_v(\theta, \dot{\theta})\dot{\theta} - \dot{\theta}^T m\dot{m}^T\dot{\theta} = \dot{\theta}^T \frac{1}{2}\dot{M}\dot{\theta}$$
$$-\dot{\theta}^T m\dot{m}^T\dot{\theta} = \dot{\theta}^T[\frac{1}{2}(\dot{m}m^T - m\dot{m}^T)]\dot{\theta} = v^T\frac{1}{2}m^{-1}(\dot{m}m^T - m\dot{m}^T)(m^{-1})^T v = 0, \quad (15)$$

because the matrix $(\dot{m}m^T - m\dot{m}^T)$ is a skew symmetric one. From the above derivation arises that $C_v(\theta, v) = \frac{1}{2}m^{-1}(\dot{m}m^T - m\dot{m}^T)(m^{-1})^T$.
The third decomposition method Herman (2005b; 2006) is an extension of the method Loduha & Ravani (1995) and it is based on the NGVC with $M(\theta) = \Phi^T\Phi$. Hence the two first-order equations (the diagonalized equation of motion and the velocity transformation equation) for rigid manipulator can be rewritten in the form:

$$\dot{\vartheta} + C_\vartheta(\theta, \vartheta)\vartheta + G_\vartheta(\theta) = \varpi, \quad (16)$$
$$\vartheta = \Phi(\theta)\dot{\theta}, \quad (17)$$

where

$$\dot{\vartheta} = \Phi\ddot{\theta} + \dot{\Phi}\dot{\theta}, \quad \Phi = N^{\frac{1}{2}}Y^{-1}, \quad (18)$$
$$C_\vartheta(\theta, \vartheta) = [(\Phi^T)^{-1}C(\theta, \dot{\theta}) - \dot{\Phi}]\Phi^{-1}, \quad (19)$$
$$G_\vartheta(\theta) = (\Phi^T)^{-1}G(\theta), \quad (20)$$
$$\varpi = (\Phi^T)^{-1}Q. \quad (21)$$

Remark 1. If the viscous damping forces are taken into account then we have the following classical equations of motion, and e.g. the equations in terms of the GVC and the NGVC:

$$M(\theta)\ddot{\theta} + C(\theta, \dot{\theta})\dot{\theta} + G(\theta) + f(\dot{\theta}) = Q, \quad (22)$$
$$N\dot{u} + C_u(\theta, u)u + G_u(\theta) + f_u(\theta, \dot{\theta}) = \pi, \quad (23)$$
$$\dot{\vartheta} + C_\vartheta(\theta, \vartheta)\vartheta + G_\vartheta(\theta) + f_\vartheta(\theta, \dot{\theta}) = \varpi, \quad (24)$$

where $f_u(\theta, \dot{\theta}) = Y^T f(\dot{\theta})$ and $f_\vartheta(\theta, \dot{\theta}) = (\Phi^T)^{-1}f(\dot{\theta})$.

2.3 Other decomposition methods

The main problem concerning the transformed equations of motion is the selection method for decomposition of the inertia matrix. There are various known methods for decomposition of the inertia matrix to obtain a diagonal matrix or the identity matrix. For this purpose the Cholesky factorization (which can be referred to Hurtado (2004); Matlab (1996); Sovinsky et al. (2005) or decomposition into the eigenvalues and the eigenvectors considered in Junkins & Schaub (1997); Matlab (1996). Moreover, using e.g. the Schur decomposition or the singular value decomposition Matlab (1996) we are able to decompose the inertia matrix. The eigenvalue and eigenvector based decomposition method, the Schur decomposition method and the singular value decomposition method for a symmetric and positive definite matrix M lead to obtaining a transformation matrix which has, in general, all nonzero elements. This fact complicates a possible controller design because the number of necessary numerical operation increase and each variable. However, sometimes the use of the appropriate method (not very much time consuming) may decide about performance of a non-adaptive sliding mode controller.

2.4 Some useful properties of IQV

Some advantages arising from the description of motion in terms of the IQV concern an insight into the manipulator dynamics. The kinetic energy of the manipulator is expressed as (compare Herman (2005a), Jain & Rodriguez (1995), and Herman (2005b), respectively):

$$K(\theta, u) = \frac{1}{2}\dot{\theta}^T M(\theta)\dot{\theta} = \frac{1}{2}u^T N u = \frac{1}{2}\sum_{k=1}^{\mathcal{N}} N_k u_k^2 = \sum_{k=1}^{\mathcal{N}} K_k, \tag{25}$$

$$K(\theta, \nu) = \frac{1}{2}\dot{\theta}^T M(\theta)\dot{\theta} = \frac{1}{2}\dot{\theta}^T m(\theta) m^T(\theta)\dot{\theta} = \frac{1}{2}\nu^T \nu = \frac{1}{2}\sum_{k=1}^{\mathcal{N}} \nu_k^2 = \sum_{k=1}^{\mathcal{N}} K_k, \tag{26}$$

$$K(\theta, \vartheta) = \frac{1}{2}\dot{\theta}^T M(\theta)\dot{\theta} = \frac{1}{2}\dot{\theta}^T \Phi^T \Phi\dot{\theta} = \frac{1}{2}\vartheta^T \vartheta = \frac{1}{2}\sum_{k=1}^{\mathcal{N}} \vartheta_k^2 = \sum_{k=1}^{\mathcal{N}} K_k. \tag{27}$$

The above given formulas allow one to determine the part of energy corresponding to each inertial quasi-velocity individually (and also concerning the appropriate link taking into account the dynamical coupling).

Additionally, it is possible to calculate elements of the matrix N (GVC and NGVC) or D (NQV) - see *Notation* - which can be understood as a rotational inertia about each joint axis or a mass shifted along the translational joint. Using the equation (1) this information is inaccessible.

3. Sliding mode controllers using inertial quasi-velocities

3.1 Control algorithms in joint space

In classical form the sliding mode controller in joint space of a manipulator can be expressed as follows Sciavicco & Siciliano (1996); Slotine & Li (1991):

$$Q = M(\theta)\ddot{\theta}_r + C(\theta, \dot{\theta})\dot{\theta}_r + G(\theta) + k_D s. \tag{28}$$

The used symbols denote: $\ddot{\theta}_r = \ddot{\theta}_d + \Lambda\dot{\tilde{\theta}}$, $\dot{\theta}_r = \dot{\theta}_d + \Lambda\tilde{\theta}$ with $\ddot{\theta}_d$ as the desired joint acceleration vector and $\tilde{\theta} = \theta_d - \theta$, $\dot{\tilde{\theta}} = \dot{\theta}_d - \dot{\theta}$ the joint velocities error, and the joint error between the desired and actual posture, respectively. The matrix Λ is constant and it has eigenvalues

strictly in the right-half complex plane and k_D is a constant positive definite control gain matrix. The vector s is defined as $s = \dot{\tilde{\theta}} + \Lambda\tilde{\theta}$.
In terms of the GVC introduced originally by Loduha & Ravani (1995) the non-adaptive sliding mode controller can be presented in the given below proposition. Recall also that from (3) arises the relationship $u = \Upsilon^{-1}\dot{\theta}$ (the matrix Υ is invertible) and the time derivative of $\dot{\theta} = \Upsilon u$ is $\ddot{\theta} = \dot{\Upsilon}u + \Upsilon\dot{u}$. It is assumed the following sliding surface of the objective point

$$\Upsilon^{-1}(\dot{\tilde{\theta}} + \Lambda\tilde{\theta}) = 0. \tag{29}$$

Proposition 1. Consider the system (2) and (3) together with the controller in terms of the GVC Herman (2005a)

$$\pi = N\dot{u}_r + C(\theta, u)u_r + G_u(\theta) + k_D s_u + \Upsilon^T k_P \tilde{\theta}, \tag{30}$$

where

$$u_r = \Upsilon^{-1}\dot{\theta}_r, \quad \dot{u}_r = \Upsilon^{-1}(\ddot{\theta}_r - \dot{\Upsilon}u_r), \tag{31}$$

$$s_u = u_r - u = \Upsilon^{-1}(\dot{\tilde{\theta}} + \Lambda\tilde{\theta}), \tag{32}$$

with positive definite k_D, k_P, Λ control gain matrices, and $\ddot{\theta}_r = \ddot{\theta}_d + \Lambda\dot{\tilde{\theta}}$, $\dot{\theta}_r = \dot{\theta}_d + \Lambda\tilde{\theta}$, $\tilde{\theta} = \theta_d - \theta$, $\dot{\tilde{\theta}} = \dot{\theta}_d - \dot{\theta}$. Using the definition (29) and if the signals θ_d, $\dot{\theta}_d$, $\ddot{\theta}_d$ are bounded, then the equilibrium point $[s_u^T, \tilde{\theta}^T]^T = 0$ is globally asymptotically stable in the sense of Lyapunov. The joint forces (which arises from (7)) are given as $Q = (\Upsilon^T)^{-1}\pi$.
Proof Herman (2005a). The closed loop system with control (30) using s_u is given as follows

$$N\dot{u} + C(\theta, u)u + G_u(\theta) = N\dot{u}_r + C(\theta, u)u_r + G_u(\theta) + k_D s_u + \Upsilon^T k_P \tilde{\theta}, \tag{33}$$

what leads to

$$N\dot{s}_u + [C(\theta, u) + k_D]s_u + \Upsilon^T k_P \tilde{\theta} = 0. \tag{34}$$

As a Lyapunov function candidate consider the following expression

$$\mathcal{L}(s_u, \tilde{\theta}) = \frac{1}{2}s_u^T N s_u + \frac{1}{2}\tilde{\theta}^T k_P \tilde{\theta}. \tag{35}$$

The time derivative of N equals (where $M = M(\theta)$)

$$\dot{N} = \frac{d}{dt}(\Upsilon^T M \Upsilon) = \dot{\Upsilon}^T M \Upsilon + \Upsilon^T \dot{M} \Upsilon + \Upsilon^T M \dot{\Upsilon}. \tag{36}$$

Next we calculate the time derivative of (35), use (2)-(7), (34), (36), and the property, e.g. Kelly & Moreno (2005); Slotine & Li (1991)

$$q^T \left[\frac{1}{2}\dot{M}(\theta) - C(\theta, \dot{\theta})\right] q = 0, \quad \forall q, \theta, \dot{\theta} \in R^{\mathcal{N}}. \tag{37}$$

After transposition of (3) one can obtain ($\dot{\mathcal{L}} = \frac{d\mathcal{L}}{dt}$):

$$\dot{\mathcal{L}}(s_u, \tilde{\theta}) = s_u^T N \dot{s}_u + \frac{1}{2} s_u^T \dot{N} s_u + \dot{\tilde{\theta}}^T k_P \tilde{\theta} = s_u^T [-C(\theta, u) s_u - k_D s_u + \frac{1}{2} \dot{N} s_u] - s_u^T Y^T k_P \tilde{\theta}$$
$$+\dot{\tilde{\theta}}^T k_P \tilde{\theta} = s_u^T [-Y^T M \dot{Y} s_u - Y^T C(\theta, \dot{\theta}) Y s_u - k_D s_u + \frac{1}{2} (\dot{Y}^T M Y + Y^T \dot{M} Y$$
$$+Y^T M \dot{Y}) s_u] - s_u^T Y^T k_P \tilde{\theta} + \dot{\tilde{\theta}}^T k_P \tilde{\theta} = -s_u^T k_D s_u + s_u^T [\frac{1}{2} Y^T M \dot{Y} - Y^T M \dot{Y} + \frac{1}{2} \dot{Y}^T M Y$$
$$+Y^T (\frac{1}{2} \dot{M} - C(\theta, \dot{\theta})) Y] s_u - s_u^T Y^T k_P \tilde{\theta} + \dot{\tilde{\theta}}^T k_P \tilde{\theta} = -s_u^T k_D s_u + \frac{1}{2} s_u^T (\dot{Y}^T M Y - Y^T M \dot{Y}) s_u$$
$$-s_u^T Y^T k_P \tilde{\theta} + \dot{\tilde{\theta}}^T k_P \tilde{\theta} = -s_u^T k_D s_u - s_u^T Y^T k_P \tilde{\theta} + \dot{\tilde{\theta}}^T k_P \tilde{\theta}. \quad (38)$$

Using (32) one can write:

$$\dot{\mathcal{L}}(s_u, \tilde{\theta}) = -s_u^T k_D s_u - (\dot{\tilde{\theta}}^T + \tilde{\theta}^T \Lambda^T) k_P \tilde{\theta} + \dot{\tilde{\theta}}^T k_P \tilde{\theta} = -s_u^T k_D s_u - \tilde{\theta}^T \Lambda^T k_P \tilde{\theta}. \quad (39)$$

Assumption that $k_P = k_D \Lambda$ leads to

$$\dot{\mathcal{L}}(s_u, \tilde{\theta}) = -s_u^T k_D s_u - \tilde{\theta}^T \Lambda^T k_D \Lambda \tilde{\theta}. \quad (40)$$

The time derivative $\dot{\mathcal{L}}$ (40) is a negative semidefinite function. Invoking Lyapunov direct method Khalil (1996); Slotine & Li (1991) the above proof is completed. Therefore, $[s_u^T, \tilde{\theta}^T]^T = 0$ is globally asymptotically stable in the sense of Lyapunov.
Remark 2. The control law (30) can be also simplified as follows: $\pi = N\dot{u}_r + C(\theta, u)u_r + G_u(\theta) + k_D s_u$. The proof, in such case, can be given basing on the Barbalat's Lemma Slotine & Li (1991). However, the performance of the simplified control algorithm is worse than if the controller (30) is used because of absence the additional position error regulation term.
The analogous tracking control problem can be considered in terms of the NQV. Consider the following surface:

$$m^T(\dot{\tilde{\theta}} + \Lambda\tilde{\theta}) = 0, \quad (41)$$

which is also a sliding surface of the objective point (the matrix m^T is invertible Jain & Rodriguez (1995)).
Proposition 2. Consider the system (8) and (9) together with the controller in terms of the NQV

$$\epsilon = \dot{v}_r + C_v(\theta, v)v_r + G_v(\theta) + k_D s_v + m^{-1} k_P \tilde{\theta}, \quad (42)$$

where

$$v_r = m^T \dot{\theta}_r, \quad \dot{v}_r = m^T(\ddot{\theta}_r - (\dot{m}^T)^{-1} v_r), \quad (43)$$
$$s_v = v_r - v = m^T(\dot{\tilde{\theta}} + \Lambda\tilde{\theta}), \quad (44)$$

with positive definite k_D, k_P, Λ control gain matrices, and $\ddot{\theta}_r = \ddot{\theta}_d + \Lambda\dot{\tilde{\theta}}$, $\dot{\theta}_r = \dot{\theta}_d + \Lambda\tilde{\theta}$, $\tilde{\theta} = \theta_d - \theta$, $\dot{\tilde{\theta}} = \dot{\theta}_d - \dot{\theta}$. Using the definition (41) and if the signals $\dot{\theta}_d$, $\dot{\theta}_d$, $\ddot{\theta}_d$ are bounded, then the equilibrium point $[s_v^T, \tilde{\theta}^T]^T = 0$ is globally asymptotically stable in the sense of Lyapunov. The input forces vector of manipulator $Q = m\epsilon$ arises from (13).
Proof. The closed loop system with control (42) using s_v is given as follows:

$$\dot{v} + C_v(\theta, v)v + G_v(\theta) = \dot{v}_r + C_v(\theta, v)v_r + G_v(\theta) + k_D s_v + m^{-1} k_P \tilde{\theta}, \quad (45)$$

which, using (44), leads to equation:

$$\dot{s}_v + [C_v(\theta,v) + k_D]s_v + m^{-1}k_P\tilde{\theta} = 0. \tag{46}$$

As a Lyapunov function candidate consider the following expression:

$$\mathcal{L} = \frac{1}{2}s_v^T s_v + \frac{1}{2}\tilde{\theta}^T k_P \tilde{\theta}. \tag{47}$$

Next, calculating the time derivative of (47), using (46), definition (44) and property (14) one obtains:

$$\begin{aligned}\dot{\mathcal{L}} &= s_v^T\dot{s}_v + \dot{\tilde{\theta}}^T k_P\tilde{\theta} = s_v^T(-C_v(\theta,v)s_v - k_D s_v - m^{-1}k_P\tilde{\theta}) + \dot{\tilde{\theta}}^T k_P\tilde{\theta} \\ &= -s_v^T C_v(\theta,v)s_v - s_v^T k_D s_v - s_v^T m^{-1}k_P\tilde{\theta} + \dot{\tilde{\theta}}^T k_P\tilde{\theta} = -s_v^T k_D s_v - (\dot{\tilde{\theta}}^T + \tilde{\theta}^T\Lambda^T)k_P\tilde{\theta} \\ &+\dot{\tilde{\theta}}^T k_P\tilde{\theta} = -s_v^T k_D s_v - \tilde{\theta}^T\Lambda^T k_P\tilde{\theta}.\end{aligned} \tag{48}$$

Choosing $k_P = k_D\Lambda$ yields

$$\dot{\mathcal{L}} = -s_v^T k_D s_v - \tilde{\theta}^T\Lambda^T k_D\Lambda\tilde{\theta}. \tag{49}$$

One can observe that $\dot{\mathcal{L}}$ (49) is a negative semidefinite function. Invoking Lyapunov direct method Khalil (1996); Slotine & Li (1991) the above proof is completed. Therefore, $[s_v^T, \tilde{\theta}^T]^T = 0$ is globally asymptotically stable in the sense of Lyapunov.

3.2 Control algorithms in operational space

Consider the sliding mode controller in the workspace of a rigid serial manipulator if the viscous damping is taken into account. In classical case the controller related to (22) can be described as follows Sciavicco & Siciliano (1996); Slotine & Li (1987):

$$Q = M(\theta)\ddot{\theta}_r + C(\theta,\dot{\theta})\dot{\theta}_r + G(\theta) + F\dot{\theta}_r + k_D s \tag{50}$$

where in order to extend the joint space controller to task space it is necessary to introduce:

$$\dot{\theta}_r = J_A^{-1}(\theta)\left[\dot{x}_d + \Lambda(x_d - x)\right] \tag{51}$$

$$\ddot{\theta}_r = J_A^{-1}(\theta)\left\{[\ddot{x}_d + \Lambda(\dot{x}_d - \dot{x})] - \dot{J}_A\dot{\theta}_r\right\} \tag{52}$$

$$s = \dot{\theta}_r - \dot{\theta} = J_A^{-1}(\theta)[\dot{x}_d - J_A(\theta)\dot{\theta} + \Lambda(x_d - x)]. \tag{53}$$

In the above equations the x_d, $\dot{x}_d$ and $\ddot{x}_d$ are the desired end-effector posture (position and orientation), velocity and acceleration, respectively. Moreover, x and $\dot{x}$ denote actual end-effector posture and velocity whereas θ_r, $\dot{\theta}_r$ and $\ddot{\theta}_r$ are reference joint position, velocity and acceleration Slotine & Li (1987). The matrix k_D is positive definite whereas the matrix Λ is a diagonal (constant) matrix whose eigenvalues are strictly in the right-half complex plane. The used symbol J_A means the analytical Jacobian because the end-effector velocity can be defined by the kinematic relationship $\dot{x} = J_A(\theta)\dot{\theta}$ Sciavicco & Siciliano (1996). In general $J_A^{-1}(\theta)$ have to be replaced by the right pseudo-inverse of J_A i.e. $J_A^{\dagger} = J_A^T(J_A J_A^T)^{-1}$. Using the controller (50) the sliding surface

$$\dot{\tilde{x}} + \Lambda\tilde{x} = 0, \tag{54}$$

where $\dot{\tilde{x}} = \dot{x}_d - \dot{x}$ and $\tilde{x} = x_d - x$ is reached which in turn implies that $\tilde{x} \to 0$ as $t \to 0$.

Proposition 3 Herman (2009a). Consider the system (23) and (3) together with the controller in terms of the GVC

$$\pi = N\dot{u}_r + C_u(\theta, u)u_r + G_u(\theta) + Y^T F Y u_r + k_D s_u + J_{Au}^T(\theta)k_P\tilde{x}, \tag{55}$$

where

$$u_r = J_{Au}^{-1}(\theta)\left[\dot{x}_d + \Lambda(x_d - x)\right], \tag{56}$$

$$\dot{u}_r = J_{Au}^{-1}(\theta)\left\{[\ddot{x}_d + \Lambda(\dot{x}_d - \dot{x})] - \dot{J}_{Au}(\theta)u_r\right\}, \tag{57}$$

$$s_u = u_r - u = J_{Au}^{-1}(\theta)[\dot{x}_d - J_{Au}(\theta)u + \Lambda(x_d - x)] = J_{Au}^{-1}(\theta)(\dot{\tilde{x}} + \Lambda\tilde{x}), \tag{58}$$

with positive definite k_D, k_P, Λ control gain matrices. Using the definition (54) (assuming that $J_{Au}^{-1}(\theta)$ is a nonsingular matrix) and if the signals $x_d, \dot{x}_d, \ddot{x}_d$ are bounded then the end-effector posture (position and orientation) error $\tilde{x} = x_d - x$ and the velocity error $\dot{\tilde{x}} = \dot{x}_d - \dot{x}$ are convergent to zero and the equilibrium point $[s_u^T, \tilde{x}^T]^T = 0$ is globally exponentially stable. The joint forces (which arises from (7)) are given as $Q = (Y^T)^{-1}\pi$.
The end-effector velocity can be defined by $\dot{x} = J_A(\theta)\dot{\theta}$ Sciavicco & Siciliano (1996) as well as by $\dot{x} = J_{Au}(\theta)u$. Comparing both relationships and taking into account (3) we conclude that $J_{Au}(\theta) = J_A(\theta)Y$. Besides, in general case instead of $J_{Au}^{-1} = J_{Au}^{-1}(\theta)$ the right pseudo-inverse matrix $J_{Au}^{\dagger} = J_{Au}^T(J_{Au}J_{Au}^T)^{-1}$ should be used. It is also assumed (basing on (3)) that $\dot{\theta}_r = Yu_r$.
Proof can be found in Herman (2009a).
In terms of the NGVC we propose the following nonlinear controller.
Proposition 4 Herman (2009c). Consider a system (16) and (17) together with the controller

$$\varpi = \dot{\vartheta}_r + C_\vartheta(\theta, \vartheta)\vartheta_r + G_\vartheta(\theta) + (\Phi^T)^{-1}F\Phi^{-1}\vartheta_r + k_D s_\vartheta + J_{A\vartheta}^T(\theta)k_P\tilde{x}, \tag{59}$$

where

$$\vartheta_r = J_{A\vartheta}^{-1}(\theta)\left[\dot{x}_d + \Lambda(x_d - x)\right], \tag{60}$$

$$\dot{\vartheta}_r = J_{A\vartheta}^{-1}(\theta)\left\{[\ddot{x}_d + \Lambda(\dot{x}_d - \dot{x})] - \dot{J}_{A\vartheta}(\theta)\vartheta_r\right\}, \tag{61}$$

$$s_\vartheta = \vartheta_r - \vartheta = J_{A\vartheta}^{-1}(\theta)\left[\dot{x}_d - J_{A\vartheta}(\theta)\vartheta + \Lambda(x_d - x)\right] = J_{A\vartheta}^{-1}(\theta)(\dot{\tilde{x}} + \Lambda\tilde{x}), \tag{62}$$

with positive definite k_D, k_P, Λ control gain matrices. Using the definition (54) (assuming that $J_{A\vartheta}^{-1}(\theta)$ is a nonsingular matrix) and if the signals $x_d, \dot{x}_d, \ddot{x}_d$ are bounded, then the end-effector position error $\tilde{x} = x_d - x$ and the velocity error $\dot{\tilde{x}} = \dot{x}_d - \dot{x}$ are convergent to zero, and the equilibrium point $[s_\vartheta^T, \tilde{x}^T]^T = 0$ is globally exponentially stable. The joint forces (which arises from (21)) are given as $Q = \Phi^T\varpi$.
Proof (based on Herman (2009c)). The closed-loop system (16) and (17) together with the controller (59) can be written as:

$$\begin{aligned}&\dot{\vartheta} + C_\vartheta(\theta, \vartheta)\vartheta + G_\vartheta(\theta) + (\Phi^T)^{-1}F\Phi^{-1}\vartheta \\ &\quad = \dot{\vartheta}_r + C_\vartheta(\theta, \vartheta)\vartheta_r + G_\vartheta(\theta) + (\Phi^T)^{-1}F\Phi^{-1}\vartheta_r + k_D s_\vartheta + J_{A\vartheta}^T k_P\tilde{x},\end{aligned} \tag{63}$$

what leads to

$$\dot{s}_\vartheta + [C_\vartheta(\theta, \vartheta) + k_D + (\Phi^T)^{-1}F\Phi^{-1}]s_\vartheta + J_{A\vartheta}^T k_P\tilde{x} = 0. \tag{64}$$

The proposed the Lyapunov function candidate is assumed as follows:

$$\mathcal{L}(s_\vartheta, \tilde{x}) = \frac{1}{2}s_\vartheta^T s_\vartheta + \frac{1}{2}\tilde{x}^T k_P\tilde{x}. \tag{65}$$

The time derivative of $\mathcal{L}(s_\vartheta, \tilde{x})$ along of the system trajectories (16) and (17) is given by:

$$\dot{\mathcal{L}}(s_\vartheta, \tilde{x}) = s_\vartheta^T \dot{s}_\vartheta + \dot{\tilde{x}}^T k_P \tilde{x} = s_\vartheta^T[-C_\vartheta - k_D - (\Phi^T)^{-1} F \Phi^{-1}] s_\vartheta - s_\vartheta^T J_{A\vartheta}^T k_P \tilde{x} + \dot{\tilde{x}}^T k_P \tilde{x}. \tag{66}$$

Consider the term $-s_\vartheta^T C_\vartheta s_\vartheta$. Calculating the time derivative of the inertia matrix $M = \Phi^T \Phi$ (see 2.2) one obtains $\dot{M} = \frac{d}{dt}(\Phi^T \Phi) = \dot{\Phi}^T \Phi + \Phi^T \dot{\Phi}$. Introducing $s_\phi = \Phi^{-1} s_\vartheta$ and using (19) one gets:

$$\begin{aligned}
-s_\vartheta^T C_\vartheta s_\vartheta &= -s_\vartheta^T[(\Phi^T)^{-1} C - \dot{\Phi}]\Phi^{-1} s_\vartheta = s_\phi^T(\Phi^T \dot{\Phi} - C) s_\phi \\
&= s_\phi^T(\frac{1}{2}\Phi^T \dot{\Phi} + \frac{1}{2}\Phi^T \dot{\Phi} - C + \frac{1}{2}\dot{\Phi}^T \Phi - \frac{1}{2}\dot{\Phi}^T \Phi) s_\phi = s_\phi^T \left[(\frac{1}{2}\dot{M} - C) + \frac{1}{2}(\Phi^T \dot{\Phi} - \dot{\Phi}^T \Phi) \right] s_\phi. \quad (67)
\end{aligned}$$

Because the matrix $(\frac{1}{2}\dot{M} - C)$ is skew-symmetric then we can use (37). Moreover, the matrix $(\Phi^T \dot{\Phi} - \dot{\Phi}^T \Phi)$ is also skew-symmetric (see Herman (2009c)). Thus, one can write:

$$\dot{\mathcal{L}}(s_\vartheta, \tilde{x}) = -s_\vartheta^T[k_D + (\Phi^T)^{-1} F \Phi^{-1}] s_\vartheta - s_\vartheta^T J_{A\vartheta}^T k_P \tilde{x} + \dot{\tilde{x}}^T k_P \tilde{x}. \tag{68}$$

Using now (62) one obtains:

$$\begin{aligned}
\dot{\mathcal{L}}(s_\vartheta, \tilde{x}) &= -s_\vartheta^T[k_D + (\Phi^T)^{-1} F \Phi^{-1}] s_\vartheta - (\dot{\tilde{x}}^T + \tilde{x}^T \Lambda^T) k_P \tilde{x} + \dot{\tilde{x}}^T k_P \tilde{x} \\
&= -s_\vartheta^T[k_D + (\Phi^T)^{-1} F \Phi^{-1}] s_\vartheta - \tilde{x}^T \Lambda^T k_P \tilde{x}. \quad (69)
\end{aligned}$$

Assuming that $k_P = \delta\Lambda$ (where δ is a positive constant serving for the position error regulation) we have:

$$\dot{\mathcal{L}}(s_\vartheta, \tilde{x}) = -s_\vartheta^T[k_D + (\Phi^T)^{-1} F \Phi^{-1}] s_\vartheta - \tilde{x}^T \delta \Lambda^T \Lambda \tilde{x}. \tag{70}$$

As a result, one can write the above equation in the following form:

$$\dot{\mathcal{L}}(s_\vartheta, \tilde{x}) = -\begin{bmatrix} s_\vartheta \\ \tilde{x} \end{bmatrix}^T \underbrace{\begin{bmatrix} k_D + (\Phi^T)^{-1} F \Phi^{-1} & 0 \\ 0 & \delta \Lambda^T \Lambda \end{bmatrix}}_{A} \begin{bmatrix} s_\vartheta \\ \tilde{x} \end{bmatrix}. \tag{71}$$

Note that the symmetric matrix A is positive definite. Thus, $\lambda_m\{A\} > 0$. Denoting now $x_s = [s_\vartheta^T, \tilde{x}^T]^T$ one can write

$$\dot{\mathcal{L}}(t, x_s) \leq -\lambda_m\{A\}||x_s||^2 \tag{72}$$

for all $t \geq 0$ and $x_s \in R^{2\mathcal{N}}$.
Therefore, basing on the Lyapunov direct method Khalil (1996); Slotine & Li (1991), the conclusion that the state space origin of the system (16) and (17) together with the controller (59)

$$\lim_{t \to \infty} \begin{bmatrix} s_\vartheta(t) \\ \tilde{x}(t) \end{bmatrix} = 0 \tag{73}$$

is globally exponentially convergent can be done.
The end-effector velocity is defined by $\dot{x} = J_A(\theta)\dot{\theta}$ Sciavicco & Siciliano (1996). Introducing the analytical Jacobian in terms of the vector ϑ we can write the relationship $\dot{x} = J_{A\vartheta}(\theta)\vartheta$. Comparing $\dot{x}$ from both relationships and taking into account Eq.(17) we conclude that

$J_{A\vartheta}(\theta) = J_A(\theta)\Phi^{-1}$. Moreover, in general case, instead of $J_{A\vartheta}^{-1} = J_{A\vartheta}^{-1}(\theta)$ the right pseudo-inverse matrix $J_{A\vartheta}^{\dagger} = J_{A\vartheta}^T(J_{A\vartheta}J_{A\vartheta}^T)^{-1}$ should be used. Based on Eq.(17) it is also assumed that $\vartheta_r = \Phi\dot{\theta}_r$. Kinematics singularities are the same as in $J_A(\theta)$ because we obtain only a new Jacobian, but the structure of the manipulator is the same.
Remark 3. Analogous proofs can be carried out regarding the controllers in the manipulator joint space considered earlier.

3.3 Advantages and disadvantages of the IQV controllers

Consider some aspects of the presented controllers in terms of the IQV.

1. The controllers expressed in terms of IQV seem complicated. Note however, that the controls algorithms can be realized using quantities arising from the spatial operators which decrease their computational complexity Jain & Rodriguez (1995). Also Kane's equations are computationally effective Kane & Levinson (1983). Thus, the algorithms seem a useful tool for simulation of serial rigid manipulators.

2. The manipulator input torque Q can be calculated from the relationship $Q = (\Upsilon^{-1})^T\pi$ and $Q = m\epsilon$, i.e for the controllers (30) and (42) it has have the form:

$$Q = M(\theta)\ddot{\theta}_r + C(\theta,\dot{\theta})\dot{\theta}_r + G(\theta) + k_P\tilde{\theta} + (\Upsilon^{-1})^T k_D \Upsilon^{-1} s, \tag{74}$$

$$Q = M(\theta)\ddot{\theta}_r + C(\theta,\dot{\theta})\dot{\theta}_r + G(\theta) + k_P\tilde{\theta} + m k_D m^T s. \tag{75}$$

Comparing (75) and (74) with (28) it can be seen that the difference relies on an additional term $k_P\tilde{\theta}$ and the use of the matrix $(\Upsilon^{-1})^T k_D \Upsilon^{-1}$ or mk_Dm^T instead of the matrix k_D. The term $k_P\tilde{\theta}$ causes that one obtains more precise trajectory tracking than using the controller (28). In spite of that in Berghuis & Nijmeijer (1993) the classical controller with the term $k_P\tilde{\theta}$ was proposed, the controllers in terms of the IQV have one more benefit. The matrix mk_Dm^T contains both kinematic and dynamical parameters which are present in the matrix $M(\theta)$. As a result, the matrices m and m^T give an additional gains and improve the controller performance (after some time their elements are almost constant). Similarly, the use of the controller (59), in comparison with the classical controller (50), has two advantages. First, after transformation $Q = \Phi^T\varpi$ (see (21)) the generalized force vector is as follows:

$$Q = M\ddot{\theta}_r + C(\theta,\dot{\theta})\dot{\theta}_r + G(\theta) + F\dot{\theta}_r + \Phi^T k_D \Phi s + J_A^T(\theta)k_P\tilde{x}. \tag{76}$$

Recalling (50) one can observe that the NGVC controller has two terms which are absent in the classical control algorithm. The first term contains instead of the matrix k_D the matrix $\Phi^T k_D \Phi$. The elements of Φ^T and Φ give an additional gain and, as a result, the desired position and velocity using the NGVC controller is achieved faster or with smaller coefficients of k_D than using the classical controller. The second term $J_A^T(\theta)k_P\tilde{x}$ ensures the position error convergence in the operational space. Lack of the term causes that the proof of the error convergence can be done based on the Barbalat's Lemma Slotine & Li (1991).

3. An important advantages of the controllers in terms of the IQV arises from the fact, that the matrices $(\Upsilon^{-1})^T k_D \Upsilon^{-1}$ or mk_Dm^T reflect dynamics of the considered system. Consequently, the elements of k_D serve for tuning of gain coefficients (in contrast in (28) the matrix k_D is selected using various methods depending on experience of the user).

4. The sliding mode control algorithm described by Slotine and Li Slotine & Li (1987) enables also adaptive trajectory control. The equation (1) can be written as follows Sciavicco & Siciliano (1996):
$$M(\theta)\ddot{\theta} + C(\theta,\dot{\theta})\dot{\theta} + G(\theta) = Y(\theta,\dot{\theta},\ddot{\theta})p = \tau \tag{77}$$
where p is an m-dimensional vector of constant parameters and Y is an $(\mathcal{N} \times m)$ matrix which is a function of joint positions, velocities and accelerations. Decomposition of the matrix M in Eq.(1) which leads to Eq.(8) (after multiplication by the matrix m^{-1}) causes that one obtains $m^{-1}Y(\theta,\dot{\theta},\ddot{\theta})p = \epsilon$. However, for dynamics equation in terms of the nominal parameters one has $\hat{m}^{-1}Y(\theta,\dot{\theta},\ddot{\theta})\hat{p} = \hat{\epsilon}$. Therefore, in terms of the NQV vector adaptation with respect to the vector of parameters p is impossible because parameters of the system are involved in matrices m^{-1} and $\hat{m}^{-1}$. Analogous conclusion can be made about other kinds of the IQV that is an disadvantage.
5. Robustness issue. In case where uncertainty of parameters occurs we should ask about robustness of the proposed controller. The appropriate case concerning the GVC controller is considered in Herman (2005a).

4. Simulation results using various controllers

4.1 Examples of serial manipulators

The DDA manipulator is characterized by the following set of manipulator parameters An et al. (1988) (see Table 1):

Link number k		1	2	3
m_k	kg	19.67	53.01	67.13
J_{xx}	kgm^2	0.1825	3.8384	23.1568
J_{xy}	kgm^2	0	0	0
J_{xz}	kgm^2	-0.0166	0	0.3145
J_{yy}	kgm^2	0.4560	3.6062	20.4472
J_{yz}	kgm^2	0	-0.0709	1.2948
J_{zz}	kgm^2	0.3900	0.6807	0.7418
c_{xk}	m	0.0158	0	0
c_{yk}	m	0	-0.0643	-0.0362
c_{zk}	m	0.0166	-0.1480	0.5337
l_k	m	0	0.462	0

Table 1. Parameters of the DDArm manipulator An et al. (1988)

The Yasukawa-like manipulator is characterized by the parameters given in Table 2. The appropriate equations of motion can be found in reference Kozlowski (1992).

4.2 GVC - joint space

4.2.1 DDArm manipulator

In order to show performance and advantages of the controller (30) consider DDArm manipulator depicted in Figure 1(a). The results are based on Herman (2005a).
The following fifth-order polynomial was chosen for tracking: initial points $\theta_{i1} = (-7/6)\pi$ rad, $\theta_{i2} = (269.1/180)\pi$ rad, $\theta_{i3} = (-5/9)\pi$ rad, and final points $\theta_{f1} = (2/9)\pi$ rad, $\theta_{f2} = (19.1/180)\pi$ rad, $\theta_{f3} = (5/6)\pi$ rad, with time duration $t_f = 1.3$ s. Starting points were different from initial points $\Delta = +0.2, +0.2, +0.2$ rad, respectively. All simulations (were

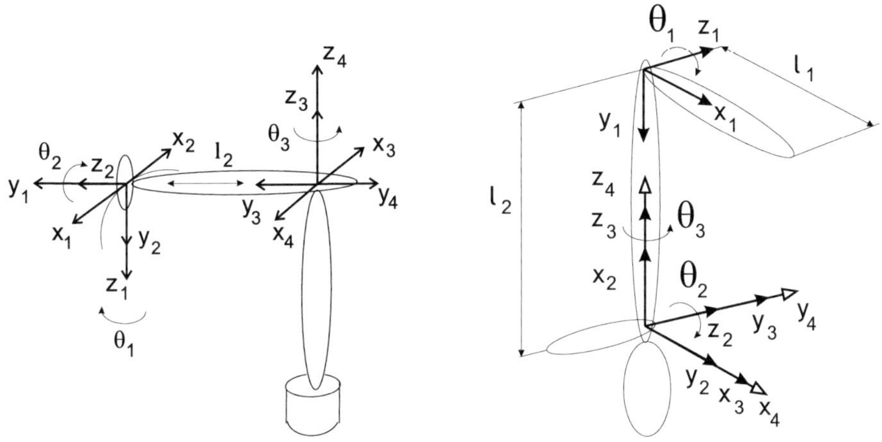

(a) (b)

Fig. 1. Examples of spatial manipulator: (a) kinematic scheme of DDArm; (b) kinematic scheme of Yasukawa

Link number k		1	2	3
m_k	kg	10	30	65
J_{xx}	kgm^2	0.4	0.2	0
J_{xy}	kgm^2	0	0.01	0
J_{xz}	kgm^2	0	−0.01	0
J_{yy}	kgm^2	0.04	0.7	0
J_{yz}	kgm^2	−0.01	−0.01	0
J_{zz}	kgm^2	0.5	1.5	1.5
c_{xk}	m	0	0.01	0
c_{yk}	m	0.1	0.01	0
c_{zk}	m	0.01	0	0
l_k	m	0.4	0.65	0

Table 2. Parameters of the Yasukawa-like manipulator

realized using MATLAB with SIMULINK). The assumed diagonal control coefficients were as follows: $k_D = \text{diag}\{10, 10, 10\}$, $\Lambda = \text{diag}\{15, 15, 15\}$, $k_P = \text{diag}\{150, 150, 150\}$ for the GVC controller and $k_D = \text{diag}\{10, 10, 10\}$, $\Lambda = \text{diag}\{30, 30, 30\}$ for the classical controller (CL). Diagonal values of the matrix Λ are two times smaller than for the classical controller in order to show some differences between both control algorithms. The set of control gains is a trade-of between acceptable position trajectory error and over-regulation.

Profiles of the desired joint position and velocity trajectories are shown in Figure 2(a). In Figures 2(b) and 2(c) the joint position errors for GVC and classical (CL) controllers are presented. The GVC controller gives similar errors for the first and the second joint. However, all errors tend to zero very quickly. For CL controller (for the third joint) position error tends slower. It can be concluded that third diagonal value of k_D and Λ are not sufficient to obtain comparable performance. But increasing of these gains may lead to over-regulation. Figure 2(d) – the error norm (in logarithmic scale) says that the position error is reduced faster if the GVC controller is used. From Figures 2(e) and 2(f) arises that joint torques obtained using GVC controller and CL have comparable values. Each element of the matrix N given in Figure 2(g)

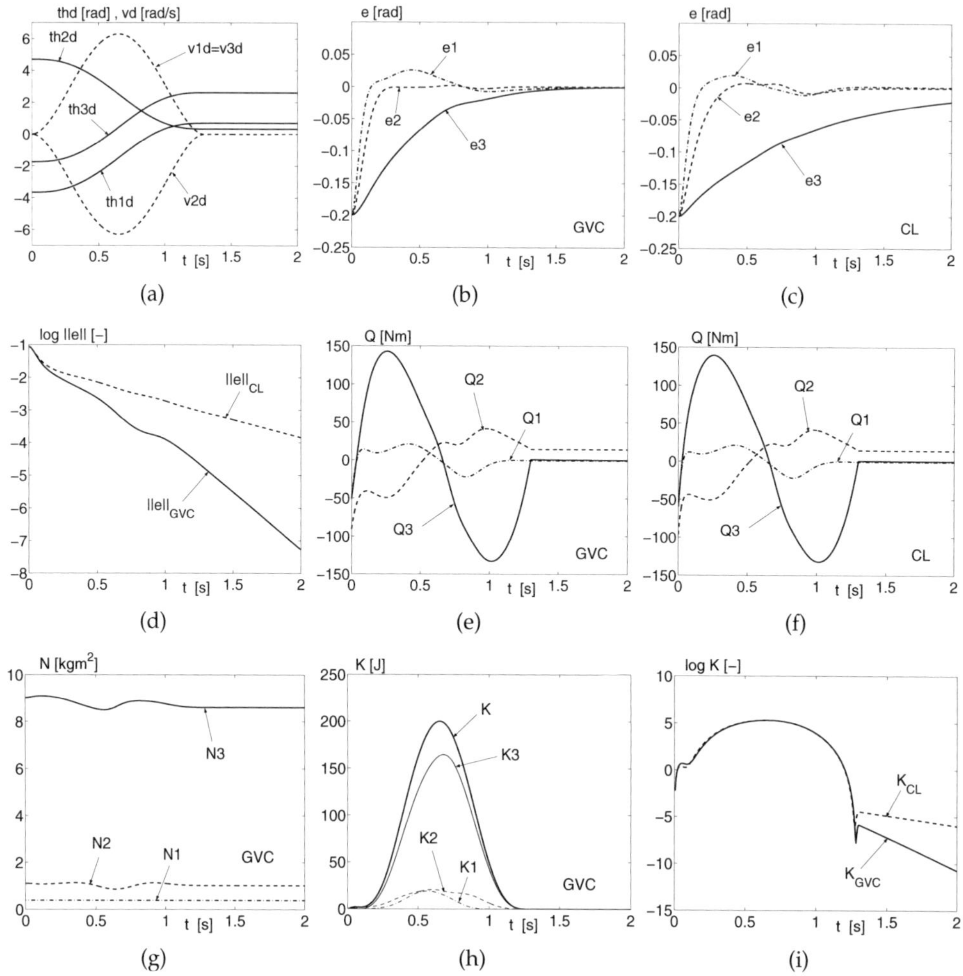

Fig. 2. Simulation results - joint space control (DDArm based on Herman (2005a)): a) profiles of desired joint position and velocity trajectory; b) joint position errors e for GVC controller; c) joint position errors e for classical (CL) controller; d) comparison between joint position error norms $||e||$ (in logarithmic scale) for both controllers; e) joint torques Q obtained using GVC controller; f) joint torques Q obtained using CL controller; g) elements of matrix N obtained from GVC controller; h) kinetic energy in all joints and for the entire manipulator (GVC controller); i) comparison between kinetic energy reduction (in logarithmic scale) for both classical (K_{CL}) and GVC (K_{GVC}) controller

represents a rotational inertia about joint axis arising from the motion of other manipulator links. Figure 2(h) compares the kinetic energy for the whole manipulator K and for all joints. The great value of $K3$ can be related to the dominant values of $N3$ (the two informations can be obtained only for the GVC controller). From Figure 2(i) it is observable that the kinetic energy is reduced faster using GVC controller than using the CL controller.

4.2.2 Yasukawa-like manipulator

The manipulator is depicted in Figure 1(b). The given below results are based on Herman (2009b).

The polynomial trajectories were described with initial points $\theta_{i1} = (1/3)\pi$ rad, $\theta_{i2} = \pi$ rad, $\theta_{i3} = (-1/2)\pi$ rad, final points $\theta_{f1} = (-2/3)\pi$ rad, $\theta_{f2} = 0$ rad, $\theta_{f3} = (1/2)\pi$ rad, and the time duration $t_f = 1$ s. The starting points were different from the initial points $\Delta = +0.2, +0.2, +0.2$ rad. It was assumed the following control coefficients set: $k_D = \text{diag}\{10, 10, 10\}$, $\Lambda = \text{diag}\{15, 15, 15\}$, $k_P = \text{diag}\{150, 150, 150\}$ for the GVC controller (30). For the classical controller (28) we assumed the set $k_D = \text{diag}\{10, 10, 10\}$, $\Lambda = \text{diag}\{30, 30, 30\}$. Diagonal elements values of the matrix Λ are two times smaller for the GVC controller than for the classical one. For the same set of coefficients performance of the classical controller are worse than for the considered case.

Profiles of the desired joint position and velocity trajectories are shown in Figure 3(a). The joint position errors for the GVC and the classical (CL) controller are shown in Figures 3(b) and 3(c), respectively. It is observable that the errors for the GVC controller tend very fast to zero and the manipulator works correctly. But for the CL controller the joint position errors tend to zero more slowly. Increasing the gain coefficients k_D or Λ could lead to better performance obviously under condition avoidance undesirable over-regulation. This observation confirms Figure 3(d) because the error norm (in logarithmic scale) has distinctly smaller values for the GVC controller than for the CL one. Figure 3(e) presents the joint torques obtained using the GVC controller (for the classical one they have almost the same values). In Figure 3(f) elements of the matrix N are shown (such information gives only for the GVC controller). These quantities represents some rotational inertia along each axis which arise from other links motion. They are characteristic for the tested manipulator and for the desired joint velocity set. Values $N3$ are dominant almost all time what says that the third joint is the most laden. Figure 3(g) a kinetic energy time history for the total manipulator K and for all joints is presented. Most of the kinetic energy is related to the second joint ($K2$) (and also to the same link). This fact may be associated with the dominant values $N2$ in the time interval $0.4 \div 0.6$ s. Figure 3(h) compares the kinetic energy reduction for the manipulator if both control algorithms are used. It can be noticed that after some time this energy is reduced much faster using the GVC controller than using the classical one.

4.3 NQV - joint space

Simulations were done for the DDArm manipulator with the parameters given in Table 1 and under the same conditions. The assumed gain coefficients set was $k_D = \text{diag}\{10, 10, 10\}$, $\Lambda = \text{diag}\{15, 15, 15\}$, $k_P = \text{diag}\{150, 150, 150\}$ for the NQV controller and $k_D = \text{diag}\{10, 10, 10\}$, $\Lambda = \text{diag}\{15, 15, 15\}$ for the CL one. This means also that the desired joint position and velocity trajectories are assumed as in Figure 2(a).

Simulation results obtained from the NQV controller (42) and the CL controller (28) are presented in Figure 4. The joint position errors for the NQV and the CL controller, are presented in Figure 4(a) and 4(b). One can observe that for the NQV controller all position

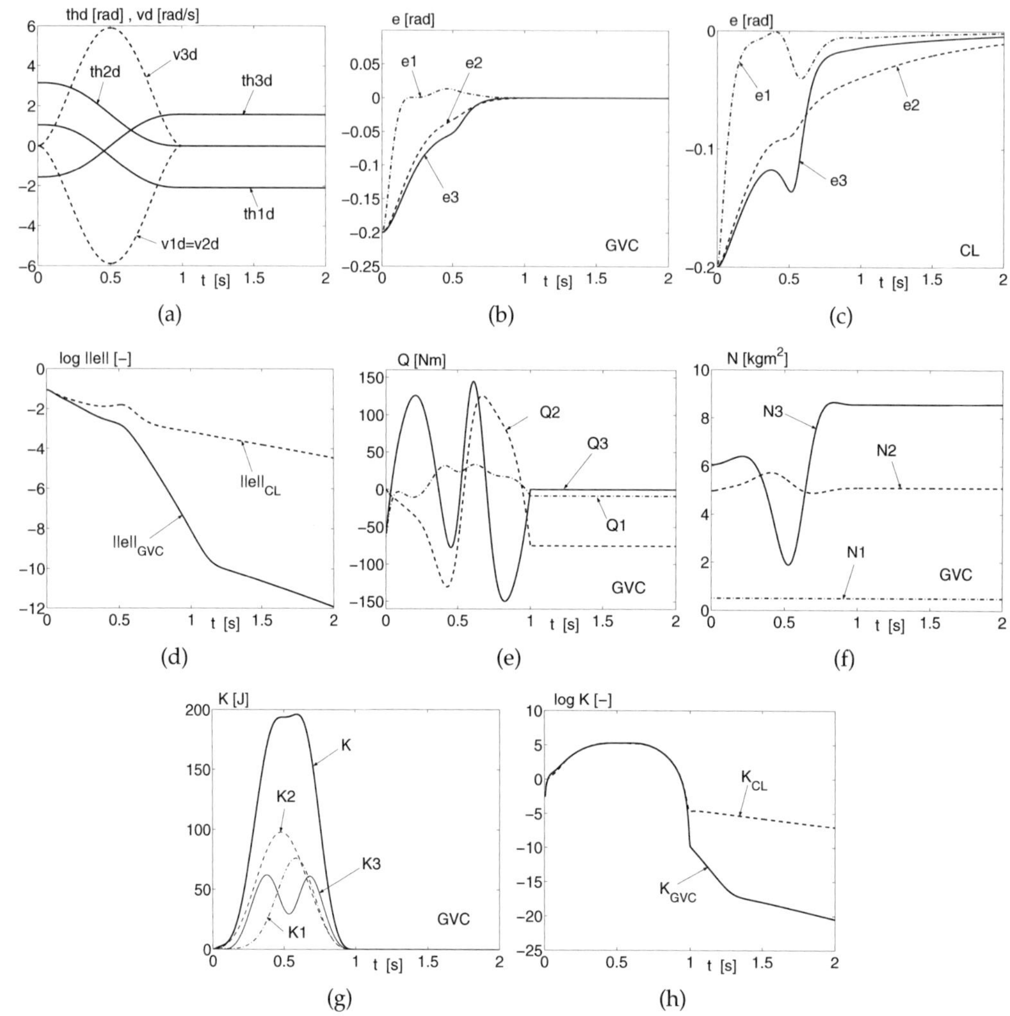

Fig. 3. Simulation results - joint space control (Yasukawa based on Herman (2009b)): a) desired joint position *thd* and joint velocity *vd* trajectory for all joints of manipulator; b) joint position errors e for GVC controller; c) joint position errors e for classical (CL) controller; d) comparison between joint position error norms $||e||$ (in logarithmic scale) for both controllers; e) joint torques Q obtained using GVC controller; f) elements of matrix N obtained from GVC controller; g) kinetic energy reduced by each joints and by the total manipulator (GVC controller); h) comparison between kinetic energy (in logarithmic scale) for classical (K_{CL}), and GVC (K_{GVC}) controller

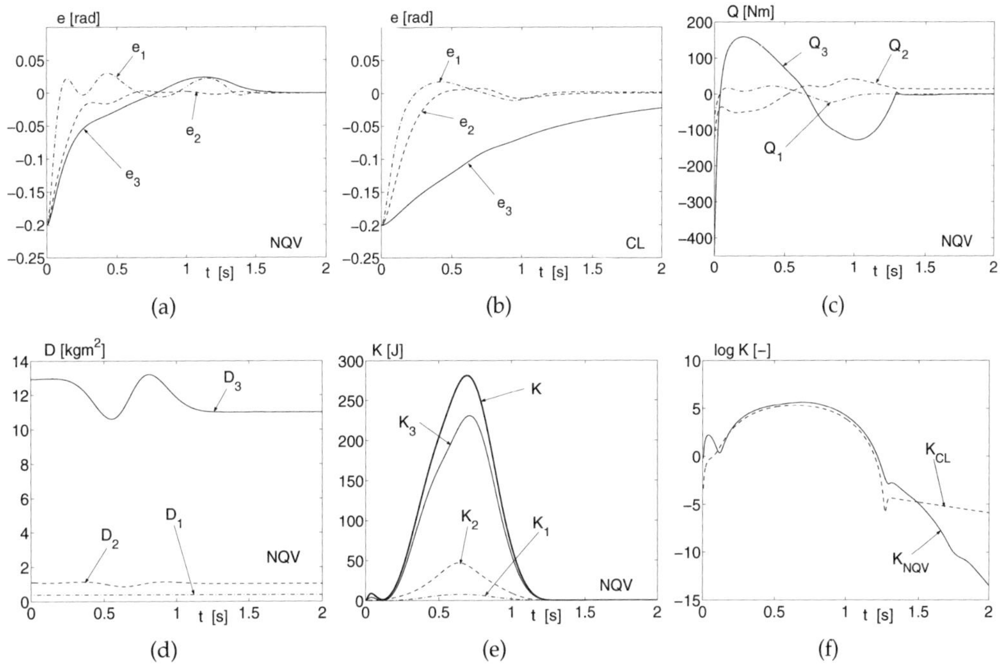

Fig. 4. Simulation results - joint space control (DDArm): a) joint position errors e for NQV controller; b) joint position errors e for CL controller; c) joint torques Q obtained using NQV controller; d) articulated inertias D_k for all joints; e) kinetic energy K_1, K_2, K_3 for all manipulator joints and entire kinetic energy K; f) comparison between kinetic energy reduction for NQV and CL controller (in logarithmic scale)

errors tend to zero after about 1.6 s. For the CL controller errors e_1, e_2 tend very fast to zero but e_3 tends to zero more slowly than for the NQV controller. Figure 3(c) shows the joint torques obtained from the NQV controller. The big initial value of the joint torque Q_3 arises from the fact that we feed back some quantity including the kinematic and dynamical parameters of the manipulator instead of the joint velocity only. However, for the tested manipulator this value is allowed as results from reference An et al. (1988). The articulated inertia D_k for each joint (Figure 4(d)) can be obtained only using the NQV controller. Each value D_k says how much inertia rotates about the k-th joint axis. Most of the rotational inertia is transfered by the third joint axis which means that dynamical interactions are great for the third joint and the third link. Figure 4(e) gives a time history of the kinetic energy for each joint and for the manipulator. Most of the energy is related to the third link which can be explained by great values of D_3. Next Figure 4(f) compares the kinetic energy (in logarithmic scale) which is reduced by the manipulator. After about 1.6 s the kinetic energy is canceled for NQV controller much faster than for CL controller.

4.4 GVC - operational space

The simulation results are obtained for a 3 D.O.F. Yasukawa-like manipulator Herman (2009a). The first objective is to show performance of the GVC controller (55) in the manipulator operational space. The following parameters are different than in Table 2:

- link masses: $m_1 = 5$ kg, $m_3 = 60$ kg;

- link inertias: $J_{xx2} = 0.6$ kgm^2, $J_{xz1} = 0.02$ kgm^2, $J_{yy1} = 0.05$ kgm^2, $J_{yy2} = 0.8$ kgm^2, $J_{zz2} = 2.0$ kgm^2, $J_{zz3} = 3.0$ kgm^2;
- distance: axis of rotation - mass center : $c_{x1} = 0.01$ m, $c_{x2} = 0.1m$, $c_{y1} = 0.01$ m;
- length of link: $l_2 = 1.3$ m.

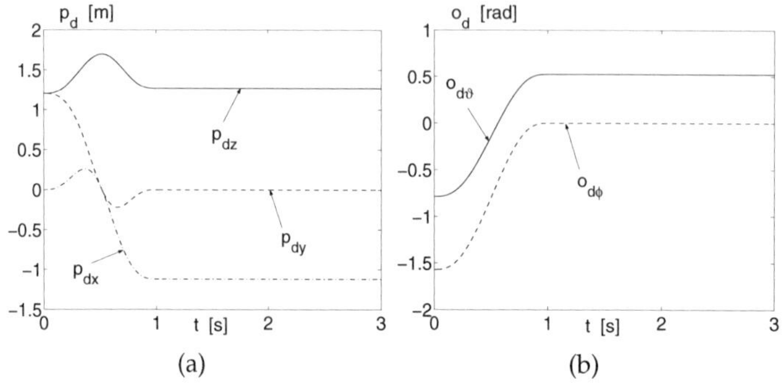

Fig. 5. Simulation results - operational space control (Yasukawa based on Herman (2009c) - the same as in Herman (2009a)): a) desired position trajectories in the operational space; b) desired orientation trajectories in the operational space (used for GVC and NGVC case)

The desired position and orientation described by the vector $x_d = \begin{bmatrix} p_{dx} & p_{dy} & p_{dz} & o_{d\phi} & o_{d\vartheta} \end{bmatrix}^T$ are shown in Figures 5(a) and 5(b) Herman (2009c).
The simulations results realized in MATLAB/SIMULINK (Figure 6) come from reference Herman (2009a).) The control gain matrices were assumed for all controllers as follows: $k_D = \text{diag}\{20, 20, 20\}$, $\Lambda = \text{diag}\{20, 20, 20, 20, 20, 20\}$, $k_P = \text{diag}\{20, 20, 20, 20, 20\}$, $\rho = 1$. Viscous damping coefficients were the same for all joints $F = \text{diag}\{2, 2, 2\}$.
Figures 6(a) and 6(b) show the position and the orientation error for the GVC controller (55) in the operational space, respectively. One can observe that both errors converge to zero after about 2 s. Next, in Figures 6(c) and 6(d) the same errors for the classical controller (50) are presented. As arises from both figures in order to achieve the steady-state the controller needs more than 3 s. At the same time the orientation errors are only close to zero. In the first phase of the manipulator motion the classical controller (CL) gives smaller orientation error than the GVC controller but after about 1 s the GVC controller gives better performance. This phenomenon results from the fact that the dynamical parameters set in the controller (55) is used. From Figure 6(e) one can observe that after 1 s the kinetic energy K_{GVC} (for the GVC controller) is reduced faster than for the classical controller K_{CL} (results are presented on logarithmic scale).
In Figure 6(f) the position error norms (on logarithmic scale) measured in the manipulator task space for the GVC controller and the classical controller (CL) are compared. It can be seen that the position error norm $||ep||_{GVC}$ is smaller than the error norm $||ep||_{CL}$. Comparison between the orientation error norms for both controllers are given in Figure 6(g). In the first phase of the manipulator motion the classical controller (CL) gives smaller orientation error than the GVC controller but after about 0.9 s the latter controller gives better performance. This behavior also results from the fact that the dynamical parameters set in the controller (55) is used. The joint torques for the GVC controller are shown in Figure 6(h). It is observable that at the start (before 0.2 s) the torque in the third joint Q_3 has great value (it is a consequence

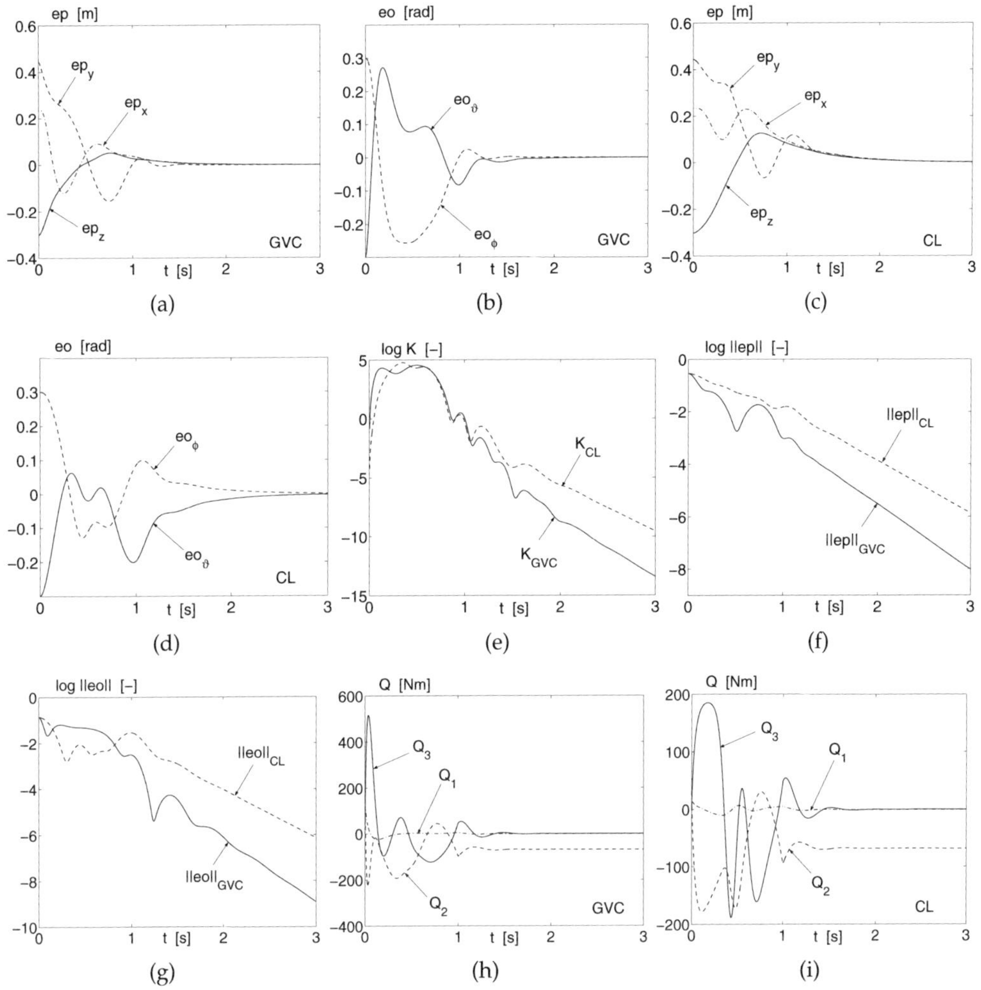

Fig. 6. Simulation results - operational space control (Yasukawa - based on Herman (2009a)): a) position errors in the operational space for GVC controller; b) orientation errors in the operational space for GVC controller; c) position errors in the operational space for classical (CL) controller; d) orientation errors in the operational space for CL controller; e) comparison between kinetic energy reduction (on logarithmic scale) for GVC and CL controller; f) comparison between position error norm on logarithmic scale for both controllers; g) comparison between orientation error norm on logarithmic scale for both controllers (GVC and CL); h) joint torques Q_k for GVC controller; i) joint torques Q_k for CL controller

of including dynamical parameters of the system in the GVC controller). As it is shown from Figure 6(i) the third joint torque for the CL controller has smaller value than using the GVC one. However, after about 0.3 s the torques for both controllers are comparable.

4.5 NGVC - operational space

Consider the Yasukawa-like manipulator again. The following parameters are different than in Table 2:

- link masses: $m_1 = 5$ kg, $m_3 = 60$ kg;
- link inertias: $J_{xx1} = 0.5$ kgm^2, $J_{xx2} = 0.6$ kgm^2, $J_{xz1} = 0.02$ kgm^2, $J_{yy1} = 0.05$ kgm^2, $J_{yy2} = 0.8$ kgm^2, $J_{zz3} = 3.0$ kgm^2;
- distance: axis of rotation - mass center : $c_{x1} = 0.01$ m, $c_{x2} = 0.1m$, $c_{y1} = 0.01$ m, $c_{z1} = 0.02$ m;
- length of link: $l_2 = 1.3$ m.

The results obtained for the NGVC (59) controller are compared with the obtained from the classical controller (50) in Figures 7 and 8 Herman (2009c).
The gain matrices were chosen as (the same for both controllers, i.e. the NGVC and the CL): $k_D = \text{diag}\{4,4,4\}$, $\Lambda = \text{diag}\{20,20,20,20,20,20\}$, $k_P = \text{diag}\{5,5,5,5,5\}$, $\delta = 0.25$ whereas the viscous damping coefficients were $F = \text{diag}\{2,2,2\}$.
In Figures 7(a) and 7(b) the position and the orientation error for the NGVC controller (59) in the operational space are shown. Both errors tend to zero after about 1.5 s. The same errors for the classical (CL) controller (50) are given in Figures 7(c) and 7(d). After 3 s (Figure 7(c)) the position steady-state is not achieved. As a result to ensure the satisfying error convergence, the CL controller needs more time than 3 s. The same conclusion can be made about the orientation error convergence (Figure 7(d)). The joint applied torques for the NGVC controller are shown in Figure 7(e). Comparing Figures 7(e) and 7(f) it can be observed that maximum values of the torques using the NGVC controller are not much larger than if the CL controller is applied.
The diagonal elements of the matrix Φ are given in Figure 8(a) whereas the off-diagonal ones in Figure 8(b). Recall that the matrices Φ^T and Φ give an additional gain in the term $\Phi^T k_D \Phi$ of the controller (59). It can be concluded that the NGVC controller uses small control coefficients k_{Dk} to ensure fast position and orientation trajectory tracking. Moreover, each element Φ^2_{kk} represents an rotational inertia corresponding to the k-th quasi-velocity, whereas Φ_{ki} (for $i \neq k$) show dynamic coupling between the joint velocities (and also between the appropriate links). Such information is available only from the NGVC controller.
From Figure 8(c) it can be seen that the kinetic energy K which must be reduced by the manipulator concerns mainly the third quasi-velocity K_3 (and also by the 3-th link). Figure 8(d) compares the kinetic energy reduction (on logarithmic scale) for both controllers. After about 1 s the kinetic energy K_{NGVC} for the NGVC controller decreases faster than for the classical controller K_{CL}. Consequently, the NGVC control algorithm gives faster error convergence than the CL control algorithm.

4.6 Discussion

From the presented simulation results arises the fact that the proposed nonlinear controllers in terms of the IQV ensures faster, than the classical controller, the position and orientation error convergence. Moreover, the kinetic energy reduction is also faster if the IQV controller is used. An disadvantage of the IQV controllers is that sometimes, at the beginning of motion, great

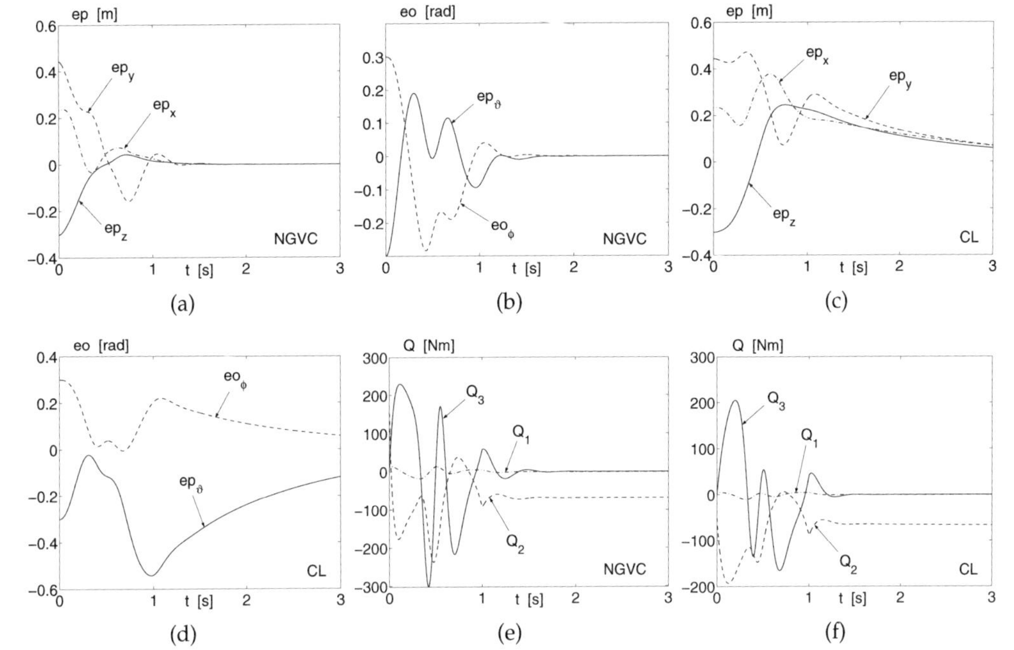

Fig. 7. Simulation results - operational space control (Yasukawa - based on Herman (2009c)): a) position errors in the in the operational space for NGVC controller; b) orientation errors in the operational space for NGVC controller; c) position errors in the operational space for classical (CL) controller; d) orientation errors in the operational space for classical (CL) controller; e) joint applied torques Q for NGVC controller; f) joint applied torques Q for CL controller

initial torque can occur. The great values come from including the manipulator parameters set into the control algorithm. Note, however that the same reason causes the benefit concerning the fast error convergence and fast kinetic energy reduction. Thus, it should be verified if for the considered manipulator the real torques are acceptable. It can be done via simulation because the expected torques are determined from the time history of Q. To obtain comparable results as for the IQV controller we have to assume for the CL controller the matrix k_D with bigger gain coefficients. However, at the same time elements of the matrix Λ should be enough great to ensure fast error convergence. From all presented cases arise that if the IQV controller is used then the gain matrix k_D has rather small values. One can say that they serve for precise tuning because the resultant gain matrix is related to the system dynamics.

5. Conclusion

In this paper, a review of a theoretical framework of non-adaptive sliding mode controllers in terms of the inertial quasi-velocities (IQV) for rigid serial manipulators was provided. The dynamics of the system using several kind of the IQV, namely: the GVC, the NQV, and the NGVC was presented. The IQV equations of motion offer some advantages which are inaccessible if the classical second-order differential equations are used. The IQV sliding mode control algorithms, based on the decomposition of the manipulator inertia matrix, can be realized both in the manipulator joint space and in its the operational space. It was shown

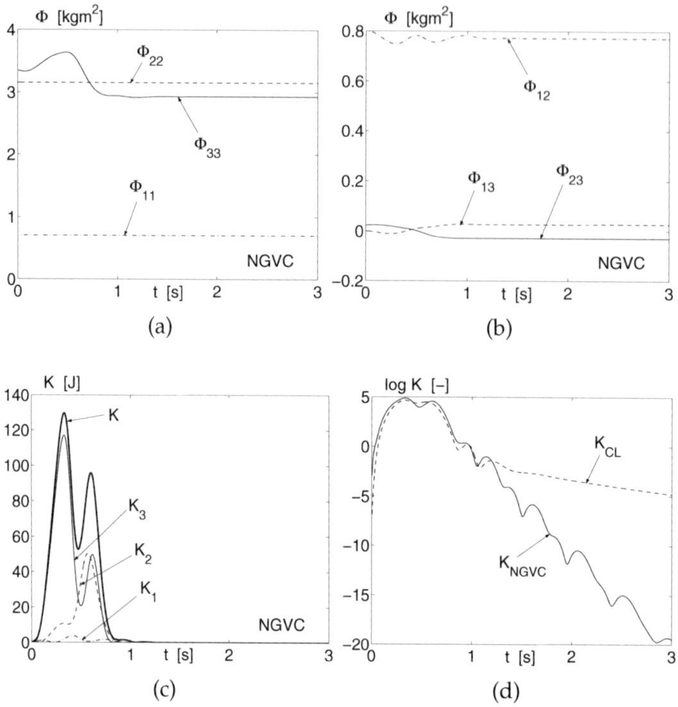

Fig. 8. Simulation results - operational space control (Yasukawa - based on Herman (2009c)): a) diagonal elements of the matrix Φ; b) other elements of the matrix Φ; c) kinetic energy time history corresponding to each quasi-velocity ϑ_k; d) comparison between kinetic energy reduction (on logarithmic scale) for the NGVC controller and the CL controller

that the considered controllers are made the equilibrium point globally asymptotically or exponentially stable in the sense of Lyapunov. Some advantages and disadvantages of the IQV controllers were also given in the work. Moreover, the proposed control schemes are also feasible if the damping forces are taken into account. Simulations results for two different 3 D.O.F. spatial manipulators have shown that the IQV controllers can give faster position and orientation error convergence and/or using smaller velocity gain coefficients than the related classical control algorithms. Faster kinetic energy reduction is also possible if the classical controller is replaced by the IQV one. It is worth noting that the discussed controllers can serve for dynamical coupling detection between the manipulator links via simulation which allows one to avoid some expensive experimental tests.

Future works should concern investigation of the IQV controllers with models of friction, especially with Coulomb friction and dynamic friction models. In order to show real performance and properties of the controllers, experimental validation is expected.

6. References

An, Ch.H.; Atkeson, Ch.G. & Hollerbach, J.M. (1988). *Model-Based Control of a Robot Manipulator*, The MIT Press.

Berghuis, H. & Nijmeijer, H. (1993). A Passivity Approach to Controller - Observer Design for Robots. *IEEE Transactions on Robotics and Automation*, Vol.9, No. 6: 740-754.

Herman, P. (2005a). Sliding mode control of manipulators using first-order equations of motion with diagonal mass matrix. *Journal of the Franklin Institute*, Vol.342, No. 4: 353-363.

Herman, P. (2005b). Normalised-generalised-velocity-component-based controller for a rigid serial manipulator. *IEE Proc. - Control Theory & Applications*, Vol.152, No.5: 581-586.

Herman, P. (2006). On using generalized velocity components for manipulator dynamics and control. *Mechanics Research Communications*, Vol.33, No. 3: 281-291.

Herman, P. (2009a). A nonlinear controller for rigid manipulators, PapersOnline: *Methods and Models in Automation and Robotics*, Vol. 14, Part 1, 10.3182/20090819-3-US-00105.

Herman, P. (2009b). Strict Lyapunov function for sliding mode control of manipulators using quasi-velocities. *Mechanics Research Communications*, Vol.36, No. 2: 169-174.

Herman, P. (2009c). A quasi-velocity-based nonlinear controller for rigid manipulators. *Mechanics Research Communications*, Vol.36, No. 7: 859-866.

Hurtado, J.E. (2004). Hamel Coefficients for the Rotational Motion of a Rigid Body. *The Journal of the Astronautical Sciences*, Vol.52, Nos.1 and 2: 129-147.

Jain, A. & Rodriguez, G. (1995). Diagonalized Lagrangian Robot Dynamics. *IEEE Transactions on Robotics and Automation*, Vol.11, No.4: 571-584.

Junkins, J.L. & Schaub H. (1997). An Instantaneous Eigenstructure Quasivelocity Formulation for Nonlinear Multibody Dynamics. *The Journal of the Astronautical Sciences*, Vol.45, No.3: 279-295.

Kane, T.R. & Levinson D.A. (1983). The Use of Kane's Dynamical Equations in Robotics. *The International Journal of Robotics Research*, Vol.2, No.3: 3-21.

Kelly R. & Moreno J. (2005). Manipulator motion control in operational space using joint velocity inner loops. *Automatica*, Vol.41, No. 8: 1423-1432.

Khalil, H. (1996). *Nonlinear Systems*, Prentice Hall, New Jersey.

Kozlowski, K. (1992). *Mathematical Dynamic Robot Models and Identi□cation of Their Parameters*, Poznan University of Technology Press, Poznan, (in Polish).

Kwatny, H.G. & Blankenship, G.L. (2000). *Nonlinear Control and Analytical Mechanics*, Birkhäuser, Boston.

Loduha T.A. & Ravani B. (1995). On First-Order Decoupling of Equations of Motion for Constrained Dynamical Systems. *Journal of Applied Mechanics - Transactions of the ASME*, Vol.62, March: 216-222.

(1996). *MATLAB, Using Matlab*, The MathWorks, Inc.

Mochiyama, H.; Shimemura, E. & Kobayashi H. (1999). Shape Control of Manipulator with Hyper Degrees of Freedom. *The International Journal of Robotics Research*, Vol.18, No.6: 584-600.

Moreno, J. & Kelly, R. (2003). Velocity control of robot manipulators: analysis and experiments. *International Journal of Control*, Vol.76, No.14: 1420-1427.

Moreno, J.; Kelly, R. & Campa, R. (2003). Manipulator velocity control using friction compensation. *IEE Proc. - Control Theory and Applications*, Vol.150, no.2: 119-126.

Moreno-Valenzuela J. & Kelly R. (2006). A Hierarchical Approach Manipulator Velocity Field Control Considering Dynamic Friction Compensation. *Journal of Dynamic Systems, Measurement, and Control - Transactions of the ASME*, Vol.128, September: 670-674.

Santibanez, V. & Kelly R. (1997). Strict Lyapunov Functions for Control of Robot Manipulators. *Automatica*, Vol.33, No.4: 675-682.

Sciavicco L. & Siciliano B. (1996). *Modeling and Control of Robot Manipulators*, The McGraw-Hill Companies, Inc., New York.

Slotine J.-J., & Li W. (1987). On the Adaptive Control of Robot Manipulators. *The International Journal of Robotics Research*, Vol.6, No.3: 49-59.
Slotine J.-J. & Li W. (1991). *Applied Nonlinear Control*, Prentice Hall, New Jersey.
Sovinsky, M.C.; Hurtado, J.E.; Griffith D.T. & Turner, J.D. (2005). The Hamel Representation: Diagonalized Poincaré Form, *Proceedings of IDETC'05 2005 ASME International Design Engineering Technical conference and Computers and Information in Engineering Conference*, DETC2005-85650, Long Beach, California, USA, September 24-28, 2005.
Spong, M.W. (1992). On the Robust Control of Robot Manipulators. *IEEE Transactions on Automatic Control*, Vol.37, No.11: 1782-1786.

Part 4

Selected Applications of Sliding Mode Control

15

Force/Motion Sliding Mode Control of Three Typical Mechanisms

Rong-Fong Fung[1] and Chin-Fu Chang[2]
[1]*Department of Mechanical & Automation Engineering*
National Kaohsiung First University of Science and Technology,
1 University Road, Yenchau, Kaohsiung 824,
[2]*Institute of Engineering Science and Technology,*
National Kaohsiung First University of Science and Technology,
1 University Road, Yenchau, Kaohsiung 824,
Taiwan

1. Introduction

A number of papers [1-7] have been presented to address the issues of multi-body mechanisms. Examples of their applications are found in gasoline and diesel engines, where the gas force acts on the slider and the motion is transmitted through the links. Whether the connecting rod is assumed to be rigid or not, the steady-state and dynamic responses of the connecting rod of the mechanism with time-dependent boundary condition were obtained by Fung *et al.* [1-3]. In addition, a number of controllers, for example, repetitive control [4], adaptive control [5], computed torque control [6], and fuzzy neural network control [7] were designed for the multi-body mechanisms.
Over the past 25 years, the SMC algorithm [8-10] has been taken into account for dynamic control problems. The main feature of the SMC is to allow the sliding mode to occur on a prescribed switching surface, so that the system is only governed by the sliding equation and remains insensitive to a class of disturbances and parameter variations [8]. It is noted that the SMC is a robust control method and has been well established in pure motion control [9]. Afterwards, in order to eliminate the chattering phenomenon, which is commonly found in simulation of discontinuous SMC systems, and to simplify a hybrid numerical method that incorporates benefits of both SMC and differential algebraic equations, the (DAE) stabilization method was developed and successfully used to simulate constrained multi-body systems (MBS) whether under holonomic constraint or not [10]. However, the development of a control law which has been induced by a constrained force has not been adequately developed consistently in the previous studies. Su *et al.* [11] attempted to use the SMC for simultaneous position and force control on a constrained robot manipulator. They asserted that the control law, along with inclusion of the constraint force error in the definition of the sliding surface, produces an asymptotically stable force tracking error. However, Grabbe and Bridges [12] addressed their formulation as being a departure from the typical definition of a sliding surface, which is a linear differential equation in one tracking error variable [13], and the errors in the separate force control law and stability analysis were presented in [11]. Recently, Lian and Lin [14] have proposed a

new sliding surface in terms of motion error and force error, and claimed that the errors in [11] are improved; therefore, the asymptotic stability of the motion-tracking error and force-tracking error can be ensured. However, Dixon and Zergeroglu [15] pointed out an error in the sliding mode control stability analysis of [14].

In this chapter, our intent is to improve the errors in [11, 14] and simplify the control design and stability proof for the three typical mechanisms, including the slider-crank mechanism, the quick-return mechanism and the toggle mechanism as shown in Figs. 1~3 respectively, which are not seen in any references addressing the force/motion SMC. Here, a separate sliding surface is proposed using the measurements of the angular position and speed of the crank, but the SMC algorithm is derived as well in a simple manner using only the force tracking error to construct the controller. In these schemes, the force tracking error is shown to be arbitrarily small by changing the force control feedback gain. Then, by exploiting the structure of its dynamics, the fundamental properties of the dynamics are obtained to facilitate controller design, whereby the asymptotic stability of motion tracking error in sliding surface and force tracking error accumulated in controller can be ensured.

The organization of this chapter is arranged as follows. In Section 2, the kinematic and dynamic analysis of the multi-body mechanism is investigated. A number of previous papers [4-7, 16-17] have shown the position and speed controllers for the regulation and tracking problems of the multi-body mechanism in the theoretical analysis and experimental results. However, control of the constrained force has not been investigated. The SMC laws are designed in Section 3. The simulated examples are shown in Section 4 and, finally, some conclusions are drawn.

2. Dynamics analysis

2.1 Dynamic equation of motion

Based on the Euler-Lagrange formulation [4], the equation of motion for a mechanism can be expressed as:

$$\mathbf{M(Q)\ddot{Q}} + \mathbf{N}\left(\mathbf{Q},\dot{\mathbf{Q}}\right) + \mathbf{\Phi}_{\mathbf{Q}}^{\mathbf{T}}\boldsymbol{\lambda} = \mathbf{Q}^{\mathbf{A}} + \mathbf{U}\,. \tag{1}$$

where $\mathbf{M(Q)}$ is an $n \times n$ inertia matrix, $\mathbf{Q} \in R^n$ is the generalized coordinate vector, $\mathbf{N(Q,\dot{Q})} \in R^n$ is the nonlinear vector, $\boldsymbol{\lambda} \in R^m$ is the vector of Lagrange multipliers, $\mathbf{\Phi}_{\mathbf{Q}} = \left[\, \partial\mathbf{\Phi}/\partial\, \mathbf{Q}\right] \in R^{m\times n}$ is the partial derivative of the constraint equation with respect to the coordinate and is called the constraint Jacobian matrix, $\mathbf{Q}^{\mathbf{A}} \in R^n$ is the vector of non-conservative forces and $\mathbf{U} \in R^n$ is the vector of applied control efforts.

In order to obtain the general form of the force/motion controller design, we rewrite the nonlinear vector as:

$$\mathbf{N(Q,\dot{Q})} = \mathbf{N}_{\mathbf{C}}\mathbf{(Q,\dot{Q})\dot{Q}} + \mathbf{N}_{\mathbf{G}}\mathbf{(Q)}\,. \tag{2}$$

where $\mathbf{N}_{\mathbf{C}}\mathbf{(Q,\dot{Q})} \in R^{n\times n}$ is the vector of coriolis and centrifugal forces; $\mathbf{N}_{\mathbf{G}}\mathbf{(Q)} \in R^n$ is the vector of gravitational force.

Then, Equation (1) becomes:

$$\mathbf{M(Q)\ddot{Q}} + \mathbf{N}_{\mathbf{C}}\mathbf{(Q,\dot{Q})\dot{Q}} + \mathbf{N}_{\mathbf{G}}\mathbf{(Q)} = \mathbf{Q}^{\mathbf{A}} + \mathbf{U} + \mathbf{F}\,. \tag{3}$$

where $\mathbf{F} = -\mathbf{\Phi}_{\mathbf{Q}}^{\mathbf{T}}\boldsymbol{\lambda}$ is the constraint force.

2.2 Dynamic properties of the mechanism

Equation (3) is similar to the motion equation of an n-link rigid constrained robot [11, 15] in the state space. Two simplifying properties should be noted about this dynamic structure:
Property 1. The individual terms on the left-hand side of Equation (3) and the whole dynamics are linear in terms of a suitably selected set of equivalent manipulator and load parameters, *i.e.*,

$$\mathbf{M(Q)\ddot{Q}} + \mathbf{N_C(Q,\dot{Q})\dot{Q}} + \mathbf{N_G(Q)} = \mathbf{Y(Q,\dot{Q},\ddot{Q})}\alpha . \tag{4}$$

where $\mathbf{Y(Q,\dot{Q},\ddot{Q})}$ is a $n \times r$ matrix; $\alpha \in R^r$ is the vector of equivalent parameters.
Property 2. From the given proper definition of the matrix $\mathbf{N_C(Q,\dot{Q})}$, $\mathbf{\dot{M}(Q)} - 2\mathbf{N_C(Q,\dot{Q})}$ is skew-symmetric. The detailed proof can be seen in Appendix A.
Due to the presence of *m* constraints, the degree of freedom of the mechanism is *(n-m)*. In this case, *(n-m)* linearly independent coordinates are sufficient to characterize the constrained motion. From the implicit function theorem, the constraint Equation (1) can always be expressed as [18]:

$$\mathbf{p} = \sigma(\mathbf{q}) . \tag{5}$$

Equation (5) is assumed that the elements of **q** are chosen to be the last *(n-m)* components of **Q.** If not the above case, Equation (1) still could always be reordered so that the last *(n-m)* equations would correspond to **q** and the first *m* equations to **p.** That is, $\mathbf{Q} = \begin{bmatrix} \mathbf{p}^T & \mathbf{q}^T \end{bmatrix}^T$.
Then, to simplify the equation form of the dynamic model, defining

$$\mathbf{L(q)} = \frac{\partial \mathbf{Q}}{\partial \mathbf{q}} = \begin{bmatrix} \frac{\partial \mathbf{p}}{\partial \mathbf{q}} & \frac{\partial \mathbf{q}}{\partial \mathbf{q}} \end{bmatrix}^T = \begin{bmatrix} \frac{\partial \sigma(\mathbf{q})}{\partial \mathbf{q}} & \mathbf{I_{n-m}} \end{bmatrix}^T \in R^{n\times(n-m)} . \tag{6}$$

and using Equation (5), we have:

$$\mathbf{\dot{Q}} = \mathbf{L(q)\dot{q}} , \tag{7}$$

$$\mathbf{\ddot{Q}} = \mathbf{L(q)\ddot{q}} + \mathbf{\dot{L}(q)\dot{q}} . \tag{8}$$

Therefore, the dynamic model of Equation (3) restricted to the constraint surface can be expressed in a reduced form as:

$$\mathbf{M(q)L(q)\ddot{q}} + \mathbf{N_1(q,\dot{q})\dot{q}} + \mathbf{N_G(q)} = \mathbf{Q^A} + \mathbf{U} + \mathbf{F} . \tag{9}$$

where

$$\mathbf{N_1(q,\dot{q})} = \mathbf{M(q)\dot{L}(q)} + \mathbf{N_C(q,\dot{q})L(q)} . \tag{10}$$

By exploiting the structure of Equation (9), three properties can be obtained as follows:
Property 3. In terms of a suitably selected set of parameters, the motion equation (9) is still linear, *i.e.*

$$\mathbf{M(q)L(q)\ddot{q}} + \mathbf{N_1(q,\dot{q})\dot{q}} + \mathbf{N_G(q)} = \mathbf{Y_1(q,\dot{q},\ddot{q})}\alpha . \tag{11}$$

Property 4. Define the matrix as

$$\mathbf{A(q)} = \mathbf{L}^{\mathrm{T}}\mathbf{(q)M(q)L(q)} \in R^{(n-m)\times(n-m)} . \tag{12}$$

Then, $\dot{\mathbf{A}} = \mathbf{2L}^{\mathrm{T}}\mathbf{(q)N_1(q,\dot{q})}$ is skew-symmetric, where
$N_1(q,\dot{q}) = M(q)\dot{L}(q) + N_C(q,\dot{q})L(q) \in R^{n\times(n-m)}$, $L(q) \in R^{n\times(n-m)}$.
Property 5. $[\mathbf{\Phi_Q L(q)}]^{\mathrm{T}} = \mathbf{L}^{\mathrm{T}}\mathbf{(q)}\mathbf{\Phi}_{\mathrm{Q}}^{\mathrm{T}} = \mathbf{0}$.
The above three properties are basic principle in designing the force/motion SMC law.

3. Design of the SMC Law

3.1 The sliding mode controller design

A number of previous papers have only shown the position and speed controller designed for the regulation and tracking problems control of the constrained mechanisms. However, control of the constrained force has not been investigated in the previous studies. In this section, a separate sliding surface is proposed using the measurements of the angular position and speed of the crank, but the SMC algorithm is derived as well in a simple manner using only the force tracking error to construct the controller.
Given a desired trajectory $\mathbf{q_d}$ and a desired constrained force $\mathbf{F_d}$, or identically a desired multiplier $\boldsymbol{\lambda}_\mathbf{d}$, which satisfy the imposed constraint, *i.e.*, $\mathbf{\Phi(q_d)} = 0$ and $\mathbf{F_d} = -\mathbf{\Phi}_{\mathrm{Q}}^{\mathrm{T}}\mathbf{(q_d)}\boldsymbol{\lambda}_\mathbf{d}$. The control objective is to determine the SMC law such that $\mathbf{q} \to \mathbf{q_d}$ and $\boldsymbol{\lambda} \to \boldsymbol{\lambda}_\mathbf{d}$ as $t \to \infty$.
From the SMC methodology, we define the tracking error $\mathbf{e_m} \in R^{n-m}$ and a sliding surface $\mathbf{s_1} \in R^{n-m}$ as:

$$\mathbf{e_m} = \mathbf{q(t)} - \mathbf{q_d(t)} . \tag{13}$$

$$\mathbf{s_1} = \dot{\mathbf{q}} - \dot{\mathbf{q}}_\mathbf{r} = \dot{\mathbf{e}}_\mathbf{m} + \mathbf{\Lambda e}_m . \tag{14}$$

where $\mathbf{q_r} \in R^{n-m}$ is the reference trajectory and $\mathbf{\Lambda} \in R^{(n-m)\times(n-m)}$ is a tunable matrix.
The sliding controller [12] is defined as:

$$\mathbf{U} = \mathbf{Y_1(q,\dot{q},\ddot{q})}\varphi - \mathbf{L(q)s_1} + \mathbf{\Phi}_{\mathrm{Q}}^{\mathrm{T}}\mathbf{(q)}\boldsymbol{\lambda}_\mathbf{c} - \mathbf{Q^A} . \tag{15}$$

where $\mathbf{Y_1}$ is a $n\times r$ matrix of known functions of $\mathbf{q,\dot{q}}$ and $\ddot{\mathbf{q}}$, $\mathbf{L(q)}$ is defined in Equation (6), $\varphi = [\varphi_1 ... \varphi_r]^T \in R^r$ is the vector of switching functions, and $\boldsymbol{\lambda}_\mathbf{c} \in R^{\mathrm{m}}$ is a force control that is defined as:

$$\boldsymbol{\lambda}_\mathbf{c} = \boldsymbol{\lambda}_\mathbf{d} - \mathbf{K e}_\lambda . \tag{16}$$

where $\mathbf{K}$ is a $\mathrm{m}\times\mathrm{m}$ constant matrix of force control feedback gains, and $\mathbf{e}_\lambda$ is the error vector of the multipliers and defined as

$$\mathbf{e}_\lambda = \boldsymbol{\lambda} - \boldsymbol{\lambda}_\mathbf{d} \in R^{\mathrm{m}} . \tag{17}$$

3.2 Stability analysis

Substituting Equation (15) into the dynamic model of Equation (9), whose order was reduced using ***property*** 3, we have:

$$\mathbf{Y}_1(\mathbf{q},\dot{\mathbf{q}},\ddot{\mathbf{q}})\varphi - \mathbf{L}(\mathbf{q})\mathbf{s}_1 + \mathbf{\Phi}_Q^T(\mathbf{q})\boldsymbol{\lambda}_c - \mathbf{\Phi}_Q^T(\mathbf{q})\boldsymbol{\lambda} = \mathbf{Y}_1(\mathbf{q},\dot{\mathbf{q}},\ddot{\mathbf{q}})\alpha . \quad (18)$$

Defining α as a constant r-dimensional vector and replacing $\mathbf{q}$ by the reference trajectory $\mathbf{q}_r$, then the linear parameterization of the dynamics (***Property*** 3) leads to:

$$\mathbf{M}(\mathbf{q}_r)\mathbf{L}(\mathbf{q}_r)\ddot{\mathbf{q}}_r + \mathbf{N}_1(\mathbf{q}_r,\dot{\mathbf{q}}_r)\dot{\mathbf{q}}_r + \mathbf{N}_G(\mathbf{q}_r) = \mathbf{Y}_1(\mathbf{q}_r,\dot{\mathbf{q}}_r,\ddot{\mathbf{q}}_r)\alpha . \quad (19)$$

Using the derivative of the sliding surface equation (14) and substituting into Equation (11), we obtain:

$$\mathbf{ML}(\dot{\mathbf{s}}_1 + \ddot{\mathbf{q}}_r) + \mathbf{N}_1(\mathbf{s}_1 + \dot{\mathbf{q}}_r) + \mathbf{N}_G = \mathbf{Y}_1\alpha . \quad (20)$$

Then combining Equation (18) with Equation (20) and using Equation (19), we obtain:

$$\mathbf{ML}\dot{\mathbf{s}}_1 = \mathbf{Y}_1\varphi - \mathbf{Y}_1\boldsymbol{\alpha} - \mathbf{N}_1\mathbf{s}_1 - \mathbf{L}\mathbf{s}_1 + \mathbf{\Phi}_Q^T\boldsymbol{\lambda}_C - \mathbf{\Phi}_Q^T\boldsymbol{\lambda} . \quad (21)$$

According to ***property*** 5, the above equation becomes:

$$\begin{aligned}\mathbf{A}\dot{\mathbf{s}}_1 &= \mathbf{L}^T\mathbf{ML}\dot{\mathbf{s}}_1 \\ &= \mathbf{L}^T\mathbf{Y}_1\varphi - \mathbf{L}^T\mathbf{Y}_1\alpha - \mathbf{L}^T\mathbf{N}_1\mathbf{s}_1 - \mathbf{L}^T\mathbf{L}\mathbf{s}_1.\end{aligned} \quad (22)$$

To derive the control algorithm, the generalized Lyapunov function is considered as:

$$\mathbf{V} = \frac{1}{2}\mathbf{s}_1^T\mathbf{A}\mathbf{s}_1 . \quad (23)$$

Differentiating V with respect to time and using ***property 4***, Equation (23) becomes:

$$\begin{aligned}\dot{\mathbf{V}} &= s_1{}^T A\dot{s}_1 + \frac{1}{2}s_1{}^T\dot{A}s_1 = s_1{}^T A\dot{s}_1 + s_1{}^T L^T N_1 s_1 \\ &= s_1{}^T(L^T Y_1\varphi - L^T Y_1\alpha - L^T N_1 s_1 - L^T L s_1) + s_1^T L^T N_1 s_1 \\ &= s_1{}^T(\mathbf{L}^\mathbf{T} Y_1\varphi - L^T Y_1\alpha - L^T L s_1).\end{aligned} \quad (24)$$

The φ is chosen as:

$$\varphi = -\bar{\alpha}_1\mathbf{sgn}\left(\sum_{j=1}^{n-m}\mathbf{s}_{1j}(\mathbf{L}^T\mathbf{Y}_1)_{ji}\right); i = 1,2,\ldots,r \quad (25)$$

such that

$$\dot{\mathbf{V}} \le \mathbf{s}_1^T\mathbf{L}^T\mathbf{L}\mathbf{s}_1 < 0 . \quad (26)$$

In the derivation of Eq. (24), it is noted that $\dot{M}(Q) - 2N_C(Q,\dot{Q})$ is skew-symmetric, and $L^T(\dot{M} - 2N_C)L$ is also skew-symmetric, which is the same as those in [11, 14]. Besides, the special cases of the three typical mechanisms in this chapter, $L^T(\dot{M} - 2N_C)L$ is always equal to zero for $n - m = 1$.

To reduce the chattering phenomenon along the sliding surface $s = 0$, we adopt the quasi-linear mode controller [13], which replaces the discontinuous term of sign function of

Equation (25) with a continuous function inside a boundary layer around the sliding surface [24]. Therefore, the sgn(S) is replaced by the saturated function:

$$sat(\frac{s}{\varepsilon}) = \begin{cases} 1 & if\ s > \varepsilon, \\ \frac{s}{\varepsilon} & if\ -\varepsilon < s < \varepsilon, \\ -1 & if\ s > \varepsilon, \end{cases}$$

where ε is the width of the boundary layer. This limits the tracking error and guarantees an accuracy of ε order while alleviating the chattering phenomenon.
From Equation (23) and Equation (26), it is evident that a sliding surface $\|\mathbf{s_1}\|$ is at last converged exponentially to zero, *i.e.*, $\mathbf{e_m} \to \mathbf{0}$ as $t \to \infty$. As if $\mathbf{q} \to \mathbf{q_d}$, the condition $\mathbf{p_d} = \boldsymbol{\sigma}(\mathbf{q_d})$ also implies that $\mathbf{p} \to \mathbf{p_d}$.
Therefore,

$$\mathbf{q} \to \mathbf{q_d} \quad as\ t \to \infty .$$

4. Simulation examples of the three typical mechanisms

4.1 The slider-crank mechanism

For more details on the kinematic and dynamic analysis of the slider-crank mechanism, refer to [19]. Using Hamilton's principle and Lagrange multipliers [20] and adopting the generalized coordinate vector $\mathbf{Q} = [\phi \quad \theta]^{\mathrm{T}}$ in Equation (1) for the slider-crank mechanism shown in Figure 1, the dynamic equation can be obtained associated with the following matrices and elements:

$$\begin{aligned}
&\mathbf{M} = \begin{bmatrix} A & E \\ E & B \end{bmatrix} \quad \mathbf{N_C} = \begin{bmatrix} K_1 & K_2 \\ P_1 & P_2 \end{bmatrix} \quad \mathbf{N_G} = \begin{bmatrix} K_3 \\ P_3 \end{bmatrix} \\
&\mathbf{\Phi_Q} = [-l\cos\phi \quad r\cos\theta] \quad \mathbf{Q^A} = \begin{bmatrix} (F_B + F_E)l\sin\phi \\ (F_B + F_E)r\sin\theta \end{bmatrix} \quad \mathbf{U} = \begin{bmatrix} u_1 \\ u_2 \end{bmatrix} \\
&A = -\frac{1}{3}m_2 l^2 - m_B l^2 \sin^2\phi \quad E = -\left(\frac{1}{2}m_2 + m_B\right) rl\sin\theta\sin\phi \\
&B = -\frac{1}{2}m_1 r^2 - (m_2 + m_B) r^2 \sin^2\theta \\
&K_1 = -m_B l^2 \dot{\phi}\sin\phi\cos\phi \quad K_2 = -\left(\frac{1}{2}m_2 + m_B\right) rl\dot{\theta}\cos\theta\sin\phi \quad K_3 = -\frac{1}{2}m_2 gl\cos\phi \\
&P_1 = -\left(\frac{1}{2}m_2 + m_B\right) rl\dot{\phi}\sin\theta\cos\phi \quad P_2 = -(m_2 + m_B) r^2 \dot{\theta}\sin\theta\cos\theta \quad P_3 = 0
\end{aligned} \tag{27}$$

where the dimensions of the slider-crank mechanism are $n = 2, m = 1$, and $r = 1$ in the dynamic analysis. For the single degree-of-freedom slider-crank mechanism, only one constraint equation exists, which can be shown as:

$$\mathbf{\Phi(Q)} = r\sin\theta - l\sin\phi = 0 . \tag{28}$$

The position of the slider **B** can be expressed as:

$$x_B = r\cos\theta + l\cos\phi \,. \tag{29}$$

Substituting Equation (28) into Equation (29) yields:

$$x_B = r\cos\theta + \left[l^2 - r^2\sin^2\theta\right]^{\frac{1}{2}} \tag{30}$$

The angular displacement of the crank can be obtained as:

$$\theta = \cos^{-1}\left[\frac{x_B{}^2 + r^2 - l^2}{2rx_B}\right]. \tag{31}$$

The result can also be obtained if the cosine law is applied.
The Jacobian matrix of the constraint equation (28) is

$$\mathbf{\Phi_Q} = \left[\frac{\partial\mathbf{\Phi(Q)}}{\partial\mathbf{Q}}\right] = \left[\frac{\partial\mathbf{\Phi(Q)}}{\partial\phi} \quad \frac{\partial\mathbf{\Phi(Q)}}{\partial\theta}\right] = \left[-l\cos\phi \quad r\cos\theta\right]. \tag{32}$$

Differentiating Equation (28) with respect to time yields the constraint velocity equation:

$$\dot{\mathbf{\Phi}}\mathbf{(Q)} = r\dot{\theta}\cos\theta - l\dot{\phi}\cos\phi = 0\,. \tag{33}$$

Therefore, the matrix defined in Equation (6) becomes

$$\mathbf{L(q)} = \frac{\partial\mathbf{Q}}{\partial\mathbf{q}} = \begin{bmatrix}\frac{\partial\phi}{\partial\theta} & \frac{\partial\theta}{\partial\theta}\end{bmatrix}^T = \begin{bmatrix}\frac{\dot{\phi}}{\dot{\theta}} & 1\end{bmatrix}^T = \begin{bmatrix}\frac{r\cos\theta}{l\cos\phi} & 1\end{bmatrix}^T \tag{34}$$

and its first time derivative becomes

$$\dot{\mathbf{L}}\mathbf{(q)} = \begin{bmatrix}\frac{rl\dot{\phi}\sin\phi\cos\theta - rl\dot{\theta}\cos\phi\sin\theta}{l^2\cos^2\phi} & 0\end{bmatrix}^T \tag{35}$$

The dynamic equation (9) of the slider-crank mechanism, when restricted to the constraint equation (28), can be expressed as:

$$\begin{bmatrix}A\frac{r\cos\theta}{l\cos\phi} + E \\ E\frac{r\cos\theta}{l\cos\phi} + B\end{bmatrix}\ddot{\theta} + \begin{bmatrix}A\frac{rl\dot{\phi}\sin\phi\cos\theta - rl\dot{\theta}\cos\phi\sin\theta}{l^2\cos^2\phi} + K_1\frac{r\cos\theta}{l\cos\phi} + K_2 \\ E\frac{rl\dot{\phi}\sin\phi\cos\theta - rl\dot{\theta}\cos\phi\sin\theta}{l^2\cos^2\phi} + P_1\frac{r\cos\theta}{l\cos\phi} + P_2\end{bmatrix}\dot{\theta} + \begin{bmatrix}-\frac{1}{2}m_2 gl\cos\phi \\ 0\end{bmatrix} = \begin{bmatrix}(F_B + F_E)l\sin\phi \\ (F_B + F_E)r\sin\theta\end{bmatrix} + \begin{bmatrix}u_1 \\ u_2\end{bmatrix} + \begin{bmatrix}l\cos\phi \\ -r\cos\theta\end{bmatrix}\lambda. \tag{36}$$

The symbols $A, E, B, K_1, K_2, P_1, P_2$ and P_2 are shown in Equation (27). It is noted that F_B is the friction force, F_E is the external force, and $f_1 = l\cos\phi\lambda$ and $f_2 = -r\cos\theta\lambda$ are the constraint forces. From the results presented above, ***property 1~property 5*** mentioned in Section 2.2 are all verified and fully satisfied in this example.

The control objective is to design a feedback controller so that the angle θ tracks the desired trajectory θ_d and maintains the constraint force $[f_1 \quad f_2]^T$ to the desired one $\mathbf{F_d}$. In here, θ_d and $\mathbf{F_d}$ are assumed to be consistent with the imposed constraint. The block diagram of the SMC algorithm is shown in Figure 4.
Since $\lambda \to \boldsymbol{\lambda}_\mathbf{d}$ means $[f_1 \quad f_2]^T \to \mathbf{F_d}$, $\boldsymbol{\theta}_\mathbf{d}$ and $\boldsymbol{\lambda}_\mathbf{d}$ are chosen as $\boldsymbol{\theta}_\mathbf{d} = 5.76(rad)$ and $\boldsymbol{\lambda}_\mathbf{d} = 15$ in the simulations. The initial values of the constraint forces are assumed to be $\mathbf{F_d(0)} = [\mathbf{f_1(0)} \quad \mathbf{f_2(0)}]^\mathbf{T} = 0$, *i.e.*, $\boldsymbol{\lambda}(0) = 0$.
Using Equation (19), the applied control effort $\mathbf{U} = [u_1 \quad u_2]^T$ can be derived as:

$$\mathbf{U} = \mathbf{Y_1}\varphi - \mathbf{L s_1} + \boldsymbol{\Phi}_\mathbf{Q}^\mathbf{T}\boldsymbol{\lambda}_\mathbf{C} - \mathbf{Q^A}. \tag{37}$$

where

$$\mathbf{Y_1} = \frac{\mathbf{1}}{\boldsymbol{\alpha}}\left[\mathbf{ML}\ddot{\boldsymbol{\theta}}_\mathbf{d} + \mathbf{N_1}\dot{\boldsymbol{\theta}}_\mathbf{d} + \mathbf{N_G}\right],\ \varphi = \begin{cases} \bar{\boldsymbol{\alpha}}_\mathbf{1}, & \mathbf{s_1}(\mathbf{L^T Y_1}) < 0 \\ -\bar{\boldsymbol{\alpha}}_\mathbf{1}, & \mathbf{s_1}(\mathbf{L^T Y_1}) > 0 \end{cases},\ \mathbf{s_1} = \dot{\mathbf{e}}_\mathbf{m} + \boldsymbol{\Lambda}\mathbf{e_m} = (\dot{\boldsymbol{\theta}} - \dot{\boldsymbol{\theta}}_\mathbf{d}) + \boldsymbol{\Lambda}(\boldsymbol{\theta} - \boldsymbol{\theta}_\mathbf{d}).$$

For numerical simulations, the parameters of the slider-crank mechanism are chosen as: $m_1 = 3.64\ kg,\ m_2 = 1.18\ kg,\ m_B = 1.8\ kg,\ r = 0.1\ m,\ l = 0.305\ m,\ lp = 0.055\ m.$ and $\alpha_1 = 1$, $\bar{\alpha}_1 = 1$ and $\Lambda = 5$.
Since the trajectory tracking on the constraint surface with a specified constraint force is of interest, the initial position and speed of the slider-crank mechanism are chosen on the desired trajectory as:

$$\begin{aligned} &\theta(0) = 4.712(rad);\quad \phi(0) = -0.334(rad);\quad \dot{\theta}(0) = 0;\quad \dot{\phi}(0) = 0; \\ &x_B(t_0) = 0.343\ m,\ x_B(t_f) = 0.443\ m. \end{aligned}$$

All the parameters in the SMC controller are chosen to achieve the best transient performance in numerical simulations under the limitation of the control effort and the requirements of stability. Furthermore, for the reason of using a single input actuation on joint 1, the control effort is only needed in the second equation of the constrained motion of Equation (36). As to the first part of Equation (36), it shows that the force equilibrium either with holonomic or nonholonomic constraints and the torque is exerted at joint 2. The responses of the crank angle, which is shown in Figure 5(a), reach the desired value in about 0.8 sec. The slider position manipulated from Equation (29) is shown in Figure 5(b). The tracking results of the crank angle θ and the slider position x_B coincide with previous studies by Fung *et al.* [16, 17]. The control effort of the applied torque τ is shown in Figure 5(c) and the sliding surface $\mathbf{s_1}$ is shown in Figure 5(d). The Lagrange multiplier λ_C is shown in Figure 6(a), and constraint forces of joints 1 and 2 are shown in Figure 6(b) and Figure 6(c), respectively. From Figures 5 and 6, the control objectives of force/motion of the slider-crank mechanism are achieved successfully.

4.2 The quick-return mechanism

To present the robustness and a well-established control method of the SMC controller, the quick-return and toggle mechanisms (see Section 4.3) will be chosen to verify the SMC algorithm, which is then adequately developed to the general case of multi-body mechanisms. The quick-return mechanism is addressed first, where the kinematic and dynamic analysis of the mechanism is found in [21] and the generalized coordinate vector $\mathbf{Q}=[\phi \ \ \beta \ \ \theta]^{\mathrm{T}}$ in Equation (1) for the quick-return mechanism shown in Figure 2 is adopted. The dynamic equation can be obtained and is associated with the following matrices and elements:

$$\mathbf{M}=\begin{bmatrix} A1 & G1 & 0 \\ H1 & B1 & 0 \\ 0 & 0 & CC \end{bmatrix} \quad \mathbf{N}_C=\begin{bmatrix} P1 & T1 & 0 \\ Y1 & R1 & 0 \\ 0 & 0 & 0 \end{bmatrix} \quad \mathbf{N}_{\mathbf{G}}=0$$

$$\mathbf{\Phi}_{\mathrm{Q}}=\begin{bmatrix} \cos\phi(D+R\cos\theta)+R\sin\theta\sin\phi & 0 & -R\sin\phi\sin\theta-R\cos\theta\cos\phi \\ -L\sin\phi & S\cos\beta & 0 \end{bmatrix}$$

$$\mathbf{Q}^{\mathbf{A}}=\begin{bmatrix} PL\cos\phi \\ PS\sin\beta \\ 0 \end{bmatrix} \quad \mathbf{U}=\begin{bmatrix} u_1 \\ u_2 \\ u_3 \end{bmatrix}$$

$$A1=-\frac{1}{3}m_1L^2-(m_2+m_C)L^2\cos^2\phi\,,\ G1=-\left(\frac{1}{2}m_2+m_C\right)SL\sin\beta\cos\phi\,,$$

$$H1=-\left(\frac{1}{2}m_2+m_C\right)SL\sin\beta\cos\phi\,,\ B1=-\left(\frac{1}{3}m_2+m_C\sin^2\beta\right)S^2\,,$$

$$CC=-\frac{1}{3}m_3R^2,\ \ P1=(m_2+m_C)L^2\dot{\phi}\cos\phi\sin\phi\,,\,,$$

$$T1=-\left(\frac{1}{2}m_2+m_C\right)SL\dot{\beta}\cos\beta\cos\phi\,,\ Y1=\left(\frac{1}{2}m_2+m_C\right)SL\dot{\phi}\sin\beta\sin\phi\,,$$

$$R1=-m_CS^2\dot{\beta}\sin\beta\cos\beta. \tag{38}$$

where $n=3$, $m=2$ and $r=1$ are employed in the dynamic analysis.

For the single degree-of-freedom quick-return mechanism, there exist two constraint equations as follows:

$$\mathbf{\Phi}(\mathbf{Q})=\begin{bmatrix} \sin\phi(D+R\cos\theta)-R\sin\theta\cos\phi \\ S\sin\beta-L(1-\cos\phi) \end{bmatrix}=0\,. \tag{39}$$

where φ can be obtained by analyzing its geometric relations

$$\phi=\tan^{-1}\frac{R\sin\theta}{D+R\cos\theta}\,. \tag{40}$$

The position of slider C can be expressed as:

$$x_C=S\cos\beta-L\sin\phi\,. \tag{41}$$

The Jacobian matrix of the constraint equations is:

$$\mathbf{\Phi_Q} = \begin{bmatrix} \cos\phi(D+R\cos\theta)+R\sin\theta\sin\phi & 0 & -R\sin\phi\sin\theta - R\cos\theta\cos\phi \\ -L\sin\phi & S\cos\beta & 0 \end{bmatrix}. \tag{42}$$

Therefore, the matrix defined in Equation (6) is:

$$\begin{aligned} \mathbf{L(q)} &= \frac{\partial \mathbf{Q}}{\partial \mathbf{q}} = \begin{bmatrix} \frac{\partial\phi}{\partial\theta} & \frac{\partial\beta}{\partial\theta} & \frac{\partial\theta}{\partial\theta} \end{bmatrix}^T = \begin{bmatrix} \frac{\dot\phi}{\dot\theta} & \frac{\dot\beta}{\dot\theta} & 1 \end{bmatrix} = \begin{bmatrix} \frac{\dot\phi}{\dot\theta} & \frac{\dot\beta}{\dot\phi}\times\frac{\dot\phi}{\dot\theta} & 1 \end{bmatrix} \\ &= \begin{bmatrix} \frac{(DR\cos\theta+R^2)}{D^2+R^2+2DR\cos\theta} & \frac{L\sin\phi(DR\cos\theta+R^2)}{S\cos\beta(D^2+R^2+2DR\cos\theta)} & 1 \end{bmatrix}^T \end{aligned} \tag{43}$$

Then, differentiating Equation (43) with respect to time yields:

$$\dot{\mathbf{L}}\mathbf{(q)} = \begin{bmatrix} \frac{DR\dot\theta\sin\theta(R^2-D^2)}{(D^2+R^2+2DR\cos\theta)^2} \\ \frac{LSDR\dot\theta\sin\phi\sin\theta\cos\beta(R^2-D^2)}{S^2\cos^2\beta(D^2+R^2+2DR\cos\theta)^2} + \\ \frac{LS(D^2+R^2+2DR\cos\theta)(DR\cos\theta+R^2)(\dot\beta\sin\phi\sin\beta+\dot\phi\cos\phi\cos\beta)}{S^2\cos^2\beta(D^2+R^2+2DR\cos\theta)^2} \\ 0 \end{bmatrix} = \begin{bmatrix} \dot L_{11} \\ \dot L_{21} \\ 0 \end{bmatrix}. \tag{44}$$

The dynamic equation of a quick-return mechanism, when restricted to the constraint equation, can be expressed as:

$$\begin{aligned} &\begin{bmatrix} A1\times\left\{\frac{(DR\cos\theta+R^2)}{D^2+R^2+2DR\cos\theta}\right\}+G1\times\left\{\frac{L\sin\phi(DR\cos\theta+R^2)}{S\cos\beta(D^2+R^2+2DR\cos\theta)}\right\} \\ H1\times\left\{\frac{(DR\cos\theta+R^2)}{D^2+R^2+2DR\cos\theta}\right\}+B1\times\left\{\frac{L\sin\phi(DR\cos\theta+R^2)}{S\cos\beta(D^2+R^2+2DR\cos\theta)}\right\} \\ CC \end{bmatrix}\ddot\theta_1 + \\ &+\begin{bmatrix} A1\times\dot L_{11}+G1\times\dot L_{21}+\frac{P1\times(DR\cos\theta+R^2)}{D^2+R^2+2DR\cos\theta}+\frac{T1\times L\sin\phi(DR\cos\theta+R^2)}{S\cos\beta(D^2+R^2+2DR\cos\theta)} \\ H1\times\dot L_{11}+B1\times\dot L_{21}+\frac{Y1\times(DR\cos\theta+R^2)}{D^2+R^2+2DR\cos\theta}+\frac{R1\times L\sin\phi(DR\cos\theta+R^2)}{S\cos\beta(D^2+R^2+2DR\cos\theta)} \\ 0 \end{bmatrix}\dot\theta_1 = \\ &= \begin{bmatrix} PL\cos\phi \\ PS\sin\beta \\ 0 \end{bmatrix} + \begin{bmatrix} u_1 \\ u_2 \\ u_3 \end{bmatrix} + \begin{bmatrix} \cos\phi(D+R\cos\theta)+R\sin\theta\sin\phi & -L\sin\phi \\ 0 & S\cos\beta \\ -R\sin\phi\sin\theta-R\cos\theta\cos\phi & 0 \end{bmatrix}\begin{bmatrix} \lambda_1 \\ \lambda_2 \end{bmatrix}. \end{aligned} \tag{45}$$

The parameters $A1, G1, B1, H1, CC, \dot{L}_{11}, \dot{L}_{21}, P1, T1, Y1, R1$ are shown in Equation (38). It is noted that P is the cutting force acting on the slider C, τ is the external force acting on rod 3, and the constraint forces are expressed as:

$$f_1 = \{\cos\phi(D + R\cos\theta) + R\sin\theta\sin\phi\}\lambda_1 - L\lambda_2\sin\phi,$$

$$f_2 = S\cos\beta\lambda_2,$$

$$f_3 = \{-R\sin\phi\sin\theta - R\cos\theta\cos\phi\}\lambda_1.$$

The control objective is to control the slider C to move periodically. Since $\lambda \to \boldsymbol{\lambda}_{\mathbf{d}}$ means $[f_1 \quad f_2]^T \to \mathbf{F_d}$, we chose $\boldsymbol{\theta}_{\mathbf{d}} = 4.712\ rad$ and $\boldsymbol{\lambda}_{\mathbf{d}} = 15$ in the simulations. The initial position of x is 2.167 m (*i.e.* $\theta_0 = 3.1416\,rad$) for the slider C and the controlled stroke of the slider C are set to be 0.85 m. Substituting the slider position x into Equations (39)-(41), the crank angle θ can be obtained.
In the simulations, the responses of the crank angle showing in Figure 7(a) reach the desired value in about 0.9 sec. The responses of the position of slider C are shown in Figure 7(b), the associated control efforts τ are shown in Figures 7(c), and the sliding surface $\mathbf{s_1}$ is shown in Figure 7(d). The Lagrange multiplier λ_C is showed in Figure 8(a). The constraint forces of joints 1-3 are shown in Figures 8(b)-(d), respectively. From Figures 7 and 8, the control objectives of force/motion of the quick-return mechanism are achieved successfully.

4.3 The toggle mechanism

For more details of the kinematic and dynamic analysis of the toggle mechanism, refer to [23] and adopt the generalized coordinate vector $\mathbf{Q} = [\theta_5 \quad \theta_2 \quad \theta_1]^{\mathrm{T}}$ in Equation (1) for the toggle mechanism shown in Figure 3. The dynamic equation can be obtained and is associated with the following matrices and elements:

$$\mathbf{M} = \begin{bmatrix} A & 0 & E \\ 0 & B & H \\ E & H & C_W \end{bmatrix} \quad \mathbf{N}_C = \begin{bmatrix} I & 0 & J \\ 0 & K_W & L \\ P_W & Q_W & R_W \end{bmatrix} \quad \mathbf{N_G} = 0$$

$$\mathbf{\Phi_Q} = \begin{bmatrix} \dfrac{\partial \mathbf{\Phi}_n(\mathbf{Q})}{\partial \mathbf{Q}_i} \end{bmatrix} = \begin{bmatrix} 0 & r_3\cos\theta_2 & r_1\cos\theta_1 \\ r_5\cos\theta_5 & 0 & r_4\cos(\theta_1 + \phi) \end{bmatrix}$$

$$\mathbf{Q^A} = \begin{bmatrix} F_\mathbf{C} r_5 \sin\theta_5 \\ (F_\mathbf{B} + F_\mathbf{E}) r_3 \sin\theta_2 \\ (F_\mathbf{B} + F_\mathbf{E}) r_1 \sin\theta_1 + F_\mathbf{C} r_4 \sin(\theta_1 + \phi) \end{bmatrix}$$

$$A = -\frac{1}{2}\left\{\left(\frac{2}{3}m_5 + 2m_C\sin^2\theta_5\right)r_5^2\right\}, \quad B = -\frac{1}{2}\left\{\left(\frac{2}{3}m_3 + 2m_B\sin^2\theta_2\right)r_3^2\right\},$$

$$C_w = -\frac{1}{2}\left\{\frac{m_2 r_1^2 r_4^2}{r_2^2}\sin^2\phi + 2(m_3 + m_B)r_1^2\sin^2\theta_1 + 2(m_5 + m_C)r_4^2\sin^2(\theta_1 + \phi)\right\},$$

$$E=-\frac{1}{2}\left\{(2m_C+m_5)r_4r_5\sin(\theta_1+\phi)\sin\theta_5\right\},\ Q_W=-\frac{1}{2}\left\{(m_3+2m_B)r_1r_3\sin\theta_1\cos\theta_2\right\}\dot{\theta}_2,$$

$$H=-\frac{1}{2}\left\{(2m_B+m_3)r_1r_3\sin\theta_1\sin\theta_2\right\},\ I=-\frac{1}{2}\left\{2m_Cr_5{}^2\sin\theta_5\cos\theta_5\right\}\dot{\theta}_5,$$

$$J=-\frac{1}{2}\left\{(m_5+2m_C)r_4r_5\cos(\theta_1+\phi)\sin\theta_5\right\}\dot{\theta}_1,\ K_W=-\frac{1}{2}\{2m_Br_3{}^2\sin\theta_2\cos\theta_2\}\dot{\theta}_2,$$

$$L=-\frac{1}{2}\left\{(m_3+2m_B)r_1r_3\cos\theta_1\sin\theta_2\right\}\dot{\theta}_1,\ P_w=-\frac{1}{2}\left\{(m_5+2m_C)r_4r_5\sin(\theta_1+\phi)\cos\theta_5\right\}\dot{\theta}_5,$$

$$R_W=-\frac{1}{2}\{2(m_3+m_B)r_1^2\sin\theta_1\cos\theta_1+2(m_5+m_C)r_4^2\sin(\theta_1+\phi)\cos(\theta_1+\phi)\}\dot{\theta}_1 \tag{46}$$

Where $n=3$, $m=2$ and $r=1$ are employed in the dynamic analysis.
For the single degree-of-freedom toggle mechanism, there exist two constraint equations as follows:

$$\mathbf{\Phi}(\mathbf{Q})=\begin{bmatrix} r_1\sin\theta_1+r_3\sin\theta_2-f \\ r_5\sin\theta_5+r_4\sin(\theta_1+\phi)-h-f\end{bmatrix}=0. \tag{47}$$

where ϕ can be obtained by analyzing its geometric relation as:

$$\phi=\cos^{-1}\left(\frac{r_1^2+r_4^2-r_2^2}{2r_1r_4}\right). \tag{48}$$

The positions of sliders B and C can be expressed as follows:

$$x_B=r_1\cos\theta_1+\left[r_3^2-(f-r_1\sin\theta_1)^2\right]^{\frac{1}{2}}, \tag{49}$$

$$x_C=r_4\cos(\theta_1+\phi)+\{r_5^2-[(h+f)-r_4\sin(\theta_1+\phi)]^2\}^{\frac{1}{2}}. \tag{50}$$

The Jacobian matrix of the constraint equation is:

$$\mathbf{\Phi_Q}=\left[\frac{\partial\,\mathbf{\Phi}_n(\mathbf{Q})}{\partial\,\mathbf{Q}_i}\right]=\begin{bmatrix} 0 & r_3\cos\theta_2 & r_1\cos\theta_1 \\ r_5\cos\theta_5 & 0 & r_4\cos(\theta_1+\phi)\end{bmatrix}. \tag{51}$$

Therefore, the matrix defined in Equation (6) is:

$$\mathbf{L(q)}=\frac{\partial\mathbf{Q}}{\partial\mathbf{q}}=\begin{bmatrix}\frac{\partial\theta_5}{\partial\theta_1} & \frac{\partial\theta_2}{\partial\theta_1} & \frac{\partial\theta_1}{\partial\theta_1}\end{bmatrix}^T=\begin{bmatrix}-\frac{r_4\cos(\theta_1+\phi)}{r_5\cos\theta_5} & -\frac{r_1\cos\theta_1}{r_3\cos\theta_2} & 1\end{bmatrix}^T. \tag{52}$$

Then, differentiating Equation (52) with respect to time yields:

$$\dot{\mathbf{L}}(\mathbf{q})=\begin{bmatrix}\dfrac{r_4r_5\dot{\theta}_1\sin(\theta_1+\phi)\cos\theta_5-r_4r_5\dot{\theta}_5\sin\theta_5\cos(\theta_1+\phi)}{r_5^2\cos^2\theta_5}\\ \dfrac{r_1r_3\dot{\theta}_1\sin\theta_1\cos\theta_2-r_1r_3\dot{\theta}_2\sin\theta_2\cos\theta_1}{r_3^2\cos^2\theta_2}\\ 0\end{bmatrix}=\begin{bmatrix}\dot{L}_{11}\\ \dot{L}_{21}\\ 0\end{bmatrix}. \tag{53}$$

The dynamic equation of the toggle mechanism, when restricted to the constraint equation (51), can be expressed as:

$$\begin{bmatrix}A\times\left\{-\dfrac{r_4\cos(\theta_1+\phi)}{r_5\cos\theta_5}\right\}+E\\ B\times\left\{-\dfrac{r_1\cos\theta_1}{r_3\cos\theta_2}\right\}+H\\ E\times\left\{-\dfrac{r_4\cos(\theta_1+\phi)}{r_5\cos\theta_5}\right\}+A\times\left\{-\dfrac{r_1\cos\theta_1}{r_3\cos\theta_2}\right\}+C_W\end{bmatrix}\ddot{\theta}_1+\begin{bmatrix}A\times\dot{L}_{11}+I\times\left\{-\dfrac{r_4\cos(\theta_1+\phi)}{r_5\cos\theta_5}\right\}+J\\ B\times\dot{L}_{21}+K_W\times T+L\\ E\times\dot{L}_{11}+H\times\dot{L}_{21}+P_W\times\left\{-\dfrac{r_4\cos(\theta_1+\phi)}{r_5\cos\theta_5}\right\}+Q_W\times\left\{-\dfrac{r_1\cos\theta_1}{r_3\cos\theta_2}\right\}+R_W\end{bmatrix}\dot{\theta}_1$$
$$=\begin{bmatrix}F_Cr_5\sin\theta_5\\ (F_{\mathbf{B}}+F_{\mathbf{E}})r_3\sin\theta_2\\ (F_{\mathbf{B}}+F_{\mathbf{E}})r_1\sin\theta_1+F_Cr_4\sin(\theta_1+\phi)\end{bmatrix}+\begin{bmatrix}u_1\\ u_2\\ u_3\end{bmatrix}+\begin{bmatrix}0 & r_5\cos\theta_5\\ r_3\cos\theta_2 & 0\\ r_1\cos\theta_1 & r_4\cos(\theta_1+\phi)\end{bmatrix}\begin{bmatrix}\lambda_1\\ \lambda_2\end{bmatrix}. \tag{54}$$

The parameters $A,E,B,H,C_W,\dot{L}_{11},\dot{L}_{21},I,J,K_W,T,L,P_W,Q_W$ and R_W are shown in Equation (46). It is noted that F_B is the friction force, F_E is the external force acting on the slider B, F_C is the applied force acting on the slider C, and $f_1=r_5\lambda_2\cos\theta_5$, $f_2=r_3\lambda_2\cos\theta_2$ and $f_3=r_1\lambda_1\cos\theta_1+r_4\lambda_2\cos(\theta_1+\phi)$ are the constraint forces.

The control objective is to regulate the position of slider B moving from the left to the right ends. The initial position of X_B is 0.104 m (*i.e.* $\theta_2(t_0)=4.712\,rad$), and its expected position is 0.114 m. The desired values are $\boldsymbol{\theta}_d(t_f)=5.76\,rad$ and $\lambda_d=15$ in the simulations. Furthermore, in order to show that the SMC is insensitive to parametric variation, the effects of friction forces in joints are considered in this toggle mechanism system by using the Lagrange multiplier method.

The comparisons between the nominal case without considering friction forces and the case with friction forces are shown in Figures 9-11. Figures 9(a)-(c) show the trajectories of angles θ_1, θ_2 and θ_5, respectively. Figures 10(a)-(b) show the positions of sliders B and C, respectively. Figures 10(c)-(d) illustrate the control effort τ and the sliding surface s_1, respectively. Finally, Figure 11(a) shows the Lagrange multiplier λ_C and Figures 11(b)-(d) address the constraint forces f_1, f_2 and f_3 acting on the joints 1, 2, and 3, respectively.

From the numerical results, it is found that the control efforts τ are almost identical for both cases whether the friction forces are considered or not. From the above figures, the force/motion control objective of a toggle mechanism using the SMC is achieved successfully and the system responses are insensitive to the effects of friction forces.

5. Conclusion

Based on the Lyapunov theorem, we successfully derived a generalized SMC algorithm in a simple manner. The algorithm used tracking the Lagrange multiplier error to facilitate controller design and proposed a separate sliding surface in terms of the displacement and velocity. Furthermore, some properties of the dynamic structure were presented and used to reduce the dynamic model equations. Finally, the slider-crank, quick-return, and toggle mechanisms were employed to illustrate and verify the methodology developed. From the numerical results, we conclude that the effectiveness in application of the developed SMC method is successfully verified in regards to the force/motion controls for these three typical mechanisms. First, fast attainment of the control objective: in the simulations, the three typical mechanisms reached the desired value in less than 1 second. Second, no overshoot in the control process: from the numerical results, the force/motion control objectives of the three mechanisms using the SMC were achieved successfully and the system responses did not overshoot in the whole control process. Third, insensitivity to parametric variation: the control efforts τ are almost identical and the system responses are insensitive to the effects of the friction forces.

6. Acknowledgment

The authors would like to thank Nation Science Council of Republic of China, Taiwan, for financial support of this research under contract no. NSC 98-2221-E-327 -010 -MY3.

Appendix A

Define the matrix $N(Q,\dot{Q}) = \dot{M}(Q) - 2N_C(Q,\dot{Q})$, then $N(Q,\dot{Q})$ is skew symmetric, i.e., the components n_{jk} of N satisfy $n_{jk} = -n_{kj}$.
Proof: Given the inertia matrix $M(Q)$, the kj^{th} component of $\dot{M}(Q)$ is given by the chain rule as

$$\dot{m}_{kj} = \sum_{i=1}^{n} \frac{\partial m_{kj}}{\partial Q_i} \dot{Q}_i, \tag{A.1}$$

and the kj^{th} component of $N(Q,\dot{Q}) = \dot{M}(Q) - 2N_C(Q,\dot{Q})$ is given by

$$\begin{aligned} \dot{n}_{kj} &= \dot{m}_{kj} - 2c_{kj} \\ &= \sum_{i=1}^{n} \left[\frac{\partial m_{kj}}{\partial Q_i} - \left\{ \frac{\partial m_{kj}}{\partial Q_i} + \frac{\partial m_{ki}}{\partial Q_i} - \frac{\partial m_{ij}}{\partial Q_k} \right\} \dot{Q}_i \right] = \sum_{i=1}^{n} \left[\frac{\partial m_{ij}}{\partial Q_k} - \frac{\partial m_{ki}}{\partial Q_j} \right] \dot{Q}_i \end{aligned} \tag{A.2}$$

Since the inertia matrix $M(Q)$ is symmetric, *i.e.*, $m_{ij} = m_{ji}$, it follows from Equation (A.2), the jk^{th} component of $N(Q,\dot{Q}))$ is

$$n_{jk} = \sum_{i=1}^{n} \left[\frac{\partial m_{ij}}{\partial Q_j} - \frac{\partial m_{ki}}{\partial Q_k} \right] \dot{Q} = -n_{kj}. \tag{A.3}$$

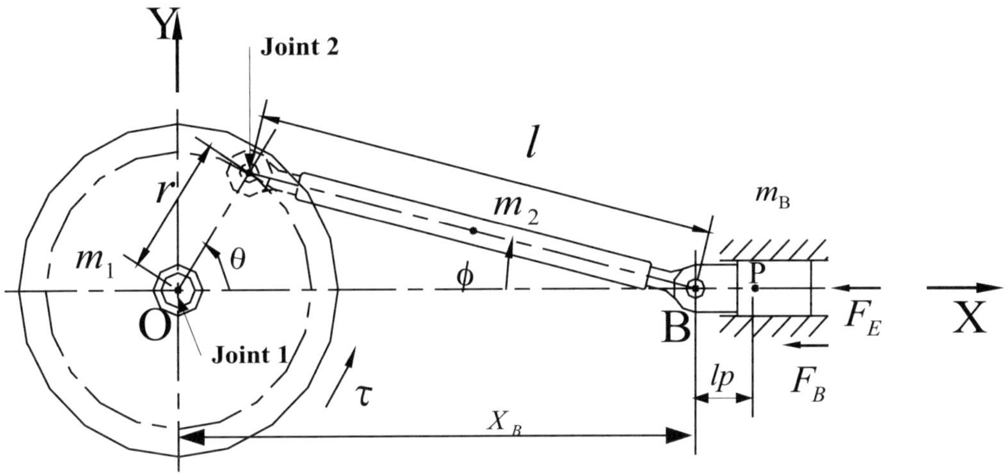

Fig. 1. The physical model of the slider-crank mechanism

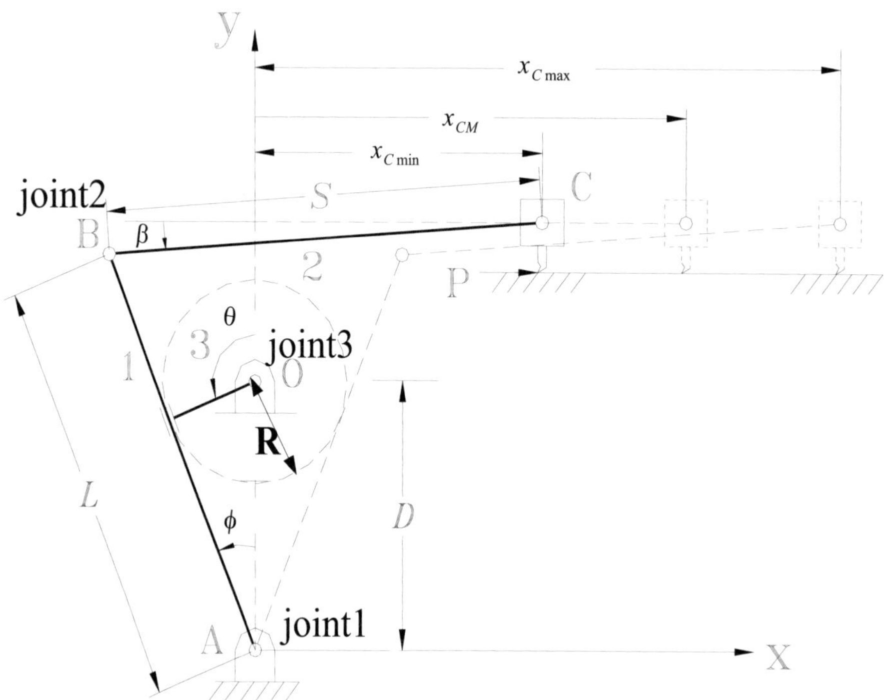

Fig. 2. The physical model of the quick-return mechanism

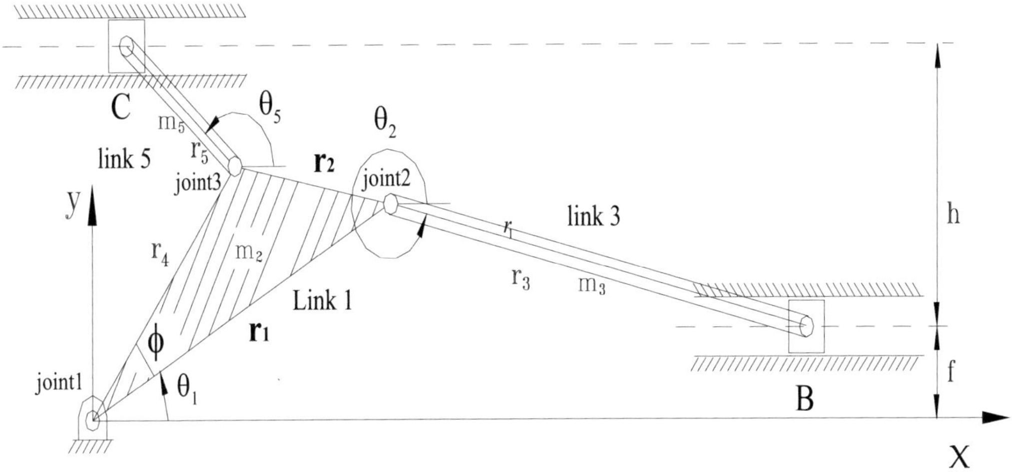

Fig. 3. The physical model of the toggle mechanism.

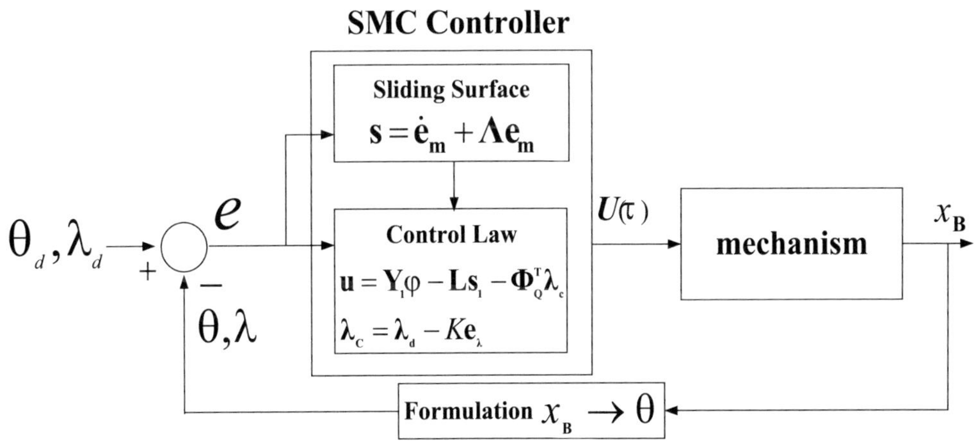

Fig. 4. The block diagram of mechanisms using the SMC controller

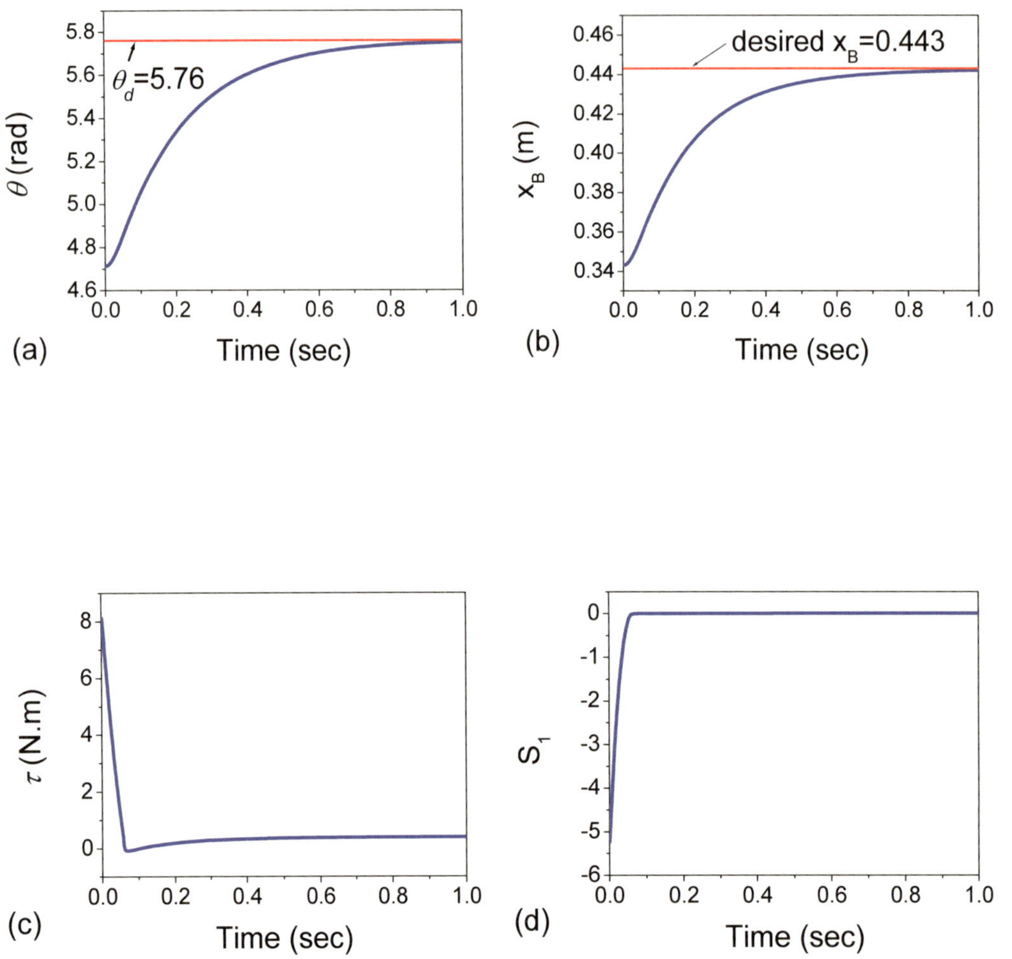

Fig. 5. The simulation results of the slider-crank mechanism. ('—'desired value; '—'actual trajectory) (a) Response trajectories of the crank angle θ. (b) Response trajectories of the slider B in position X_B. (c) Response trajectories of the control effort τ. (d) Response trajectories of the sliding surface s_1.

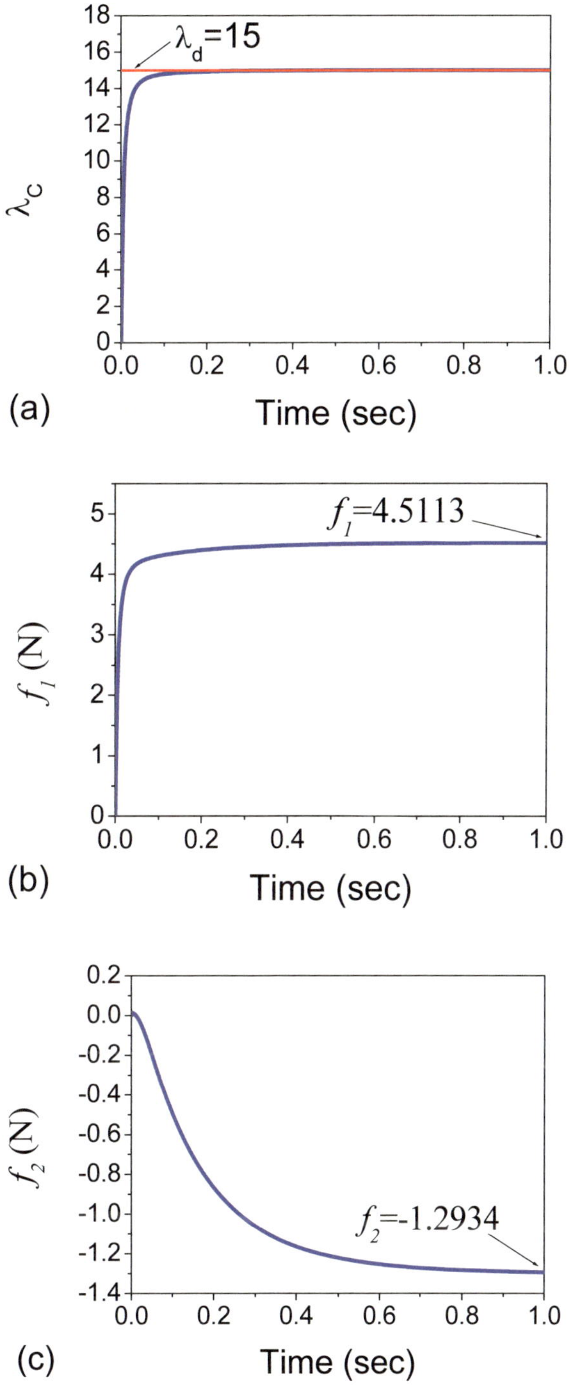

Fig. 6. The simulation results of the slider-crank mechanism. ('—'desired value; '—'actual trajectory) (a) Response trajectories of the Lagrange multiplier λ_C . (b) Response trajectories of the constraint force f_1 . (c) Response trajectories of the constraint force f_2 .

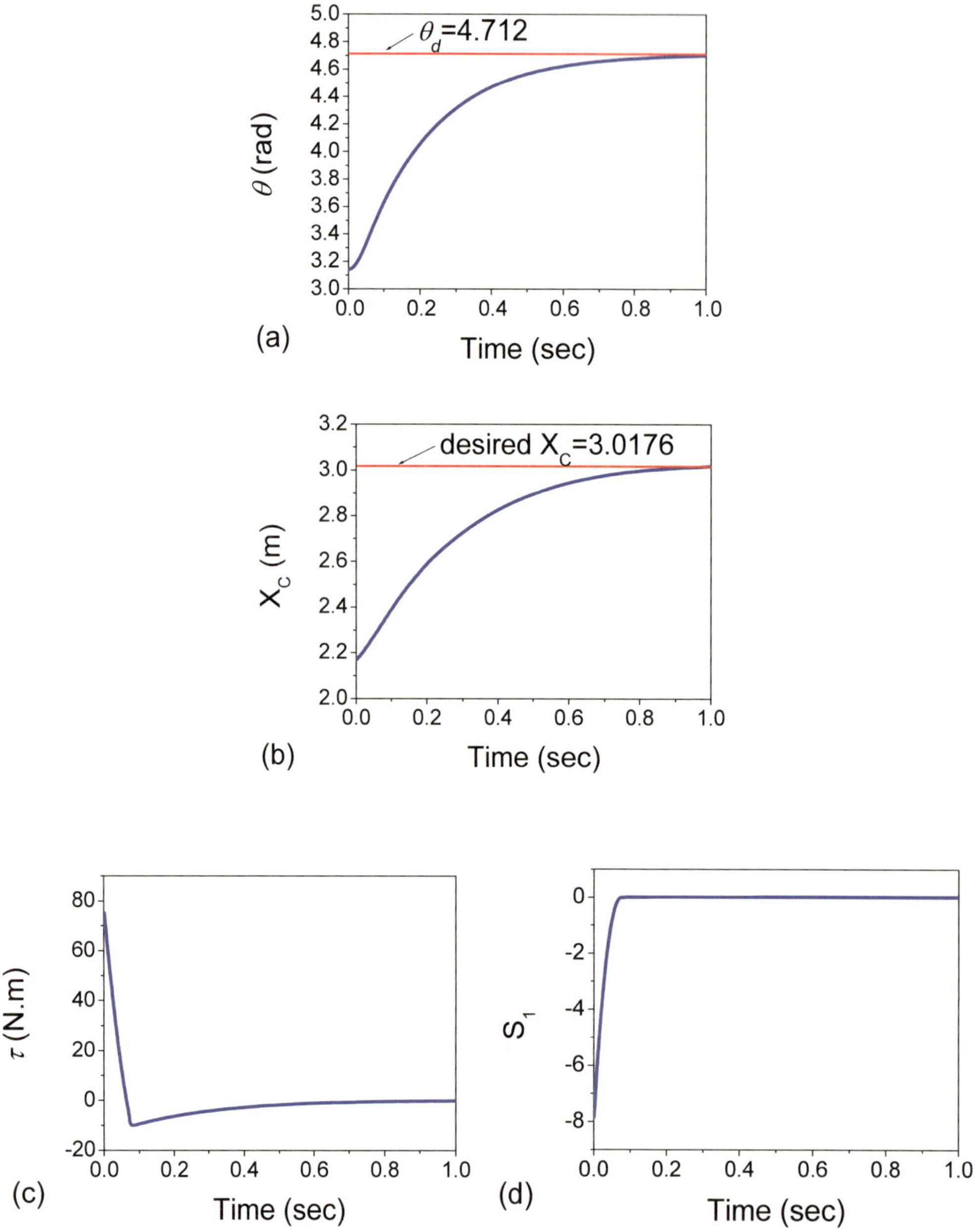

Fig. 7. The simulation results of the quick-return mechanism. ('—'desired value; '—'actual trajectory) (a) Response trajectories of the crank angle displacement θ . (b) Response trajectories of the slider C in position X_C . (c) Response trajectories of the control effort τ . (d) Response trajectories of the sliding surface s_1 .

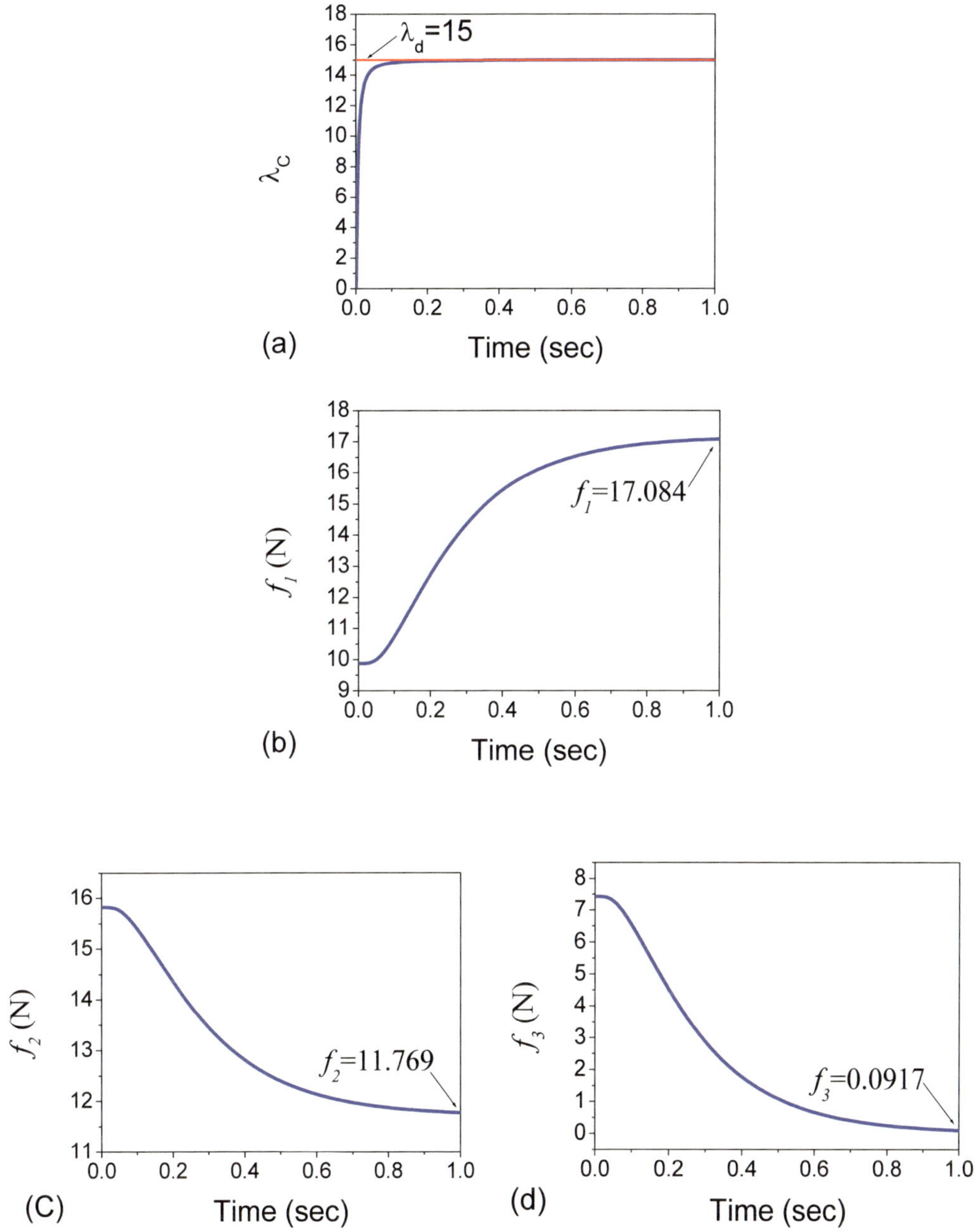

Fig. 8. The simulation results of the quick-return mechanism. ('—'desired value; '—'actual trajectory) (a) Response trajectories of the Lagrange multiplier λ_C. (b) Response trajectories of the constraint force f_1. (c) Response trajectories of the constraint force f_2. (d) Response trajectories of the constraint force f_3.

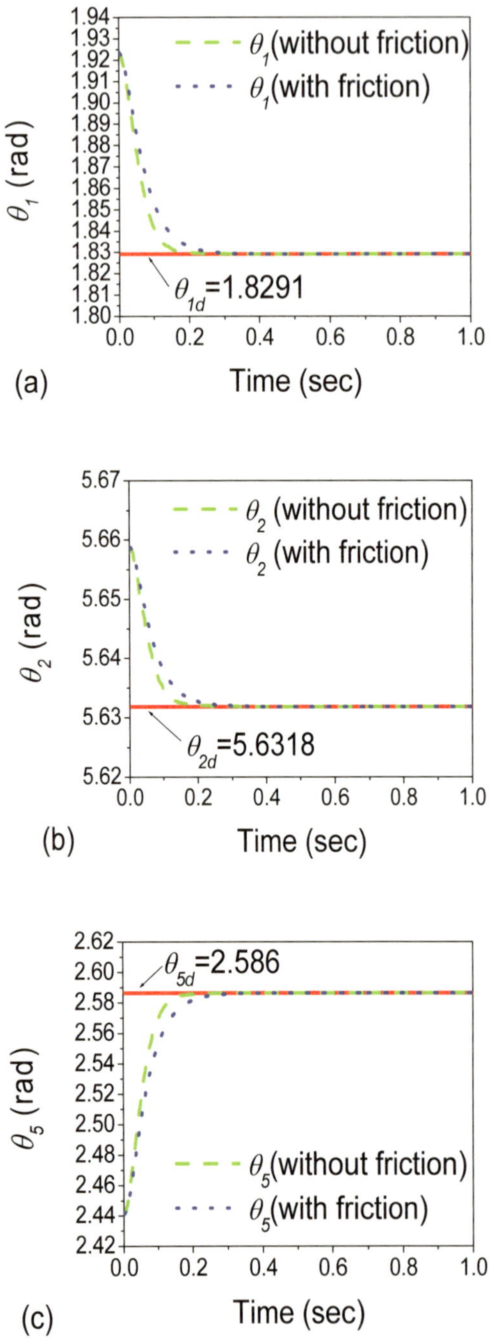

Fig. 9. The simulation results of the toggle mechanism. ('—'desired curve; '---'actual trajectory (without friction), '---'actual trajectory (with friction and $f_r = 0.3$)) (a) Response trajectories of the angle displacement θ_1 . (b) Response trajectories of the angle displacement θ_2 . (c) Response trajectories of the angle displacement θ_5 .

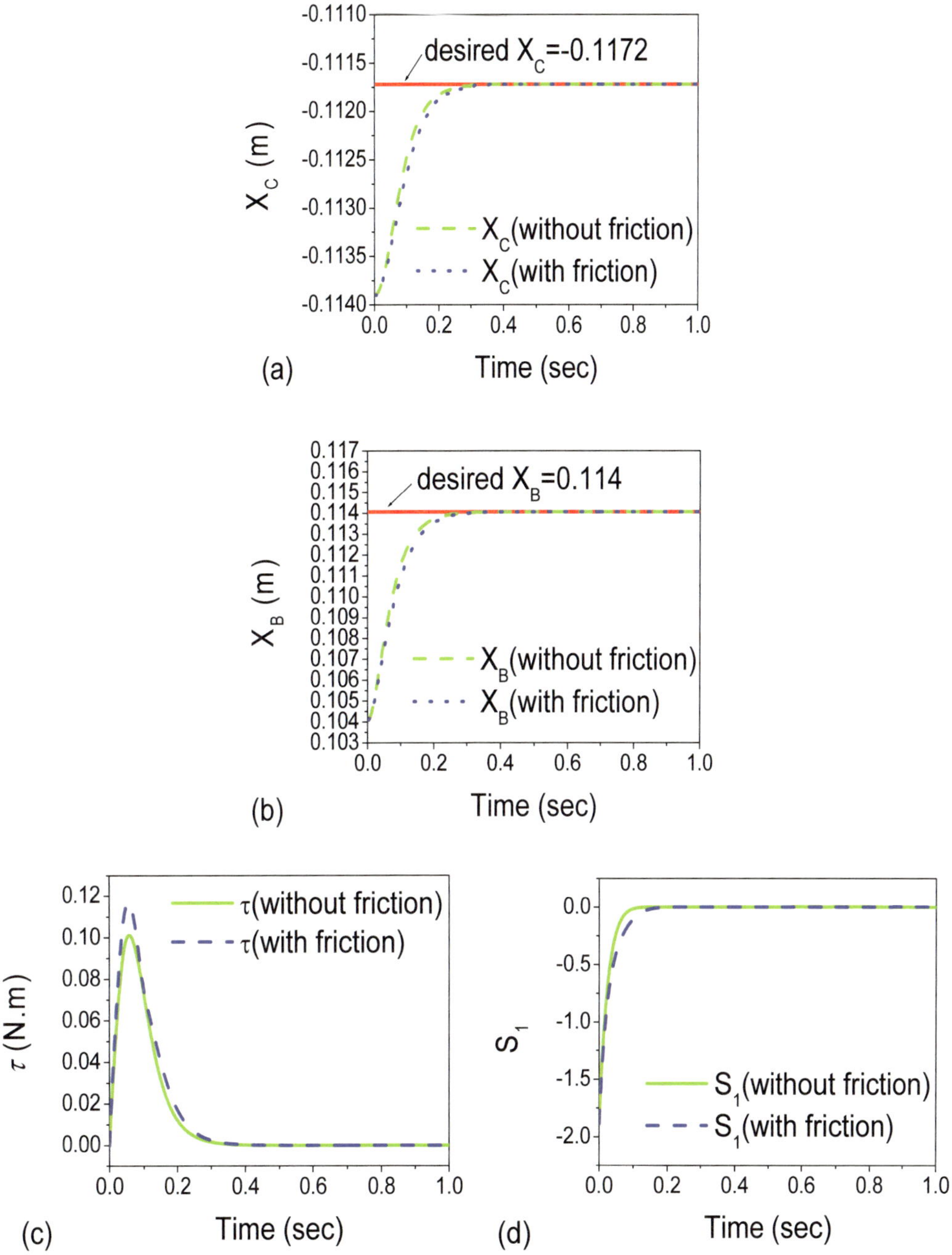

Fig. 10. The simulation results of the toggle mechanism. ('—'desired curve; '---'actual trajectory (without friction), '---'actual trajectory (with friction and $f_r = 0.3$)) (a) Response trajectories of the slider C in position X_C . (b) Response trajectories of the slider B in position X_B . (c) Response trajectories of the control effort τ . (d) Response trajectories of the sliding surface s_1 .

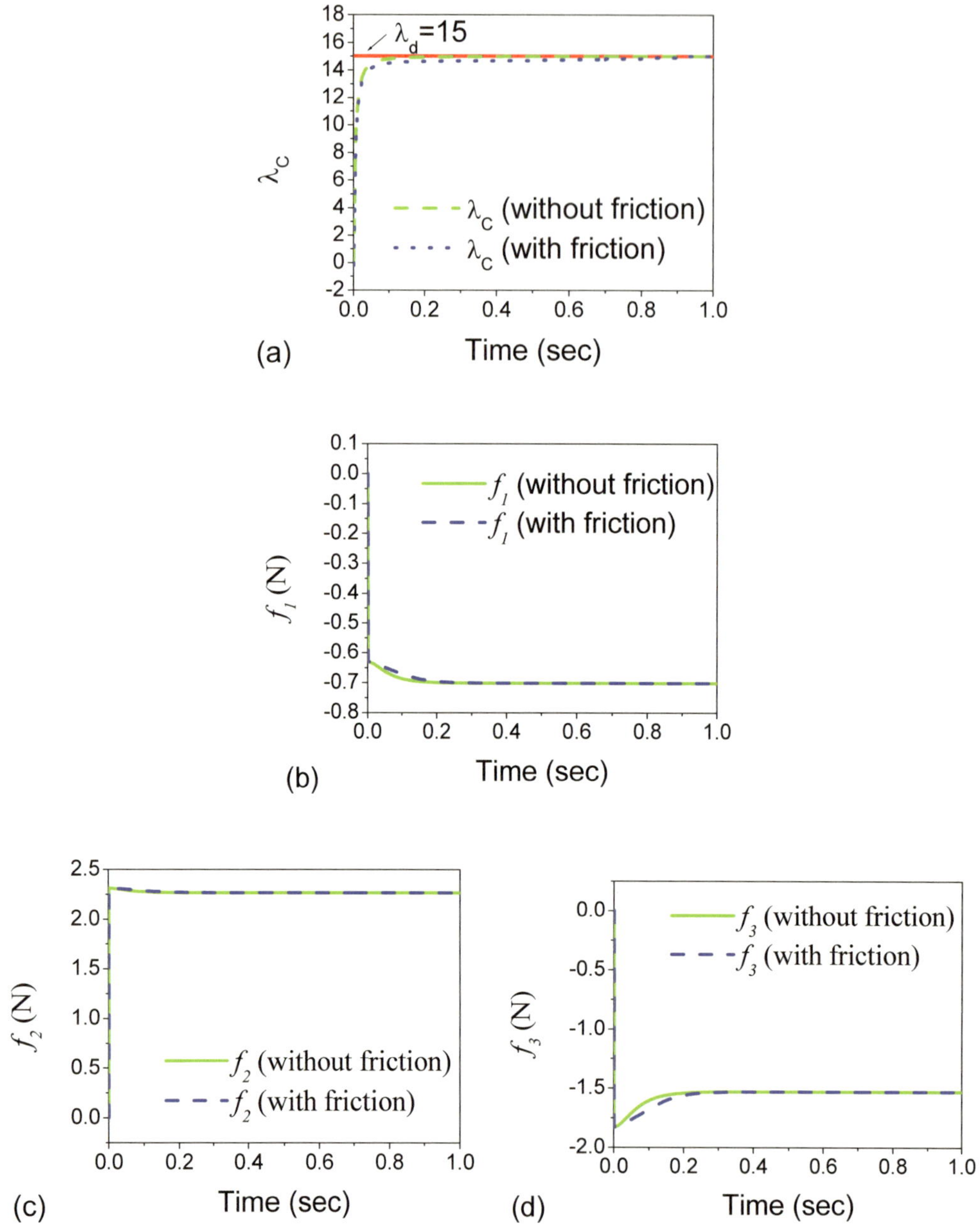

Fig. 11. The simulation results of the toggle mechanism. ('—'desired curve; '---'actual trajectory (without friction), '---'actual trajectory (with friction and $f_r = 0.3$)) (a) Response trajectories of the Lagrange multiplier λ_C. (b) Response trajectories of the constraint force f_1. (c) Response trajectories of the constraint force f_2. (d) Response trajectories of the constraint force f_3

7. References

[1] Fung, R. F., "Dynamic Analysis of the Flexible Connecting Rod of a Slider-Crank Mechanism," *ASME Journal of Vibration and Acoustic*, Vol. 118, No. 4, pp. 687-689(1996).
[2] Fung, R. F., and Chen, H. H., "Steady-State Response of the Flexible Connecting Rod of a Slider-Crank Mechanism with Time-Dependent Boundary Condition," *Journal of Sound and Vibration*, Vol. 199, No. 2, pp. 237-251(1997).
[3] Fung, R. F., "Dynamic Response of the Flexible Connecting Rod of a Slider-Crank Mechanism with Time-Dependent Boundary Effect," *Computer & Structure*, Vol. 63, No. 1, pp. 79-90(1997).
[4] Fung, R. F., Huang, J. S., Chien, C. C., and Wang, Y. C., "Design and Application of a Continuous Repetitive Controller for Rotating Mechanisms," *International Journal of Mechanical Sciences*, Vol. 42, pp. 1805-1819(2000).
[5] Lin, F. J., Fung, R. F., and Lin Y. S., "Adaptive Control of Slider-Crank Mechanism Motion: Simulations and Experiments," *International Journal of Systems Science*, Vol. 28, No. 12, pp. 1227-1238(1997).
[6] Lin, F. J., Lin, Y. S. and Chiu, S. L., "Slider-Crank Mechanism Control using Adaptive Computed Torque Technique," *Proceedings of the IEE Control Theory Application*, Vol. 145, No. 3, pp. 364-376(1998).
[7] Lin, F. J., Fung, R. F., Lin, H. H., and Hong, C. M., "A Supervisory Fuzzy Neural Network Controller for Slider-Crank Mechanism," *Proceedings of the IEEE Control Applications Conferences*, pp. 1710-1715(1999).
[8] Utkin, V. I., *Sliding Modes and Their Applications*, Mir: Moscow (1978).
[9] Utkin, V. I., "Discontinuous Control System: State of the Art in Theory and Application," *Preprint 10th IFAC World Congress*, Vol. 1, pp. 75(1987).
[10] Compere, M. D. and Longoria, R. G., "Combined DAE and Sliding Mode Control Methods for Simulation of Constrained Mechanical System," *ASME Journal of Dynamic System, Measurement, and Control*, Vol. 122, pp. 691-698(2000).
[11] Su, C. Y., Leung, T. P., and Zhou, Q. J., "Force/Motion Control of Constrained Robots Using Sliding Mode," *IEEE Transactions on Automatic Control*, Vol. 37, No. 5, pp. 668-672(1992).
[12] Grabbe, M. T., and Bridges, M. M., "Comments on "Force/Motion Control of Constrained Robots Using Sliding Mode"," *IEEE Transactions on Automatic Control*, Vol. 39, No. 1, pp. 179(1994).
[13] Slotine, J. J. E. and Li, W., *Applied Nonlinear Control*. Englewood Cliffs, NJ: Prentice-Hall (1991).
[14] Lian, K. Y. and Lin, C. R., "Sliding Mode Motion/Force Control of Constrained Robots," *IEEE Transactions on Automatic Control*, Vol. 43, No. 8, pp. 1101-1103(1998).
[15] Dixon, W. E. and Zergeroglu, E., "Comments on "Sliding Mode Motion/Force Control of Constrained Robots"," *IEEE Transactions on Automatic Control*, Vol. 45, No. 8, pp. 1576(2000).
[16] Fung, R. F., Shue, L. C, "Regulation of a Flexible Slider–Crank Mechanism by Lyapunov's Direct Method," *Mechatronics*, Vol. 12, pp. 503-509(2002).
[17] Fung, R. F., Sun, J. H, "Tracking Control of the Flexible Slider-Crank Mechanism System Under Impact," *Journal of Sound and Vibration*, Vol. 255, pp. 337-355(2002).

[18] McClamroch, N. H., and Wang, D. W., "Feedback Stabilization and Tracking of Constrained Robots," *IEEE Transactions on Automatic Control*, Vol. 33, No. 5, pp. 419-426(1988).

[19] Fung, R. F., Lin, F. J., Huang, J. S., and Wang, Y. C., "Application of Sliding Mode Control with A Low Pass Filter to the Constantly Rotating Slider-Crank Mechanism," *The Japan Society of Mechanical Engineering*, Series C, Vol. 40, No. 4, pp. 717-722(1997).

[20] Parviz, E. N., *Computer-Aided Analysis of Mechanical System*. Prentice-Hall, Englewood Cliffs NJ (1988).

[21] Fung, R. F. and Chen, K. W., "Constant Speed Control of the Quick-return Mechanism," *The Japan Society of Mechanical Engineering*, Series C, Vol. 40, No. 3, pp. 454-461(1997).

[22] Fung, R. F. and Yang, R. T., "Motion control of an electrohydraulic actuated toggle mechanism," *Mechatronics*, Vol. 11, pp. 939-946(2001).

[23] Fung, R. F., Wu, J. W. and Chen, D. S., "A variable structure control toggle mechanism driven by a linear synchronous motor with joint coulomb friction," *Journal of sound and vibration*, Vol. 274, No. 4, pp. 741-753(2001).

[24] Slotine, J. J. E. and Sastry, S. S., "Tracking control of nonlinear system using sliding surface with application to robot manipulators," *International journal of control*, Vol. 38, pp. 465-492(1983).

16

Automatic Space Rendezvous and Docking Using Second Order Sliding Mode Control

Christian Tournes[1], Yuri Shtessel[2] and David Foreman[3]
[2]*University of Alabama Huntsville*
[1,3]*Davidson Technologies Inc*
USA

1. Introduction

This chapter presents a Higher Order Sliding Mode (HOSM) Control for automatic docking between two space vehicles. The problem considered requires controlling the vehicles' relative position and relative attitude. This type of problem is generally addressed using optimal control techniques that are, unfortunately, not robust. The combination of optimum control and Higher Order Sliding Mode Control provides quasi-optimal robust solutions. Control of attitude includes a receiver vehicle passive mode option where the pursuing vehicle controls the relative attitude using the active pixels of a camera viewing a network of lights placed on the receiving vehicle, which by sharing considerable commonality with manual operations allows possible human involvement in the docking process.

2. Problem description

The complexity of satellite formation and automatic space docking arises from the formulation of Wilshire equations. These equations are nonlinear and exhibit coupling of normal and longitudinal motions. The problem is compounded by the characteristics of the on/off thrusters used. Typical solutions to the problem involve application of optimal control. The problem with optimal control is that it is not robust and it only works well when a perfectly accurate dynamical model is used. This subject has been investigated extensively by the research community (Wang, 1999), (Tournes, 2007). Since this is a navigation and control problem involving two bodies, one question is how to obtain the measurements to be used. Of course a data link from the receiving vehicle to inform the pursuer about its state can be used, whereby the pursuer receives the current position velocity and attitude state of the receiving vehicle. One could also mount distance measurement equipment on the vehicles such as a Lidar to provide accurate range and range rate measurements. The exchange of attitude represents a larger challenge, as the relative motion will be the difference of the measurements/estimations by separate Inertial Measurement Units (IMU) of their attitude. Such a difference will contain the drift and the noise of two IMUs.
The transversal aspect of this chapter presents lateral and longitudinal guidance algorithms, based on measurements of range and range rate without regard to the source of these

measurements which could be provided by a Lidar system (Tournes, 2007) or interpreted from visible cues using a pattern of reference lights.

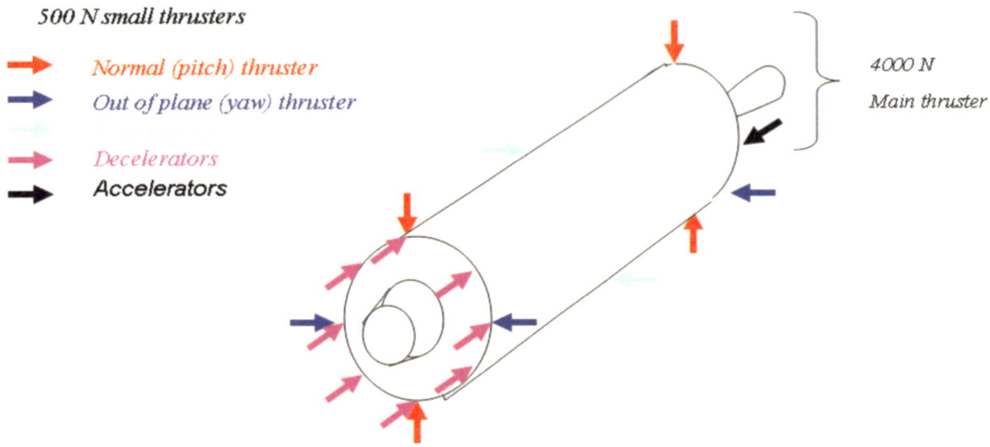

Fig. 1. Notional vehicle.

The attitude aspect presents a workable solution that does not require any reporting by the receiving unit and is based on a pattern of reference lights, that when viewed by the pursuer would allow the latter to evaluate the relative attitude orientation error. The quaternion representing the relative attitude is estimated in real time by a nonlinear curvefit algorithm and is used as the feedback of a second order sliding mode attitude control algorithm.
For simulation purposes, we assumed the pursuing vehicle (as shown in Fig. 1) to be similar in characteristics to ESA's Automated Transfer Vehicle (ESA 2006). Its initial mass is 10000 kg. It is equipped with a main / sustainer orientable thruster providing 4000 N thrust. Twenty small thrusters of 500 N are used by pairs to steer roll, pitch, and yaw attitude as well as lateral and normal motion. Regarding axial dynamics, we assume that several axial thrusters could be used to achieve axial deceleration. We assume that using all of them would provide a "maximum" braking; using half would provide a "medium" breaking; and using a quarter would provide "small" braking. A major goal in the study was to obtain extremely small velocity, position and attitude errors at the docking interface.

3. Governing equations and problem formulation

Equations governing the relative motion of the pursuer with respect to the pursued vehicle are along in-track, out of plane and normal axis represented by Wilshire equations (Chobotov, 2002).

$$\begin{aligned}
\mathbf{r}_{sv} &= \mathbf{r}_T + \boldsymbol{\rho} \\
\ddot{\mathbf{r}}_{sv} &= \ddot{\mathbf{r}}_T + \ddot{\boldsymbol{\rho}} + 2(\boldsymbol{\omega} \times \dot{\boldsymbol{\rho}}) + \dot{\boldsymbol{\omega}} \times \boldsymbol{\rho} + \boldsymbol{\omega} \times (\boldsymbol{\omega} \times \boldsymbol{\rho}) \\
\ddot{\mathbf{r}}_{sv} &= \mathbf{g} + \frac{\mathbf{F}}{m} = \mathbf{g} + \boldsymbol{\Gamma} \\
\ddot{\boldsymbol{\rho}} &= \boldsymbol{\Gamma} + \mathbf{f(t)}
\end{aligned} \tag{1}$$

Where $\mathbf{r}_{sv}, \mathbf{r}_T, \boldsymbol{\rho}$ represent respectively the space vehicle position pursued vehicle position and relative position vectors; Γ, g are the thrust and gravity accelerations.

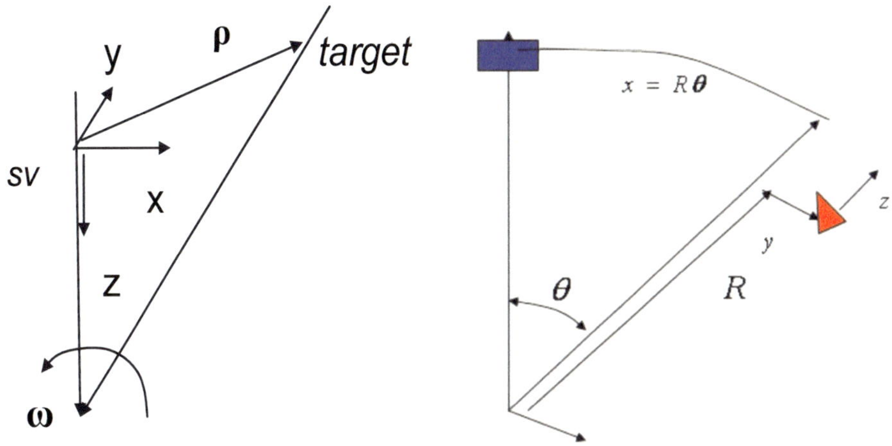

Fig. 2. System of axes used.

3.1 Translational dynamics

The system of axes used is shown in Fig. 2. Equation (1) is linearized, assuming that the thrust **F** is aligned with the pursuer longitudinal axis. Expressing the three components of gravity vector **g** as function of the pursuer position vector, one obtains

$$
\begin{aligned}
&\ddot{x} = f_x(.) + \frac{F_x}{m}; \quad f_x(.) = -\mu\frac{x}{r^3} + 2\omega\dot{z} + \dot{\omega}z + \omega^2 x \\
&\ddot{y} = f_y(.) + \frac{F_y}{m}; \quad f_y(.) = -\mu\frac{y}{r^3} \\
&\ddot{z} = f_z(.) + \frac{F_z}{m}; \quad f_z(.) = -\mu\frac{z + r_T}{r^3} + \frac{\mu}{r_T^2} - 2\omega\dot{x} - \dot{\omega}x + \omega^2 z
\end{aligned}
\tag{2}
$$

Where x, y, z are relative coordinates; ω is a rotational speed of a frame connected to the pursued vehicle, μ represents the gravitational constant. Functions: $f_x(.)$, $f_y(.)$, $f_z(.)$ represent the effects in Eq. (1) other than caused by thrust and are treated as disturbances. They are smooth functions which tend to zero as the vehicles get closer. When variable attitude mode is in effect, Eq. (2) is generalized to a form

$$
\ddot{x} = \Gamma\sqrt{1-\delta_y^2}\sqrt{1-\delta_z^2}\,\delta_x + f_x(.);\ \ddot{y} = \Gamma\sqrt{1-\delta_z^2}\,\delta_y + f_y(.);\ \ddot{z} = \Gamma\delta_z + f_z(.) \tag{3}
$$

Here, $\Gamma = F/m$; F (the magnitude of the thrust) can take three discrete values, the vehicle mass m varies slowly with time, δ_x can take discrete values 1,-0, 1. Pursuer pitch and yaw attitude angles are defined as $\theta = asin(\delta_z)$ and $\psi = atan2(\delta_x\sqrt{1-\delta_y^2}, \delta_y)$ respectively.
When fixed attitude mode is in effect, Eq. (2) is written as:

$$
\ddot{z} = f_z(.) + \frac{F}{m}u_z;\ \ u_z = \{-1,0,1\};\ \ddot{y} = f_y(.) + \frac{F}{m}u_y;\ \ u_y = \{-1,0,1\};\ \ddot{x} = f_x(.) + \frac{F}{m}u_x;\ \ u_x = \{-1,0,1\} \tag{4}
$$

3.2 Attitude dynamics

The body attitude is represented by quaternion $\mathbf{Q}_{(.)}^{body}$ the dynamics of which is governed by

$$\dot{\mathbf{Q}}_{(.)}^{body} = -\frac{1}{2}\begin{bmatrix} 0 & p & q & rr \\ -p & 0 & -rr & q \\ -q & rr & 0 & -p \\ -rr & -q & p & 0 \end{bmatrix}\mathbf{Q}_{(.)}^{body} \tag{5}$$

Where (.) represents some non rotating reference, i.e. Earth Centered Inertial and Where p, q, rr represent the body rates expressed in the body frame. An alternate notation, using quaternion multiplication (Kuipers, 1999) is:

$$\dot{\mathbf{Q}}_{(.)}^{body} = \mathbf{Q}_{(.)}^{body}\,\boldsymbol{\Omega}$$

The dynamics p, q, rr are governed by

$$\dot{\boldsymbol{\Omega}} = \begin{bmatrix} \dot{p} \\ \dot{q} \\ \dot{rr} \end{bmatrix} = -\mathbf{I}^{-1}\boldsymbol{\Omega}\times\mathbf{I} + \mathbf{I}^{-1}\begin{bmatrix} \frac{l}{2} r_p F_p \delta_p \\ (x_{\delta q} - x_{cg})F_q\delta_q \\ -(x_{\delta q} - x_{cg})F_q\delta_{rr} \end{bmatrix} \tag{6}$$

Where $\mathbf{I}$ represent the vehicle matrix of inertia, $\boldsymbol{\Omega}$ the rotation matrix in body axes and $F_p, F_q, r_p, x_q, x_{cg}, \delta_p, \delta_q, \delta_{rr}$ represent respectively roll, pitch/yaw thruster maximum force, roll thrusters radial position, pitch/yaw thruster axial position, and corresponding normalized control amplitudes in roll, pitch and yaw.

3.3 Problem formulation

3.3.1 Lateral control: The control must steer the vehicle position to the prescribed orbital plane and orbit altitude. For that matter during the *initial rendezvous,* out-of-plane and relative orbit positions with respect to pursued vehicle are calculated at the onset of the maneuver. The HOSM lateral trajectory control calculates required acceleration to follow the desired approach profile and calculates the required body attitude represented by quaternion $\mathbf{Q}\,*_{(.)}^{body}$. During subsequent drift, braking and final docking phases the pursuer is maintained in the orbital plane and at the correct altitude by means of on-off HOSM control applied by the corresponding thrusters.

3.3.2 Longitudinal control: During initial *rendezvous* the pursuer accelerates using the main thrust/sustainer. Corresponding thrust is shut down when the pursuer is in the orbital plane, has attained the pursued vehicle's orbit altitude and desired closing rate. During the drift segment no longitudinal control is applied. The braking segment begins at a range function of the range rate. Following coast, braking is applied until reaching the terminal sliding mode condition. On-off deceleration pulses are then commanded by the HOSM longitudinal control.

3.3.3 Attitude control: During the initial rendezvous, continuous HOSM controls the attitude such that $\mathbf{Q}_{(.)}^{body} \to \mathbf{Q}\,*_{(.)}^{body}$ where $\mathbf{Q}_{(.)}^{body}$ represents current body attitude. During following segments the pursuing vehicle regulates its body attitude so that $\mathbf{Q}_{(.)}^{body} \to \mathbf{Q}\#_{(.)}^{body}$ where $\mathbf{Q}\#_{(.)}^{body}$ represents the attitude of the pursued vehicle.

4. Why higher order sliding mode control

HOSM control is an emerging (less than 10 years old) control technique (Shtessel, 2003), (Shkolnikov, 2000), (Shtessel, 2000), (Shkolnikov, 2005), (Tournes, 2006), (Shtessel, 2010) which represents a game changer. It should not be confused with first order sliding mode control which has been used for the last 30 years. Its power resides in four mathematically demonstrated properties:

1. **Insensitivity to matched disturbances:** Consider a system of relative degree n, with its output tracking error dynamics represented as:

$$x^{(n)} = f(x,t) - u \tag{7}$$

where $f(x,t)$ represents some unknown disturbance. A convergence function $u = C(x,\dot{x},..x^{(n-1)})$ is selected so that the output tracking error x in Eq. (7) and its consecutive derivatives up to degree $n-1$ converge to zero in finite time in the presence of the disturbance $f(x,t)$ provided that $|f(x,t)| < M$ is bounded. In this application, such a bound exists (Chobotov, 2002), (Wang, 1999). This property of HOSM control is inherited from classical sliding mode control (SMC). Being implemented in discrete time, the output tracking error is not driven to precisely zero but is ultimate bounded in the sliding mode with sliding accuracy proportional to the k^{ith} power of time increment Δt. This property makes HOSM an enhanced-accuracy robust control technique applicable to controllers and to observer design.

2. **Dynamical collapse:** Unlike traditional control techniques that seek asymptotic convergence, HOSM achieves finite time convergence in systems with arbitrary relative degree, just as classical SMC achieves the same result for the system with relative degree one. This is much more than an academic distinction; it means that when the sliding mode is reached the effective transfer function of inner loops with relative degree greater than one becomes an identity.
3. **Continuous / smooth guidance laws:** HOSM controllers can yield continuous and even smooth controls that are applicable in multiple-loop integrated guidance/autopilot control laws.
4. **Continuous / Discontinuous actuators:** HOSM techniques are nonlinear robust control techniques. When discontinuous actuators such as on-off thrusters must be used, all linear control laws require a re-design into a discontinuous control law that approximates the effects of the initial control law. HOSM design produces directly, when need arises, a discrete pulse width modulated control law that achieves the same level of accuracy as a linear control law.

5. Docking strategy

It is assumed in Fig. 3 that the automatic docking starts at a relatively large distance (>40-50 km). The pursuer, during *Initial Rendezvous* manages using its main thrust / sustainer to get in a coplanar circular orbit with altitude equal to that of the receiving vehicle, but with a slightly higher longitudinal velocity. Maintaining this altitude will require infrequent thruster firings by the pursuer. Alternately, one could place the pursuer on a circular coplanar orbit consistent with its longitudinal velocity and design the control law to track the orbit associated to its current velocity which "in time" will end up being the same as the

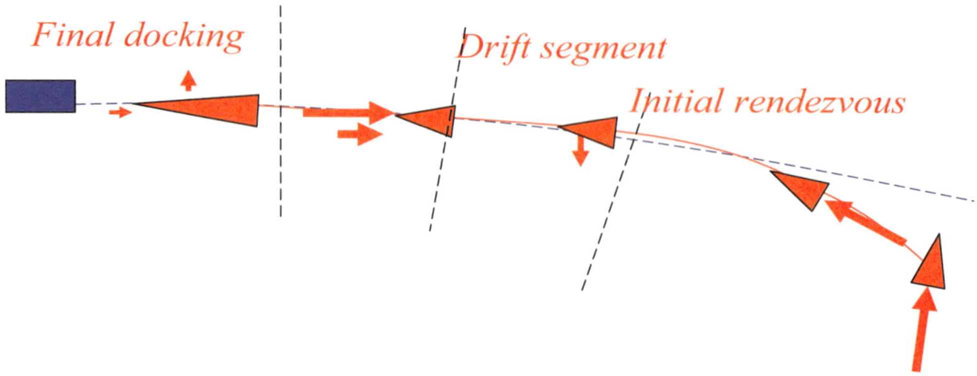

Fig. 3. Docking strategy.

pursued vehicle altitude. During the initial rendezvous, the pursuing vehicle is set to the desired drift velocity relative to the pursued vehicle. This maneuver is represented by trajectory 0-1-2 in the phase portrait of Fig. 4. During this initial segment, a varying attitude mode is applied. The transition from variable attitude to fixed attitude takes place when the normal and out-of plane errors become lower than a prescribed threshold defined as

$$V1 = (y^2 + z^2 + \dot{y}^2 + \dot{z}^2); \quad V1 < \varepsilon 3 \tag{9}$$

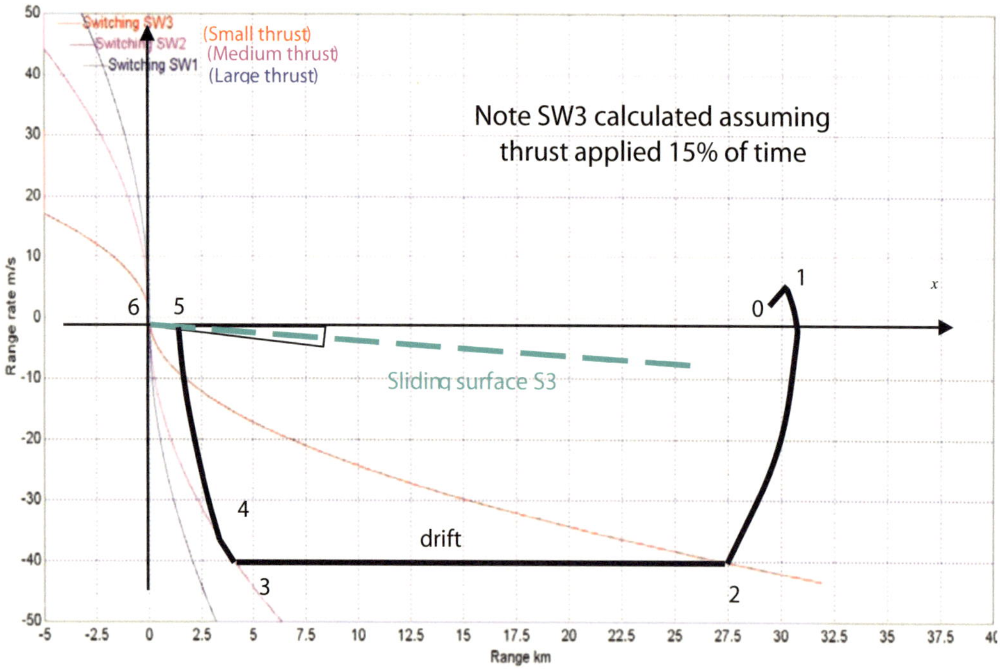

Fig. 4. Longitudinal control strategy.

During the drift segment, normal and lateral control is applied to keep the pursuer vehicle at the prescribed altitude and in the prescribed plane. The drift motion (2-3) begins with

$$V = (x^2 + y^2 + z^2 + \dot{x}^2 + \dot{y}^2 + \dot{z}^2); \quad V < \varepsilon \tag{10}$$

The end of the drift segment is calculated using Pontyagyn's Principle of Maximum. Three switching surfaces are defined as:

$$SW1 = x + \frac{sign(\dot{x})\dot{x}^2 m(t)}{2F_1}; SW2 = x + \frac{sign(\dot{x})\dot{x}^2 m(t)}{2F_2}; SW3 = x + \frac{sign(\dot{x})\dot{x}^2 m(t)}{2\alpha F_3} \tag{11}$$

Large, medium, or small thrust is applied as thresholds $SW1, SW2, SW3$ are reached depending on the braking strategy used and this thrust is applied until the distance from the terminal switching surface becomes small enough. At that point, the terminal thrust is shut down. The termination of the decelerating maneuver is governed by

$$\sigma_x = 2x + \dot{x}; \quad |\sigma_x| > \varepsilon 2 \tag{12}$$

Once (12) is satisfied, terminal docking begins: radial and out-of-plane errors are almost null and the only disturbance left is radial with a magnitude $f_z(.) = -2\omega\dot{x}$ and this has already been greatly reduced by previous in-track braking.

6. HOSM design of the relative navigation

6.1 Normal / Lateral control during initial rendezvous

During the initial phase of the rendezvous, the pursuing vehicle is steered by the continuous orientation of its main thruster/sustainer. We select the relative normal / lateral positions as the sliding variables. Given that the ultimate objective of this initial rendezvous is to set the pursuing vehicle in an orbit coplanar to the pursued vehicle's orbit and at the same altitude, we define $z^*(t)_{(.)}; (.) = radial, out\ of\ plane$ to be a profile joining initial pursuer vehicle with its terminal objective, this profile is designed to be terminally tangent to pursued vehicle orbit. The *initial rendezvous* objective is thus, to steer the pursuer trajectory so that $z(t) \rightarrow z^*(t)_{(.)}$. Sliding variable is chosen as:

$$\sigma_{(.)} = z_{(.)}{}^* - z_{(.)} \tag{13}$$

Applying the relative degree procedure, we differentiate twice the sliding variable before the control appears, with Eqs. (4, 13) we obtain a dynamics of sliding variable of relative degree two.

$$\begin{gathered} \ddot{\sigma}_{(.)} = d - bu_{(.)}; (.)z, y \\ d_{(.)} = \ddot{z}_{(.)} - f_{(.)}(.); b_{(.)} = \frac{F_{(.)}}{m} \end{gathered} \tag{14}$$

Consider sliding variable dynamics given by a system with a relative degree two.

$$\ddot{\sigma} = h(\sigma, \dot{\sigma}, t) + k(t)u_{\delta}, \quad k(t) > 0 \tag{15}$$

In the considered case, the controls are continuous. Define auxiliary sliding surfaces $s_{(.)}$ as dynamical sliding manifolds

$$s_{(.)} = \dot{\sigma}_{(.)} + \varpi \left|\sigma_{(.)}\right|^{1/2} sign(\sigma_{(.)})$$
$$s_{(.)} \to 0 \Rightarrow \sigma_{(.)}, \dot{\sigma}_{(.)} \to 0 \tag{16}$$

As the sliding manifolds are relative degree 1 with respect to the system, the controller is now relative degree 1 with respect to the sliding manifold. The corresponding Super-Twist controllers are given by:

$$u_{(.)} = Limit[-0.5, 0.5](-\alpha sign(s_{(.)})\left|s_{(.)}\right|^{1/2} - \beta \int_0^t sign(s_{(.)})d\tau \tag{17}$$

Where the Limit [,] is imposed because the relative attitude with respect to the trajectory must be bounded such as to leave enough longitudinal control authority to steer the longitudinal relative motion.

6.2 Normal / Lateral control during fixed attitude mode

After reaching the prescribed altitude and the prescribed orbital plane, normal/lateral on-off thrusters are used to keep the pursuing vehicle at the proper altitude and in the orbital plane.

With $k_m < k(t) < k_M$ and $\left|h(\sigma, \dot{\sigma}, t)\right| \le L$; it is shown (Edwards, 1998), (Utkin, 1999), (Levant, 2001), (Shtessel, 2003), (Shkolnikov, 2000), (Shtessel, 2000) that a sliding variable σ given by (10) is stabilized at zero altogether with its derivative $\dot{\sigma}$ in finite time by means of the SOSM controller

$$u = -\bar{\rho} \cdot sign(\dot{\sigma} + \lambda \left|\sigma\right|^{0.5} sign(\sigma)), \quad \lambda > 0, \quad \bar{\rho} > 0 \tag{18}$$

where $\bar{\rho} > \left(0.5\lambda^2 + L\right) / k_M$. This controller is called a *second order sliding mode controller with prescribed convergence law.* It is worth noting that the high frequency switching SOSM controller (18) achieves the finite time stabilization of σ and $\dot{\sigma}$ at zero in the presence of a bounded disturbance $h(\sigma, \dot{\sigma}, t)$.

Controller (18) yields on-off control that can be applied directly to the on-off thrusters. Here we choose $\lambda = 8\ rad / \sec$, and $\bar{\rho} = 0.1\ m / s^2$ is imposed by the acceleration achieved by the on-off thrusters.

6.3 Simulation

The Six Degrees of Freedom simulation was ran in Earth Centered Inertial Coordinates over rotating spherical Earth[1]. Attitude motion was calculated using Quaternions representing the body attitude with respect to ECI frame[2]. The simulation was calculated in normalized units with unit of length being the equatorial radius, the unit of velocity the circular velocity at the surface level, and the time unit the ratio of previous quantities. The results are presented in SI units and the gains used in normalized units converted to SI units.

---

[1] The simulation could be easily extended to work over oblate Earth. However since the problem is a problem of relative motion, this easy extension was not considered

[2] The problem to solve is a problem of relative attitude, and for that matter any other reference could have been chosen such as North East Down.

Integration step used was 10^{-6} normalized time units that is about 0.000806 sec. The integrations were performed using Runge-Kutta 4 algorithm build in the Vissim simulation software.

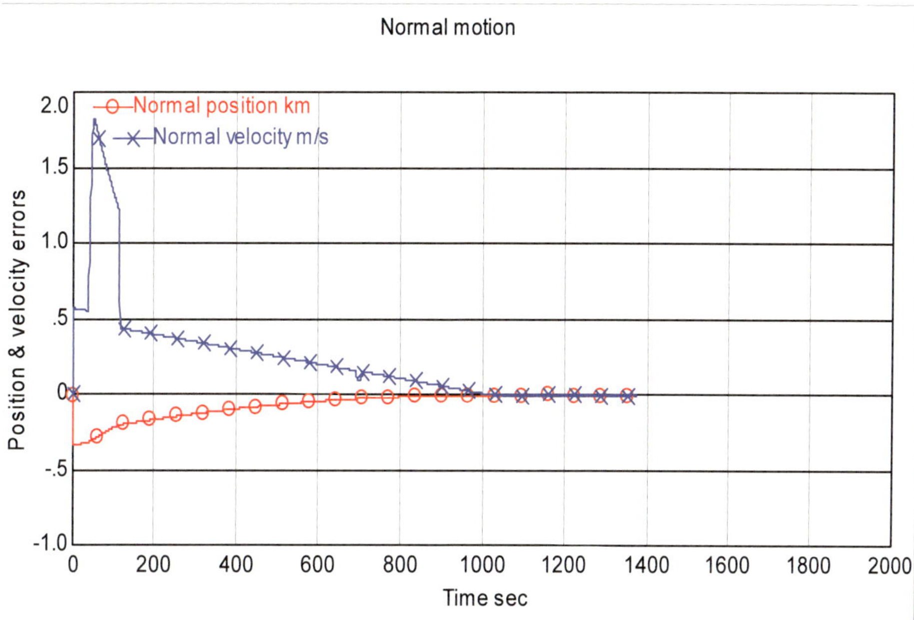

Fig. 5. Normal position and velocity error.

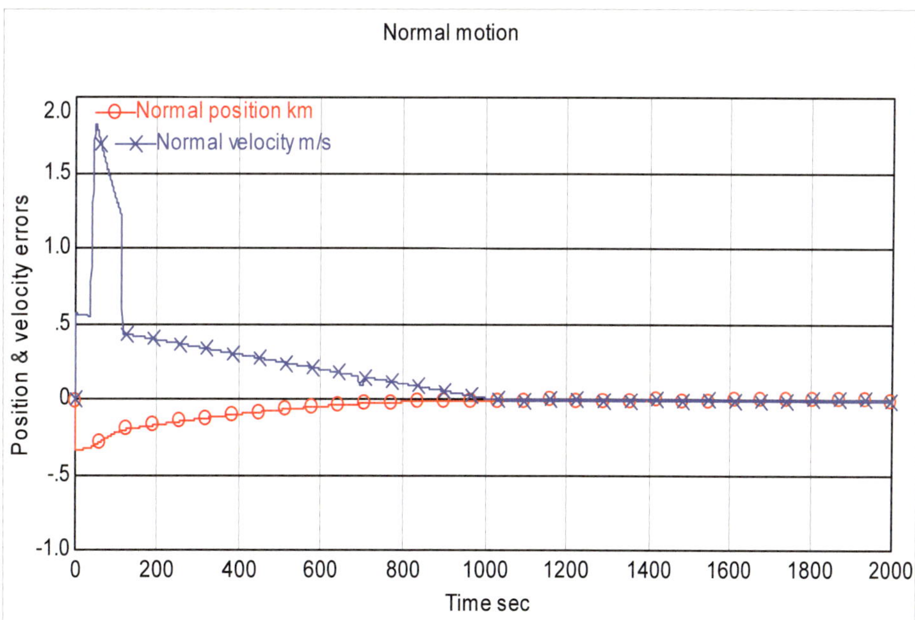

Fig. 6. Vehicle relative pitch attitude error.

The results Fig. 5 show that after the initial rendezvous normal/lateral distances to the receiving vehicle's orbit are kept within millimeters, millimeters /sec. Figure 6. depicts the corresponding vehicle attitude.

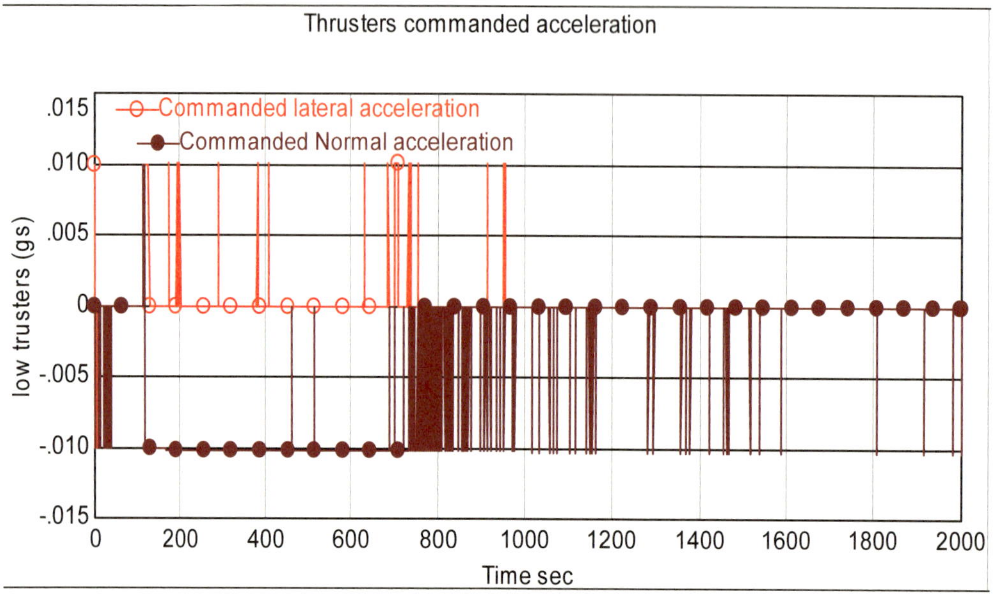

Fig. 7. Activity of the small thrusters.

The result Fig. 7 exhibits thruster commands during an important interval of activity in the segment 114-930 sec. The interval 114-537 corresponds to the drift segment during which the pursuing vehicle is at the same altitude that the pursued vehicle but has larger velocity by approximately 40 m/s. The interval 537-936 records deceleration to a much smaller longitudinal relative velocity. From there, as the longitudinal velocity is constantly reduced, the firing of normal thrusters becomes more and more infrequent. Conversely the activity of transversal thrusts reduces much more rapidly as this error is driven to zero.

6.3 Longitudinal control during terminal sliding mode phase

The prescribed longitudinal relative motion is defined by sliding variable
Figure 6. displays the corresponding vehicle normal and lateral (out-of-plane) thrusters' activity.

$$\sigma_x = \dot{x} + cx \tag{20}$$

When the longitudinal sliding surface is reached (when $\sigma_x \approx 0$), this forces the longitudinal velocity to reduce as the range becomes smaller. Using this surface the pulse width controller is given by

$$\begin{aligned} &w_x = -Asign(\sigma_x)|\sigma_x|^{1/2} - B\int_0^t sign(\sigma_x)d\tau;\ u_x = PWM(w_x) \\ &PWM(u) = DeadBand(1-\nu, u) + 0.5 + Triangle(1, \varepsilon) \\ &Triangle(A, f) = triangular\ wave\ amplitude = A,\ frequency = f \end{aligned} \tag{21}$$

6.4 Longitudinal breaking strategies and gates

Several control strategies have been analyzed which use braking maneuvers of different intensity and duration. We present hereafter the medium breaking strategy.

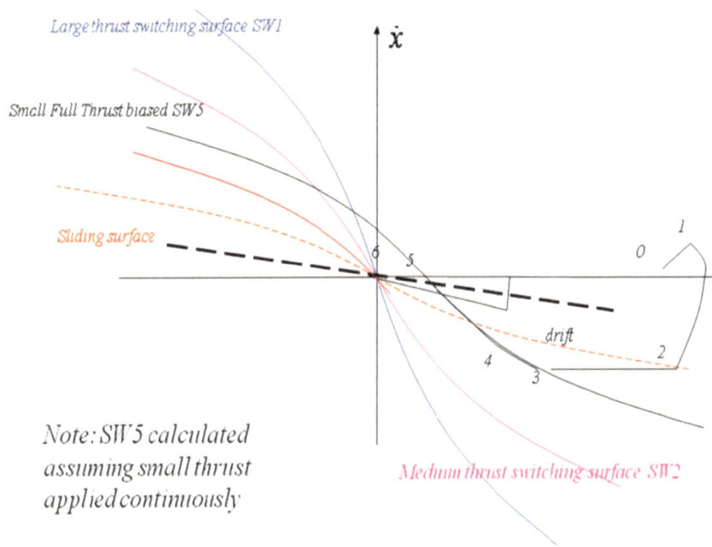

Fig. 8. Longitudinal control strategy 2 medium breaking.

Longitudinal control starts at point 1, the beginning of initial rendezvous. The pursuing vehicle accelerates using the main thruster / sustainer until point 2 when the relative prescribed closing velocity is reached. This point is selected such that a 15% duty cycle of small thruster deceleration would be required to steer the relative position and velocity approximately to zero. It is followed by a drift segment until reaching the second breaking curve at point 3, represented by a medium breaking stategy biased by some positive range.
The medium deceleration is applied from 3-5 until reaching the sliding surface. From 5-6 the longitudinal motion is governed by the linear manifold Eq. (12).
Results in Fig 9 show the variation of longitudinal range and range rate as functions of time. One can note that after significant initial variations in range and range rate, their values decrease asymptotically after reaching the sliding surface at t=914.

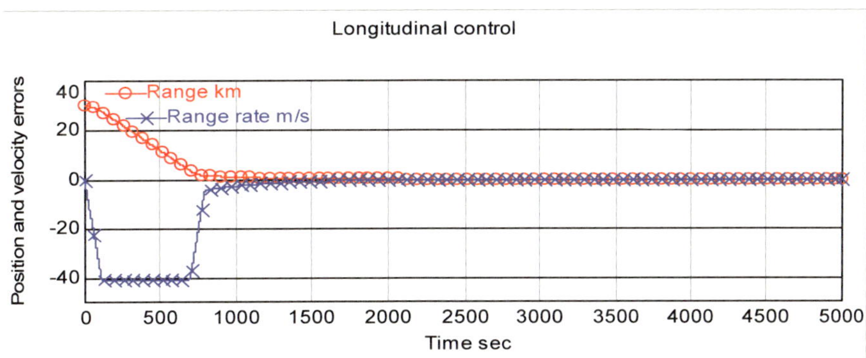

Fig. 9. Longitudinal control.

Results Fig. 10 show the absence of longitudinal control during the "drift" segment and also the continuous application of the "medium" deceleration from 700-796 sec. Results in Fig.10 show the pattern of longitudinal thrust. Starting on the left, one can note the sustainer thrust followed by the drift segment where no longitudinal thrust is applied, the deceleration pulse, then the deceleration segment where braking thrust is applied continuously;

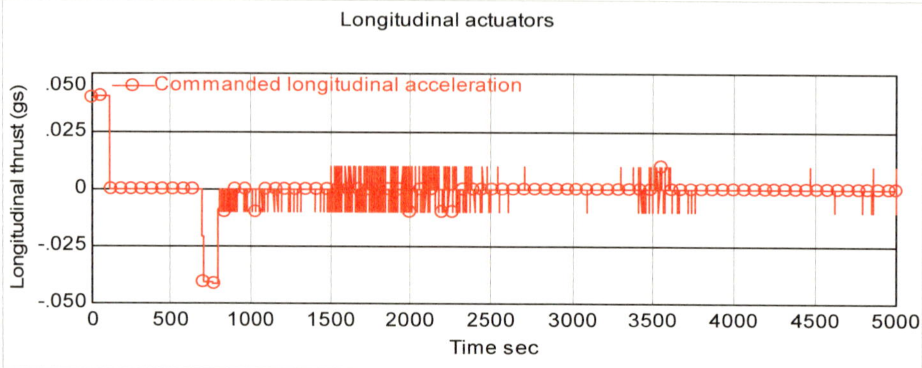

Fig. 10. Longitudinal thruster activity.

Results Fig. 10 also show the absence of longitudinal control during the "drift" segment and the continuous application of "medium" deceleration from 700-796 sec. Results in Fig. 10 show the pattern of longitudinal thrust. Starting on the left, one can note the sustainer thrust followed by the drift segment where no longitudinal thrust is applied, the deceleration pulse, then the deceleration segment where braking thrust is applied continuously; thereafter, the firing becomes sparser and the durations of the thrust pulses smaller, and reaches "soft kiss" conditions with range and range rate in the sub-millimeter and millimeter / sec. It is possible to make the docking faster by modifying parameter c in Eq. (20) and to interrupt it sooner as docking tolerances are reached. Another factor that may be considered in the automatic docking is the incorporation of cold gas thrusters to provide small and clean propulsive increments for final docking.

Three gateways are designed to check that the automatic docking is on track; equivalently, that provided the interceptor position is within the gate, docking can be pursued safely; specifically, that the margin of error they define can be corrected safely with available control authority.

For that matter we are going to present the gates from final to initial.

The third gateway is defined at the beginning of the deceleration The outer range is the minimum range such that if small thrusters are applied continously, the deceleration will achieve a zero velocity and distance from the receiving station. The deceleration must begin at the latest when intersecting the outside elliptical contour. The inner contour represents the minimum time for driving the longitudinal sliding variable to zero. The terminal deceleration in sliding mode must be initiated before reaching the inner contour.

At point 3 of Fig. 11, the pursuing vehicle begins medium braking, segment 3-5. Point 4 is at the intersection with the contour where there is enough stopping power to overcome the disturbances and stop at the origin using the small break. The breaking maneuver with small break must begin at the latest at point 4. The point 5 is designed to be on the intersection of the sliding manifold Eq. (12), with the small braking biased contour.

Evidently, the point 5 must be outside the inner elliptical contour that defines the minimum time needed to drive the terminal sliding surface to the origin.

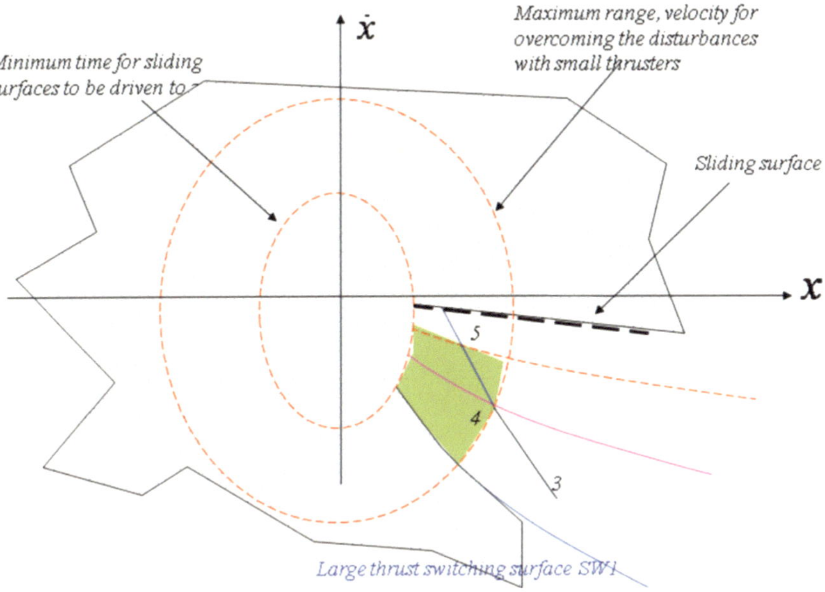

Fig. 11. Third gate.

The second gate Fig. 12 defines the drift segment. It begins at point 2; the intersection of the drift segment with SW3 and it ends at point 3 the beginning of the braking maneuver on biased SW5.

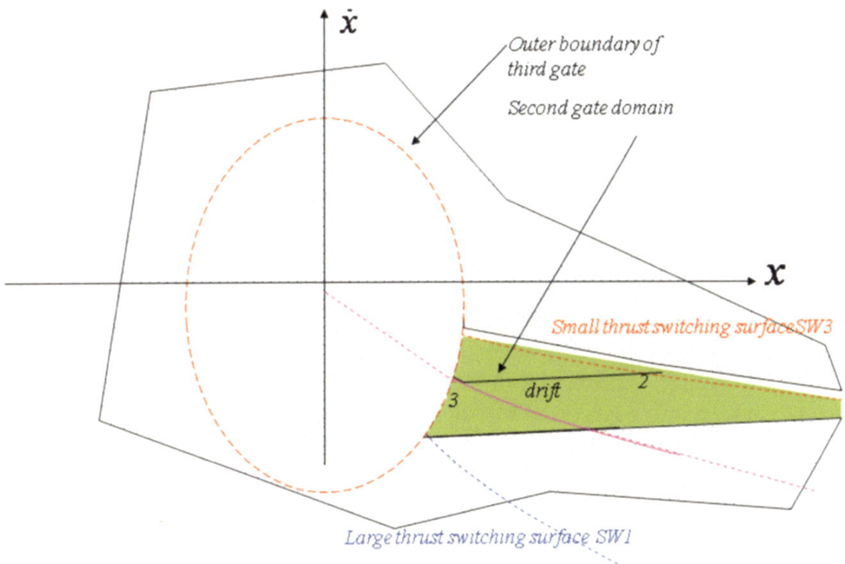

Fig. 12. Second gate.

The first gate (Fig. 13) defines the initial contour where the interceptor must be in the phase plane to intersect the small partial thrust SW3 with a viable drift velocity value and suffcient drift time. In any case the initial point 1 must be above SW3 and there is some latitude regarding the initial velocity and range.

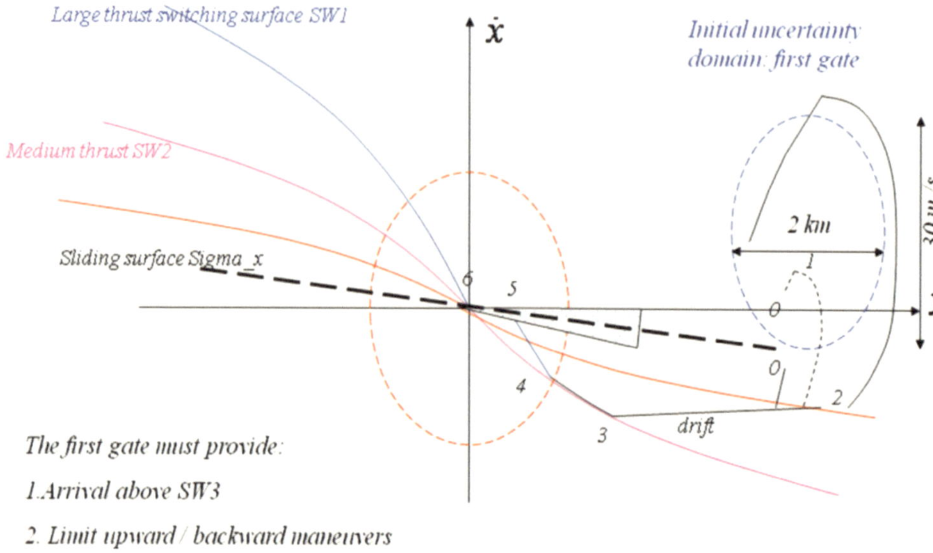

Fig. 13. First gate.

7. Use of active bitmap pixels to control relative attitude

Regulation of pursuer attitude for automated docking can be broken into two functional segments. While the objects are far apart, the pursuer's attitude is controlled to align its axial direction with the relative line of sight and to place its normal direction in the orbital plane. Control during this segment has been done many times and is not the subject of this discussion. When the objects are very close, and before docking can occur, the pursuer must align its mating surface with that of the pursued vessel. In this section, we discuss one practical method that this alignment can be performed efficiently, reliably and automatically.

Any geometry will do, but suppose that both mating surfaces are circular and that the target object is fitted with a series of detectable objects (i.e. lights) equally spaced around the mating surface. Suppose further that the pursuer is fitted with an array of suitable detectors which we shall call the Focal Plane Array (FPA) and that this FPA can be considered to lie in the center of its mating surface. As described in figure 14, if the surfaces are ready for docking, the pursuer will perceive a circular ring of lights in the center of the FPA. If the surfaces are offset, then the ring will be offset on the FPA. If the surfaces are misaligned, the ring will be elliptical rather than circular. The apparent size of this perceived ring of lights will indicate separation distance; the center will indicate normal and lateral error; the eccentricity of the ellipse will indicate the degree of angular error; and the orientation of the ellipse will indicate the relative axis about which the pursuer must rotate for successful docking. Although we will not address relative roll in this chapter, if one of the lights is

distinct from the others, a roll error could also be deduced. This is nearly equivalent to the information a human pilot would use to accomplish the same task.
Automated control of attitude for docking is thus reduced to two necessary tasks. First, information from the FPA must be interpreted (in the presence of noise) to yield a real-time measure of attitude error. Second, that error must be used to correctly orient the vehicle. We will apply a nonlinear least-squares curvefit and multidimensional search to the corrupted pattern of lights in order to estimate the equation of the perceived ellipse. The relative magnitude and orientation of the semi-major and semi-minor axes of this ellipse are used to generate a necessary angle of rotation and the unit vector we must rotate about, respectively. The relative degree approach will be used to generate a second-order sliding mode controller of the type described in (Levant, 2003). Finally, these methods will be implemented and tested using simulation.

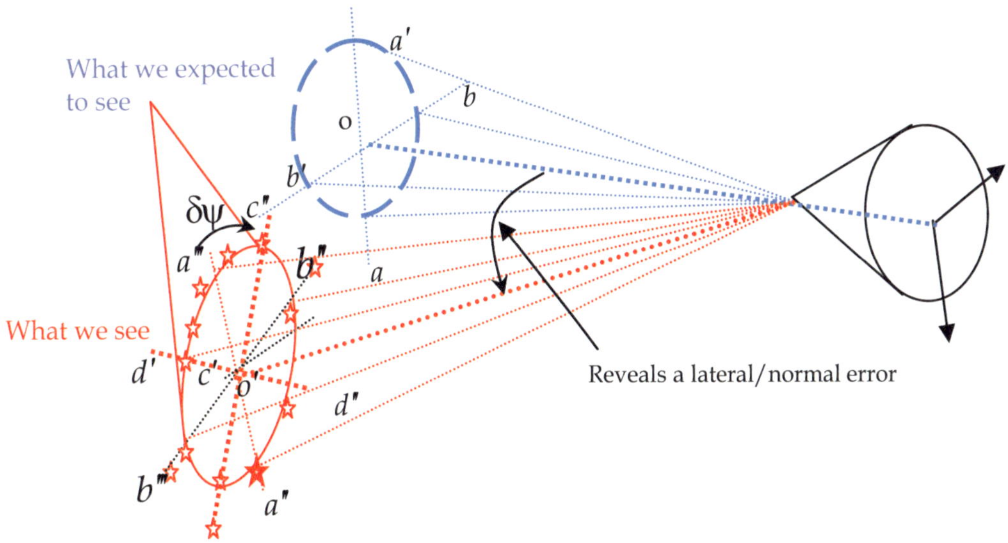

Fig. 14. Use of light patterns to extract relative position and attitude.

7.1 Mathematical background: Quaternions

The idea of relating two oriented surfaces (equivalently, two reference frames) by a single rotation about a specified axis is precisely the motivation behind the concept of quaternions. Since many readers will not be familiar with quaternions, we introduce a few important concepts here. Those wishing to understand quaternions in greater depth are referred to Dr. Kuipers' excellent book (Kuipers, 1999) on the subject.
Let us describe the relationship between two right-hand coordinate systems as a single rotation about a specified axis. Let us package this description into a 4-vector as follows:

$$\mathbf{Q} = \begin{bmatrix} \cos\eta \\ [\hat{\mathbf{u}}\sin\eta] \end{bmatrix} = \begin{bmatrix} \mathbf{q_0} \\ \mathbf{q} \end{bmatrix} \quad \begin{matrix} \eta = \textit{the magnitude of rotation} \\ \hat{\mathbf{u}} = \textit{the (unit) vector to rotate about} \end{matrix} \tag{22}$$

It is easily verified that this construct has (Euclidian) norm 1. If we define multiplication of these objects in a particular way, they exhibit several useful traits. Define:

$$\mathbf{PQ} = \mathrm{p_0 q_0} - \mathbf{p} \bullet \mathbf{q} + \mathrm{p_0}\mathbf{q} + \mathrm{q_0}\mathbf{p} + \mathbf{p} \times \mathbf{q} \tag{23}$$

The following four useful and remarkable properties hold:

1. For any quaternion Q, $\mathbf{QQ}^* = \begin{bmatrix} q_0 \\ \mathbf{q} \end{bmatrix} \begin{bmatrix} q_0 \\ -\mathbf{q} \end{bmatrix} = \mathbf{Q}^*\mathbf{Q} = \begin{bmatrix} 1 & 0 & 0 & 0 \end{bmatrix}^t$ and this is the quaternion relating any coordinate frame to itself.
2. Given a vector v in the initial reference frame, the vector part of $\mathbf{Q}\begin{bmatrix} 0 \\ \mathbf{v} \end{bmatrix}\mathbf{Q}^*$ is the equivalent vector in the rotated frame.
3. Given quaternion P relating frame 1 to frame 2 and quaternion Q relating frame 2 to frame 3, the product QP is the quaternion relating 1 to frame 3.
4. If Q is the quaternion relating frame 1 to frame 2 and $\boldsymbol{\Omega} = \begin{bmatrix} \mathbf{pp} \\ \mathbf{qq} \\ \mathbf{rr} \end{bmatrix}$ represents the turning rate of frame 1 relative to frame 2 (i.e. the body rates) then $\dot{\mathbf{Q}} = \mathbf{Q}\begin{bmatrix} 0 \\ \boldsymbol{\Omega} \end{bmatrix}$

7.2 Attitude error from FPA measurements

In this exercise we are assuming that the pursuer's on-board sensor is the only source of attitude feedback. Specifically, this information takes the form of a set of Cartesian positions on the FPA corresponding to the location of the docking lights; with the detectors on the FPA working in the same way as rods on a human retina. Our challenge is to interpret, from this list of positions, the relative orientation of the pursuer and target docking surfaces.
As discussed in the introduction, if the docking surfaces are not perfectly adjusted, a circular pattern of indistinguishable lights (Fig. 15a) will appear as an offset ellipse (15b).

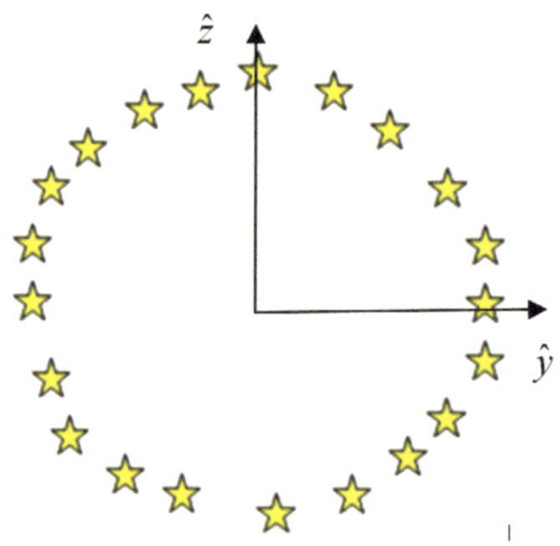

Fig. 15a. Circular pattern of docking lights.

Lateral and longitudinal guidance was described earlier in this chaper; thus we are only concerned that the pursuer's attitude be modified such that the percieved ellipse become circular. We proceed in two steps: first determine the equation of the ellipse that most nearly fits the measurements; then compute attitude error from this equation.

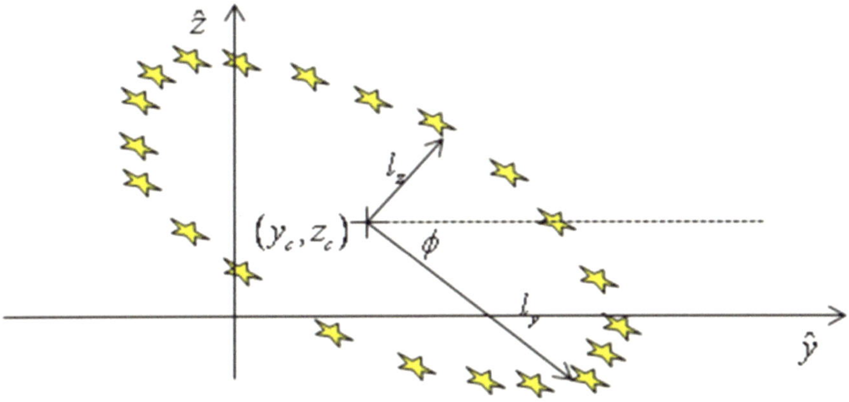

Fig. 15b. Docking lights as seen by pursuer.

The formal equation of an offset, rotated ellipse is:

$$\frac{\left((y-y_c)\cos\phi+(z-z_c)\sin\phi\right)^2}{l_y^{\,2}}+\frac{\left(-(y-y_c)\sin\phi+(z-z_c)\cos\phi\right)^2}{l_z^{\,2}}=1 \tag{24}$$

To perform least-squares curvefit from a set of measured points (x,z), define a function:

$$E=\sum_{all\,lights}\left\{1-\frac{\left((y-y_c)\cos\phi+(z-z_c)\sin\phi\right)^2}{l_y^{\,2}}+\frac{\left(-(y-y_c)\sin\phi+(z-z_c)\cos\phi\right)^2}{l_z^{\,2}}\right\}^2 \tag{25}$$

We will find a local minimum value of E with respect to the parameters $\{y_c, z_c, l_y, l_z, \phi\}$ using the steepest descent method:

$$\begin{bmatrix} y_c \\ z_c \\ l_y \\ l_z \\ \phi \end{bmatrix}_+ = \begin{bmatrix} y_c \\ z_c \\ l_y \\ l_z \\ \phi \end{bmatrix}_- - \frac{\rho}{\sqrt{\left(\partial E/\partial y_c\right)^2+\left(\partial E/\partial z_c\right)^2+\left(\partial E/\partial l_y\right)^2+\left(\partial E/\partial l_z\right)^2+\left(\partial E/\partial \phi\right)^2}}\begin{bmatrix} \partial E/\partial y_c \\ \partial E/\partial z_c \\ \partial E/\partial l_y \\ \partial E/\partial l_z \\ \partial E/\partial \phi \end{bmatrix} \tag{26}$$

with the ad-hoc addition that, if $E_+ > E_-$ then $\rho \leftarrow \frac{\rho}{10}$. This iteration is allowed to continue until the function E converges to a constant value at which the parameters describing the "best-fit" ellipse are established[3].

[3] The multivariate search described above requires an initial guess for each parameter. Convergence rate is sensitive to this guess and to the initial step size ρ. Furthermore, if care is not exercised, this search may converge to a local (and not global) minimum. An extensive discussion of multivariate search isoutside the scope of this chapter. outside the scope of this chapter.

The magnitude of rotation necessary for the ellipse to appear circular is described by:

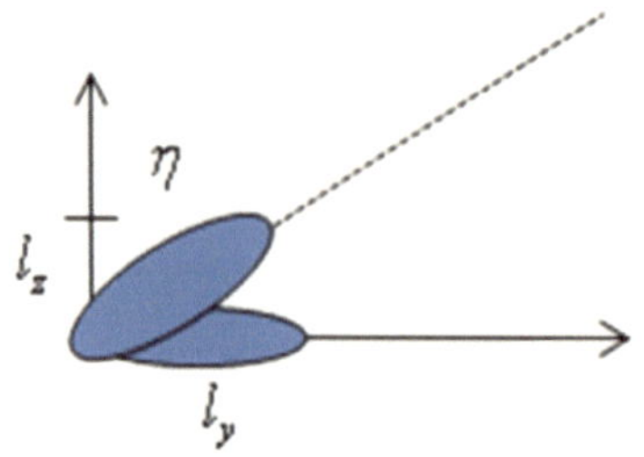

$$l_z = l_y \cos\eta \quad \Rightarrow \quad \eta = a\cos\frac{l_z}{l_y}$$

The required axis of rotation is the ellipse semi-major axis, which is described by:

$$\hat{\mathbf{u}} = \cos\phi\,\hat{\mathbf{y}} + \sin\phi\,\hat{\mathbf{z}} \tag{27}$$

The quaternion relating the pursuer's attitude to that necessary for docking is, therefore:

$$\mathbf{Q} = \begin{bmatrix} \cos\frac{\eta}{2} \\ 0 \\ \sin\frac{\eta}{2}\cos\phi \\ \sin\frac{\eta}{2}\sin\phi \end{bmatrix} = \begin{bmatrix} \sqrt{\frac{1}{2}\left(1 - {}^{l_z}\!/_{l_y}\right)} \\ 0 \\ \sqrt{\frac{1}{2}\left(1 + {}^{l_z}\!/_{l_y}\right)}\cos\phi \\ \sqrt{\frac{1}{2}\left(1 + {}^{l_z}\!/_{l_y}\right)}\sin\phi \end{bmatrix} \tag{28}$$

7.3 Derivation of the attitude control law

The relative degree approach to derivation of a control law consists of a sequence of general steps. First, establish an approximate mathematical model for the object to be controlled. If (as is always the case) this model is imperfect, we include an unknown "disturbance" function into which all of the uncertainties, approximations and unknowable quantities are swept. Second, the feedback error is defined. This error must be generated from measured quantities and must be positive definite. In the third step, a mathematical relationship is established between the feedback error and the actual control. This relationship is made to fit a template equation that is well-behaved in the presence of the expected disturbance. Finally, the relationship is solved to describe the necessary control in terms of the feedback error, possibly other measured quantities and the disturbance, which is discarded.

Let Q represent the quaternion relating the pursuer body frame to the required attitude for docking as computed in (24). In practice, the pursued vehicle may be rotating, but because we derive all our information from the pattern of docking lights, the pursued vehicle's rotation is confounded with the pursuing vehicle's rotation and is thus unknowable. Therefore we shall consider the desired attitude to be an inertial frame and consider any error resulting from this supposition to be part of the disturbance function. Further define:

$\mathbf{Q},\boldsymbol{\omega} = \begin{bmatrix} pp \\ qq \\ rr \end{bmatrix}$ is the vector of the pursuer's body rates

$\mathbf{I} \in \mathbb{R}^{3x3}$ is the pursuer's matrix of inertia, which is considered nonsingular

$\begin{cases} \mathbf{B} \in \mathbb{R}^{3x3} \\ \mathbf{u} \in \mathbb{R}^{3} \end{cases}$ such that Bu represents the moment contribution of control in the body axis

The equations of state may be described as:

$$\begin{aligned} \dot{\mathbf{Q}} &= \mathbf{Q}\boldsymbol{\omega} + \boldsymbol{\Delta}_1 \\ \dot{\boldsymbol{\omega}} &= -\mathbf{I}^{-1}\boldsymbol{\omega} \times \mathbf{I}\boldsymbol{\omega} + \mathbf{I}^{-1}\mathbf{B}\mathbf{u} + \boldsymbol{\Delta}_2 \end{aligned} \tag{29}$$

For docking, we want the pursuer's body frame to align with the desired frame; this is equivalent to driving $\mathbf{Q} \to \begin{bmatrix} 1 & 0 & 0 & 0 \end{bmatrix}^t$. Because Q has norm 1, driving the vector part to zero will accomplish this desire. If we consider desired rotation about the body x-axis to be zero and restrict the remaining axis of rotation to quadrants 1 and 2 (accounting for the direction of rotation by other means) taking feedback error to be the vector part of Q results in a positive definite function. Therefore, with obvious notation, let:

$$\boldsymbol{\sigma} = [\mathbf{Q}]_{123} \tag{30}$$

ignoring disturbances and differentiating: $\dot{\boldsymbol{\sigma}} = [\mathbf{Q}\boldsymbol{\omega}]_{123}$

$$\ddot{\boldsymbol{\sigma}} = [\mathbf{Q}\mathbf{Q}\boldsymbol{\omega} + \mathbf{Q}\dot{\boldsymbol{\omega}}]_{123} = \left[\mathbf{Q}\mathbf{Q}\boldsymbol{\omega} + \mathbf{Q}(-\mathbf{I}^{-1}\boldsymbol{\omega} \times \mathbf{I}\boldsymbol{\omega} + \mathbf{I}^{-1}\mathbf{B}\mathbf{u})\right]_{123} \tag{31}$$

Before proceeding, we will need the following theorem:

Theorem: For quaternions $\mathbf{P} = \begin{bmatrix} \mathrm{p}_0 \\ [\mathbf{p}] \end{bmatrix}, \mathbf{Q} = \begin{bmatrix} \mathrm{q}_0 \\ [\mathbf{q}] \end{bmatrix}, \; \left\{\mathbf{P}^* [\mathbf{PQ}]_{123}\right\}_{123} = \mathrm{p}_0\mathbf{q} + \mathrm{p}_0\mathrm{q}_0\mathbf{p}$ (32)

Proof: from Kuipers (p.108):

$$\mathbf{PQ} = \begin{bmatrix} \mathrm{p}_0\mathrm{q}_0 - \mathbf{p}\bullet\mathbf{q} \\ [\mathrm{p}_0\mathbf{q} + \mathrm{q}_0\mathbf{p} + \mathbf{p}\times\mathbf{q}] \end{bmatrix}$$

$$\Rightarrow \; \mathbf{P}^*[\mathbf{PQ}]_{123} = \begin{bmatrix} \mathrm{p}_0(0) + \mathbf{p}\bullet(\mathrm{p}_0\mathbf{q} + \mathrm{q}_0\mathbf{p} + \mathbf{p}\times\mathbf{q}) \\ [\mathrm{p}_0(\mathrm{p}_0\mathbf{q} + \mathrm{q}_0\mathbf{p} + \mathbf{p}\times\mathbf{q}) - (0)\mathbf{p} - \mathbf{p}\times(\mathrm{p}_0\mathbf{q} + \mathrm{q}_0\mathbf{p} + \mathbf{p}\times\mathbf{q})] \end{bmatrix}$$

$$\begin{aligned} \left\{\mathbf{P}^*[\mathbf{PQ}]_{123}\right\}_{123} &= \mathrm{p}_0^2\mathbf{q} + \mathrm{p}_0\mathrm{q}_0\mathbf{p} + \mathrm{p}_0(\mathbf{p}\times\mathbf{q}) - \mathrm{p}_0(\mathbf{p}\times\mathbf{q}) - \mathrm{q}_0(\mathbf{p}\times\mathbf{p}) - \mathbf{p}\times\mathbf{p}\times\mathbf{q} \\ &= \mathrm{p}_0^2 q + \mathrm{p}_0\mathrm{q}_0\mathbf{p} \end{aligned}$$

□

Define:

$\mathbf{S}(\dot{\sigma},\sigma)_{1,2,3} = -\rho\,\mathbf{SIGN}\left(\dot{\sigma}_{1,2,3} + \mu|\sigma_{1,2,3}|^{1/2} sign(\sigma_{1,2,3})\right)$ where ρ and μ are positive constants. (33)

It is shown [26] that the equation: $\ddot{\sigma}_{1,2,3} - \mathbf{S}(\dot{\sigma},\sigma)_{1,2,3} = \Delta_{1,2,3}$ is finite-time stable and displays "good" transient behavior in each of its three elements so long as elements of the disturbance Δ are bounded by the proportionality constant ρ. Substituting for the second derivative in (31):

$$\mathbf{S}=\left[\mathbf{QQ}\omega+\mathbf{Q}(-\mathbf{I}^{-1}\omega\times\mathbf{I}\omega+\mathbf{I}^{-1}\mathbf{Bu})\right]_{123}$$

Pre-multiply both sides by $\mathbf{Q}^*$ and apply the theorem:

$$\mathbf{Q}^*\mathbf{S}=\mathbf{Q}^*\left[\mathbf{QQ}\omega\right]_{123}+\mathbf{Q}^*\left[\mathbf{Q}\left(-\mathbf{I}^{-1}\omega\times\mathbf{I}\omega+\mathbf{I}^{-1}\mathbf{Bu}\right)\right]_{123}$$

$$\left[\mathbf{Q}^*\mathbf{S}\right]_{123}=\mathbf{q}_0^{\ 2}\left[\mathbf{Q}\omega\right]_{123}+\mathbf{q}_0\left[\mathbf{Q}\omega\right]_0\left[\mathbf{Q}^*\right]_{123}-\mathbf{q}_0^{\ 2}(\mathbf{I}^{-1}\omega\times\mathbf{I}\omega)+\mathbf{q}_0^{\ 2}(\mathbf{I}^{-1}\mathbf{Bu})$$

Solve for the control u:

$$\mathbf{u}=\mathbf{B}^{-1}\mathbf{I}\left\{\frac{1}{\mathbf{q}_0^{\ 2}}\left[\mathbf{Q}^*\mathbf{S}\right]_{123}-\left[\mathbf{Q}\omega\right]_{123}+\frac{\left[\mathbf{Q}\omega\right]_0}{q_0}\left[\mathbf{Q}\right]_{123}-(\mathbf{I}^{-1}\omega\times\mathbf{I}\omega)\right\} \tag{34}$$

7.4 Simulation results

In order to demonstrate this method of attitude control for automated docking, a ten-second interval near the end of a docking mission was simulated. The initial separation is 11 m and the closing velocity is 1 m/sec. Lateral and longitudinal control are not included in this exercise, nor is roll attitude. Initially, the docking surfaces are misaligned by .1 radian (~6 degrees) in the pitch direction and .25 radians(14 degrees) in yaw. Additionally, we have initial body rates equal to .05 rad/sec away from zero in the pitch and .1 rad/sec towards zero in yaw. Realistically, seeker error would decrease as the surfaces approach, but for demonstration purposes, a uniformly-distributed 5% error was added to the y- and z-positions of each docking light.

The gains ρ and μ of Eq. (33) were empirically set to 5 and 0.25, respectively; these gains were intentionally not fine-tuned and it was observed that acceptable behavior is exhibited when either or both of these are halved or doubled.

Results are summarized in Figs. 16 - 18.

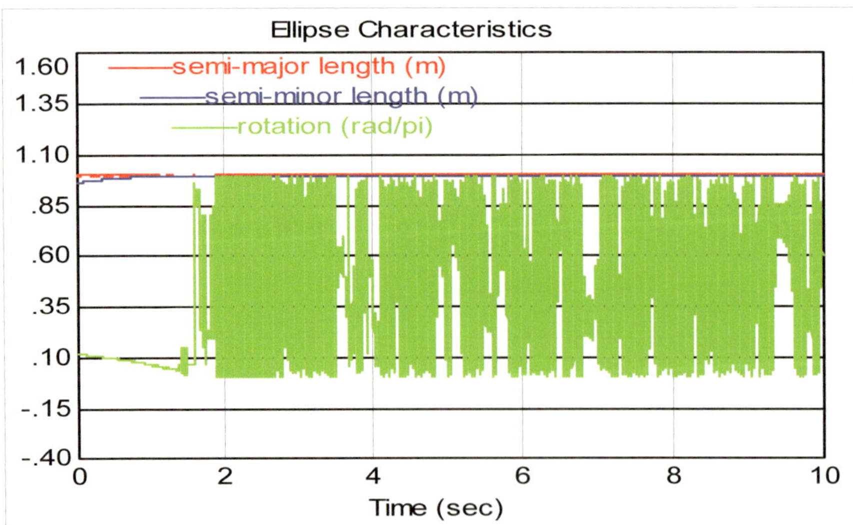

Fig. 16. Characteristics of the curvefit ellipse.

In Fig. 16 we observe that the (normalized) semi-major axis length is constant at unity. This is necessary, as the apparent length (adjusting for changes in proximity) does not change with aspect. The semi-minor axis length is initially somewhat less, but quickly converges to one; this is an indication that the percieved ellipse becomes a percieved circle. At about the time the semi-minor axis approaches unity, the apparent rotation of the ellipse becomes chaotic. This is expected - as the FPA image becomes more circular, definition of the semi-major and semi-minor axes is largely determined by noise.

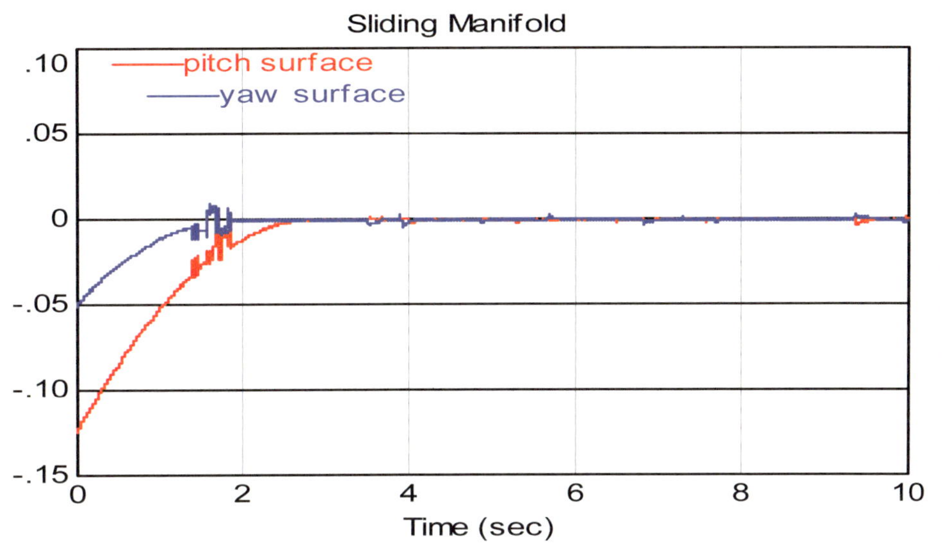

Fig. 17. Quaternion elements 2 and 3.

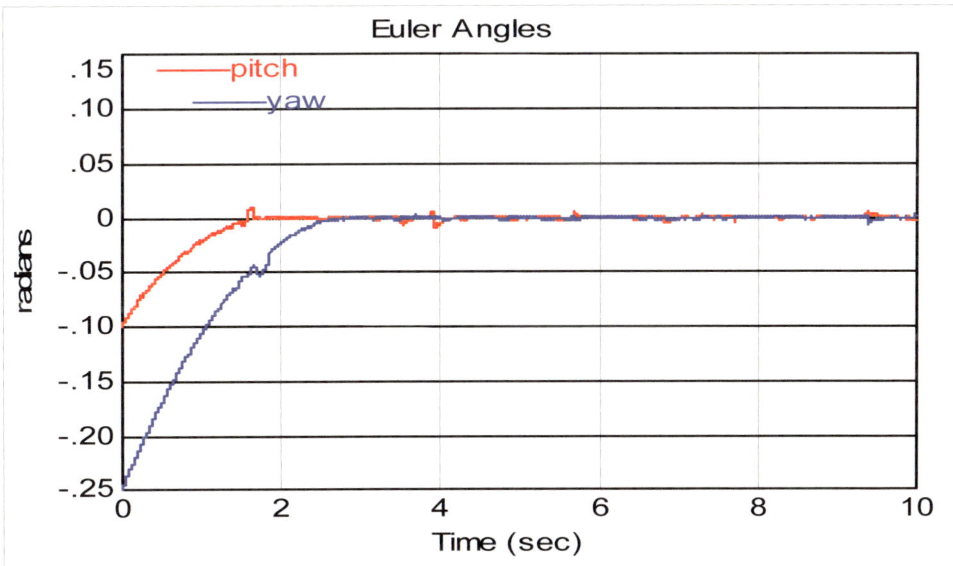

Fig. 18. Corresponding pitch and yaw angles.

In Fig. 17 we observe that the sliding variables are driven into a narrow band about zero in finite time and remain within that band thereafter. Note that actual convergence to the sliding surface occurs significantly after the quaternion axis (green line of Fig. 16) becomes chaotic. It is apparent that the averaged reaction to extremely noisy feedback is still useful for control. If the seeker noise was correlated in time, we might expect to see a small and persistent error away from zero.

Euler angles are easily extracted from the quaternion elements. In Fig. 18 we see the pitch misalignment, which started nearer to zero converge first, followed by yaw. After the transient, both angles are constrained to within about 3 or 4 milliradians (0.2 degrees). Speed of convergence and ultimate boundary are largely dictated by the gains ρ and μ of Eq. (33), subject to limitations on thruster force and the need to dominate the sum of all disturbances.

7.5 Observations

Before concluding this section, let us make some interesting and important observations concerning the demonstrated method for automatic control of attitude for docking.

First, this automated method is very similar to the approach taken by a human pilot; rather than assembling position and attitude information from a variety of sources, computing a time profile and inverting the physical model to produce attitude commands, this method "sees" that the ring of docking lights is slightly out of round and nudges the controls in response. This not only increases confidence in the robustness of our method, but introduces the possibility of Human Assisted Control (HAC) for docking attitude.

Second, there is no Inertial Measuring Device (IMU) input involved in this method. This means no IMU errors, no acquisition and processing of IMU data, no synchronization of IMUs between the pursuer and pursued and no provisioning for loss of data. All feedback is from a single, reliable on-board source. On a related note, there is no participation required on the part of the pursued object and no communication requirement. This is extremely favorable because communication increases risk and always introduces delay. Delay is extremely detrimental to sliding mode control, which is fundamentally based on high-frequency switching.

Finally, the reader may have spotted a significant flaw in our method. When interpreting the ring of docking lights as an ellipse on the FPA, the magnitude of rotation and the axis of rotation can be determined, but there is no inherent way to determine the direction of rotation. In other words, we cannot tell if the ellipse is tipped "towards" us or "away". This perceptive reader is correct; some other method such as Doppler ranging or a comparison of the relative brightness on each side of the semi-major axis must be used to supply this final bit of information. While generating the results of Figs. 16-18, we assumed that the directionality was known and correct.

7.6 Conclusion: Attitude control

It is possible to control relative attitude by simply constructing a quaternion error function of the pattern of lights. One must note that the algorithm process is very similar to the human control processes in that the idea is to drive errors to zero. These solutions are enabled by the property that sliding mode controllers are perfectly insensitive to matched disturbances. Using this property it is possible to not represent explicitly in the design some dynamical terms of the sliding variable dynamics and to treat them simply as disturbance terms.

8. Chapter conclusion

The simplicity and elegance of the solution is a unique attribute of this emerging technique which makes it a *game changer.* Proposed design could conciliate the optimality of bang-bang solutions which are not robust with the robustness of HOSM which is not optimum. The result is a very simple design that conciliates a quasi-optimality with a perfect robustness. The insensitivity property of HOSM controllers to matched disturbances allowed to treat all the dynamical effects other than caused by the control to be treated as disturbances and compensated implicitly. Likewise for the attitude motion where by treating all dynamical effects other than the torques created by attitude command thrusters the three attitude motions could be treated as explicitly de-coupled[4] which greatly simplified the design of the control. Finally by running the simulation for a very long duration we showed that final results of extreme accuracy could be achieved.

9. References

W. A. Chobotov. (2002). *Orbital Mechanics (3rd Ed.).* AIAA Educational Series, Reston VA. pp. 155-158.

C. Edwards. And S. Spurgeon S. (1998). *Sliding Mode Control: Theory and Applications.*

Taylor & Francis, Bristol. ESA (2006) Document No. EUC-ESA-FSH-003 REV-1.2.

L. Fridman, Y. Shtessel, C. Edwards, and X. G. Yan. (2008). Higher Order Sliding Mode Observer for State Estimation and Input Reconstruction in Nonlinear Systems. *International Journal of Robust and Nonlinear Control, Special Issue on Advances in Higher Order Sliding Mode Control,* Vol. 18, Issue 4-5 (March). pp. 399-412.

C. Hall and Y. Shtessel. (2006). Sliding Mode Disturbance Observers-based Control for a Reusable Launch Vehicle. *AIAA Journal on Guidance, Control, and Dynamics,* Vol. 29, No. 6, (November-December). pp. 1315-1329.

C. D. Karlgard. (2006). Spacecraft. *AIAA Journal of Guidance, Control, and Dynamics,* Vol. 29, No. 1. pp. 495-499.

J. Kuipers. (1999). *Quaternions and Rotation Sequences.* Princeton University Press, Princeton NJ

A. Levant. (2001). Universal SISO sliding-mode controllers with finite-time convergence. IEEE Transactions on Automatic Control, Vol. 46, No. 9. pp. 1447-1451.

A. Levant. (2003). Higher-order sliding modes, differentiation and output-feedback control. *International Journal of Control.* Vol. 76, No. 9/10. pp. 924-941.

T. Massey and Y. Shtessel. (2005). Continuous Traditional and High Order Sliding Modes for Satellite Formation Control. *AIAA Journal on Guidance, Control, and Dynamics,* Vol. 28, No. 4, (July-August). pp. 826-831.

I. A. Shkolnikov, Y. B. Shtessel, M. Whorton, and M. Jackson. (2000). Robust to Noise Microgravity Isolation Control System Design via High-Order Sliding Mode Control. *Proceedings of the Conference on Guidance, Navigation, and Control,* Denver, CO. AIAA paper No. 2000-3954.

---

[4] The coupling between attitude channels is treated as disturbance and is thus, compensated implicitly by the controller

I. Shkolnikov, Y.B. Shtessel, and D. Lianos. (2005). The effect of sliding mode observers in the homing guidance loop,. *ImechE Journal on Aerospace Engineering,* Part G, 219, 2. pp. 103-111.

Y. Shtessel, C. Hall, and M. Jackson. (2000). Reusable Launch Vehicle Control in Multiple Time Scale Sliding Modes. *AIAA Journal on Guidance, Control, and Dynamics,* Vol. 23, No. 6. pp. 1013-1020

Y.B. Shtessel, I. Shkolnikov, and M. Brown. (2003). An Asymptotic Second-Order Smooth Sliding Mode Control. *Asian Journal of Control,* Vol. 4, No. 5. pp. 498-504.

Y. Shtessel, I. Shkolnikov and A. Levant. (2007). Smooth Second Order Sliding Modes: Missile Guidance Application. *Automatica,* Vol. 43, No.8. pp. 1470-1476.

Y. Shtessel, S. Baev, C. Edwards, and S. Spurgeon. (2010). HOSM observer for a class of non-minimum phase causal nonlinear MIMO systems. *IEEE Transactions on Automatic Control,* Vol. 55, No. 2. pp. 543-548.

P. Singla, K. Subbarao, and J. L. Junking. (2006) Adaptive Output Feedback Control for Spacecraft Rendezvous and Docking under Measurement Uncertainty. *AIAA Journal of Guidance, Control, and Dynamics,* Vol. 22, No. 4. pp. 892-902.

A. Sparks. (2000). Satellite Formation Keeping in the Presence of Gravity Perturbations, *Proceedings of the American Control Conference* (June).

C. Tournes, and Y.B. Shtessel. (2006). Autopilot for Missiles Steered by Aerodynamic Lift and Divert Thrusters using Second-Order Sliding Mode. *AIAA Journal of Guidance Control and Dynamics* Vol. 29, No. 3. pp. 617-623.

C. Tournes, Y.B. Shtessel. (2007). Automatic Docking using Second Order Sliding Mode Control. *Proceedings of the 2007 IEEE American Control Conference.*

V. Utkin, J. Guldner, and J. Shi. (1999). *Sliding Modes in Electromechanical Systems.* Taylor and Francis, London.

P. K. Wang, F. Y. Hadaegh and K. Lau. (1999) Synchronized Formation Rotation and Attitude Control of Multiple Free-Flying Spacecrafts. *AIAA Journal of Guidance, Control, and Dynamics,* Vol. 29, No. 1. pp. 28-35.

H. Wong, V, Karpila. (2001). Adaptive Output Feedback Tracking Control of Multiple Spacecraft. *Proceedings of the American Control Conference,* (June).

17

High Order Sliding Mode Control for Suppression of Nonlinear Dynamics in Mechanical Systems with Friction

Rogelio Hernandez Suarez[1], America Morales Diaz[2], Norberto Flores Guzman[3], Eliseo Hernandez Martinez[4] and Hector Puebla[4]
[1]*Instituto Mexicano del Petróleo, México D.F.*
[2]*Centro de Investigación y Estudios Avanzados del IPN, Saltillo Coahuila*
[3]*Centro de Investigación en Matemáticas, Guanajuato Guanajuato, Saltillo, Coahuila,*
[4]*Universidad Autónoma Metropolitana, México D.F.*
México

1. Introduction

Friction occurs in all mechanical systems, (e.g. bearings, transmissions, hydraulic and pneumatic cylinders, valves, brakes and wheels). Friction is the tangential reaction force between two surfaces in contact. There is a wide range of physical phenomena that cause friction, this includes elastic and plastic deformations, fluid mechanics and wave phenomena, and material sciences (Bowden & Tabor, 1950; Armstrong-Hélouvry, 1994; Rabinowicz, 1995). In mechanical systems, friction can limit the performance in terms of increasing tracking errors and, under certain conditions, friction leads to oscillatory behavior, including simple periodic oscillations and chaos (Feeny & Moon, 1994; Hikihara & Moon, 1994; Ibrahim, 1994). In many practical situations it may be desirable that a given system originally undergoing complicated behavior should be forced to display regular motions (e.g., suppression of oscillatory dynamics). For instance, it could be desirable to induce regular dynamics in mechanical oscillators to avoid errors (as in the case of precise position mechanisms) lead by external vibrations and magnetic fields (Chatterjee, 2007; Fradkov & Pogromsky, 1998; Southward et al., 1991).

To deal with systems with friction, it is necessary to have a good characterization of the structure of the friction model and then to design appropriate compensation techniques. As a friction phenomenon has not yet been completely understood, friction modeling is not an easy task. Indeed, uncertainty exists on most models that contain a friction component (Hinrichs et al., 1998; Feeny, 1998; Armstrong-Hélouvry et al., 1994; Olsson et al., 1998). Thus, for control design purposed for systems with friction it is necessary to consider the uncertainty of models that includes friction.

Different control approaches for friction compensation have been proposed. For instance, linear control system with type PI controllers (Puebla & Alvarez-Ramirez, 2008), adaptive compensation (Canudas de Wit & Lischinsky, 1997; Huang et al., 2000; Tomei, 2000), neural networks (Lin & Wai, 2003), and others nonlinear model-based methods (Alvarez-Ramirez et al., 1995; Chatterjee, 2007; Xie, 2007; Zeng & Sepehri, 2008). The use of linear techniques is very limiting due the highly nonlinearity of mechanical systems with friction, whereas nonlinear control design needs, in general, too much information about the process. The performance of a model-based adaptive control is limited by the accurateness of the model used to describe the various friction-related effects.

Sliding mode control is a robust control method that has long been applied in simple mechanical systems (Hangos et al., 2004). A drawback of this methodology, which usually limits its applicability to control mechanical systems, is the high-frequency switching of the control action which induces the so-called chattering phenomenon, i.e. undesired oscillations of the relevant signals causing vibrations and unacceptable mechanical wear (Hangos et al., 2004). A possibility to overcome this problem is to rely on high order sliding mode control (Laghrouche et al., 2007; Levant, 2005; Levant, 2001). In (Hernandez-Suarez et al., 2009) we have applied the integral high order sliding mode control (IHOSMC), to suppress stick-slip oscillation in an oil-drillstring. The resulting feedback control approach leads to a robust feedback control scheme that deals with uncertainties in the friction model and drillstrings parameters. The IHOSMC approach consider the integration of a fractional power of the absolute value of the tracking error, coupled with the sign function (Laghrouche et al., 2007; Aguilar-Lopez et al., 2010). This control structure provides simplification of the control law and good robustness properties. In this chapter we extend the application of IHOSMC approaches to control a general class of mechanical systems with friction. The proposed controller yield to a robust performance in presence of external disturbances and uncertainty on the parameters of systems with friction.

This chapter is organized as follows: In Section 2, for the sake of clarity in presentation, we briefly provide some issues on the friction phenomenology and modeling, and the class of mechanical systems with friction is also introduced. In Section 3 we present the IHOSMC approach and introduce a recursive cascade control scheme for the control of the class of mechanical systems considered in this chapter. Numerical benchmark examples are used to illustrate the control performance of the proposed control approach. Finally, in Section 6 we close this chapter with some concluding remarks.

2. Mechanical systems with friction

In this section we briefly discuss the friction phenomenology. Next we present some classical models of friction phenomena. Finally, we introduce the class of mechanical systems under consideration in this chapter.

2.1 Friction phenomenology

Friction involves two solid surfaces sliding against each other. The friction force is affected by many factors such as the properties of bulk and surface layer materials, the roughness of the surfaces in contact, the stress levels, the sliding speed, the temperature, the environment, the properties of the lubricants and the lubrication conditions. Lubrication has the main purpose of creating a fluid film between the two contacting surfaces, avoiding solid-to-solid contact (Bowden & Tabor, 1950; Bowden & Tabor, 1964; Armstrong-Hélouvry et al., 1994; Rabinowicz, 1995).

Friction is a torque, or a force, that depends on the relative velocity of the moving surfaces. Although there is disagreement on the character of the functionality of the friction forces with the velocity, experiments have confirmed that, for moderate and low velocities, the main components in the friction forces are mainly caused by the following phenomena (Bowden & Tabor, 1950; Bowden & Tabor, 1964; Armstrong-Hélouvry et al., 1994; Rabinowicz, 1995):

1. Coulomb and sticktion: Coulomb friction is due to sticking effects. At zero speed, the friction torque is equal and opposite of the applied torque, unless the latter one is larger than the stiction torque. In this case, the friction torque is equal to the stiction torque, F_s. The stiction torque is a torque at the moment of breakaway and is larger than the Coulomb torque. Kinetic friction F_r is the resisting force which acts on a body after the force of static friction has been overcome.
2. Stribeck (downward bends): After the sticktion force has been overcame, the friction force decreases exponentially, reaching a minimum, and then increases proportionally with the velocity. These bends occur at velocities close to zero. The friction forces are due to a partial lubrication, where the velocity is adequate to entrain some fluid in the junction but not enough to fully separate the surfaces.
3. Viscous: Here the surfaces are fully separated by fluid film. In this regime the viscosity of the lubricant dominates and friction increases with velocity.
4. Asymmetries and position dependence: Imperfections and unbalances in the mechanism induce asymmetries and position dependence of the friction forces.

2.2 Friction models

Many friction models have been developed and reflect different aspects of the friction phenomena (Armstrong-Hélouvry et al., 1994; Hinrichs et al., 1998; Olsson et al., 1998). In general, existing friction models are usually classified as static and dynamic, where the fundamental difference between them is the frictional memory. Static models usually have a form of direct dependence between the friction force and relative velocity. Dynamic friction models where memory effect is described with a complimentary dynamics between the velocity and the friction force. Typical friction models are the Columb, Dahl, and the LuGre, friction models.

1. Columb model: Coulomb proposed the first model for the physical origin of friction, which explained some of the important properties of dry friction in a simple way. Coulomb sliding friction is given by,
$$f = F_N f_c sign(v) \quad v \neq 0 \tag{1}$$
where f is a friction force, F_N the normal load, f_c the coefficient of Coulomb friction and v is relative velocity. The indeterminate and discontinuous nature of the Coulomb model makes it extremely difficult to simulate the dynamics of the mechanical systems (Olsson et al, 1998).
2. Dahl model: Dahl developed a simple model, which can be considered as a generalization of Coulomb friction. The frictional hysteresis during pre-sliding is approximated by a generalized first order equation of the position depending only on the sign of the velocity. Dahl proposed the following equation,
$$\frac{df}{dt} = \sigma_0 (1 - sign(v)\frac{F}{F_s})^{\delta_d} v \tag{2}$$
where σ_0 denotes the initial stiffness of the contact at velocity reversal and δ_d denotes a model parameter determining the shape of the hysteresis. v is the relative moving speed.

F_s is the highest steady state friction. The Dahl model produces a smooth transition around zero velocity (Olsson et al, 1998; Dahl, 1976).

3. Stribeck friction: The Stribeck friction may be represented by,

$$f = F_0 sign(v) + F_s \exp(-(v/v_s)^2) - 1) sign(v) \tag{3}$$

Where F_0 is the static friction, F_s is the magnitude of the Stribeck effect, v_s is the critical velocity of the Stribeck effect, and v is the velocity. This model is empirical and in most cases has a good fit to data (Olsson et al, 1998).

4. LuGre model: The LuGre model describe major features of dynamic friction, including presliding displacement, varying break-away force and Stribeck effect. The LuGre model is given by,

$$\begin{aligned} f &= \sigma_0 z + \sigma_1 \dot{z} + \alpha_2 v \\ \dot{z} &= v - \frac{|v|}{g(v)} z \\ g(v) &= \alpha_0 + \alpha_1 \exp(-(v/v_s)^2) \end{aligned} \tag{4}$$

where z represents the unmeasurable internal friction state, σ_0, σ_1, α_2 are parameters associated to the stiffness of the elastic bristle, damping coefficient in elastic range, and the viscous friction coefficient, respectively, v is the relative velocity, and v_s is the Stribeck velocity. The function $g(v)$ is positive and it describes the Stribeck effect. Direct use of the above LuGre model for friction compensation may have some implementation problems. Namely, as the internal friction state z is unmeasurable, it is necessary to construct observers to estimate z for dynamic friction compensation (Canudas de Wit et al., 1995; Olsson et al, 1998).

2.2.1 Stick-slip

This phenomenon consists on the sudden and successive change from "stick" state to "sliding" state, provoking the apparition of vibration and noise. Stick-slip friction is present in any elements involving relative motion, such as gears, pulleys, bearings, DC motors. Stick-slip friction is generally described as a composite of two different processes: the static process when an object is stationary (no sliding is involved) and likely to move under certain applied torque, and the dynamic process when sliding is involved (Fidlin, 2006; Denny, 2004).
The static process is characterized by the maximum static torque (or breakaway torque), under which static state remains and the magnitude of the static friction force is equal to that of the applied force. The slipping process is relatively complicated. Slipping torque is usually modeled as a linear combination of Coulomb torque, viscous torque, exponential torque used to represent Stribeck effect, and position dependent components (Fidlin, 2006; Denny, 2004).
A mathematical formulation of the stick-slip friction (denoted by τ_{fm}) with some commonly used friction components is (Fidlin, 2006; Denny, 2004),

$$\begin{aligned} \tau_{fm} &= \tau_{stm} sign(v) + k_{vis} v - \tau_{cm}(1 - \exp(-T_0 |v|)) sign(v) \\ &+ \tau_{pm} sign(v) + (1 - sign(v)) \tau_i \end{aligned} \tag{5}$$

where k_{vis} is the coefficient of viscous friction, τ_{stm} represents the maximum static friction torque, the third term on the right side represents the Stribeck effect, where T_0 is a positive constant, τ_i is the input torque, so that the last term on the right side stands for the static friction force whose magnitude is equal to the applied force, and τ_{pm} is the position dependent friction torque, which can be modeled as,

$$\tau_{pm} = \beta_1 \sin(\beta_2 q + \beta_3) \tag{6}$$

where β_i are constants.

2.3 The class of mechanical systems with friction

We consider a n order generic simple model of a mechanical system with friction, which, possibly after a change of coordinates, can be described by,

$$\begin{aligned} \dot{x}_i &= f_i(x) + g_i(x)x_{i+1} \quad 1 \leq i \leq j-1 \\ \dot{x}_j &= f_j(x) + g_j(x)u \quad j \leq n \\ \dot{x}_{j+l} &= f_{j+l}(x) \quad j+l = n \quad for\, j < n \end{aligned} \tag{7}$$

where x are the states of the system and u is the control input. Note that the j first equations in dynamical system (7) are in the so called chained form. Its not hard to see that several model of mechanical systems with friction can be described by (7). For instance, a basic model formulation, that covers various mechanical models reported in literature is described by,

$$\ddot{x} = -F(\dot{x}, t) + u \tag{8}$$

where x is the position, $\dot{x}$ is the speed, F is a nonlinear function including friction components, and u is an external input. This system can be described by (7) under a simple coordinates change.

3. Integral high order sliding model control of mechanical systems with friction

In this section, the IHOSMC is presented to control the class of mechanical systems described in the above section. By exploiting the chained form of model (7) we use a recursive cascade control configuration, where a virtual control input is introduced for the control design, and a single control input u, related to the electrical properties of the motor and consequently, the torque supplied by the motor is employed.

3.1 Control problem

The control objective is the regulation or tracking of an intermediate state $(x_1 \leq x_i \leq j-1)$ of the class of mechanical systems with friction in the form (7) about a given reference, *i.e.* $x_i \rightarrow x_{i,ref}$ under the following assumptions,

A1 States x_1 to x_j are available for control design purposes.

A2 Nonlinear terms f and g and model parameters are uncertain, and can be available rough estimates of these terms.

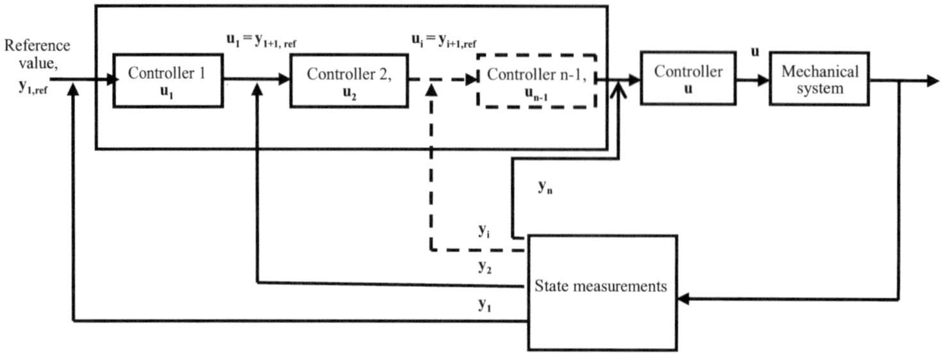

Fig. 1. Cascade control for mechanical systems with friction.

The following comments are in order:

- Dynamical and design problems in mechanical systems can be addressed using large finite element models, which give quantitative information and can help to give practical recommendations to circumvent mechanical problems. However, in practice, control and optimization techniques tend to be based on simple models. Indeed, for control systems design purposes, both low dimensional and less complex models can provide (to some degree) qualitative insight on the dominant complex phenomenon occurring in mechanical systems with friction.
- Assumption $A1$ is a reasonable assumption in several applications. For instance, high precision optical encoders for position measurement. On the other hand, even in the absence of such measurements, a state estimator can be designed.
- $A2$ considers that nonlinear terms, including friction, can contain uncertain parameters, or in the worst case the whole terms are unknown. From a practical viewpoint, obtaining a model that can embody all such characteristics of the friction force is not an easy task. Having selected an appropriate model, the parameters of the model are needed to be experimentally identified to implement the model. Friction identification is another challenging part of the friction compensation process.

3.2 A cascade control scheme

We can exploit the structure of the model given by (7) to design a cascade procedure to control mechanical systems with friction. Cascade control is a common control configuration in process control, which can be thought of as partial state feedback. A typical cascade control structure has two feedback controllers with the output of the primary (master) controller changing the set point of the secondary (slave) controller (Alvarez-Ramirez et al., 2002; Krishnaswamy et al., 1990).

Figure 1 shows the recursive cascade control configuration for the class of mechanical systems given by (7). The cascade control configuration is based on the design of an intermediate virtual control function $u_{vi} = u_i$. The design is recursive because the computation of u_{i+1} requires the computation of u_i. For instance, for the simple class of mechanical system (8), the master controller regulates the mechanical position x_1 to a desired reference $x_{1,ref}$ with the virtual input $x_2 = u_{vi}$, the slave controller regulates the velocity state variable x_2 to the reference $x_{2,ref} = u_{vi}$ with the real control input u. In other words, the master controller provides reference values $x_{2,ref}$ to the slave controller, which is driven by the real control input u.

3.3 Integral high order sliding mode control

Sliding mode control techniques have long been recognized as a powerful robust control method (Hangos et al., 2004; Levant, 2001; Sira-Ramirez, 2002). Sliding mode control is a nonlinear controller due of the switching control action. Sliding-mode control schemes, have shown several advantages like allowing the presence of matched model uncertainties and convergence speed over others existing techniques as Lyapunov-based techniques, feedback linearization and extended linearization. However standard sliding-mode controllers have some drawbacks: the closed-loop trajectory of the designed solution is not robust even with respect to the matched disturbances on a time interval preceding the sliding motion, the classical sliding-mode controllers are robust in the case of matched disturbances only, the designed controller ensures the optimality only after the entrance point into the sliding mode. To try to avoid the above a relatively new kind of sliding-mode structures have been proposed as the named high-order sliding-mode technique, these techniques consider a fractional power of the absolute value of the tracking error coupled with the sign function, this structure provides several advantages as simplification of the control law, higher accuracy and chattering prevention (Hangos et al., 2004; Levant, 2001; Sira-Ramirez, 2002).

3.3.1 Control design

Sliding mode control designs consists of two phases. In the first phase the sliding surface is to be reached (reaching mode), while in the second the system is controlled to move along the sliding surface (sliding mode). In fact, these two phases can be designed independently from each other. Reaching the sliding surface can be realized by appropriate switching elements (Hangos et al., 2004).
Defining,

$$\sigma(e) = e_i = y - y_{ref} \tag{9}$$

as the sliding surface, we have that the continuous part of the sliding mode controller is given by,

$$u_{eq,i} = -g_i(x)^{-1}(f_i(x) - \dot{y}_{i,ref}) \tag{10}$$

such that,

$$\dot{\sigma}(e_i) = 0 \tag{11}$$

where $\dot{y}_{i,ref}$is the time-derivative of the desired trajectory signal. In sliding mode, the controlled system satisfies the condition $de_i/dt = 0$, such that the tracking error will be driven to zero.
To force the system trajectory to converge to the sliding surface in the presence of both model uncertainties and disturbances, with chattering minimization and finite-time convergence, the sliding trajectory is proposed as (Levant, 2001; Aguilar-Lopez et al., 2010),

$$u_{sld,i} = -g_i(x)^{-1}[\delta_{1,i}e_i + \delta_{2,i}\int_0^t \left\{sign(e_i)\,|e_i|^{1/p}\right\} d\tau] \tag{12}$$

where $\delta_{1,i}$ and $\delta_{2,i}$ are control design parameters. In essence, to achieve a zero tracking error all system trajectory must be forced to converge to $\sigma(e_i)$ in finite time and to remain on $\sigma(e_i)$ afterwards.
The complete IHOSMC is given by,

$$u_i = u_{eq,i} + u_{sld,i} = -g_i(x)^{-1}(f_i(x) - \dot{y}_{i,ref} + \delta_{1,i}e_i + \delta_{2,i}\int_0^t \left\{sign(e_i)\,|e_i|^{1/p}\right\} d\tau) \tag{13}$$

The synthesis of the above control law requires accurate knowledge of both $f_i(x)$ and $dy_{i,ref}/dt$ to be realizable.
To enhance the robust performance of the above control laws, the uncertain terms $f_i(x)$ are lumped in single terms and compensated with a reduced-order observer (Alvarez-Ramirez, 1999).
Then, we define the modelling error function as follows,

$$\eta_i = f_i(x) \tag{14}$$

In order to get estimated values of the modelling error functions $\widetilde{\eta}$, we introduce the following reduced-order observer,

$$\dot{\widetilde{\eta}}_i = \lambda_i(\eta_i - \widetilde{\eta}_i) \tag{15}$$

where λ_i are observer design parameters. From (7) and (15), and after of some direct algebraic manipulation we get,

$$\dot{w}_i = -g_i(x)u_i - \widetilde{\eta}_i \qquad \widetilde{\eta}_i = \lambda_i(w_i + y_i) \tag{16}$$

The robust IHOSMC law is written as,

$$u_i = -g_i(x)(\widetilde{\eta}_i - \dot{y}_{i,ref} + \delta_{1,i}e_i + \delta_{2,i}\int_0^t \left\{sign(e_i)\,|e_i|^{1/p}\right\} d\tau) \tag{17}$$

By exploiting the properties of the sliding part of the sliding-mode type controllers to compensates uncertain nonlinear terms, the knowledge of nonlinear terms $f_i(x)$ can be avoided. On the order hand, extensive simulation examples show that the derivative of the set point variable can be eliminated without affecting the closed loop system performance.
Summarizing, the IHOSMC is composed by a proportional action, which has stabilizing effects on the control performance, and a high order sliding surface, which compensates the uncertain nonlinear terms to provide robustness to the closed-loop system. This behavior is exhibited because, once on the sliding surface, system trajectories remain on that surface, so the sliding condition is taken and make the surface and invariant set. This implies that some disturbances or dynamic uncertainties can be compensated while still keeping the surface an invariant set.
The following comments are in order:

- The IHOSMC approach has two control design parameters, namely, δ_1 and δ_2 which will be chosen as $\delta_1 > \delta_2 > 0$. On the other hand, parameter λ, of the uncertainty observer should be chosen as $0 < \lambda < \omega_c$, where ω_c is the open-loop dominant frequency of the systems oscillations.

- Although a rigorous robustness analysis is beyond the scope of this study, numerical examples will show that the feedback controller is able of yielding robust control performance despite significant parameter departures from parameter nominal values.
- Stability must be preserved in the context of both structured uncertainties in the parameters as well as unstructured errors in modeling. A stability analysis for the proposed control configurations should parallel the steps reported in (Aguilar-Lopez et al., 2010) and (Aguilar-López & Martínez-Guerra, 2008).

4. Applications

In this section, simulation results are presented for both position regulation and tracking of mechanical systems with friction (7) with the IHOSMC approach described above. The control performance is evaluated considering set point changes and typical disturbances of mechanical systems. We consider the following five examples: (i) Mechanical system with Coulomb friction, (ii) an inverted pendulum, (iii) an AC induction motor, and (vi) a levitation magnetic system.

4.1 Mechanical system with Coulomb friction

We consider a mechanical system described in (Alvarez-Ramirez et al., 1995) with a Coulomb friction law. The dimensionless equation of motion is,

$$\begin{aligned}\frac{dx_1}{dt} &= x_2 \\ \frac{dx_2}{dt} &= \frac{1}{m}\{-F(x_1,x_2) - \alpha x_1 + \tau_l + u\}\end{aligned} \tag{18}$$

where m is the mass of the system, τ_l is an unknown external force, which way be due to loads and/or noise acting in the mechanism, u is a manipulated forced used to control the system and the term $F(x_1, x_2)$ includes all friction effects and is determined by the following expression,

$$F(x_1,x_2) = \phi f(x_1)$$

where ϕ is the coefficient of friction and $f(x_1)$ is the normal load which vary with displacement,

$$\begin{aligned}f(x_1) &= -\mu_k \quad x_1 < 0 \\ -\mu_s \leq f(x_1) &\leq \mu_s x_1 = 0 \\ f(x_1) &= \mu_k \quad x_1 > 0\end{aligned} \tag{19}$$

Control objetive is the position tracking to the periodic reference,

$$y_{ref} = x_{1,ref} = 0.3\sin(0.5t)$$

The parameters of the controller are set as $\delta_{1,i} = [25, 10]$, and $\delta_{2,i} = [2.3, 1]$. Model simulation parameters are taken from (Alvarez-Ramirez et al., 1995). The control law is turned on the $t = 50$ time units and $\tau_l = A\sin(1.25t)$. At $t = 75$ the amplitude A of the external force τ_l is

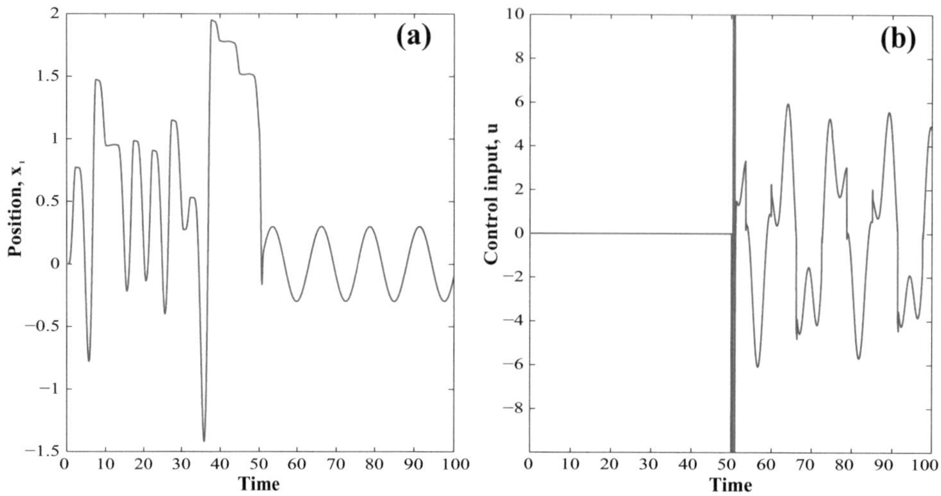

Fig. 2. (a) Cascade control for mechanical system, (18) and (b) control input.

changed in 20 %. Figure 2 shows the position trajectory before and after the control activation. In Figure 2 the control input is also displayed. It can be seen that the proposed cascade control scheme is able to track the desired reference and rejects the applied perturbation. After that the control input reach the saturation levels ($-10 < u < 10$) the control inputs displays a complex oscillatory behavior.

4.2 Inverted pendulum

The inverted pendulum has been used as a classical control example for nearly half a century because of its nonlinear, unstable, and nonminimum-phase characteristics. In this case we consider a single inverted pendulum.

The equation of motion for a simple inverted pendulum with Coulomb friction and external perturbation is (Poznyak et al., 2006),

$$\begin{aligned}\frac{dx_1}{dt} &= x_2 \\ \frac{dx_2}{dt} &= -g\sin(x_1)/l - v_s x_2/J - p_s sign(x_2)/J + \tau_d + u/J\end{aligned} \tag{20}$$

where g is the gravitational acceleration, l is the distance between the rotational axis and center of gravity of the pendulum, $J = ml^2$ is the inertial moment, where m is the mass of the system, $\tau_d = 0.5\sin(2t) + 0.5\cos(5t)$ is an external disturbance, which may be due to loads and/or noise acting in the mechanism, u is a manipulated forced used to control the system. Let $y_{ref} = x_{1,ref} = \sin(t)$ be the desired orbit of the pendulum position. Figure 3 shows the control performance using the control parameters $\delta_{1,i} = [12, 7]$, and $\delta_{2,i} = [1, 0.5]$. In this case the IHOSMC controller is activated at $t = 15$ and from 0 to 15 time units the pendulum is driven by the twisting controller introduced by Poznyak et al. (2006). It can be seen from Figure 3 that the IHOSMC controller is able to follow the periodic orbit with a better closed loop behavior that the twisting controller.

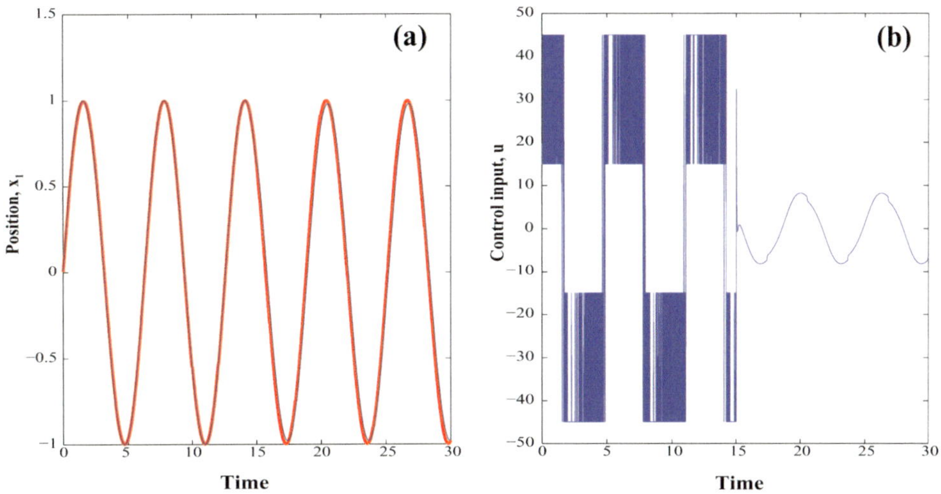

Fig. 3. (a) Control performance for inverted pendulum system and (b) control input.

4.3 Induction AC motors

Induction motors have found considerable applications in industry due to their reliability, ruggedness and relatively low cost. Their mechanical reliability is due to the fact that there is no mechanical commutation as in most DC motors. Furthermore, induction motors can also be used in volatile environments because no sparks are produced. An induction motor is composed of three stator windings and three rotor windings.

A simple mathematical model of an induction motor, under field-oriented control with a constant rotor flux amplitude, which was presented in (Tan et al., 2003), is the following,

$$\begin{aligned} \frac{dx_1}{dt} &= x_2 \\ \frac{dx_2}{dt} &= \frac{K_T}{J}x_3 - \frac{F}{J} - \frac{\tau_l}{J} \\ \frac{dx_3}{dt} &= a_1x_2 + a_2x_3 + bu \end{aligned} \tag{21}$$

where x_1 is the rotor angle, x_2 is the rotor angular velocity, x_3 is the component of stator current, u is the component of stator voltage, J is the rotor inertia, τ_l is the load torque, and F is the friction force.

Friction force is modeled by the LuGre friction model with friction force variations,

$$\begin{aligned} \frac{dz}{dt} &= x_2 - \frac{|x_2|}{g(x_2)}z \\ F &= \sigma_0 z + \sigma_1\frac{dz}{dt} + \sigma_2 x_2 \end{aligned} \tag{22}$$

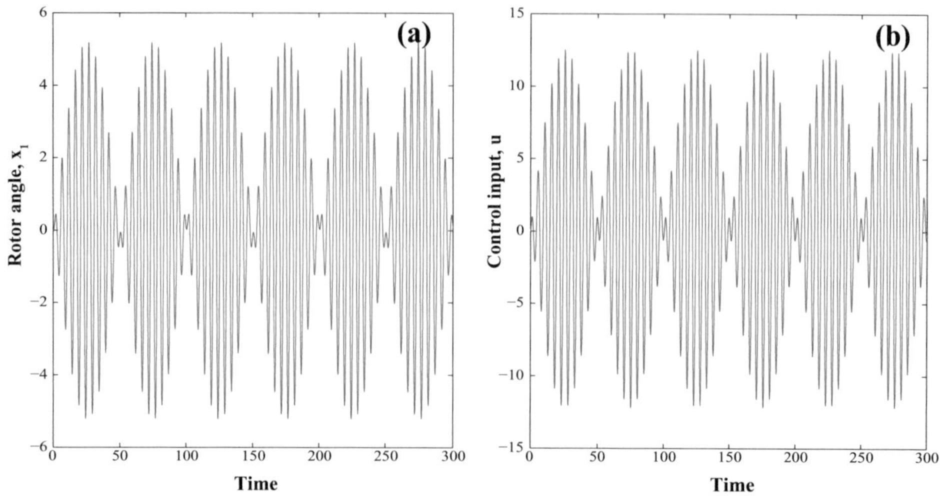

Fig. 4. (a) Cascade control for induction AC motors system and (b) control input.

where z is the friction state that physically stands for the average deflection of the bristles between two contact surfaces. The nonlinear function is used to describe different friction effects and can be parameterized to characterize the Stribeck effect,

$$g(x_2) = F_c + (F_s - F_c)\exp(-\frac{x_2}{v_s})^2 \tag{23}$$

where F_c is the Coulomb friction value, F_s is the stiction force value, and v_s is the Stribeck velocity.

The control objective is to asymptotically track a given bounded reference signal $y_{ref} = x_{1,ref}$ given by,

$$y_{ref} = 5.6\sin(0.4\pi t)\sin(0.02\pi t) \tag{24}$$

A load disturbance $\tau_l = 0.8\ N \cdot m$ is injected into the induction motor simulation model. The position of the rotor angle and the corresponding control input are shown in Figure 4. It can be seen that the controller is able to track the desired reference (24) using a periodic input of the control input. The external disturbance is also rejected without an appreciable degradation of the closed-loop system.

4.4 Levitation system

Magnetic levitation systems have been receiving considerable interest due to their great practical importance in many engineering fields (Hikihara & Moon, 1994). For instance, high-speed trains, magnetic bearings, coil gun and high-precision platforms. We consider the control of the vertical motion in a class of magnetic levitation given by a single degree of freedom (specifically, a magnet supported by a superconducting system). In particular, we consider a magnet supported by superconducting system which can be represented by a second-order differential equation with a nonlinear term which involves hysteresis and periodic external excitation force. Without loss of generality, one can consider that the model of the levitation system is modelled by the following equation (Femat, 1998),

$$\frac{dx_1}{dt} = x_2 \quad (25)$$
$$\frac{dx_2}{dt} = -\delta x_2 - x_1 + x_3 + \tau_l + u$$
$$\frac{dx_3}{dt} = -\gamma(x_3 - F)$$

x_1 is defined as a displacement from the surface of a high *Tc* superconductor (HTSC) surface, x_2 is the velocity, x_3 is a dynamical force between the HTSC and the magnet, which includes hysteresis effects, δ represents a mechanical damping coefficient, γ is a relaxation coefficient, τ_l is an external excitation force, and u is the control force.
The nonlinear function F is given by (Femat, 1998),

$$F = F_{x_1} \exp(-x_1)(1 - F_{x_2}) \quad (26)$$
$$F_{x_1} = F_0 \exp(-x_1)$$
$$F_{x_1} = \begin{cases} -\mu_1 - x_2 & \epsilon \leq x_2 \\ \frac{-x_2(\mu_1 - \mu_2)}{2\epsilon} & -\epsilon \leq x_2 < \epsilon \\ \mu_2 & x_2 < -\epsilon \end{cases}$$

where the exponential term F_{x_1} shows the force-displacement relation without hysteresis, F_0 denotes the maximum force between the HTSC and the magnet, μ_1 and μ_2 are constants. The control problem is the regulation to the origin of the vertical motion, i.e. $y_{ref} = x_{1,ref} = 0.0$. In the Figure 6 the controlled position and the corresponding control input are presented (control action is turn on a $t = 100.0$ time units). It can be seen from Figure 5 that the controller can regulate the vertical position of the levitation system via a simple periodic manipulation of the control force. The control input reaches saturation levels in the first 20 time units, which can be related to high values of the controller parameters.

5. Conclusions

In mechanical systems, the control performance is greatly affected by the presence of several significant nonlinearities such as static and dynamic friction, backlash and actuator saturation. Hence, the productivity of industrial systems based on mechanical systems depend upon how control approaches are able to compensate these adverse effects. Indeed, fiction in mechanical systems can lead to premature degradation of highly expensive mechanical and electronic components. On the other hand, due to uncertainties and variations in environmental factors a mathematical model of the friction phenomena present significant uncertainties.
In this chapter, by means of an IHOSMC approach and a cascade control configuration we have derived a robust control approach for both regulation and tracking position in mechanical systems. The underlying idea behind the control approach is to force the error dynamics to a sliding surface that compensates uncertain parameters and unknown term. The sliding mode control law is enhanced with an uncertainties observer. We have show via numerical simulations how the motion can be regulate and tracking to a desired reference in presence of uncertainties in the control design and changes in model parameters. Although

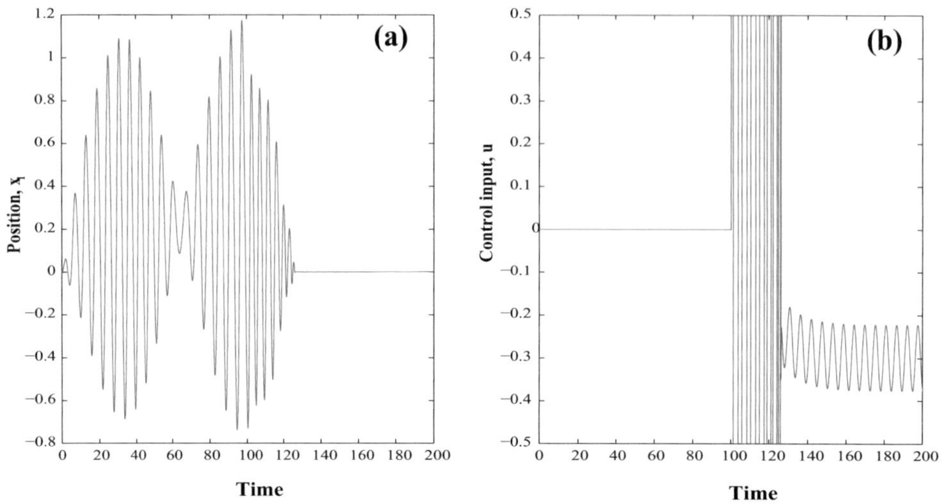

Fig. 5. Levitation system: (a) motion vertical control and (b) control input.

the control design is restricted to certain class of mechanical systems with friction, the concepts presented in our work should find general applicability in the control of friction in other systems.

6. References

Aguilar-Lopez, R.; Martinez-Guerra, R.; Puebla, H. & Hernadez-Suarez, R. (2010). High order sliding-mode dynamic control for chaotic intracellular calcium oscillations. *Nonlinear Analysis B: Real World Applications*, 11, 217-231.

Aguilar-Lopez, R. & Martinez-Guerra, R. (2008). Control of chaotic oscillators via a class of model free active controller: suppression and synchronization. *Chaos Solitons Fractals*, 38, No. 2, 531-540.

Alvarez-Ramirez, J. (1999). Adaptive control of feedback linearizable systems: a modelling error compensation approach. *Int. J. Robust Nonlinear Control*, 9, 361.

Alvarez-Ramirez, J.; Alvarez, J. & Morales, A. (2002). An adaptive cascade control for a class of chemical reactors, *International Journal of Adaptive Signal and Processing*, Vol. 16, 681-701.

Alvarez-Ramirez, J.; Garrido, R. & Femat, R. (1995). Control of systems with friction. *Physics Review E*, 51, No 6, 6235-6238.

Armstrong-Hélouvry, B.; Dupont, P. & Canudas de Wit, C. (1994). A survey of models, analysis tools and compensations methods for the control of machines with friction. *Automatica*, 30, No. 7, 1083-1138.

Bowden, F.P. & Tabor, D. (1950). *The Friction and Lubrication of Solids*. Oxford Univ. Press, Oxford.

Bowden, F.P. & Tabor, D. (1964). *The Friction and Lubrication of Solids: Part II*. Oxford Univ. Press, Oxford.

Canudas de Wit, C. & Lischinsky, P. (1997). Adaptive friction compensation with partially known dynamic friction model. *Int. J. Adaptive Control Signal Processing*, Vol. 11, 65-80.

Canudas de Wit, C.; Olsson, H.; Astrom, K.J. & Lischinsky, P. (1995). A new model for control of systems with friction. *IEEE Tran. Automatic Control*, 40, No. 3, 419-425.

Chatterjee, S. (2007). Non-linear control of friction-induced self-excited vibration. *Int. J. Nonlinear Mech.*, 42, No. 3, 459-469.

Dahl, P.R. (1976). Solid friction damping of mechanical vibrations. *AIAA J.*, 14, No. 12, 1675-1682.

Denny, M. (2004). Stick-slip motion: an important example of self-excited oscillation. *European J. Physics*, 25, 311-322.

Feeny, B. & Moon, F.C. (1994). Chaos in a forced dry-friction oscillator: experimental and numerical modelling. *J. Sound Vibration*, 170, No. 3, 303-323.

Feeny, B. (1998). A nonsmooth coulomb friction oscillator. *Physica D*, 59, 25-38.

Femat, R. (1998). A control scheme for the motion of a magnet supported by type-II superconductor. *Physica D*, 111, 347-355.

Fidlin, A. (2006). *Nonlinear Oscillations in Mechanical Engineering*. Springer, Berlin, Heidelberg, 2006.

Fradkov, A.L. & Pogromsky, A.Y. (1998). *Introduction to Control of Oscillations and Chaos*. World Scientific Publishing.

Hangos, K.M.; Bokor, J. & Szederkényi, G. (2004). *Analysis and Control of Nonlinear Process Systems*. Springer-Verlag London.

Hernandez-Suarez, R.; Puebla, H.; Aguilar-Lopez, R. & Hernandez-Martinez, E. (2009). An integral high-order sliding mode control approach for controlling stick-slip oscillations in oil drillstrings. *Petroleum Science Technology*, 27, 788-800.

Hikihara, T. & Moon, F.C. (1994), Chaotic levitated motion of a magnet supported by superconductor. *Phys. Lett. A*, 191, 279.

Hinrichs, N.; Oestreich, M. & Popp, K. (1998). On the modelling of friction oscillators. *Journal Sound Vibration*, 216, No. 3, 435-459.

Huang, S.N.; Tan, K.K. & Lee, T.H. (2000). Adaptive friction compensation using neural network approximations. *IEEE Trans. Syst., Man and Cyber.-C*, 30, No. 4, 551–557.

Ibrahim, R.A. (1994). Friction-induced vibration, chatter, squeal and chaos: part I: mechanics of contact and friction. *Applied Mechanics Reviews*, 47, No. 7, 209–226.

Krishnaswamy, P.R.; Rangaiah, G.P.; Jha, R.K. & Deshpande, P.B. (1990). When to use cascade control. *Industrial Engineering Chemistry Research*, 29, 2163-2166.

Laghrouche, S.; Plestan, F. & Glumineau, A. (2007). Higher order sliding mode control based on integral sliding mode. *Automatica*, 45, 531-537.

Levant, A. (2005). Homogeneity approach to high-order sliding mode design. *Automatica*, 41, 823- 830.

Levant, A. (2001). Universal SISO sliding-mode controllers with finite-time convergence. *IEEE Trans. Automat. Contr.*, 46, 1447-1451.

Lin, F.J. & Wai, R.J. (2003). Robust recurrent fuzzy neural network control for linear synchronous motor drive system. *Neurocomputing*, 50, 365-390.

Olsson, H.; Åström, K. J.; Canudas de Wit, C.; Gäfvert, M. & Lischinsky, P. (1998). Friction models and friction compensation. *Eur. J. Control*, 4, No. 3, 176–195.

Poznyak, A.; Shtessel, Y.; Fridman, L.; Davila, J. & Escobar, J. (2006). Identification of parameters in dynamic systems via sliding-mode techniques. In: *Advances in Variable Structure and Sliding Mode Control*, LNCIS, Vol. 334, 313-347.

Puebla, H.; Alvarez-Ramirez, J. & Cervantes, I. (2003). A simple tracking control for chuas circuit. *IEEE Trans. Circ. Sys.* , 50, 280.

Puebla, H. & Alvarez-Ramirez, J. (2008). Suppression of stick-slip in oil drillstrings: a control approach based on modeling error compensation. *Journal Sound Vibration*, 310, 881-901,

Rabinowicz, E. (1995). *Friction and Wear of Materials*. New York, Wiley, Second edition.

Sira-Ramirez, H. (2002). Dynamic second order sliding-mode control of the Hovercraft vessel. *IEEE Trans. Control Syst. Tech.*, 10, 860-865.

Southward, S.C.; Radcliffe, C.J. & MacCluer, C.R. (1991). Robust nonlinear stick-slip friction compensation. *Journal Dynamic Systems, Measurement, and Control*, Vol. 113, 639-645.

Tan, Y.; Chang, J. & Hualin, T. (2003). Adaptive backstepping control and friction compensation for AC servo with inertia and load uncertainties. *IEEE Tran. Industrial Electronics*, 50, No. 5, 944-952.

Tomei, P. (2000). Robust adaptive friction compensation for tracking control of robot manipulators. *IEEE Trans. Automat. Contr.*, 45, No. 11, 2164–2169.

Xie, W.F. (2007). Sliding-mode-observer-based adaptive control for servo actuator with friction. *IEEE Trans. Ind. Elec.*, 54, No. 3.

Zeng, H. & Sepehri, N. (2008). Tracking control of hydraulic actuators using a LuGre friction model compensation. *Journal Dynamic Systems Measurement Control*, 30.

18

Control of ROVs using a Model-free 2nd-Order Sliding Mode Approach

Tomás. Salgado-Jiménez, Luis G. García-Valdovinos and Guillermo Delgado-Ramírez
Center for Engineering and Industrial Development – CIDESI
Mexico

1. Introduction

Remotely Operated Vehicles (ROVs) have had significant contributions in the inspection, maintenance and repair of underwater structures, related to the oil industry, especially in deep waters, not easily accessible to humans. Two important capabilities for industrial ROVs are: position tracking and the dynamic positioning or station-keeping (the vehicle's ability to maintain the same position with respect to the structure, at all times).

It is important to remember that underwater environment is highly dynamic, presenting significant disturbances to the vehicle in the form of underwater currents, interaction with waves in shallow water applications, for instance. Additionally, the main difficulties associated with underwater control are the parametric uncertainties (as added mass, hydrodynamic coefficients, etc.). Sliding mode techniques effectively address these issues and are therefore viable choices for controlling underwater vehicles. On the other hand, these methods are known to be susceptible to chatter, which is a high frequency signal induced by control switches. In order to avoid this problem a High Order Sliding Mode Control (HOSMC) is proposed. The HOSMC principal characteristic is that it keeps the main advantages of the standard SMC, thus removing the chattering effects.

The proposed controller exhibits very interesting features such as: *i.* a model-free controller because it does neither require the dynamics nor any knowledge of parameters, *ii.* It is a smooth, but robust control, based on second order sliding modes, that is, a chattering-free controller is attained. *iii.* The control system attains exponential position tracking and velocity, with no acceleration measurements.

Simulation results reveal the effectiveness of the proposed controller on a nonlinear 6 degrees of freedom (DOF) ROV, wherein only 4 DOF (x, y, z, ψ) are actuated, the rest of them are considered intrinsically stable. The control system is tested under ocean currents, which abruptly change its direction. Matlab-Simulink, with Runge-Kutta ODE45 and variable step, was used for the simulations. Real parameters of the KAXAN ROV, currently under construction at CIDESI, Mexico, were taken into account for the simulations. In Figure 1 one can see a picture of KAXAN ROV.

For performance comparison purposes, numerical simulations, under the same conditions, of a conventional PID and a model-based first order sliding mode control are carried out and discussed.

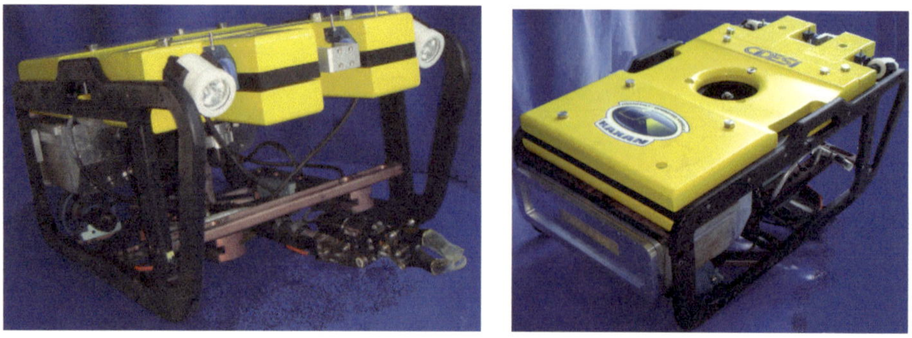

Fig. 1. ROV KAXAN; frontal view (left) and rear view (right).

1.1 Background

In this section an analysis of the state of the art is presented. This study aims at reviewing ROV control strategies ranging from position trajectory to station-keeping control, which are two of the main problems to deal with. There are a great number of studies in the international literature related to several control approaches such as PID-like control, standard sliding mode control, fuzzy control, among others. A review of the most relevant works is given below:

Visual servoing control

Some approaches use vision-based control (Van Der Zwaan & Santos-Victor, 2001)(Quigxiao et al., 2005)(Cufi et al., 2002)(Lots et al., 2001). This strategy uses landmarks or sea bed images to determine the ROV's actual position and to maintain it there or to follow a specific visual trajectory. Nevertheless, underwater environment is a blurring place and is not a practical choice to apply neither vision-based position tracking nor station-keeping control.

Intelligent control

Intelligent control techniques such as Fuzzy, Neural Networks or the combined Neuro-Fuzzy control have been proposed for underwater vehicle control, (Lee et al., 2007)(Kanakakis et al., 2004)(Liang et al., 2006). Intelligent controllers have proven to be a good control option, however, normally they require a long process parameter tuning, and they are normally used in experimental vehicles; industrial vehicles are still an opportunity area for these control techniques.

PID Control

Despite the extensive range of controllers for underwater robots, in practice most industrial underwater robots use a Proportional-Derivative (PD) or Proportional-Integral-Derivative (PID) controllers (Smallwood & Whitcomb, 2004)(Hsu et al., 2000), thanks to their simple structure and effectiveness, under specific conditions. Normally PID-like controllers have a good performance; however, they do not take into account system nonlinearities that eventually may deteriorate system's performance or even lead to instability.
The paper (Lygouras, 1999) presents a linear controller sequence (P and PI techniques) to govern x position and vehicles velocity u. Experimental results with the THETIS (UROV) are shown. The paper (Koh et al., 2006) proposes a linearizing control plus a PID technique for depth and heading station keeping. Since the linearizing technique needs the vehicle's model, the robot parameters have to be identified. Simulation and swimming pool tests show that the control is able to provide reasonable depth and heading station keeping control. An adaptive

control law for underwater vehicles is exposed in (Antonelli et al., 2008)(Antonelli et al., 2001). The control law is a PD action plus a suitable adaptive compensation action. The compensation element takes into account the hydrodynamic effects that affect the tracking performance. The control approach was tested in real time and in simulation using the ODIN vehicle and its 6 DOF mathematical model. The control shows asymptotic tracking of the motion trajectory without requiring current measurement and *a priori* exact system dynamics knowledge. Self-tuning autopilots are suggested in (Goheen & Jefferys, 1990), wherein two schemes are presented: the first one is an implicit linear quadratic on-line self-tuning controller, and the other one uses a robust control law based on a first-order approximation of the open-loop dynamics and on line recursive identification. Controller performance is evaluated by simulation.

Model-based control (Linearizing control)

Other alternative to counteract underwater control problems is the model-based approach. This control strategy considers the system nonlinearities. On the other hand it is important to notice that the system's mathematical model is needed as well as the exact knowledge of robot parameters. Calculation and programming of a full nonlinear 6 DOF dynamic model is time consuming and cumbersome. In (Smallwood & Whitcomb, 2001) a preliminary experimental evaluation of a family of model-based trajectory-tracking controllers for a full actuated underwater vehicle is reported. The first experiments were a comparison of the PD controller versus fixed model-based controllers: the Exact Linearizing Model-Based (ELMB) and the Non Linear Model-Based (NLMB) while tracking a sinusoidal trajectory. The second experiments were followed by a comparison of the adaptive controllers: adaptive exact Linearizing model-based and adaptive non linear model-based versus the fixed model-based controllers ELMB and NLMB, tracking the same trajectory. The experiments corroborate that the fixed model-based controllers outperformed the PD Controller. The NLMB controller outperforms the ELMB. The adaptive model-based controllers all provide more accurate trajectory tracking than the fixed model-based. However, notice that in order to implement such model-based controllers, at least the vehicle's dynamics is required, and in some cases the exact knowledge of the parameters as well, which is difficult to achieve in practice. In paper (Antonelli, 2006) a comparison between six controllers was performed, and four of them are model-based type; the others are a non model-based and a Jacobian-transpose-based. Numerical simulations using the 6 DOF mathematical model of ODIN were carried out. The paper concludes that the controller's effort is very similar; however the model-based approaches have a better behavior. In paper (McLain et al., 1996), real-time experiments were conducted at the Monterey Bay Aquarium Research Institute (MBARI) using the OTTER vehicle. The control strategy was a model-based linearizing control. Additionally interaction forces acting on the vehicle due to arm motion were predicted and fed into the vehicle's controller. Using this method, station-keeping capability was greatly enhanced. Finally, other exact linearizing model-based control has been used in (Ziani-Cherif, 19998).

First order Sliding Mode Control (SMC)

Sliding mode techniques effectively address underwater control issues and are therefore viable choices for controlling underwater vehicles. However, it is well known that these methods are susceptible to chatter, which is a high frequency signal induced by the switching control. Some relevant studies that use SMC are described next. The paper (Healey & Lienard, 1993) used a sliding mode control for the combined steering, diving and speed control. A series of simulations in the NPS-AUV 6 DOF mathematical model are conducted. (Riedel, 2000) proposes a new Disturbance Compensation Controller (DCC), employing on board vehicles sensors that allow the robot to learn and estimate the seaway dynamics. The estimator is based on a Kalman filter and the control law is a first order sliding mode, which induces harmful high frequency signals on the actuators. The paper (Gomes et al., 2003) shows some control techniques tested in PHANTON 500S simulator. The control laws are: conventional PID, state feedback linearization and first order sliding modes control. The author presented a comparative analysis wherein the sliding mode has the best performance, at the expense of a high switching on the actuators. Work (Hsu et al., 2000) proposes a dynamic positioning system for a ROV based on a mechanical passive arm, as a measurement system. This measurement system was selected from a group of candidate systems, including long base line, short baseline, and inertial system, among others. The selection was based on several criteria: precision, construction cost and operational facilities. The position control laws were a conventional P-PI linear control. Last, the other position control law was the variable structure model-reference adaptive control (VS-MRAC). Finally, in the paper (Sebastián, 2006) a model-based adaptive fuzzy sliding mode controller is reported.

Adaptive first order Sliding Mode Control (ASMC)

SMC have a good performance when the controller is well tuned, however if the robot changes its mass or its center of mass, for instance, because of the addition of a new arm or a tool, the system dynamics changes and the control performance may be affected; similarly, if a change in the underwater disturbances occurs (current direction, for instance), a new tuning should be done. In order to reduce chattering problems, ASMC have been proposed. These controllers are excellent alternative to counteract changes in the system dynamics and environment, nevertheless design and tuning time could be longer, and robot model is required. Following, some relevant works are enumerated. In (Da Cunha, 1995), an adaptive control scheme for dynamic positioning of ROVs, based on a variable structure control (first order sliding mode), is proposed. This sliding mode technique is compared with a P-PI controller. Their performances are evaluated by simulation and in pool tests, proving that the sliding mode approach has a better result. The paper (Bessa, 2007) describes a depth SMC for remotely operated vehicles. The SMC is enhanced by an adaptive fuzzy algorithm for uncertainties/disturbances compensation. Numerical simulations in 1 DOF (depth) are presented to show the control performance. This SMC also uses the vehicle estimated model.
Paper (Sebastián & Sotelo, 2007) proposes the fusion of a sliding mode controller and an adaptive fuzzy system. The main advantage of this methodology is that it relaxes the required exact knowledge of the vehicle model, due to parameter uncertainties are compensated by the fuzzy part. A comparative study between; PI controller, classic sliding mode controller and the adaptive fuzzy sliding mode is carried out. Experimental results demonstrate the good performance of the proposed controller. (Song & Smith, 2006) combine sliding mode control with fuzzy logic control. The combination objective is to reduce chattering effect due to model parameter uncertainties and unknown perturbations. Two control approaches are tested: Fuzzy Sliding Mode Controller (FSMC) and Sliding Mode Fuzzy Controller (SMFC). In the FSMC uses a simple fuzzy logic control to fuzzify the relationship of the control command and the distance between the actual state and the sliding surface. On the other hand, at the SMFC each rule is a sliding mode controller. The boundary layer and the coefficients of the sliding surface become the coefficients of the rule output function. Open water experiments were conducted to test AUV's depth and heading controls. The better behavior was detected in the

SMFC. Finally, an adaptive first order sliding mode control for an AUV for the diving maneuver was implemented in (Cristi et al., 1990). This control technique combines the adaptivity of a direct adaptive control algorithm with the robustness of a sliding mode controller. The control is validated by numerical simulations.

High Order Sliding Mode Control (HOSMC)

In order to avoid chattering problem and system model requirement a new methodology called High Order Sliding Mode Control (HOSMC) is proposed in (Garcia-Valdovinos, 2009). HOSMC principal characteristic is that it keeps the main advantages of the standard SMC, removing the chattering effects (Perruquett & Barbot, 1999).

The methodology proposed in this chapter was firstly reported in (Garcia-Valdovinos, 2009), where it is proposed a second order sliding-PD control to address the station keeping problem and trajectory tracking under disturbances. The control law is tested in an under-actuated 6-DOF ROV under Matlab-Simulink simulations, considering unknown and abrupt changing currents direction.

2. General 6 DOF underwater system model

Following standard practice (Fossen, 2002), a 6 DOF nonlinear model of an underwater vehicle is obtained. By using a global reference *Earth-fixed frame* and *Body-fixed frame*, see Figure 2. The *Body-fixed frame* is attached to the vehicle. Its origin is normally on the center of gravity. The motion of the *Body-fixed frame* is described relative to the *Earth-fixed frame*.

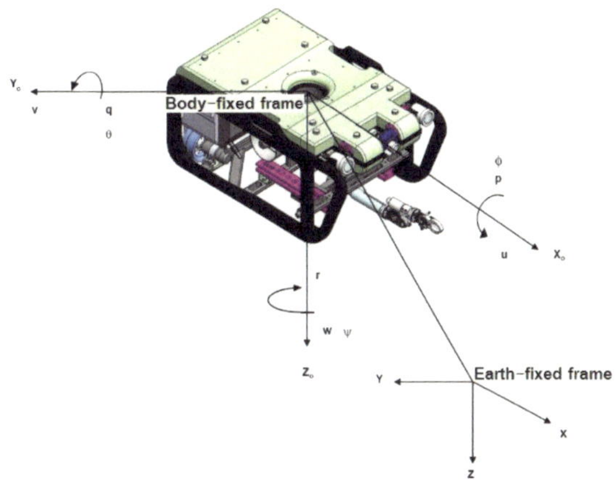

Fig. 2. Reference Earth-fixed frame and Body-fixed frame.

The notation defined by SNAME (Society of Naval Architects and Marine Engineers) established that the *Body-fixed frame* has components of motion given by the linear velocities vector $\nu_1 = \begin{bmatrix} u & v & w \end{bmatrix}^T$ and angular velocities vector $v_2 = \begin{bmatrix} p & q & r \end{bmatrix}^T$ (Fossen, 2002).. The general velocity vector is represented as:

$$\nu = \begin{bmatrix} \nu_1 & \nu_2 \end{bmatrix}^T = \begin{bmatrix} u & v & w\, p & q & r \end{bmatrix}^T$$

where u is the linear velocity in surge, v the linear velocity in sway, w the linear velocity in heave, p the angular velocity in roll, q the angular velocity in pitch and r the angular velocity in yaw.

The position vector $\eta_1 = [x \quad y \quad z]^T$ and orientation vector $\eta_2 = [\phi \quad \theta \quad \psi]^T$ coordinates expressed in the Earth-fixed frame are:

$$\eta = [\eta_1 \quad \eta_2]^T = [x \quad y \quad z \quad \phi \quad \theta \quad \psi]^T$$

where *x, y, z* represent the Cartesian position in the *Earth-fixed frame* and φ represent the roll angle, θ the pitch angle and ψ the yaw angle.

Kinematic model. It is the transformation matrix between the Body and Earth frames, expressed as (Fossen, 2002):

$$\dot{\eta} = J(\eta)\nu$$

$$\begin{bmatrix} \dot{\eta}_1 \\ \dot{\eta}_2 \end{bmatrix} = \begin{bmatrix} J_1(\eta_2) & 0_{3x3} \\ 0_{3x3} & J_2(\eta_2) \end{bmatrix} \begin{bmatrix} \nu_1 \\ \nu_2 \end{bmatrix} \tag{1}$$

where $J_1(\eta_2)$ is the rotation matrix that gives the components of the linear velocities ν_1 in the *Earth-fixed frame* and $J_2(\eta_2)$ is the matrix that relates angular velocity ν_2 with vehicle's attitude in the global reference frame.

Well-posed Jacobian: The transformation (1) is ill-posed when $\theta = \pm 90^o$. To overcome this singularity, a quaternion approach might be considered. However, the vehicle KAXAN is not required to be operated on $\theta = \pm 90^o$. In addition, the ROV is completely stable in roll and pitch coordinates.

Hydrodynamic model: The equation of motion expressed in the *Body-fixed frame* is given as follows (Fossen, 2002):

$$M\dot{\nu} + C(\nu)\nu + D(\nu)\nu + g(\eta) = \tau \tag{2}$$

where $\nu \in R^{n6\,x\,1}$, $\eta \in R^{nx1}$, and $\tau \in R^{p\,x\,1}$. τ denotes the control input vector. Matrix $M \in R^{nxn}$, is the inertia matrix including hydrodynamic added mass, $C \in R^{n\,x\,n}$, is a nonlinear matrix including Coriolis, centrifugal and added terms, $D \in R^{n\,x\,n}$, denotes dissipative influences, such as potential damping, viscous damping and skin friction, finally vector $g \in R^{n\,x\,1}$, denotes restoring forces and moments.

Ocean currents. Some factors that generate current are: tide, local wind, nonlinear waves, ocean circulation, density difference, etc. It's not the objective of this work to make a deeply study of this phenomena, but only to study the current model proposed by (Fossen, 2002). This methodology proposes that the equations of motion can be represented in terms of the relative velocity:

$$\nu_r = \nu - V_c \tag{3}$$

where $V_c = [u_c \quad v_c \quad w_c \quad 0 \quad 0 \quad 0]^T$ is a vector of irrotation Body-fixed current velocities. The average current velocity V_c is related to Earth-fixed current velocity components $\begin{bmatrix} u_c^E & v_c^E & w_c^E \end{bmatrix}$ by the following expression:

$$\begin{aligned} u_c^E &= V_c \cos(\alpha_c)\cos(\beta_c) \\ v_c^E &= V_c \cos(\beta_c) \\ w_c^E &= V_c \sin(\alpha_c)\cos(\beta_c) \end{aligned} \tag{4}$$

where α_c is the angle of attack and β_c the sideslip angle.
Finally, the Earth-fixed current velocity could be computed at the Body-fixed frame, by using

$$\begin{bmatrix} u_c \\ v_c \\ w_c \end{bmatrix} = J_1(\eta_2) \begin{bmatrix} u_c^E \\ v_c^E \\ w_c^E \end{bmatrix} \tag{5}$$

In order to simulate the current and their effect on the ROV, the following model will be applied

$$M\dot{\nu} + C_{RB}(\nu)\nu + C_A(\nu_r)\nu_r + D(\nu_r)\nu_r + g(\eta) = \tau \tag{6}$$

where C_{RB} is the Coriolis from rigid body inertia, and C_A is the Coriolis from added mass.
Assuming that Body-fixed current velocity is constant or at least slowly varying, $\dot{v}_c = 0 \quad \Rightarrow \quad \dot{v}_r = \dot{v}$.

Control input vector. The τ_η comprises the thruster force applied to the vehicle. KAXAN has four thrusters, whose forces and moments are distributed as:

- F_1 Thruster located at rear (left).
- F_2 Thruster located at rear (right).
- F_3 Lateral thruster.
- F_4 Vertical thruster.

F_1 and F_2 propel the vehicle in the x direction and generates the turn in ψ when $F_1 \neq F_2$, F_3 propels the vehicle sideways (y direction) and F_4 allows the vehicle to move up and down (z direction). Then the control signal τ_η must be multiplied by a $\boldsymbol{B}$ matrix comprising forces and moments according to the force application point to the center of mass.

$$\tau_\eta = \begin{bmatrix} F_1 + F_2 \\ F_3 \\ F_4 \\ -F_3 d_{3z} + F_4 d_{4y} \\ F_1 d_{1z} + F_2 d_{2z} - F_4 d_{4x} \\ -F_1 d_{1y} + F_2 d_{2y} - F_3 d_{3x} \end{bmatrix} \tag{7}$$

Rewriting (7) gives rise to

$$\tau_\eta = \underbrace{\begin{bmatrix} X \\ Y \\ Z \\ K \\ M \\ N \end{bmatrix}}_{Control\ Force} = \underbrace{\begin{bmatrix} 1 & 1 & 0 & 0 \\ 0 & 0 & 1 & 0 \\ 0 & 0 & 0 & 1 \\ 0 & 0 & -d_{3z} & d_{4y} \\ d_{1z} & d_{2z} & 0 & -d_{4x} \\ -d_{1y} & d_{2y} & d_{3x} & 0 \end{bmatrix}}_{B} \begin{bmatrix} F_1 \\ F_2 \\ F_3 \\ F_4 \end{bmatrix} \tag{8}$$

3. Control systems

In this section the PID control and model-based first order SMC laws are reminded, later the model-free 2-order sliding mode control technique is introduced (hereafter called HOSMC). These control laws behavior are shown in the next section.

3.1 PID control

The Proportional-Integral-Derivative control law is (Ogata, 1995):

$$\tau = K_P e(k\Delta_T) + \frac{K_P \Delta_T}{T_I} \sum_{h=1}^{k} \frac{e(h\Delta_T) + e((h-1)\Delta_T)}{2} + K_P T_D \left[e(k\Delta_T) - e((k-1)\Delta_T) \right] \tag{9}$$

where Δ_T is the sample time, $e(k\Delta_T)$ is the error measured at the sample time $k\Delta_T$. K_P is the proportional gain, T_I is the integral time and T_D is the derivative time. The PID control gains are shown in Table 1.

	Gains					
	x	y	z	ϕ	θ	ψ
K_p	1600	1800	1300	0	0	18000
T_d	3000	15000	3000	0	0	70000
T_i	0.5	10	0.5	0	0	0.25

Table 1. PID control gains.

3.2 Model-based first order sliding mode control (SMC)

Using the methodology given in (Slotine & Li, 1991), the sliding surface is defined as

$$s = \dot{\tilde{\eta}} - \alpha \tilde{\eta} \tag{10}$$

where $\tilde{\eta} = \eta - \eta_d$.

The SMC control law is given by

$$\tau = \tau_{eq} + K_s sign(\beta s) \tag{11}$$

where τ_{eq} is the equivalent control given by the system estimated dynamic. Parameters β and K_s are constants, *sign* denotes the sign function. Table 2 lists the control gains used in the simulation.

	Gains					
	x	y	Z	ϕ	θ	ψ
K_s	530	700	10	0	0	40
α	530	500	25	0	0	15

Table 2. SMC control gains.

3.3 Model-free 2nd-order sliding mode control (HOSMC)

To analyze the proposed controller is necessary to introduce the following preliminaries. Let the nominal reference $\dot{\eta}_r$ be:

$$\dot{\eta}_r = \dot{\eta}_d - \alpha\tilde{\eta} + S_d - K_i \int_0^t sign\left(S_q\right) d\sigma \tag{12}$$

where α, K_i are diagonal positive definite $n \times n$ gain matrices, function *sign(x)* stands for sign function of $x \in \Re^n$, and

$$\begin{aligned} S_q &= S - S_d \\ S &= \dot{\tilde{\eta}} - \alpha\tilde{\eta} \\ S_d &= S(t_0)\ e^{-\kappa\ t} \end{aligned} \tag{13}$$

for $\kappa > 0$. $S(t_0)$ stands for $S(t)$ at $t=0$.
Now, let the extended error variable be defined as follows:

$$S_r = \dot{\eta} - \dot{\eta}_r \tag{14}$$

and substituting (12) into (14) yields,

$$S_r = S_q + K_i \int_0^t sign\left(S_q\right) d\sigma \tag{15}$$

Notice that the task is defined in the *Earth-fixed frame* for the sake of simplicity.

Controller definition

The control design and some structural properties are now given.
Theorem. Consider the vehicle dynamics (2) in closed loop with the control law given by

$$\tau_\eta = -K_d S_r \tag{16}$$

where K_d is a positive $n \times n$ feedback gain matrix. Exponential tracking is guaranteed, provided that K_i in (15) and K_d are large enough, for small initial error condition.
Proof. A detailed analysis shows that the above Theorem fulfills, see (Garcia-Valdovinos et al. 2006) and (Parra-Vega et al., 2003) for more details ■.
Remark 1. Since the control (15) is computed in the Earth-fixed frame it is necessary to map it into the Body- fixed frame by using the transpose Jacobian (1) as follows:

$$\tau = J^T \tau_\eta \tag{17}$$

Remark 2. Expanding the control law (16) can be rewritten as follows:

$$\tau_\eta = -\underbrace{K_d\alpha\tilde{\eta}}_{P} - \underbrace{K_d\dot{\tilde{\eta}}}_{D} - \underbrace{K_d K_i \int_0^t sign\left(S_q\right) d\sigma}_{Sliding\ part} \tag{18}$$

which gives rise to a sliding PD-like controller.

3.3.1 Comments on HOSMC

How to tune the controller: The stability proof (see (Garcia-Valdovinos et al. 2006) and (Parra-Vega et al., 2003) for more details) suggests that arbitrary small K_i and small α can be set as a starting point. Increase feedback gain K_d until acceptable boundedness of S_r appears.

Then, increase gradually K_i until the sliding mode arises. Finally, increase α to achieve a better position tracking performance. Notice that K_i is not a high gain result since a larger K_i does not mean a larger domain of stability.

Robustness: The system has inherent robustness of typical variable structure systems, since the invariance property is attained for all time, whose convergence is governed solely by (13) when $S_q(t)=0$ for all time, independently of bounded disturbances.

Smooth Controller: Higher-order sliding modes, in this case second order sliding mode (SOSM), have emerged to solve the problem of chattering, which is induced by first order sliding modes (FOSM). Besides preserving the advantages of FOSM, the scheme SOSM totally removes the chattering effect of FOSM, and provides for even higher accuracy. In our case, SOSM is induced, and chattering is circumvented by integrating the sign function of S_q.

Finite time convergence: Since sliding mode exists for all time, it is possible to attain finite time convergence of position tracking errors by means of well-posed terminal attractors. Finite time convergence can be tuned arbitrarily via a time-varying gain $\alpha(t)$ so as to drive smoothly $\Delta x(t)$ toward its equilibrium $\Delta x(t)=0$. Gain $\alpha(t)$ is tailored with a Time Base Generator (TBG), which may be a fifth order polynomial that smoothly goes from $0 \to 1$, for more details see (Garcia-Valdovinos et al. 2006).

4. Numerical simulations

Performance of the controllers is verified through some simulations with a 6 DOF underwater vehicle (2), where only 4 DOF are actuated, that is (x, y, z, ψ). Evidently, ϕ and θ are not actuated, though these are bounded (stable). Position tracking simulations are presented. Matlab-Simulink has been used to perform the simulations with ODE Runge-Kutta 45, variable step.

4.1 Controller's gains

Feedback gains for the controller are show in Table 3.

	Gains					
	x	y	z	ϕ	θ	ψ
α	30	30	30	0	0	50
K_d	1000	1000	1000	0	0	1000
K_i	0.05	0.05	0.05	0	0	0.05
κ			5			

Table 3. Model-free 2-order sliding mode contol gains.

4.2 Ocean current parameters

The current starts flowing to the north and after some time, it suddenly changes to east. In all cases the current is $V_c=1.1\ m/s$. According to (2) and (4) one has the following:

1. North: When flowing to the north, parameters are the following: $\alpha_c = 0\ rad$ and $\beta_c = 0\ rad$.
2. East: When flow is in the east direction, parameters are the following: $\alpha_c = 0\ rad$ and $\beta_c = \pi/2\ rad$.

4.3 Position tracking

Now, the proposed controller is evaluated for tracking tasks, under ocean currents. The task is divided into two stages. First stage consists of moving the vehicle smoothly from an initial

position $[x_i, y_i, z_i, \psi_i] = [0, 0, 0, 0]$ to a final position $[x_f, y_f, z_f, \psi_i]=[1, 0, 0.5, \pi/2]$, see the linear path in figures 3, 7 and 11, for the PID, SMC and HOSMC, respectively. This stage lasts 15 seconds, from $t=0$ s to $t=15$ s.
Second stage, once the vehicle is correctly oriented, it is requested to follow a circumference of radio $r=1$ m, centered at $(h, k) = (0, 0)$. The circumference is executed at a rate given by $\omega=0.628$ rad/s, that is, in $t=10$ s. Notice that the circumference is designed in plane x, y, and ψ is always tangential to the circumference, see the circular path in figures (3, 7 and 11, for the PID, SMC and 2-order sliding mode control, respectively). This stage lasts 10 seconds, from $t>15$ s to $t=25$ s.
From $t=10$ s to $t<15$ s (first stage) the ocean current flows to the north (u_c). The lasts 15 seconds, from $t>10$ s to $t=25$ s, current flows to the east (υ_c).

4.4 Description of results

Figures 3 (PID), 7 (SMC) and 11 (HOSMC), depict the complete trajectory tracking by the system.
Figures 4 (PID), 8 (SMC) and 12 (HOSMC), show the system position tracking comparison x vs x_d, y vs y_d and z vs z_d.
Figures 5 (PID), 9 (SMC) and 13 (HOSMC), give the robot inclination behavior; notice that the angular position tracking in ψ is attained (even under currents influence). As it was mentioned ϕ and θ are not actuated, however they are stable, they present a slight deviation from zero, due to the changing current.
The control signal behavior is described in Figures 6 for the PID control, 10 for the SMC and 14 for the HOSMC. The figures show the propulsion force in the x, y and z directions (from top to bottom), and the last box represent the momentum around in the ψ angle.
Finally the control performance is compared by using the Mean Square Error (MSE). Figure 15 represents the MSEs values for the three control techniques. The figure show two bars for each control technique, the first represents the MSE_R and the second is the MSE_ψ. Where the MSE_R is defined by:

$$MSE_R = \sqrt{(MSE_x)^2 + (MSE_y)^2 + (MSE_z)^2} \tag{19}$$

4.5 PID control

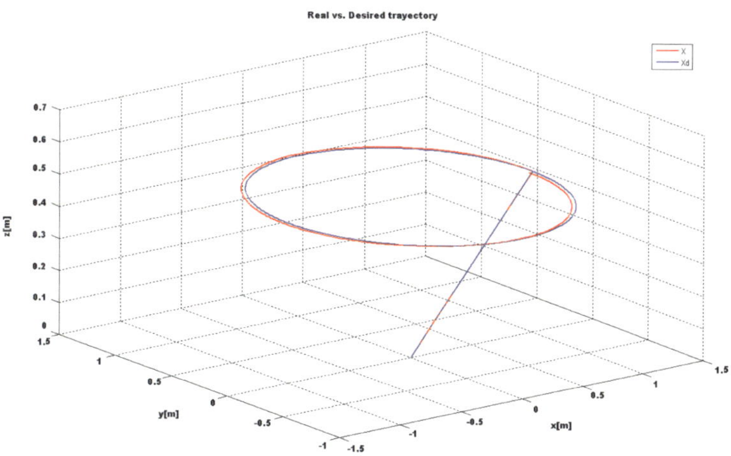

Fig. 3. Position tracking performance under PID control.

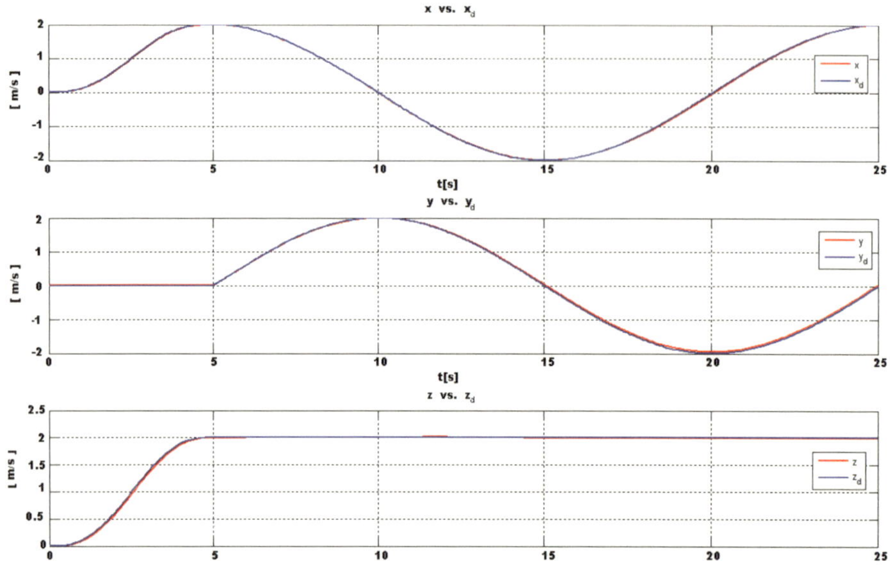

Fig. 4. Position tracking performance (x vs x_d, y vs y_d and z vs z_d) under the PID control.

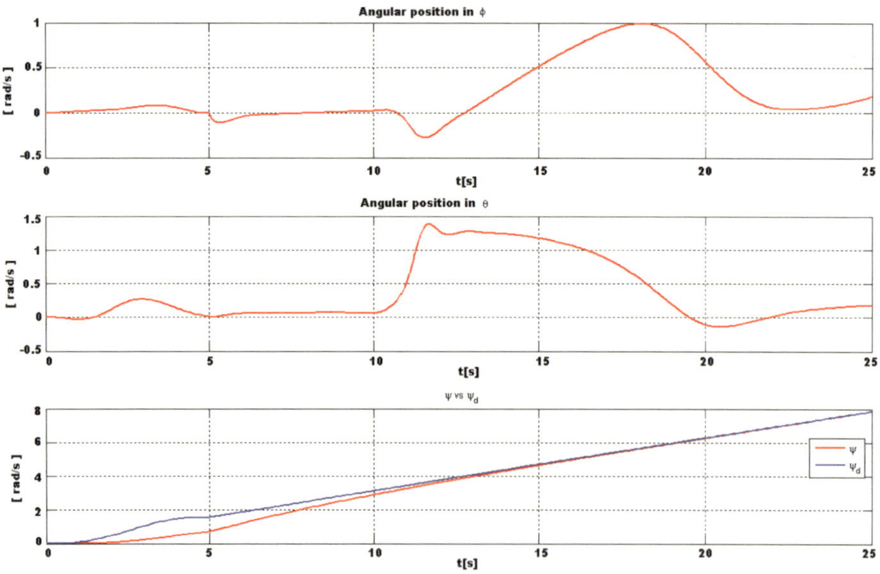

Fig. 5. Angular inclinations behavior (ϕ, θ and ψ vs ψ_d) under the PID control.

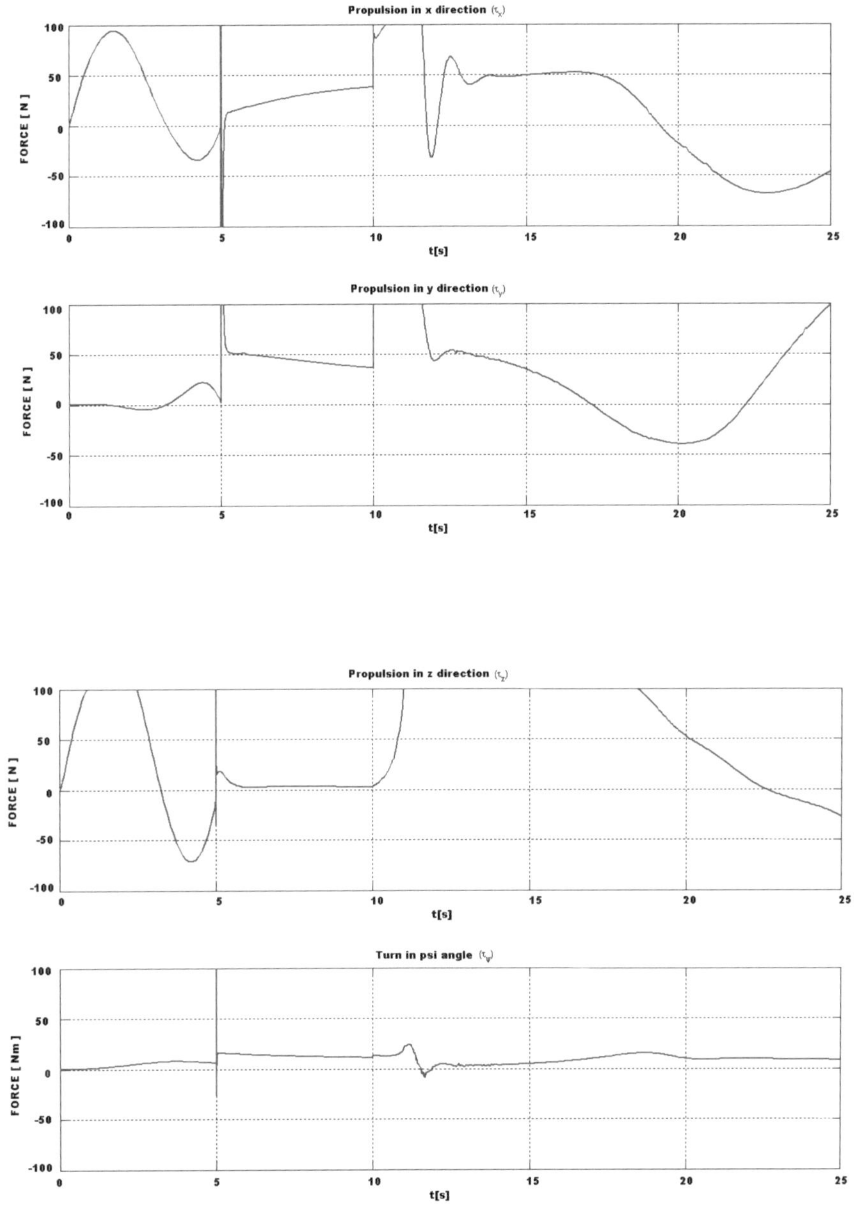

Fig. 6. Control signal behavior. From top to bottom propulsion force in the x, y and z directions, and the last box represent the momentum around the ψ angle (PID control).

4.6 Model-based first order mode control (SMC)

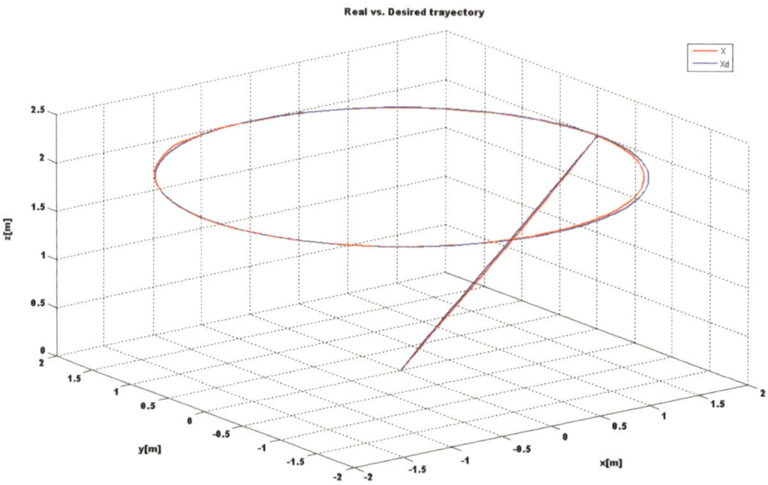

Fig. 7. Position tracking performance with SMC.

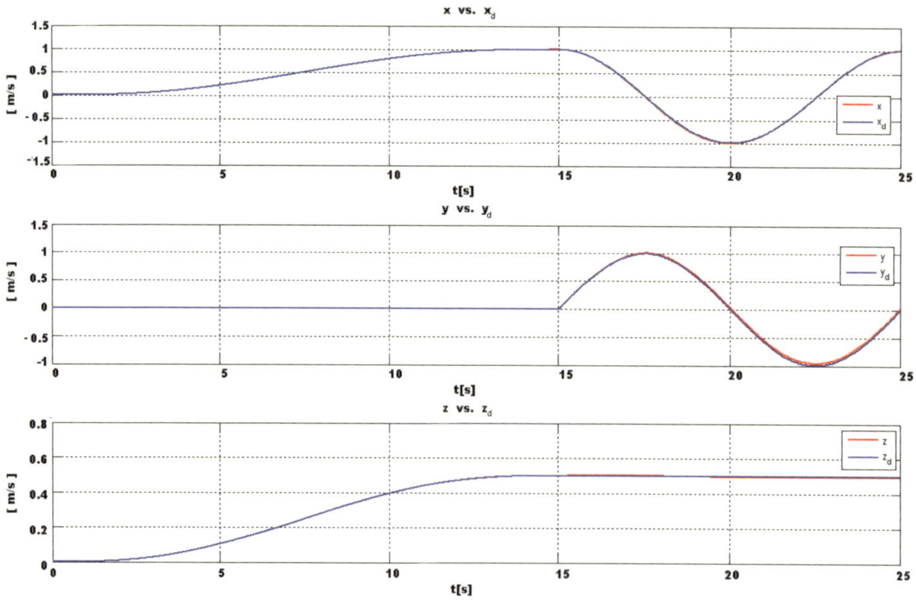

Fig. 8. Position tracking performance (x *vs* x_d, y *vs* y_d and z *vs* z_d) with the SMC.

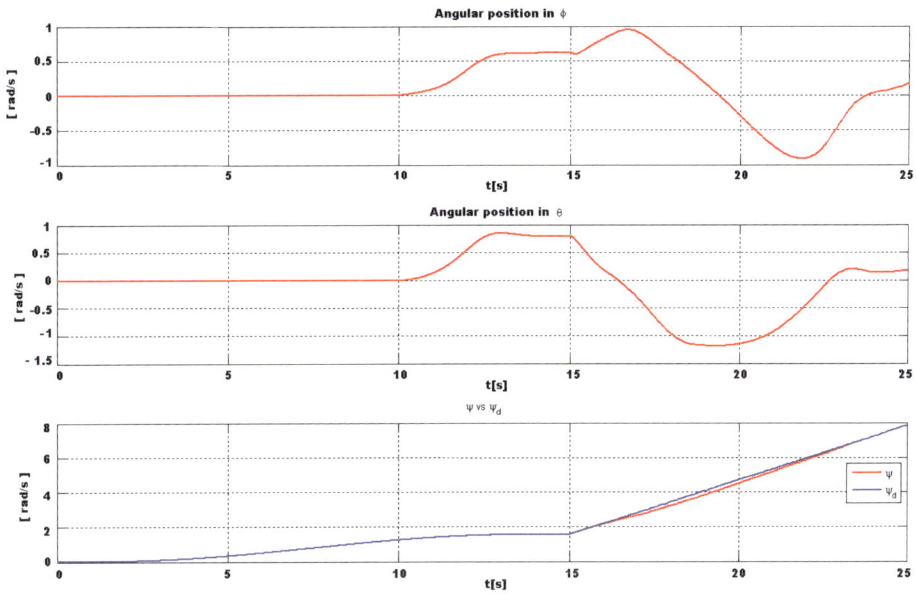

Fig. 9 Angular inclinations behavior (ϕ, θ and ψ *vs* ψ_d) with the SMC control.

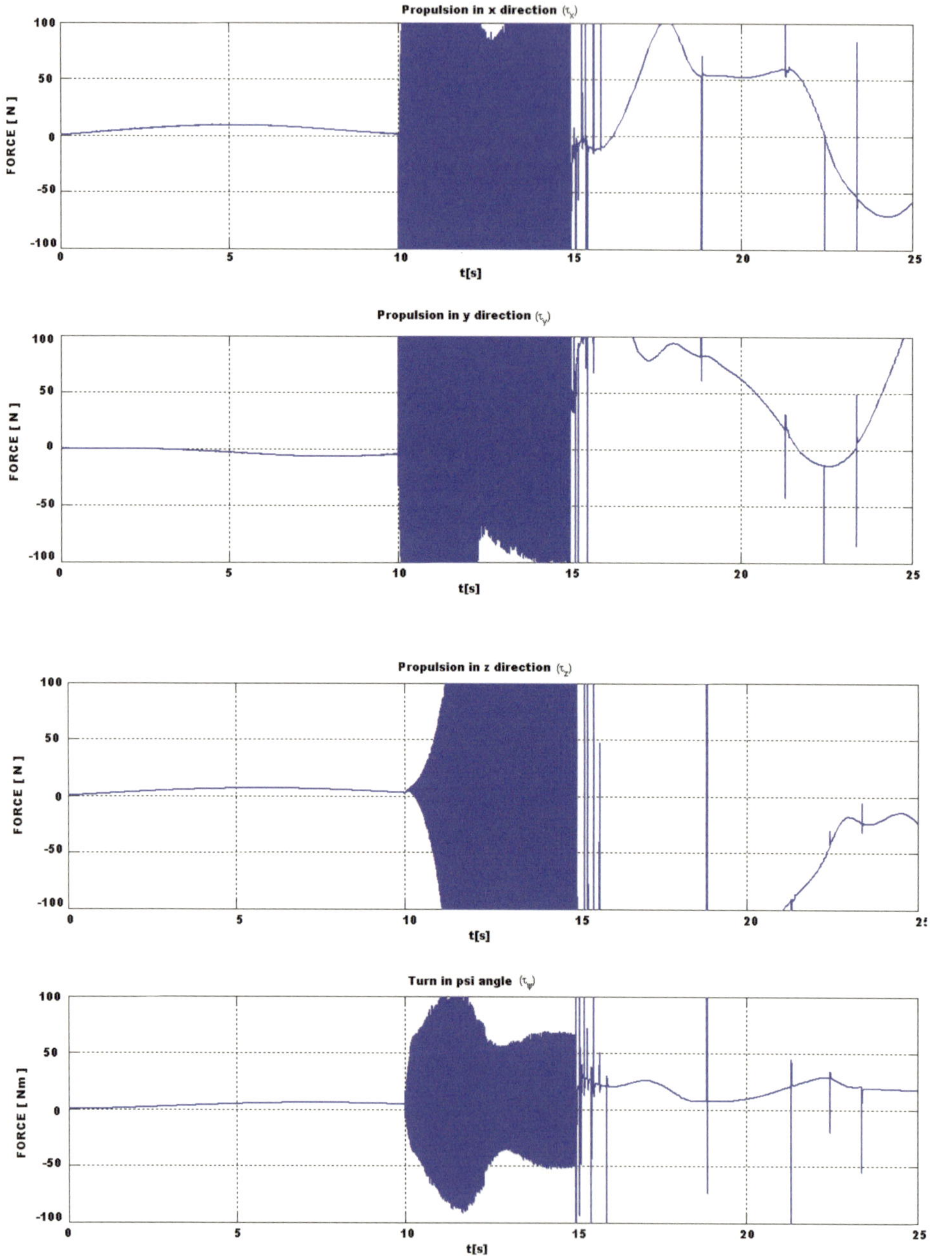

Fig. 10. Control signal behavior. From top to bottom propulsion force in the x, y and z directions, and the last box represent the momentum around in the ψ angle (SMC).

4.7 Model-free 2nd-Order sliding mode control

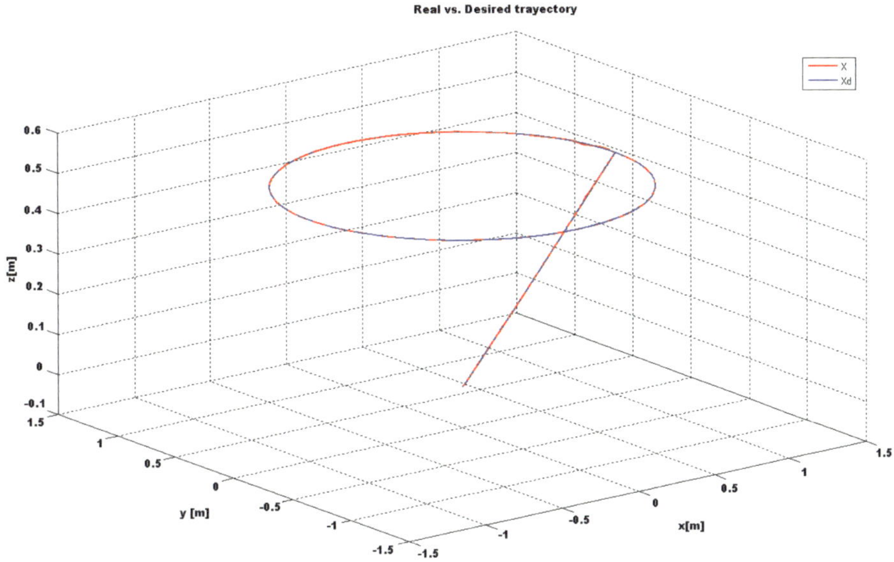

Fig. 11. Position tracking performance with HOSMC.

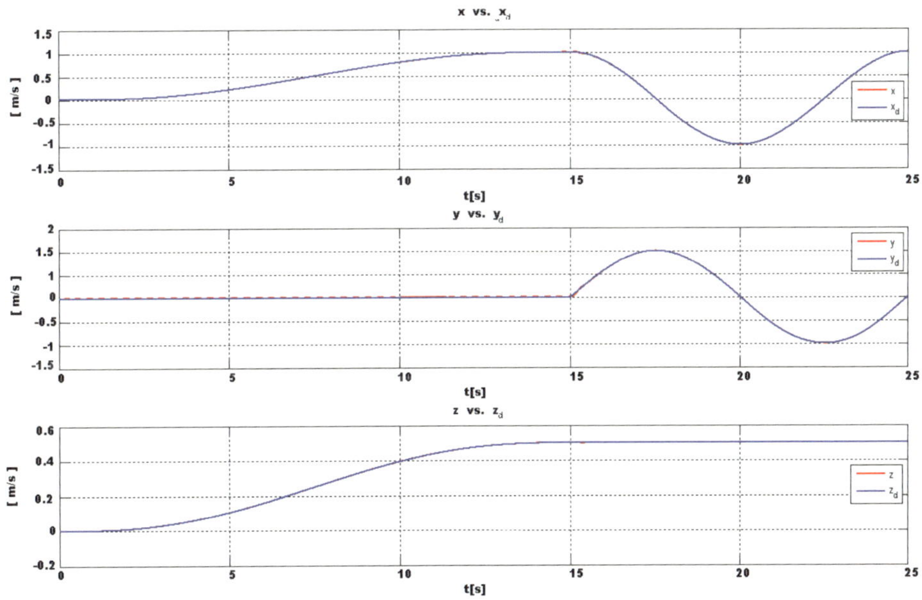

Fig. 12. Position tracking performance (x *vs* x_d, y *vs* y_d and z *vs* z_d) with the HOSMC.

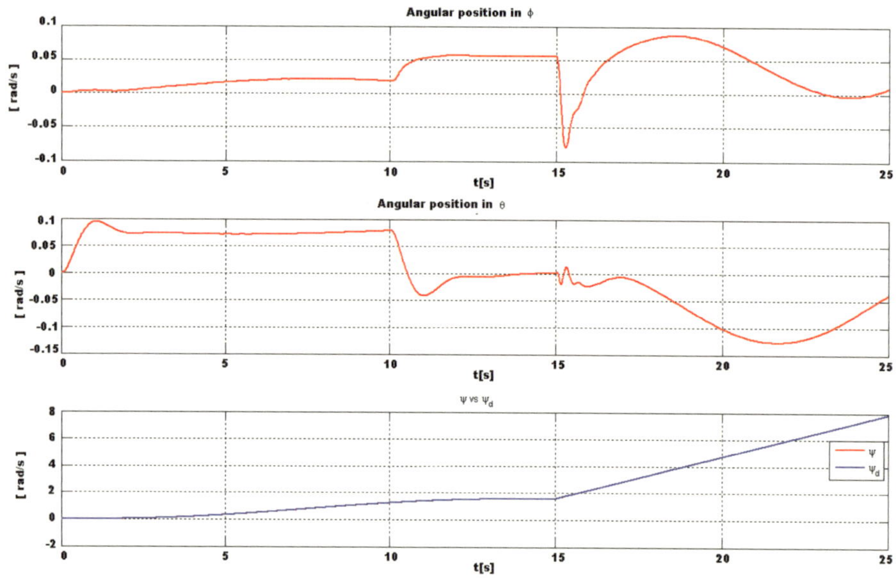

Fig. 13. Angular inclinations behavior (ϕ, θ and ψ *vs* ψ_d) with the HOSMC.

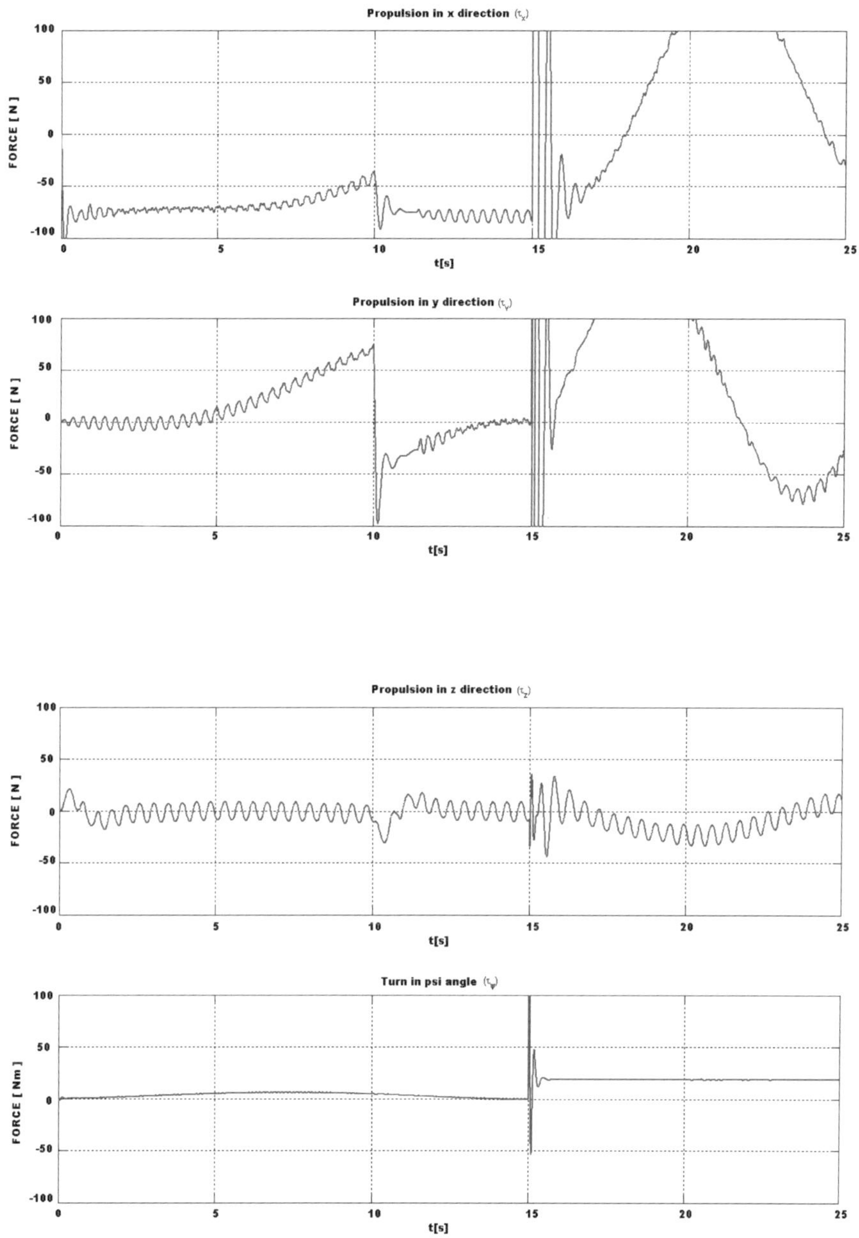

Fig. 14. Control signal behavior. From top to bottom propulsion force in the x, y and z directions, and the last box represent the momentum around in the ψ angle (HOSMC).

4.8 Control performance comparison by Mean Square Error (MSE)

An MSE study reveals that the proposed controller (HOSMC) exhibits the best performance in terms of position tracking.

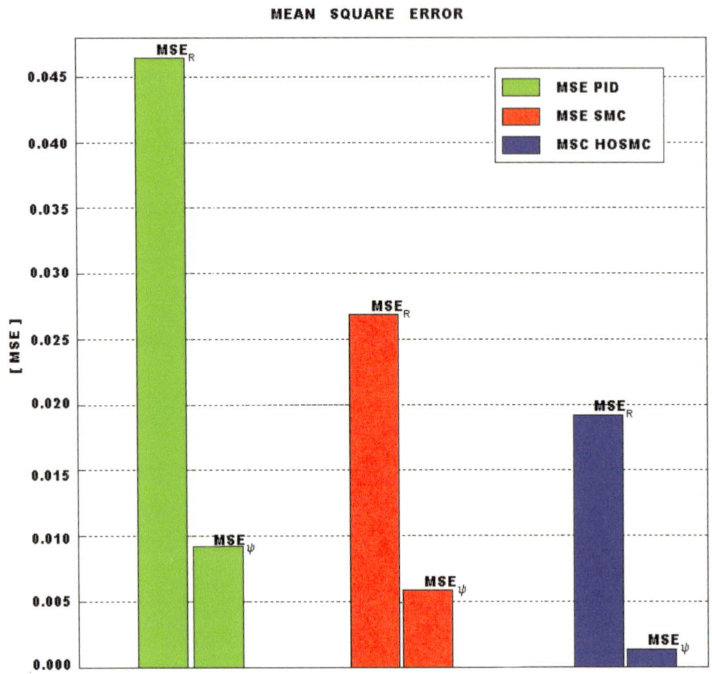

Fig. 15. Mean Square Error (MSE) values for the three control techniques.

5. References

Antonelli, G. (2006). Dynamic Control of 6-DOF AUVs. *Springer Tracts in Advanced Robotics,* Volume 2.

Antonelli, G.; Chiaverini, S.; Sarkar, N. & West, M. (2001). Adaptive control of an autonomous underwater vehicle: experimental results on ODIN. *Transactions on Control Systems Technology, IEEE, Sep 2001.*

Antonelli, G.; Fossen, T. I. & Yoerger, D. R. (2008). Underwater Robotics, *Springer Handbook of Robotics, Springer Berlin Heidelberg.*

Bessa, W. M.; Dutra, M. S. & Kreuzer, E. (2007). Depth control of remotely operated underwater vehicles using an adaptive fuzzy sliding mode controller. *Robotics and Autonomous system, Elsevier.*

Cristi, R.; Papoulias, F. A. & Healey, A. J. (1990). Adaptive sliding mode control of autonomous underwater vehicles in the dive plane. *IEEE Journal of Oceanic Engineering,* Vol. 15, No. 15 July 1990.

Cufi, X.; Garcia, R. & Ridao. P. (2002). An approach to vision-based station keeping for an unmanned underwater vehicle. IEEE/RSJ International Conference on Intelligent Robots and System.

Da Cunha, J.P.V.S.; Costa, R.R. & Liu, Hsu. (1995). Design of a high performance veriable structure position control of ROV's. *IEEE Journal of Oceanic Engineering,* Volume 20, No. 1, Page(s):42 - 54, Jan.

Fossen, T. I. (2002). *Marine control systems; guidance, navigation and control of ships. Rigs and underwater vehicles.* Marine Cybernetics.

García-Valdovinos, L. G.; Parra-Vega, V. & Arteaga, M.A. (2006). Bilateral Cartesian Sliding PID Force/Position Control for Tracking in Finite Time of Master-Slave Systems, *Proceedings of the IEEE ACC'06,* pp. 369-375, Minneapolis, Minnesota, EUA.

Garcia-Valdovinos, L. G.; Salgado-Jimenez. T. & Torres-Rodríguez, H. (2009). Model-free high order sliding mode control for ROV: Station-keeping approach, *Proceedings of OCEANS MTS/IEEE.*

Goheen K. R. & Jefferys E.R. (1990). Multivariable self-tuning autopilots for autonomous and remotely operated underwater vehicles. *IEEE Journal of Oceanic Engineering.* Jul.

Gomes, R. M. F.; Sousa, J. B. & Pereira, F. L. (2003). Modelling and control of the IES project ROV. *Proceedings of European control conference,* Cambridge, UK.

Healey, A. J. & Lienard, D. (1993). Multivariable sliding mode control for autonomous diving and steering of unmanned underwater vehicles. *IEEE Journal of Oceanic Engineering,* Vol. 18, No. 3, July.

Hsu L.; Costa, R. R. & Lizarralde, F. (2000). Dynamic positioning of remotely operated underwater vehicles. *IEEE Robotics and Automation Magazine.* September, 2000.

Kanakakis, V.; Valavanis, K. P. & Tsourveloudis, N. C. (2004). Fuzzy-Logic Based Navigation of Underwater Vehicles. *Journal of Intelligent and Robotic Systems,* Springer Netherlands, Volume 40, Number 1 / May.

Koh, T. H.; Lau, M. W. S.; Seet, G. & Low, E., A. (2006). Control Module Scheme for an Underactuated Underwater Robotic Vehicle. *Journal of Intelligent and Robotic Systems,* Springer Netherlands. Volume 46, Number 1,/ May.

Lee, J.; Roh, M.; Lee, J. & Lee, D. (2007). Clonal Selection Algorithms for 6-DOF PID Control of Autonomous Underwater Vehicles. *Lecture Notes in Computer Science,* Springer Berlin / Heidelberg. Volume 4628/2007.

Liang, X.; Li, Y.; Xu, Y.; Wan, L. & Qin, Z. (2006). Fuzzy neural network control of underwater vehicles based on desired state programming. *Journal of Marine Science and Application,* Harbin Engineering University. Volume 5, Number 3 / September.

Lots, J.-F.; Lane, D.M.; Trucco, E. & Chaumette, F. (2001). A 2D visual servoing for underwater vehicle station keeping. *Proceedings of the IEEE International Conference on Robotics and Automation,* ICRA2001.

Lygouras, J. N. (1999). DC Thruster Controller Implementation with Integral Anti-wind up Compensator for Underwater ROV. *Journal of Intelligent and Robotic Systems,* Springer Netherlands. Volume 25, Number 1 / May.

McLain, T. W.; Rock, S. M. & Lee, M. J. (1996). Experiments in the coordinated control of an underwater arm/vehicle system. *Autonomous Robots,* Springer Netherlands. Vol. 3, Numbers 2-3 / June.

Ogata, K. (1995). *Discrete-Time Control Systems.* 2nd Edition, Prentice-Hall.

Parra-Vega, V.; Arimoto, S.; Liu, Y. H.; Hirzinger, G. & Akella, P. (2003). Dynamic sliding PID control for tracking of robot manipulators: Theory and experiments. *IEEE Trans. on Rob. and Autom.,* 19(6):967--976, December.

Perruquett, W. & Barbot, J. P. (1999). *Sliding Modes control in Engineering.* Marcel Dekker, Inc.

Qingxiao, W.; Shuo, L., Yingming, H. & Feng, Z. (2005). A model-based monocular vision system for station keeping of an underwater vehicle. *Proceedings of the IEEE International Conference on Robotics and Biomimetics (ROBIO).*

Sebastián, E. (2006). Adaptive Fuzzy Sliding Mode Controller for the Snorkel Underwater Vehicle. *Lecture Notes in Computer Science,* Springer Berlin / Heidelberg, Volume 4095/2006.

Sebastián, E. & Sotelo, M. A. (2007). Adaptive Fuzzy Sliding Mode Controller for the Kinematic Variables of an Underwater Vehicle. *Journal of Intelligent and Robotic Systems,* Springer Netherlands, Volume 49, Number 2, June.

Slotine J.-J. and W. Li. (1991). *Applied Nonlinear Control.* Prentice-Hall.

Smallwood, D. A. & Whitcomb, L. L. (2001). Toward Model Based dynamic Positioning of underwater robotics Vehicles. *Proceedings of the OCEANS MTS/IEEE.*

Smallwood, D.A. & Whitcomb, L. L. (2004). Model-based dynamic positioning of underwater robotic vehicles: theory and experiment. *IEEE Journal of Oceanic Engineering,* Volume 29, Issue 1, Page(s):169 - 186, January.

Song, F. & Smith, S. M. (2006). Combine Sliding Mode Control and Fuzzy Logic Control for Autonomous Underwater Vehicles. *Lecture Notes in Advanced Fuzzy Logic Technologies in Industrial Applications,* Publisher Springer London.

Riedel, J. S. (2000). Shallow water station-keeping of an autonomous underwater vehicle: The experimental Results of a disturbance compensation controller. *Proceedings of the OCEANS MTS/IEEE.*

Van Der Zwaan, S. & Santos-Victor, J. (2001). Real-time vision-based station keeping for underwater robots. *Proceedings of the OCEANS MTS/IEEE.*

Ziani-Cherif, S. (1998). *Contribution à la modélisation, l'estimation des paramètres dynamiques et la commande d'un engin sous-marin.* Ph. D. thesis in French, Thèse de docteur de l'Ecole Central de Nantes, France.

19

Sliding Mode Control Applied to a Novel Linear Axis Actuated by Pneumatic Muscles

Dominik Schindele and Harald Aschemann
Chair of Mechatronics, University of Rostock
18059 Rostock, Germany

1. Introduction

Pneumatic muscles are innovative tensile actuators consisting of a fiber-reinforced vulcanised rubber tubing with appropriate connectors at both ends. The working principle is based on a rhombical fibre structure that leads to a muscle contraction in longitudinal direction when the pneumatic muscle is filled with compressed air. This contraction can be used for actuation purposes. Pneumatic muscles are low cost actuators and offer several further advantages in comparison to classical pneumatic cylinders: significantly less weight, no stick-slip effects, insensitivity to dirty working environment, and a higher force-to-weight ratio. A major advantage of pneumatic drives as compared to electrical drives is their capability of providing large maximum forces for a longer period of time. In this case electrical drives are in risk of overheating and may result in increasing errors due to thermal expansion. For these reasons, different researchers have investigated pneumatic muscles as actuators for several applications, e.g. a planar elbow manipulator in Lilly & Yang (2005), a 2-DOF serial manipulator in Van-Damme et al. (2007) or a parallel manipulator in Zhu et al. (2008).

Pneumatic muscles are characterised by dominant nonlinearities, namely the force and volume characteristics. Hence, these nonlinearities have to be considered by suitable control approaches such as sliding mode control. In this contribution the sliding mode technique is applied to a novel linear drive actuated by four pneumatic muscles. This pneumatic linear drive allows for maximum velocities of approximately 1.3 m/s in a workspace of approximately 1 m. In Aschemann & Hofer (2004) and Aschemann et al. (2006) the authors presented the implementation of a trajectory control for a carriage with a pair of pneumatic muscles arranged at opposite sides of a carriage. Unfortunately, this direct actuation by pneumatic muscles suffers from two main drawbacks: On the one hand, the maximum velocity of the carriage is limited to approx. 0.3 m/s, on the other hand the workspace is constrained to the maximum contraction length of the pneumatic muscles, in the given case to approx. 0.25 m. To increase the available workspace as well as the maximum carriage velocity, a new test-rig has been built up. At this test-rig, a rocker transmits the drive force of the pneumatic muscles to the carriage, see Aschemann & Schindele (2008) or Schindele & Aschemann (2010). One disadvantage of this setup is the required height, necessary for the kinematics considered there. To reduce the overall size of the drive mechanism, now, the muscle force is transmitted to the carriage by a pulley tackle consisting of a wire rope and several deflection pulleys, see Fig. 1. The mentioned components are installed such that the required muscle force as well as the maximum workspace and velocity of the carriage are

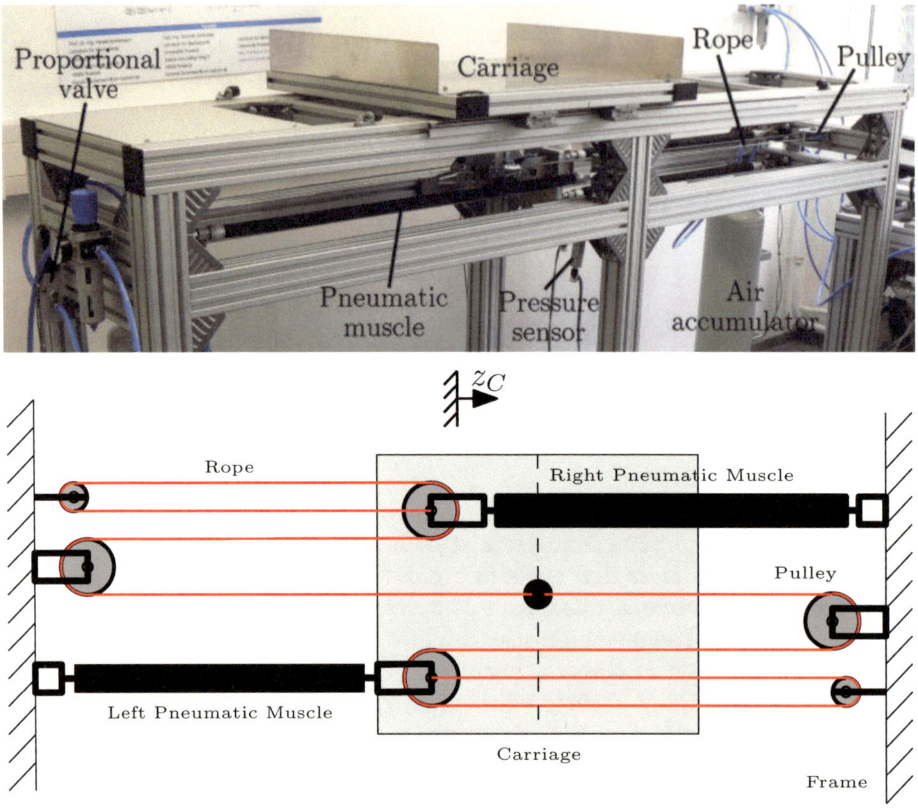

Figure 1. Experimental setup.

increased by a factor of three, in comparison to a directly driven configuration. For actuation of the carriage, four pneumatic muscles are employed, whereas two muscles are used for each direction of tension, respectively. The mass flow rate of compressed air in and accordingly out of each pneumatic muscle is controlled by means of two separate proportional valves. One proportional valve is employed for the two left pneumatic muscles and the other proportional valve is utilised for the two right pneumatic muscles. Pressure declines in the case of large mass flow rates are avoided by using an air accumulator for each valve.
In the paper, first, a control-oriented model of the pneumatically driven high-speed linear axis is derived in section 2 as the basis of control design. At this, polynomial descriptions are utilised to describe the nonlinear characteristics of the pneumatic muscle, i.e., the muscle volume and the muscle force as functions of both contraction length and internal muscle pressure. Second, in sections 3 and 4, sliding mode control techniques are employed to design a nonlinear cascade control. For this purpose the differential flatness-property of the system is exploited. The inner control loops involve a fast pressure control for each muscle, respectively. The outer control loop achieves a decoupling of the carriage position and the mean muscle pressure as controlled variables and provides the reference pressures for the inner pressure control loops. As an alternative to the standard sliding mode technique, additionally, a second-order sliding mode controller and a proxy-based sliding mode controller has been designed for the outer control loop. Proxy-based sliding mode control is a modification of sliding mode control as well as an extension of PID-control, see Kikuuwe & Fujimoto

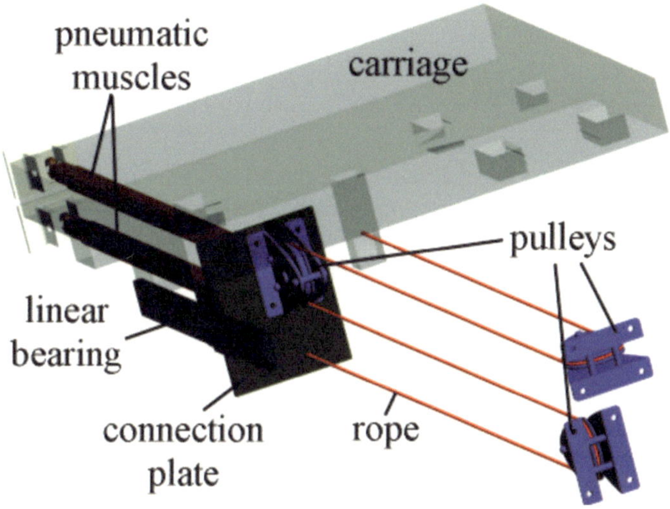

Figure 2. Drawing of the left pulley tackle.

(2006), Van-Damme et al. (2007). The basic idea is to introduce a virtual carriage, called proxy, which is controlled using sliding mode techniques, whereas the proxy is connected to the real carriage by a PID-type coupling force. The goal is to achieve precise tracking during normal operation and smooth, overdamped recovery in the presence of large position errors, which leads to an inherent safety property. In sections 5 and 6, nonlinear friction and remaining model uncertainties in the equations of motion are considered by a feedforward friction compensation module, based on the LuGre model in combination with a nonlinear reduced-order disturbance observer. Finally, in section 8, the proposed control strategy has been implemented at the test rig of the Chair of Mechatronics, University of Rostock. Thereby, desired trajectories for the carriage position can be tracked with high accuracy.

2. System modelling

The modelling of the pneumatically driven high-speed linear axis involves the mechanical subsystem and the pneumatic subsystem, which are coupled by the tension forces of the pneumatic muscles.

2.1 Modelling of the mechanical subsystem

The mechanical model of the high-speed linear axis consists of the carriage and two pulley tackles, at which one pulley tackle transmits the tension force of two pneumatic muscles to the carriage in each case. In this way two pneumatic muscles as well as one pulley tackle is employed for each moving direction of the carriage, see Fig. 2. For modelling the mechanical subsystem is divided into the following elements (Fig. 1 and Fig. 2): a lumped mass for the carriage (mass m_C), the two connection plates, which are also modelled as lumped masses (mass m_{MFi}, $i = \{l,r\}$) and the six pulleys (mass moment of inertia J_{ij}, $i = \{l,r\}$, $j = \{1,2,3\}$). The motion of the linear axis is completely described by the generalised coordinate $z_C(t)$, which denotes the carriage position. The equation of motion directly follows from Lagrange's equations in form of a second-order differential equation

$$m \cdot \ddot{z}_C = \frac{a_M}{k} \left(F_{Mr} - F_{Ml} \right) - F_U \,, \tag{1}$$

with the reduced mass

$$m = \frac{1}{k^2}\left(k^2 \cdot m_C + m_{MFl} + m_{MFr} + \sum_{j=1}^{3} J_{lj}\left(\frac{j}{r}\right)^2 + \sum_{j=1}^{3} J_{rj}\left(\frac{j}{r}\right)^2\right). \tag{2}$$

The parameter $k = 3$ denotes the number of pulleys (radius r) employed for each pulley tackle, and the parameter $a_M = 2$ stands for the two muscles, used for actuation in the left or right direction, respectively. All remaining model uncertainties are taken into account by the disturbance force F_U. On the one hand, these uncertainties stem from approximation errors concerning the static muscle force characteristics and non-modelled viscoelastic effects of the vulcanised rubber material. On the other hand, time-varying damping and friction acting on the carriage, the connection plates and the pulleys depend in a complex manner on lots of influence factors and cannot be accurately represented by a simple friction model.

2.2 Modelling of the pneumatic subsystem

Under the assumption, that the dynamic behaviour of the internal muscle pressure is identically for the two left and right muscles, for modelling and control of the pneumatic subsystem only one muscle for each drive direction is considered. The larger force obtained by utilising two muscles for each pulley tackle is regarded by the factor α_M in equation (1). A mass flow $\dot{m}_{Mi}$, $i = \{l, r\}$ into the pneumatic muscle leads to an increase in internal pressure p_{Mi}, and a contraction $\Delta\ell_{Mi}$ of the muscle in longitudinal direction due to specially arranged fibers. The maximum contraction length $\Delta\ell_{M,max}$ is given by 25% of the uncontracted length. This contraction effect can be exploited to generate forces. The force F_{Mi} and the volume V_{Mi} of a pneumatic muscle depend nonlinear on the according internal pressure p_{Mi} and the contraction length $\Delta\ell_{Mi}$. Given the length of the uncontracted muscle ℓ_M, the contraction length of a pneumatic muscle is related to the carriage position by the following equations

$$\Delta\ell_{Ml} = \ell_M - \frac{1}{k} z_C \,, \tag{3}$$

$$\Delta\ell_{Mr} = \ell_M + \frac{1}{k} z_C \,. \tag{4}$$

The dynamics of the internal muscle pressure follows directly from a mass flow balance in combination with the energy equation for the compressed air in the muscle. As the internal muscle pressure is limited by a maximum value of $p_{Mi,max} = 7$ bar, the ideal gas equation represents an accurate description of the thermodynamic behaviour of the air in muscle $i = \{l, r\}$ (Smith et al. (1996))

$$\frac{p_{Mi}}{\rho_{Mi}} = R_L \cdot T_{Mi} \,. \tag{5}$$

Here, the density ρ_{Mi}, the gas constant of air R_L and the thermodynamic temperature T_{Mi} are introduced. The thermodynamic process is modelled as a polytropic change of state (Smith et al. (1996))

$$\frac{p_{Mi}}{\rho_{Mi}^n} = const. \tag{6}$$

with $n = 1.26$ as identified polytropic exponent. The polytropic exponent is in between $n = 1$ for an isothermal process, and $n = \kappa$ for an isentropic process. Thus, the relationship between the time derivative of the pressure and the time derivative of the density is given by

$$\dot{p}_{Mi} = n \cdot R_L \cdot T_{Mi} \cdot \dot{\rho}_{Mi} \,. \tag{7}$$

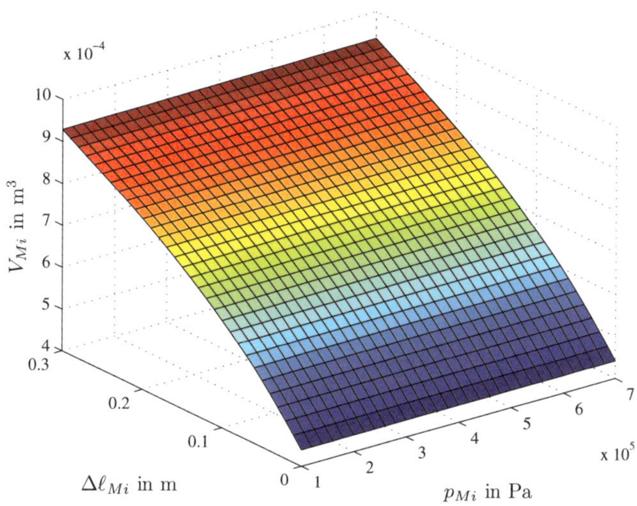

Figure 3. Identified volume characteristic of the pneumatic muscle.

The mass flow balance for the pneumatic muscle is governed by

$$\dot{\rho}_{Mi} \cdot V_{Mi} = \dot{m}_{Mi} - \rho_{Mi} \cdot \dot{V}_{Mi} \tag{8}$$

The identified volume characteristic (Fig. 3) of the pneumatic muscle can be described by a polynomial function of both contraction length $\Delta\ell_{Mi}$ and the muscle pressure p_{Mi}

$$V_{Mi}\left(\Delta\ell_{Mi}, p_{Mi}\right) = \sum_{j=0}^{3} a_j \cdot \Delta\ell_{Mi}^{j} \cdot \sum_{k=0}^{1} b_k \cdot p_{Mi}^{k}. \tag{9}$$

By inserting (7) and (9), the pressure dynamics (8) for the muscle i results in

$$\begin{aligned} \dot{p}_{Mi} &= \frac{n}{V_{Mi} + n \cdot \frac{\partial V_{Mi}}{\partial p_{Mi}} \cdot p_{Mi}} \left[u_{Mi} - \frac{\partial V_{Mi}}{\partial \Delta\ell_{Mi}} \cdot \frac{\mathrm{d}\Delta\ell_{Mi}}{\mathrm{d}z_C} \cdot p_{Mi} \cdot \dot{z}_C \right] \\ &= k_{ui}\left(\Delta\ell_{Mi}, p_{Mi}\right) u_{Mi} - k_{pi}\left(\Delta\ell_{Mi}, \Delta\dot{\ell}_{Mi}, p_{Mi}\right) p_{Mi}, \end{aligned} \tag{10}$$

where $u_{Mi} = R_L \cdot T_{Mi} \cdot \dot{m}_{Mi}$ denotes the input variable. The internal temperature T_{Mi} can be approximated with good accuracy by the constant temperature T_{amb} of the ambiance. In this way, temperature measurements are avoided, and the implementational effort is significantly reduced.

The force characteristic $F_{Mi}\left(p_{Mi}, \Delta\ell_{Mi}\right)$ of a pneumatic muscle states the resulting tension force for given internal pressure p_{Mi} as well as given contraction length $\Delta\ell_{Mi}$ and represents the connection of the mechanical and the pneumatic system part. The nonlinear force characteristic (Fig. 4) has been identified by static measurements and, then, approximated by the following polynomial description

$$F_{Mi}(p_{Mi}, \Delta\ell_{Mi}) = \begin{cases} \bar{F}_{Mi}(p_{Mi}, \Delta\ell_{Mi}), & \bar{F}_{Mi} > 0 \\ 0 & , \text{ else} \end{cases}, \tag{11}$$

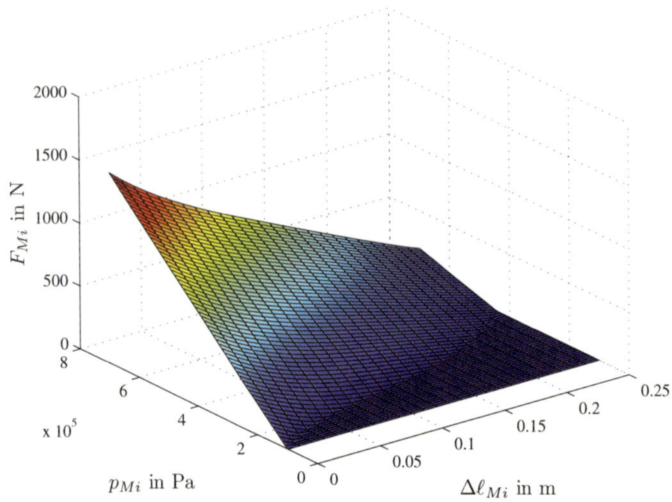

Figure 4. Identified force characteristic of the pneumatic muscle.

with

$$\bar{F}_{Mi}(p_{Mi}, \Delta l_{Mi}) = \underbrace{\sum_{m=0}^{3} \left(a_m \cdot \Delta\ell_{Mi}^m\right) p_{Mi}}_{f_{1i}} - \underbrace{\sum_{n=0}^{4} \left(b_n \cdot \Delta\ell_{Mi}^n\right)}_{f_{2i}} . \tag{12}$$

3. Control of the carriage position

The different sliding mode controllers for the carriage position are designed by exploiting the differential flatness property of the system under consideration (Fliess et al. (1995), Sira-Ramirez & Llanes-Santiago (2000)). For the mechanical system the carriage position z_C and the mean muscle pressure $p_M = 0.5\,(p_{Ml} + p_{Mr})$ are chosen as flat output candidates. The trajectory control of the mean pressure allows for increasing stiffness concerning disturbance forces acting on the carriage (Bindel et al. (1999)). As the inner controls have been assigned a high bandwidth, these underlying controlled muscle pressures can be considered as ideal control inputs of the outer control

$$\boldsymbol{u} = \begin{bmatrix} u_l \\ u_r \end{bmatrix} = \begin{bmatrix} p_{Ml} \\ p_{Mr} \end{bmatrix} . \tag{13}$$

Subsequent differentiation of the first flat output candidate until one of the control inputs appears leads to

$$y_1 = z_C , \tag{14a}$$

$$\dot{y}_1 = \dot{z}_C , \tag{14b}$$

$$\ddot{y}_1 = \frac{a_M}{k \cdot m}\left(F_{Mr} - F_{Ml}\right) - \frac{1}{m} F_U = \ddot{z}_C\left(z_C, \dot{z}_C, p_{Ml}, p_{Mr}, F_U\right) , \tag{14c}$$

whereas the second variable directly depends on the control inputs

$$y_2 = p_M = 0.5\,(p_{Ml} + p_{Mr}) . \tag{15}$$

The disturbance force F_U is estimated by a disturbance observer and used for disturbance compensation. Due to the differential flatness of the system, the inverse dynamics can be obtained by solving the equations (14) and (15) for the input variables

$$\boldsymbol{u} = \frac{1}{a_M\left(f_{1l}+f_{1r}\right)}\begin{bmatrix} a_M f_{2l} - a_M f_{2r} - km\ddot{z}_C - kF_U + 2a_M p_M f_{1r} \\ a_M f_{2r} - a_M f_{2l} + km\ddot{z}_C + kF_U + 2a_M p_M f_{1l} \end{bmatrix}. \tag{16}$$

3.1 Sliding mode control

Now, the tracking error $e_z = z_{Cd} - z_C$ can be stabilised by sliding mode control. For this purpose, the following sliding surface s_z is defined for the outer control loop in the form

$$s_z = \dot{z}_{Cd} - \dot{z}_C + \alpha\left(z_{Cd} - z_C\right). \tag{17}$$

At this, the coefficient α must be chosen positive in order to obtain a Hurwitz-polynomial. The convergence to the sliding surfaces in face of model uncertainty can be achieved by specifying a discontinuous signum-function

$$\dot{s}_z = -W_z \cdot sign(s_z), \;\; W_z > 0. \tag{18}$$

With a properly chosen positive coefficient W_z dominating the corresponding model uncertainties, the sliding surface $s_z = 0$ is reached in finite time depending on the initial conditions. This leads to the stabilising control law for each crank angle

$$v_z = \ddot{q}_{id} + \alpha \cdot \left(\dot{z}_{Cd} - \dot{z}_C\right) + W_z \cdot sign(s_z). \tag{19}$$

Here, the carriage position z_C, the carriage velocity $\dot{z}_C$, the desired trajectory for the carriage position z_{Cd} and their first two time derivatives have to be provided. For the second stabilising control input v_p, the desired trajectory for the mean pressure p_{Md} is directly utilised in a feedforward manner, i.e., $v_p = p_{Md}$. Inserting these new defined inputs into (16), the inverse dynamics becomes

$$\boldsymbol{u} = \frac{1}{a_M\left(f_{1l}+f_{1r}\right)}\begin{bmatrix} a_M f_{2l} - a_M f_{2r} - kmv_z - kF_U + 2a_M v_p f_{1r} \\ a_M f_{2r} - a_M f_{2l} + kmv_z + kF_U + 2a_M v_p f_{1l} \end{bmatrix}. \tag{20}$$

Having once reached the sliding surfaces, the final sliding mode is maintained during trajectory tracking provided that the tracking error $e_z = z_{Cd} - z_C$ is governed by an asymptotically stable first-order error dynamics

$$\dot{e}_z + \alpha \cdot e_z = 0. \tag{21}$$

Then, a globally asymptotically stable tracking of desired trajectories for the carriage position is guaranteed leading to

$$\lim_{t\to\infty} e_z(t) = 0. \tag{22}$$

For reduction of high frequency chattering the switching function $sign(s_z)$ in (19) can be replaced by the smooth function $\tanh\left(\frac{s_z}{\epsilon}\right)$, $\epsilon > 0$

$$v_z = \ddot{z}_{Cd} + \alpha \cdot \left(\dot{z}_{Cd} - \dot{z}_C\right) + W_z \cdot \tanh\left(\frac{s_z}{\epsilon}\right). \tag{23}$$

This regularisation, however, implicates a non-ideal sliding mode within a resulting boundary layer determined by the parameter ϵ in the switching function.

3.2 Higher-order sliding mode control

An alternative method to reduce high frequency chattering effects is to employ higher-order sliding mode techniques for control design, Levant (2008). For this approach the control derivative is considered as a new control input. Containing an integrator in the dynamic feedback law, real discontinuities in the control input are avoided at higher-order sliding mode. In this contribution a quasi-continuous second-order sliding mode controller as proposed in Levant (2005) is utilised. Then the tracking error is stabilised by the following control law

$$v_z = \alpha \frac{\dot{s}_z + \beta \left|s_z\right|^{\frac{1}{2}} sign\left(s_z\right)}{\left|\dot{s}_z\right| + \beta \left|s\right|^{\frac{1}{2}}} . \tag{24}$$

In Pukdeboon et al. (2010) a slightly modified version of this controller is introduced. For a reduction of the chattering phenomena, a small positive scalar ν is added to the denominator of (24). Then the smoothed control law is given by

$$v_z = \alpha \frac{\dot{s}_z + \beta \left|s_z\right|^{\frac{1}{2}} sign\left(s_z\right)}{\left|\dot{s}_z\right| + \beta \left|s\right|^{\frac{1}{2}} + \nu} . \tag{25}$$

For further reduction of the chattering phenomena, similar to the first-order sliding mode control law (23) the discontinuous function $sign\left(s_z\right)$ in (25) can be replaced by the smooth function $\tanh\left(\frac{s_z}{\epsilon}\right)$, $\epsilon > 0$. Again, the new control input v_z has to be inserted in the inverse dynamics (16), at which the second control input v_p remains the same.

3.3 Proxy-based sliding mode control

Proxy-based sliding mode control is a modification of sliding mode control as well as an extension of PID-control, see Kikuuwe & Fujimoto (2006), Van-Damme et al. (2007). The basic idea is to introduce a virtual carriage, called proxy, which is controlled using sliding mode techniques, whereas the proxy is connected to the real carriage by a PID-type coupling force, see Fig. 5. The goal of proxy-based sliding mode is to achieve precise tracking during normal operation and smooth, overdamped recovery in case of large position errors. The sliding mode control law for the virtual carriage results from equation (19) with z_S denoting the carriage position of the proxy

$$v_a = \ddot{z}_{Cd} + \alpha \cdot \left(\dot{z}_{Cd} - \dot{z}_S\right) + W_z \cdot tanh\left(\frac{\dot{z}_{Cd} - \dot{z}_s + \alpha\left(z_{Cd} - z_S\right)}{\epsilon}\right) . \tag{26}$$

The PID-type virtual coupling between the proxy and the real carriage is given by

$$v_c = K_I \int \left(z_S - z_C\right) \mathrm{d}t + K_P\left(z_S - z_C\right) + K_D\left(\dot{z}_s - \dot{z}_C\right) . \tag{27}$$

Assuming a proxy with vanishing mass, the condition $v_a = v_c$ holds. By introducing the new variable a as integrated difference between the real and the virtual carriage position $a = \int \left(z_S - z_C\right) \mathrm{d}t$, the virtual coupling (27) and the stabilising proxy-based sliding mode control law (26) result in (Kikuuwe & Fujimoto (2006))

$$v_c = K_I a + K_P \dot{a} + K_D \ddot{a} \, , \tag{28}$$

$$v_a = \ddot{z}_{Cd} + \alpha \dot{e}_z - \alpha \ddot{a} + W_z tanh\left(\frac{\dot{e}_z + \alpha e_z - \alpha \dot{a} - \ddot{a}}{\epsilon}\right) . \tag{29}$$

The implementation of the control law is shown in the right part of Fig. 5.

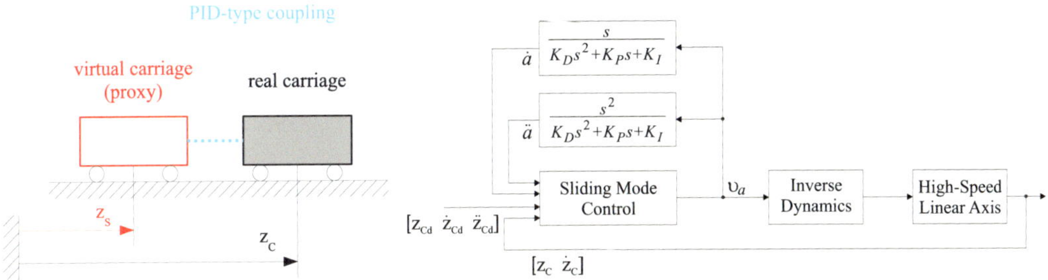

Figure 5. Coupling between virtual and real carriage (left). Implementation of the proxy-based sliding mode control (right).

4. Control of internal muscle pressure

The internal pressures of the pneumatic muscles are controlled separately with high accuracy in fast underlying control loops. The pneumatic subsystem represents a differentially flat system with the internal muscle pressure as flat output, see Aschemann & Schindele (2008). Hence, equation (10) can be solved for the input variable

$$u_{Mi} = \frac{1}{k_{ui}\left(\Delta\ell_{Mi}, p_{Mi}\right)}\left[\dot{p}_{Mi} + k_{pi}\left(\Delta\ell_{Mi}, \Delta\dot{\ell}_{Mi}, p_{Mi}\right) p_{Mi}\right] . \tag{30}$$

The contraction length $\Delta\ell_{Mi}$ as well as its time derivative $\Delta\dot{\ell}_{Mi}$ can be considered as scheduling parameters in a gain-scheduled adaptation of k_{ui} and k_{pi}. With the internal pressure as flat output, its first time derivative $\dot{p}_{Mi} = v_i$ is introduced as new control input. The error dynamics of each muscle pressure $p_{Mi}, i = \{l, r\}$, can be asymptotically stabilised by the following control law

$$v_i = \dot{p}_{Mid} + a_i \cdot \left(p_{Mid} - p_{Mi}\right) , \tag{31}$$

where the constant a_i is determined by pole placement. By introducing the definition $e_i = p_{Mid} - p_{Mi}$ for the control error w.r.t. the internal muscle pressure, the corresponding error dynamics is governed by the following first order differential equation

$$\dot{e}_i + a_i \cdot \dot{e}_i = 0 . \tag{32}$$

5. Feedforward friction compensation

The main part of the friction is considered by a dynamical friction model in a feedforward manner. For this purpose, the LuGre friction model, introduced by de Wit et al. (1995), is employed. This friction model is capable of describing the Stribeck effect, hysteresis, stick-slip limit cycling, presliding displacement as well as rising static friction

$$\dot{z} = \dot{z}_{Cd} - \frac{|\dot{z}_{Cd}|}{g\left(\dot{z}_{Cd}\right)} z , \tag{33}$$

$$F_{Fr} = \sigma_0 z + \sigma_1 \dot{z} + \sigma_2 \dot{z}_{Cd} , \tag{34}$$

where the function $g\left(\dot{z}_{Cd}\right)$ is given by

$$g\left(\dot{z}_{Cd}\right) = F_C + \left(F_S - F_C\right) e^{-\left(\frac{\dot{z}_{Cd}}{v_S}\right)^2} . \tag{35}$$

Here, the internal state variable z describes the deflection of the contact surfaces. The model parameters are given by the static friction F_S, the Coulomb friction F_C and the Stribeck velocity v_S. The parameter σ_0 is the stiffness coefficient, σ_1 the damping coefficient and σ_2 the viscous friction coefficient. All parameters have been identified using nonlinear least square techniques.

6. Reduced nonlinear disturbance observer

Disturbance behaviour and tracking accuracy in view of model uncertainties can be significantly improved by introducing a compensating control action provided by a nonlinear reduced-order disturbance observer as described in Friedland (1996). The observer design is based on the equation of motion. The key idea for the observer design is to extend the state equation with integrators as disturbance models

$$\begin{aligned} \dot{\mathbf{y}} &= \mathbf{f}\left(\mathbf{y}, F_U, \mathbf{u}\right) \,, \\ \dot{F}_U &= 0 \,, \end{aligned} \tag{36}$$

where $\mathbf{y} = \begin{bmatrix} \mathbf{q} \ \dot{\mathbf{q}} \end{bmatrix}^T$ denotes the measurable state vector. The estimated disturbance force $\hat{F}_U$ is obtained from $\hat{F}_U = \mathbf{h}^T\mathbf{y} + z$ with the chosen observer gain vector $\mathbf{h}^T$.

$$\mathbf{h}^T = \begin{bmatrix} h_1 \ h_1 \end{bmatrix} \,. \tag{37}$$

The state equation for z is given by

$$\dot{z} = \Phi\left(\mathbf{y}, \hat{F}_U, \mathbf{u}\right) \,. \tag{38}$$

The observer gain vector $\mathbf{h}$ and the nonlinear function Φ have to be chosen such that the steady-state observer error $e = F_U - \hat{F}_U$ converges to zero. Thus, the function Φ can be determined as follows

$$\dot{e} = 0 = \dot{F}_U - h^T\mathbf{f}\left(\mathbf{y}, \hat{F}_U, \mathbf{u}\right) - \Phi\left(\mathbf{y}, F_U, \mathbf{u}\right) \,. \tag{39}$$

In view of $\dot{F}_U = 0$, equation (39) yields

$$\Phi\left(\mathbf{y}, F_U, \mathbf{u}\right) = -\mathbf{h}^T\mathbf{f}\left(\mathbf{y}, \hat{F}_U\mathbf{u}\right) \,. \tag{40}$$

The linearised error dynamics $\dot{e}$ has to be made asymptotically stable. Accordingly, all eigenvalues of the Jacobian

$$\mathbf{J}_e = \frac{\partial\Phi\left(\mathbf{y}, F_U, \mathbf{u}\right)}{\partial F_U} \tag{41}$$

must be located in the left complex half-plane. This can be achieved by proper choice of the observer gain h_1. The stability of the closed-loop control system has been investigated by thorough simulations.

7. Control implementation

For the implementation at the test rig the control structure as depicted in Fig. 6 has been used. Fast underlying pressure control loops achieve an accurate tracking behaviour for the desired pressures stemming from the outer control loop. The nonlinear valve characteristic (VC) has been identified by measurements, see Aschemann & Schindele (2008), and is compensated by

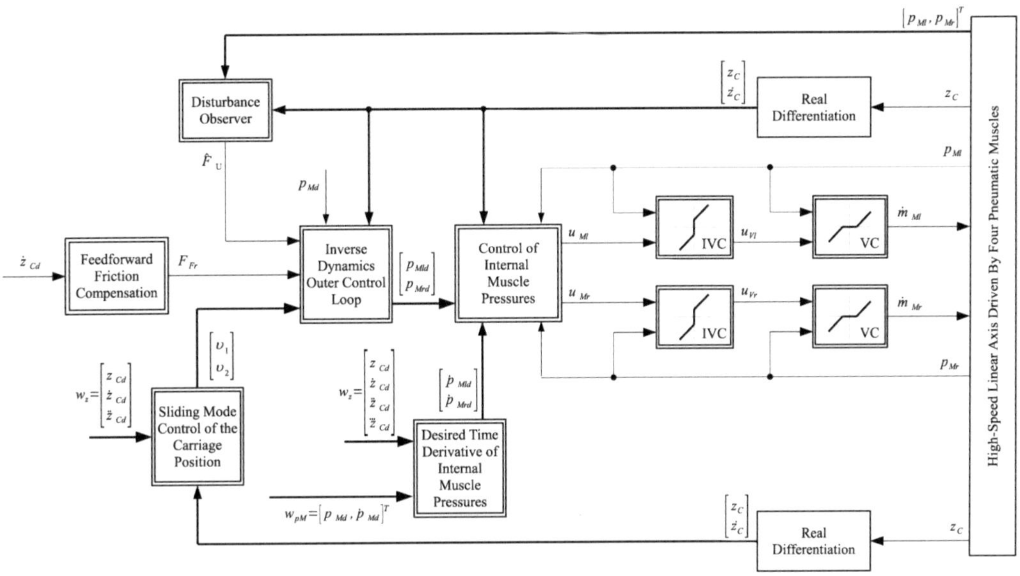

Figure 6. Implementation of the cascaded control structure.

its approximated inverse valve characteristic (IVC) in each input channel. For each pulley tackle one pneumatic muscle is equipped with a piezo-resistive pressure sensor mounted at the connection flange that connects the muscle with the connection plate. The carriage position z_C is obtained by a linear incremental encoder providing high resolution. The carriage velocity $\dot{z}_C$ is derived from the carriage position z_C by means of real differentiation using a DT_1-System with the corresponding transfer function $G_{DT1}(s) = \frac{s}{T_1 s+1}$. The desired value for the time derivative of the internal muscle pressure can be obtained either by real differentiation of the corresponding control input p_{Mi} in (16) or by model-based calculation using only desired values, i.e.

$$\dot{p}_{Mid} = \dot{p}_{Mid}\left(z_{Cd}, \dot{z}_{Cd}, \ddot{z}_{Cd}, \dddot{z}_{Cd}, p_{Md}, \dot{p}_{Md}, \hat{F}_U, \dot{\hat{F}}_U\right). \tag{42}$$

The corresponding desired trajectories are obtained from a trajectory planning module that provides synchronous time optimal trajectories according to given kinematic and dynamic constraints. It becomes obvious that a continuous time derivative $\dot{p}_{Mid}$ requires a three times continuously differentiable desired carriage trajectory. In (42) the time derivative of $\hat{F}_U$ is needed. Considering equation (38) and the first time derivatives of the system states, the value of $\dot{\hat{F}}_U$ can be obtained as follows

$$\dot{\hat{F}}_U = \mathbf{h}^T \dot{\mathbf{y}} + \dot{z}. \tag{43}$$

8. Experimental results

Both tracking performance and steady-state accuracy w.r.t. the carriage position z_C have been investigated by experiments at the test rig of the Chair of Mechatronics, University of Rostock. It is equipped with four pneumatic muscles DMSP-20 from FESTO AG. The control algorithm has been implemented on a dSpace real time system. For the experiments the trajectory shown in Fig. 7 have been used. Here the desired carriage position varies in an interval between

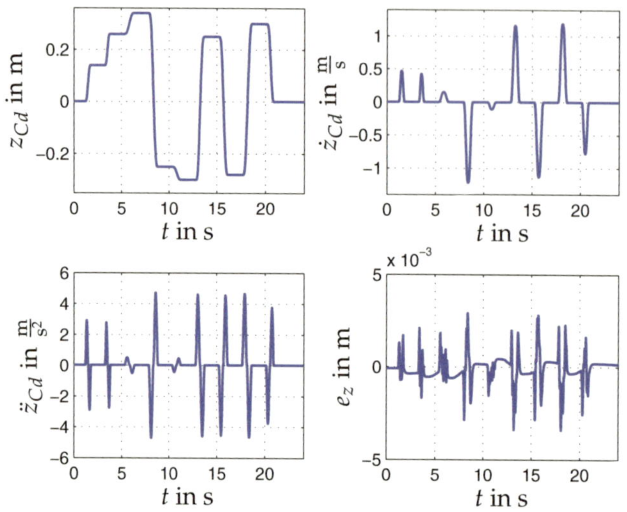

Figure 7. Desired values for the carriage position, velocity, and acceleration. Corresponding control error $e_z = z_{Cd} - z_C$ for standard sliding mode control.

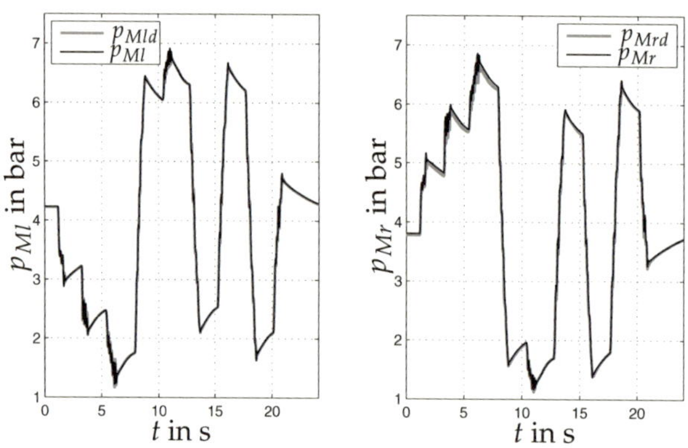

Figure 8. Comparison of desired and actual values for the left and right muscle pressure.

−0.35 m and 0.35 m. The maximum velocities are approximately 1.3 m/s and the maximum accelerations are about 5 m/s^2. The resulting tracking errors for the carriage $e_z = z_{Cd} - z_C$ are shown in the right lower part of Fig. 7. As for the carriage position, the maximum tracking error during the acceleration and deceleration intervals is approximately 3.5 mm. The maximum steady-state error is approximately 0.6 mm. Fig. 8 shows the corresponding desired and actual values of the internal muscle pressure. Obviously, the underlying fast control loops achieve a precise tracking of the desired values, which stem from the outer decoupling control loop. Due to a time-optimal trajectory planning using desired ansatzfunctions with limited jerk as described in Aschemann & Hofer (2005), the admissible range of the internal muscle pressure is not exceeded. In Fig. 9 the different control approaches, introduced in this contribution, are compared concerning the control error e_z. The higher-order sliding mode (HOSM) control approach results in a slightly larger maximum tracking error than

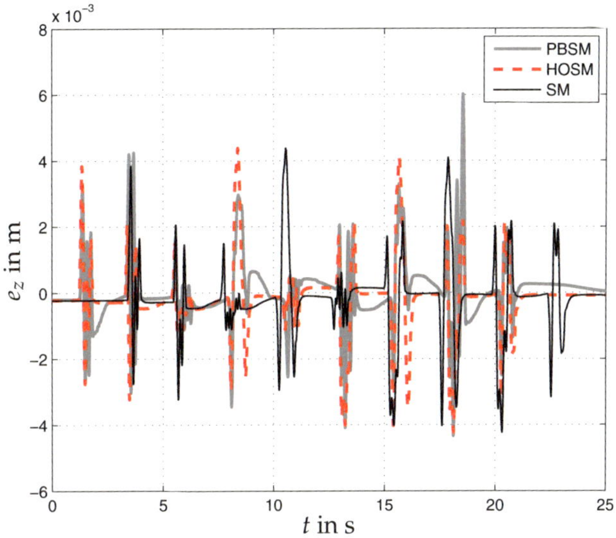

Figure 9. Comparison of different control approaches concerning the corresponding control errror e_z: Proxy-based sliding mode control (PBSM), Higher-order sliding mode control (HOSM) and standard sliding mode control (SM).

with the standard sliding mode technique (SM). Nevertheless, the steady-state accuracy of the HOSM approach is superior to the other approaches. As the chattering phenomena is reduced by HOSM control the parameter ϵ in equation (25) can be chosen very small, so that the hyperbolic tangent function is very close to the ideal switching-function. The parameter ϵ in (23) have to be chosen about 100 times larger as compared to the value in HOSM, to avoid the high-frequency chattering, which is critical for the proportional valves and results in a reduced lifetime of the valves. The largest tracking errors occur with proxy-based sliding mode (PBSM) control, which represents a PID-controller at normal operation. The benefits of the PBSM control are its high robustness and its slow and safe recovery from unexpected disturbances and abnormal events, which leads to an inherent safety property. In Fig. 10 the impact of the feedforward friction compensation and the nonlinear reduced disturbance observer is demonstrated. Here the tracking errors of SM control with feedforward friction compensation (f.f.c.) and disturbance observer (d.o.), SM control only with f.f.c and SM control without f.f.c. and d.o. are depicted. As can be seen the tracking errors can be significantly reduced by employing the proposed disturbance compensation strategy. The sum of the feedforward friction force F_{Fr} and the disturbance force estimated by the disturbance observer $\hat{F}_U$ is depicted in Fig. 11. The robustness of the proposed solution is shown by a unmodelled additional mass of 25 kg, which represents almost the double of the nominal value. In the corresponding force, the increase due to the higher inertial forces becomes obvious. The corresponding tracking errors are shown in Fig. 12. All three control approaches show similar results. Whereas the steady-state errors remain almost unchanged, the maximum tracking errors are now approximately 8 mm due to the inertia forces during the acceleration and deceleration phases. The closed-loop stability is not affected by this parametric uncertainty.

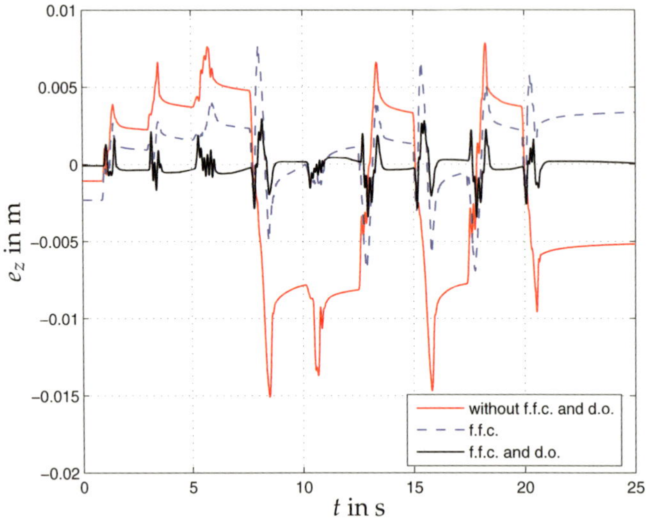

Figure 10. Tracking errors of SM control without disturbance compensation, SM control with feedforward friction compensation (f.f.c.) and SM control with f.f.c. and disturbance observer (d.o.).

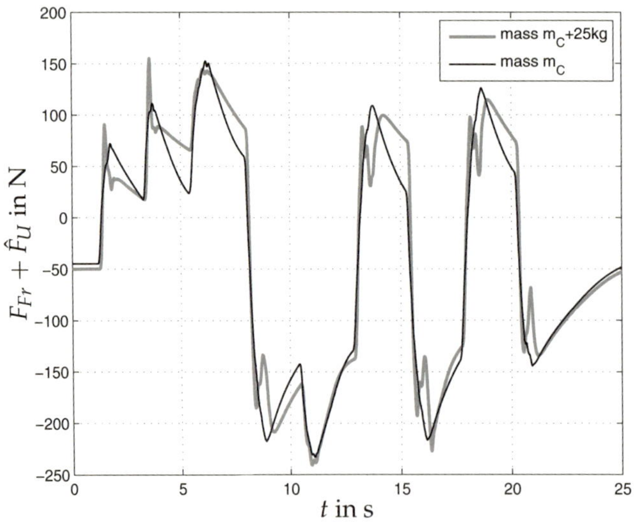

Figure 11. Estimated disturbance force with and without additional mass of 25 kg.

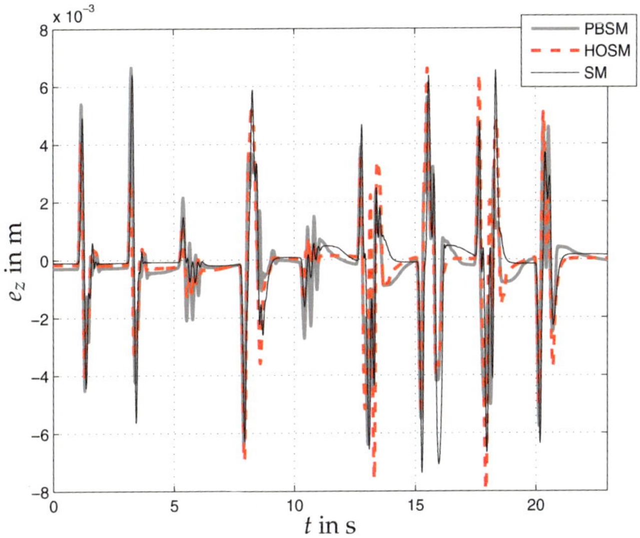

Figure 12. Tracking errors with an additional mass of 25 kg.

9. Conclusions

In this paper, a nonlinear cascaded trajectory control was presented for a new linear axis driven by pneumatic muscles that offers a significant increase in both workspace and maximum velocity as compared to a directly actuated solution. Furthermore, the proposed setup requires a relativ small overall size in comparison to a drive concept with an rocker as in Aschemann & Schindele (2008). The modelling of this mechatronic system leads to nonlinear system equations of fourth order containing identified polynomial descriptions of the main nonlinearities of the pneumatic subsystem: the characteristic of the pneumatic valve and the characteristics of the pneumatic muscle. The inner control loops of the cascade involve a decentralised control of the internal muscle pressures with high bandwidth. For the outer control loop different sliding mode control approaches have been investigated leading to a decoupling of the carriage position and the mean pressure as controlled variables. Thereby, critical high frequency chattering can be avoided either by a regularisation of the switching function or by using a second-order sliding mode controller. Model uncertainties in the muscle force characteristic as well as nonlinear friction are directly taken into account by a compensation scheme consisting of a feedforward friction compensation and a nonlinear reduced disturbance observer. Experimental results emphasise the excellent closed-loop performance with maximum position errors of approximately 4 mm. The robustness of the proposed control is shown by measurements with an almost doubled carriage mass.

10. References

Aschemann, H. & Hofer, E. (2004). Flatness-based trajectory control of a pneumatically driven carriage with uncertainties, *Proceedings of NOLCOS 2004, Stuttgart, Germany* pp. 239–244.

Aschemann, H. & Hofer, E. (2005). Flatness-based trajectory planning and control of a parallel robot actuated by pneumatic muscles, *CD-Proceedings of ECCOMAS Thematic Conf. Multibody Dyn., Madrid, Spain* .

Aschemann, H. & Schindele, D. (2008). Sliding-mode control of a high-speed linear axis driven by pneumatic muscle actuators, *IEEE Trans. Ind. Electronics* 55(11): 3855–3864.

Aschemann, H., Schindele, D. & Hofer, E. (2006). Nonlinear optimal control of a mechatronic system with pneumatic muscle actuators, *CD-Proceedings of MMAR 2006, Miedzyzdroje, Poland* .

Bindel, R., Nitsche, R., Rothfuß, R. & Zeitz, M. (1999). Flatness based control of two valve hydraulic joint actuator of a large manipulator, *CD-Proceedings of ECC 1999, Karlsruhe, Germany* .

de Wit, C. C., Olsson, H., Åström, K. & Lischinsky, P. (1995). A new model for control of systems with friction, *IEEE Transactions on Automatic Control* 40(3): 419–425.

Fliess, M., Levine, J., Martin, P. & Rouchon, P. (1995). Flatness and defect of nonlinear systems: Introductory theory and examples, *Int. J. Control* 61: 1327–1361.

Friedland, B. (1996). *Advanced Control System Design*, Prentice-Hall.

Kikuuwe, R. & Fujimoto, H. (2006). Proxy-based sliding mode control for accurate and safe position control, *IEEE Trans. on Industr. Electr.* 53(5): 25–30.

Levant, A. (2005). Quasi-continuous high-order sliding-mode controllers, *IEEE Transactions on Automatic Control* 50(11): 1812–1816.

Levant, A. (2008). Homogeneous high-order sliding modes, *Proceedings of the 17th IFAC World Congress, Seoul, Korea* pp. 3799–3810.

Lilly, J. & Yang, L. (2005). Sliding mode control tracking for pneumatic muscle actuators in opposing pair configuration, *IEEE Trans. on Contr. Syst. Techn.* 13(4): 550–558.

Pukdeboon, C., Zinober, A. S. I. & Thein, M.-W. L. (2010). Quasi-continuous higher order sliding-mode controllers for spacecraft-attitude-tracking maneuvers, *IEEE Transactions on Industrial Electronics* 57(4): 1436–1444.

Schindele, D. & Aschemann, H. (2010). Norm-optimal iterative learning control for a high-speed linear axis with pneumatic muscles, *Proc. of NOLCOS 2010, Bologna, Italy.* to be published.

Sira-Ramirez, H. & Llanes-Santiago, O. (2000). Sliding mode control of nonlinear mechanical vibrations, *J. of Dyn. Systems, Meas. and Control* 122(12): 674–678.

Smith, J., Ness, H. V. & Abott, M. M. (1996). *Introduction to Chemical Engineering Thermodynamics*, McGraw-Hill, New York.

Van-Damme, M., Vanderborght, R., Ham, R. V., Verrelst, B., Daerden, F. & Lefeber, D. (2007). Proxy-based sliding-mode control of a manipulator actuated by pleated pneumatic artificial muscles, *Proc. IEEE Int. Conf. on Robotics and Automation, Rome, Italy* pp. 4355–4360.

Zhu, X., Tao, G., Yao, B. & Cao, J. (2008). Adaptive robust posture control of a parallel manipulator driven by pneumatic muscles, *Automatica* 44(9): 2248–2257.

Adaptive Sliding Mode Control of Adhesion Force in Railway Rolling Stocks

Jong Shik Kim, Sung Hwan Park,
Jeong Ju Choi and Hiro-o Yamazaki
School of Mechanical Engineering, Pusan National University
Republic of Korea

1. Introduction

Studies of braking mechanisms of railway rolling stocks focus on the adhesion force, which is the tractive friction force that occurs between the rail and the wheel (Kadowaki, 2004). During braking, the wheel always slips on the rail. The adhesion force increases or decreases according to the slip ratio, which is the difference between the velocity of the rolling stocks and the tangential velocity of each wheel of the rolling stocks normalized with respect to the velocity of the rolling stocks. A nonzero slip ratio always occurs when the brake caliper holds the brake disk, and thus the tangential velocity of the wheel so that the velocity of the wheel is lower than the velocity of the rolling stocks. Unless an automobile is skidding, the slip ratio for an automobile is always zero. In addition, the adhesion force decreases as the rail conditions change from dry to wet (Isaev, 1989). Furthermore, since it is impossible to directly measure the adhesion force, the characteristics of the adhesion force must be inferred based on experiments (Shirai, 1977).

To maximize the adhesion force, it is essential to operate at the slip ratio at which the adhesion force is maximized. In addition, the slip ratio must not exceed a specified value determined to prevent too much wheel slip. Therefore, it is necessary to characterize the adhesion force through precise modeling.

To estimate the adhesion force, observer techniques are applied (Ohishi, 1998). In addition, based on the estimated value, wheel-slip brake control systems are designed (Watanabe, 2001). However, these control systems do not consider uncertainty such as randomness in the adhesion force between the rail and the wheel. To address this problem, a reference slip ratio generation algorithm is developed by using a disturbance observer to determine the desired slip ratio for maximum adhesion force. Since uncertainty in the traveling resistance and the mass of the rolling stocks is not considered, the reference slip ratio, at which adhesion force is maximized, cannot always guarantee the desired wheel slip for good braking performance.

In this paper, two models are developed for the adhesion force in railway rolling stocks. The first model is a static model based on a beam model, which is typically used to model automobile tires. The second model is a dynamic model based on a bristle model, in which the friction interface between the rail and the wheel is modeled as contact between bristles (Canudas de Wit, 1995). The validity of the beam model and bristle model is verified through an adhesion test using a brake performance test rig.

We also develop wheel-slip brake control systems based on each friction model. One control system is a conventional PI control scheme, while the other is an adaptive sliding mode control (ASMC) scheme. The controller design process considers system uncertainties such as the traveling resistance, disturbance torque, and variation of the adhesion force according to the slip ratio and rail conditions. The mass of the rolling stocks is also considered as an uncertain parameter, and the adaptive law is based on Lyapunov stability theory. The performance and robustness of the PI and adaptive sliding mode wheel-slip brake control systems are evaluated through computer simulation.

2. Wheel-slip mechanism for rolling stocks

To reduce braking distance, automobiles are fitted with an anti-lock braking system (ABS) (Johansen, 2003). However, there is a relatively low adhesion force between the rail and the wheel in railway rolling stocks compared with automobiles. A wheel-slip control system, which is similar to the ABS for automobiles, is currently used in the brake system for railway rolling stocks.
The braking mechanism of the rolling stocks can be modeled by

$$F_a = \mu(\lambda)N \tag{1}$$

$$\lambda = \frac{v - r\omega}{v} \tag{2}$$

where F_a is the adhesion force, $\mu(\lambda)$ is the dimensionless adhesion coefficient, λ is the slip ratio, N is the normal force, v is the velocity of the rolling stocks, and ω and r are the angular velocity and radius of each wheel of the rolling stocks, respectively. The velocity of the rolling stocks can be measured (Basset, 1997) or estimated (Alvarez, 2005). The adhesion force F_a is the friction force that is orthogonal to the normal force. This force disturbs the motion of the rolling stocks desirably or undesirably according to the relative velocity between the rail and the wheel. The adhesion force F_a changes according to the variation of the adhesion coefficient $\mu(\lambda)$, which depends on the slip ratio λ, railway condition, axle load, and initial braking velocity, that is, the velocity at which the brake is applied. Figure 1 shows a typical shape of the adhesion coefficient $\mu(\lambda)$ according to the slip ratio λ and rail conditions.
To design a wheel-slip control system, it is useful to simplify the dynamics of the rolling stocks as a quarter model based on the assumption that the rolling stocks travel in the longitudinal direction without lateral motion, as shown in Fig. 2 the equations of motion for the quarter model of the rolling stocks can be expressed as

$$J\dot{\omega} = -B\omega + T_a - T_b - T_d \tag{3}$$

$$M\dot{v} = -F_a - F_r \tag{4}$$

where B is the viscous friction torque coefficient between the brake pad and the wheel, $T_a = rF_a$ and T_b are the adhesion and brake torques, respectively, T_d is the disturbance torque due to the vibration of the brake caliper, J and r are the inertia and radius, respectively, of each wheel of the rolling stocks, and M and F_r are the mass and traveling resistance force of the rolling stocks, respectively.

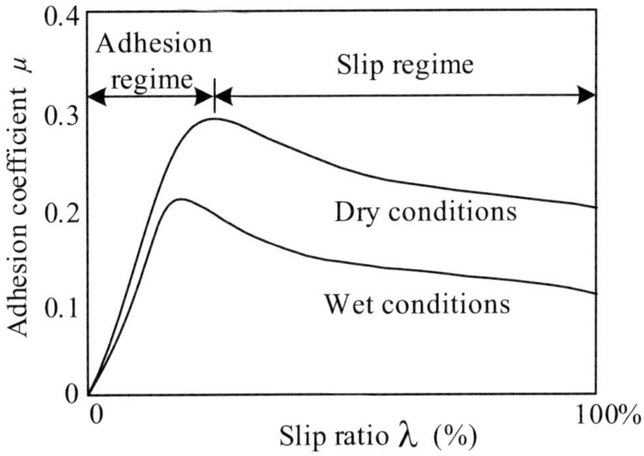

Fig. 1. Typical shape of the adhesion coefficient according to the slip ratio and rail conditions.

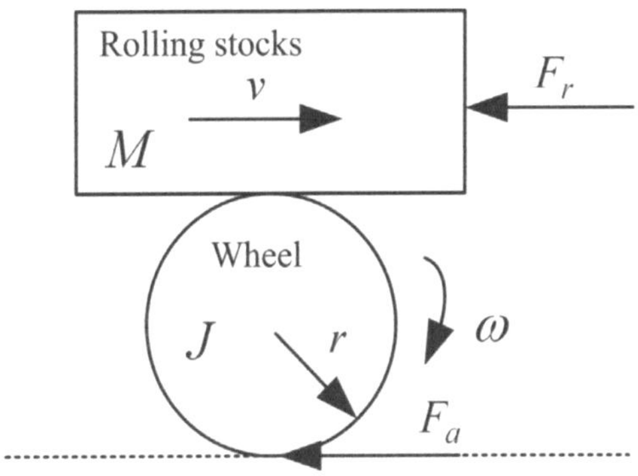

Fig. 2. Quarter model of the rolling stocks.

From (3) and (4), it can be seen that, in order to achieve sufficient adhesion force, a large brake torque T_b must be applied. When T_b is increased, however, the slip ratio increases, which causes the wheel to slip. When the wheel slips, it may develop a flat spot on the rolling surface. This flat spot affects the stability of the rolling stocks, the comfort of the passengers, and the life cycle of the rail and the wheel. To prevent this undesirable braking situation, a desired wheel-slip control is essential for the brake system of the rolling stocks.

In addition, the adhesion force between the wheel and the contact surface is dominated by the initial braking velocity, as well as by the mass M and railway conditions. In the case of automobiles, which have rubber pneumatic tires, the maximum adhesion coefficient changes from 0.4 to 1 according to the road conditions and the materials of the contact surface (Yi, 2002). In the case of railway rolling stocks, where the contact between the wheel and the rail is that of steel on steel, the maximum adhesion force coefficient changes from approximately 0.1 to 0.4 according to the railway conditions and the materials of the contact surface (Kumar, 1996). Therefore, railway rolling stocks and automobiles have significantly different adhesion force coefficients because of different materials for the rolling and contact surfaces. However, the brake characteristics of railway rolling stocks (Jin, 2004) and automobiles (Li, 2006) are similar.

According to adhesion theory, the maximum adhesion force occurs when the slip ratio is approximately between 0.1 and 0.4 in railway rolling stocks. Therefore, the slip ratio at which the maximum adhesion force is obtained is usually used as the reference slip ratio for the brake control system of the rolling stocks. Figure 3 shows an example of a wheel-slip control mechanism based on the relationship between the slip ratio and braking performance.

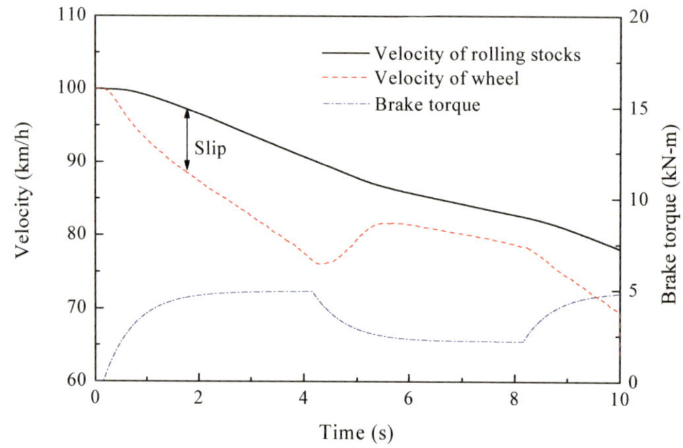

Fig. 3. Example of a wheel-slip control mechanism based on the relationship between the slip ratio and braking performance.

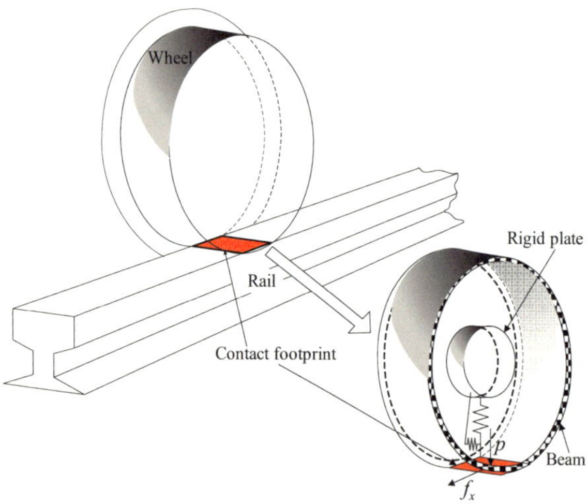

Fig. 4. Simplified contact model for the rail and wheel.

3. Static adhesion force model based on the beam model

To model the adhesion force as a function of the slip ratio, we consider the beam model, which reflects only the longitudinal adhesion force. Figure 4 shows a simplified contact model for the rail and wheel, where the beam model treats the wheel as a circular beam supported by springs. The contact footprint of an automobile tire is generally approximated as a rectangle by the beam model (Sakai, 1987). In a similar manner, the contact footprint between the rail and the wheel is approximated by a rectangle as shown in Fig. 5.

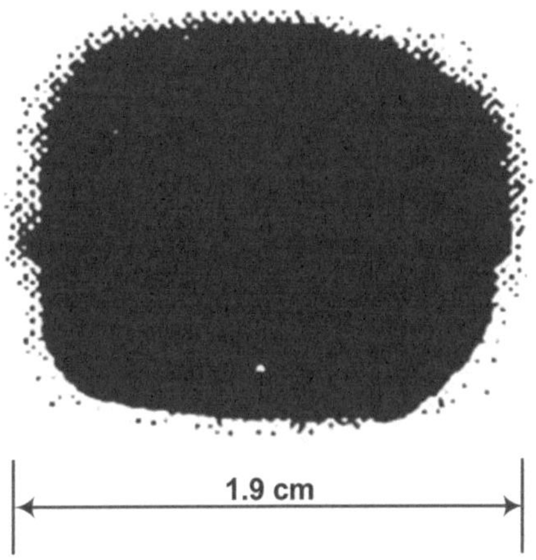

Fig. 5. Contact footprint between the rail and the wheel.

The contact pressure p between the rail and the wheel at the displacement x_c from the tip of the contact footprint in the longitudinal direction is given by (Sakai, 1987)

$$p = \frac{6N}{l^3 w}\left[\left(\frac{l}{2}\right)^2 - \left(x_c - \frac{l}{2}\right)^2\right] \tag{5}$$

where N is the normal force, and l and w are the length and width of the contact footprint, respectively. Figure 6 shows a typical distribution of the tangential force coefficient in a contact footprint (Kalker, 1989).

In Fig. 6, the variable f_x, which is the derivative of the adhesion force F_a with respect to the displacement x_c from the tip of the contact footprint, is given by

$$f_x = \begin{cases} C_x \lambda w x_c & for \;\; 0 \le x_c \le l_h, \\ \mu_d p & for \;\; l_h < x_c \le l, \end{cases} \tag{6}$$

where C_x is the modulus of transverse elasticity, l_h is the displacement from the tip of the contact footprint at which the adhesion-force derivative f_x changes rapidly, and μ_d is the dynamic friction coefficient. In particular, μ_d is defined by

$$\mu_\text{d} = \mu_\text{max} - \frac{a\lambda v l}{(l - l_h)} \tag{7}$$

where μ_max is the maximum adhesion coefficient, a is a constant that determines the dynamic friction coefficient in the slipping regime, and l_h is expressed as (Sakai, 1987)

$$l_h = l\left(1 - \frac{K_x \lambda}{3\mu_\text{max} N}\right) \tag{8}$$

where K_x is the traveling stiffness calculated by

$$K_x = \frac{1}{2} C_x l^2 \tag{9}$$

The wheel load, which is the normal force, is equal to the integrated value of the contact pressure between the rail and the wheel over the contact footprint. Therefore, the adhesion force F_a between the rail and the wheel can be calculated by integrating (6) over the length of the contact footprint and substituting (7) and (8) into (6), which is expressed as

$$\begin{aligned} F_a = & \frac{1}{2} C_x \lambda w l^2 \left(1 - \frac{K_x \lambda}{3\mu_{\max} N}\right)^2 + \frac{1}{2} K_x \lambda - \frac{3}{2} Na(v - r\omega) \\ & - \frac{1}{2} \mu_{\max} N \left[1 - 3\frac{Na(v - r\omega)}{K_x \lambda}\right]\left[1 - \left(1 - \frac{2K_x \lambda}{3\mu_{\max} N}\right)^3\right]. \end{aligned} \tag{10}$$

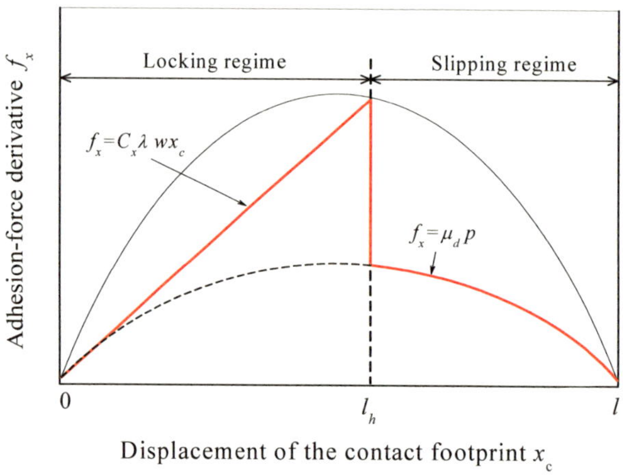

Fig. 6. A typical distribution of the tangential force coefficient in a contact footprint.

4. Dynamic adhesion force model based on bristle contact

As a dynamic adhesion force model, we consider the Dahl model given by (Dahl, 1976)

$$\frac{dz}{dt} = \sigma - \frac{\alpha|\sigma|}{F_c} z \tag{11}$$

$$F = \alpha z \tag{12}$$

where z is the internal friction state, σ is the relative velocity, α is the stiffness coefficient, and F and F_c are the friction force and Coulomb friction force, respectively. Since the steady-state version of the Dahl model is equivalent to Coulomb friction, the Dahl model is a generalized model for Coulomb friction. However, the Dahl model does not capture either the Stribeck effect or stick-slip effects. In fact, the friction behavior of the adhesion force according to the relative velocity σ for railway rolling stocks exhibits the Stribeck effect, as shown in Fig. 7. Therefore the Dahl model is not suitable as an adhesion force model for railway rolling stocks.

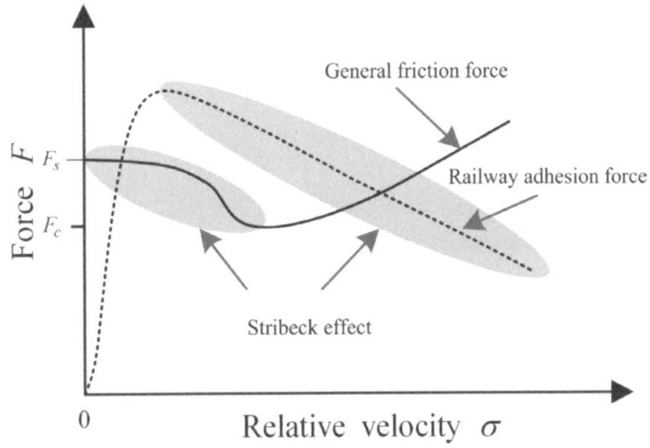

Fig. 7. Typical shape of the general friction force and adhesion force in railway rolling stocks according to the relative velocity.

However, the LuGre model (Canudas de Wit, 1995), which is a generalized form of the Dahl model, can describe both the Stribeck effect and stick-slip effects. The LuGre model equations are given by

$$\frac{dz}{dt} = \sigma - \frac{\alpha|\sigma|}{g(\sigma)} z \tag{13}$$

$$g(\sigma) = F_c + (F_s - F_c)e^{-(\sigma/v_s)^2} \tag{14}$$

$$F = \alpha z + \alpha_1 \dot{z} + \alpha_2 \sigma \tag{15}$$

where z is the average bristle deflection, v_s is the Stribeck velocity, and F_s is the static friction force. In addition, α, α_1, and α_2 are the bristle stiffness coefficient, bristle damping coefficient, and viscous damping coefficient, respectively.

The functions $g(\sigma)$ and F in (14) and (15) are determined by selecting the exponential term in (14) and coefficients α, α_1, and α_2 in (15), respectively, to match the mathematical model with the measured friction. For example, to match the mathematical model with the measured friction, the standard LuGre model is modified by using $e^{-|\sigma/v_s|^{\frac{1}{2}}}$ in place of the term $e^{-(\sigma/v_s)^2}$ in (14). Furthermore, for the tire model for vehicle traction control, the function F given by (15) is modified by including the normal force. Thus, (13)-(15) are modified as (Canudas de Wit, 1999)

$$\frac{dz}{dt} = \sigma - \frac{\alpha'|\sigma|}{g(\sigma)} z \tag{16}$$

$$g(\sigma) = \mu_c + (\mu_s - \mu_c)e^{-|\sigma/v_s|^{\frac{1}{2}}} \tag{17}$$

$$F = (\alpha' z + \alpha_1' \dot{z} + \alpha_2' \sigma) N \tag{18}$$

where μ_s and μ_c are the static friction coefficient and Coulomb friction coefficient, respectively, $N = mg$ is the normal force, m is the mass of the wheel, and $\alpha' = \frac{\alpha}{N}$, $\alpha_1' = \frac{\alpha_1}{N}$, and $\alpha_2' = \frac{\alpha_2}{N}$ are the normalized wheel longitudinal lumped stiffness coefficient, normalized wheel longitudinal lumped damping coefficient, and normalized viscous damping coefficient, respectively.

In general, it is difficult to measure and identify all six parameters, α, α_1, α_2, F_s, F_c, and v_s in the LuGre model equations. In particular, identifying friction coefficients such as α and α_1 requires a substantial amount of experimental data (Canudas de Wit, 1997). We thus develop a dynamic model for friction phenomena in railway rolling stocks, as shown in Fig. 7. The dynamic model retains the simplicity of the Dahl model while capturing the Stribeck effect.

As shown in Fig. 8 (Canudas de Wit, 1995), the motion of the bristles is assumed to be the stress-strain behavior in solid mechanics, which is expressed as

$$\frac{dF_a}{dx} = \alpha\left[1 - h(\sigma)F_a\right] \tag{19}$$

where F_a is the adhesion force, α is the coefficient of the dynamic adhesion force, and x and σ are the relative displacement and velocity of the contact surface, respectively. In addition, the function $h(\sigma)$ is selected according to the friction characteristics.

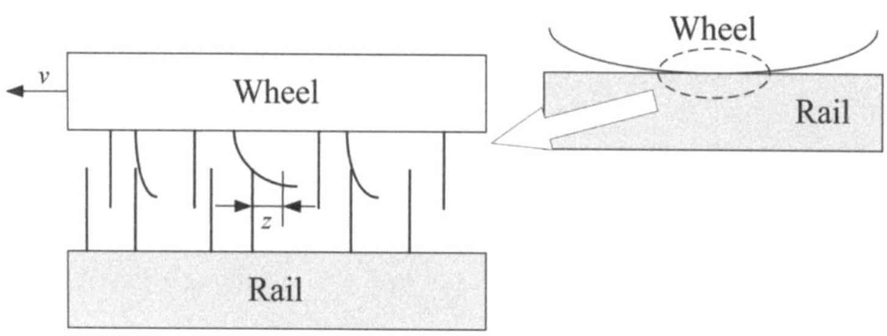

Fig. 8. Bristle model between the rail and the wheel.

Defining z to be the average deflection of the bristles, the adhesion force F_a is assumed to be given by

$$F_a = \alpha z \tag{20}$$

The derivative of F_a can then be expressed as

$$\frac{dF_a}{dt} = \frac{dF_a}{dx}\frac{dx}{dt} = \frac{dF_a}{dx}\sigma = \alpha\left[1 - h(\sigma)F_a\right]\sigma = \alpha\frac{dz}{dt} \tag{21}$$

It follows from (20) and (21) that the internal state z is given by

$$\dot{z} = \sigma\left[1 - h(\sigma)F_a\right] = \sigma\left[1 - \alpha h(\sigma) z\right]. \tag{22}$$

To select the function $h(\sigma)$ for railway rolling stocks, the term $e^{-\sigma/v_s}$ is used in place of $e^{-(\sigma/v_s)^2}$ in (14). This term is simplified by executing the Taylor series expansion for $e^{-\sigma/v_s}$ and by taking only the linear term $1-\frac{\sigma}{v_s}$. In addition, neglecting the coefficients α_1 and α_2 in (15) for simplicity yields

$$g(\sigma)=F_c+(F_s-F_c)\left(1-\frac{\sigma}{v_s}\right)=F_s-(F_s-F_c)\frac{\sigma}{v_s}. \tag{23}$$

By comparing (13) and (23) with (22) and by considering the relative velocity σ, which is positive in railway rolling stocks, $h(\sigma)$ in (22) can be derived as

$$h(\sigma)=\frac{\beta}{\gamma-\sigma} \tag{24}$$

where $\beta=\frac{1}{F_s-F_c}v_s$ and $\gamma=\frac{F_s}{F_s-F_c}v_s$. In general, β and γ are positive tuning parameters because F_s is larger than F_c as shown in Fig. 7 In the dynamic model, the parameter α is the coefficient for the starting point of the slip regime, where the adhesion force decreases according to the relative velocity, and the parameters β and γ are the coefficients for the slope and shift in the slip regime, respectively.

5. Verification of the adhesion force models

To verify the adhesion force models, experiments using a braking performance test rig in the Railway Technical Research Institute in Japan and computer simulations are carried out under various initial braking velocity conditions. Figure 9 shows the test rig for the braking performance test. The conceptual schematic diagram is shown in Fig. 10. This test rig consists of a main principal axle with a wheel for rolling stocks on a rail, flywheels, a main motor, a sub-axle with a wheel, and a brake disk. After accelerating to the target velocity by the main motor, the brake caliper applies a brake force to the wheel. The inertia of the flywheels plays the role of the inertia of the running railway rolling stocks.

Fig. 9. Test rig for the brake performance test.

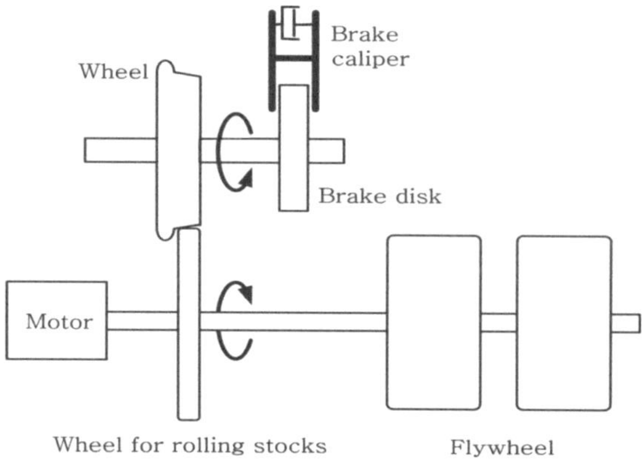

Fig. 10. Conceptual schematic diagram of the test rig for the brake performance test.

The test conditions are shown in Table 1. During the experiments, the brake torque T_b, the wheel load N, the angular velocity of the wheel ω, and the velocity of the rolling stocks v are measured simultaneously. The adhesion torque T_a between the rail and the wheel used in the calculation of the adhesion coefficient is also estimated in real time. As in the case of running vehicles, it is impossible to measure the adhesion torque directly on the brake performance test rig.

Test Condition	Value
Initial braking velocity	30, 60, 100, 140 km/h
Slip ratio	0 - 50%
Wheel load	34.5 kN
Wheel inertia	60.35 kg-m^2
Viscous friction torque coefficient	0.25 N-m-s

Table 1. Test conditions of the test rig for the brake performance test

It is essential that knowledge of the adhesion torque be available for both ABS in automobiles and wheel-slip control of rolling stocks. However, it is difficult to directly acquire this information. While an optical sensor, which is expensive (Basset, 1997), can be used to acquire this information, the adhesion force between the wheel and the rail is estimated through the application of a Kalman filter (Charles, 2006). By using this scheme, the adhesion force can be estimated online during the normal running of the vehicle before the brake is applied. A disturbance observer considering the first resonant frequency of the rolling stocks is designed in order to avoid undesirably large wheel slip, which causes damage to the rail and wheel (Shimizu, 2007). A sliding mode adhesion-force observer using the estimation error of the wheel angular velocity and based on a LuGre model can be used for this purpose (Patel, 2006).

We now consider an adhesion-torque observer for estimation. In (3), we neglect the unknown disturbance torque of the wheel T_d because the dominant disturbance torque caused by the vibration of the brake caliper acts only for a moment in the initial braking time. Then the adhesion torque T_a is expressed as

$$T_a = J\dot{\omega} + B\omega + T_b \tag{25}$$

Taking Laplace transforms yields

$$T_a(s) = Js\omega(s) + B\omega(s) + T_b(s) \tag{26}$$

Since a differential term is included in (26), we implement a first-order lowpass filter of the form

$$\hat{T}_a(s) = \frac{Js}{\tau s + 1}\omega(s) + B\omega(s) + T_b(s) \tag{27}$$

or

$$\hat{T}_a(s) = \left(B + \frac{J}{\tau} - \frac{J/\tau}{\tau s + 1} \right)\omega(s) + T_b(s) \tag{28}$$

where τ is the time constant of the lowpass filter in the adhesion-torque observer, which is illustrated in Fig. 11. The estimated adhesion coefficient $\hat{\mu}$ can now be obtained by

$$\hat{\mu} = \frac{\hat{T}_a}{Nr} \tag{29}$$

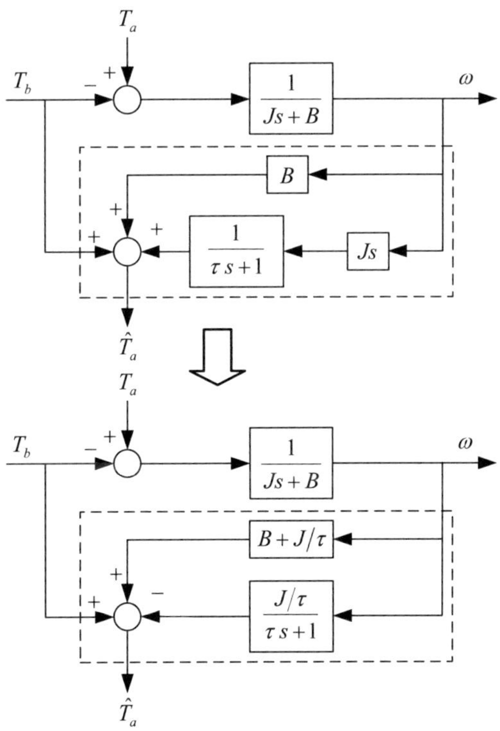

Fig. 11. Adhesion-torque observer.

As shown by the experimental wheel-slip results in Fig. 12, before 4.5 s, the velocity v of the rolling stocks matches the tangential velocity $v_w = r\omega$ of the wheel, where r and ω are the radius and angular velocity of the wheel, respectively, while a large difference occurs between the velocity of the rolling stocks and the tangential velocity of the wheel at 4.5 s when a large brake torque is applied. This difference means that large wheel slip occurs as a result of braking. The controller ceases the braking action at 6.1 s when the slip ratio exceeds 50%. Henceforth, the tangential velocity of the wheel recovers, and the slip ratio decreases to zero by the adhesion force between the rail and the wheel. In the experiment, to prevent damage due to excessive wheel slip, the applied brake torque is limited so that the slip ratio does not exceed 50%.

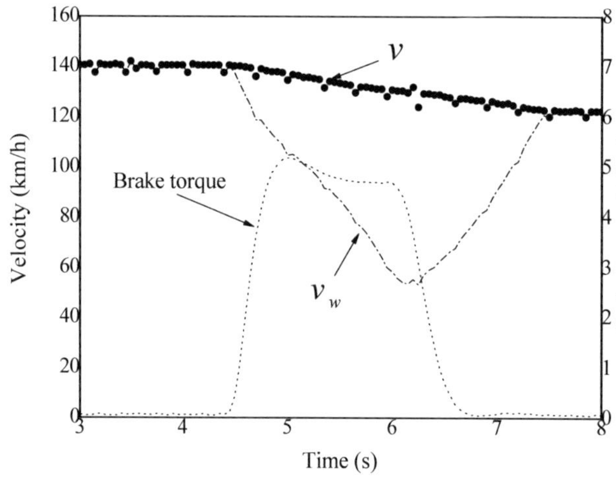

Fig. 12. Experimental wheel-slip results.

Table 2 shows the parameters of the adhesion force models for computer simulation. In Table 2, the parameter values for the length l and the width w of the contact footprint are taken from (Uchida, 2001). The constant a in (7) for the beam model is determined as 0.0013 h/km based on the adhesion experimental results at the initial braking velocity of 140 km/h.

Parameter	Notation	Value
Modulus of transverse elasticity	C_x	1.52×10^9 N/m^2
Length	l	0.019 m
Width	w	0.019 m
Wheel load	N	34.5 kN
Maximum adhesion coefficient for v_0 = 30, 60, 100, 140 km/h	μ_{max}	0.360, 0.310, 0.261, 0.226
Radius of the wheel	r	0.43 m

Table 2. Parameters of the beam and bristle models for computer simulation.

Figure 13 shows experimental and simulation results of the adhesion coefficient according to the slip ratio and initial braking velocity. As shown in Fig. 13, the variation of the adhesion coefficients obtained by the experiments is large. It is therefore difficult to determine a precise mathematical model for the adhesion force. In spite of these large variations, it is found that the experimental results of the mean value of the adhesion coefficient according to the slip ratio are consistent with the simulation results based on the two kinds of adhesion force models. Table 3 shows the mean values of the absolute errors between the experimental results for the mean value of the adhesion coefficient and the simulation results for the beam and bristle models according to the initial braking velocity of the rolling stocks. Mean values of the absolute errors in the relevant range of the initial braking velocity for the beam and bristle models are 0.011 and 0.0083, respectively. Using the bristle model in place of the beam model yields 24.5% improvement in accuracy.

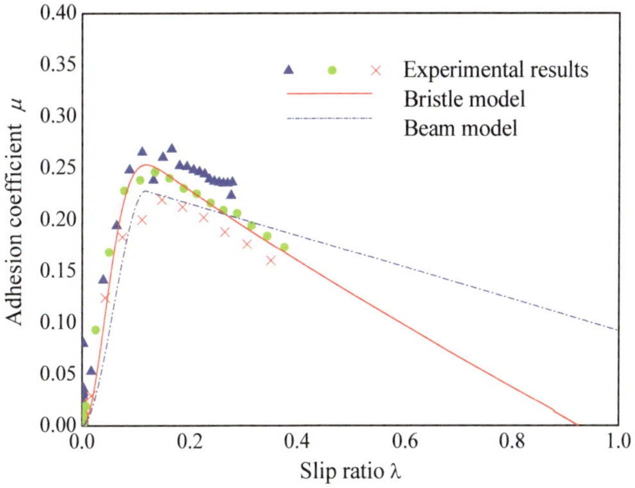

(a) Initial braking velocity $v_0 = 140$ km/h

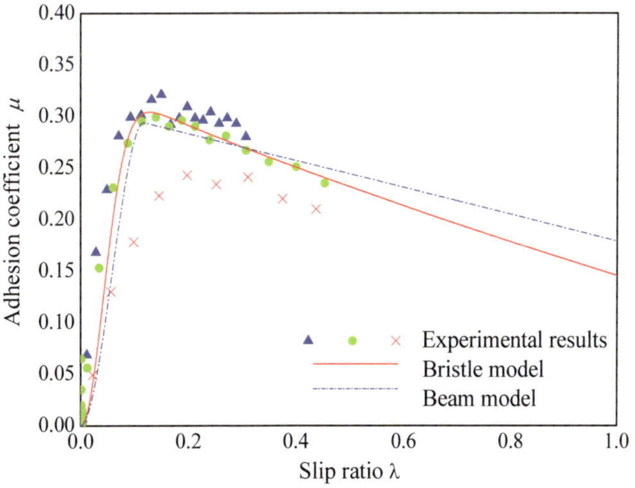

(b) Initial braking velocity $v_0 = 100$ km/h

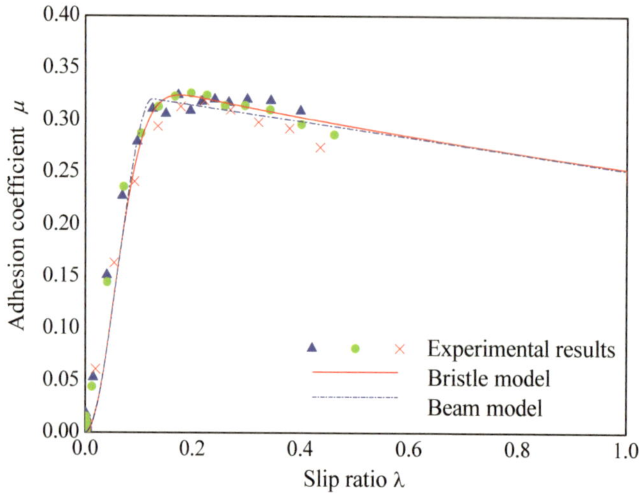

(c) Initial braking velocity v_0 = 60 km/h

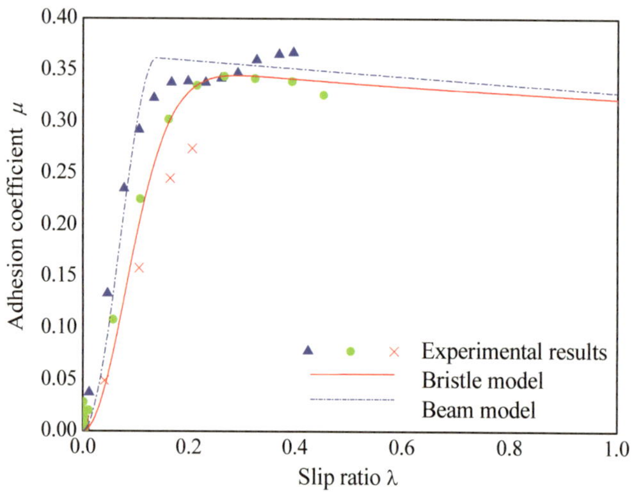

(d) Initial braking velocity v_0 = 30 km/h

Fig. 13. Experimental and simulation results of the adhesion coefficient.

Adhesion model \ Initial braking velocity	30 km/h	60 km/h	100 km/h	140 km/h
Beam model	0.0130	0.0085	0.0132	0.0093
Bristle model	0.0080	0.0080	0.0102	0.0077

Table 3. Mean values of the absolute errors between the experimental results for the mean value of the adhesion coefficient and the simulation results for the beam and bristle models.

From the experimental results in Fig. 13, the parameters α , β , and γ of the bristle model (19) - (22), (24) can be expressed as

$$\alpha = 5.455 \times 10^{4} - 3.641 \times 10^{2} v_0 + 3.798 \times 10^{-1} v_0^2 \quad (30)$$

$$\beta = 1.873 \times 10^{-2} - 6.059 \times 10^{-5} v_0 + 5.500 \times 10^{-8} v_0^2 \quad (31)$$

$$\gamma = 2.345 \times 10^{2} - 8.620 \times 10^{-1} v_0 + 1.053 \times 10^{-4} v_0^2 \quad (32)$$

where v_0 is the initial braking velocity of the rolling stocks. The coefficients in (30), (31), and (32) are obtained by curve fitting for the values of the parameters according to the initial braking velocity.

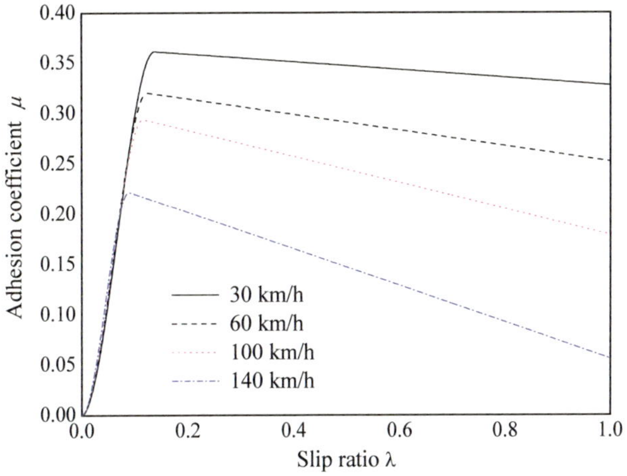

Fig. 14. Simulation results of the mean value of the adhesion coefficient for the beam model.

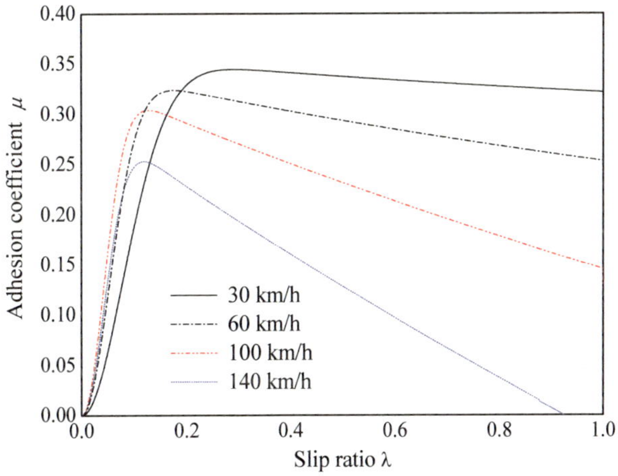

Fig. 15. Simulation results of the mean value of the adhesion coefficient for the bristle model.

Simulation results of the mean value of the adhesion coefficients for the beam model and bristle model according to the slip ratio and initial braking velocity, respectively, are shown in Fig. 14 and 15. These results show a similar tendency for the change in the initial braking velocity conditions. However, the adhesion force model based on the beam model cannot represent the dynamic characteristics of friction. The beam model is obtained by curve fitting the experimental results on the adhesion force, while the bristle model, which includes the friction dynamics, describes the effect of the initial braking velocity accurately in the adhesion regime, where the adhesion force increases according to the slip ratio, as shown in Fig. 15. Therefore, the bristle model is more applicable than the beam model for the desired wheel-slip controller design.

6. Desired wheel slip using adaptive sliding mode control

The desired wheel-slip brake control system is designed by using an adaptive sliding mode control (ASMC) scheme to achieve robust wheel-slip brake control. In the controller design process, the random value of adhesion torque, the disturbance torque due to the vibration of the brake caliper, and the traveling resistance force of the rolling stocks are considered as system uncertainties. The mass of the rolling stocks and the viscous friction torque coefficient are also considered as parameters with unknown variations. The adaptive law for the unknown parameters is based on Lyapunov stability theory.
The sliding surface s for the design of the adaptive sliding mode wheel-slip brake control system is defined as

$$s = e + \rho \int_0^t e\,dt \tag{33}$$

where $e = \sigma_d - \sigma$ is the tracking error of the relative velocity, $\sigma = \lambda v = v - r\omega$ is the relative velocity, σ_d is the reference relative velocity, and ρ is a positive design parameter.
The sliding mode control law consists of equivalent and robust control terms, that is,

$$T_b = U_{eq} + U_r \tag{34}$$

where U_{eq} and U_r are the equivalent and robust control terms. To obtain U_{eq} and U_r, we combine (3), (4), with the derivative of the sliding surface in (33), and include random terms in the adhesion force $F_{ar} = F_a + F_r$ and the adhesion torque $T_{ar} = T_a + T_r$, where F_r and T_r are the random terms of the adhesion force and adhesion torque, respectively. Then, the derivative of the sliding surface can be written as

$$\dot{s} = \dot{\sigma}_d + \left(\frac{1}{rM} + \frac{r}{J}\right)T_a + \left(\frac{1}{rM} + \frac{r}{J}\right)T_r + \frac{1}{M}F_r - \frac{r}{J}T_b - \frac{rB}{J}\omega - \frac{r}{J}T_d + \rho e \tag{35}$$

To determine the equivalent control term U_{eq}, uncertainties such as random terms in the adhesion force and adhesion torque F_r and T_r, as well as the disturbance torque T_d in (35) are neglected, and it is assumed that the sliding surface s is at steady state, that is, $\dot{s} = 0$, then the equivalent control law can be determined as

$$U_{eq} = \frac{J}{r}\left[\dot{\sigma}_d + \left(\frac{1}{rM} + \frac{r}{J}\right)T_a - \frac{rB}{J}\omega + \rho e\right] \tag{36}$$

Thus, $\dot{s}$ can be rewritten as

$$\dot{s} = \left(\frac{1}{rM} + \frac{r}{J}\right)T_r + \frac{1}{M}F_r - \frac{r}{J}T_d - \frac{r}{J}U_r \tag{37}$$

In the standard sliding mode control, to satisfy the reachability condition that directs system trajectories toward a sliding surface where they remain, the derivative of the sliding surface is selected as

$$\dot{s} = -K\,\mathrm{sgn}(s) \tag{38}$$

In this case, chattering occurs in the control input. To attenuate chattering in the control input, the derivative of the sliding surface is selected as (Gao, 1993)

$$\dot{s} = -Ds - K\,\mathrm{sgn}(s) \tag{39}$$

where the parameters D and K are positive.
To determine a control term U_r that achieves robustness to uncertainties such as random terms in the adhesion force and adhesion torque, as well as the disturbance torque, it is assumed that

$$D_0|s| + K_0 > \left|\left(\frac{1}{rM}+\frac{r}{J}\right)T_r + \frac{1}{M}F_r - \frac{r}{J}T_d\right| + \eta \tag{40}$$

where the parameters $D_0 = \frac{r}{J}D$, $K_0 = \frac{r}{J}K$, and η are positive. Then, the robust control law can be determined as

$$U_r = Ds + K\,\mathrm{sgn}(s) \tag{41}$$

and using (40), the reachability condition is satisfied as

$$\begin{aligned} s\dot{s} &= s\left[\left(\frac{1}{rM}+\frac{r}{J}\right)T_r + \frac{1}{M}F_r - \frac{r}{J}T_d - D_0 s - K_0\,\mathrm{sgn}(s)\right] \\ &\le |s|\left[\left|\left(\frac{1}{rM}+\frac{r}{J}\right)T_r + \frac{1}{M}F_r - \frac{r}{J}T_d\right| - D_0|s| - K_0\right] \\ &< -\eta|s|. \end{aligned} \tag{42}$$

Finally, the sliding mode control law is selected as

$$\begin{aligned} T_b &= U_{eq} + U_r \\ &= \frac{J}{r}\left[\dot{\sigma}_d + \left(\frac{1}{rM}+\frac{r}{J}\right)T_a - \frac{rB}{J}\omega + \rho e\right] + Ds + K\,\mathrm{sgn}(s), \end{aligned} \tag{43}$$

where the reference slip acceleration $\dot{\sigma}_d$ and the adhesion torque T_a cannot be measured during operation. Therefore, to implement the control system, the reference slip acceleration $\dot{\sigma}_d = \lambda_d \dot{v}$ must be estimated by $\lambda_d \hat{\dot{v}}$, where $\hat{\dot{v}}$ is the estimated acceleration of the rolling stocks, which can be obtained by the measured velocity of the rolling stocks through the first-order filter $G_f(s) = \frac{s}{\tau s + 1}$. In addition, the adhesion torque $T_a = rF_a$ must be replaced by the calculated value given by (20) and (22) with the measured relative velocity σ.
If the mass of the rolling stocks M and the viscous friction torque coefficient B are considered as parameters with variation, that is, $M = M_n + M_p$ and $B = B_n + B_p$, where the

subscripts n and p denote the nominal and perturbation values, respectively, then the uncertainty ψ in the mass of the rolling stocks and the viscous friction torque coefficient is defined as

$$\psi = \frac{1}{rM_p}T_m - \frac{rB_p}{J}\omega = \theta^T\phi \tag{44}$$

where $\theta^T = \begin{bmatrix} \frac{1}{rM_p} & -\frac{rB_p}{J} \end{bmatrix}$ and $\phi = \begin{bmatrix} T_m \\ \omega \end{bmatrix}$. The parameter vector θ is considered as an unknown parameter vector, which can be estimated by using the update law. From (43) and the estimated unknown parameter vector $\hat{\theta}$, the estimated sliding model control law can be selected as

$$\hat{T}_b = \frac{J}{r}\left[\dot{\sigma}_d + \left(\frac{1}{rM_n} + \frac{r}{J}\right)T_m - \frac{rB_n}{J}\omega + \hat{\theta}^T\phi + \rho e\right] + Ds + K\,\mathrm{sgn}(s) \tag{45}$$

In order to obtain the update law for the unknown parameters, we consider the Lyapunov candidate

$$V = \frac{1}{2}s^2 + \frac{1}{2k}\tilde{\theta}^T\tilde{\theta} \tag{46}$$

where $\tilde{\theta} = \theta - \hat{\theta}$, θ and $\hat{\theta}$ are the nominal and estimated parameter vectors, respectively, and k is a positive parameter. The derivative of the Lyapunov candidate including sliding dynamics is expressed as

$$\dot{V} = s\left[\dot{\sigma}_d + \left(\frac{1}{rM_n} + \frac{r}{J}\right)T_m - \frac{rB_n}{J}\omega - \frac{r}{J}\hat{T}_b + \theta^T\phi + \rho e\right] - \frac{1}{k}\tilde{\theta}^T\dot{\hat{\theta}} \tag{47}$$

Substituting the estimated brake torque $\hat{T}_b$ given by (45) into (47) yields

$$\dot{V} = -Ds^2 - Ks\,\mathrm{sgn}(s) + \tilde{\theta}^T\left(s\phi - \frac{1}{k}\dot{\hat{\theta}}\right) \tag{48}$$

By using the update law for the unknown parameters given by

$$\dot{\hat{\theta}} = ks\phi \tag{49}$$

the derivative of the Lyapunov candidate (48) is nonpositive. The invariant set theorem then guarantees asymptotic stability of the wheel-slip brake control system (Khalil, 1996).

7. Performance evaluation of the desired wheel-slip control system

The characteristics of the wheel-slip control system shown in Fig. 16 are evaluated by simulation. The performance and robustness of the wheel-slip control system using the ASMC scheme are evaluated for railway rolling stocks, while considering system uncertainties such as parameter variation, railway conditions, disturbances, and unmodeled dynamics.

For simulation, the bristle model is used for the adhesion force model because the bristle model is relatively close to the actual adhesion force compared with the beam model. In addition, it is assumed that the brake torque is applied when the velocity of the rolling stocks is 100 km/h. From the experimental results in Fig. 13, it is assumed that the random adhesion force F_r is a white noise signal with a Gaussian distribution that has a standard deviation of 0.431 kN. Since the actual brake force is applied to the wheel disk by the brake caliper, the vibration occurs on the brake caliper at the initial braking moment. Therefore, the disturbance torque $T_d = 0.05T_b e^{-4t} \sin 10\pi t$, caused by the vibration of the brake caliper, is considered in the simulation. In addition, the traveling resistance force $F_r = 0.63v^2$ of the rolling stocks and the viscous friction $B = 0.25^{-0.01t}$ is considered, which causes overheating between the wheel disk and the brake pad. Finally, the unmodeled dynamics $G_a(s) = \frac{e^{-0.15s}}{0.6s+1}$ of the pneumatic actuator of the brake control system are considered.

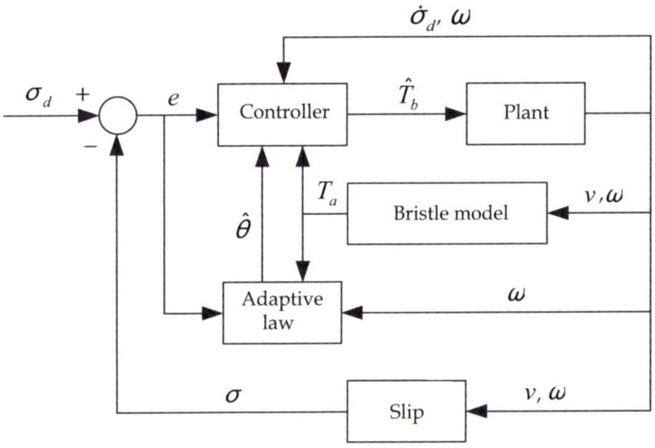

Fig. 16. Wheel-slip control system.

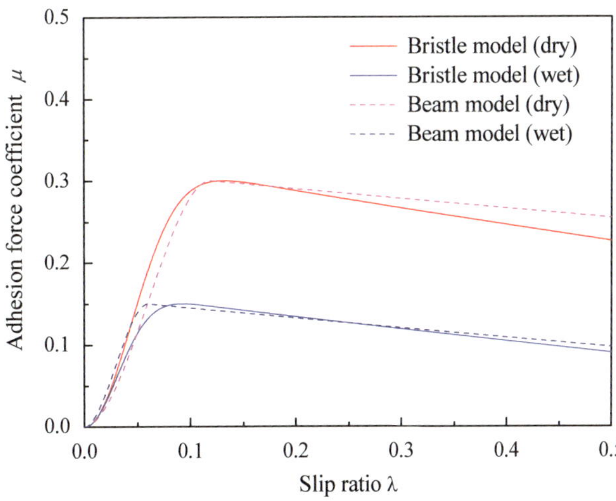

Fig. 17. Relationship between the adhesion force coefficient and the slip ratio according to the change in rail conditions from dry to wet based on the beam and bristle models.

To assess the braking performance in the presence of parameter variations, the simulation is carried out under the assumption that the mass of the rolling stocks changes according to the number of passengers and that the rolling stocks travel in dry or wet rail conditions. It is assumed that the mass of the rolling stocks changes from 3517 to 5276 kg at 25 s. It is also assumed that the maximum adhesion force under wet rail conditions is approximately half of the maximum adhesion force under dry rail conditions and that the rail conditions change from dry to wet at 25 s. Figure 17 shows the relationship between the adhesion force coefficient and the slip ratio according to the change in rail conditions from dry to wet based on the beam and bristle models, which are considered in the simulation. As shown in Fig. 17, the reference slip ratio is assumed to be 0.119 and 0.059 under dry and wet rail conditions, respectively, for the beam model, and 0.132 and 0.092 under dry and wet rail conditions, respectively, for the bristle model.

In order to verify the performance and robustness of the ASMC system, the desired wheel-slip control system using the ASMC scheme is compared with a PI control system through simulation. Control gains of the PI and ASMC systems are selected by trial and error by considering various constraints for each case, such as the maximum brake torque and the maximum slip ratio allowed until the desired performance and robustness are obtained, which are summarized in Table 4. In controller design, the bristle model and beam model are considered for the adhesion force model.

Control scheme	Control gain	Beam model	Bristle model
PI	K_p	400 N-h	650 N-h
	K_i	54 N	27 N
ASMC	D	1.62 h^{-1}	1.53 h^{-1}
	K	70 km/h^2	70 km/h^2
	ρ	1.65 h^{-1}	1.54 h^{-1}
	k	2.1×10^{-8}	2.1×10^{-8}

Table 4. Control gains of the PI and ASMC systems.

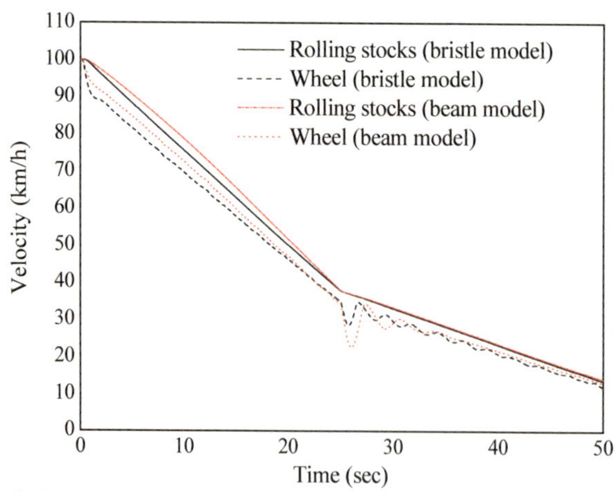

Fig. 18. Velocities of the wheel and rolling stocks for the PI control systems based on the beam and bristle models.

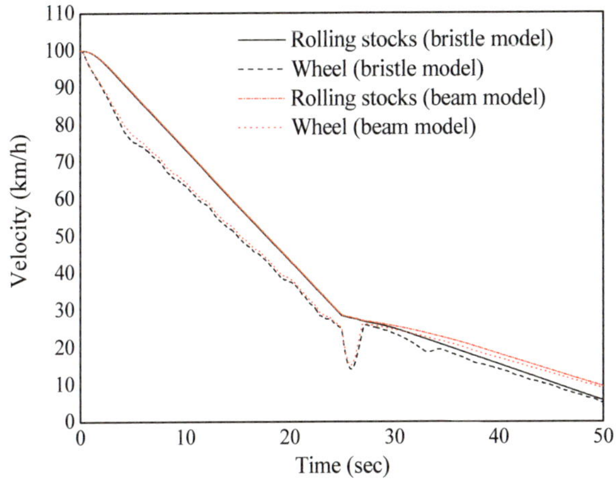

Fig. 19. Velocities of the wheel and rolling stocks for the adaptive sliding mode control (ASMC) systems based on the beam and bristle models.

Figures 18 and 19 show the velocities of the wheel and rolling stocks for the PI and ASMC systems based on the beam and bristle models, respectively. As shown in Fig. 18, the braking distance and time of the PI control system until the velocity of the rolling stocks reaches 5 km/h are 700 m and 59.5 s, respectively, for the PI control based on the beam model, and 682 m and 58.9 s, respectively, for the PI control based on the bristle model. By using the PI control based on the bristle model in place of the PI control based on the beam model, the braking distance and time are improved by 2.6% and 1%, respectively. However, the PI control system cannot effectively compensate for system uncertainties such as the mass of the rolling stocks, railway conditions, the traveling resistance force, and variations of the viscous friction coefficient.

As shown in Fig. 19 for the ASMC system, the braking distance and time are 607 m and 55.3 s, respectively, for the ASMC based on the beam model and 581 m and 50.7 s, respectively, for the ASMC based on the bristle model. Figure 19 shows that the ASMC system provides robust velocity regulation of the rolling stocks in the presence of variations in the mass of the rolling stocks and rail conditions. In this case, the braking distance and time are improved by 4.3% and 8.3%, respectively, by using the ASMC based on the bristle model in place of the ASMC based on the beam model.

Figure 20 shows the brake torques for the PI and ASMC systems based on the beam and bristle models. The expended braking energies of the PI and ASMC systems during braking time are 1.77×10^7 N-m and 1.71×10^7 N-m, respectively. Therefore, by using the adaptive sliding mode control system, it is possible to effectively reduce the braking time and distance using a relatively small braking energy consumption.

The operation of the PI and ASMC wheel-slip control systems can also be demonstrated through the slip ratios. Figure 21 shows the slip ratios of the PI and ASMC systems based on the beam and bristle models. Figure 21 shows that the PI control system has a large tracking error of slip ratio compared with the ASMC system. However, the wheel-slip control system using the ASMC scheme can maintain the slip ratio near the reference slip ratio during the braking time although the slip ratios fluctuate slightly after 25s when the system uncertainties are applied. Therefore, it is appropriate to use the adaptive sliding mode control system to obtain the maximum adhesion force and a short braking distance. Using the ASMC based on the bristle model in place of the PI control based on the beam model yields 28% improvement in the wheel slip. Table 5 summarizes the performance of the PI and ASMC systems based on the beam and bristle models.

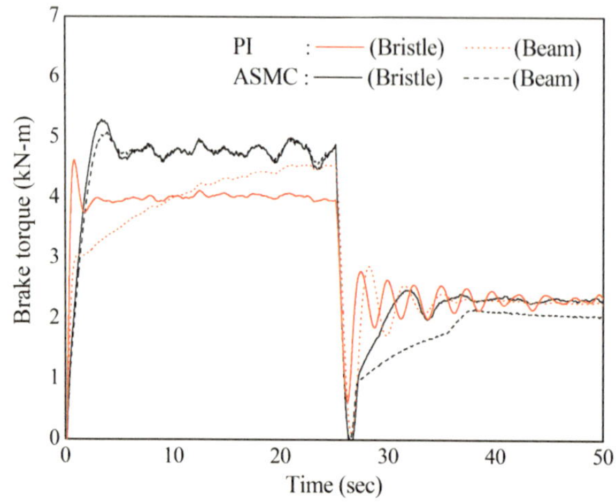

Fig. 20. Brake torques for the PI and ASMC systems based on the beam and bristle models.

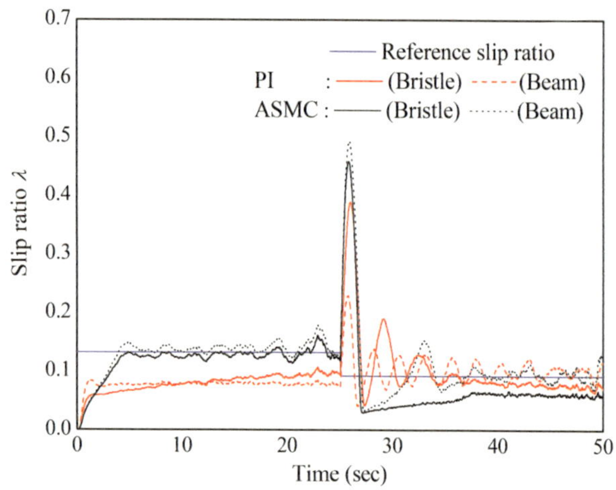

Fig. 21. Slip ratios of the PI and ASMC systems based on the beam and bristle models.

Performance	PI		ASMC	
	Beam model	Bristle Model	Beam model	Bristle model
Braking distance (m)	700	682	607	581
Braking time (s)	59.5	58.9	55.3	50.7
Expended braking energy(kN-m)	1.77×10^7	1.77×10^7	1.77×10^7	1.77×10^7
Mean value of the absolute error between λ and λ_d	0.0378	0.0351	0.0354	0.0272

Table 5. The performances of the PI and ASMC systems based on the beam and bristle models.

8. Conclusions

Two kinds of models, namely, the beam and bristle models, for the adhesion force in railway rolling stocks are developed. The validity of the beam and bristle models is obtained through an adhesion test using a brake performance test rig. By comparing the simulation results of the two kinds of adhesion force models with the experimental results, it is found that the two kinds of adhesion force models can effectively represent the experimental results. However, the adhesion force model based on the beam model cannot represent the dynamic characteristics of friction, while the bristle model can mathematically include the dynamics on friction and can precisely consider the effect of the initial braking velocity in the adhesion regime. Therefore, the bristle model is more appropriate than the beam model for the design of the wheel-slip controller.

In addition, based on the beam and bristle models, the PI and ASMC systems are designed to control wheel slip in railway rolling stocks. Through simulation, we evaluate the performance and robustness of the PI and ASMC systems based on the beam and bristle models for railway rolling stocks. It is verified from the simulation study that, among the four types investigated according to control schemes and adhesion force models, the adaptive sliding mode control system based on the bristle model is the most suitable system for the wheel slip in the rolling stocks with system uncertainties such as the mass and traveling resistance force of the rolling stocks, rail conditions, random adhesion torque, disturbance torque due to the vibration of the brake caliper, and unmodeled actuator dynamics.

9. References

L. Alvarez, J. Yi, R. Horowitz, and L. Olmos (2005), Dynamic friction model-based tire-road friction estimation and emergency braking control, *ASME Journal of Dynamic Systems, Measurement, and Control*, vol. 127, no. 1, pp. 22-32.

M. Basset, C. Zimmer, and G. L. Gissinger (1997), Fuzzy approach to the real time longitudinal velocity estimation of a FWD car in critical situations, *Vehicle System Dynamics*, vol. 27, no. 5 & 6, pp. 477-489.

C. Canudas de Wit, and P. Tsiotras (1999), Dynamic tire friction models for vehicle traction control, *Proceedings of the 38th IEEE Conference on Decision and Control*, Phoenix, Arizona, pp. 3746-3751.

C. Canudas de Wit, and P. Lischinsky (1997), Adaptive friction compensation with partially known dynamic model, *International Journal of Adaptive Control Signal Processing*, vol. 11, pp. 65-80.

C. Canudas de Wit, K. J. Astrom, and P. Lischinsky (1995), A new model for control of systems with friction, *IEEE Transaction on Automatic Control*, vol. 40, no. 3, pp. 419-425.

G. Charles, and R. Goodall (2006), Low adhesion estimation, *Proceedings of the Institution of Engineering and Technology International Conference on Railway Condition Monitoring*, Birmingham, UK, pp. 96-101.

P. Dahl (1976), Solid friction damping of mechanical vibrations, *AIAA Journal*, vol. 14, no. 12, pp. 1675-1682.

W. Gao, and J. C. Hung (1993), Variable structure control of nonlinear systems, *IEEE Transaction on Industrial Electronics*, vol. 40, no. 1, pp. 45-55.

I. P. Isaev, and A. L. Golubenko (1989), Improving experimental research into adhesion of the locomotive wheel with the rail, *Rail International*, vol. 20, no. 8, pp. 3-10.

X. S. Jin, W. H. Zhang, J. Zeng, Z. R. Zhou, Q. Y. Liu, and Z. F. Wen (2004), Adhesion experiment on a wheel/rail system and its numerical analysis, *IMechE J. of Engineering Tribology*, vol. 218, pp. 293-303.

T. A. Johansen, I. Petersen, J. Kalkkuhl, and J. Lüdemann, "Gain-scheduled wheel slip control in automotive brake systems," IEEE Transactions on Control Systems Technology, vol. 11, no. 6, pp. 799-811.

S. Kadowaki, K. Ohishi, S. Yasukawa, and T. Sano (2004), Anti-skid re-adhesion control based on disturbance observer considering air brake for electric commuter train, *The 8th* IEEE *International Workshop on Advanced Motion Control, Kawasaki, Japan*, pp. 607-612.

J. J. Kalker, and J. Piotrowski (1989), Some new results in rolling contact, *Vehicle System Dynamics*, vol. 18, no. 4, pp. 223-242.

H. K. Khalil (1996), *Nonlinear systems*. Upper Saddle River, NJ: Prentice Hall

S. Kumar, M. F. Alzoubi, and N. A. Allsayyed (1996), Wheel/rail adhesion wear investigation using a quarter scale laboratory testing facility, *Proceedings of the 1996 ASME/IEEE Joint Railroad Conference*, Oakbrook, IL, pp. 247-254.

K. J. Lee, H. M. Kim, and J. S. Kim (2004), Design of a chattering-free sliding mode controller with a friction compensator for motion control of a ball-screw system, *IMechE J. of Systems and Control Engineering*, vol. 218, pp. 369-380.

L. Li, F. Y. Wang, and Q. Zhou (2006), Integrated longitudinal and lateral tire/road friction modeling and monitoring for vehicle motion control, *IEEE Transactions on Intelligent Transportation Systems*, vol. 7, no. 1, pp. 1-19.

K. Ohishi, K. Nakano, I. Miyashita, and S. Yasukawa (1998), Anti-slip control of electric motor coach based on disturbance observer, *The 5th IEEE International Workshop on Advanced Motion Control*, Coimbra, Portugal, pp. 580-585.

N. Patel, C. Edwards, and S. K. Spurgeon (2006), A sliding mode observer for tyre friction estimation during braking, *Proceedings of the 2006 American Control Conference*, Minneapolis, Minnesota, pp. 5867-5872.

H. Sakai (1987), *Tire Engineering*. Tokyo, Japan: Grand Prix Publishing, Japanese

Y. Shimizu, K. Ohishi, T. Sano, S. Yasukawa, and T. Koseki (2007), Anti-slip/skid re-adhesion control based on disturbance observer considering bogie vibration, *Proceedings of the Fourth Power Conversion Conference*, Nagoya, Japan, pp. 1376-1381.

S. Shirai (1977), Adhesion phenomena at high-speed range and performance of an improved slip-detector, *Quarterly Report of Railway Technical Research Institute*, vol. 18, no. 4, pp. 189-190.

S. Uchida, "Brake of railway III (2001), *Japan Train Operation Association Journal, Brake of Railway Vehicles*, vol. 3, pp. 25-28.

T. Watanabe, and M. Yamashita (2001), A novel anti-slip control without speed sensor for electric railway vehicles, *The 27th Annual Conference of the IEEE Industrial Electronics Society*, Denver, CO, vol. 2, pp. 1382-1387.

J. Yi, L. Alvarez, and R. Horowitz (2002), Adaptive emergency braking control with underestimation of friction coefficient, *IEEE Transactions on Control Systems Technology*, vol. 10, no. 3, pp. 381-392.

21

A Biomedical Application by Using Optimal Fuzzy Sliding-Mode Control

Bor-Jiunn Wen
Center for Measurement Standards, Industrial Technology Research Institute Hsinchu, Taiwan, R.O.C.

1. Introduction

The development of biochips is a major thrust of the rapidly growing biotechnology industry. Research on biomedical or biochemical analysis miniaturization and integration has made explosive progress by using biochips recently. For example, capillary electrophoresis (CE), sample preconcentration, genomic DNA extraction, and DNA hybridization have been successfully miniaturized and operated in a single-step chip. However, there is still a considerable technical challenge in integrating these procedures into a multiple-step system. In biometric and biomedical applications, the special transporting mechanism must be designed for the μTAS (micro total analysis system) to move samples and reagents through the microchannels that connect the unit procedure components in the system. Therefore, an important issue for this miniaturization and integration is microfluid management technique, i.e., microfluid transportation, metering, and mixing. This charter introduced a method to achieve the microfluidic manipulated implementation on biochip system with a pneumatic pumping actuator and a feedback-signal flowmeter by using an optimal fuzzy sliding-mode control (OFSMC) design based on the 8051 microprocessor.

However, the relationships of the pumping mechanisms, the operating conditions of the devices, and the transporting behavior of the multi-component fluids in these channels are quite complicated. Because the main disadvantages of the mechanical valves utilized moving parts are the complexity and expense of fabrication, and the fragility of the components. Therefore, a novel recursively-structured apparatus of valveless microfluid manipulating method based on a pneumatic pumping mechanism has been utilized in this study. The working principle of this pumping design on this device should not directly relate to the nature of the fluid components. The driving force acting on the microliquid drop in the microchannel of this device is based on the pneumatic pumping which is induced by a blowing airflow. Furthermore, the pneumatic pumping actuator should be independent of the actuation responsible for the biochemical analysis on the chip system, so the contamination of pneumatic pumping source can be avoided. The total biochip mechanism consists of an external pneumatic actuator and an on-chip planar structure for airflow reception.

In order to achieve microfluidic manipulation in the microchannel of the biochip system, pneumatic pumping controller plays an important role. Therefore, a design of the controller has been investigated numerically and experimentally in the present charter. In the control structure of biochip system, at first, the mathematical model of the biochip mechanism is identified by ARX model. Second, according to the results of the biochip-mechanism identification, the control-algorithm design is developed. By the simulation results of the biochip system with a feedback-signals flowmeter, they show the effectiveness of the developed control algorithm. Third, architecture of the control algorithm is integrated on a microprocessor to implement microfluidic manipulation. Since the mathematical model of the flow control mechanism in the biochip microchannels is a complicated nonlinear plant, the fuzzy logic control (FLC) design of the controller will be utilized. Design of the FLC based on the fuzzy set theory has been widely applied to consumer products or industrial process controls. In particular, they are very effective techniques for complicated, nonlinear, and imprecise plants for which either no mathematical model exists or the mathematical model is severely nonlinear. The FLC can approximate the human expert's control behaviors to work fine in such ill-defined environments. For some applications, the FLC can be divided into two classes 1) the general-purpose fuzzy processor with specialized fuzzy computations and 2) the dedicated fuzzy hardware for specific applications. Because the general-purpose fuzzy processor can be implemented quickly and applied flexibly, and dedicated fuzzy hardware requires long time for development, the general-purpose fuzzy processor-8051 microcontroller can be used. Nevertheless, there are also systemic uncertainties and disturbance in FLC controller. Because sliding-mode control (SMC) had been known as an effective approach to restrain the systemic uncertainties and disturbance, SMC algorithm was utilized. In order to achieve a robust control system, the microcontroller of the biochip system combining FLC and SMC algorithms optimally has been developed. Therefore, an OFSMC based on an 8051 microcontroller has been investigated numerically and experimentally in this charter. Hence, microfluidic manipulation in the microchannel of the biochip system based on OFSMC has been implemented by using an 8051 microcontroller.

The microfluidic manipulation based on the microcontroller has successfully been utilized to improve the reaction efficiency of molecular biology. First, it was used in DNA hybridization. There are two methods to improve the efficiency of the nucleic acid hybridization in this charter. The first method is to increase the velocity of the target nucleic acid molecules, which increases the effective collision into the probe molecules as the target molecules flow back and forth. The second method is to introduce the strain rates of the target mixture flow on the hybridization surface. This hybridization chip was able to increase hybridization signal 6-fold, reduce non-specific target-probe binding and background noises within 30 minutes, as compared to conventional hybridization methods, which may take from 4 hours to overnight. Second, it was used in DNA extraction. When serum existed in the fluid, the extraction efficiency of immobilized beads with solution flowing back and forth was 88-fold higher than that of free-beads. Third, it could be integrated in lab-on-a-chip. For the Tee-connected channels, it demonstrated the ability of manipulating the liquid drop from a first channel to a second channel, while simultaneously preventing flow into the third channel. Because there is a continuous airflow at the "outlet" during fluid manipulation, it can avoid contamination of the air source similar to the "laminar flow hook" in biological experiments.

The charter is organized as follows. In Section 2, we introduce the structure of the biochip control system. In Section 3, the fundamental knowledge of OFSMC and the model of the biochip system are introduced, and we address the OFSMC scheme and the associated simulations. In Section 4, the OFSMC IC based on 8051 microprocessor is designed, and the results of the real-time experiment are presented. In Section 5, the efficiency improvement for the molecular biology reaction and DNA extraction by using OFSMC method are presented. Finally, the conclusion is given in Section 6.

2. Structure of the biochip control system

The structure of the biochip control system (Fig. 1) contains six parts: an air compressor, two flow controllers and two flowmeters, a flow-control chip, a biochip, photodiodes system, and a control-chip circuit system. One had designed a pneumatic device with planar structures for microfluidic manipulation (Chung, Jen, Lin, Wu & Wu, 2003). Pneumatic devices without any microfabricated electrodes or heaters, which will have a minimal effect on the biochemical properties of the microfluid by not generating electrical current or heat, are most suitable for μTAS. A pneumatic structure possessing the ability of bi-directional pumping should be utilized in order to implement a pneumatic device which can control the movement of microfluid without valves or moving parts.

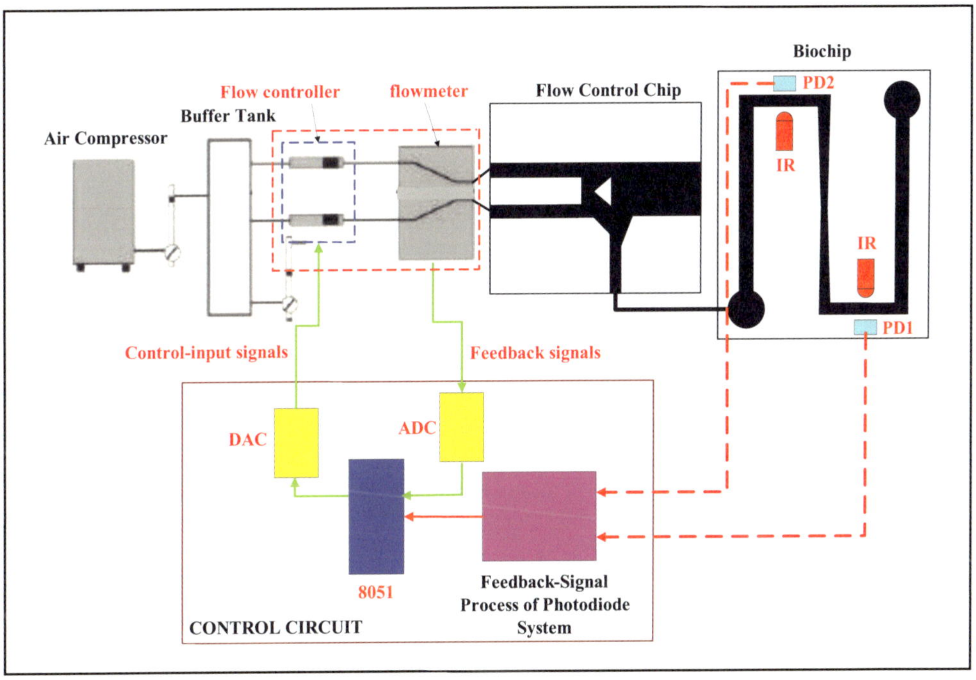

Fig. 1. Structure of the biochip control system.

The schematic diagram of the single pneumatic structure, which provides suction and exclusion by two inlets, is depicted in Fig. 2. When the air flows through inlet A only, it causes a low-pressure zone behind the triangular block and suction occurs in the vertical microchannel. Furthermore, when the air flows through inlet B only, the airflow is induced into the vertical microchannel to generate exclusion. The numerical and experimental results of the pressure and the stream tracers for the condition of the flow-control chip have been demonstrated (Marquardt, 1963). According to the principle of the flow-control chip, the microfluidic manipulation on the biochip is presented in this study by using OFSMC rules with two flow controllers and two flowmeters. Since the biochip in the biochip system is a consumer, the photodiodes system should be utilized for sensing the feedback signals of the position of the reagent in the microchannel of the biochip. Hence, DNA extraction can be achieved in this study.

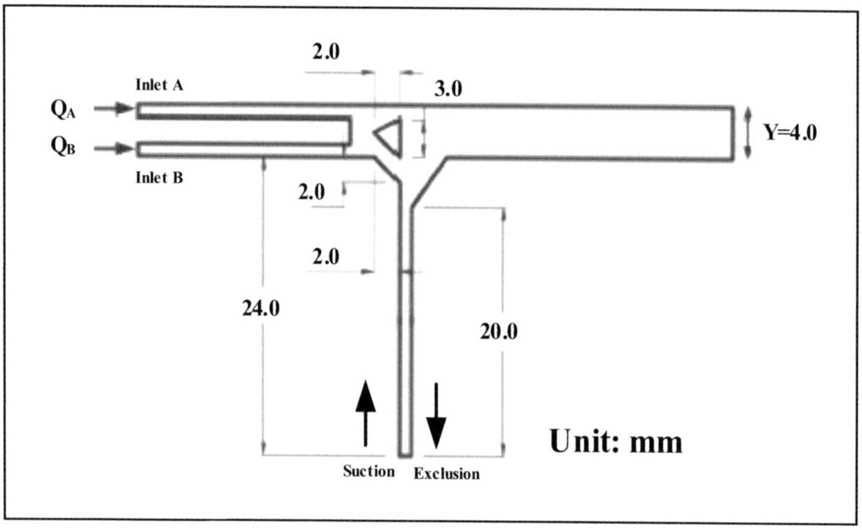

Fig. 2. Single pneumatic structure.

3. Design of the biochip control system

3.1 Design of optimal fuzzy sliding mode control

The biochip system of this design is shown in Fig. 1. If the biochip is DNA extraction chip, the extraction beads are immobilized on the channels. When the bio-fluidics does not flow the place without beads, the time of not extracting DNA can be reduced, and the extraction efficiency will also be improved. So the control of bio-fluidics' location is critical to DNA extraction (or hybridization) efficiency.

The biochip system depicted in Fig. 1 is a nonlinear system. Since the mathematical model of the flow-control mechanism and the microchannels in the biochip is a complicated nonlinear model, FLC design of the controller was utilized. The basic idea behind FLC is to incorporate the expert experience of a human operator in the design of the controller in controlling a process whose input-output relationship is described by a collection of fuzzy control rules (Altrock, Krause & Zimmermann, 1992). The heart of the fuzzy control rules is a knowledge base consisting of the so-called fuzzy IF-THEN rules involving linguistic variables rather than a complicated dynamic model. The typical architecture of a FLC, shown in Fig. 3, is comprised of four principal components: a fuzzification interface, a knowledge base, an inference engine, and defuzzification interface. The fuzzification interface has the effect of transforming crisp measured data into suitable linguistic values; it was designed first so that further fuzzy inferences could be performed according to the fuzzy rules (Polkinghorne, Roberts, Burns & Winwood, 1994). The heart of the fuzzification interface is the design of membership function. There are many kinds of membership functions - Gaussian, trapezoid, triangular and so on - of the fuzzy set. In this paper, a triangular membership function was utilized, as shown in Figs. 4-5.

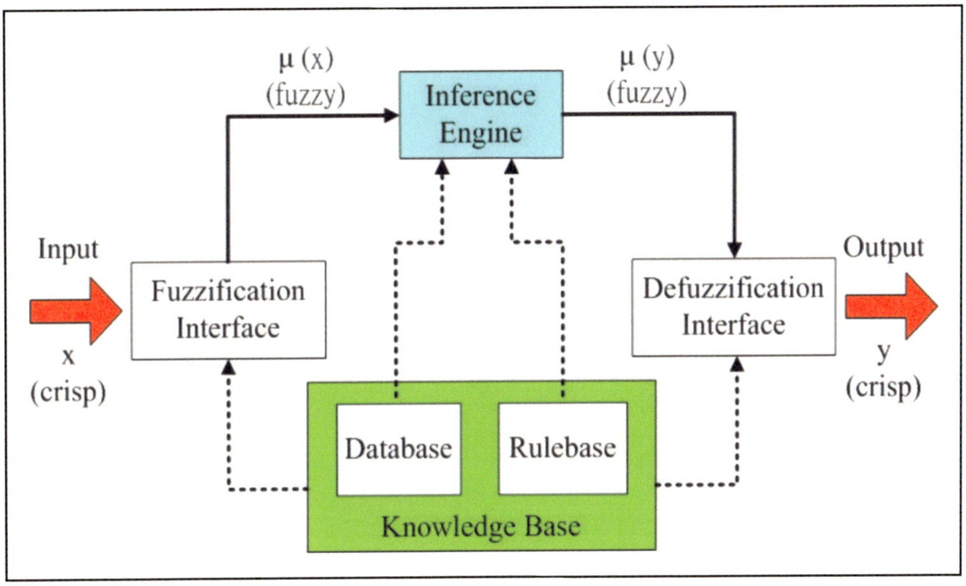

Fig. 3. Architecture of a fuzzy logic controller.

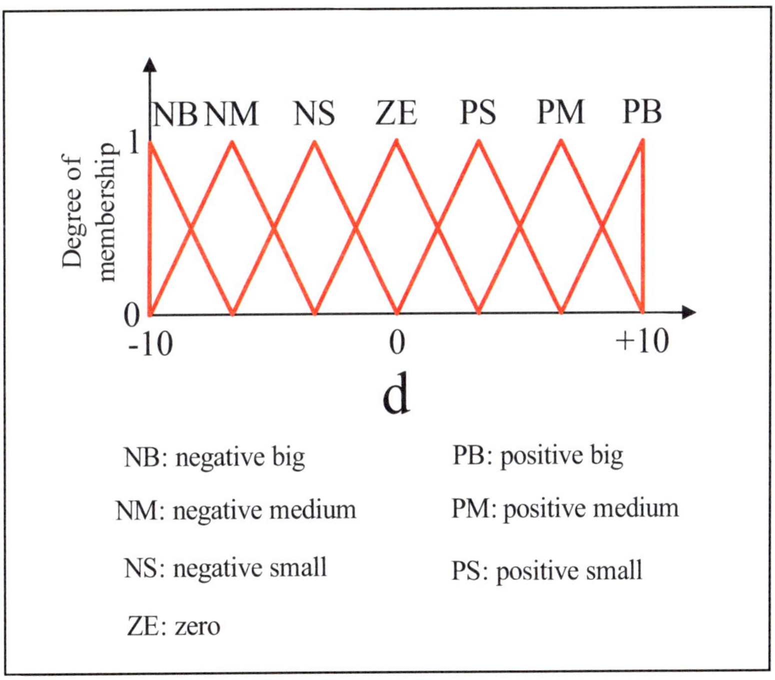

Fig. 4. Membership function-input variable (d) of photodiode detector.

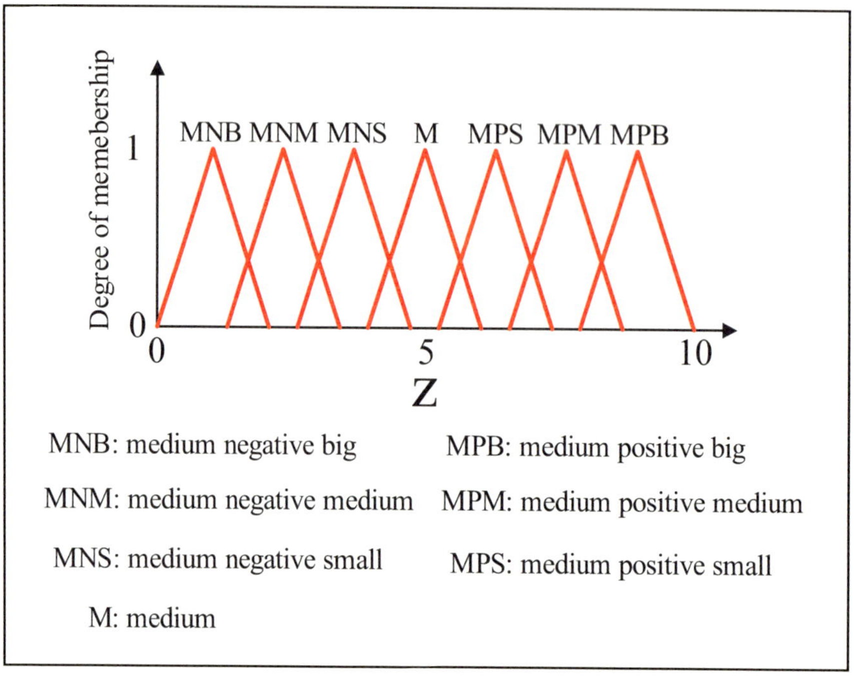

Fig. 5. Membership function-output variable (z) of photodiode detector.

The overall fuzzy rules for the biochip system are defined as the following:

IF d is NB then z_j is MPB

IF d is NM then z_j is MPM

IF d is NS then z_j is MPS

IF d is ZE then z_j is M

IF d is PS then z_j is MNS

IF d is PM then z_j is MNM

IF d is PB then z_j is MNB

where d is input variable of the photodiode signal, and z is output variable of the photodiode signal.

The inference engine is based on the compositional rule of inference with knowledge base for approximate reasoning suggested by Zadeh (Zadeh, 1965; Zadeh, 1968). An inference engine is the kernel of the FLC in modeling human decision making within the conceptual framework of fuzzy logic and reasoning. Hence, the fuzzification interface and fuzzy rules are designed completely before fuzzy reasoning. In this paper, since there are many structures of inference engine, fuzzy reasoning-Mamdani's minimum fuzzy implication rule (MMFIR) method (Mamdani, 1977; Lee, 1990; Altrock, Krause & Zimmermann, 1992; Lin and Lee, 1999) was utilized. For simplicity, assume two fuzzy rules as follows:

R1: IF x is A1 and y is B1, then z is C1,

R2: IF x is A2 and y is B2, then z is C2.

Then the firing strengths α_1 and α_2 of the first and second rules may be expressed as

$$\alpha_1 = \mu_{A_1}(x_0) \wedge \mu_{B_1}(y_0) \text{ and } \alpha_2 = \mu_{A_2}(x_0) \wedge \mu_{B_2}(y_0),$$

where $\mu_{A_1}(x_0)$ and $\mu_{B_1}(y_0)$ indicate the degrees of partial match between the user-supplied data and the data in the fuzzy rule base.
In MMFIR fuzzy resoning, the ith fuzzy control rule leads to the control decision

$$\mu_{C_i'}(w) = \alpha_i \wedge \mu_{C_i}(w).$$

The final inferred consequent C is given by

$$\mu_C(w) = \mu_{C_1'} \vee \mu_{C_2'} = [\alpha_1 \wedge \mu_{C_1}(w)] \vee [\alpha_2 \wedge \mu_{C_2}(w)].$$

The fuzzy reasoning process is illustrated in Fig. 6.

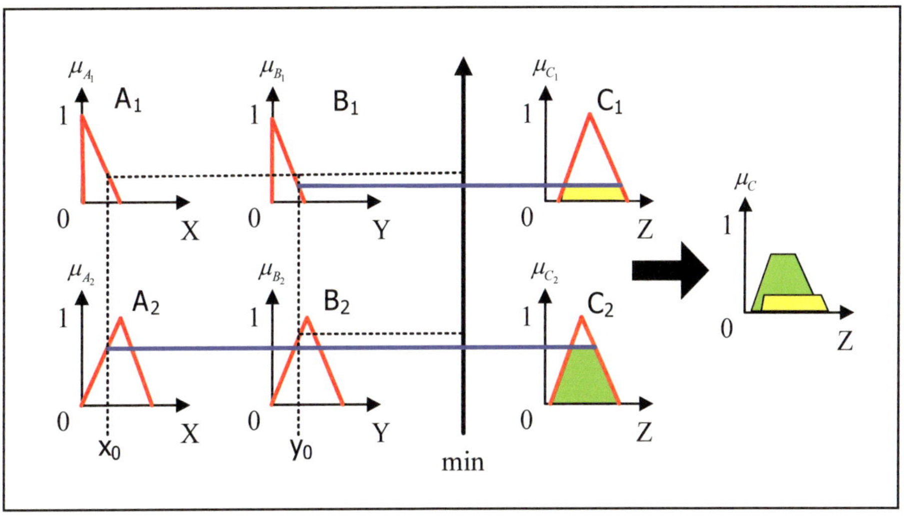

Fig. 6. Fuzzy reasoning of MMFIR method.

Defuzzification is a mapping from a space of fuzzy control actions defined over an output universe of discourse into a space of crisp control actions. This process is necessary because fuzzy control actions cannot be utilized in controlling the plant for practical applications. Hence, the widely used center of area (COA) method, which generates the center of gravity of the possibility distribution of a control action, was utilized. In the case of a discrete universe, this method yields

$$z_{COA} = \frac{\sum_{j=1}^{n} \mu_C(z_j) z_j}{\sum_{j=1}^{n} \mu_C(z_j)} \tag{1}$$

where n is the number of quantization levels of the output, z_j is the amount of control output at the quantization level j, and $\mu_C(z_j)$ represents its membership degree in the output fuzzy set C.

The biochip system depicted in Fig. 1 is a nonlinear system that has been used as an application to study real world nonlinear control problems by different control techniques (Cheng & Li, 1998; Li & Shieh, 2000). The model of the biochip system is identified by ARX model, as

$$\begin{cases} X(k+1) = A_z X(k) + B_z u(k) \\ y(k) = C_z X(k) \end{cases} \tag{2}$$

where $X(k) \in R^n$ is the state variables of system, $u(k) \in R^m$ is the input voltage of the flow controller and $y(k) \in R^r$ is the assumed model output related to the position of the reagent in the microchannel of the biochip. The system is controllable and observable.

Sliding mode control's robust and disturbance-insensitive characteristics enable the SMC controller to perform well in systems with model uncertainty, disturbances and noises. In this paper, in addition to FLC controller, SMC controller was utilized to design the control input voltage of the flow controller. To design SMC controllers, a sliding function was designed first, and then enforced a system trajectory to enter sliding surface in a finite time. As soon as the system trajectory entered the sliding surface, they moved the sliding surface to a control goal. To sum up, there are two procedures of sliding mode, as shown in Fig. 7.

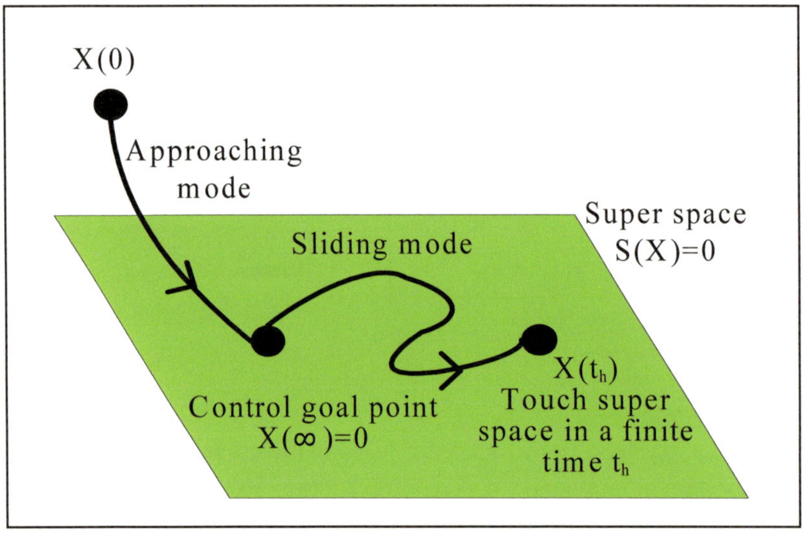

Fig. 7. Generation of sliding mode.

The proposed SMC controller was based on pole placement (Chang, 1999), since the sliding function could be designed by pole placement. Some conditions were set for the sliding vector design in the proposed sliding mode control:

1. $\text{Re}\{\lambda_i\} < 0$, $\alpha_j \in R$, $\alpha_j < 0$, $\alpha_j \neq \lambda_i$.
2. Any eigenvalue in $\{\alpha_1, ..., \alpha_m\}$ is not in the spectrum of A_z.
3. The number of any repeated eigenvalues in $\{\lambda_1, ..., \lambda_{n-m}, \alpha_1, ..., \alpha_m\}$ is not greater than m, the rank of B_z.

where $\{\lambda_1, \lambda_2, ..., \lambda_{n-m}\}$ are sliding-mode eigenvalues and $\{\alpha_1, \alpha_2, ..., \alpha_m\}$ are virtual eigenvalues.

As proved by Sinswat and Fallside (Sinswat & Fallside, 1977), if the condition (3) in the above is established, the control system matrix $A_z - B_z K$ can be diagonalized as

$$A_z - B_z K = \begin{bmatrix} V \\ F \end{bmatrix}^{-1} \begin{bmatrix} \Phi_V & 0 \\ 0 & \Gamma_F \end{bmatrix} \begin{bmatrix} V \\ F \end{bmatrix} \tag{3}$$

where $\Phi_V = diag[\lambda_1, \lambda_2, ..., \lambda_{n-m}]$, $\Gamma_F = diag[\alpha_1, \alpha_2, ..., \alpha_m]$, and V and F are left eigenvectors with respect to Φ_V and Γ_F, respectively. Hence, Eq. (3) can be rewritten as

$$\begin{cases} V(A_z - B_z K) = \Phi_V V \\ F(A_z - B_z K) = \Gamma_F F \end{cases} \tag{4}$$

Rearrangement of Eq. (4) yields

$$FA_z - \Gamma_F F = (FB_z)K \tag{5}$$

According to Chang (Chang, 1999),

$$rank(FA_z - \Gamma_F F) = rank(F) \tag{6}$$

Since F contains m independent left eigenvectors, one has $rank(F) = m$. From Eqs. (5) and (6), it is also true that $rank(FA_z - \Gamma_F F) = rank((Fb_z)K) = rank(F) = m$. In other words, FB_z is invertible. With the designed left eigenvector F above, the sliding function $S(k)$ is designed as

$$S(k) = FX(k) \tag{7}$$

The second step is the discrete-time switching control design. A different and much more expedient approach than that of Gao et al. (Gao, Wang & Homaifa, 1995) is adopted here. This approach is called the reaching law approach that has been proposed for continuous variable structure control (VSC) systems (Gao, 1990; Hung, Gao & Hung, 1993; Gao & Hung, 1993). This control law is synthesized from the reaching law in conjunction with a plant model and the known bounds of perturbations. For a discrete-time system, the reaching law is (Gao, Wang & Homaifa, 1995)

$$S(k+1) - S(k) = -qTS(k) - \varepsilon T \operatorname{sgn}(S(k)) \tag{8}$$

where $T > 0$ is the sampling period, $q > 0$, $\varepsilon > 0$ and $1 - qT > 0$. Therefore, the switching control law for the discrete-time system is derived based on this reaching law. From Eq. (7) and pole-placement method, $S(k)$ and $S(k+1)$ can be obtained in terms of sliding vector F as,

$$\begin{cases} S(k) = FX(k) \\ S(k+1) = FX(k+1) = F(A_z - B_z K)X(k) + FB_z u(k) \end{cases} \tag{9}$$

where $K \in R^n$ is a gain matrix obtained by assigning n desired eigenvalues $\{\lambda_1, ..., \lambda_{n-m}, \alpha_1, ..., \alpha_m\}$ of $A - BK$.
It follows that

$$S(k+1)-S(k)=F(A_z-B_zK)X(k)+FB_zu(k)-FX(k) \tag{10}$$

From Eqs. (8) and (10),

$$S(k+1)-S(k)=-qTS(k)-\varepsilon T\,\mathrm{sgn}(S(k))=F(A_z-B_zK)X(k)+FB_zu(k)-FX(k)$$

Solving for $u(k)$ obtains the switching control law

$$u(k)=-(FB_z)^{-1}[F(A_z-B_zK)X(k)+(qT-1)FX(k)+\varepsilon T\,\mathrm{sgn}(FX(k))] \tag{11}$$

In order to achieve the output tracking control, a reference command input $r(k)$ is introduced into the system by modifying the state feedback control law $u_p(k)=-KX(k)$ with pole-placement design method (Franklin, Powell & Workman, 1998) to become

$$u_p(k)=N_ur(k)-K(X(k)-N_xr(k)) \tag{12}$$

where

$$\begin{bmatrix} N_u \\ N_x \end{bmatrix}=\begin{bmatrix} A_z-I & B_z \\ C_z & 0 \end{bmatrix}^{-1}\begin{bmatrix} 0 \\ I \end{bmatrix} \tag{13}$$

The proposed SMC input, based on Eq. (13), is assumed to be

$$u_s(k)=u_p(k)+u=N_ur(k)-K(X(k)-N_xr(k))+u \tag{14}$$

Substituting Eq. (11) into (14) gives the proposed SMC input as

$$\begin{aligned} u_s(k)=&N_ur(k)-K(X(k)-N_xr(k)) \\ &-(FB_z)^{-1}[F(A_z-B_zK)X(k)+(qT-1)FX(k)+\varepsilon T\,\mathrm{sgn}(FX(k))] \end{aligned} \tag{15}$$

The pole-placement SMC design method utilizes the feedback of all the state variables to form the desired sliding vector. In practice, not all the state variables are available for direct measurement. Hence, it is necessary to estimate the state variables that are not directly measurable.

In practice, a discrete linear time-invariant system sometimes has system disturbances and measurement noise. Hence, linear quadratic estimator (LQE) will be applied here to estimate optimal states in having system disturbances and measurement noise.

According to Eq. (2), consider a system model as

$$\begin{cases} X(k+1)=A_zX(k)+B_zu(k)+G\nu(k) \\ y(k)=C_zX(k)+\omega(k) \end{cases} \tag{16}$$

where $X(k)\in R^n$ is the state variable, $u(k)\in R^m$ is the control input voltage , $y'(k)\in R^r$ is the assumed plant output related to the XY stage position, and $\nu(k)\in R^n$ and $\omega(k)\in R^r$ are system disturbances and measurement noise with covariances $E[\omega\omega^T]=Q$, $E[\nu\nu^T]=R$ and $E[\omega\nu^T]=0$.

The objective of LQE is to find a vector $\hat{X}(k)$ which is an optimal estimation of the present state $X(k)$. Here "optimal" means the cost function

$$J = \lim_{T \to \infty} E\left\{ \int_0^T (X^T Q X + u^T R u) dt \right\} \tag{17}$$

is minimized. The solution is the estimator as

$$\begin{cases} \hat{X}(k+1) = A_z \hat{X}(k) + B_z u(k) + K_f (y(k) - C_z \hat{X}(k)) \\ \hat{y}(k) = C_z \hat{X}(k) \end{cases} \tag{18}$$

where K_f is the "optimal Kalman" gain $K_f = PC_z^T R^{-1}$ and P is the solution of the algebraic Riccati equation

$$A_z P + P A_z^T - P C_z^T R^{-1} C_z P + Q = 0 \tag{19}$$

FLC, SMC, and LQE were combined into the so called optimal fuzzy sliding-mode control (OFSMC) and utilized to control input voltage of the flow controller. The OFSMC block diagram with LQE is shown in Fig. 8.

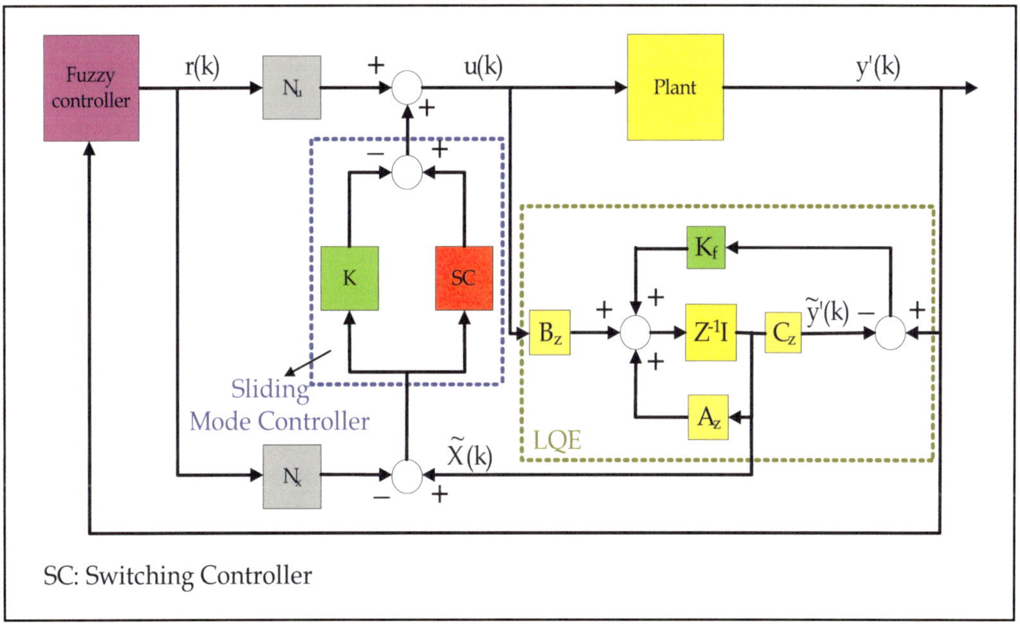

Fig. 8. OFSMC block diagram.

In the biochip system, the photodiode system provided the position feedback signal for FLC and LQE. Then, the FLC could use the position feedback signals to generate the input signals for SMC. And the LQE could estimate optimal states in having system disturbances and measurement noise for SMC by the position feedback signals. Hence, the SMC with FLC and LQE could implement the microfuildic manipulation very well and robustly. The performance of the OFSMC would be explained in detail by simulation and experimental results, which are presented in Section 3.2.

3.2 Simulation of OFSMC

This section deals with a system model described by Eq. (2) and defines a reference command input $r(k)$, which is an input voltage of the flow controller by fuzzy controller with designed photodiode signals. The pole-placement algorithm described in Section 3.1 is utilized to determine a sliding vector. In this study, Ackermann's Formula is used to determine the pole-placement feedback gain matrix K. In practice, the fact that not all state variables are available for direct measurement results in the necessity to estimate the state variables that are not directly measurable. Hence, the full-order state observer designed by Ackermann's Formula and LQE will be utilized in this study.

In order to achieve the biochip application, the microfluidic reagent has to be manipulated to flow back and forth in the central zone of the microchannel between the PD1 and PD2, shown in Fig. 1. During the simulations, the external disturbance would be added in system plant. Figs. 9 and 10 show the simulations of the biochip system model at 2 Hz of back and forth flowing based on FLC, and fuzzy sliding mode control (FSMC), respectively, with the full-order estimator (FOE), and OFSMC by using the MATLAB and Simulink. In Figure 9, the blue solid lines represent reference command input whereas the red dotted lines, the green dash-dot lines and the magenta dashed line are the system output based on FLC, FSMC with FOE and OFSMC respectively. Every turn of the curve represents a reversal of the flowing reagent during its back and forth flow in the microchannel on the biochip. Fig. 10 is the error performance of the simulation results of biochip system model based on the three controllers.

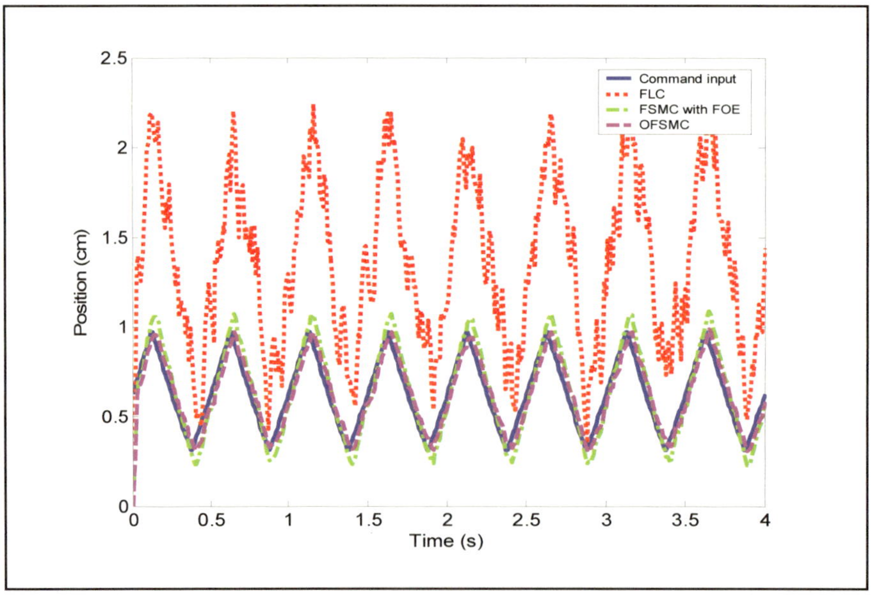

Fig. 9. Simulation results of biochip system model based on FLC, FSMC with the FOE and OFSMC at 2 Hz. The blue solid lines represent reference command input whereas the red dotted lines, the green dash-dot lines and the magenta dashed line are the system output based on FLC, FSMC with FOE and OFSMC respectively.

Increasing emphasis on the mathematical formulation and measurement of control system performance can be found in recent literature on modern control. Therefore, as an always-positive number or zero, the performance index that can be calculated or measured and used to evaluate the system's performance is usually utilized. The best system is defined as the system that minimizes this index. In this study, integrated absolute error (IAE) that is often of practical significance is used as the performance index and is expressed as

$$IAE = \int_0^T |e(t)|\,dt \tag{20}$$

where $e(t)$ is a error function of the plant and T is a finite time. In addition to IAE, integral of time multiplied by absolute error (ITAE) that provides the performance index of the best sensitivity is expressed as

$$ITAE = \int_0^T t|e(t)|\,dt \tag{21}$$

where $e(t)$ is a error function of the plant and T is a finite time. Using the above two methods, the performance of the system will be evaluated exactly.

In molecular biology applications, increasing the velocity of the target nucleic acid molecules increases the number of effective collision into the probe molecules as the target molecules flow back and forth, which will ultimately increase the efficiency of biochemical reaction obviously. Therefore, according to the issue, the performance of the simulation results with the three control rules as the target molecules flowing back and forth at 0.2, 0.5, 1 and 2 Hz would be presented in Table 1.

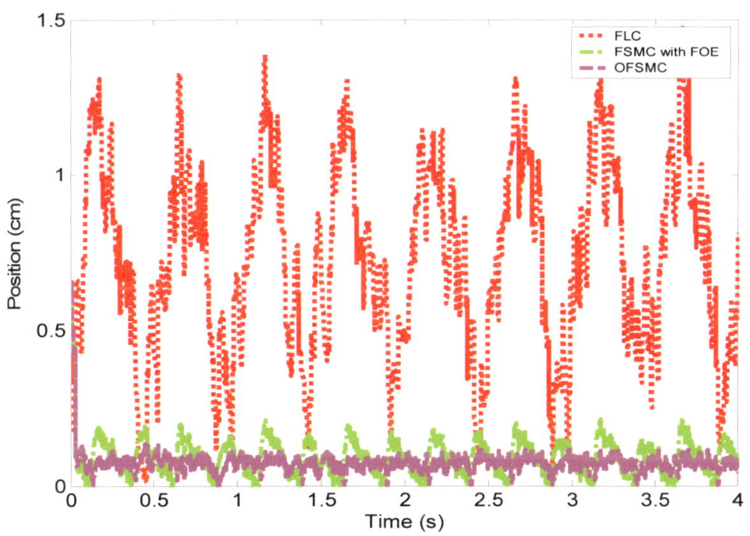

Fig. 10. Error performance of simulation results of biochip system model based on FLC, FSMC with the FOE and OFSMC at 2 Hz. The red dotted lines, the green dash-dot lines and the magenta dashed line are the system output based on FLC, FSMC with FOE and OFSMC respectively.

The overshoot means the reagent is out of the central zone as it is manipulated. Here, out of the length of the central zone is defined as overshoot value. And the error performance of the simulation results were also evaluated by using IAE and ITAE indices and the results are shown in Tables 2 and 3. The following conclusions can be arrived at from the analysis of the simulation results from Figs. 9 and 10, and Tables 1 to 3.

Frequency (Hz) \ Overshot (mm)	FLC	FSMC	OFSMC
0.2	8.0	1.1	0.5
0.5	10.0	1.3	0.7
1	11.5	1.7	0.9
2	12.0	1.8	1.1

Table 1. Performances of the simulation results with FLC, FSMC, and OFSMC control rules

Frequency (Hz) \ IAE	FLC	FSMC	OFSMC
0.2	3061	1106	73
0.5	2888	1084	100
1	2889	1279	162
2	2892	1824	301

Table 2. IAE index of the control systems with FLC, FSMC, and OFSMC control rules

Frequency (Hz) \ ITAE	FLC	FSMC	OFSMC
0.2	5438	2063	118
0.5	5514	2023	170
1	5712	2473	303
2	5814	3529	570

Table 3. ITAE index of the control systems with FLC, FSMC, and OFSMC control rules

1. According to Figs. 9, 10, and the values of IAE and ITAE (Table 1), the biochip system model based on OFSMC controller at 2 Hz performs better than that based on FLC, or FSMC controller with FOE. In addition, the performances of the biochip system model based on OFSMC controller at 0.2, 0.5, and 1 Hz are obviously better than the other two, according to Tables 1 to 3. Therefore, the OFSMC controller can perform well in the biochip system with disturbances.
2. It is certain that the OFSMC control method is capable of manipulating the position of the reagent in the microchannel on the biochip robustly and successfully. The experimental results of microfluidic manipulation on biochip system with OFSMC controller based on 8051 microprocessor are shown in Section 4.

4. Experimental results of OFSMC

The control block diagram of the biochip system with OFSMC controller described in Section 2 and 3 is shown in Fig. 8. In order to provide a quick and useful product for non PC-based systems, the microfluidic manipulation is implemented by 8051 microprocessor in this study. And the A/D and D/A chips were utilized to convert the photodiode or flowmeter feedback analog signals into digital signals for the microprocessor as well as to convert digital signals into analog signals for the flow controller. Then, the circuit of the photodiode-signal process should be designed. Assembly language was utilized to program the OFSMC control rules to embed into 8051 microprocessor with flow chart of the program shown in Fig. 11. The experimental results of microfluidic manipulation on biochip system with OFSMC controller based on 8051 microprocessor are shown in Fig. 12, where the volume of reagent used is 94 μL. The reagent on the biochip system was controlled excellently to flow back and forth at 2 Hz, because the overshoot of the control performance was very small and the control system was very stable. The experimental results of the control performance with FLC, FSMC, and OFSMC control rules are shown in Fig. 13.

According to Fig. 13, the microfluidic manipulation with FLC control rule can only be implemented to flow back and forth at 0.2 Hz, and the overshoot of the performance is -10, which means the reagent could not be manipulated between the PD1 and PD2. Either it was pushed out of the biochip, or it was manipulated under 1 cm length of the undershoot at 0.2 Hz of back and forth flowing. In addition, according to the results of the performance with FSMC control rule, the overshoot became larger and larger by increasing the frequency of back and forth flowing. Compared to FLC and FSMC control rule, the overshoots of the performance with OFSMC control rule were the least of the three control rules and the performance was the most stable and the best of the three at all frequencies of back and forth flowing. The microfluidic manipulation on biochip system with OFSMC rules can keep flowing back and forth at 2 Hz within 1 h while the other two can not.

Since the experimental and simulation results are in good agreement, it could be concluded that the control performance with OFSMC control rule was better than that with FLC and FSMC. Compared to FLC and FSMC, it was more successful to overcome the variable parameters and nonlinear model to achieve a better microfluid management with OFSMC control rule when using different biochip for every time. Therefore, it is certain that the OFSMC control method is capable of manipulating the position of the reagent in the microchannel on the biochip robustly and successfully.

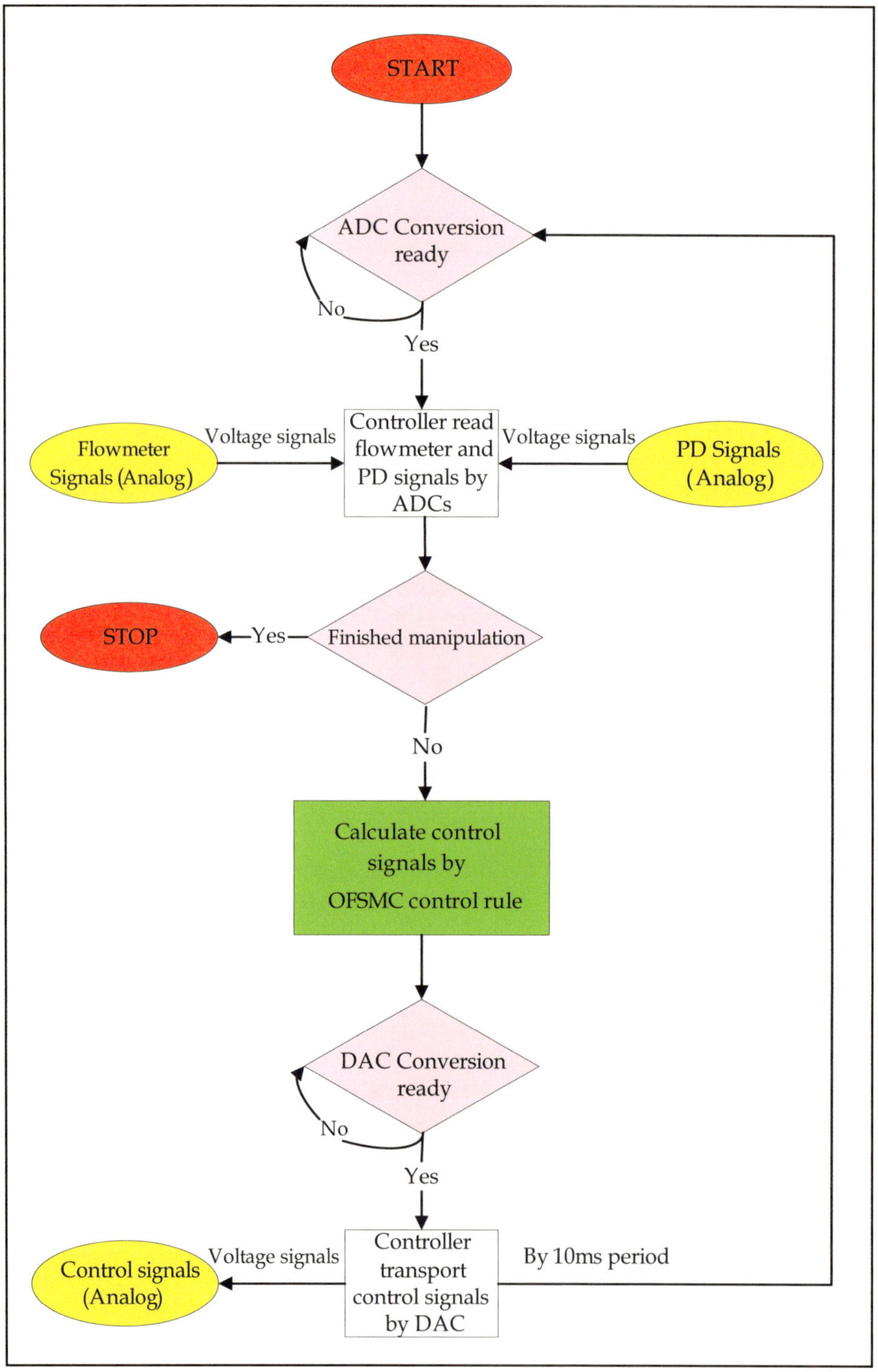

Fig. 11. Flow chart of the program.

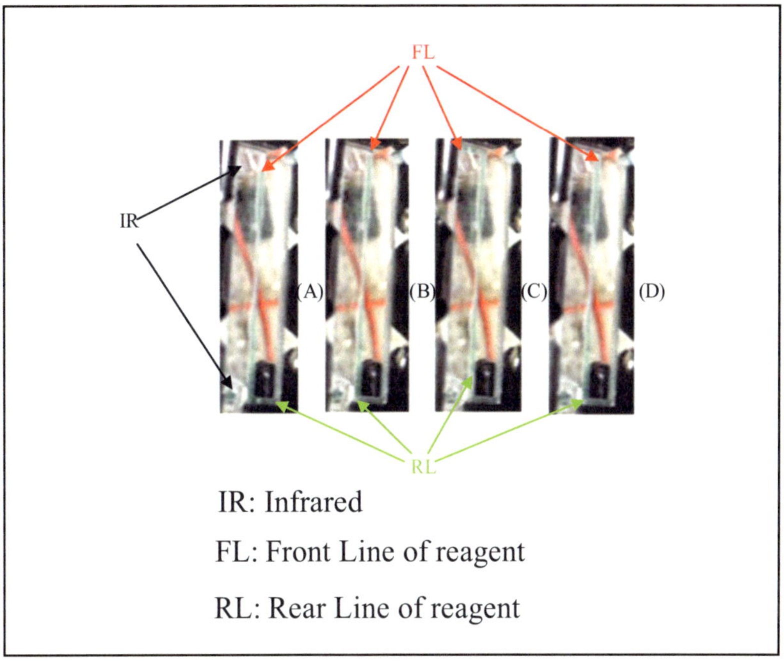

Fig. 12. Experimental results of microfluidic manipulation at 2 Hz (A period from (A) to (D)).

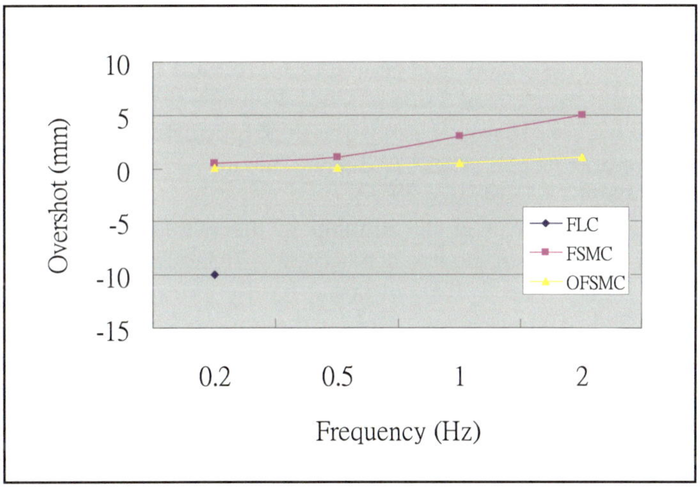

Fig. 13. Experimental results of microfluidic manipulation with FLC, FSMC, and OFSMC control rules.

5. Biomedical application results

According to the experimental results given in Section 4, the microfluidic manipulation based on the microcontroller could be utilized in biotechnology, as it successfully improved the efficiency of the biochemical reaction. First, it was used in DNA hybridization. There are two methods to improve the efficiency of the nucleic acid hybridization in this charter. The first method is to increase the velocity of the target nucleic acid molecules, which increases the effective collision into the probe molecules as the target molecules flow back and forth. The second method is to introduce the strain rates of the target mixture flow on the hybridization surface. This hybridization chip was able to increase hybridization signal 6-fold, reduce non-specific target-probe binding and background noises within 30 minutes, as compared to conventional hybridization methods, which may take from 4 hours to overnight. Second, it was used in DNA extraction. In this section, DNA extraction from Chung, Wen and Lin (Chung, Wen & Lin, 2007) is introduced.

The microfluidic DNA extraction chip was designed and fabricated onto a poly-methylmethacrylate (PMMA) substrate with flow channels. Machined and immobilized-beads PMMA substrate and blank PMMA were bonded together to form the device. The beads used for DNA extraction was obtained from Magic Bead Inc. (USA) A plasma generator was used to perform the surface treatment. Plasma source gas consisting of a mixture of ammonia and oxygen was used to activate the surface of PMMA substrate.

Escherichia coli (E. coli) was cultured in 5 ml of LB medium (NaCl: 10 g/l, Tryptone: 10 g/l, yeast extract: 5 g/l) in 15 ml tubes at 37°C and 225 rpm. After 16 h, the optical density (OD) of the culture was measured in a spectrophotometer (U-2100, Hitachi, Japan). The number of E. coli cells or the amount of DNA was calculated from an OD versus cell number. The culture was then diluted by distilled water to obtain varying numbers (102 -105) of E. coli cells per micro liter. DNA was extracted from the blood of one of the members of the group using the microchip. Whole blood was directly used without any pretreatment.

The sample flowed forward and backward with the immobilized beads at a frequency of 1 Hz inside the channel. E. coli cells were treated with a buffer (B1+B2, Magic Bead, USA) to lyse the cells and to release the DNA. The DNA extracted using the microchips was amplified by PCR (Polymerase Chain Reaction). The forward and reverse primers were 5′-CAGGATTAGATACCCTGGTAG-3′ and 5′-TTCCCCTACGGTTACCTTGTT-3′, respectively. The PCR condition was: one cycle of 5 min at 95°C, 40 cycles of 30 s at 95°C, 40 s at 58°C and 40 s at 72°C, and one cycle of 10 min at 72°C. The PCR products were analyzed qualitatively in a Mupid-2 electrophoresis equipment (Advance, Japan) and quantitatively in an Agilent 2100 bioanalyzer (Agilent Technologies, USA).

The extraction efficiencies of E. coli cell number in the whole blood were tested and are shown in Fig. 18a. It showed that the free beads could efficiently extract DNA as the number of E. coli cells was higher than 5×10^4, but could hardly extract DNA as the number was smaller than 10^4 in 25 μl of whole blood. And for the immobilized beads, the corresponding boundary number of E. coli cells for efficient and hard extractions of DNA were 2×10^2 and 10^2, respectively. After the analysis in the bioanalyzer, the results were as shown in Fig. 18b. When the number of E. coli cells was 2×10^2 to 10^4 in 25 μl of whole blood, the extraction efficiency of immobilized beads with solution flowing back and forth was about 600-fold larger than that of free beads.

6. Conclusions

In biometric and biomedical applications, an important issue for miniaturization and integration is microfluid management. This charter introduced the optimal fuzzy sliding-mode control (OFSMC) design based on the 8051 microprocessor and the complete microfluidic manipulated system implementation of biochip system with a pneumatic pumping actuator, two feedback-signal photodiodes and flowmeters for better microfluid management. The newly developed microfluid management technique was successfully utilized to improve the reaction and extraction efficiency of a biochemical reaction.

7. Acknowledgements

Thanks for Prof. Chung to provide the flow control chip and biochip.

8. References

Altrock, C., Krause V. B., & Zimmermann, H. J. (1992). Advanced Fuzzy Logic Control of a Model Car in Extreme Situation. *Fuzzy Sets and Systems*, Vol. 48, 41-52.

Alfaro-Cid, E., McGookin, E. W., Murray-Smith, D. J. & Fossen, T. I. (2005), Genetic algorithms optimisation of decoupled Sliding Mode controllers: simulated and real results, *Control Engineering Practice*, Vol. 13, No.6, 739-748.

Bart, S. F., Tavrow, L. S., Mehregany, M., & Lang, J. H. (1990). Microfabricated electrohydrodynamic pumps. *Sens. Actuators*, A21-A23, 193-197.

Bousse, L., & Minalla. A. (1999). Optimization of sample injection components in electrokinetic microfluid system. *Technical Digest of Twelfth IEEE International Conference on Micro Electro Mechanical Systems*, 309-314, Orlando, Florida, U.S.A..

Cheng, C. C., & Li, T. H. S. (1998). Parallel Fuzzy Sliding-mode Control of a spring-linked Cart-pole system. *IEEE IECON'98. Aachen.*

Chang, J. L. (1999). *Sliding-Mode Control Design based on Conventional Pole-Assignment Method.* Ph D dissertation, National Chiao Tung University.

Chung, Y. C., Jen, C. P., Lin, Y. C., Wu, C. Y., & Wu, T. C. (2003). Design of a Recursively-Structured Valveless Device for Microfluidic Manipulation. *Lab on a Chip*, Vol. 3, No. 3, 168-172.

Chung, Y. C., Wen, B. J. & Lin, Y. C. (2007), Optimal Fuzzy Sliding-mode Control for Bio-microfluidic Manipulation. *Control Engineering Practice*, Vol. 15, 1093 - 1105.

Edwards, C and Tan, C. P. (2006), Sensor fault tolerant control using sliding mode observers, *Control Engineering Practice*, Vol. 14, No. 8, 897-908.

Fuhr, G., Hagedorn, R., Muller, T, Benecke, W., & Wagner, B. (1992). Pumping of water solution in microfabricated electrohydrodynamic systems. *Proceedings of the IEEE Micro Electro Mechanical Systems*, 25-29, Travemünde, Germany.

Franklin, G. F., Powell, J. D., & Workman, M. L. (1998). *Digital Control of Dynamic Systems*, Addison Wesley Longman, Inc., 3rd ed..

Gao, W. B. (1990). *Foundation of Variable Structure Control*. Beijing: China Press of Science and Technology.

Gao, W. B., & Hung, J. C. (1993). Variable Structure Control of Nonlinear System: A New Approach. *IEEE Transactions on Industrial Electronics*, Vol. 40, No. 1, 45-56.

Gao, W. B., Wang, Y., & Homaifa, A. (1995). Discrete-Time Variable Structure Control Systems. *IEEE Transactions on Industrial Electronics*, Vol. 42, No. 2.

Harrison, D. J., Seiler, K., Manz, A., & Fan, Z. (1992). Chemical analysis and electrophoresis systems integrated on glass and silicon chips. *Technical Digest of IEEE Solid-State Sensor and Actuators Workshop*, 110-113, Hilton Head Island, South Carolina, U.S.A..

Hung, J. Y., Gao, W. B., & Hung, J. C. (1993). Variable Structure Control: A Survey. *IEEE Transactions on Industrial Electronics*, Vol. 40, No. 1, 2-22.

Hashimoto, Y., Farkas, I., Murase, H., Carson, E. R. and Sano, A. (2004), Control approaches to bio- and ecological systems, *Control Engineering Practice*, Vol. 12, No.5, Fuzzy System Applications in Control, 595-603.

Jen, C. P., & Lin, Y. C. (2002). Design and simulation of bi-directional microfluid driving systems. *J. Micromech. Microeng.*, Vol. 17, 115-121.

Lintel, H. T. G. van, Pol, F. C. M. van de, Bouwstra, S. (1988). A piezoelectric micropump based on micromachining of silicon. *Sens. Actuators*, A 15, 153-167.

Lee, C. C. (1990). Fuzzy Logic in Control System: Fuzzy Logic Controller-Part I, and II. *IEEE Trans. Systems, Mans, and Cybernetics*, Vol. 20, No.2, 404-435.

Lammerink, T. S. J., Elwenspoek, M., & Fluitman, J. H. J. (1993). Integrated micro-liquid dosing system. *Proceedings of the IEEE Micro Electro Mechanical Systems*, 254-259, Fort Lauderdale, U.S.A..

Lin, C. T., & Lee, C. S. (1999). *Neural Fuzzy Systems*. Prentice-Hall Pte Ltd.

Li, T. H. S., & Shieh, M. Y. (2000). Switching-type Fuzzy Sliding Mode Control of a Cart-pole System. *Mechatronics*, Vol. 10, 91-109.

Marquardt, D. (1963). An algorithm for least-squares estimation of nonlinear parameters. *SIAM Journal of Applied Mathematics*, Vol. 11, 431–441.

Mamdani, E. H. (1977). Application of Fuzzy Logic to Approximate Reasoning using Linguistic Synthesis. *IEEE Trans. Computers*, C-26.

Miyazaki, S., Kawai, T., & Araragi, M. (1991). A piezo-electric pump driven by a flexural progressive wave. *Proceedings of the IEEE Micro Electro Mechanical Systems*, 283-288.

Moroney, R. M., White, R. M., & Howe, R. T. (1991). Ultrasonic induced microtransport. *Proceedings of the IEEE Micro Electro Mechanical Systems*, 277-282.

Polkinghorne, M. N., Roberts, G. N., Burns, R. S., & Winwood, D. (1994). The implementation of fixed rulebase fuzzy logic to the control of small surface ships. *Control Engineering Practice*, Vol. 3, No. 3, 321–328.

Peng, H., Ozaki, T., Toyoda, Y., & Oda, K. (2001). Exponential ARX model based long-range predictive control strategy for power plants. *Control Engineering Practice*, Vol. 9, 1353–1360.

Richter, A., Plettner, A., Hofmann, K. A., & Sandmaier H. (1991). Electrohydrodynamic pumping and flow measurement. *Proceedings of the IEEE Micro Electro Mechanical Systems*, 271-276, Nara, Japan.

Sinswat, V., & Fallside, F. (1977). Eigenvalue/Eigenvector Assignment by State Feedback. *Int. J. Control*, Vol. 23, 183-196.

Sentoni, G., Agamennoni, O., Desages, A., & Romagnoli, J. (1996). Approximate models for nonlinear process control. *A.I.Ch.E. Journal*, Vol. 42, 2240–2250.

Tas, N. R., Berenschot, J. W., Lammerink, T. S. J., Elwenspoek M., & Berg, A. van-der. (2002). Nanofluidic bubble pump using surface tension directed gas injection. *Anal. Chem.*, Vol. 74, 2224-2227.

Wen, B. J. (2003). *Sliding- Mode Control for Nanoparticle Manipulation Using an Atomic Force Microscopy*. Master Thesis, National Chiao Tung University.

Zadeh, L. A. (1965). Fuzzy Sets. *Inform. Control*, Vol. 8, 338-353.

Zadeh, L. A. (1968). Fuzzy Algorithm. *Inform. Control*, Vol. 12, 94-102.

Zengerle, R., Richer, A., & Sandmaier, H. (1992). A micro membrane pump with electrostatic actuation. *Proceedings of the IEEE Micro Electro Mechanical Systems*, 19-24, Travemünde, Germany.

Part 5

New Trends in the Theory of Sliding Mode Control

22

Sliding Mode Control of Second Order Dynamic System with State Constraints

Aleksandra Nowacka-Leverton and Andrzej Bartoszewicz
Technical University of Łódź, Institute of Automatic Control
18/22 Stefanowskiego St. 90-924 Łódź,
Poland

1. Introduction

In recent years much of the research in the area of control theory focused on the design of discontinuous feedback which switches the structure of the system according to the evolution of its state vector. This control idea may be illustrated by the following example.
Example 1. Let us consider the second order system

$$\begin{aligned} \dot{x}_1 &= x_2 \\ \dot{x}_2 &= x_2 + u_i \qquad i = 1,2, \end{aligned} \tag{1}$$

where $x_1(t)$ and $x_2(t)$ denote the system state variables, with the following two feedback control laws

$$u_1 = f_1(x_1, x_2) = -x_2 - x_1 \tag{2}$$

$$u_2 = f_2(x_1, x_2) = -x_2 - 4x_1 \tag{3}$$

The performance of system (1) controlled according to (2) is shown in Fig. 1, and Fig. 2 presents the behaviour of the same system with feedback control (3). It can be clearly seen from those two figures that each of the feedback control laws (2) and (3) ensures the system stability only in the sense of Lyapunov.
However, if the following switching strategy is applied

$$i = \begin{cases} 1 & \text{for} \quad \min\{x_1, x_2\} < 0 \\ 2 & \text{for} \quad \min\{x_1, x_2\} \geq 0 \end{cases} \tag{4}$$

then the system becomes asymptotically stable. This is illustrated in Fig. 3. Moreover, it is worth to point out that system (1) with the same feedback control laws may exhibit completely different behaviour (and even become unstable). For example, if the switching strategy (4) is modified as

$$i = \begin{cases} 1 & \text{for} \quad \min\{x_1, x_2\} \geq 0 \\ 2 & \text{for} \quad \min\{x_1, x_2\} < 0 \end{cases} \tag{5}$$

then the system output increases to infinity. The system dynamic behaviour, in this situation, is illustrated in Fig. 4.

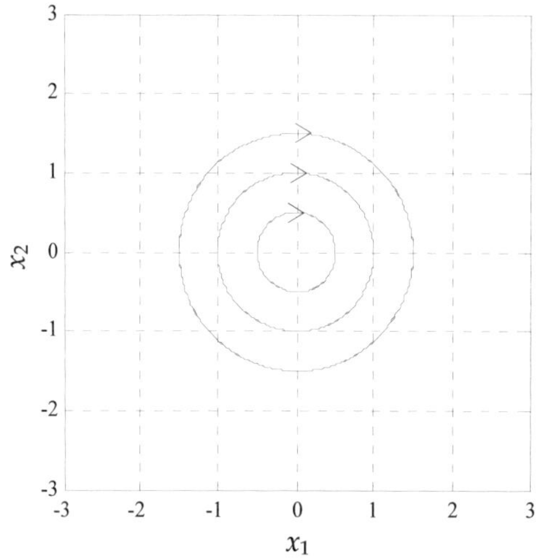

Fig. 1. Phase portrait of system (1) with controller (2).

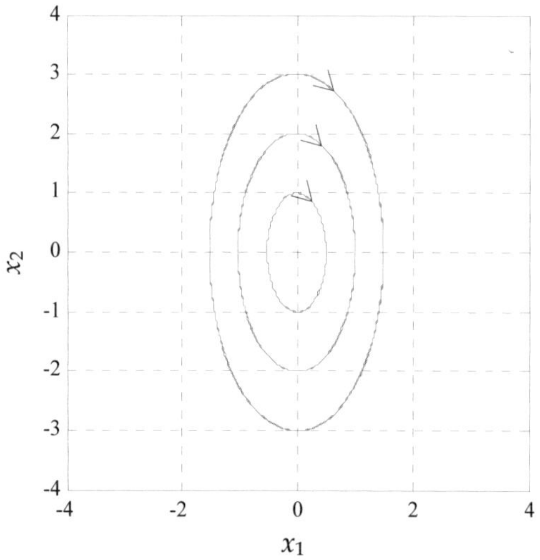

Fig. 2. Phase portrait of system (1) with controller (3).

This example presents the concept of variable structure control (VSC) and stresses that the system dynamics in VSC is determined not only by the applied feedback controllers but also, to a large extent, by the adopted switching strategy. VSC is inherently a nonlinear technique and as such, it offers a variety of advantages which cannot be achieved using conventional linear controllers. Our next example shows one of those favourable features - namely it demonstrates that VSC may enable finite time error convergence.

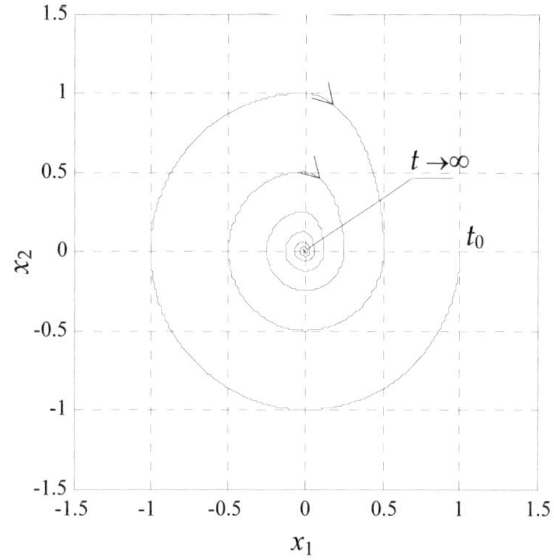

Fig. 3. Phase portrait of system (1) when switching strategy (4) is applied.

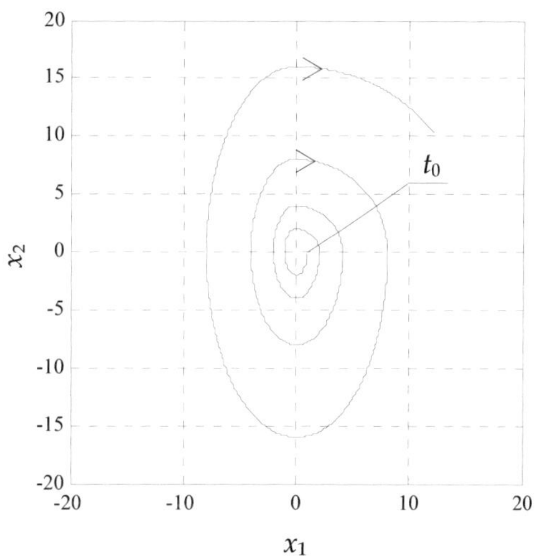

Fig. 4. Phase portrait of system (1) when switching strategy (5) is applied.

Example 2. In this example, again we consider system (1), however now we apply the following controller

$$u = -x_2 - a\,\mathrm{sgn}(x_1) - b\,\mathrm{sgn}(x_2) \tag{6}$$

where $a > b > 0$. Closer analysis of the behaviour of system (1) with control law (6) demonstrates that, in this example, the system error converges to zero in finite time which can be expressed as

$$T = \frac{a}{b}\sqrt{2x_{10}}\left(\frac{1}{\sqrt{a-b}} + \frac{1}{\sqrt{a+b}}\right) \tag{7}$$

where x_{10} and $x_{20} = 0$ represent initial conditions of system (1). Even though the error converges to zero in finite time, the number of oscillations in the system tends to infinity, with the period of the oscillations decreasing to zero. This is illustrated in Figs. 5 and 6. In the simulation example presented in the figures, the following parameters are used a = 7, b = 3, $x_{10} = 20$ and $x_{20} = 0$. Consequently, the system error is nullified at the time instant T = 12.045 and remains equal to zero for any time greater than T. Clearly these favourable properties are achieved using finite control signal. This controller, due to the way the phase trajectory - shown in Fig. 5 - is drawn, is usually called "twisting controller". It also serves as a good, simple example of the second order sliding mode controllers.

The two examples presented up to now demonstrate the principal properties of VSC systems. However, the main advantage of the systems is obtained when the controlled plant exhibits the sliding motion (DeCarlo et al., 1988; Hung et al., 1993; Slotine & Li, 1991; Utkin, 1977). The idea of sliding mode control (SMC) is to employ different feedback controllers acting on the opposite sides of a predetermined surface in the system state space. Each of those controllers pushes the system representative point (RP) towards the surface, so that the RP approaches the surface, and once it hits the surface for the first time it stays on it ever after. The resulting motion of the system is restricted to the surface, which graphically can be interpreted as "sliding" of the system RP along the surface. This idea is illustrated by the following example.

***Example* 3.** Let us consider another second order plant

$$\begin{aligned} \dot{x}_1 &= x_2 \\ \dot{x}_2 &= b\cos(m\,x_1) + u \qquad |b| < 1, \end{aligned} \tag{8}$$

where b and m are possibly unknown constants. We select the following line in the state space

$$s = x_2 + c x_1 = 0 \tag{9}$$

(c = const.) and apply the controller

$$u = -c\,x_2 - \mathrm{sgn}(s) \tag{10}$$

In this equation sgn(.) function represents the sign of its argument, i.e. sgn(s < 0) = –1 and sgn(s > 0) = +1. With this controller the system representative point moves towards line (9) always when it does not belong to the line. Then, once it hits the line, the controller switches the plant input (in the ideal case) with infinite frequency. Therefore, line (9) is called the switching line. Furthermore, since after reaching the line, the system RP slides along it, then the line is also called the sliding line. This example is illustrated in Fig. 7. The system parameters used in the presented simulation are c = 0.5, b = 0.75, m = 10 and the simulation is performed for the following initial conditions $x_{10} = 5$ and $x_{20} = 1$. Notice that when the plant remains in the sliding mode, its dynamics is completely determined by the switching line (or in general the switching hypersurface) parameters. This implies that neither model uncertainty nor matched external disturbance affects the plant dynamics (Draženović, 1969) which is a highly desirable system property. This property can also be justified geometrically, if one notices that in our example the slope of line (9) fully governs the plant motion in the sliding mode. Therefore, in SMC systems we usually make the distinction between two phases: the first one – called the reaching phase – lasts until the controlled plant RP hits the switching surface, and the second one - the sliding phase - begins when the RP reaches the surface. In the latter phase the plant insensitivity to a class of modeling inaccuracies and external disturbances is ensured. Let us stress that the system robustness with respect to unmodeled dynamics, parameter uncertainty and external disturbances is guaranteed only in the sliding mode. Therefore, shortening or (if possible) even complete elimination of the reaching phase is an important and timely research issue (see for example Bartoszewicz & Nowacka-Leverton, 2009; Pan & Furuta, 2007; Sivert, 2004; Utkin & Shi, 1996) in the field of SMC.

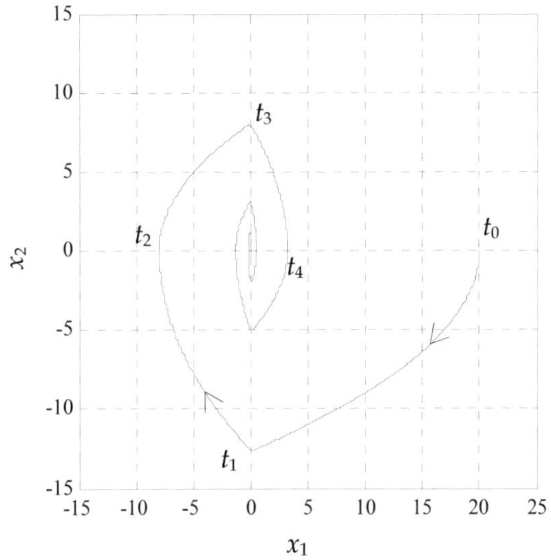

Fig. 5. Phase portrait of system (1) controlled according to (6).

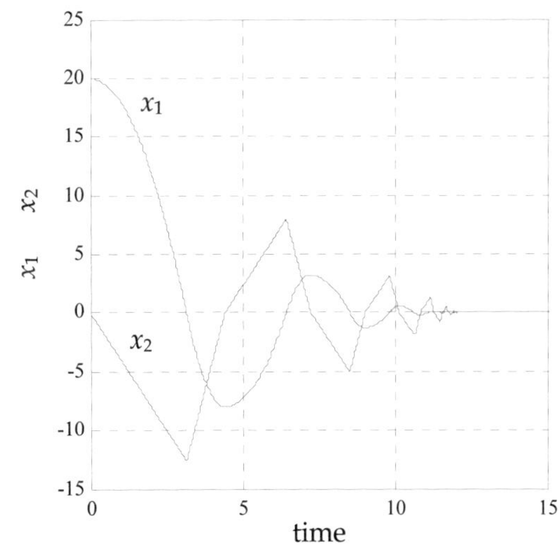

Fig. 6. State variables of system (1) controlled according to (6).

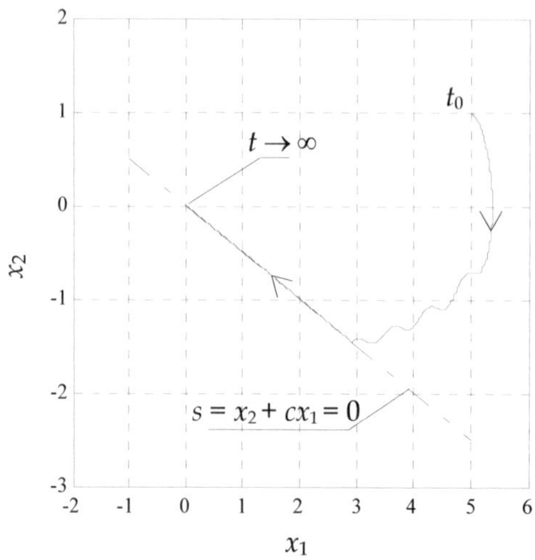

Fig. 7. Phase trajectory of system (1) controlled according to (10).

Another immediate consequence of the fact that in the sliding mode, the system RP is restricted to the switching hypersurface (which is a subset of the state space) is reduction of the system order. If the system of the order n has m independent inputs, then the sliding mode takes place on the intersection of m hypersurfaces and the reduced order of the system is equal to the difference n - m. To be more precise, in multi-input systems the sliding mode may take place either independently on each switching hypersurface or only on the intersection of the surfaces. In the first case the system RP approaches each surface at any time instant and once it hits any of the surfaces it stays on this surface ever after. In the latter case, however, the system RP does not necessarily approach each of the surfaces, but it always moves towards their intersection. In this case the system RP may hit a surface and move away from it (might possibly cross a switching surface), but once it reaches the intersection of all the surfaces, then the RP never leaves it.

One of the major tasks in the SMC system design is the selection of an appropriate control law. This can be achieved either by assuming a certain kind of the control law (usually motivated by some previous engineering experience) and proving that this control satisfies one of the so-called reaching conditions or by applying the reaching law approach. The reaching conditions (Edwards & Spurgeon, 1998) ensure stability of the sliding motion and therefore they are naturally derived using Laypunov stability theory. On the other hand, if the reaching law approach is adopted for the purpose of a sliding mode controller construction (Bartoszewicz, 1998; Bartoszewicz, 1996; Gao et al, 1995; Golo & Milosavljević, 2000; Hung et al., 1993), then a totally different design philosophy is employed. In this case the desired evolution of the switching variable s is specified first, and then a control law ensuring that s changes according to the specification is determined.

Sliding mode controllers guarantee system insensitivity with respect to matched disturbance and model uncertainty (Draženović, 1969), and cause reduction of the plant order. Moreover, they are computationally efficient, and may be applied to a wide range of various, possibly nonlinear and time-varying plants. However, often they also exhibit a serious drawback which essentially hinders their practical applications. This drawback - high frequency oscillations which inevitably appear in any real system whose input is supposed to switch infinitely fast - is usually called chattering. If system (8) exhibits any, even arbitrarily small, delay in the input channel, then control strategy (10), will cause oscillations whose frequency and amplitude depend on the delay. With the decreasing of the delay time, the frequency rises and the amplitude is getting smaller. This is a highly undesirable phenomenon, because it causes serious wear and tear on the actuator components. Therefore, a few methods to eliminate chattering have been proposed. The most popular of them uses function

$$\mathrm{sat}(s)=\begin{cases} -1 & \text{for } s<-\rho \\ \frac{1}{\rho}s & \text{for } |s|\le\rho \\ 1 & \text{for } s>\rho \end{cases} \tag{11}$$

instead of sgn(s) in the definition of the discontinuous control term. With this modification the term becomes continuous and the switching variable does not converge to zero but to the closed interval $[-\rho, \rho]$. Consequently, the system RP after the reaching phase termination, belongs to a layer around the switching hyperplane and therefore this strategy is called boundary layer controller (Slotine & Li, 1991).

Other approaches to the chattering elimination include:

- introduction of other nonlinear approximations of the discontinuous control term, for example the so called fractional approximation defined as

$$\mathrm{approx}(s)=\frac{s}{\varepsilon+|s|} \tag{12}$$

where ε is a small positive constant (Ambrosino et al., 1984; Xu et al., 1996);
- replacing the boundary layer with a sliding sector (Shyu et al., 1992; Xu et al., 1996);
- using dynamic sliding mode controllers (Sira-Ramirez, 1993a; Sira-Ramirez, 1993b; Zlateva, 1996);
- using fuzzy sliding mode controllers (Palm, 1994; Palm et al., 1997);
- using second (or higher) order sliding mode controllers (Bartolini et al., 1998; Levant, 1993).

The phenomenon of chattering has been extensively analyzed in many papers using describing function method and various stability criteria (Shtessel & Lee, 1996).

As it has already been mentioned, the switching surface completely determines the plant dynamics in the sliding mode. Therefore, selecting this surface is one of the two major tasks in the process of the SMC system design. In order to stress this issue let us point out that the same controller which has been considered in the last example may result in a very different system performance, if the sliding line slope c is selected in another way. This can be easily noticed if one takes into account any negative c. Then, controller (10) still ensures stability of the sliding motion on line (9), i.e. the system RP still converges to the line, however the system is unstable since both state variables x_1 and x_2 tend to (either plus or minus) infinity while the system RP slides away from the origin of the phase plane along line (9).

Since sliding mode control is well known to be a robust and computationally efficient regulation technique which may be applied to nonlinear and possibly time-varying plants, then the proper design of the sliding mode controllers has recently become one of the most extensively studied research topics within the field of control engineering. This design process usually breaks into two distinct parts: in the first part the switching surface is selected, and in the second one the control signal which always makes the system representative point approach the surface is chosen. Once the representative point hits the surface, then under the same control signal, the point remains on the surface. Thus, the switching surface fully determines the system dynamics in the sliding mode and should be carefully selected by the system designer.

In this chapter we consider the second order, nonlinear, time-varying system subject to the acceleration and velocity constraints. We introduce a continuously time-varying switching line adaptable to the initial conditions of the system which guarantees the existence of a sliding mode on this line. At the time $t = t_0$ the line passes through the representative point, specified by the initial conditions of the system, in the error state space. Afterwards, the line moves smoothly, with a constant deceleration and a constant angle of inclination, to the origin of the space and having reached the origin the line remains fixed. Thus the proposed control algorithm eliminates the reaching phase and forces the representative point of the system to always stay on the switching line. Consequently, our control is robust with respect to the external disturbance and model uncertainty from the very beginning of the control action. Furthermore, in order to obtain good dynamic performance of the considered system, the switching line is designed in such a way that the integral absolute error (IAE) over the whole period of the control action is minimised and state constraints are satisfied at the same time. The presented method is verified by the simulation example.

The control algorithm proposed in this chapter may be regarded as an alternative solution to the elegant and currently widely accepted integral sliding mode control technique (Utkin & Shi, 1996). The main advantage of our approach is explicit consideration of state constraints in the controller design process. Furthermore, the novelty of our work demonstrates itself also in the IAE optimal performance and error convergence without oscillations or overshoots.

2. Problem formulation

In this chapter we consider the time-varying and nonlinear, second order system described by the following equations

$$\begin{aligned} \dot{x}_1 &= x_2 \\ \dot{x}_2 &= f(x,t) + \Delta f(x,t) + b(x,t)u + d(t) \end{aligned} \tag{13}$$

where x_1, x_2 are the state variables of the system and $x(t) = [x_1(t)\ x_2(t)]^T$ is the state vector, t denotes time, u is the input signal, b, f - are a priori known, bounded functions of time and the system state, Δf and d are functions representing the system uncertainty and external disturbances, respectively. Further in this chapter, it is assumed that there exists a strictly positive constant δ which is the lower bound of the absolute value of $b(x, t)$, i.e. $0 < \delta = \inf\{|b(x, t)|\}$. Furthermore, functions Δf and d are unknown and bounded. Therefore, there exists a constant μ which for every pair (x, t) satisfies the following inequality $|\Delta f(x, t) + d(t)| \leq \mu$. The initial conditions of the system are denoted as x_{10}, x_{20} where $x_{10} = x_1(t_0)$, $x_{20} = x_2(t_0)$. System (13) is supposed to track the desired trajectory given as a function of time $x_d(t) = [x_{1d}(t)\ \ x_{2d}(t)]^T$ where $x_{2d}(t) = \dot{x}_{1d}(t)$ and $x_{2d}(t)$ is a differentiable function of time. The trajectory tracking error is defined by the following vector

$$e(t) = [e_1(t) \ \ e_2(t)]^T = x(t) - x_d(t) \tag{14}$$

Hence, we have $e_1(t) = x_1(t) - x_{1d}(t)$ and $e_2(t) = x_2(t) - x_{2d}(t)$.
In this chapter it is assumed that at the initial time $t = t_0$, the tracking error and the error derivative can be expressed as

$$e_1(t_0) = e_0 \neq 0, e_2(t_0) = 0 \tag{15}$$

where e_0 is an arbitrary real number different from zero. This condition is indeed satisfied in many practical applications such as position control or set point change of second order systems. An example of these applications is point to point (PTP) control of robot manipulators, that is moving the manipulator arm from its initial location where it is originally at a halt, to another predefined position at which the arm stops and again is expected to remain at rest.
Further in this chapter, we present a detailed description of the sliding mode control strategy which ensures optimal performance of the system and its robustness with respect to both the system uncertainty $\Delta f(\mathbf{x}, t)$ and external disturbance d(t).

3. Sliding mode controller

In order to effectively control system (13), i.e. to eliminate the reaching phase and to obtain system insensitivity with respect to both external disturbance d(t) and the model uncertainty $\Delta f(x, t)$ from the very beginning of the system motion, we introduce a time-varying switching line. The line slope does not change during the control process, which implies that the line moves on the phase plane without rotating. In other words, the line is shifted in the state space with a constant angle of inclination. At the beginning the line moves with a constant deceleration in the state space and then it stops at a time instant $t_f > t_0$. Consequently, the switching line can be described by the following equation

$$s(e,t) = 0 \ \text{ where } \ s(e,t) = e_2(t) + ce_1(t) + \left(Ct^2 + Bt + A\right)\delta \tag{16}$$

where

$$\delta = \begin{cases} 1 & \text{for } t \in [0, t_f) \\ 0 & \text{for } t \in [t_f, \infty) \end{cases} \tag{17}$$

and c, A and B are constants. The selection of these constants will be considered further in this chapter.
In order to ensure system (13) stability in the sliding motion on the line described by equations (16) parameter c in this equation must be strictly positive, i.e. $c > 0$. Furthermore, in order to actually eliminate the reaching phase, and consequently to ensure insensitivity of the considered system from the very beginning of its motion, constants A, B, C and c should be chosen in such a way that the representative point of the system at the initial time $t = t_0$ belongs to the switching line. For that purpose, the following condition must be satisfied

$$s\left[e(t_0), t_0\right] = e_2(t_0) + ce_1(t_0) + Ct_0^2 + Bt_0 + A = 0 \tag{18}$$

Notice that the input signal

$$u = \left\{ -f(x,t) - ce_2(t) + \dot{x}_{2d}(t) - (2Ct + B)\delta - \gamma \operatorname{sgn}\left[s(e,t) \right] \right\} / b(x,t) \tag{19}$$

where $\gamma = \eta + \mu$ and η is a strictly positive constant, ensures the stability of the sliding motion on the switching line (16). In order to verify this property we consider the product $s(e,t)\dot{s}(e,t) = s(e,t)\left[\dot{e}_2(t) + ce_2(t) + (2Ct + B)\delta\right]$. Taking into account (13) and (19), we obtain

$$\begin{aligned} s(e,t)\dot{s}(e,t) &= s(e,t)\left[f(x,t) + \Delta f(x,t) + b(x,t)u + d(t) - \dot{x}_2(t) + ce_2(t) + (2Ct + B)\delta\right] = \\ &= s(e,t)\left\{\Delta f(x,t) - \gamma \operatorname{sgn}\left[s(e,t)\right] + d(t)\right\} \le -\eta\left|s(e,t)\right| \end{aligned} \tag{20}$$

which proves the stability of the sliding motion on the switching line (16). In order to find the system tracking error we solve equation (16). First we consider the following equation

$$e_2(t) + ce_1(t) + Ct^2 + Bt + A = 0 \tag{21}$$

which determines the considered switching line for any time $t \le t_f$, i.e. when the line moves and δ=1. Solving equation (21) with initial condition (15) and assuming for the sake of clarity that $t_0 = 0$, we can calculate the tracking error and its derivative for the time $t \in \langle 0, t_f)$. Furthermore, taking into account condition (18) and the assumption that $t_0 = 0$ we obtain

$$A = -ce_0 \tag{22}$$

Then, the tracking error and its derivative can be written as

$$e_1(t) = \left(-\frac{B}{c^2} + \frac{2C}{c^3} \right) e^{-ct} - \frac{C}{c} t^2 + \frac{2C - cB}{c^2} t + e_0 + \frac{B}{c^2} - \frac{2C}{c^3} \tag{23}$$

$$e_2(t) = \left(\frac{B}{c} - \frac{2C}{c^2} \right) e^{-ct} - \frac{2C}{c} t + \frac{2C - cB}{c^2} \tag{24}$$

Now we solve equation

$$e_2(t) + ce_1(t) = 0 \tag{25}$$

which determines the considered switching line for any time $t > t_f$ i.e. for the time when the line does not move which is equivalent to the case δ=0. For this purpose we calculate values of (23) and its derivative (24) for $t = t_f$

$$e_1(t_f) = \left(-\frac{B}{c^2} + \frac{2C}{c^3} \right) e^{-ct_f} - \frac{C}{c} t_f^2 + \frac{2C - cB}{c^2} t_f + e_0 + \frac{B}{c^2} - \frac{2C}{c^3} \tag{26}$$

$$e_2(t_f) = \left(\frac{B}{c} - \frac{2C}{c^2} \right) e^{-ct_f} - \frac{2C}{c} t_f + \frac{2C - cB}{c^2} \tag{27}$$

Then, after some calculations, we obtain the evolution of the tracking error

$$e_1(t) = \left[-\frac{B}{c^2} + \frac{2C}{c^3} + \left(-\frac{C}{c} t_f^2 + \frac{2C - cB}{c^2} t_f + e_0 + \frac{B}{c^2} - \frac{2C}{c^3} \right) e^{ct_f} \right] e^{-ct} \tag{28}$$

$$e_2(t) = \left[\frac{B}{c} - \frac{2C}{c^2} + \left(Ct_f^2 - \frac{2C - cB}{c} t_f - ce_0 - \frac{B}{c} + \frac{2C}{c^2} \right) e^{ct_f} \right] e^{-ct} \tag{29}$$

Notice that the error described by (23) and (28) converges to zero monotonically. Next, we show the procedure for finding the optimal switching line.

4. Switching line design

Now we present how to choose the optimal switching line under the assumption that the line moves with a constant deceleration to the origin of the error state space. It means that we consider the line defined by (16) where $C \neq 0$. Notice that for the time $t > t_f$, switching line (16) is fixed and passes through the origin of the error state space. This leads to the condition

$$Ct_f^2 + Bt_f + A = 0 \tag{30}$$

Furthermore, in order to avoid rapid input changes, the velocity of the introduced line should change smoothly. Thus, the following condition should hold

$$2Ct_f + B = 0 \tag{31}$$

Using relations (30), (31) and (22), we obtain the formula expressing the time when the line stops moving

$$t_f = \frac{2e_0 c}{B} \tag{32}$$

In order to choose the switching line parameters, the integral of the absolute error (IAE)

$$J = \int_0^\infty |e_1(t)| dt \tag{33}$$

is minimised subject to the system velocity

$$|e_2(t)| \leq v_{max} \tag{34}$$

and the system acceleration

$$|\dot{e}_2(t)| \leq a_{max} \tag{35}$$

constraints, where v_{max}, a_{max} represent the maximum admissible velocity and maximum admissible acceleration of the considered system, respectively. In order to facilitate further minimisation procedure, we define the following positive constant

$$k = \frac{e_0 c^2}{B} \tag{36}$$

From (36), we get

$$c = \sqrt{\frac{Bk}{e_0}} \tag{37}$$

We begin the procedure for finding optimal switching line parameters with calculating the IAE criterion. Substituting equations (23) and (28) into (33), calculating appropriate integrals and considering relation (37), we obtain

$$J(k,B) = \frac{|e_0|^{3/2}}{\sqrt{|B|}}\left(\frac{1}{\sqrt{k}} + \frac{2}{3}\sqrt{k}\right) \tag{38}$$

This criterion will be minimised with constraints (34) and (35). Since the considered criterion decreases with increasing value of B, the minimisation procedure of two variable function J(k, B) can be replaced by the minimisation of a single variable function. This remark will be very useful further in the chapter. Considering constraints, firstly we take into account each of the two constraints separately, and then we require both of them to be satisfied simultaneously.

4.1 Velocity constraint

In this section we will consider system (13) subject to velocity constraint (34). For any time $t \le t_f$ the system velocity is described by equation (24) and for the time $t \ge t_f$ by relation (29). Calculating the maximum value of $|e_2(t)|$ we get

$$\max|e_2(t)| = \left|\frac{B}{c}\left[\frac{\ln(1+2k)}{2k} - 1\right]\right| \tag{39}$$

Then using relations (34), (39) and taking into account condition (37), we obtain the following inequality

$$|B| \le \frac{v_{max}^2 k}{|e_0|}\left[\frac{\ln(1+2k)}{2k} - 1\right]^{-2} \tag{40}$$

As it was mentioned, because criterion (38) decreases with increasing value of | B | the minimisation of criterion J as a function of two variables (k, B) with the velocity constraint may be replaced by the minimisation of the following single variable function

$$J_v(k) = \frac{e_0^2}{v_{max}}\left|\frac{\ln(1+2k)}{2k} - 1\right|\left(\frac{1}{k} + \frac{2}{3}\right) \tag{41}$$

This function, for any fixed k expresses the minimum value of criterion J(k, B) which can be achieved when the velocity constraint is satisfied. Closer analysis of this criterion as a single variable function shows that (41) reaches its minimum for numerically found argument $k_{v\,opt} \approx 13.467$. Then, the optimal parameter B can be calculated from

$$B = \frac{v_{max}^2 k}{|e_0|}\left[\frac{\ln(1+2k)}{2k} - 1\right]^{-2} \operatorname{sgn}(e_0) \tag{42}$$

Substituting $k_{v\,opt}$ into (42), we obtain

$$B_{v\,opt} = \frac{v_{max}^2 k_{v\,opt}}{|e_0|}\left[\frac{\ln\left(1+2k_{v\,opt}\right)}{2k_{v\,opt}} - 1\right]^{-2} \operatorname{sgn}(e_0) \tag{43}$$

The other switching line parameters can be derived from (22), (31), (32) and (37) , and they are given below

$$A_{v\,opt} = v_{max} k_{v\,opt} \left| \frac{\ln\left(1 + 2k_{v\,opt}\right)}{2k_{v\,opt}} - 1 \right|^{-1} \text{sgn}(e_0) \tag{44}$$

$$t_{f\,v\,opt} = \frac{2|e_0|}{v_{max}} \left| \frac{\ln\left(1 + 2k_{v\,opt}\right)}{2k_{v\,opt}} - 1 \right| \tag{45}$$

$$C_{v\,opt} = -\frac{v_{max}^3 k_{v\,opt}}{4e_0^2} \left| \frac{\ln\left(1 + 2k_{v\,opt}\right)}{2k_{v\,opt}} - 1 \right|^{-3} \text{sgn}\left(e_0\right) \tag{46}$$

$$c_{v\,opt} = \frac{v_{max} k_{v\,opt}}{|e_0|} \left| \frac{\ln\left(1 + 2k_{v\,opt}\right)}{2k_{v\,opt}} - 1 \right| \tag{47}$$

That concludes the analysis of the velocity constraint taken into account separately.

4.2 Acceleration constraint

Now we consider the system acceleration constraint given by (35). Let us calculate the greatest value of $|\dot{e}_2(t)|$. The maximum absolute value of this signal, achieved at the initial time $t_0 = 0$ is equal to $|\dot{e}_2(0)| = |B|$. Then, the acceleration constraint can be expressed as follows

$$|B| \le a_{max} \tag{48}$$

Now we will analyse the criterion J minimisation task. Notice that for any given value of k, the minimum of criterion (38) is obtained for the greatest value of $|B|$ satisfying constraint (48). Therefore, the solution of the considered minimisation task can be found as a minimum of the following single variable function J

$$J_a(k) = \frac{|e_0|^{3/2}}{\sqrt{a_{max}}} \left(\frac{1}{\sqrt{k}} + \frac{2}{3}\sqrt{k} \right) \tag{49}$$

In order to analyse the minimisation task we calculate the derivative of expression (49) with respect to k. Then, we conclude that function (49) reaches its minimum for $k_{a\,opt} = 1.5$ and the optimal parameter B can be calculated from

$$B = a_{max} \text{sgn}\left(e_0\right) \tag{50}$$

The other optimal switching line parameters can be calculated from relations

$$A_{a\,opt} = -\sqrt{\frac{3a_{max}|e_0|}{2}} \text{sgn}(e_0) \tag{51}$$

$$t_{f\,a\,opt} = \sqrt{\frac{6|e_0|}{a_{max}}} \tag{52}$$

$$C_{a\,opt} = -\frac{a_{max}^{3/2}}{2\sqrt{6|e_0|}}\operatorname{sgn}(e_0) \tag{53}$$

$$c_{a\,opt} = \sqrt{\frac{3a_{max}}{2|e_0|}} \tag{54}$$

That ends our presentation of the algorithm for switching line design with the acceleration constraint.

4.3 Velocity and acceleration constraint

Finally, we consider both of constraints, i.e. the system velocity and the system acceleration and we require that they are satisfied at the same time. In order to minimise the considered criterion with constraints (34) and (35), we will minimise the following function of a single variable k

$$J_{v\,a}(k) = \max[J_v(k), J_a(k)] \tag{55}$$

This minimisation task can be solved by considering three cases (which are illustrated in Figs. 8-10):

1. $J_v(k_{a\,opt}) \le J_a(k_{a\,opt})$
2. $J_v(k_{v\,opt}) \ge J_a(k_{v\,opt})$
3. $J_v(k_{a\,opt}) > J_a(k_{a\,opt})$ and $J_v(k_{v\,opt}) < J_a(k_{v\,opt})$

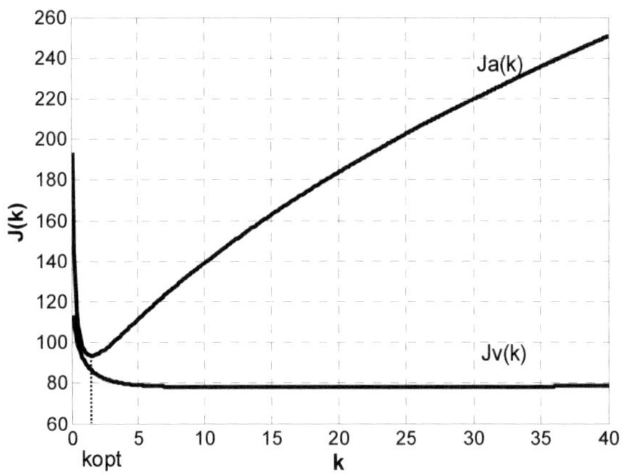

Fig. 8. Criteria $J_v(k)$ and $J_a(k)$ - case 1.

In the first case, the optimal value of k is given by $k_{opt} = k_{a\,opt} = 1.5$, and then parameter B_{opt} is given by formula (50). In the second case we obtain that $k_{opt} = k_{v\,opt} \approx 13.467$ and B_{opt} can be calculated from equation (42). In the last case, in order to find the optimal solution, we solve (numerically) equation $J_v(k) - J_a(k) = 0$ in the interval $(k_{a\,opt}, k_{v\,opt})$. Substituting numerically found value k_{opt} into either (42) or (50), we get the optimal value of B. The other optimal switching line parameters can be derived from (22), (31), (32) and (37). In this way we design the switching line which is optimal in the sense of the IAE criterion and guarantees that the state constraints are satisfied.

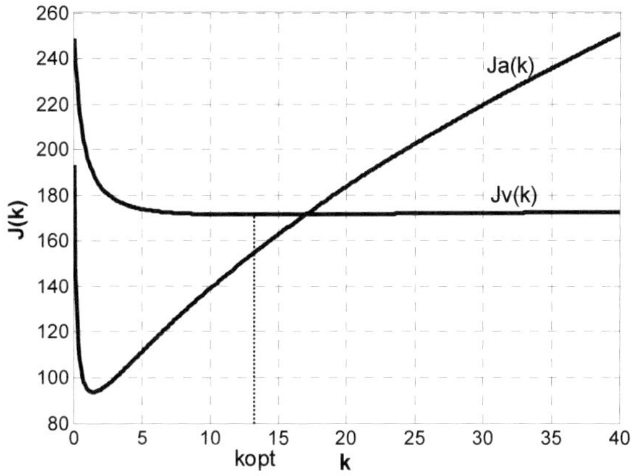

Fig. 9. Criteria $J_v(k)$ and $J_a(k)$ - case 2.

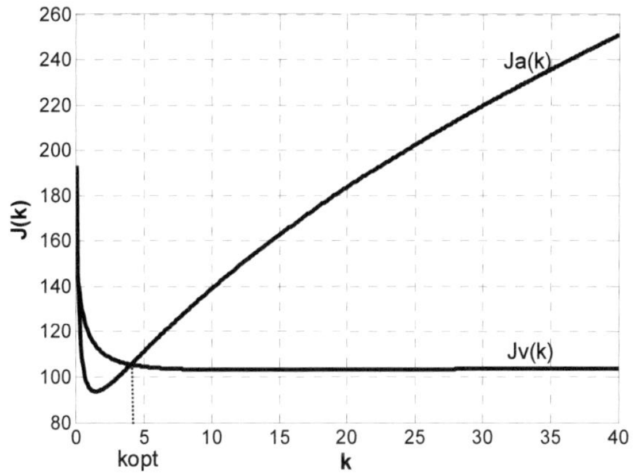

Fig. 10. Criteria $J_v(k)$ and $J_a(k)$ - case 3.

5. Simulation examples

In order to illustrate and verify the proposed method of the switching line design, we consider a suspended load described as follows

$$\dot{x}_1 = x_2, \quad \dot{x}_2 = \left[-0.15x_2 + F - f(x_1, x_2)\right]/m \tag{56}$$

where $m = 1$ kg and $f(x_1, x_2) = 0.1\text{sgn}(x_2) + 0.049x_2/(\pi|x_2| + 0.1)$ represents model uncertainty, i.e. unknown friction in the system. Consequently, $\gamma = 0.15$. The initial condition $x_0 = 0.1$ m. The demand position of system (56) is $x_d = 7$ m. We require that $v_{max} = 0.3$ m/s and $a_{max} = 0.1$ m/s^2. Then, using the presented algorithm, we obtain that

$J_v(k_{a\,opt}) > J_a(k_{a\,opt})$ and $J_v(k_{v\,opt}) < J_a(k_{v\,opt})$, and the optimal value of k can be found numerically. In the considered example it is equal to $k_{opt} \approx 4.0612$. Consequently, we obtain the following set of the optimal parameters $A_{opt} \approx$ 1.674 m/s, $B_{opt} = -0.1$ m/s^2, $c_{opt} \approx 0.2426$ 1/s and $C_{opt} \approx$ 0.0015 m/s^3. The line stops moving at the time instant $t_{f\,opt}$ equal to 33.48s.

Simulation results for the system with this line are shown in Figures 11 – 14. From Figure 11 it can be seen that the load reaches its demand position without oscillations or overshoots. Figure 12 presents the system velocity. The system acceleration is illustrated in Figure 13. The plots confirm that the required constraints are always satisfied. Furthermore, the system is insensitive from the very beginning of the control process. Figure 14 illustrates the phase trajectory of the controlled plant.

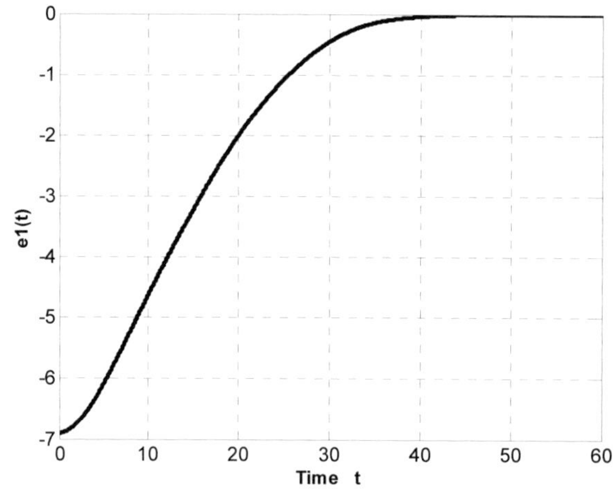

Fig. 11. System error evolution.

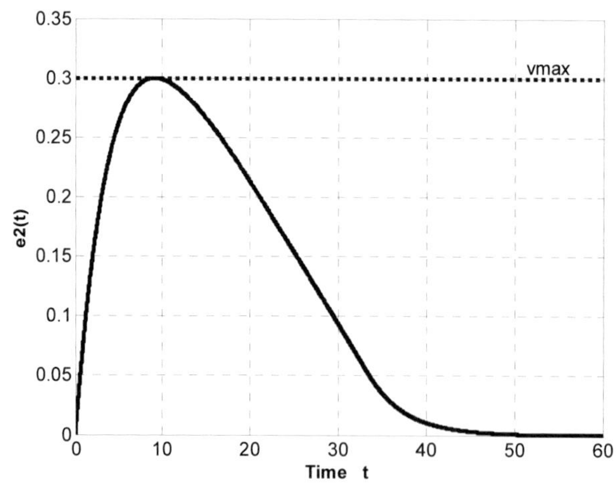

Fig. 12. System velocity.

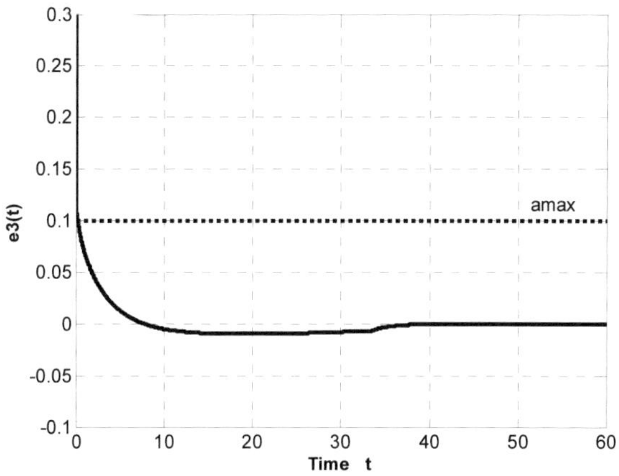

Fig. 13. System acceleration.

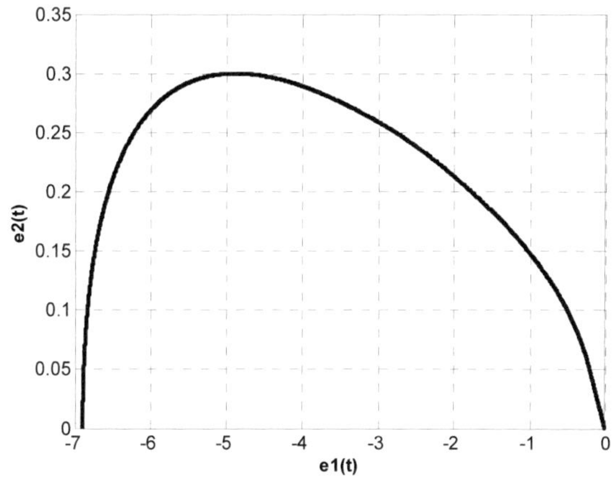

Fig. 14. Phase trajectory.

6. Conclusion

In this chapter, we proposed a method of sliding mode control. This method employs the time-varying switching line which moves with a decreasing velocity and a constant angle of inclination to the origin of the error state space. Parameters of this line are selected in such a way that integral the absolute error (IAE) is minimised with the system acceleration and the system velocity constraints. Furthermore, the tracking error converges to zero monotonically and the system is insensitive with respect to external disturbance and the model uncertainty from the very beginning of the control action.

7. Acknowledgement

This work was financed by the Polish State budget in the years 2010–2012 as a research project N N514 108638 "Application of regulation theory methods to the control of logistic processes".

8. References

Ambrosino, G.; Celentano, G. & Garofalo F. (1984). Variable structure model reference adaptive control systems. *Int. J. of Contr.* Vol., 34, 1339-1349

Bartolini, D.; Ferrara, A. & Usai, E. (1998). Chattering avoidance by second-order sliding mode control. *IEEE Trans. on Automatic Contr.* Vol., 43, 241–246

Bartoszewicz, A. (1998). Discrete-time quasi-sliding-mode control strategies. *IEEE Trans. on Ind. Electron.* Vol., 45, 633–637

Bartoszewicz, A. (1996). Remarks on 'Discrete-time variable structure control systems'. *IEEE Trans. on Ind. Electron.* Vol., 43, 235–238

Bartoszewicz, A. & Nowacka-Leverton, A. (2009) *Time-varying sliding modes for second and third order systems.* LNCIS, vol. 382, Springer-Verlag, Berlin Heidelberg

DeCarlo, R S.; Żak S. & Mathews G. (1988). Variable structure control of nonlinear multivariable systems: a tutorial. *Proceedings of the IEEE.* Vol., 76, 212–232

Draženović, B. (1969). The invariance conditions in variable structure systems. *Automatica.* Vol., 5, 287–295

Edwards C. & Spurgeon, S. K. (1998). *Sliding mode control: theory and applications.* Taylor and Francis Eds

Gao, W.; Wang Y. & Homaifa, A. (1995). Discrete-time variable structure control system. *IEEE Trans. on Ind. Electron.* Vol., 42, 1995, 117–122

Golo G. & Milosavljević, C. (2000). Robust discrete-time chattering free sliding mode control. *Syst. & Contr. Letters.* Vol., 41, 19–28

Hung, J. Y.; Gao W. & Hung, J. C. (1993). Variable structure control: a survey. *IEEE Trans. on Ind. Electron.* Vol., 40, 2–22

Levant, A. (1993).Sliding order and sliding accuracy in sliding mode control. *Int. J. of Contr.* Vol., 58, 1247–1263

Palm, R. (1994). Robust control by fuzzy sliding mode. *Automatica.* Vol., 30, 1429–1437

Palm, R.; Driankov D. & Hellendoorn, H. (1997). *Model based fuzzy control.* Springer, Berlin

Pan, Y. & Furuta, K. (2007). Variable structure control with sliding sector based on hybrid switching law. *Int. J. of Adaptive Contr. and Signal Processing.* Vol., 21, 764–778

Shtessel, Y. & Lee, Y. J. (1996). New approach to chattering analysis in systems with sliding modes. *Proceedings of the IEEE Int. Conf. on Decision and Contr.* 4014–4019

Shyu, K.; Tsai, Y. & Yung, C. (1992). A modified variable structure controller. *Automatica.* Vol., 28, 1209–1213

Sira-Ramirez, H. (1993a) A dynamical variable structure control strategy in asymptotic output tracking problems. *IEEE Trans. on Automatic Contr.* Vol., 38, 615–620

Sira-Ramirez, H. (1993b) On the dynamical sliding mode control of nonlinear systems. *Int. J. of Contr.* Vol., 57, 1039–1061

Sivert, A. et al., (2004). Robust control of an induction machine drive using a time-varying sliding surface. *Proceedings of the IEEE Int. Symposium on Ind. Electron.* 1369–1374

Slotine, J. J. & Li, W. (1991) *Applied nonlinear control.* Prentice-Hall Int. Editions

Utkin, V. (1977).Variable structure systems with sliding modes. *IEEE Trans. on Automatic Contr.* Vol., 22, 212–222

Utkin, V. & Shi, J. (1996). Integral sliding mode in systems operating under uncertainty conditions. *Proceedings of the 35th IEEE Conf. on Decision and Contr.*, 4591–4596

Xu, J. et al.(1996). Design of variable structure controllers with continuous switching control. *Int. J. of Contr.* Vol., 65, 409–431

Zlateva, P. (1996). Variable-structure control of nonlinear systems. *Contr. Engineering Practice.* Vol., 4, 1023–1028

23

Sliding Mode Control System for Improvement in Transient and Steady-state Response

Takao Sato, Nozomu Araki, Yasuo Konishi, Hiroyuki Ishigaki
University of Hyogo
Japan

1. Introduction

This chapter discusses design methods for improving sliding mode control system (Chern & Wu, 1992b; Sato, 2010; Utkin, 1977). Variable structure control (VSC) can be easily applied to nonlinear systems and is robust to plant parameter variation or load disturbance because of the existence of a sliding mode. Hence, it has been applied to various systems (e.g., an inverted pendulum system, a magnetic levitation system and robot manipulators (Ashrefiuon & Whitman, 2010; Bandal & Vernekar, 2010; Zergeroglu & Tatlicioglu, 2010)).

VSC methods employing integral compensation have been proposed to achieve servo tracking in the presence of load disturbance or plant parameter variation (Chern & Wu, 1991; 1992a;b). Robust tracking servo can be attained with a controller using integral compensation but the integral action causes phase lag, which deteriorates control performance. However, proportional compensation can adjust the gain property without changing the phase property. Hence, if control systems are designed to use proportional compensation as well as integral compensation, control performance can be further improved. Therefore, this chapter discusses a method for designing a sliding mode controller using both proportional and integral compensations. Hence, this method has higher potential than conventional methods (Chern & Wu, 1991; 1992a;b). In particular, robust servo tracking in steady state is achieved by using integral compensation, and transient response is enhanced by using proportional compensation. Hence, both responses are improved. In conventional methods, to determine the switching plane and the integral gain, a quadratic function is minimized by using the optimal linear regulator technique (Chern & Wu, 1992b) or the characteristics equation of a closed-loop system is assigned to have desired eigenvalues (Chern & Wu, 1991; 1992a). The design methods discussed in this chapter employ the optimal linear regulator technique to determine an optimal switching plane, proportional gain and integral gain to stabilize a closed-loop system.

To demonstrate the potential of these design methods, the designed variable structure controllers are applied to an inverted pendulum system that has been developed to study bifurcations and chaos (Kameoka, 2003; Sato et al., 2005; 2006). Because of the existence of unknown disturbances and unmodeled factors, its exact dynamic characteristics cannot be obtained. Hence, desired control performance cannot be attained if the system is controlled by using a controller based on a variable structure configuration. The potential of the design methods is confirmed by applying these methods to this system, as shown by simulation and

experimental results. Note that the main purpose of this chapter is not to control chaos but to develop a new method for designing a variable structure controller for improving control performance in the presence of load disturbance or plant parameter variation.
This chapter is organized as follows. In Section 2, three control systems are designed: a design method using integral compensation (2.1), proportional compensation (2.2) and both proportional and integral compensations (2.3). Section 3 gives simulation and experimental results to evaluate three method methods. Finally, concluding remarks and future works are given.

2. Design of Sliding Mode Control Systems

Consider a controlled system described as

$$\dot{x}_i = x_{i+1} \quad (i = 1, \cdots, n-1) \tag{1}$$

$$\dot{x}_n = -\sum_{i=1}^{n} a_i x_i + bu - f_d \tag{2}$$

where $x_i (i = 1, \cdots, n)$, u and f_d are the state variable, the control input and the disturbance, respectively. x_1 is the plant output, and $a_i (i = 1, \cdots, n)$ and b are the plant parameters.
To have the plant output converge to its reference input without steady-state error, a method with integral compensation (Chern & Wu, 1992b) is designed as described in 2.1, and a design method using proportional compensation and a method using both proportional and integral compensations (Sato, 2010) are designed as described in 2.2 and 2.3, respectively. For the simplicity of description, this study deals with the case of $n = 2$.

2.1 Control with integral compensation

Chern & Wu (1992b) proposed an integral variable structure controller to achieve servo tracking.

2.1.1 Design of control law with integral compensation

Error variable z is defined as:

$$\dot{z} = r - x_1 \tag{3}$$

where r is the desired state of x_1 and is set by a user. Switching function σ is chosen as:

$$\sigma = S_1(x_1 - K_I z) + x_2 \tag{4}$$

where S_1 is a constant, and constant K_I is referred to as an integral gain. Equation (4) is differentiated with respect to t, and $\dot{\sigma}$ is calculated as:

$$\dot{\sigma} = S_1(\dot{x}_1 - K_I \dot{z}) + \dot{x}_2 \tag{5}$$

Substituting equations (1) and (2) into equation (5), the next equation is obtained as:

$$\dot{\sigma} = S_1(x_2 - K_I(r - x_1)) - a_1 x_1 - a_2 x_2 + bu - f_d \tag{6}$$

The dynamic characteristics of the switching function are assigned by the differential equation:

$$\dot{\sigma} = -Q_s \mathrm{sat}(\sigma) - K_s f(\sigma) \tag{7}$$

where Q_s and K_s are arbitrary positive integers, and sat means saturation and is defined as:

$$\mathrm{sat}(\sigma) = \begin{cases} 1 & (\sigma > L) \\ \dfrac{\sigma}{L} & (|\sigma| \leq L) \\ -1 & (\sigma < -L) \end{cases} \tag{8}$$

$\sigma f(\sigma) > 0$ is required because the condition for existence of a sliding mode is $\lim_{\sigma \to 0} \sigma\dot{\sigma} < 0$ (Utkin, 1977). Hence, $f(\sigma)$ is set as $f(\sigma) = \sigma$. Then, equation (7) is rewritten as:

$$\dot{\sigma} = -Q_s \mathrm{sat}(\sigma) - K_s \sigma \tag{9}$$

Based on equations (6) and (9), a control law is derived as:

$$u = \left[-S_1(x_2 - K_I(r - x_1)) + a_1 x_1 + a_2 x_2 + f_d - Q_s \mathrm{sat}(\sigma) - K_s \sigma\right]/b \tag{10}$$

2.1.2 Design of switching surface and integral gain

While in the sliding mode, the use of $\sigma = 0$ yields:

$$x_2 = -S_1(x_1 - K_I z) \tag{11}$$

Equation (11) is substituted into equation (1), and the following equation is obtained.

$$\dot{x}_1 = -S_1(x_1 - K_I z) \tag{12}$$

Then,

$$x = Ax + Bv + Er$$
$$v = Sx$$

where

$$x = \begin{bmatrix} z \\ x_1 \end{bmatrix}, A = \begin{bmatrix} 0 & -1 \\ 0 & 0 \end{bmatrix}, B = \begin{bmatrix} 0 \\ 1 \end{bmatrix}, E = \begin{bmatrix} 1 \\ 0 \end{bmatrix}, S = \begin{bmatrix} S_1 K_I & -S_1 \end{bmatrix}$$

The optimal gain of S is found by means of the optimal linear regulator technique (Chern & Wu, 1992b), and it is derived by minimizing quadratic index I given as:

$$I = \frac{1}{2} \int_{t_s}^{\infty} (x^T Q^T x + vRv) dt \tag{13}$$

where $Q = Q^T > 0$ and $R > 0$ are a weighting matrix and a weighting parameter, and t_s is the time from when the sliding mode begins (Anderson & Moore, 1971). Weighting matrix Q

can be chosen as:

$$Q = D^T D$$

where D is a $1 \times n$ vector and pair (A, D) is observable. Then, the solution that minimizes the quadratic index is given as:

$$S = -R^{-1}B^T P$$

where P is the solution of the Riccati equation given as:

$$PA + A^T P - PBR^{-1}B^T P + Q = 0 \tag{14}$$

2.2 Control with proportional compensation

A controller employing proportional compensation is designed as described herein before a variable structure controller employing both proportional and integral compensations to be discussed in 2.3 (Sato, 2010).

The controller designed in this section cannot achieve robust servo tracking, but in comparison to the controllers employing integral compensation designed as described in 2.1 and 2.3, the effectiveness of proportional compensation in variable structure control can be confirmed.

2.2.1 Design of control law with proportional compensation

Switching function σ is defined as:

$$\sigma = S_1(x_1 - r) + x_2 \tag{15}$$

Equation (15) can be differentiated. Hence,

$$\dot{\sigma} = S_1\dot{x}_1 + \dot{x}_2$$

Based on equations (1) and (2), the equation given above is rewritten as:

$$\dot{\sigma} = S_1 x_2 - a_1 x_1 - a_2 x_2 + bu - f_d$$

Using equations (9) and the above equation, a control law is obtained as:

$$S_1 x_2 - a_1 x_1 - a_2 x_2 + bu - f_d = -Q_s \text{sat}(\sigma) - K_s \sigma \tag{16}$$

2.2.2 Design of switching surface and proportional gain

While in the sliding mode ($\sigma = 0$), equation (15) is rewritten as:

$$x_2 = -S_1(x_1 - r) \tag{17}$$

Using equation (17), equation (1) is rewritten as:

$$\dot{x}_1 = -S_1(x_1 - r)$$

Then,

$$\dot{x} = Ax + Bv + Er$$
$$v = Sx$$

where

$$x = x_1, \; A = 0, \; B = -1, \; E = S_1, \; S = S_1$$

Using the Riccati equation (14), control parameter S_1 is decided.

2.3 Control with both proportional and integral compensations

A controller is designed using both proportional and integral compensations as described in this section (Sato, 2010).

2.3.1 Design of control law with both proportional and integral compensation

Switching function σ is defined as:

$$\sigma = S_1(x_1 - r - K_I z) + x_2 \tag{18}$$

and

$$\dot{\sigma} = S_1(\dot{x}_1 - K_I \dot{z}) + \dot{x}_2 \tag{19}$$

where equation (18) can be differentiated with respect to t. Because equation (19) is equivalent to equation (5), a control law is derived as:

$$u = \left[-S_1(x_2 - K_I(r - x_1)) + a_1 x_1 + a_2 x_2 + f_d - Q_s \text{sat}(\sigma) - K_s \sigma\right] / b \tag{20}$$

2.3.2 Design of switching surface and proportional and integral gains

Using $E = [1 \;\; S_1]^T$, the control parameters of this law are decided in the same way as 2.1.2.

3. Application

3.1 Controlled plant and controller design

The controlled object is an inverted pendulum, which is a nonlinear system (Kameoka, 2003). The model of the inverted pendulum system is illustrated in Fig. 1, and its motion equation is given as:

$$J\ddot{\theta} + C\dot{\theta} + K\theta - mgh \sin\theta = mha\omega^2 \cos\theta \sin\omega t + u \tag{21}$$

where θ and u are expressed as functions of t. The system parameters in the motion equation are shown in Table 1. In particular, the damping coefficient C depends on room air temperature and is sensitive to slight changes in surroundings because the damper is an air damper. The control objective is to control the pendulum rod at a specified angle. To this end, controllers were designed using sliding mode control, as described in 2.1, 2.2 and 2.3, respectively.

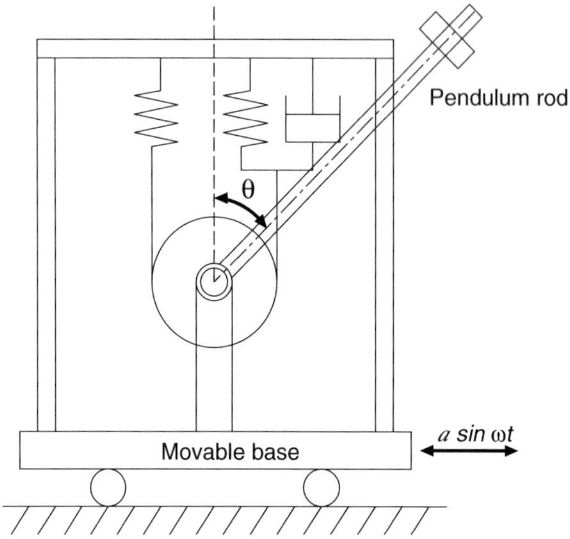

Fig. 1. Model of an inverted pendulum system

θ $[rad]$	angle of a pendulum rod
$\dot{\theta}$ $[rad/s]$	angular velocity of a pendulum rod
J $[kgm^2]$	moment of inertia
C $[Ns \cdot m/rad]$	damping coefficient
K $[Nm/rad]$	spring modulus
m $[kg]$	mass of a pendulum rod
g $[m/s^2]$	gravity acceleration
h $[m]$	distance between the center of gyration and the center of gravity of a pendulum rod
a $[m]$	amplitude of oscillation
ω $[rad/s]$	angular frequency
u $[Nm]$	input torque

Table 1. System parameters in motion equation (21)

Three control methods derived in 2.1, 2.2 and 2.3 are applied to the inverted pendulum system, and their control results are compared. These control methods are designed to control a pendulum rod at a specified angle. Their control parameters are calculated on the basis of system parameters shown in Table 2 given by pre-experiments. The true parameter of a dumper is $K = 0.587$, but assuming that there is modeling error, three control laws are designed as $K = 0.5$.
The design parameters of the control law employing just proportional compensation derived in 2.2 are set as:

$$Q = 10, \ \ R = 1, \ \ Q_s = 0.5, \ \ K_s = 1 \tag{22}$$

System parameter	Physical quantity
J	0.022 [kgm^2]
C	0.01 [Nsm/rad]
K	0.587 [Nm/rad]
m	0.547 [kg]
g	9.8 [m/s^2]
h	0.113 [m]
a	0.045 [m]
ω	1.34 [rad/s]

Table 2. Physical quantity of system parameters

Compensation method	S_1	K_I
Proportional (15)	3.2	-
Integral (4)	3.3	0.097
Proportional and Integral (18)	3.3	0.097

Table 3. Switching surface and proportional and integral compensators

The design parameters of the method employing just integral compensation derived in 2.1 are set as:

$$Q = \begin{bmatrix} 0.1 & 0 \\ 0 & 10 \end{bmatrix}, \quad R = 1, \quad Q_s = 0.5, \quad K_s = 1 \tag{23}$$

The design parameters of the control law with both the proportional and integral compensations derived in 2.3 are set to be the same as those obtained from equation (23). The calculated parameters of the switching surface and proportional and integral compensators are shown in Table 3. The reference angle of a pendulum rod is set to 10[degree], and parameter L of the saturation function (8) is set to 0.01. Control is started after 30[s].
An experimental setup is illustrated in Fig. 2. Because of the capability of a DC motor, the control input is limited as:

$$|u| < 0.0245$$

The resolution of an encoder is 0.18[degree].
To compare the control results of three control methods, a performance index is defined as:

$$E = \sum_{k=30/T_s}^{100/T_s} (r[k] - y[k])^2 \tag{24}$$

where T_s denotes the sampling interval and is set to 50[ms].

3.2 Simulation

Simulations have been conducted using the design parameters (22) and (23). The simulation results are shown in Fig. 3. The result for proportional compensation is shown in Fig. 3(a), and it is shown that the pendulum rod could not precisely follow the reference angle and steady-state error remains because of the modeling error, although its response is quick.

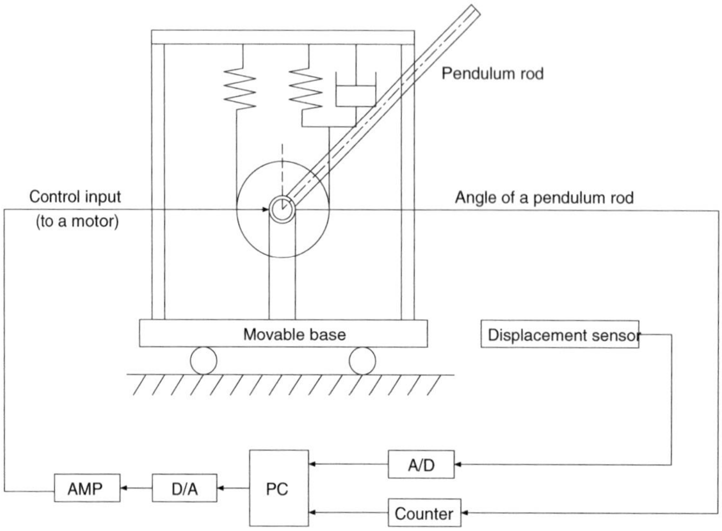

Fig. 2. Experimental setup

The angle for integral compensation is shown in Fig. 3(b), and that for proportional & integral compensation is shown in Fig. 3(c). It can be seen that the steady-state error can be eliminated in the case that integral compensation is employed. However, the transient response of Fig. 3(c) is superior to that of Fig. 3(b) since the control error is quickly improved by proportional compensation. The scores of (24) are summarized in Table 4, and E_P, E_I and E_{PI} show the scores of the control methods employing proportional, integral, and proportional-integral compensations, respectively. It can be seen that E_P is worst because steady-state error remains. In the case that integral compensation is employed, steady-state error is eliminated by using integral compensation, and E_I is better than E_P. In the case that both proportional and integral compensations are employed, steady-state error can be eliminated by using integral compensation, and control error can be quickly improved by the proportional action. Hence, E_{PI} is the smallest.

Employed compensator	Control error E
Proportional	3.3×10^3 (E_P)
Integral	1.1×10^4 (E_I)
Proportional and Integral	5.4×10^2 (E_{PI})

Table 4. Control error of simulation results

3.3 Experiment

Experiments have been conducted using the same control laws as the simulation. In the experimental setup shown in Fig. 2, angular velocity $\dot{\theta}$ cannot be directly obtained. Hence, instead of its true value, the control input is calculated using an estimated value. The sampling interval T_s is set to be the same as that of the simulation. Experimental results are shown in Fig. 4. As the performance index (24), the control results are summarized in Table 5. The experimental results are similar to the simulation results. However, the obtained plant output, that is, the angle of a pendulum rod, is quantized, and furthermore,

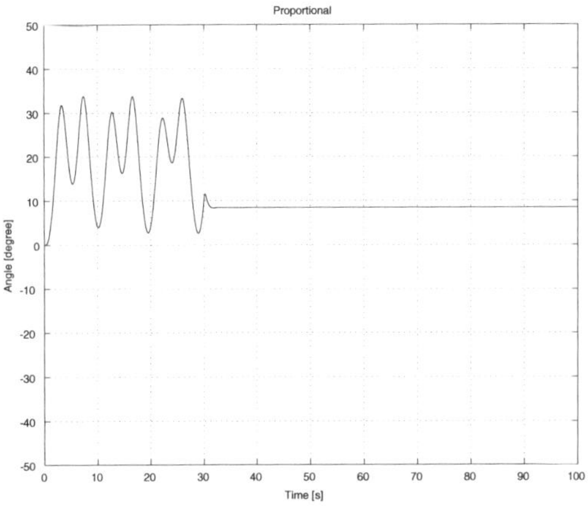

(a) Proportional compensation

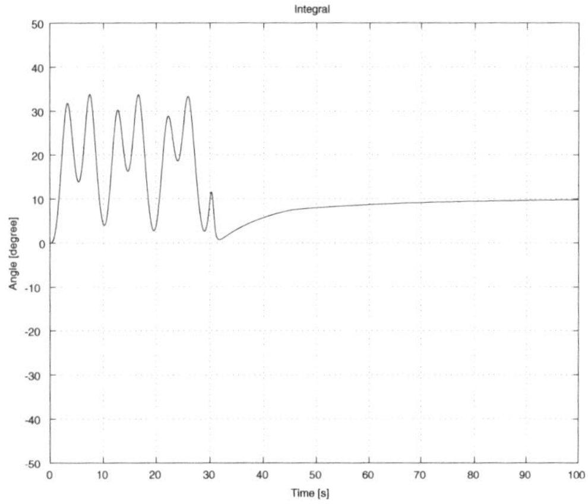

(b) Integral compensation

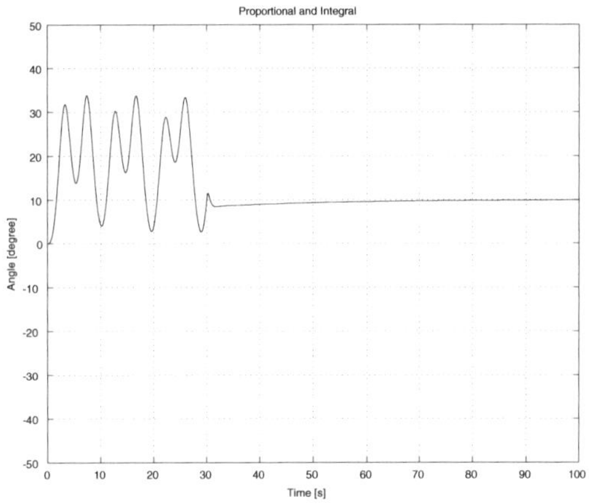

(c) Proportional and integral compensation

Fig. 3. Simulation results: angle

because its angular velocity cannot be directly obtained, an approximated value is employed instead of its true value. Hence, an obtained angular velocity is not accurate, and it usually vibrates. Therefore, a pendulum rod cannot be completely converged to a specified angle even if integral compensation is employed. However, the method using both proportional and integral compensations (E_{PI}) is better than the method employing just proportional compensation (E_P). Therefore, the effectiveness of the method using both proportional and integral compensations is confirmed.

Employed compensator	Control error E
Proportional	6.9×10^3 (E_P)
Integral	1.6×10^4 (E_I)
Proportional and Integral	6.3×10^3 (E_{PI})

Table 5. Control error of experimental results

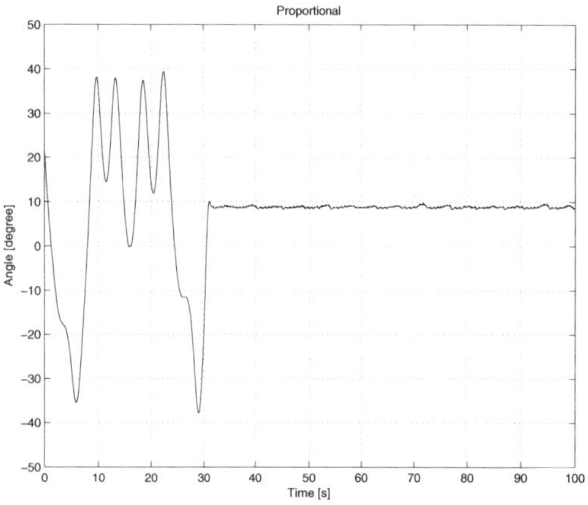

(a) Proportional compensation

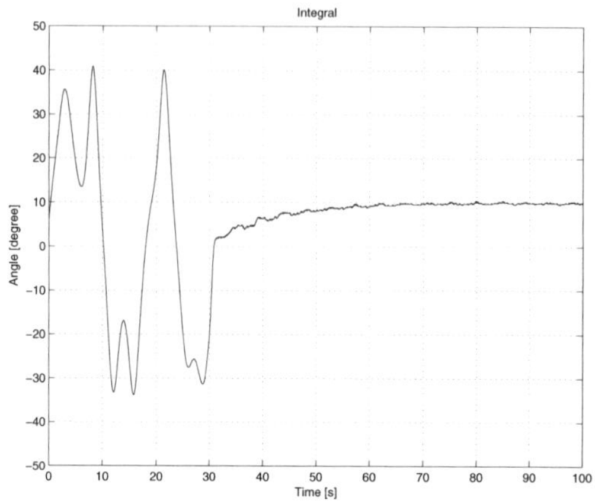

(b) Integral compensation

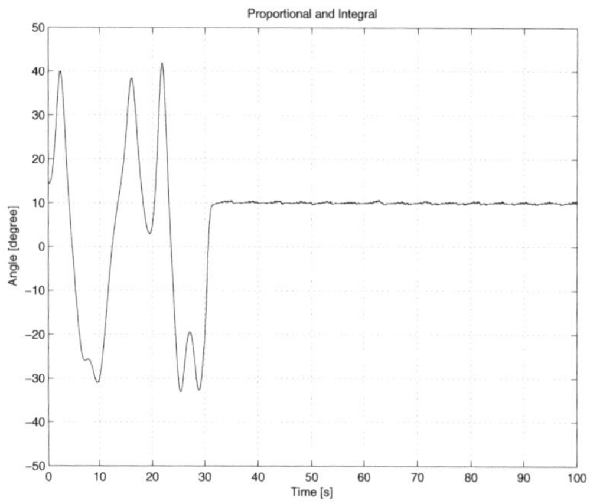

(c) Proportional and integral compensation

Fig. 4. Experimental results: angle

4. Conclusion

This chapter has discussed new methods for designing a variable structure control. Chern & Wu (1992b) have designed a sliding mode controller employing integral compensation to achieve servo tracking. To improve the transient response, a sliding mode controller has been designed using proportional compensation as well as integral compensation. If a sliding mode controller is designed using proportional compensation, the transient response can be improved but the steady-state error remains. However, the problem can be resolved by the controller using both proportional and integral compensations, which is designed employing both proportional and integral compensations. Three methods have been applied to an inverted pendulum system, these control results have been compared compared.
In this chapter, the method using proportional and integral compensations has been derived, but approaches employing derivative compensation have the potential to improve control performance if the accurate angular velocity can be obtained because the derivative action leads phase in the whole frequency domain although it increases gain in the high-frequency

domain. However, in control of an inverted pendulum system, the derivative action cannot well work because an obtained angular velocity vibrates due to the quantization of the obtained output signal.

In this study, a controlled plant is assumed to be a single-rate control system, where both the plant output and the control input are sampled or updated at the same rate. However, if a control system is extended into a multirate system, where the sampling interval of the control input differs from the hold interval of the control input, the control performance can be enhanced (Bandal & Vernekar, 2010; Bandyopadhyay & Janardhanan, 2006; Inoue et al., 2007).

5. Acknowledgment

The authors would like to express his sincere gratitude to Dr. Koichi Kameoka, Professor Emeritus at University of Hyogo, for his collaboration in this work.

6. References

Anderson, B. D. & Moore, J. B. (1971). *Linear Optimal Control Systems*, Englewood Cliffs, NJ: Prentice-Hall.

Ashrefiuon, H. & Whitman, A. (2010). Closed-loop reponse analysis of an inverted pendulum, *Proc. of ACC*, Baltimore, pp. 646–651.

Bandal, V. & Vernekar, N. (2010). Design of a discrete-time sliding mode controller for a magnetic levitation system using multirate output feedback, *Proc. of ACC*, Baltimore, pp. 4289–4294.

Bandyopadhyay, B. & Janardhanan, S. (2006). *Discrete-time Sliding Mode Control, A Multirate Output Feedback Approach*, Springer, Berlin, Gemany.

Chern, T. & Wu, Y. (1991). Design of integral variable structure controller and application to electrohydraulic velocity servosystems, *IEE Proc. Pt. D* 138(5): 439–444.

Chern, T. & Wu, Y. (1992a). Integral variable structure control approach for robot manipulators, *IEE Proc. Pt. D* 139(2): 161–166.

Chern, T. & Wu, Y. (1992b). An optimal variable structure control with integral compensation for electrohydraulic position servo control systems, *IEEE Trans., Industrial Electronics* 39(5): 460–463.

Inoue, A., Deng, M., Matsuda, K. & Bandyopadhyay, B. (2007). Design of a robust sliding mode controller using multirate output feedback, *Proc. of 16th IEEE CCA*, Singapore, pp. 200–203.

Kameoka, K. (2003). Development of a machine for studying bifurcations and chaos, *Trans. of the Society of Instrument and Control Engineers* 39(8): 786–788. (in Japanese).

Sato, T. (2010). Sliding mode control with proportional-integral compensation and application to an inverted pendulum system, *Int. J. of Innovative Computing, Information and Control* 6(2): 519–528.

Sato, T., Kondo, K. & Kameoka, K. (2005). Control experiments of an inverted pendulum system, *Proc. of SICE Annual Conference*, pp. 2671–2676.

Sato, T., Kondo, K. & Kameoka, K. (2006). Control of a machine for studying bifurcations and chaos with adaptive backstepping method, *preprints of IFAC Conference on Analysis and Control of Chaotic Systems*, Reims,France, pp. 417–422.

Utkin, V. I. (1977). Variable structure systems with sliding modes, *IEEE Trans. AC* 22: 212–222.

Zergeroglu, E. & Tatlicioglu, E. (2010). Observer based output feedback tracking control of robot manipulators, *2010 IEEE MSC*, Yokohama, pp. 602–607.

24

A New Design for Noise-Induced Chattering Reduction in Sliding Mode Control

Min-Shin Chen[1] and Ming-Lei Tseng[2]
[1]*National Taiwan University*
[2]*Chinese Culture University*
Taiwan, Republic of China

1. Introduction

The sliding mode control (Utkin, 1977; Hung *et al.*, 1993) is robust with respect to certain structured system uncertainties and disturbances. However, the early version of sliding mode control adopts a switching function in its design, and this results in high-frequency oscillations (the so-called chattering) in the control signal. Such control chattering is undesirable since it can damage the actuator and the system. Among the various solutions to reducing chattering, the boundary layer design (Burton & Zinober, 1986; Slotine & Sastry, 1983) is probably the most common approach. In the boundary layer design, a smooth continuous function is used to approximate the discontinuous sign function in a region called the boundary layer around the sliding surface. As a result, the control signal in a boundary layer design will contain no chattering in a noise free environment. However, the boundary layer design has two disadvantages. First, chattering reduction of the control signal is achieved at the sacrifice of control accuracy. To obtain smoother control signals, one must adopt a larger boundary layer width. But a larger boundary layer width results in larger errors in control accuracy. Second, when there is high-level measurement noise, the boundary layer design becomes ineffective in chattering reduction.

One of the purposes of this chapter is to show that contrary to the common belief, the boundary layer design does not completely solve the chattering problem in practical applications. Essentially this is due to the fact that the boundary layer control design is still a high gain design, and as a result, its control signal is very sensitive to high-frequency measurement noise. Control chattering may still take place due to the excitations of measurement noise. This fact will be demonstrated via the frequency domain analysis.

The other purpose of this chapter is to present a new design for chattering reduction by low-pass filtering the control signal. The new design will be shown to be able to avoid the disadvantages of conventional boundary layer design while effectively reduce chattering. The new design adopts a special control structure, in which an integrator is placed in front of the system to be controlled. A sliding mode control w is then constructed for the extended system (the original system plus the integrator). The control signal w hence has chattering, but the true control signal u going into the system is smooth since the high frequency chattering in w will be filtered out by the integrator, which acts as a low pass filter. With such a design, the chattering reduction is achieved by low pass filtering, and at the same time the control accuracy can be maintained. Another advantage of the new design is that in noisy environment, measurement noise causes severe chattering in the signal w, but the integrator can still effectively filter out the chattering. Hence, the new design has better noise immunity than the conventional boundary layer design. Previous literature (Sira-Ramirez, 1993; Sira-Ramirez *et al.*, 1996) contains no stability analysis or performance analysis, and does not address noise-induced chattering. In (Xu *et al.*, 2004), two first-order filters are employed but again noise-induced chattering is not addressed.

The new low-pass-filtering design for chattering reduction is nevertheless non-trivial. As is known, the sliding variable in sliding mode control design must be chosen such that control input shows up in the time derivative of sliding variable. In this way, the control input can influence how the sliding variable evolves. Such a design guideline must also be observed in the new design. Hence, the time derivative of the new sliding variable for the extended system should contain the sliding mode control w. This in turn suggests that the new sliding variable itself for the extended system contains the integration of w which is the true control signal u. Since the unknown disturbance d enters the system in the same place as the control signal u (the so-called matching condition (Corless & Leitmann, 1981)), the new sliding variable will inevitably contains the unknown disturbance d, and this makes evaluation of the sliding variable difficult. This is a problem that is unique to the low-pass-filtering design. Previous literature (Bartolini, 1989; Bartolini & Pydynowski, 1996) has attempted to solve this problem only with partial success. In (Bartolini, 1989), a variable structure estimator is proposed to estimate the sliding variable, but it must assume *a priori* that the system state is uniformly bounded before proving the system stability. In (Bartolini & Pydynowski, 1996), a one-dimensional observer is proposed to estimate the sliding variable, but stability is guaranteed only if a differential inequality with bounded coefficients is satisfied. This chapter will propose a complete solution by using the disturbance estimator proposed in (Chen & Tomizuka, 1989) for sliding variable estimation. A rigorous stability proof of the new sliding mode control will also be presented.

This chapter is organized as follows. Section 2 reviews the boundary layer design for the sliding mode control of a linear uncertain system. A simulation example is given to reveal the weakness of boundary layer design. Section 3 introduces the new chattering reduction control design. A second simulation example is given to confirm the advantage of new design. Finally, Section 4 gives the conclusions.

2. Boundary layer control

The purpose of this section is to review the boundary layer design in sliding mode control for a linear system with matching disturbance.

2.1 Noise-free boundary layer control

Consider a linear system with matching disturbance :

$$\dot{x}_0 = Ax_0 + B(u_0 + d), \tag{1}$$

where $x_0 \in R^n$ is the system state available from noise-free measurement, $u_0 \in R^1$ is the scalar control input, and d is an unknown disturbance with known upper bounds $|d| \leqslant D_0$, $|\dot{d}| \leqslant D_1$. The system matrices $A \in R^{n\times n}$ and $B \in R^{n\times 1}$. The pair (A, B) is controllable. The control objective is to eliminate the interference of the disturbance d with the control u_0.

To achieve this objective, a sliding mode control with boundary layer design has previously been proposed as :

$$u_0 = -Kx_0 + v_0, \tag{2}$$

where K is the state feedback gain that places the poles of $(A - BK)$ to the left half plane so that there exists a positive definite matrix P satisfying the Lyapunov equation

$$(A - BK)^T P + P(A - BK) = -I, \tag{3}$$

and v_0 is the boundary layer control :

$$v_0 = -\rho \frac{s_0}{|s_0| + \epsilon} \quad , \qquad \rho > D_0 \tag{4}$$

with the sliding variable s_0 given by

$$s_0 = 2B^T P x_0, \tag{5}$$

and ϵ a small positive number specifying the boundary layer width. Since the above boundary layer control is a continuous function of the system state, the resultant control signal (2) will have no chattering phenomenon if there is no measurement noise and unmodeled dynamics. Note that close to the sliding surface ($s_0 \approx 0$), the boundary layer control (4) reduces to a proportional control with high control gain: $v_0 = -(\rho/\epsilon)s_0$. This high gain characteristics is the cause of noise-induced chattering introduced in the next section.

2.2 Noise-corrupted boundary layer control

In order to analyze how the conventional boundary layer control responds to measurement noise, a zero-mean stochastic noise n is introduced into the measurement of system state x. The state equation (1) thus becomes :

$$\dot{x}_1 = Ax_1 + B(u_1 + d), \tag{6}$$

where x_1 is the noise-affected system state, u_1 is the noise-affected control input :

$$u_1 = -K(x_1 + n) + v_1, \qquad v_1 = -\rho \frac{s_1}{|s_1| + \epsilon} \quad , \tag{7}$$

where n is the stochastic measurement noise, and s_1 the sliding variable

$$s_1 = 2B^T P(x_1 + n). \tag{8}$$

Define the perturbed control input $\delta u = u_1 - u_0$ as the difference between the noise-free u_0 in the previous section and noise-affected u_1 in this section. Similarly, the perturbed state $\delta x = x_1 - x_0$ is the difference between the noise-free x_0 and noise-affected x_1. Since the measurement noise n is assumed to be of small magnitude, so are δu and δx. As a result, one can apply linearization technique to the nonlinear boundary layer control system; in particular, one can derive the linear transfer function $T_n(s)$ from the measurement noise n to the perturbed control δu. From this transfer function $T_n(s)$, one can learn how the high-frequency measurement noise n affects the perturbed control δu and the noise-affected input $u_1 = u_0 + \delta u$. If the high-frequency gain of $T_n(s)$ is large, it suggests that measurement

noise n can induce high-frequency chattering in the control u_1. The derivation procedure of $T_1(s)$ is as follows.
It follows from (1) and (6)

$$\delta\dot{x} = A\delta x + B\delta u, \tag{9}$$

and from (2) and (7),

$$\delta u = -K(\delta x + n) + v_1 - v_0. \tag{10}$$

If one defines $f(x) = \rho\frac{2B^TPx}{|2B^TPx|+\epsilon}$, according to (4) and (7),

$$v_1 - v_0 = -f(x_1 + n) + f(x_0) \doteq -\frac{\partial f}{\partial x}_{x=x_0} \cdot (x_1 + n - x_0) = -N(\delta x + n), \tag{11}$$

where the second equality results from the Taylor series expansion of $f(x_1 + n)$ at x_0, and

$$N = \frac{\partial f}{\partial x}_{x=x_0} = \rho\frac{2B^TP\epsilon}{(|s_0| + \epsilon)^2} \in R^{1\times n} \tag{12}$$

Note that in the above Taylor series expansion of the nonlinear function $f(\cdot)$, one can neglect all high-order terms and retains only the first order term because $\delta x + n$ is small.
Combining equations (10) and (11) gives

$$\delta u = -(K + N)(\delta x + n). \tag{13}$$

Substituting the above equation into (9) results in the closed-loop transfer function from n to δx:

$$\delta x = -\left[sI - A + B(K + N)\right]^{-1}B(K + N)n, \tag{14}$$

where s represents the Laplace transform operator. Finally, the transfer function from $n \in R^n$ to $\delta u \in R^1$ can be deduced from (13) and (14),

$$\delta u = T_n(s)\, n, \qquad T_n(s) = \{(K + N)\left[sI - A + B(K + N)\right]^{-1}B - I\}(K + N). \tag{15}$$

One may now use Equation (15) to study how the stochastic measurement noise n affects the control input $u_1 = u_0 + \delta u$ in the boundary layer control. In particular, one is interested in knowing whether the high-frequency measurement noise n will contribute to the chattering (high-frequency oscillations) of control signals in a boundary layer design. Note that control chattering occurs only after the sliding variable s_0 approaches almost zero. When this occurs, the vector N in (12) may be approximated by

$$N \doteq \rho\frac{2B^TP}{\epsilon}.$$

One may now plot the Bode diagram of $T_n(s)$ in (15) with the row vector N given as above to check how sensitive the boundary layer control is to the measurement noise.
A simulation example is given below to show that even if a boundary layer design has been used, control chattering may still take place due to the measurement noise.

Example 1: Consider the system (1) with

$$A = \begin{bmatrix} 0 & 1 & 0 \\ 0 & 0 & 1 \\ 7 & -1 & 2 \end{bmatrix}, \qquad B = \begin{bmatrix} 0 \\ 0 \\ 1 \end{bmatrix},$$

and a disturbance $d = cos(t)$. The sliding mode control (2) and (4) has design parameters: boundary layer width $\epsilon = 1, 0.01$, and 0.001 respectively, control gain $\rho = 1.2$, and state feedback gain $K = \begin{bmatrix} 67 \ 46 \ 14 \end{bmatrix}$. From (15), the singular value of transfer function $T_n(s)$ from n to $\delta u \in R^1$ is plotted in Figure 1.

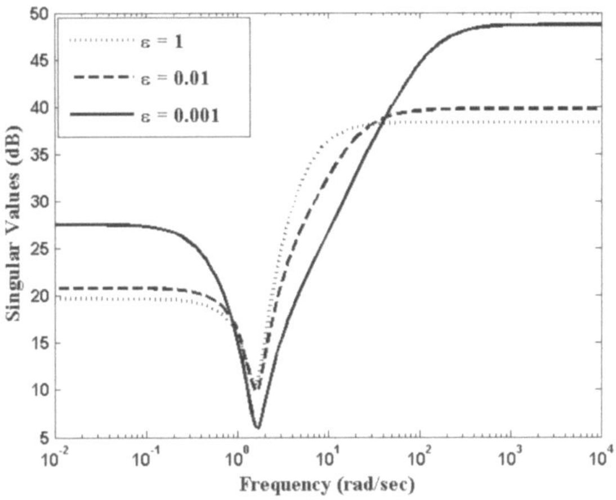

Fig. 1. Singular values of $T_n(s)$ with different ϵ

The high gain of $T_n(s)$ at high frequency suggests that the sliding mode control signal is very sensitive to high-frequency measurement noise. The smaller the boundary layer width ϵ, the more sensitive the control input to the measurement noise. As a result, the high frequency measurement noise n will create substantial high frequency oscillations (chattering) in the perturbed control δu, and hence in the noise-affected input $u_1 = u_0 + \delta u$. Figure 2 shows the time response of control input u_1, which confirms the existence of high frequency chattering even if a boundary layer of width $\epsilon = 0.005$ has been introduced into the sliding mode control design.

3. Filtered sliding mode control

3.1 Sliding variable design

As is demonstrated in the simulation example 1, sliding mode control with the boundary layer design still exhibits the chattering phenomenon when there is a high level of measurement noise. Hence, a solution better than the boundary layer design is required to reduce the chattering in sliding mode control. To this end, one will introduce the Filtered Sliding Mode Control in this section, whose control structure is depicted in Figure 3. In Figure 3, an integrator is intentionally placed in front of the system, and $w = \dot{u}$ is treated as the control variable for the extended system. A switching sliding mode control law is chosen for w to suppress the effects of disturbance d. Even though w is chattering, the control input u to the system will be smooth because the high-frequency chattering will be filtered out by the

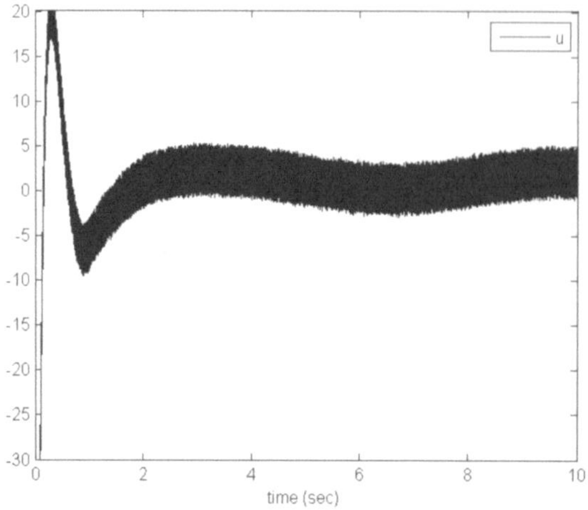

Fig. 2. Time history of control input

integrator, which acts as a low-pass filter. In other words, the new control design removes chattering by filtering the control signal, hence, the control structure in Figure 3 is called *Filtered* Sliding Mode Control.

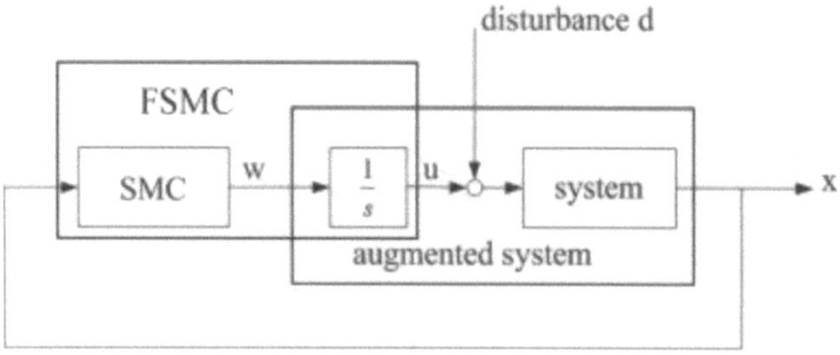

Fig. 3. Filtered sliding mode control

Consider a linear system with disturbance:

$$\dot{x} = Ax + B(u + d). \tag{16}$$

For the design of filtered sliding mode control, one chooses the sliding variable as follows.

$$s_2 = \dot{z} + \lambda z, \qquad z = Cx, \tag{17}$$

where λ is a positive constant, and the row vector $C \in R^{1\times n}$ is chosen such that (A, B, C) is of relative degree one, and the $(n-1)$ zeros of the system (A, B, C) are in the stable locations. It will be shown in the proof of Theorem 3 below that when s_2 is driven to zero, the system state x will also be convergent to zero.
Using (17) and (16), one finds

$$s_2 = CAx + CB(u + d) + \lambda Cx, \tag{18}$$

and, by taking the time derivative of s_2,

$$\begin{aligned}\dot{s}_2 &= (CA^2 + \lambda CA)x + (CAB + \lambda CB)u + CBw \\ &+ (CAB + \lambda CB)d + CB\dot{d}.\end{aligned}$$

Note that the control variable $w = \dot{u}$ appears in the time derivative of the sliding variable s_2, suggesting that one can control the evolution of s_2 by properly choosing the control variable w. However, there is a problem that according to (18), the expression of s_2 contains the unknown disturbance term d. Therefore, it is difficult to evaluate the sliding variable s_2.
To solve this problem, one will use the Disturbance Estimator proposed in (Chen & Tomizuka, 1989) to estimate the disturbance d. With an estimate of d, one can obtain an estimate of the sliding variable s_2 via (18). In the sequel, an estimator for the unknown disturbance d will be constructed based on the scalar variable z defined in (17). Note that z satisfies the following differential equation,

$$\dot{z} = CAx + CB(u + d), \tag{19}$$

Call $\hat{z}$ an estimate of z, and denote the estimation error as

$$e = z - \hat{z}.$$

Construct the governing equation of $\hat{z}$ as follows.

$$\dot{\hat{z}} = CAx + \beta e + CB(u + v), \qquad v = \rho \frac{e}{|e| + \epsilon}, \tag{20}$$

where β is a positive constant, ρ an estimator gain larger than the disturbance upper bound D_0, and ϵ is a positive constant close to zero. With the above estimator (20), an estimate of the disturbance d will be provided by

$$\hat{d} = \frac{\beta}{CB} e + v = \frac{\beta e}{CB} + \frac{\rho e}{|e| + \epsilon}. \tag{21}$$

Once one has obtained an estimate of d, one can approximate s_2 in (18) by

$$\hat{s}_2 = CAx + CB(u + \hat{d}) + \lambda Cx. \tag{22}$$

The following theorem proves the effectiveness of the above disturbance estimator (20) and (21).
Theorem 2: The disturbance estimation error $d - \hat{d}$, where $\hat{d}$ is given by (21), will become arbitrarily small if the estimator gain ρ in (20) is sufficiently large.
Proof: One can refer to the original paper on disturbance estimator (Chen & Tomizuka, 1989). For completeness of this chapter, a simple proof will be given below. From (19) - (20), one can easily obtain

$$\begin{aligned}\dot{e} &= -\beta e - CB(v - d) \\ &= -\beta e - CB(\rho \frac{e}{|e| + \epsilon} - d).\end{aligned} \tag{23}$$

It will be shown that both e and $\dot{e}$ will become arbitrarily small if ρ is sufficiently large. Notice from (23) and (21) that $\dot{e} = CB(d - \hat{d})$. Therefore, the smallness of $\dot{e}$ implies the smallness of $d - \hat{d}$ and hence, the success of disturbance estimation.

Let Lyapunov function $V_1 = \frac{1}{2}e^2$, and take its time derivative,

$$\begin{aligned}\dot{V}_1 = e\,\dot{e} \;&=\; e(-\beta e - CB(v - d)) \\ &\leq -\beta e^2 - |CBe|(\rho\frac{|e|}{|e|+\epsilon} - D_0) \\ &\leq -\beta {n_1}^2, \qquad \text{for all } e \notin N_1,\end{aligned}$$

where $N_1 = \{e : |e| < n_1 = (\epsilon D_0)/(\rho - D_0)\}$. With the last inequality, one can prove (see (Chen & Tomizuka, 1989)) that $|e(t)| < n_1$ for all $t > T_1$ for some finite time T_1. Since $n_1 = (\epsilon D_0)/(\rho - D_0)$ becomes arbitrarily small as the disturbance estimator gain ρ becomes sufficiently large, one concludes that e becomes arbitrarily small within a finite time if ρ is sufficiently large.
To check the behavior of $\dot{e}$, one chooses $V_2 = \frac{1}{2}\dot{e}^2$, and take its time derivative,

$$\begin{aligned}\dot{V}_2 = \dot{e}\,\ddot{e} \;&=\; \dot{e}(-\beta\dot{e} - CB(\frac{\rho\epsilon\dot{e}}{(|e|+\epsilon)^2} - \dot{d})) \\ &\leq -\beta\dot{e}^2 - \frac{\rho\epsilon|CB||\dot{e}|^2}{(n_1+\epsilon)^2} + |CB\dot{e}|D_1, \;\; t \geq T_1, \\ &\leq -\beta\dot{e}^2 - \frac{\rho\epsilon|CB\dot{e}|}{(n_1+\epsilon)^2}(|\dot{e}| - \frac{D_1(n_1+\epsilon)^2}{\rho\epsilon}), \;\; t \geq T_1, \\ &\leq -\beta {n_2}^2, \qquad \text{for all } \dot{e} \notin N_2,\end{aligned}$$

where $N_2 = \{\dot{e} : |\dot{e}| < n_2 = D_1(n_1+\epsilon)^2/(\rho\epsilon)\}$. From the last inequality, one can prove (Chen & Tomizuka, 1989) that $|\dot{e}(t)| < n_2$ for all $t > T_2$ for some finite time T_2. Since $n_2 = D_1(n_1 + \epsilon)^2/(\rho\epsilon)$ becomes arbitrarily small as the disturbance estimator gain ρ becomes sufficiently large, one concludes that $\dot{e} = CB(d - \hat{d})$ becomes arbitrarily small within a finite time if ρ is sufficiently large. End of proof.

3.2 Control variable design

In the filtered sliding mode control, the objective of the control variable $w = \dot{u}$ is to drive the sliding variable s_2 to (almost) zero in the face of unknown disturbance. For this purpose, one chooses

$$\begin{aligned}\dot{u} &= w \\ &= -(CA^2 + \lambda CA)x - (CAB + \lambda CB)u \\ &\quad -\sigma s_2 - \delta\,\mathrm{sgn}(s_2),\end{aligned}$$

where $\sigma > 0$, $\mathrm{sgn}(\cdot)$ is the sign function, and δ is an upper bound of the uncertainty $|\Delta p|$ with

$$\Delta p = (CAB + \lambda CB)d + \dot{d}. \tag{24}$$

As explained in the previous section, it is impossible to evaluate the sliding variable s_2 due to the disturbance d involved. Hence, to implement the proposed control, one uses the estimate $\hat{s}_2$ in place of s_2,

$$\begin{aligned}\dot{u} &= w \\ &= -(CA^2 + \lambda CA)x - (CAB + \lambda CB)u \\ &\quad -\sigma\hat{s}_2 - \delta\,\mathrm{sgn}(\hat{s}_2),\end{aligned} \tag{25}$$

where $\hat{s}_2$ comes from (22). Finally, it is commented that in the above filtered sliding mode control law one can replace the sign function by other smooth approximations such as the saturation function or other boundary layer design.

From (25), the true control input to the system is given by

$$u = H(s)w, \quad H(s) = \frac{1}{s}. \tag{26}$$

Even though the switching control w in (25) contains high-frequency chattering, the high-frequency chattering will be filtered out by the low-pass filter $H(s)$. The control input u to the real system can be obtained by direct integration and then becomes chattering free.

The following theorem, which is the main result of this chapter, proves that the proposed control (25) is practically stabilizing.

Theorem 3: The proposed filtered sliding mode control (25) *practically* stabilizes the system (16) with bounded control u, in the sense that the system state is asymptotically driven into a residual set around the origin, with the size of residual set becoming arbitrarily small when the estimator gain ρ in the disturbance estimator (20) becomes sufficiently large.

Proof: Denote $\tilde{s}_2 = s_2 - \hat{s}_2$, where s_2 and $\hat{s}_2$ are as given by (18) and (22) respectively. It is easy to check that $\tilde{s}_2 = CB(d - \hat{d})$. To study the evolution of s_2, choose Lyapunov function $V = \frac{1}{2}s_2^2$ and check its time derivative under the proposed control w in (25),

$$\begin{aligned}
\dot{V} &= s_2[(CA^2 + \lambda CA)x + (CAB + \lambda CB)u + CBw \\
&\quad + (CAB + \lambda CB)d + CB\dot{d}] \\
&= s_2[-\sigma \hat{s}_2 - \delta\, \mathrm{sgn}(\hat{s}_2) + \Delta p] \\
&= -\sigma s_2^2 + \sigma s_2 \tilde{s}_2 + s_2[-\delta\, \mathrm{sgn}(\hat{s}_2) + \Delta p],
\end{aligned} \tag{27}$$

where Δp is as given in (24), and one has used $\hat{s}_2 = s_2 - \tilde{s}_2$ to obtain the third equality. There are two possible cases for the square brackets in the above equation.

Case 1. $|s_2| > |\tilde{s}_2|$: In this case, $\mathrm{sgn}(\hat{s}_2) = \mathrm{sgn}(s_2 - \tilde{s}_2) = \mathrm{sgn}(s_2)$. Equation (27) then becomes

$$\begin{aligned}
\dot{V} &\leq -\sigma s_2^2 + \sigma s_2 \tilde{s}_2 - |s_2|(\delta - |\Delta p|) \\
&\leq -\sigma s_2^2 + \sigma s_2 \tilde{s}_2 \\
&\leq -\sigma |s_2|^2 + \sigma |s_2| v,
\end{aligned}$$

where the second inequality results from the design choice $\delta > |\Delta p|$, and the third inequality (with v an arbitrarily small number) comes from Theorem 2 that $\tilde{s}_2 = CB(d - \hat{d})$ becomes arbitrarily small asymptotically. From the last inequality, it is not difficult to show that asymptotically one has $\lim_{t\to\infty} |s_2| \leq v$; that is, s_2 becomes arbitrarily small asymptotically.

Case 2. $|s_2| \leq |\tilde{s}_2|$: Since $\tilde{s}_2 = CB(d - \hat{d})$, it follows from Theorem 2 that $|s_2|$ becomes arbitrarily small asymptotically.

Judging from conclusions of both Case 1 and 2, one can say that the sliding variable s_2 becomes arbitrarily small asymptotically. One next shows that the system state x will also become arbitrarily small as s_2 does. To this end, introduce a state transformation (Isidori, 1989),

$$x = T\begin{bmatrix} z \\ \eta \end{bmatrix}, \quad T \in R^{n\times n}. \tag{28}$$

where the external state $z \in R^1$ is as defined in (17), and the *internal state* $\eta \in R^{n-1}$ satisfies

$$\dot{\eta} = Q\eta + Lz, \tag{29}$$

for some matrices Q, L, in which Q is a square matrix whose eigenvalues are open-loop zeros of the triple (A, B, C) (Isidori, 1989). Since, in the design of sliding variable in (17), $z = Cx$ is chosen such that (A, B, C) has only stable zeros, Q is stable.
When s_2 becomes arbitrarily small, it follows from (17) that the external state z also becomes arbitrarily small since $\dot{z} + \lambda z = s_2$ can be regarded as a stable system z subject to small input signal s_2. Similarly, (29) can be regarded as a stable system η subject to small input signal z. Hence, its state η will also become arbitrarily small asymptotically. Finally, since both z and η become arbitrarily small, so does the original system state x according to the state transformation (28). End of proof.
To show the efficacy of the proposed filtered sliding mode control in noisy environments, a simulation example is presented below.
Example 2: Filtered sliding mode control.
The same system as in Example 1 is tested again for the proposed filtered sliding mode control (25). Here one has chosen

$$C = [2,\ 3,\ 1]\,.$$

The disturbance $d = \cos(t)$ and the state measurement is contaminated with a uniform noise with zero-mean and standard deviation 0.05. One tests the proposed filtered sliding mode control (25). The parameters are chosen such that $\lambda = 2$ in (17) and $\beta = 100$, $\rho = 1.2$, $\epsilon = 0.005$ in (20). Other design parameters are $\sigma = 30$ and $\delta = 9.6$ in (25). The plot of Figure 4(a) shows the time history of system state, which achieves almost the same performance as that with the boundary layer control. However, note from the plot of Figure 4(b) that the filtered sliding mode design has successfully removed chattering in the control input u even in this noisy environment.

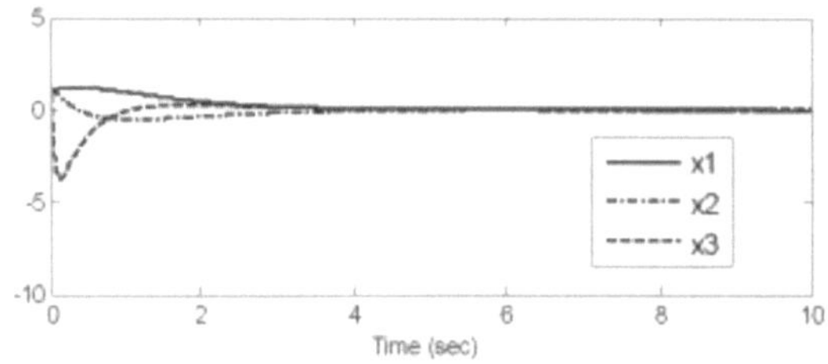

(a) System state

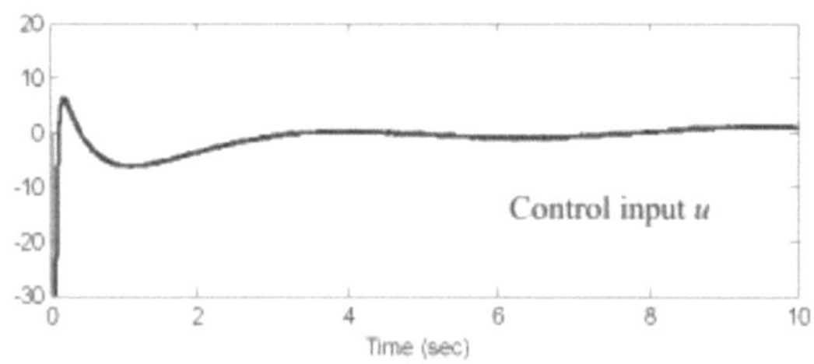

(b) Control input

Fig. 4. Filtered sliding mode control with noise

4. Conclusions

This chapter first shows via the linearization technique and the frequency domain analysis that the boundary layer design in the sliding mode control can still exhibit control chattering due to the excitations of measurement noise. Hence, other solutions to the chattering reduction such as those in (Chen *et al.*, 2002; 2007) should be searched.

Second, a new design is proposed to reduce control chattering in sliding mode control by low-pass filtering the control signal. The new design requires estimation of the sliding variable, and this is achieved by the use of a disturbance estimator. The unique feature of this new design is that chattering reduction is achieved by low-pass filtering the control signal, and control accuracy can be maintained by a sufficiently large disturbance estimator gain. This is contrary to the conventional boundary layer design, where chattering reduction is achieved at the price of sacrificing the control accuracy. This chapter further shows via simulation examples that when there is high-level measurement noise, the boundary layer design can no longer reduce chattering, but the new design in this chapter can effectively reduce chattering even in noisy environments.

5. References

Bartolini, G. (1989). Chattering phenomena in discontinuous control systems, *Int. J. Systems Sci.*, Vol. 20, 2471-2481.

Bartolini, G., & Pydynowski, P. (1996). An improved, chattering free, V.S.C scheme for uncertain dynamical systems, *IEEE Trans. Autom. Control*, 41, 1220-1226.

Burton, J. A. & Zinober, A. S. I. (1986). Continuous approximation of variable structure control. *Int. J. System Science*, 17, 875-885.

Chen, M. S. & Tomizuka, M. (1989). Disturbance estimator and its application in estimation of system output derivatives, *Proceedings of Conference on Decision and Control*, Tampa, pp. 452-457.

Chen, M. S., Hwang, Y. R. & Tomizuka, M. (2002). A state-dependent boundary layer design for sliding mode control. *IEEE Trans. Autom. Control*, 47, 1677-1681.

Chen, M. S., Chen, C. H. & Yang, F. Y. (2007). An LTR-observer-based dynamic sliding mode control for chattering reduction. *Automatica*, 43, 1111-1116.

Corless, M. J., & Leitmann, G. (1981). Continuous state feedback guaranteeing uniform ultimate boundedness for uncertain dynamic systems, *IEEE Trans. Autom. Control*, 26, 1139-1143.

Hung, J. Y., Gao, W. B. & Hung, J. C. (1993). Variable structure control: a survey. *IEEE Trans. Ind. Electron.*, 40, 2-22.

Isidori, A. (1989) *Nonlinear Control Systems*, Springer-Verlag, New York.

Sira-Ramirez, H. (1993). On the dynamical sliding mode control of nonlinear systems, *Int. J. Control*, 57, 1039-1061.

Sira-Ramirez, H., Llanes-Santiago, O. & Fernadez, N. A. (1996). On the stabilization of nonlinear systems via input-dependent sliding surfaces, *Int. J. of Robust and Nonlinear Control*, 6, 771-780.

Slotine, J. J. E. & Sastry, S. S. (1983). Tracking control of nonlinear systems using sliding surfaces with application to robot manipulator. *Int. J. Contr.*, 38, 931-938.

Utkin, V. I. (1977). Variable structure systems with sliding modes. *IEEE Trans. Autom. Control*, AC-22, 212-222.

Xu, J.-X., Pan, Y.J. & Lee, T.H. (2004). A new sliding mode control with closed-loop filtering architecture for a class of nonlinear systems, *IEEE Transactions on Circuits and Systems II: Express Briefs*, 51 (4), 168-173.

25

Multimodel Discrete Second Order Sliding Mode Control: Stability Analysis and Real Time Application on a Chemical Reactor

Mohamed MIHOUB, Ahmed Said NOURI and Ridha BEN ABDENNOUR
National School of Engineers of Gabes, University of Gabes
Tunisia

1. Introduction

The variable structure control is principally characterized by its robustness with respect to the system's modeling uncertainties and external disturbances (Decarlo et al. (1988); Filippov (1960); Lopez & Nouri (2006); Utkin (1992)). Sliding Mode Systems are a particular case of the Variable Structure Systems (VSS). They are feedback systems with discontinuous gains switching the system' s structure according to the state evolution, in order to maintain the trajectory within some specified subspace called the sliding surface (Utkin (1992)). However, the application of this control law is confronted to a serious problem. In fact, sliding mode necessitates an infinite switching frequency which is impossible to realize in numerical applications because of the calculation time and of the sensors dynamics that can not be neglected. The discontinuous control generates in that case oscillations on the state and on the switching function (Utkin (1992)). Owing to the many advantages of the digital control strategy (Ben Abdennour et al. (2001)), the discretization of the sliding mode control (SMC) has become an interesting research field. Unfortunately, the chattering phenomenon is more obvious in this case, because the sampling rate is more reduced.
Many approaches have been suggested in order to resolve this last problem. Most of them propose a reduction in the oscillation amplitude at cost of the robustness of the control law (Utkin et al. (1999)). In the eighties, a new control technique, called high order sliding mode control, have been investigated. Its main idea is to reduce to zero, not only the sliding function, but also its high order derivatives. In the case of the r-order sliding mode control, the discontinuity is applied on the (r-1) derivative of the control. The effective control is obtained by (r-1) integrations and can, then, be considered as a continuous signal. In other words, the oscillations generated by the discontinuous control are transferred to the higher derivatives of the sliding function. This approach permits to reduce the oscillations amplitude, the notorious sliding mode systems robustness remaining intact (Levant (1993)).
Another problem of the SMC is its vulnerability to external disturbances, parametric variations and non linearity, essentially, during the reaching phase. A solution to this problem, based on the multimodel approach, was proposed by the authors in (Mihoub et al. (2009a)).
The combination of the multimodel approach and the second order discrete sliding mode control (2-DSMC) allows resolving both the chattering problem and the vulnerability during

the reaching phase. A stability analysis of the multimodel discrete second order sliding mode control (MM-2-DSMC) is proposed in this work. The performances offered by the second order approach and by the multimodel approach are illustrated by a comparison between the experimental results on a chemical reactor of the first order DSMC, the 2-DSMC and the MM-2-DSMC (Mihoub et al. (2009a;b)).

2. Discrete second order sliding mode control

2.1 High order sliding mode approach

The high order sliding mode control concept have been introduced in the eighties at the aim of resolving the chattering phenomenon. Levantovsky (Levantovsky (1985)) and Emelyanov (Emelyanov et al. (1986)) proposed to transfer it on the higher derivatives of the control law. Therefore, the system's input becomes continuous.
Let's consider the non linear system defined by:

$$\dot{x} = f(t, x, u) \tag{1}$$

where :

- $x(t) = [x_1(t), ..., x_n(t)]^T \in X$ state vector, $X \subset R^n$.
- $u(t, x)$ is the control.
- $f(t, x, u)$ is a function supposed sufficiently differentiable.

We denote by $S(t, x)$ the sliding function. It is a differentiable function with its $(r-1)$ first derivatives relatively to the time depending only on the state $x(t)$ (that means they contain no discontinuities).
Definition **(Salgado (2004))**
A sliding mode is said "first order sliding mode" if and only if $S(t, x) = 0$ and $S(t, x)\dot{S}(t, x) < 0$
A sliding mode is said "r^{th} order sliding mode" if and only if:

$$S(t, x) = \dot{S}(t, x) = ... = S^{(r-1)}(t, x) = 0 \tag{2}$$

The aim of first order sliding mode control is to force the state to move on the switching surface $S(t, x) = 0$. In high order sliding mode control, the purpose is to force the state to move on the switching surface $S(t, x) = 0$ and to keep its $(r-1)$ first successive derivatives null (Salgado (2004)).
In the case of second order sliding mode control, we must verify:

$$S(t, x) = \dot{S}(t, x) = 0 \tag{3}$$

We introduce here the equivalent control approach for second order sliding mode control (Salgado (2004)).
The derivative of the sliding function is:

$$\frac{d}{dt}S(t, x) = \frac{\partial}{\partial t}S(t, x) + \frac{\partial}{\partial x}S(t, x)\frac{\partial x}{\partial t} \tag{4}$$

Considering the relation (1), we can write:

$$\dot{S}(t,x,u) = \frac{\partial}{\partial t}S(t,x) + \frac{\partial}{\partial x}S(t,x)f(t,x,u) \tag{5}$$

The second order derivative of $S(t,x)$ is:

$$\tfrac{d}{dt}\dot{S}(t,x,u) = \tfrac{\partial}{\partial t}\dot{S}(t,x,u) + \tfrac{\partial}{\partial x}\dot{S}(t,x,u)\tfrac{\partial x}{\partial t} + \tfrac{\partial}{\partial u}\dot{S}(t,x,u)\tfrac{\partial u}{\partial t} \tag{6}$$

This last equation can be written as follow:

$$\frac{d}{dt}\dot{S}(t,x) = \theta(t,x) + \varsigma(t,x)\dot{u}(t) \tag{7}$$

with:

$$\begin{aligned} \theta(t,x) &= \tfrac{\partial}{\partial t}\dot{S}(t,x,u) + \tfrac{\partial}{\partial x}\dot{S}(t,x,u)f(t,x,u) \\ \varsigma(t,x) &= \tfrac{\partial}{\partial u}\dot{S}(t,x,u) \end{aligned}. \tag{8}$$

Let's consider now the new system whose state variables are the sliding function $S(t,x)$ and its derivative $\dot{S}(t,x)$:

$$\begin{cases} y_1(t,x) = S(t,x) \\ y_2(t,x) = \dot{S}(t,x) \end{cases} \tag{9}$$

By using the equations (8) and (9), we can write:

$$\begin{cases} \dot{y}_1(t,x) = y_2(t,x) \\ \dot{y}_2(t,x) = \theta(t,x) + \varsigma(t,x)\dot{u}(t) \end{cases} \tag{10}$$

The system described by (10) is a second order system. For this new system a new sliding function can be proposed:

$$\begin{aligned} \sigma(t,x) &= y_2(t,x) + \alpha y_1(t,x) \\ &= \dot{S}(t,x) + \alpha S(t,x) \end{aligned} \tag{11}$$

The system whose input is $\dot{u}(t)$ and output $\sigma(t,x)$ has got a relative order equal to one and a sliding mode can be involved on $\sigma(t,x) = 0$ (Sira-Ramirez (1988)). The correspondent control law can be of the form:

$$\dot{u}(t) = \dot{u}_{eq(t)} - Msign(\sigma(t,x)) \tag{12}$$

The term $\dot{u}_{eq}(t)$ is deduced from:

$$\dot{\sigma}(t,x) = \dot{y}_2(t,x) + \alpha\dot{y}_1(t,x) = \ddot{S}(t,x) + \alpha\dot{S}(t,x) = 0 \tag{13}$$

with: $\ddot{S}(t,x) = C^T\ddot{x}(t)$.

The vector $\ddot{x}(t)$ can be deduced from the considered system:

$$\ddot{x}(t) = \frac{\partial}{\partial t}f(t,x,u) + \frac{\partial}{\partial x}f(t,x,u)\dot{x}(t) + \frac{\partial}{\partial u}f(t,x,u)\dot{u}(t) \tag{14}$$

The equivalent control for the new system is, then, written:

$$\begin{aligned}\dot{u}_{eq}(t) = & -\frac{1}{C^T\frac{\partial}{\partial u}f(t,x,u)}\Big(C^T\frac{\partial}{\partial}f(t,x,u) \\ & +C^T\frac{\partial}{\partial x}f(t,x,u)\dot{x}(t) + \alpha\dot{S}(t,x)\Big)\end{aligned} \tag{15}$$

The control input for the new system is:

$$\dot{u}(t) = \dot{u}_{eq}(t) + u_{dis}(t) \tag{16}$$

with $\quad u_{dis}(t) = -\ M\,sign\,(\sigma(t,x))$
The effective control to apply to the system (1) is obtained by integration:

$$u(t) = \int \dot{u}_{eq}(t)dt - \int u_{dis}(t)dt \tag{17}$$

If we consider a system whose output is the sliding function $S(t,x)$ with a relative order equal to one, the control algorithm, described above, is convergent if there exist positive constants Γ_m, Γ_M, Φ and s_0 such that, in a neighborhood $|S(t,x)| \leq s_0$, the following conditions are verified(Salgado (2004)):

$$\begin{aligned}& 0 < \Gamma_m \leq \varsigma(t,x) \leq \Gamma_M \\ & |\theta(t,x)| \leq \Phi\end{aligned} \tag{18}$$

This approach requires the knowledge of a model of the system, and guaranties an asymptotic convergence of the sliding function to zero according to a desired dynamic.

2.2 Discrete second order sliding mode approach

Let's consider the following system :

$$\begin{aligned}& x(k+1) = Ax(k) + Bu(k) \\ & y(k) = Hx(k)\end{aligned} \tag{19}$$

The sliding function relative to this system is taken in this linear form:

$$S(k) = C^T(x(k) - x_d(k)) \tag{20}$$

with $x_d(k)$ is the desired state vector and C is the sliding function's parameters' vector. A discrete first order sliding mode control can be given by the following expression (Gao et al. (1995)):

$$u(k) = (C^TB)^{-1}[\varphi S(k) - C^TAx(k) - M\text{sign}(S(k))] \tag{21}$$

In order to develop a second order sliding mode controller, a fictive system whose state variables are $S(k+1)$ and $S(k)$ is considered. The new sliding function $\sigma(k)$ is defined by:

$$\sigma(k) = S(k+1) + \beta S(k) \tag{22}$$

with:

$$\begin{aligned}S(k+1) &= C^T(x(k+1) - x_d(k+1)) \\ &= C^T\left(Ax(k) + Bu(k) - x_d(k+1)\right)\end{aligned} \tag{23}$$

We note that β is chosen in the interval [0, 1[, in order to ensure the convergence of $\sigma(k)$.

By analogy with the case of the first order discrete sliding mode control law (1-DSMC), the equivalent control that forces the system to evolute on the sliding function is deduced from :

$$\sigma(k+1) = \sigma(k) = 0 \tag{24}$$

The equations (22), (23) and (24) give:

$$S(k+1) + \beta S(k) = 0 \tag{25}$$

and

$$\begin{aligned} S(k+1) &= \sigma(k+1) - \beta S(k) \\ &= C^T(x(k+1) - x_d(k+1)) \\ &= C^T\left(Ax(k) + Bu_{eq}(k) - x_d(k+1)\right) \end{aligned} \tag{26}$$

Then:

$$u_{eq}(k) = (C^T B)^{-1}[-\beta S(k) - C^T Ax(k) + C^T(x_d(k+1)] \tag{27}$$

The robustness is ensured by the addition of a discontinuous term (sign of the new sliding function $\sigma(k)$). By analogy with the continuous-time case, we apply to the system (19) the integral of the discontinuous term which will be approximated by a first order transformation.

$$u_{dis}(k) = u_{dis}(k-1) - T_e M\, sign\,(\sigma(k)) \tag{28}$$

The control at the instant k is then (Mihoub et al. (2009b)):

$$u(k) = u_{eq}(k) + u_{dis}(k) \tag{29}$$

The integration of the discontinuous term of the control allows its use in the case of many applications where actuators can be damaged by the discontinuity of the 1-DSMC (gates, motoring...). However, this approach does not ameliorate the robustness of the system during the reaching phase (Mihoub et al. (2008)). To resolve this problem, the multimodel approach is exploited in the following paragraph.

3. A multimodel for the 2-DSMC

3.1 Multimodel approach

Instead of exploiting one global model of the system for the equivalent control calculation, the multimodel approach suggests the use of some partial models that express the process dynamics. Two problems must be resolved in this case: the construction of the partial models and the choice of the right one at the right time (Ltaief et al. (2003a;b; 2004); Mihoub et al. (2008; 2009a); Talmoudi et al. (2002a;b; 2003)). If the final model is built by the fusion technique, we must, of course, compute partial models validities.

3.1.1 Construction of the partial models

Some approaches have been proposed for the systematic determination of a generic models base. In (Lahmari (1999)), Ksouri L. proposed a models' base based on the Kharitonov's algebric approach. Four extreme models and a medium one can be exploited by the multimodel strategy. Ben Abdennour et al. (Ltaief et al. (2003a;b; 2004); Talmoudi et al.

(2002a;b; 2003)) have proposed two contributions for the systematic determination of the models'base. The first is based on the Chiu's approach for fuzzy classification (Chiu (1994)) and the second exploites the classification strategy based on the Kohenen's Neural Network.

3.1.2 The validities computing

The validities estimation can be insured, classically, by the residue approach:

$$v_i(k) = \frac{1 - \frac{r_i(k)}{\sum_{c=1}^{md} r_c(k)}}{md - 1}, \quad i \in [1, md] \tag{30}$$

$$r_i(k) = |y(k) - y_i(k)| \tag{31}$$

with $y(k)$ is the system's output, $y_i(k)$ is the output of the ith model and md is the models number.

In order to reduce the perturbation phenomenon due to the inadequate models, we reinforce the validities as follow:

$$v_i^{renf}(k) = v_i(k) \prod_{\substack{c=1 \\ i \neq c}}^{md} \left(1 - e^{-\left(\frac{r_c(k)}{g}\right)^2}\right) \tag{32}$$

with g is a positive coefficient. The normalized reinforced validities are given by:

$$v_{in}^{renf}(k) = \frac{v_i^{renf}(k)}{\sum_{c=1}^{md} v_c^{renf}(k)} \tag{33}$$

3.2 The Multimodel 2-DSMC

As already mentioned, the 2-DSMC helps to reduce the chattering phenomenon by the integration of the discontinuous term which is used to guaranty the robustness of the control law. Unfortunately, this discontinuous term does not switch during the reaching phase (because the system has not reached the sliding surface yet). Consequently, during this phase the robustness is not guaranteed. A solution for this problem was proposed in (Mihoub et al. (2008; 2009a)) by combining the second order discrete sliding mode control and the multimodel approach.

The multimodel discrete second order sliding mode control (MM-2-DSMC) structure is given by the figure 1.

In our case, the partial models can be represented as follows :

$$\begin{aligned} &\text{Modèle 1} \; : \begin{cases} x(k+1) = A_1 x(k) + B_1 u(k) \\ y(k) = Hx(k) \end{cases} \\ &\vdots \\ &\text{Modèle } md : \begin{cases} x(k+1) = A_{md} x(k) + B_{md} u(k) \\ y(k) = Hx(k) \end{cases} \end{aligned} \tag{34}$$

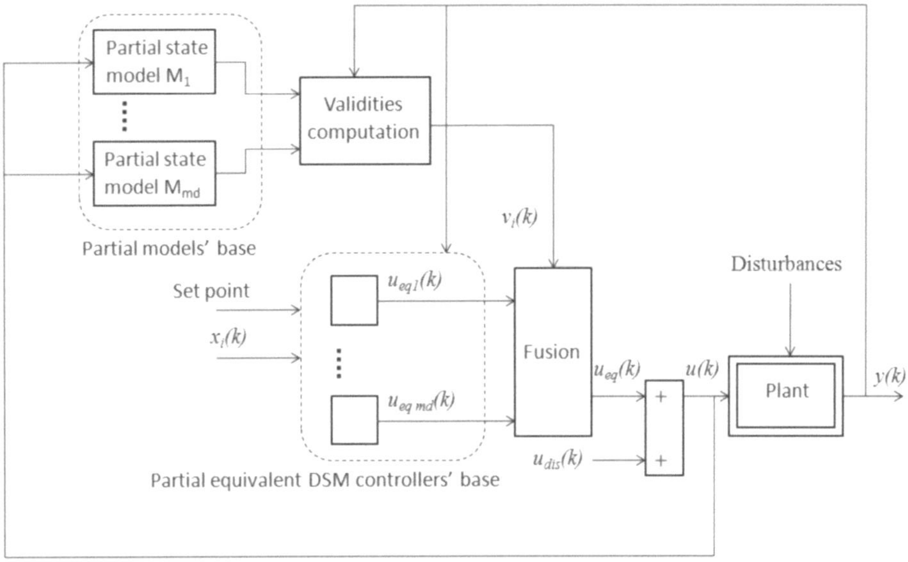

Fig. 1. The structure of a multimodel discrete second order sliding mode control (MM-2-DSMC).

where *md* is the number of the partial models. The control applied to the system is given by the following relation:

$$u(k) = v_1(k)u_{1eq}(k) + v_2(k)u_{2eq}(k) + ... + v_{md}(k)u_{mdeq}(k) + u_{dis}(k); \tag{35}$$

with

- $v_i(k)$: the validity of the ith local state model,
- $u_{ieq}(k)$: the partial equivalent term of the 2-DSMC calculated using the ith local state model,
- $u_{dis}(k)$: the discontinuous term of the control.

$$u_{eqi}(k) = (C^T B_i)^{-1} \left(\alpha\, S(k) - C^T A_i x(k) + C^T (x_d(k+1)\right)$$

A_i et B_i are the matrixes of the ith partial state model. The discontinuous term is given by the following expression:

$$u_{dis}(k) = u_{dis}(k-1) - M\, sign\, (\sigma(k))$$

The multimodel discrete second order sliding mode control (MM-2-DSMC) is, then, given by:

$$u(k) = \sum_{i=1}^{m_d} v_i(k) u_{eqi}(k) + u_{dis}(k) \tag{36}$$

A stability analysis of this last control law is proposed in the following paragraph.

3.3 Stability analysis of the MM-2-DSMC

Let's consider the following non stationary system:

$$\begin{aligned} x(k+1) &= A_d x(k) + B_d u(k) + \Gamma(k) \\ y(k) &= Hx(k) \end{aligned} \tag{37}$$

$\Gamma(k)$ represents eventual non linearities and external disturbances.

Considering the following notations: $A_m = \sum_{i=1}^{md} v_i A_i$ and $B_m = \sum_{i=1}^{md} v_i B_i$,
we obtain the following model:

$$\begin{aligned} x(k+1) &= A_m x(k) + B_m u(k) \\ y(k) &= Hx(k) \end{aligned} \tag{38}$$

which is the multimodel approximation of the system (37). This last system can be, then, written in the following form:

$$\begin{aligned} x(k+1) &= (A_m + \Delta A_m)\, x(k) + (B_m + \Delta B_m)\, u(k) + \Gamma(k) \\ y(k) &= Hx(k) \end{aligned} \tag{39}$$

such that:

$$\begin{aligned} A_d &= A_m + \Delta A_m \\ B_d &= B_m + \Delta B_m \end{aligned} \tag{40}$$

We note:

$$\Delta(k) = \Delta A_m x(k) + \Delta B_m u(k) + \Gamma(k) \tag{41}$$

The system (37) can, in this case, be written as follows:

$$\begin{aligned} x(k+1) &= A_m x(k) + B_m u(k) + \Delta(k) \\ y(k) &= Hx(k) \end{aligned} \tag{42}$$

The control law given by (36) is applied to the system. In the case of the multimodel, the equivalent term $\sum_{i=1}^{m_d} v_i(k) u_{eqi}(k)$ of (36) is written as follow:

$$u_{eq}(k) = (C^T B_m)^{-1}[-\beta S(k) - C^T A_m x(k)] \tag{43}$$

In this case, the sliding function dynamics are given by the following expression:

$$C^T x(k+1) = S(k+1) = -\beta S(k) + C^T \Delta(k) + (C^T B_m) u_{dis}(k) \tag{44}$$

The sliding function variation $[S(k+1) - S(k)]$ is given by the following relation:

$$\begin{aligned} S(k+1) - S(k) = &- \beta(S(k) - S(k-1) + C^T(\Delta(k) - \Delta(k-1)) \\ &- (C^T B_m) M sign\,(S(k) + \beta S(k-1)) \end{aligned} \tag{45}$$

In what follows, the quantity $(C^T B_m)M$ will be noted M^*.

Which gives:

$$S(k+1) + \beta S(k) = (S(k) + \beta S(k-1) + C^T (\Delta(k) - \Delta(k-1)) - M^* sign\,(S(k) + \beta S(k-1)) \quad (46)$$

The relation (46) can be written:

$$\sigma(k+1) = \sigma(k) + C^T (\Delta(k) - \Delta(k-1)) - M^* sign\,(\sigma(k)) \quad (47)$$

In discrete time sliding mode control, instead of the sliding mode, a quasi sliding-mode is considered in the vicinity of the sliding surface, such that $|\sigma(k)| < \varepsilon$, where $\sigma(k)$ is the sliding function and ε is a positive constant called the quasi-sliding-mode band width.
Bartoszewicz, in (Bartoszewicz (1998)), gave the following sufficient and necessary condition for a system to satisfy a convergent quasi sliding mode:

$$\begin{cases} \sigma(k) > \varepsilon \Rightarrow -\varepsilon \leq \sigma(k+1) < \sigma(k) \\ \sigma(k) < -\varepsilon \Rightarrow \sigma(k) < \sigma(k+1) \leq \varepsilon \\ |\sigma(k)| < \varepsilon \Rightarrow |\sigma(k+1)| \leq \varepsilon \end{cases} \quad (48)$$

$\forall k$, $C^T (\Delta(k) - \Delta(k-1))$ is supposed to be bounded such that:

$$\left| C^T (\Delta(k) - \Delta(k-1)) \right| < \Delta_0 \quad (49)$$

with Δ_0 being a positive constant.

Théorème 0.1. *Let's consider the system (37) to which the MM-2-DSMC given by (36) is applied. If the discontinuous term amplitude M is chosen such that:*

$$M^* > \Delta_0 \quad (50)$$

where $M^* = (C^T B_m)M$ and Δ_0 is the external disturbances and system's parameters' variation bound given by (49), then, the MM-2-DSMC of (36) results in a convergent quasi sliding mode.

Proof.
ε is chosen equal to $M^* + \Delta_0$.
To prove the convergence of the proposed control technique, we must, then, check the following three conditions:

$$\sigma(k) > M^* + \Delta_0 \Rightarrow -(M^* + \Delta_0) \leq \sigma(k+1) < \sigma(k) \quad (51)$$

$$\sigma(k) < -(M^* + \Delta_0) \Rightarrow \sigma(k) < \sigma(k+1) \leq M^* + \Delta_0 \quad (52)$$

$$|\sigma(k)| < M^* + \Delta_0 \Rightarrow |\sigma(k+1)| \leq M^* + \Delta_0 \quad (53)$$

1. Let's begin by the condition (51):

$$\sigma(k) > M^* + \Delta_0 \Rightarrow -M^* - \Delta_0 < \sigma(k+1) < \sigma(k)$$

   * The inequality

$$\sigma(k+1) < \sigma(k) \quad (54)$$

can be written as follows:

$$\sigma(k) + C^T(\Delta(k) - \Delta(k-1)) - M^* sign\,(\sigma(k)) < \sigma(k) \quad (55)$$

Knowing that $\sigma(k) > 0$, the inequality (55) becomes:

$$\sigma(k) + C^T(\Delta(k) - \Delta(k-1)) - M^* < \sigma(k) \quad (56)$$

By substructing $\sigma(k)$ from both sides of this last inequality, we obtain:

$$C^T(\Delta(k) - \Delta(k-1)) - M^* < 0 \quad (57)$$

This last inequality is true because M^* is chosen such that $M^* > \Delta_0$

* The inequality

$$-M^* - \Delta_0 < \sigma(k+1) \quad (58)$$

can be written as follows:

$$-M^* - \Delta_0 < \sigma(k) + C^T(\Delta(k) - \Delta(k-1)) - M^* \quad (59)$$

which gives:

$$-\Delta_0 - C^T(\Delta(k) - \Delta(k-1)) < \sigma(k) \quad (60)$$

This last inequality is true, knowing that $\sigma(k) > M^* + \Delta_0 > 0$ and $-\Delta_0 - C^T(\Delta(k) - \Delta(k-1)) < 0$

2. Let's consider condition (52):

$$\sigma(k) < -M^* - \Delta_0 \Rightarrow \sigma(k) < \sigma(k+1) < M^* + \Delta_0$$

By replacing $\sigma(k+1)$ by its expression, we obtain:

$$\sigma(k) < \sigma(k) + C^T(\Delta(k) - \Delta(k-1)) + M^* < M^* + \Delta_0 \quad (61)$$

* The inequality

$$\sigma(k) + C^T(\Delta(k) - \Delta(k-1)) + M^* < M^* + \Delta_0 \quad (62)$$

can be written as follows:

$$\sigma(k) + C^T(\Delta(k) - \Delta(k-1)) < \Delta_0 \quad (63)$$

which gives:

$$\sigma(k) < \Delta_0 - C^T\,(\Delta(k) - \Delta(k-1)) \quad (64)$$

This last inequality is true because $\Delta_0 - C^T\,(\Delta(k) - \Delta(k-1)) > 0$ and $\sigma(k) < 0$.

* Besides, it is evident that $\sigma(k) < \sigma(k) + C^T(\Delta(k) - \Delta(k-1)) + M^*$, knowing that:

$$M^* > \Delta_0 > C^T(\Delta(k) - \Delta(k-1)) \quad (65)$$

3. Let's consider condition (53):

$$|\sigma(k)| < M^* + \Delta_0 \Rightarrow |\sigma(k+1)| < M^* + \Delta_0$$

* If $\sigma(k) > 0$, then, the inequality

$$|\sigma(k)| < M^* + \Delta_0 \tag{66}$$

becomes:

$$0 < \sigma(k) < M^* + \Delta_0 \tag{67}$$

which gives:

$$\begin{aligned}
C^T(\Delta(k) - \Delta(k-1)) - M^* &< \sigma(k) + C^T(\Delta(k) - \Delta(k-1)) - M^* \\
&< M^* + \Delta_0 + C^T(\Delta(k) - \Delta(k-1)) - M^* \\
\Rightarrow -\Delta_0 - M^* &< \sigma(k) + C^T(\Delta(k) - \Delta(k-1)) - M^* < M^* + \Delta_0 + \Delta_0 - M^* \\
\Rightarrow -\Delta_0 - M^* &< \sigma(k) + C^T(\Delta(k) - \Delta(k-1)) - M^* < \Delta_0 + \Delta_0 \\
\Rightarrow -\Delta_0 - M^* &< \sigma(k) + C^T(\Delta(k) - \Delta(k-1)) - M^* < M^* + \Delta_0 \\
\Rightarrow -\Delta_0 - M^* &< \sigma(k+1) < M^* + \Delta_0
\end{aligned}$$

Then,

$$|\sigma(k+1)| < M^* + \Delta_0 \tag{68}$$

* If $\sigma(k) < 0$, then,

$$|\sigma(k)| < M^* + \Delta_0 \tag{69}$$

becomes

$$-M^* - \Delta_0 < \sigma(k) < 0 \tag{70}$$

then,

$$\begin{aligned}
C^T(\Delta(k) - \Delta(k-1)) + M^* - M^* - \Delta_0 &< \sigma(k) + C^T(\Delta(k) - \Delta(k-1)) + M^* \\
&< C^T(\Delta(k) - \Delta(k-1)) + M^* \\
\Rightarrow -\Delta_0 - \Delta_0 &< \sigma(k) + C^T(\Delta(k) - \Delta(k-1)) + M^* < M^* + \Delta_0 \\
\Rightarrow -\Delta_0 - M^* &< \sigma(k) + C^T(\Delta(k) - \Delta(k-1)) + M^* < M^* + \Delta_0 \\
\Rightarrow -\Delta_0 - M^* &< \sigma(k+1) < M^* + \Delta_0
\end{aligned}$$

so,

$$|\sigma(k+1)| < M^* + \Delta_0 \tag{71}$$

The verification of the three conditions (51), (52) and (53) proves the existence of the convergent quasi sliding mode. Therefore, the controller given by (29) is stable. □

Note

The bound of $\Delta(k)$ can be determined by studying the uncertainties of the different partial models, how much they cover the different real system's dynamics and the method used for the calculation of the validities degrees. The linear matrixes inequalities (LMI) approach can be used in these conditions.

4. Experimentation on a chemical reactor

4.1 Process description

The semi-batch reactor control provides a very challenging problem for the process control engineer, due to the high non linearity that characterizes its dynamic behavior. Therefore, we choose to apply the proposed control laws for temperature control of the chemical reactor presented in this section (a photo of the considered reactor is given by figure 2).

This process is used to esterify olive oil. The produced ester is widely used for the

Fig. 2. The esterification reactor used for the experiments.

manufacture of cosmetic products. A specific temperature profile sequence must be followed in order to guarantee an optimal exploitation of the involved reagents' quantities. The olive oil contains, essentially, a mono-unsaturated fatty acid that react with alcohol to give water and ester as shown by the following reaction equation:

$$Acid + Alcohol \underset{2}{\overset{1}{\rightleftharpoons}} Ester + Water \tag{72}$$

The final solution contains all the reagents and products in certain proportions. To drive the reaction equilibrium in the way 1 and, consequently, increase the ester's proportion, we should take away water from the solution. This is done by vaporization. The fatty acid (oleic acid) and the ester ebullition temperatures are approximately $300°C$. The chosen alcohol (1-butanol) is characterized by an ebullition temperature of $118°C$. Consequently, heating the reactor to a temperature slightly over $100°C$ will result in the vaporization of water only (which is evacuated through the condenser).

The reactor is heated by circulating a coolant fluid through the reactor jacket. This fluid is, in turn, heated by three resistors located in the heat exchanger (Figure 3). The reactor temperature control loop monitors temperature inside the reactor and manipulates the power delivered to the resistors. It is, also, possible to cool the coolant fluid by circulating cold water through a coil in the heat exchanger. Cooling is, normally, done when the reaction is over, in order to accelerate the reach of ambient temperature.

The process can be considered as a single input - single output system. The input is the heating power $P(W)$. The output is the reactor temperature $TR(°C)$. The interface between the process and the calculator is ensured by a data acquisition card of the type RTI 810. The data acquisition card ensures the conversion of the analog measures of the temperature to digital values and the conversion of the digital control value to an analog electric signal proportional the power applied to the heating resistors.

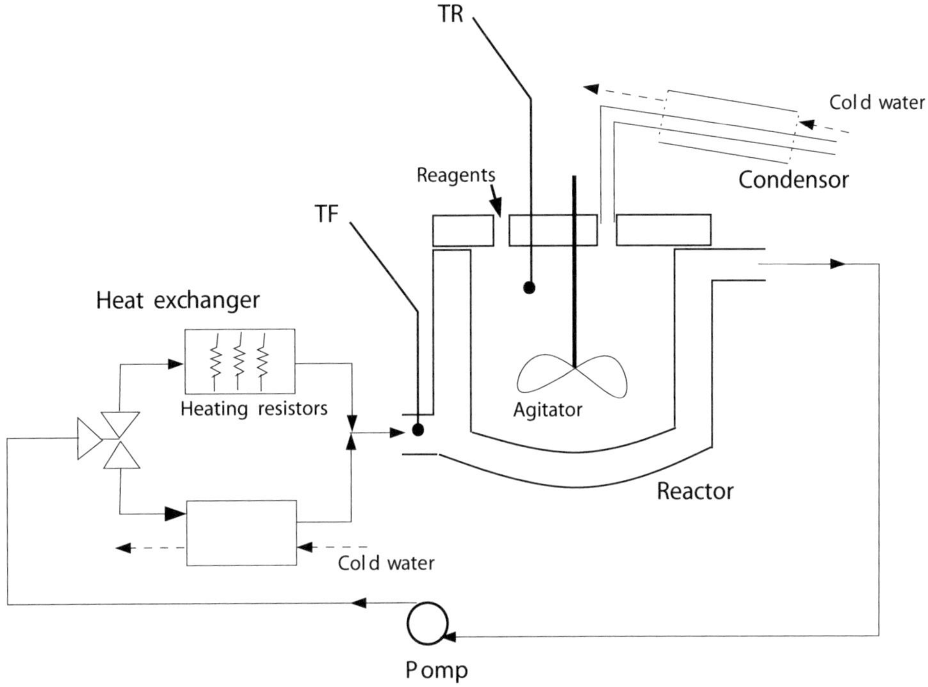

Fig. 3. Synoptic scheme of the real process.

The control law must carry out the following three stages :

- Bring the reactor's temperature TR to $105°C$.
- Keep the reactor's temperature to this value until the reaction is over (no more water dripping out of the condenser).
- Lower the reactor's temperature.

We chose, therefore, the set point given by figure 4.

We represent, in figure 5, the static characteristic of the system. The different coordinates are taken relatively to the three stages of the process. We notice that the system can be considered as a linear one, though, with some approximations. According to the step responses of the system, the retained sampling step is equal to $180s$. A Pseudo-Random, Binary input Signal (PRBS) is applied to the real system. An identification of the system structure, based on the instrumental determinants ratio method (Ben Abdennour et al. (2001)), led to a discrete second order linear model.

Due to the nature of the control law to be applied to the reactor, the needed model is a state model. The considered state variables are the reactor's temperature $TR(°C)$ (noted $x_1(k)$), which is at the same time the system's output, and the coolant fluid temperature $TF(°C)$ (noted $x_2(k)$).

The state variables sequences $x_1(k)$ and $x_2(k)$ relative to the PRBS excitation input are measured and used for the parametric identification of the system. The least square method leads to the following nominal model:

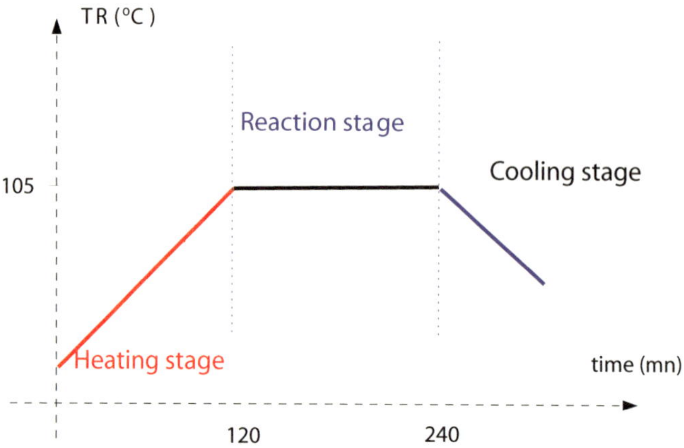

Fig. 4. Desired reactor temperature.

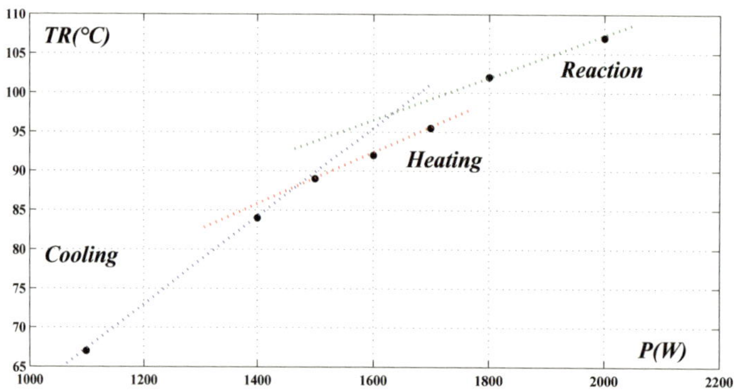

Fig. 5. Static characteristic of the reactor.

$$\begin{aligned} x(k+1) &= \begin{bmatrix} 0.6158 & 0.3600 \\ 0.0409 & 0.9118 \end{bmatrix} x(k) + \begin{bmatrix} 0 \\ 0.0033 \end{bmatrix} u(k) \\ y(k) &= \begin{bmatrix} 1 & 0 \end{bmatrix} x(k) \end{aligned} \tag{73}$$

The application of the multimodel approach and by using the least square method applied on the input/states sequence relative to each reaction stage leads to three partial models of the form:

$$\begin{aligned} x(k+1) &= A_i x(k) + B_i u(k) \\ y(k) &= H x(k) \qquad\qquad i \in [1,3] \end{aligned} \tag{74}$$

with for:

- the heating stage:

$$\begin{aligned} A_1 &= \begin{bmatrix} 0.4712 & 0.4953 \\ -0.1296 & 1.0651 \end{bmatrix} \; ; \; B_1 = \begin{bmatrix} 0 \\ 0.0036 \end{bmatrix} \\ H &= \begin{bmatrix} 1 & 0 \end{bmatrix} \end{aligned} \tag{75}$$

- the reaction stage:

$$A_2 = \begin{bmatrix} 0.4114 & 0.5482 \\ -0.0358 & 0.9787 \end{bmatrix} \; ; \; B_2 \begin{bmatrix} 0 \\ 0.0034 \end{bmatrix}$$
$$H = \begin{bmatrix} 1 \; 0 \end{bmatrix} \tag{76}$$

- the cooling stage:

$$A_3 = \begin{bmatrix} 0.6914 & 0.2877 \\ -0.0386 & 0.0.9912 \end{bmatrix} \; ; \; B_3 = \begin{bmatrix} 0 \\ 0.0032 \end{bmatrix}$$
$$H = \begin{bmatrix} 1 \; 0 \end{bmatrix} \tag{77}$$

The control performance and robustness of the previously mentioned control laws, with respect to the model-system mismatch and external disturbance, are illustrated and compared through the experimental results given in the following paragraph.

4.2 Experimental results

In this paragraph, the performance of the MM-2-DSMC is shown by an experimentation on the chemical reactor. Firstly, the chattering reduction, obtained by exploiting the second order sliding mode control, is illustrated by a comparison between the results obtained by the first order discrete sliding mode control with those realized by the 2-DSMC (Mihoub et al. (2009b)). The nominal model (73) is used for both the DSMC and the 2-DSMC.

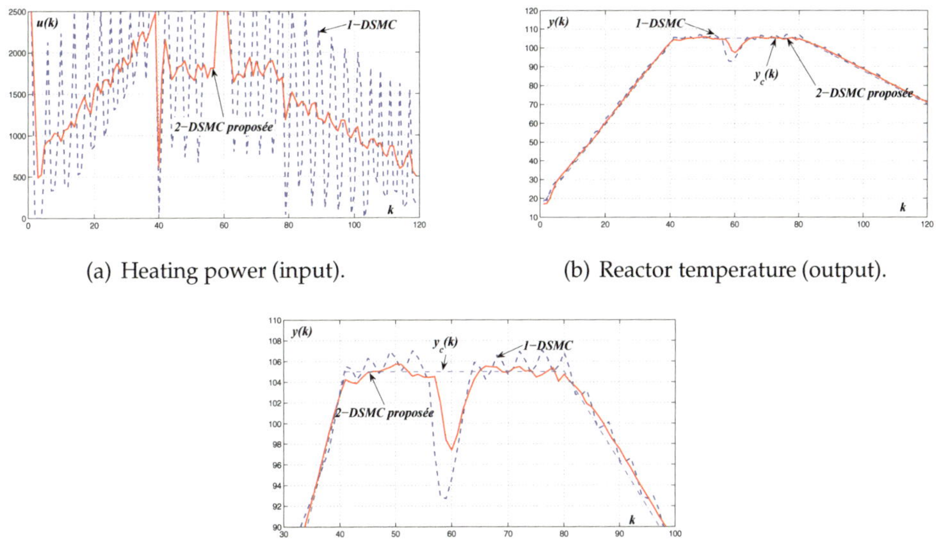

(a) Heating power (input).

(b) Reactor temperature (output).

(c) Zoom on the reactor temperature evolution.

Fig. 6. Comparison between DSMC and 2-DSMC

We observe that the chattering of the control ($u(k)$) is remarkably reduced (figure 6.a). A better set point tracking is, consequently, obtained as shown by figures 6.b and 6.c, which represent, respectively, the evolution of the reactor temperature and a zooming of this last one in the neighborhood of 105° C. As mentioned above, the reaction takes place essentially during this phase. If the temperature reactor overshoots 105° C, a large amount of alcohol is evaporated and wasted and if it does not reach 105° C, the reaction kinetics are slowed down. So, the 2-DSMC results in a better efficiency relatively to the first order DSMC.

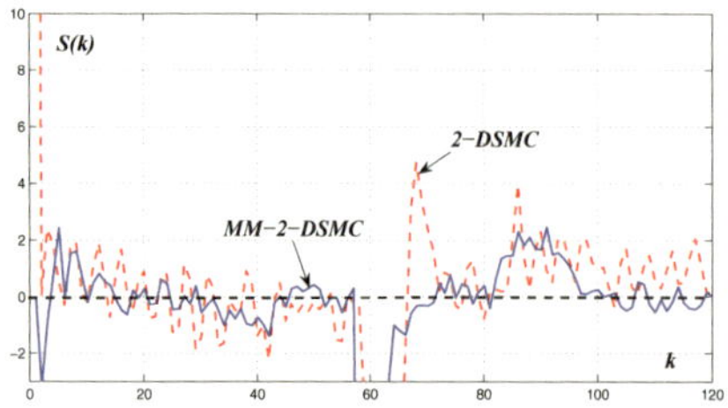

(a) Sliding function.

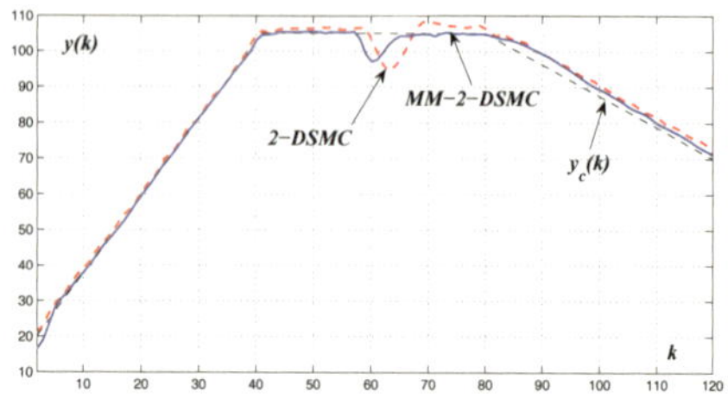

(b) Reactor temperature (output).

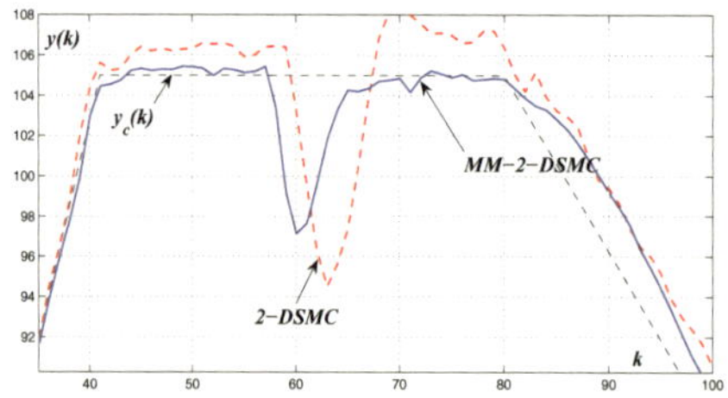

(c) Zoom of the reactor temperature evolution.

Fig. 7. Comparison between 2-DSMC and MM-DSMC

Secondly, the multimodel approach is combined with the 2-DSMC in order to enhance the reaching phase. The MM-2-DSMC and the 2-DSMC are represented together in figure 7 (Mihoub et al. (2009a)). It can be observed that the sliding function overshoots due to a bad reaching phase in the case of the 2-DSMC are reduced thanks to the multimodel approach (see figure 7.a). A better set point tracking is then obtained, as shown by figures 7.b and 7.c. An amelioration of the efficiency of the chemical reactor is, consequently, obtained.

5. Conclusion

In this work, the problems of the discrete sliding mode control are discussed. A solution to the chattering problem can be given by the second order sliding mode. To enhance the reaching phase, the multimodel approach is exploited. A combination of the 2-DSMC and the multimodel approach is, then, used. A stability analysis of the multimodel second order discrete sliding mode control is proposed in this work. An experimentation on a chemical reactor is considered. On the one hand, a comparison between the results obtained by the first order DSMC and those obtained by the 2-DSMC showed the chattering reduction offered by the second order approach. On the other hand, a comparison between the results of the 2-DSMC and those of the MM-2-DSMC, illustrated both an enhancement of the reaching phase and a notable reduction of the chattering phenomenon. A better efficiency of the reactor is, therefore, obtained.

6. References

Bartoszewicz., A. (1998) "Discrete-time quasi-sliding-mode control strategies". *IEEE Trans. Ind. Electronics*, 45(4):633–637.

Ben Abdennour, R., Borne, P., Ksouri, M. & M'sahli, F. (2001) *"Identification et comm&e numérique des procédés industriels"*. Editions Technip, Paris.

Chiu., S. L. (1994) "Fuzzy model identification based on cluster estimation". *Journal of Intelligent & Fuzzy Systems*, 2:267–278.

Decarlo, R. A., Zak, H. S. & Mattews, G.P. (1988) "Variable structure control of nonlinear multivariable systems: A tutorial". *Proceeding IEEE*, 73:212–232.

Emelyanov, S. V., Korovin, S. K. & Levantovsky, L.V. (1986) "Higher order sliding modes in the binary control systems". *Soviet Physics, Doklady*, 31:291–293.

Filippov, A. (1960) "Equations différentielles à second membre discontinu". *Journal de Mathématiques*, 51(1):99–128.

Gao, W., Wang, Y. & Homaifan, H. (1995) "Discrete-time variable structure control systems". *IEEE Trans. Ind. Electronics*, 42(2):117–122.

Jimenez, T. S. (2004) *"Contribution à la comm&e d'un robot sous-marin autonome de type torpille"*. Thèse de doctorat, Université Montpellier II.

Ksouri Lahmari, M. (1999) *"Contribution à la comm&e multimodèle des processus complexes"*. PhD thesis, USTL, Lille.

Levant, A. (1993) "Sliding order & sliding accurcy in sliding mode control". *International Journal of Control*, 58(6):1247–1263.

Levantovsky, L. V. (1985) "Second order sliding algorithms: their realization". In *Dynamics of Heterogeneous Systems. Materials of the seminar*, 32–43, Moscow.

Lopez, P. & Nouri, A.S. (2006) *"Théorie élémentaire et pratique de la comm&e par les régimes glissants"*. Mathématiques et applications 55, SMAI, Springer - Verlag.

Ltaief, M.,Ben Abdennour, R., Abderrahim, K. & Ksouri, M. (2003) "A new systematic determination approach of a models base for the representation of uncertain systems". In *SSD'03*, Sousse, Tunisie.

Ltaief, M., Abderrahim, K., Ben Abdennour, R. & Ksouri, M (2003) "A fuzzy fusion strategy for the multimodel approach application". *WSEAS Trans. on circuits & systems*, 2(4):686–691.

Ltaief, M., Abderrahim, K., Ben Abdennour, R. & Ksouri, M. (2004) "Systematic determination of a models base for the multimodel approach: Experimental validation". *WSEAS Trans. on Electronics*, 1(2): 331–336.

Mihoub, M., Nouri, A. S. & Ben Abdennour, R. (2008) "The multimodel approach for a numerical second order sliding mode control of highly non stationary systems". In *ACC'08*, American Control Conference, Seattle, W.A. USA.

Mihoub, M., Nouri, A.S. & Ben Abdennour, R. (2009) "A real time application of discrete second order sliding mode control to a semi-batch reactor: a multimodel approach". *Int. Journal of Modelling, Identification & Control*, 6(2):156–163.

Mihoub, M. , Nouri, A.S. & Ben Abdennour, R. (2009) "Real time application of discrete second order sliding mode control to a chemical reactor". *Control Engineering Practice*, 17:1089–1095.

Sira-Ramirez, H. (1988) "Structure at infinity, zero dynamics & normal forms of systems undergoing sliding motion". *Int. J. Systems Sci.*, 21(4):665–674.

Talmoudi, S., Ben Abdennour, R., Abderrahim, K. & Ksouri, M. (2002) "Multimodèle et multi-comm&e neuronaux pour la conduite numérique des systèmes non linéaires et non stationnaires". In *CIFA'02*, France.

Talmoudi, S., Ben Abdennour, R., Abderrahim, K. & Borne, P. (2002) "A systematic determination approach of a models'base for uncertain systems: Experimental validation". In *SMC'02*, Conference on Systems, Man & Cybernetics, Tunisie.

Talmoudi, S., Abderrahim, K., Ben Abdennour, R. & Ksouri, M. (2003) "A new technique of validities'computation for multimodel approach". In *WSEAS03 (ICOSMO03)*, Greece.

Utkin, V. I. (1992) *"Sliding Mode in Control Optimisation"*. Springer-Verlag, Berlin.

Utkin, V.I., David Young, K. & Ozguner, U. (1999) "A control engineer's guide to sliding mode control". *IEEE Transactions on Control Systems Technology*, 7:328–342.

26

Two Dimensional Sliding Mode Control

Hassan Adloo[1], S.Vahid Naghavi[2],
Ahad Soltani Sarvestani[2] and Erfan Shahriari[1]
[1]*Shiraz University, Department of Electrical and Computer Engineering*
[2]*Islamic Azad University, Zarghan Branch*
Iran

1. Introduction

In nature, there are many processes, which their dynamics depend on more than one independent variable (e.g. thermal processes and long transmission lines (Kaczorek, 1985)). These processes are called multi-dimensional systems. Two Dimensional (2-D) systems are mostly investigated in the literature as a multi-dimensional system. 2-D systems are often applied to theoretical aspects like filter design, image processing, and recently, Iterative Learning Control methods (see for example Roesser, 1975; Hinamoto, 1993; Whalley, 1990; Al-Towaim, 2004; Hladowski et al., 2008). Over the past two decades, the stability of multi-dimensional systems in various models has been a point of high interest among researchers (Anderson et al., 1986; Kar, 2008; Singh, 2008; Bose, 1994; Kar & Singh, 1997; Lu, 1994). Some new results on the stability of 2-D systems have been presented - specifically with regard to the Lyapunov stability condition which has been developed for RM (Lu, 1994). Then, robust stability problem (Wang & Liu, 2003) and optimal guaranteed cost control of the uncertain 2-D systems (Guan et al., 2001; Du & Xie, 2001; Du et al., 2000) came to be the area of interest. In addition, an adaptive control method for SISO 2-D systems has been presented (Fan & Wen, 2003). However, in many physical systems, the goal of control design is not only to satisfy the stability conditions but also to have a system that takes its trajectory in the predetermined hyperplane. An interesting approach to stabilize the systems and keep their states on the predetermined desired trajectory is the sliding mode control method. Generally speaking, SMC is a robust control design, which yields substantial results in invariant control systems (Hung et al., 1993). The term invariant means that the system is robust against model uncertainties and exogenous disturbances. The behaviour of the underlying SMC of systems is indeed divided into two parts. In the first part, which is called reaching mode, system states are driven to a predetermined stable switching surface. And in the second part, the system states move across or intersect the switching surface while always staying there. The latter is called sliding mode. At a glance in the literature, it is understood that there are many works in the field of SMC for 1-D continuous and discrete time systems. (see Utkin, 1977; Asada & Slotine, 1986; Hung et al., 1993; DeCarlo et al., 1988; Wu and Gao, 2008; Furuta, 1990; Gao et al., 1995; Wu & Juang, 2008; Lai et al., 2006; Young et al., 1999; Furuta & Pan, 2000; Proca et al., 2003; Choa et al., 2007; Li & Wikander, 2004; Hsiao et al., 2008; Salarieh & Alasty, 2008) Furthermore SMC has been contributed to various control methods (see for example Hsiao et al., 2008; Salarieh & Alasty, 2008) and several experimental works (Proca et al., 2003). Recently, a SMC design for a 2-D system in RM model has been presented (Wu & Gao, 2008) in which the idea of a 1-D quasi-sliding mode (Gao et al., 1995) has been extended for the 2-D system. Though the sliding surfaces design problem and the conditions for the existence of an ideal quasi-sliding mode has been solved in terms of LMI.

In this Chapter, using a 2-D Lyapunov function, the conditions ensuring the rest of horizontal and vertical system states on the switching surface and also the reaching condition for designing the control law are investigated. This function can also help us design the proper switching surface. Moreover, it is shown that the designed control law can be applied to some classes of 2-D uncertain systems. Simulation results show the efficiency of the proposed SMC design. The rest of the Chapter is organized as follows. In Section two, Two Dimensional (2-D) systems are described. Section three discusses the design of switching surface and the switching control law. In Section four, the proposed control design for two numerical examples in the form of 2-D uncertain systems is investigated. Conclusions and suggestions are finally presented in Section five.

2. Two dimensional systems

As the name suggests, two-dimensional systems represent behaviour of some processes which their variables depend on two independent varying parameters. For example, transmission lines are the 2-D systems where whose currents and voltages are changed as the space and time are varying. Also, dynamic equations governed to the motion of waves and temperatures of the heat exchangers are other examples of 2-D systems. It is interesting to note that some theoretical issues such as image processing, digital filter design and iterative processes control can be also used the 2-D systems properties.

2.1 Representation of 2-D systems

Especially, a well-known 2-D discrete systems called Linear Shift-Invariant systems has been presented which is described by the following input-output relation

$$\sum_{m}\sum_{n} b_{m,n} y_{i-m,j-n} = \sum_{m}\sum_{n} a_{m,n} u_{i-m,j-n} \tag{1}$$

Also, this input-output relation can be transformed into frequency-domain using 2-D Z transformation.

$$H(z_1, z_2) = \frac{Z\{y(i,j)\}}{Z\{u(i,j)\}} = \frac{P(z_1, z_2)}{Q(z_1, z_2)} \tag{2}$$

Similar to the one-dimensional systems, the 2-D systems are commonly represented in the state space model but what is makes different is being two independent variables in the 2-D systems so that this resulted in several state space models.

A well-known 2-D state space model was introduced by Roesser, 1975 which is called Roesser Model (RM or GR) and described by the following equations

$$\begin{aligned} \begin{bmatrix} x^h(i+1,j) \\ x^v(i,j+1) \end{bmatrix} &= \begin{bmatrix} A_1 & A_2 \\ A_3 & A_4 \end{bmatrix} \begin{bmatrix} x^h(i,j) \\ x^v(i,j) \end{bmatrix} + \begin{bmatrix} B_1 & B_2 \\ B_3 & B_4 \end{bmatrix} \begin{bmatrix} u^h(i,j) \\ u^v(i,j) \end{bmatrix} \\ y(i,j) &= \begin{bmatrix} C_1 & C_2 \end{bmatrix} \begin{bmatrix} x^h(i,j) \\ x^v(i,j) \end{bmatrix} \end{aligned} \tag{3}$$

where $x^h(i, j) \in R^n$ and $x^v(i, j) \in R^m$ are the so called *horizontal* and *vertical* state variables respectively. Also $u(i, j) \in R^p$ is an input and $y(i, j) \in R^q$ is an output variable. Moreover, i and j represent two independent variables. A_1, A_2, A_3, A_4, B_1, B_2, C_1 and C_2 are constant matrices with proper dimensions. To familiar with other 2-D state space models (Kaczorek, 1985).

2.2 Stability of Rosser Model

One of the important topics in the 2-D systems is stability problem. Similar to 1-D systems, the stability of 2-D systems can be represent in two kinds, BIBO and Internally stability. First, a BIBO stability condition for RM is stated.

Theorem 1: A zero inputs 2-D system in RM (3) is BIBO stable if and only if one of the following conditions is satisfied

1. I. A_1 is stable, II. $A_4 + A_3\left[I_n z_1 - A_1\right]^{-1} A_2$ is stable for $|z_1| = 1$.
2. I. A_2 is stable, II. $A_1 + A_2\left[I_m z_2 - A_4\right]^{-1} A_3$ is stable for $|z_2| = 1$.

Note that, in the discrete systems, a matrix is stable if all whose eigenvalues are in the unite circle. Thus, from Theorem 1, it can be easily shown that a 2-D system in RM is unstable if A_1 or A_2 is not stable.

Similar to 1-D case, the Lyapunov stability for 2-D systems has been developed such that we represented in the following theorem.

Theorem 2: Zero inputs 2-D system (3) is asymptotically stable if there exist two positive definite matrices $P_1 \in R^n$ and $P_2 \in R^m$ such that

$$A^T P A - P = -Q \tag{4}$$

where Q is a positive matrix and

$$P = \begin{bmatrix} P_1 & 0 \\ 0 & P_2 \end{bmatrix}, \quad A = \begin{bmatrix} A_1 & A_2 \\ A_3 & A_4 \end{bmatrix} \tag{5}$$

Remark 1: Note that the 2D system (3) is asymptotically stable if the state vector norms $x^h(i,j)$ and $x^v(i,j)$ converge to zero when $i+j \to \infty$.

Remark 2: The equality (3) is commonly called 2-D Lyapunov equation. As stated in the theorem 2, the condition for stability of 2-D systems in RM model is only sufficient not necessary and the Lyapunov matrix, P, is a block diagonal while in the 1-D case, the stability conditions is necessary and sufficient and the Lyapunov matrix is a full matrix.

However, it is worthy to know that the Lyapunov equation (3) can be used to define the 2-D Lyapunov function as shown below.

$$V_{00}(i,j) = X^T \begin{bmatrix} P_1 & 0 \\ 0 & P_2 \end{bmatrix} X \tag{6}$$

where $X = \left[x^h(i,j) \quad x^v(i,j)\right]^T$. Regarding (5), define delayed 2-D Lyapunov function as follows

$$V_{11}(i,j) = X_{11}{}^T \begin{bmatrix} P_1 & 0 \\ 0 & P_2 \end{bmatrix} X_{11} \tag{7}$$

where $X = \left[x^h(i+1,j) \quad x^v(i,j+1) \right]^T$. Now, we can state following fact.

Theorem 3: 2-D system (3) is asymptotically stable if there are the Lyapunov function, (6) and the delayed function (7) such that

$$\Delta V(i,j) = V_{11}(i,j) - V_{00}(i,j) < 0 \tag{8}$$

As a result, the Theorem 3 can be used to design a 2-D control system.

3. Sliding mode control of 2-D systems

In this section, we review some prominence of the 1-D sliding mode control and then present the 2-D sliding mode control for RM.

3.1 One dimensional (1-D) Sliding Mode Control

Generally speaking, Sliding Mode Control (SMC) method is a robust control policy in which the control input is designed based on the reaching and remaining on the predetermined state trajectory. This state trajectory is commonly called switching surface (or manifold). Usually, first the switching surface is determined as a function of the state and/or time, and

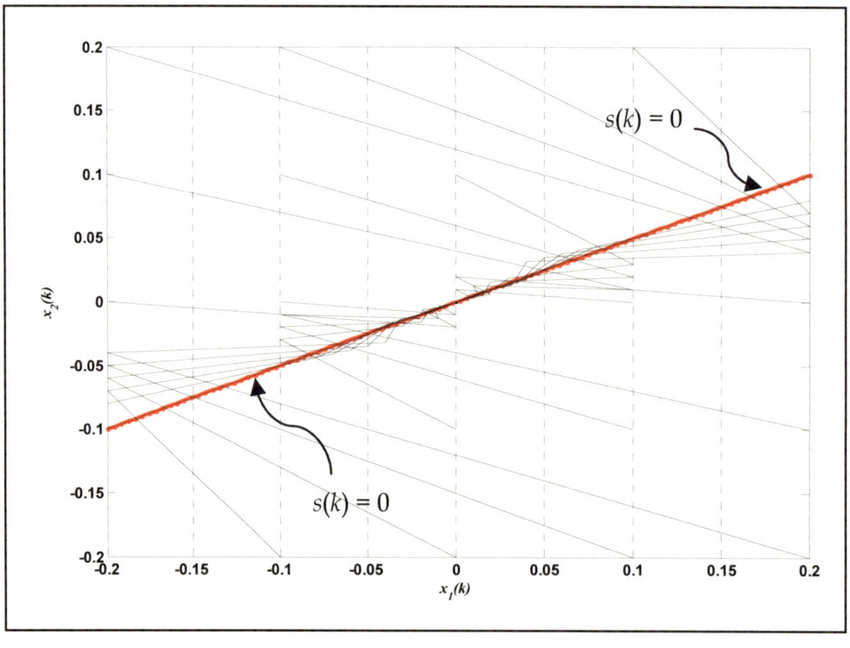

Fig. 1. State trajectory for some different initial conditions

then the control action is designed to reach and remain the state trajectory on the switching surface and move to the origin. Therefore, it can be appreciated that the switching surface should be contained the origin and designed such that the system is stabile when remaining on it. Three main advantages of the SMC method are low sensitivity to the uncertainty (high robustness), dividing the system trajectory in two sections with low degree and also easily in implementation and applicability to various systems.

To make easier understanding the 1-D SMC, let consider a simple example in which a discrete time system is given as follows

$$\begin{cases} x_1(k+1) = x_2(k) \\ x_1(k+1) = x_1(k) + x_2(k) + u(k) \end{cases} \tag{9}$$

Consider that the switching surface is

$$s(k) = x_1(k) - 2x_2(k) \tag{10}$$

also let us the control input is given as

$$u(k) = u_e(x,k) + 0.6\,s(k) \tag{11}$$

where $u_e(x,k) = -x_1(k) - 0.5x_2(k) - 0.5s(k)$. It is clear that the control input is $u_e(x, k)$ when the system remain on the surface (in other word when $s(k) = 0$). Fig. 1 illustrates the state trajectories of the system for some different initial conditions such that they converge to the surface and move to the origin in the vicinity of it. As it is shown in Fig. 1, the state trajectories switch around the surface when they reach the vicinity of it. The main reason of this phenomenon comes from the fact that the system dynamic equation is not exactly matched to the switching surface (Gao, 1995). In fact, the control policy in the SMC method is to reduce the error of the state trajectory to the switching surface using the switching surface feedback control. It is worthy to note that in the SMC method, the system trajectory is divided to two sections that are called reaching phase and sliding phase. Thus, the control input design is commonly performed in two steps, which named equivalent control law and switching are control law design. We want to use this strategy to present 2-D SMC design.

3.2 Two dimensional (2-D) sliding mode control

Consider the 2-D system in RM model as stated in (3).
In this chapter it is assumed that the 2-D system (3) starts from the boundary conditions that are satisfied following condition

$$\sum_{k=0}^{\infty} \left|x^h(0,k)\right|^2 + \left|x^v(k,0)\right|^2 < \infty \tag{12}$$

where $x^h(0, k)$ and $x^v(k, 0)$ are horizontal and vertical boundary conditions. Before introducing 2-D SMC method, some definitions are represented.
Definition 1: The horizontal and vertical linear switching surfaces denoted by $s^h(i, j)$ and $s^v(i, j)$, are defined as the linear combination of the horizontal and vertical state of the 2-D system respectively as shown below

$$\begin{aligned} s^h(i,j) &= C^h x^h(i,j) \\ s^v(i,j) &= C^v x^v(i,j) \end{aligned} \tag{13}$$

where C^h and C^v are the proper constant matrices with proper dimensions.

Definition 2: Consider the 2-D system (1) starts from $(i, j) = (i_0, j_0)$. The boundary conditions can be out of or on the switching surface. So, if the system state trajectory moves toward the switching surface (11), this case is called reaching phase (or mode). After this, if it intersects switching surface at $(i, j) = (i_1, j_1)$ and remains there for all $(i, j) > (i_1, j_1)$ then this is called sliding motion or sliding phase for 2-D systems in RM.

As it is mentioned previously, a common approach to design SMC method contains two steps. First step is determination of the proper switching surface and second step is to design a control action to reach the state trajectory the surface and after it move toward the origin.

3.3 Two dimensional switching surface design

In order to design the 2-D switching surface, we want to extend a well-known method in 1-D case to 2-D case that is equivalent control approach. The equivalent control approach is based on the fact that the system state equation should be stable when it stays on the surface. In this method two points have to be considered, one is to find condition that assures staying on the surface and other is related to the stability of the system when is laid on the surface. It can be shown that two problems can be solved by 2-D Lyapunov stability presented in the theorem 3. For this purpose, let us define following Lyapunov functions

$$\begin{aligned} V_{00}(i,j) &= \frac{1}{2}[s^h(i,j)]^2 + \frac{1}{2}[s^v(i,j)]^2 \\ V_{11}(i,j) &= \frac{1}{2}[s^h(i+1,j)]^2 + \frac{1}{2}[s^v(i,j+1)]^2 \end{aligned} \tag{14}$$

According to theorem 3, the stability condition in the sense of reaching to the switching surface is occurred when the difference of two functions V_{11} and V_{00} is negative. Consequently, the condition that presents staying on the switching surface can be

$$V_{11}(i,j) - V_{00}(i,j) = 0 \tag{15}$$

Therefore, it can be concluded that

$$\begin{aligned} s^h(i+1,j) &= s^h(i,j) \\ s^v(i,j+1) &= s^v(i,j) \end{aligned} \tag{16}$$

Let define following functions as

$$S_{00}(i,j) = \begin{bmatrix} s^h(i,j) \\ s^v(i,j) \end{bmatrix}, \quad S_{11}(i,j) = \begin{bmatrix} s^h(i+1,j) \\ s^v(i,j+1) \end{bmatrix} \tag{17}$$

Thus, equality (17) can be written as

$$S_{11}(i,j) = S_{00}(i,j) \tag{18}$$

From the equations (18) we can derive the control input equivalent to the case that 2-D system (3) stay on the switching surfaces as shown below

$$\begin{aligned}
S_{11}(i,j) &= \begin{bmatrix} C^h & 0 \\ 0 & C^v \end{bmatrix} \begin{bmatrix} x^h(i+1,j) \\ x^v(i,j+1) \end{bmatrix} \\
&= \begin{bmatrix} C^h & 0 \\ 0 & C^v \end{bmatrix} \begin{bmatrix} A_1 & A_2 \\ A_3 & A_4 \end{bmatrix} \begin{bmatrix} x^h(i,j) \\ x^v(i,j) \end{bmatrix} + \begin{bmatrix} C^h & 0 \\ 0 & C^v \end{bmatrix} \begin{bmatrix} B_1 & B_2 \\ B_3 & B_4 \end{bmatrix} \begin{bmatrix} u_{eq}^h(i,j) \\ u_{eq}^v(i,j) \end{bmatrix} \\
&= S_{00}(i,j)
\end{aligned} \tag{19}$$

So we have

$$\begin{bmatrix} u_{eq}^h(i,j) \\ u_{eq}^v(i,j) \end{bmatrix} = -F(C^h, C^v) \begin{bmatrix} x^h(i,j) \\ x^v(i,j) \end{bmatrix} \tag{20}$$

Where

$$F(C^h, C^v) = \begin{bmatrix} C^h B_1 & C^h B_2 \\ C^v B_3 & C^v B_4 \end{bmatrix}^{-1} \begin{bmatrix} C^h (A_1 - I) & C^h A_2 \\ C^v A_3 & C^v (A_4 - I) \end{bmatrix} \tag{21}$$

and it is assumed that $\begin{bmatrix} C^h B_1 & C^h B_2 \\ C^v B_3 & C^v B_4 \end{bmatrix}$ is invertible. The control input (20) is called equivalent control law. Now, we should also guarantee the stability of the system when is laid on the surfaces. To perform this, it is sufficient that the following augmented system is stable.

$$\begin{aligned}
&\begin{bmatrix} x^h(i+1,j) \\ x^v(i,j+1) \end{bmatrix} = \begin{bmatrix} A_1 & A_2 \\ A_3 & A_4 \end{bmatrix} \begin{bmatrix} x^h(i,j) \\ x^v(i,j) \end{bmatrix} + \begin{bmatrix} B_1 & B_2 \\ B_3 & B_4 \end{bmatrix} \begin{bmatrix} u_{eq}^h(i,j) \\ u_{eq}^v(i,j) \end{bmatrix} \\
&\begin{bmatrix} s^h(i,j) \\ s^v(i,j) \end{bmatrix} = 0
\end{aligned} \tag{22}$$

Aforementioned state updating equations (22) represents the 2-D system in the case that it is laid on the surface. By replacing the equivalent control law (20) we have

$$\begin{aligned}
&\begin{bmatrix} x^h(i+1,j) \\ x^v(i,j+1) \end{bmatrix} = \left(\begin{bmatrix} A_1 & A_2 \\ A_3 & A_4 \end{bmatrix} - \begin{bmatrix} B_1 & B_2 \\ B_3 & B_4 \end{bmatrix} F(C^h, C^v) \right) \begin{bmatrix} x^h(i,j) \\ x^v(i,j) \end{bmatrix} \\
&\begin{bmatrix} s^h(i,j) \\ s^v(i,j) \end{bmatrix} = 0
\end{aligned} \tag{23}$$

ith respect to the stability of the system (8), the switching surfaces can be designed.

3.4 Two dimensional control law design

After designing the proper horizontal and vertical switching surfaces, it has to be shown that the 2-D system in RM (3) with any boundary conditions, will move toward the surfaces and reach and also sliding on them toward the origin. This purpose can be interpreted as a regulating and/or tracking control strategies. To perform this purpose, consider that the control inputs are assigned as follows

$$\begin{bmatrix} u^h(i,j) \\ u^v(i,j) \end{bmatrix} = \begin{bmatrix} u_s^h(i,j) \\ u_s^v(i,j) \end{bmatrix} + \begin{bmatrix} u_{eq}^h(i,j) \\ u_{eq}^v(i,j) \end{bmatrix} \tag{24}$$

where $u_{eq}^h(i,j)$ and $u_{eq}^v(i,j)$ were designed as (20). $u_s^h(i,j)$ and $u_s^v(i,j)$ which are called switching control laws, has to be designed such that the control inputs ensure the reaching condition. In this method, it is shown that the duties of the switching control laws are to move the state trajectories toward the surfaces. Therefore, we will first determine the condition that guarantees the reaching phase. It is interesting to note that the reaching condition is also obtained in the sense of 2-D Lyapunov functions (6) and (7) using theorem 3 such that if we have

$$S_{11}^2(i,j) < S_{00}^2(i,j) \tag{25}$$

Then the state trajectories move to the surfaces. Now let us define $\Delta S = S_{11}(i,j) - S_{00}(i,j)$ and applying the equivalent control laws (20) we have

$$\begin{aligned} \Delta S = & \left(\begin{bmatrix} C^h A_1 & C^h A_2 \\ C^v A_3 & C^v A_4 \end{bmatrix} - \begin{bmatrix} C^h (A_1 - I) & C^h A_2 \\ C^v A_3 & C^v (A_4 - I) \end{bmatrix} \right) \begin{bmatrix} x^h(i,j) \\ x^v(i,j) \end{bmatrix} \\ & + \begin{bmatrix} C^h B_1 & C^h B_2 \\ C^v B_3 & C^v B_4 \end{bmatrix} \begin{bmatrix} u_s^h(i,j) \\ u_s^v(i,j) \end{bmatrix} - \begin{bmatrix} s^h(i,j) \\ s^v(i,j) \end{bmatrix} \end{aligned} \tag{26}$$

So, this results in

$$\Delta S = \begin{bmatrix} C^h B_1 & C^h B_2 \\ C^v B_3 & C^v B_4 \end{bmatrix} \begin{bmatrix} u_s^h(i,j) \\ u_s^v(i,j) \end{bmatrix} \tag{27}$$

Theorem 4: For the 2-D system in RM (3) if the switching control law is designed as

$$\begin{bmatrix} u_s^h(i,j) \\ u_s^v(i,j) \end{bmatrix} = \begin{bmatrix} k^h s^h(i,j) \\ k^v s^v(i,j) \end{bmatrix} \tag{28}$$

where k^h and k^v are the positive constant numbers and also

$$2\begin{bmatrix} C^h B_1 & C^h B_2 \\ C^v B_3 & C^v B_4 \end{bmatrix} + \begin{bmatrix} k^h C^h B_1 & C^h B_2 \\ C^v B_3 & k^v C^v B_4 \end{bmatrix}^T \begin{bmatrix} k^h C^h B_1 & C^h B_2 \\ C^v B_3 & k^v C^v B_4 \end{bmatrix} < 0 \tag{29}$$

then the reaching condition (25) is satisfied.

Proof:

As it is mentioned, to ensure the reaching phase it is sufficient that the (25) is satisfied. It is well-known we can write (25) as below

$$\frac{1}{2}(S_{00} + \Delta S)^2 < \frac{1}{2} S_{00}^2 \tag{30}$$

By replacing (20) into (30) we have

$$\frac{1}{2}\left(S_{00}+\begin{bmatrix}C^hB_1 & C^hB_2\\ C^vB_3 & C^vB_4\end{bmatrix}\begin{bmatrix}u_s^h(i,j)\\ u_s^v(i,j)\end{bmatrix}\right)^2 = \\ =\frac{1}{2}S_{00}^2+S_{00}^T\begin{bmatrix}C^hB_1 & C^hB_2\\ C^vB_3 & C^vB_4\end{bmatrix}\begin{bmatrix}u_s^h(i,j)\\ u_s^v(i,j)\end{bmatrix} \\ +\frac{1}{2}\left(\begin{bmatrix}C^hB_1 & C^hB_2\\ C^vB_3 & C^vB_4\end{bmatrix}\begin{bmatrix}u_s^h(i,j)\\ u_s^v(i,j)\end{bmatrix}\right)^2<\frac{1}{2}S_{00}^2 \tag{31}$$

Therefore,

$$S_{00}^T\begin{bmatrix}C^hB_1 & C^hB_2\\ C^vB_3 & C^vB_4\end{bmatrix}\begin{bmatrix}u_s^h(i,j)\\ u_s^v(i,j)\end{bmatrix}<-\frac{1}{2}\left(\begin{bmatrix}C^hB_1 & C^hB_2\\ C^vB_3 & C^vB_4\end{bmatrix}\begin{bmatrix}u_s^h(i,j)\\ u_s^v(i,j)\end{bmatrix}\right)^2 \tag{32}$$

Now with respect to the switching control laws (28) we can write

$$S_{00}^T\begin{bmatrix}k^hC^hB_1 & C^hB_2\\ C^vB_3 & k^vC^vB_4\end{bmatrix}S_{00}+\frac{1}{2}S_{00}^T\begin{bmatrix}k^hC^hB_1 & C^hB_2\\ C^vB_3 & k^vC^vB_4\end{bmatrix}^T\begin{bmatrix}k^hC^hB_1 & C^hB_2\\ C^vB_3 & k^vC^vB_4\end{bmatrix}S_{00}<0 \tag{33}$$

This completes the proof.

3.5 Robust control design

In this section, assume that the 2-D system in RM (3) is not given exactly and we have

$$\begin{bmatrix}x^h(i+1,j)\\ x^v(i,j+1)\end{bmatrix}=(A+\Delta A)\begin{bmatrix}x^h(i,j)\\ x^v(i,j)\end{bmatrix}+(B+\Delta B)\begin{bmatrix}u^h(i,j)\\ u^v(i,j)\end{bmatrix} \tag{34}$$

where ΔA and ΔB are denoted as the uncertainties in the system. Assume that

$$\Delta A = pA \\ \Delta B = pB \tag{35}$$

where p is an unknown constant number and there exists a known positive real number, α, such that

$$|p|<\alpha \tag{36}$$

In this case we present following theorem.

Theorem 5: The state trajectories of the uncertain 2-D system (34) is converged to the switching surfaces (13) if

$$\alpha\begin{bmatrix}C^hB_1 & C^hB_2\\ C^vB_3 & C^vB_4\end{bmatrix}+\alpha^2\begin{bmatrix}k^hC^hB_1 & C^hB_2\\ C^vB_3 & k^vC^vB_4\end{bmatrix}^T\begin{bmatrix}k^hC^hB_1 & C^hB_2\\ C^vB_3 & k^vC^vB_4\end{bmatrix}<0 \tag{37}$$

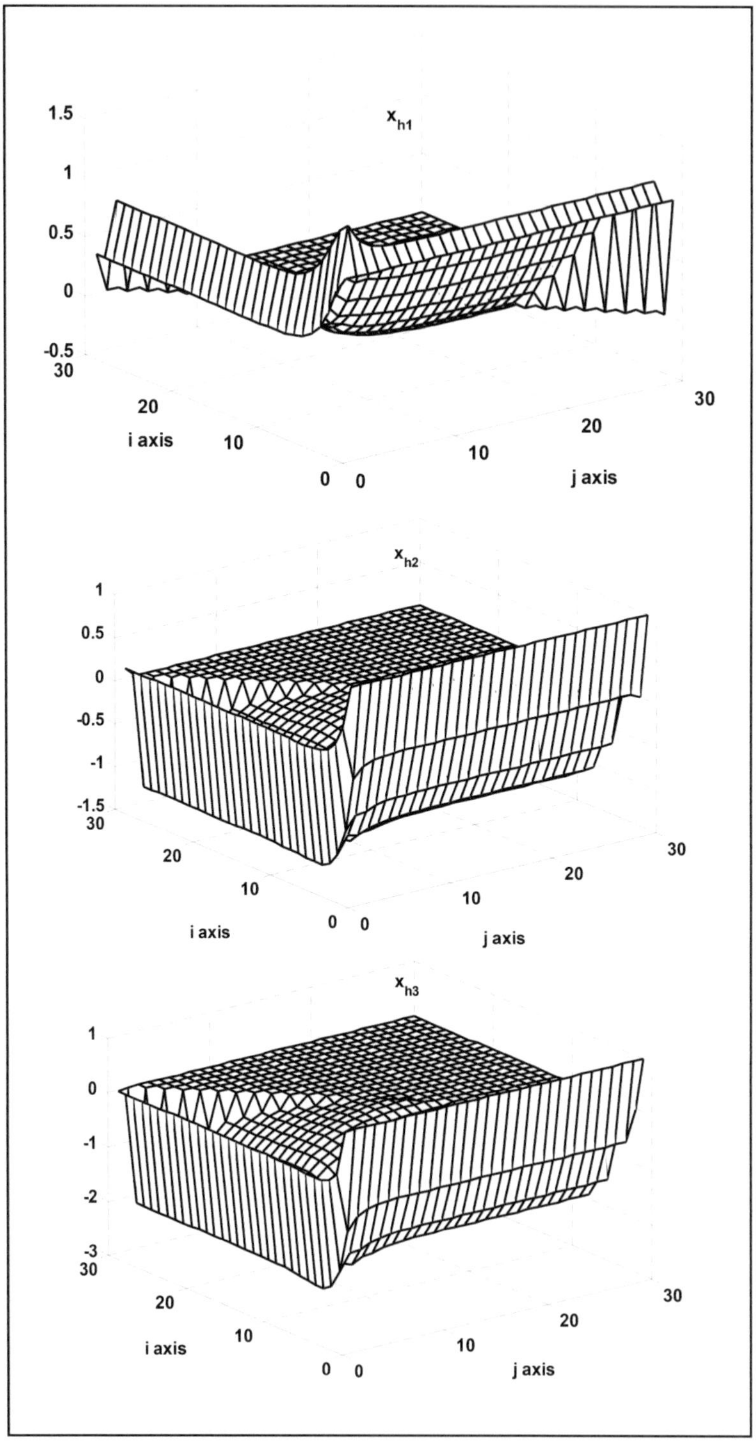

Fig. 2. The horizontal states of the system

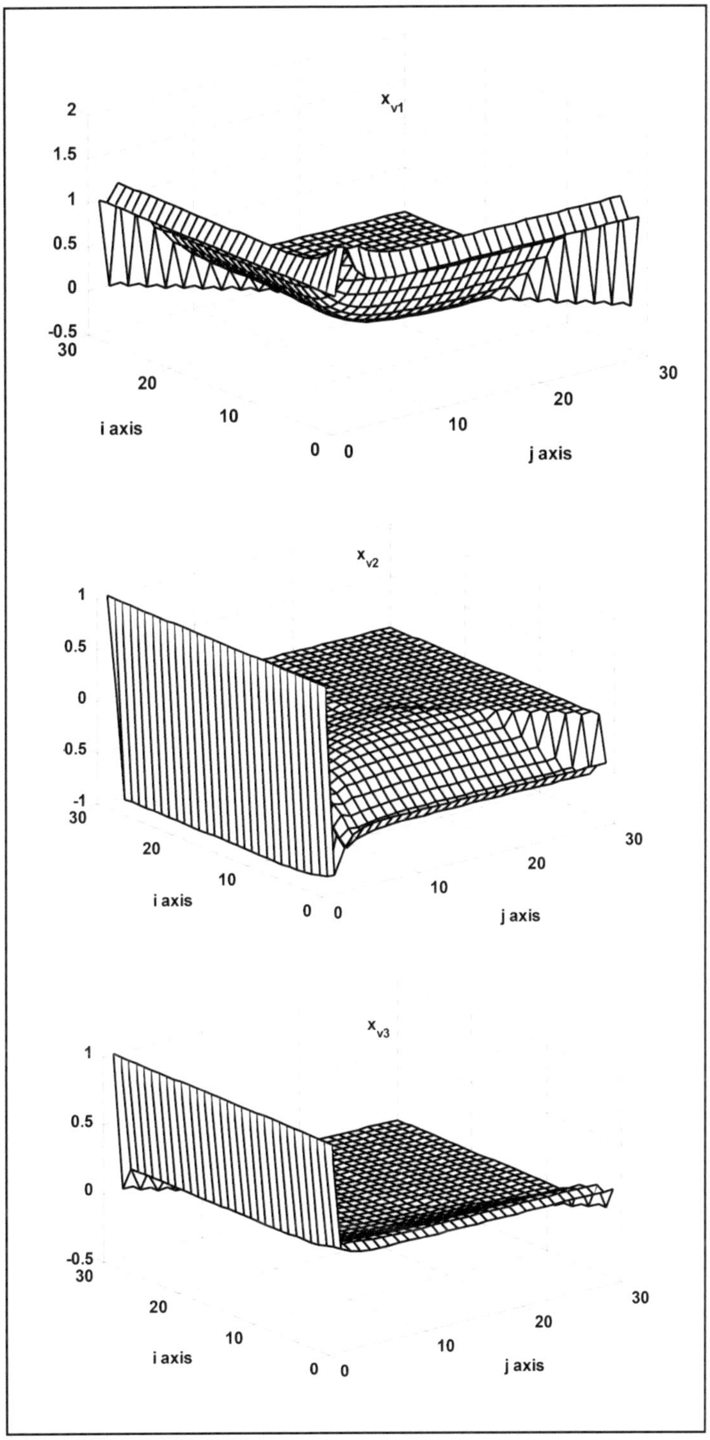

Fig. 3. The vertical states of the system

4. Numerical examples

4.1 As a first numerical example, consider a discretization of the partial differential equation of darboux equation as a 2-D system in RM (Wu & Gao, 2008) that is

$$\begin{bmatrix} x^h(i+1,j) \\ x^v(i,j+1) \end{bmatrix} = \begin{bmatrix} A_1 & A_2 \\ A_3 & A_4 \end{bmatrix} \begin{bmatrix} x^h(i,j) \\ x^v(i,j) \end{bmatrix} + \begin{bmatrix} B_1 \\ B_2 \end{bmatrix} \begin{bmatrix} u^h(i,j) \\ u^v(i,j) \end{bmatrix} \tag{38}$$

Where $u^v(i,j) \in R$, $u^h(i,j) \in R$, $x^h(i,j) \in R^3$, $x^v(i,j) \in R^3$ and

$$A_1 = \begin{bmatrix} 0.65 & -0.25 & 0.32 \\ -0.20 & 0.75 & -0.15 \\ 0.26 & 0.34 & 0.80 \end{bmatrix},\ A_2 = \begin{bmatrix} 0.25 & -0.30 & 0.20 \\ -0.30 & 0.15 & 0.24 \\ 0.15 & 0.36 & -0.48 \end{bmatrix}$$
$$A_3 = \begin{bmatrix} 0.45 & 0.20 & -0.15 \\ 0.25 & -0.30 & 0.20 \\ -0.20 & 0.65 & 0.25 \end{bmatrix},\ A_4 = \begin{bmatrix} 0.60 & 0.25 & 0.18 \\ -0.75 & -0.40 & 0.14 \\ 0.20 & 0.15 & -0.37 \end{bmatrix} \tag{39}$$

And

$$B_1 = \begin{bmatrix} 0 \\ 0 \\ 2 \end{bmatrix},\ B_2 = \begin{bmatrix} 0 \\ 0 \\ 3 \end{bmatrix} \tag{41}$$

As discussed in previous section, the switching surfaces are designed as the system equation in (22) is stable that is

$$\begin{bmatrix} x_r^h(i+1,j) \\ x_r^v(i,j+1) \end{bmatrix} = \begin{bmatrix} \frac{13}{20}-\frac{8}{25}c_1 & -\frac{1}{4}-\frac{8}{25}c_2 & \frac{1}{4}-\frac{1}{5}c_3 & -\frac{3}{10}-\frac{1}{5}c_4 \\ -\frac{1}{5}+\frac{3}{20}c_1 & \frac{3}{4}+\frac{3}{20}c_2 & -\frac{3}{10}-\frac{6}{25}c_3 & \frac{3}{20}-\frac{6}{25}c_4 \\ \frac{9}{20}+\frac{3}{20}c_1 & \frac{1}{5}+\frac{3}{20}c_2 & \frac{3}{5}-\frac{9}{50}c_3 & \frac{1}{4}-\frac{9}{50}c_4 \\ \frac{1}{4}-\frac{1}{5}c_1 & -\frac{3}{10}-\frac{1}{5}c_2 & -\frac{3}{4}-\frac{7}{50}c_3 & -\frac{2}{5}-\frac{7}{50}c_4 \end{bmatrix} \begin{bmatrix} x_r^h(i,j) \\ x_r^v(i,j) \end{bmatrix} \tag{42}$$

Where $x_r^h(i,j) \in R^2$, $x_r^v(i,j) \in R^2$ are reduced state in and

$$c_1 = \frac{c_1^h}{c_3^h},\ c_2 = \frac{c_2^h}{c_3^h},\ c_3 = \frac{c_1^v}{c_3^v}\ c_4 = \frac{c_2^v}{c_3^v} \tag{43}$$

It is easily shown that if we choose

$$C^h = \begin{bmatrix} 40.3735 & -99.1097 & 75.3160 \end{bmatrix}$$
$$C^v = \begin{bmatrix} 43.6978 & 1.3936 & -290.8205 \end{bmatrix} \tag{44}$$

Then the reduced system (42) is stable. Therefore, the equivalent control laws are

$$\begin{bmatrix} u_{eq}^{h}(i,j) \\ u_{eq}^{v}(i,j) \end{bmatrix} = -(CB)^{-1} CA \begin{bmatrix} x^{h}(i,j) \\ x^{v}(i,j) \end{bmatrix} \tag{45}$$

Where $A = \begin{bmatrix} A_1 & A_2 \\ A_3 & A_4 \end{bmatrix}$, $B = \begin{bmatrix} B_1 & 0 \\ 0 & B_2 \end{bmatrix}$, $C = \begin{bmatrix} C^h & 0 \\ 0 & C^v \end{bmatrix}$. Also, according to theorem 5 we can obtain the switching laws that are

$$\begin{bmatrix} u_{s}^{h}(i,j) \\ u_{s}^{v}(i,j) \end{bmatrix} = \begin{bmatrix} 0.0001 & 0 \\ 0 & -0.0004 \end{bmatrix} \begin{bmatrix} s^{h}(i,j) \\ s^{v}(i,j) \end{bmatrix} \tag{46}$$

The simulation results are shown in Figs. 2 - 5.

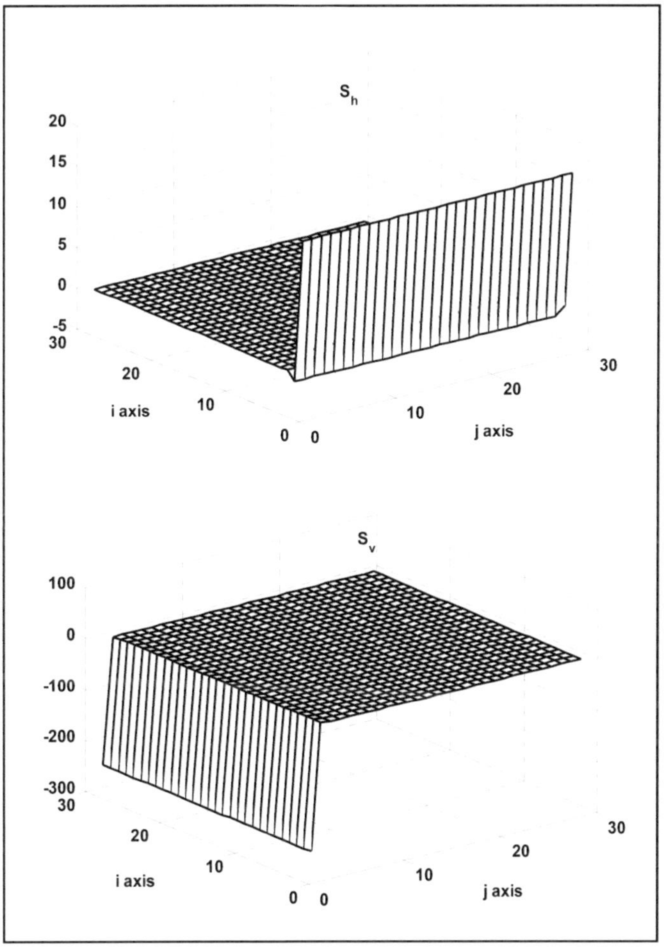

Fig. 4. The horizontal and vertical switching surfaces

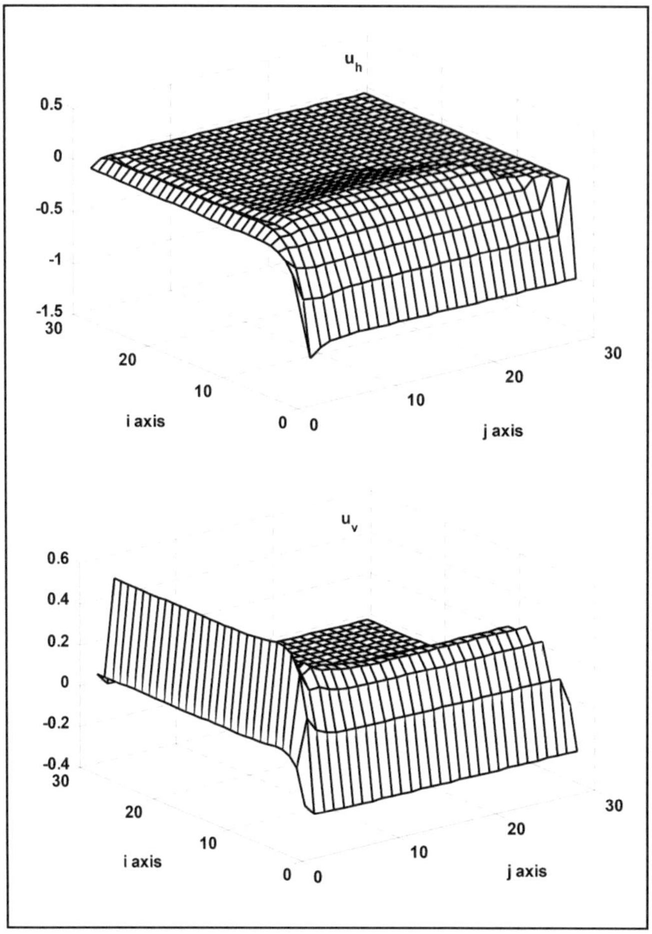

Fig. 5. The horizontal and vertical control inputs

4.2 Let a 2-D uncertain system in RM be given as follows

$$\begin{bmatrix} x_1^h(i+1,j) \\ x_2^h(i+1,j) \\ x_1^v(i,j+1) \\ x_2^v(i,j+1) \end{bmatrix} = (\mathrm{A}+\Delta\mathrm{A}) \begin{bmatrix} x_1^h(i,j) \\ x_2^h(i,j) \\ x_1^v(i,j) \\ x_2^v(i,j) \end{bmatrix} + (\mathrm{B}+\Delta\mathrm{B}) \begin{bmatrix} u^h(i,j) \\ u^v(i,j) \end{bmatrix} \tag{47}$$

Where

$$A = \begin{bmatrix} 0.7020 & 0.7846 & 1.1666 & 0.4806 \\ -1.6573 & -0.7190 & -1.7257 & -1.7637 \\ -1.0272 & -0.6165 & 1.6654 & -1.1104 \\ 0.1917 & -0.4467 & 1.0959 & -0.0200 \end{bmatrix} \quad B = \begin{bmatrix} -1.2632 & -0.3524 \\ -1.0438 & -0.2503 \\ 0.5016 & 0.8912 \\ 0.1348 & -0.0587 \end{bmatrix}$$

Suppose $\alpha = 0.5$. For this system, the switching surface is chosen as

$$\begin{bmatrix} s^h(i,j) \\ s^v(i,j) \end{bmatrix} = C \begin{bmatrix} x_1^h(i,j) \\ x_2^h(i,j) \\ x_1^v(i,j) \\ x_2^v(i,j) \end{bmatrix} \tag{48}$$

where

$$C = \begin{bmatrix} c_1^h & c_2^h & 0 & 0 \\ 0 & 0 & c_1^v & c_2^v \end{bmatrix}$$

The constant parameters c_1^h, c_2^h, c_1^v and c_2^v have to be selected such that the augmented system (22) be stable. It can be easily shown that by choosing C as

$$C = \begin{bmatrix} c_1^h & c_2^h & 0 & 0 \\ 0 & 0 & c_1^v & c_2^v \end{bmatrix} = \begin{bmatrix} -0.3608 & -0.2825 & 0 & 0 \\ 0 & 0 & 1.3173 & 0.2140 \end{bmatrix} \tag{49}$$

the augmented system (19) is stable such that

$$\begin{bmatrix} x_1^h(i+1,j) \\ x_2^h(i+1,j) \\ x_1^v(i,j+1) \\ x_2^v(i,j+1) \end{bmatrix} = \begin{bmatrix} 1.6344 & 1.1038 & 1.2997 & 0.9881 \\ -0.8101 & -0.4095 & -1.6595 & -1.2617 \\ 0.0144 & 0.0956 & 0.8075 & 0.2061 \\ -0.0887 & -0.5883 & 1.1849 & -0.2687 \end{bmatrix} \begin{bmatrix} x_1^h(i,j) \\ x_2^v(i,j) \\ x_1^v(i,j) \\ x_2^v(i,j) \end{bmatrix}$$
$$\begin{bmatrix} s^h(i,j) \\ s^v(i,j) \end{bmatrix} = \begin{bmatrix} 0 \\ 0 \end{bmatrix} \tag{50}$$

By simplifying (50), we have a reduced stable 2-D system as

$$\begin{bmatrix} x_1^h(i+1,j) \\ x_1^v(i,j+1) \end{bmatrix} = \begin{bmatrix} 0.2248 & -4.7821 \\ -0.1076 & -0.4612 \end{bmatrix} \begin{bmatrix} x_1^h(i,j) \\ x_1^v(i,j) \end{bmatrix} \tag{51}$$

So the control action that has been described in previous section is

$$\begin{bmatrix} u^h(i,j) \\ u^v(i,j) \end{bmatrix} = F \begin{bmatrix} x_1^h(i,j) \\ x_2^h(i,j) \\ x_1^v(i,j) \\ x_2^v(i,j) \end{bmatrix} - \frac{1}{2} \begin{bmatrix} s^h(i,j) \\ s^v(i,j) \end{bmatrix} \tag{52}$$

by selecting $k = 0.5$ the condition in (37) is satisfied such that

$$\left(1+\alpha^2\right)D^T D - 1 = \begin{bmatrix} -0.3636 & -0.258 \\ -0.2580 & -0.768 \end{bmatrix} \tag{53}$$

It is clear that the above matrix is a negative definite matrix. Simulation results of this example have been illustrated in Fig 6 - 8.

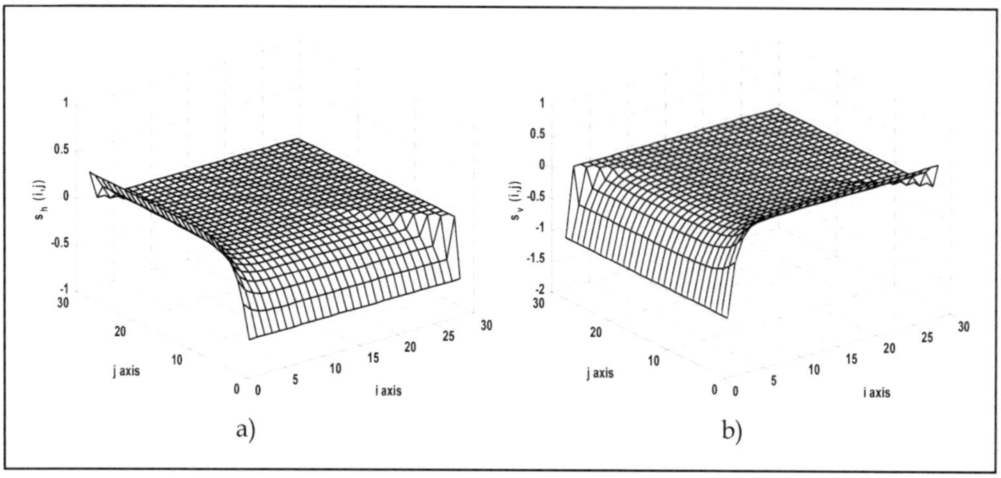

Fig. 6. a) Horizontal sliding surface $s^h(i, j)$ b) Vertical sliding surface $s^v(i, j)$

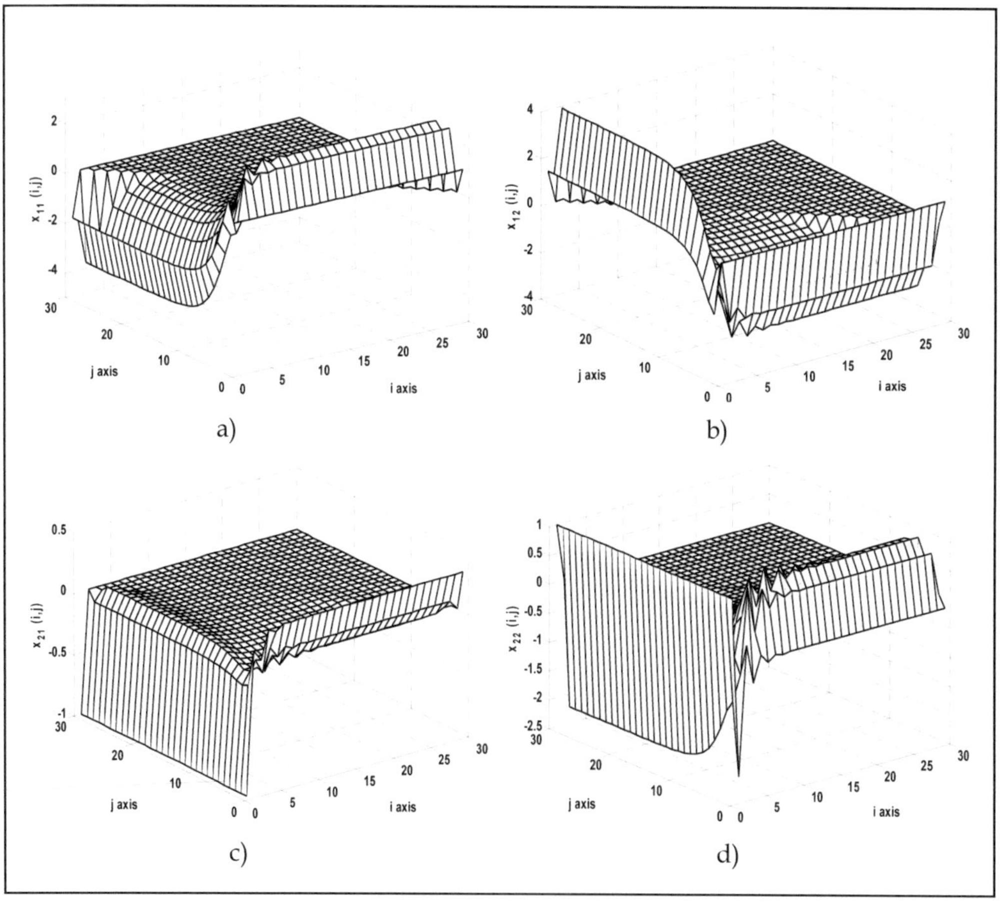

Fig. 7. System states a) $x_1^h(i,j)$, b) $x_2^h(i,j)$, c) $x_1^v(i,j)$ and d) $x_2^v(i,j)$

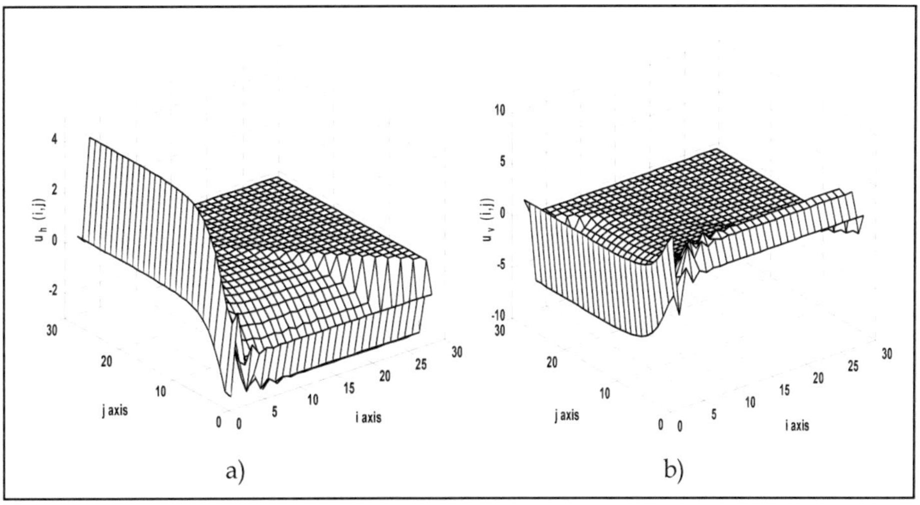

Fig. 8. a) Horizontal input control u^h (i, j), b) Vertical input control u^v (i, j)

5. Conclusion

In this Chapter, an extension of 1-D SMC design to the 2-D system in Roesser model has been proposed. Using a 2-D Lyapunov function, we first designed a linear switching surface, and then a feedback control law that satisfies reaching condition was obtained. This method can also be applied to 2-D uncertain systems with matching uncertainty.

6. References

Al-Towaim, T.; Barton, A. D.; Lewin, P. L.; Rogers E. & Owens, D. H.(2004). Iterative learning control-2D control systems from theory to application, *International Journal of Control*, Vol. 77, 877-893.

Anderson, B. D. O.; Agathoklis, P. A.; Jury, E. I. & Mansour, M. (1986). Stability and the matrix Lyapunov equation for discrete 2-dimensional systems, *IEEE Trans. Circuits Sys*, Vol. 33, No. 3, 261-267.

Asada, H. & Slotine, J. E.(1994). *Robot Analysis and Control*. New York:Wiley, 140–157.

Bose, T.(1994). Asymptotic stability of two-dimensional digital filters under quantization. *IEEE Trans. Signal Processing*, Vol. 42, 1172–1177.

Choa, J.; Principea, J. C.; Erdogmusb, D. & Motter, M. A.(2007). Quasi-sliding mode control strategy based on multiple-linear models. *Neurocomputing Elsevier*, Vol. 70, 960–974.

DeCarlo, R. A.; Zak, S. H. & Matthews, G. P.(1988). Variable structure control of nonlinear multivariable systems: A tutorial. Proc. IEEE, Vol. 76, 212-232.

Dhawan, A. & Kar, H.(2007). Optimal guaranteed cost control of 2-D discrete uncertain systems: An LMI approach. *Signal Processing Elsevier*, Vol. 87, 3075–3085.

Du, C. & Xie L.(2001). H_∞ control and robust stabilization of two-dimensional systems in Roesser models. *Automatica*, Vol. 37, 205–211.

Du, C.; Xie, L. & Soh, Y.C.(2000). H_∞ filtering of 2-D discrete systems. *IEEE Trans. Signal Process*, Vol. 48, 1760–1768.

Fan, H. & Wen C.(2003). Adaptive Control of a Class of 2-D Discrete Systems. *IEEE Trans. Circuits and Systems*, Vol. 50, 166-172.

Hinamoto, T.(1993). 2-D Lyapunov equation and filter design based on the Fornasini-Marchesini second model. *IEEE Trans. Circuits Syst. I*, Vol. 40, 102-110.
Hladowski, L. ; Galkowski, K. ; Cai, Z. ; Rogers, E. ; Freeman, C. T. & Lewin, P. L.(2008). A 2D Systems Approach to Iterative Learning Control with Experimental Validation. IFAC World Congress, vol. 17, 2832-2837, Seoul.
Hsiao, M. -Y. ; Li, T. H. S. ; Lee J. Z. ; Chao C.H. & Tsai S.H.(2008). Design of interval type-2 fuzzy sliding-mode controller. *Information Sciences Elsevier*, Vol. 178, 1696–1710.
Hung, J. Y. ; Gao, W. & Hung, J. C.(1993). Variable structure control: A survey. *IEEE Trans. Ind. Electron.*, Vol. 40, 2-22.
Gao, W.; Wang, Y. & Homaifa, A.(1995). Discrete-time variable structure control systems. *IEEE Trans. Ind. Electron.*, Vol. 42, 117–122.
Guan, X.; Long C. & Duan, G.(2001). Robust optimal guaranteed cost control for 2D discrete systems. IEEE Proc. Control Theory and Applications, Vol. 148, 355-361.
Furuta, K.(1990). Sliding mode control of a discrete system. *Syst. Contr Lett.*, Vol. 14, No. 2, 145–152.
Furuta, K. & Pan Y.(2000). Variable structure control with sliding sector. *Automatica*, Vol. 36, 211-228.
Kaczorek, (1985). *Two-dimensional Linear Systems*. Berlin: Springer-Verlag.
Kar, H.(2008). A new sufficient condition for the global asymptotic stability of 2-D state-space digital filters with saturation arithmetic. *Signal Processing Elsevier*, Vol. 88, 86–98.
Kar, H. & Singh, V.(1997). Stability analysis of 2-D state-space digital filters using Lyapunov function: a caution. *IEEE Trans. Signal Process.*, Vol. 45, 2620–2621.
Lu, W.S.(1994). Some New Results on Stability Robustness of Two-Dimensional Discrete Systems. *Multidimensional Systems and Signal Processing*, Vol. 5, 345-361.
Lai, N. O.; Edwards Ny C. & Spurgeon, S. K.(2006). Discrete output feedback sliding-mode control with integral action. *Int. J. Robust Nonlinear Control*, Vol. 16, 21–43.
Li, Y.F. & Wikander, J.(2004). Model reference discrete-time sliding mode control of linear motor precision servo systems. *Mechatronics Elsevier*, Vol. 14, 835–851.
Proca, A. B.; Keyhani, A. & Miller, J. M.(2003). Sensorless sliding-mode control of induction motors using operating condition dependent models. *IEEE Trans. Energy Conversion*, Vol. 18.
Roesser, R. P.(1975). A discrete state-space model for linear image processing. *IEEE Trans. Automat. Control*, Vol. 20, 1–10.
Salarieh, H. & Alasty, A.(2008). Control of stochastic chaos using sliding mode method. *Journal of Computational and Applied Mathematics*, 1-24.
Singh,V.(2008). On global asymptotic stability of 2-D discrete systems with state saturation. *Physics Letters A Elsevier*, Vol. 372, 5287–5289.
Utkin, V. I.(1977). Variable structure systems with sliding modes. *IEEE Trans. Automat. Contr.*, Vol. AC-22, 212–222.
Whalley, R.(1990). Two-dimensional digital filters. *Appl. Math. Modelling*, Vol. 14.
Wang, Z. & Liu X.(2003). Robust stability of Two-Dimensional uncertain discrete systems. *IEEE Signal Processing. lett*, Vol. 10, 133-136.
Wu, L. & Gao H.(2008). Sliding mode control of two-dimensional systems in Roesser model. IEEE Proc. Of Control Theory and Applications, Vol. 2, 352–364.
Wu, T. Z. & Juang, Y. T.(2008). Design of variable structure control for fuzzy nonlinear systems. *Expert Systems with Applications*, Vol. 35, 1496–1503.
Young, K. D.; Utkin, V. I. & Ozguner, U.(1999). A Control Engineer's Guide to Sliding Mode Control. *IEEE Trans. Control Systems Technology*, Vol. 7.

Sliding Mode Control Using Neural Networks

Muhammad Yasser, Marina Arifin and Takashi Yahagi
Research Institute of Signal Processing
Japan

1. Introduction

Variable structure control with sliding mode, which is commonly known as sliding mode control (SMC), is a nonlinear control strategy that is well known for its robust characteristics (Utkin, 1977). The main feature of SMC is that it can switch the control law very fast to drive the system states from any initial state onto a user-specified sliding surface, and to maintain the states on the surface for all subsequent time (Utkin, 1977), (Phuah et al., 2005 a).

The conventional SMC has two disadvantages (Ertugrul & Kaynak, 2000), (Slotine & Sastry, 1983), which are the chattering phenomenon (Slotine & Sastry, 1983), (Young et al., 1999) and the difficulty in calculating the equivalent control law of SMC that requires a thorough knowledge of the parameters and dynamics of the nominal controlled plant (Ertugrul & Kaynak, 2000), (Slotine & Sastry, 1983), (Hussain & Ho, 2004).

Many methods of SMC using neural networks (NN) have been proposed (Phuah et al., 2005 a), (Ertugrul & Kaynak, 2000), (Hussain & Ho, 2004), (Phuah et al., 2005 b), (Yasser et al., 2007), (Topalov et al., 2007).

In this paper, sliding mode controls using NN are proposed to deal with the problem of eliminating the chattering effect and the difficulty in calculating the equivalent control law of SMC that requires a thorough knowledge of the parameters and dynamics of the nominal controlled plant. The first method of this method applies a method using a simplified form of the distance function proposed in (Phuah et al., 2005 a), (Phuah et al., 2005 b). Furthermore, the simplified distance function of our method uses a sliding surface in the space of the output error and its derivations, as proposed in (Yasser et al., 2006 a), (Yasser et al., 2006 c), instead of the space of the states error to construct a corrective control input. Thus, no observer is required in the proposed method. Moreover, we also propose the application of an NN to construct the equivalent control input of SMC. The weights of the NN are adjusted using a backpropagation algorithm as in (Yasser et al., 2006 b). Hence, a thorough knowledge of the parameters and dynamics of the nominal controlled plant is not required for calculating the equivalent control law. Finally, a stability analysis is carried out, and the effectiveness of this first control method is confirmed through computer simulations. This first method has been previously discussed in (Yasser et al., 2007).

The second method of this paper applies an NN to produce the gain of the corrective control of SMC. Furthermore, the output of the switching function the corrective control of SMC is applied for the learning and training of the NN. There is no equivalent control of SMC is used in this second method. As in the first method, this second method applies a method using a sliding surface in the space of the output error and its derivations, as proposed in

(Yasser et al., 2006 a), (Yasser et al., 2006 c). The weights of the NN are adjusted using a sliding mode backpropagation algorithm, that is a backpropagation algorithm using the switching function of SMC for its plant sensitivity. Thus, this second method does not use the equivalent control law of SMC, instead it uses a variable corrective control gain produced by the NN for the SMC. Hence, a thorough knowledge of the parameters and dynamics of the nominal controlled plant is not required for calculating the control law. Finally, a stability analysis is carried out, and the effectiveness of this first control method is confirmed through computer simulations.

2. Sliding mode control

In designing a standard sliding mode controller, first we are required to construct a sliding surface that represents a desired system dynamics, and then to develop a switching control law such that a sliding mode exists on every point of the sliding surface. Any states outside the surface are driven to reach the surface in a finite time.

Let us consider an SISO nonlinear plant with BIBO described as

$$\begin{aligned} \dot{\mathbf{x}}_p(t) &= \mathbf{f}(\mathbf{x}_p(t)) + B_p u_p(t) \\ y_p(t) &= C_p \mathbf{x}_p(t) + h(\mathbf{x}_p(t)) \end{aligned} \tag{1}$$

where $\mathbf{x}_p(t)$ is an n_p th-order plant state vector, $u_p(t)$ is the control input, $y_p(t)$ is a plant output, $\mathbf{f}(\cdot)$ is a nonlinear vector function $\in R^{n_p}$, $h(\cdot)$ is a scalar nonlinear function, and B_p and C_p are matrices with appropriate dimensions. We assume that the system in (1) is controllable and observable.

The control objective is to determine a control law $u_p(t)$ such that the state vector $\mathbf{x}_p(t)$ tracks a given bounded desired state vector $\hat{\mathbf{x}}_p(t) \in R^{n_p}$. Therefore, the states error can be obtained as

$$\begin{aligned} \mathbf{e}_{x_p}(t) &= \hat{\mathbf{x}}_p(t) - \mathbf{x}_p(t) \\ &= \left[e_{x_p}(t), \dot{e}_{x_p}(t), \cdots, e_{x_p}^{(n_p-1)}(t) \right]^T. \end{aligned} \tag{2}$$

Then the sliding surface in the space of the state error can be obtained as

$$S_{x_p}(t) = \mathbf{c}_{x_p}^T \mathbf{e}_{x_p}(t) \tag{3}$$

where $\mathbf{c}_{x_p} = \left[c_{x_p 1}, c_{x_p 2}, \cdots, c_{x_p n_p} \right]^T$ is a slope of sliding surface. Generally $\mathbf{c}_{x_p}$ is chosen to force the state error converge to zero when the state is on the sliding surface.

Meanwhile, the process of SMC can be divided into two phases: the approaching phase with $S_{x_p}(t) \neq 0$ and the sliding phase with $S_{x_p}(t) = 0$. Therefore, two types of control law: an equivalent control and a corrective control can be derived separately corresponding to those two phases.

In the sliding phase, we have $S_{x_p}(t) = 0$ and $\dot{S}_{x_p}(t) = 0$, then the equivalent control term $u_{eq}(t)$ will force the system dynamics to stay on sliding surface. The equivalent control $u_{eq}(t)$ can be obtained as

$$u_{eq}(t) = \left[\mathbf{c}_{x_p}^T B_p\right]^{-1}\left[\mathbf{c}_{x_p}^T \dot{\mathbf{x}}_p(t) - \mathbf{c}_{x_p}^T \mathbf{f}(\mathbf{x}_p(t))\right]. \tag{4}$$

In the approaching phase, where $S(t) \neq 0$, a corrective control term $u_c(t)$ will force the state error outside the surface to reach the surface. The corrective control term $u_c(t)$ is defined as

$$u_c(t) = k_s sign\left(S_{x_p}(t)\right) \tag{5}$$

where k_s is a positive gain constant, and $sign\left(S_{x_p}(t)\right)$ is a sign function defined as

$$sign\left(S_{x_p}(t)\right) = \begin{cases} +1, \text{ if } S_{x_p}(t) > 0 \\ 0, \text{ if } S_{x_p}(t) = 0 \\ -1, \text{ if } S_{x_p}(t) < 0 \end{cases}. \tag{6}$$

Then, the control law of SMC will be expressed as

$$u_p(t) = u_{eq}(t) + u_c(t). \tag{7}$$

3. Sliding mode control using Neural Networks and a simplified distance function

The first method of this method applies a method using a simplified form of the distance function proposed in (Phuah et al., 2005 a), (Phuah et al., 2005 b). An NN is applied to construct the equivalent control input of SMC. The weights of the NN are adjusted using a backpropagation algorithm as in (Yasser et al., 2006 b).

3.1 Chattering elimination using a simplified distance function

Based on the concept of point to hyperplane distance, an alternative control method to calculate the corrective control term $u_c(t)$ has been proposed in (Phuah et al., 2005 a), (Phuah et al., 2005 b) to suppress the chattering phenomenon which is caused by high frequency oscillations exhibited by the corrective control law $u_c(t)$ in (5). This method uses a distance function $h(t)$ to calculate the distance between the trajectory of the state error and the sliding surface to generate the corrective control law. The distance function $h(t)$ is defined as (Phuah et al., 2005 a), (Phuah et al., 2005)

$$h(t) = \left\|\mathbf{c}_{x_p}\right\|^{-1} S_{x_p}(t) \tag{8}$$

where $\|\cdot\|$ is the usual Euclidean norm in R^{n_p}. The corrective control law is defined as (Phuah et al., 2005 a), (Phuah et al., 2005)

$$u_c(t) = kh(t) \tag{9}$$

where k is a positive constant.

To construct the corrective control law, the distance function (8) can be simplified to minimize the calculation process, and modified by applying the sliding surface in the space of the output error and its derivations, proposed in (Yasser et al., 2006 a), (Yasser et al., 2006 c), instead of the state error. For that, first, we consider a linear reference model to which the plant output required to follow in the form

$$\begin{aligned}\dot{\mathbf{x}}_{\mathbf{m}}(t) &= A_m\mathbf{x}_m(t) + B_m u_m(t)\\ y_m(t) &= C_m\mathbf{x}_m(t)\end{aligned} \tag{10}$$

where $\mathbf{x}_m(t)$ is an n_m th-order reference model state vector, $u_m(t)$ is a reference model input, $y_m(t)$ is a reference model output, A_m, C_m are matrices with appropriate dimensions, and B_m is a scalar value. The reference model can be independent of the controlled plant, and it is permissible to assume $n_m \ll n_p$. Then, we define the output error $e_{y_p}(t)$ as

$$e_{y_p}(t) = y_m(t) - y_p(t) \tag{11}$$

Thus, the simplified distance function $h_{sim}(t)$ can be described as

$$h_{sim}(t) = k_{sim} S_{y_p}(t) \tag{12}$$

where k_{sim} is a positive constant, and $S_{y_p}(t)$ is a sliding surface in the space of the output error and its derivations described as (Yasser et al., 2006 a), (Yasser et al., 2006 c)

$$\begin{aligned}S_{y_p}(t) &= \mathbf{c}_{y_p}^T \mathbf{e}_{y_p}(t)\\ &= \left[c_{y_p1}, c_{y_p2}, \cdots, c_{y_pn_s}\right]\left[e_{y_p}(t), \dot{e}_{y_p}(t), \cdots, e_{y_p}^{(n_s-1)}(t)\right]^T\end{aligned} \tag{13}$$

where $n_s > 2$.

Then, by replacing $h(t)$ in (9) with $h_{sim}(t)$ from (12), a new corrective control law can be defined as

$$u_c(t) = kh_{sim}(t) = k_{y_p} S_{y_p}(t) \tag{14}$$

where $k_{y_p} = k \cdot k_{sim}$ is a positive constant.

3.2 Neural Networks for equivalent control

To avoid the requirement of the thorough knowledge of the parameters and dynamics of the nominal plant (1), we use a feedforward NN, which consists of an input layer, a hidden layer, and an output layer as in (Yasser et al., 2006 b), to construct the equivalent control input $u_{eq}(t)$ of the SMC in (7). The equivalent control input $u_{eq}(t)$ is described as

$$\begin{aligned}u_{eq}(t) &= \alpha u_{NN}(t)\\ &= \alpha f_{ZOH}\left(u_{NN}(k)\right)\end{aligned} \tag{15}$$

where α is a positive constant, $u_{NN}(t)$ is a continuous-time output of the NN, $u_{NN}(k)$ is a discrete-time output of the NN, and $f_{ZOH}(\cdot)$ is a zero-order hold function.

As in (Yasser et al., 2006 b), we implement a sampler in front of the NN with an appropriate sampling period to obtain the discrete-time input of the NN, and a zero-order hold is

implemented to transform the discrete-time output $u_{NN}(k)$ of the NN back to the continuous-time output $u_{NN}(t)$ of the NN.
The input $i(k)$ of the NN is given as

$$i(k) = \left[e_{y_p}(k-1), \cdots, e_{y_p}(k-n) \right] \tag{16}$$

where $e_{y_p}(k)$ is the discrete-time form of $e_{y_p}(t)$ in (11). And the dynamics of the NN are given as (Yasser et al., 2006 b)

$$h_q(k) = \sum_i i_i(k) m_{iq}(k) \tag{17}$$

$$\begin{aligned} u_{NN}(k) &= o(k) \\ &= \sum_i S_1(h_q(k)) m_{qj}(k) \end{aligned} \tag{18}$$

where $i_i(k)$ is the input to the i-th neuron in the input layer ($i = 1, \cdots, n_i$), $h_q(k)$ is the input to the q-th neuron in the hidden layer ($q = 1, \cdots, n_q$), $o(k)$ is the input to the single neuron in the output layer, n_i and n_q are the number of neurons in the input layer and the hidden layer, respectively, $m_{iq}(k)$ are the weights between the input layer and the hidden layer, $m_{qj}(k)$ are the weights between the hidden layer and the output layer, and $S_1(\cdot)$ is a sigmoid function. The sigmoid function is chosen as

$$S_1(X) = \frac{2}{1 + \exp(-\mu X)} - 1 \tag{19}$$

where $\mu > 0$.
The objective of the NN training is to minimize the error function $E_{y_p}(k)$ described as

$$E_{y_p}(k) = \frac{1}{2} e_{y_p}^2(k) = \frac{1}{2} \sum_j^{n_j} \left[y_m(k) - y_p(k) \right] \tag{20}$$

where $e_{y_p}(k)$ is the discrete-time form of $e_{y_p}(t)$ in (11). The NN training is done by adapting $m_{iq}(k)$ and $m_{qj}(k)$ using the method in (Yasser et al., 2006 b) as follows

$$\begin{aligned} \Delta m_{qj}(k) &= -c \cdot \frac{\partial E(k)}{\partial m_{qj}(k)} \\ &= c \cdot \left[y_m(k) - y_p(k) \right] \cdot J_{plant} \cdot S_1(h_q(k)) \end{aligned} \tag{21}$$

$$\begin{aligned} \Delta m_{iq}(k) &= -c \cdot \frac{\partial E(k)}{\partial m_{iq}(k)} \\ &= c \cdot \left[y_m(k) - y_p(k) \right] \cdot J_{plant} \cdot m_{qj}(k) \cdot \frac{\mu}{2} (1 - S_1^2(X)) \cdot i_i(k) \end{aligned} \tag{22}$$

where c is a learning parameter, and J_{plant} represents the plant Jacobian estimated using

$$J_{plant} = sign\left(\frac{\partial y_p(k)}{\partial u_{NN}(k)}\right) \tag{23}$$

as in (Yasser et al., 2006 b).

3.3 Stability

For the stability analysis of our method, we start by defining its Lyapunov function and its derivation as follows

$$\begin{aligned} V_{SMCNN}(t) &= V_{NN}(t) + V_{SMC}(t) \\ \dot{V}_{SMCNN}(t) &= \dot{V}_{NN}(t) + \dot{V}_{SMC}(t) \end{aligned} \tag{24}$$

where $V_{NN}(t)$ is the Lyapunov function of the NN of our method, and $V_{SMC}(t)$ is the Lyapunov function of SMC of our method.

For $\dot{V}_{NN}(t)$, we assume that it can be approximated as

$$\dot{V}_{NN}(t) \cong \frac{\Delta V_{NN}(k)}{\Delta T} \tag{25}$$

where $\Delta V_{NN}(k)$ is the derivation of a discrete-time Lyapunov function, and ΔT is a sampling time. According to (Yasser et al., 2006 b), $\Delta V_{NN}(k)$ can be guaranteed to be negative definite if the learning parameter c satisfies the following conditions

$$0 < c < \frac{2}{n_q} \tag{26}$$

for the weights between the hidden layer and the output layer, $m_{qj}(k)$, and

$$0 < c < \frac{2}{n_q}\left[\max_k \left\|m_{qj}(k)\right\| \cdot \max_k \left\|i_i(k)\right\|\right]^{-2} \tag{27}$$

for the weights between the input layer and the hidden layer, $m_{iq}(k)$. Furthermore, if the conditions in (26) and (27) are satisfied, the negativity of $\dot{V}_{NN}(t)$ can also be increased by reducing ΔT in (25).

For $V_{SMC}(t)$, it is defined as

$$\begin{aligned} V_{SMC}(t) &= \frac{S_{y_p}^2(t)}{2} \\ \dot{V}_{SMC}(t) &= S_{y_p}(t)\dot{S}_{y_p}(t). \end{aligned} \tag{28}$$

Then we the following assumption.

Assumption 1: The sliding surface in (13) can approximate the sliding surface in (3) (Yasser et al., 2006 c)

$$S_{y_p}(t) \cong S_{x_p}(t). \tag{29}$$

$\dot{V}_{SMC}(t)$ in (28) can be assured to be negative definite if

$$\dot{S}_{y_p}(t) \cong \dot{S}_{x_p}(t) \\ = -k'_{y_p} S_{y_p}(t) \quad (30)$$

where k'_{y_p} is a positive constant. Following the stability analysis method in (Phuah et al., 2005 a), we apply (1)−(3), (7), (14), (15), (29) and (30) to (28) and assume that (15) can approximate (4). Thus, $\dot{V}_{SMC}(t)$ can be described as

$$\begin{aligned}\dot{V}_{SMC}(t) &= S_{y_p}(t)\dot{S}_{x_p}(t) \\ &= S_{y_p}(t)\mathbf{c}_{x_p}^T \dot{\mathbf{e}}_{x_p}(t) \\ &= S_{y_p}(t)\mathbf{c}_{x_p}^T \left[\dot{\hat{\mathbf{x}}}_p(t) - \dot{\mathbf{x}}_p(t)\right] \\ &= S_{y_p}(t)\left[\mathbf{c}_{x_p}^T \dot{\hat{\mathbf{x}}}_p(t) - \mathbf{c}_{x_p}^T \mathbf{f}(\mathbf{x}_p(t)) - \mathbf{c}_{x_p}^T B_p u_p(t)\right] \\ &= S_{y_p}(t)\left[\mathbf{c}_{x_p}^T \dot{\hat{\mathbf{x}}}_p(t) - \mathbf{c}_{x_p}^T \mathbf{f}(\mathbf{x}_p(t)) - \mathbf{c}_{x_p}^T B_p \left[u_{eq}(t) + u_c(t)\right]\right] \\ &= S_{y_p}(t)\left[-\bar{k}_{y_p} S_{y_p}(t)\right] \\ &= -\bar{k}_{y_p} S_{y_p}^2(t)\end{aligned} \quad (31)$$

which is negative definite, where $\bar{k}_{y_p} = \mathbf{c}_{x_p}^T B_p k_{y_p}$. The reaching condition (Phuah et al., 2005 a) can be achieved if

$$-\bar{k}_{y_p} S_{y_p}(t) \le \eta sign(S_{y_p}(t)). \quad (32)$$

where η is a small positive constant.

3.4 Simulation

Let us consider an SISO nonlinear plant described by (Yasser et al., 2006 b)

$$\begin{bmatrix}\dot{x}_1 \\ \dot{x}_2\end{bmatrix} = \begin{bmatrix} x_2 \\ 2x_1 \sin(x_1)\end{bmatrix} + \begin{bmatrix}0 \\ 1\end{bmatrix} u_p \\ y_p = x_1 + \sin(x_1) \quad (33)$$

and the parameters $\mathbf{c}_{y_p} = \begin{bmatrix}12 & 1\end{bmatrix}^T$ in (13), $k_{y_p} = 20$ in (14), $\alpha = 0.1$ in (15), $n_i = 1$ in(17), $n_q = 5$ in (18), $\mu = 2$ in (19), $c = 0.001$ in (21) and (22), $J_{plant} = +1$ in (21), and $\Delta T = 0.01$ in (25) are all fixed. The switching speed for the corrective control of SMC is set to 0.02 seconds. We assume a first-order reference model in (10) with parameters $A_m = -10$, $B_m = -10$, and $C_m = 1$.

Fig. 1 and Fig. 2 show the outputs of the reference model $y_m(t)$ and the plant output $y_p(t)$ using the conventional method of SMC with an NN and a sign function. These figures show that the plant output $y_p(t)$ can follow the output of the reference model $y_m(t)$ closely but not smoothly, as chattering occurs as seen in Fig. 2.

Fig. 3 and Fig. 4 show the outputs of the reference model $y_m(t)$ and the plant output $y_p(t)$ using our proposed method. It can be seen that the plant output $y_p(t)$ can follow the output of the reference model $y_m(t)$ closely and smoothly, as chattering has been eliminated as seen in Fig. 4.

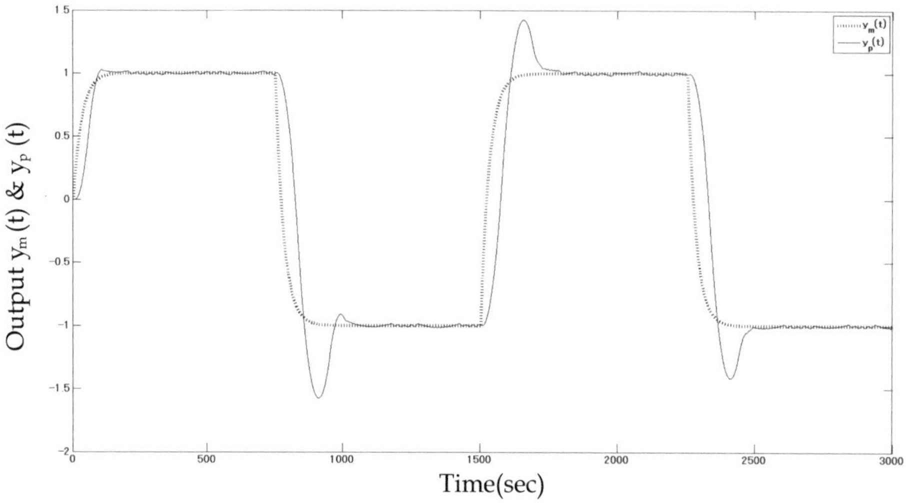

Fig. 1. $y_m(t)$ and $y_p(t)$ using SMC with NN and a sign function

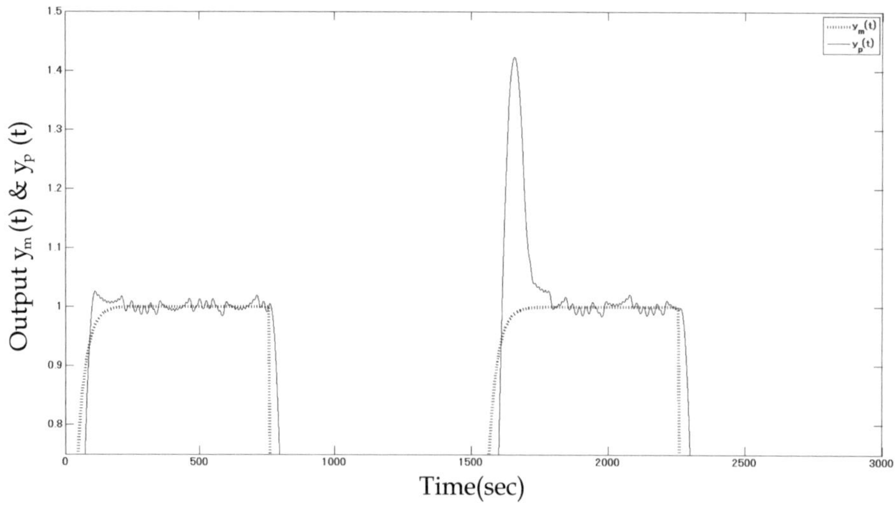

Fig. 2. Magnified upper parts of the curves in Fig. 1

4. Sliding mode control with a variable corrective control gain using Neural Networks

The method in this subsection applies an NN to produce the gain of the corrective control of SMC. Furthermore, the output of the switching function the corrective control of SMC is applied for the learning and training of the NN. There is no equivalent control of SMC is used in this second method.

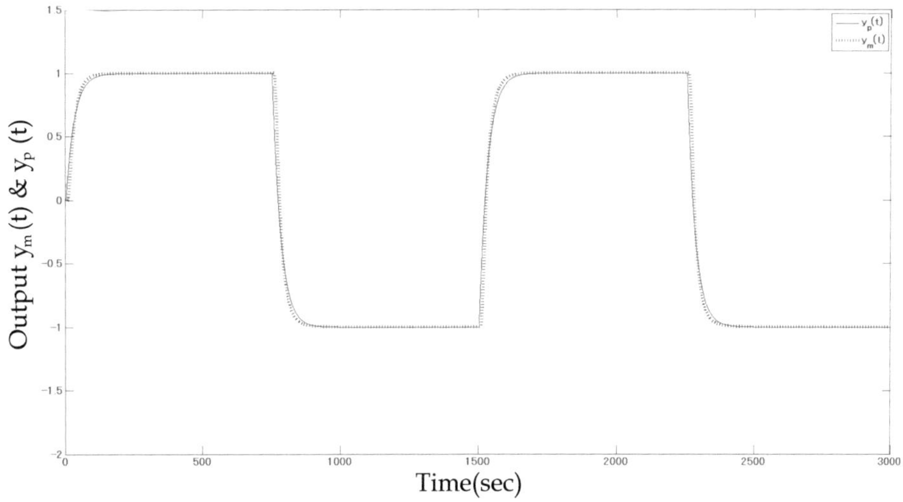

Fig. 3. $y_m(t)$ and $y_p(t)$ using SMC with NN and the simplified distance function

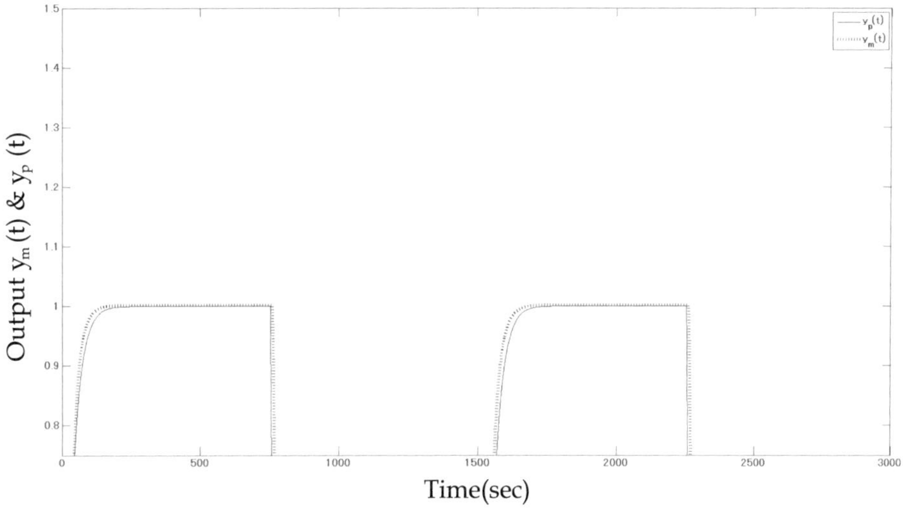

Fig. 4. Magnified upper parts of the curves in Fig. 3

4.1 A variable corrective control gain using Neural Networks for chattering elimination

Using NN to produce a variable gain for a corrective control gain of SMC, instead of using a fixed gain in the conventional SMC, can eliminate the chattering. The switching function of the corrective control is used in the sliding mode backpropagation algorithm to adjust the weight of the NN. This method of SMC does not use any equivalent control of (7) in its control law. For the SISO nonlinear plant with BIBO described in (1), the control input of SMC with a variable corrective control gain using NN is given as

$$u_p(t) = u_{cV}(t) \tag{34}$$

$$u_{cV}(t) = k_{y_p V}(t) sign\left(S_{y_p}(t)\right) \tag{35}$$

where $u_{cV}(t)$ is the corrective control with variable gain using NN, and $k_{y_p V}(t)$ is the variable gain produced by NN described as

$$\begin{aligned} k_V(t) &= \alpha \left| u_{NNV}(t) \right| \\ &= \alpha f_{ZOH}\left(u_{NNV}(k) \right) \end{aligned} \tag{36}$$

where α is a positive constant, $u_{NNV}(t)$ is a continuous-time output of the NN, $u_{NNV}(k)$ is a discrete-time output of the NN, $|\cdot|$ is an absolute function, and $f_{ZOH}(\cdot)$ is a zero-order hold function.

As in subsection 3.2, we implement a sampler in front of the NN with an appropriate sampling period to obtain the discrete-time input of the NN, and a zero-order hold is implemented to transform the discrete-time output $u_{NNV}(k)$ of the NN back to the continuous-time output $u_{NNV}(t)$ of the NN.

The input $i(k)$ of the NN is given as in (16), and the dynamics of the NN are given as

$$h_{Vq}(k) = \sum_i i_i(k) m_{Viq}(k) \tag{37}$$

$$\begin{aligned} u_{NNV}(k) &= o_V(k) \\ &= \sum_i S_1(h_{Vq}(k)) m_{Vqj}(k) \end{aligned} \tag{38}$$

where $i_i(k)$ is the input to the i-th neuron in the input layer ($i = 1, \cdots, n_{Vi}$), $h_{Vq}(k)$ is the input to the q-th neuron in the hidden layer ($q = 1, \cdots, n_{Vq}$), $o_V(k)$ is the input to the single neuron in the output layer, n_{Vi} and n_{Vq} are the number of neurons in the input layer and the hidden layer, respectively, $m_{Viq}(k)$ are the weights between the input layer and the hidden layer, $m_{Vqj}(k)$ are the weights between the hidden layer and the output layer, and $S_1(\cdot)$ is a sigmoid function. The sigmoid function is chosen as in (19).

4.2 Sliding mode backpropagation for Neural Networks training

In the sliding mode backpropagation, the objective of the NN training is to minimize the error function $E_{y_p}(k)$ described in (20). The NN training is done by adapting $m_{Viq}(k)$ and $m_{Vqj}(k)$ as follows

$$\begin{aligned} \Delta m_{Vqj}(k) &= -c \cdot \frac{\partial E(k)}{\partial m_{Vqj}(k)} \\ &= c \cdot \left[y_m(k) - y_p(k) \right] \cdot J_{V plant} \cdot S_1(h_{Vq}(k)) \end{aligned} \tag{39}$$

$$\begin{aligned} \Delta m_{Viq}(k) &= -c \cdot \frac{\partial E(k)}{\partial m_{Viq}(k)} \\ &= c \cdot \left[y_m(k) - y_p(k) \right] \cdot J_{V plant} \cdot m_{Vqj}(k) \cdot \frac{\mu}{2}(1 - S_1^2(X)) \cdot i_i(k) \end{aligned} \tag{40}$$

where c is the learning parameter, and $J_{V\,plant}$ is described as

$$\begin{aligned} J_{V\,plant} &= sign\left(\frac{\partial y_p(k)}{\partial u_{NNV}(k)}\right)\cdot sign\left(S_{y_p}(k)\right) \\ &= \bar{J}_{plant}\cdot sign\left(S_{y_p}(k)\right) \end{aligned} \tag{41}$$

where $S_{y_p}(k)$ is the time-sampled form of $S_{y_p}(t)$ in (13).

4.3 Stability

For the stability analysis of our method, we start by defining its Lyapunov function and its derivation as follows

$$\begin{aligned} V_{SMCNN_V}(t) &= V_{NN_V}(t) + V_{SMC_V}(t) \\ \dot{V}_{SMCNN_V}(t) &= \dot{V}_{NN_V}(t) + \dot{V}_{SMC_V}(t) \end{aligned} \tag{42}$$

where $V_{NNV}(t)$ is the Lyapunov function of the NN of our method, and $V_{SMCV}(t)$ is the Lyapunov function of SMC of our method.

For $\dot{V}_{NN_V}(t)$, we assume that it can be approximated as

$$\dot{V}_{NN_V}(t) \cong \frac{\Delta V_{NNV}(k)}{\Delta T} \tag{43}$$

where $\Delta V_{NNV}(k)$ is the derivation of a discrete-time Lyapunov function, and ΔT is a sampling time. According to (Yasser et al., 2006 b), $\Delta V_{NNV}(k)$ can be guaranteed to be negative definite if the learning parameter c satisfies the following conditions

$$0 < c < \frac{2}{n_{Vq}} \tag{44}$$

for the weights between the hidden layer and the output layer, $m_{Vqj}(k)$, and

$$0 < c < \frac{2}{n_{Vq}}\left[\max_k \left\|m_{Vqj}(k)\right\| \cdot \max_k \left\|i_i(k)\right\|\right]^{-2} \tag{45}$$

for the weights between the input layer and the hidden layer, $m_{Viq}(k)$. Furthermore, if the conditions in (44) and (45) are satisfied, the negativity of $\dot{V}_{VNN}(t)$ can also be increased by reducing ΔT in (43).

For $V_{SMCV}(t)$, it is defined as

$$\begin{aligned} V_{SMC_V}(t) &= \frac{S_{y_p}^2(t)}{2} \\ \dot{V}_{SMC_V}(t) &= S_{y_p}(t)\dot{S}_{y_p}(t). \end{aligned} \tag{46}$$

Then we again use assumption 1. Thus, $\dot{V}_{SMC_V}(t)$ in (46) can be assured to be negative definite if

$$\begin{aligned}\dot{S}_{y_p}(t) &\cong \dot{S}_{x_p}(t)\\ &= -k'_{y_{pV}} S_{y_p}(t)\end{aligned} \tag{47}$$

where $k'_{y_{pV}}$ is a positive constant. Based on the stability analysis method in subsection 3.3, we apply (1) – (3), (34), (35), (29) and (30) to (28). Thus, $\dot{V}_{SMC}(t)$ can be described as

$$\begin{aligned}\dot{V}_{SMC_V}(t) &= S_{y_p}(t)\dot{S}_{x_p}(t)\\ &= S_{y_p}(t)\mathbf{c}_{x_p}^T \dot{\mathbf{e}}_{x_p}(t)\\ &= S_{y_p}(t)\mathbf{c}_{x_p}^T\left[\hat{\dot{\mathbf{x}}}_p(t) - \dot{\mathbf{x}}_p(t)\right]\\ &= S_{y_p}(t)\left[\mathbf{c}_{x_p}^T\hat{\dot{\mathbf{x}}}_p(t) - \mathbf{c}_{x_p}^T\mathbf{f}(\mathbf{x}_p(t)) - \mathbf{c}_{x_p}^T B_p u_p(t)\right]\\ &= S_{y_p}(t)\left[\mathbf{c}_{x_p}^T\hat{\dot{\mathbf{x}}}_p(t) - \mathbf{c}_{x_p}^T\mathbf{f}(\mathbf{x}_p(t)) - \mathbf{c}_{x_p}^T B_p\left[k_{y_{pV}}(t) sign\left(S_{y_p}(t)\right)\right]\right]\\ &= S_{y_p}(t)\left[\mathbf{c}_{x_p}^T\hat{\dot{\mathbf{x}}}_p(t) - \mathbf{c}_{x_p}^T\mathbf{f}(\mathbf{x}_p(t))\right] - \bar{k}_{y_{pV}}(t)S_{y_p}(t) sign\left(S_{y_p}(t)\right)\end{aligned} \tag{48}$$

where $\bar{k}_{y_{pV}}(t) = \mathbf{c}_{x_p}^T B_p k_{y_p V}(t)$. $\dot{V}_{SMC}(t)$ in (48) is negative definite if $k_{y_p V}(t)$ produced by the NN is large enough. The reaching condition (Phuah et al., 2005 a) can be achieved if

$$S_{y_p}(t)\left[\mathbf{c}_{x_p}^T\hat{\dot{\mathbf{x}}}_p(t) - \mathbf{c}_{x_p}^T\mathbf{f}(\mathbf{x}_p(t))\right] - \bar{k}_{y_{pV}}(t)S_{y_p}(t) sign\left(S_{y_p}(t)\right) \le \eta sign(S_{y_p}(t)). \tag{49}$$

where η is a small positive constant.

4.4 Simulation

Let us consider an SISO nonlinear plant described in (33) and the parameters $\mathbf{c}_{y_p} = [9 \quad 1]^T$ in (13), $\alpha = 1$ in (36), $n_{Vi} = 2$ in (37), $n_q = 5$ in (38), $\mu = 2$ in (19) and (40), $c = 0.01$ in (39) and (40), $\bar{J}_{plant} = +1$ in (41), and $\Delta T = 0.01$ in (43) are all fixed. The switching speed for the corrective control of SMC is set to 0.02 seconds. We assume a first-order reference model in (10) with parameters $A_m = -10$, $B_m = -10$, and $C_m = 1$.

Fig. 5 and Fig. 6 show the outputs of the reference model $y_m(t)$ and the plant output $y_p(t)$ using our proposed method. It can be seen that the plant output $y_p(t)$ can follow the output of the reference model $y_m(t)$ closely and smoothly, as chattering has been eliminated as seen in Fig. 6.

5. Conclusion

In this chapter, we proposed two new SMC strategies using NN for SISO nonlinear systems with BIBO has been proposed to deal with the problem of eliminating the chattering effect.

In the first method, to eliminate the chattering effect, it applied a method using a simplified distance function. Furthermore, we also proposed the application of an NN using the backpropagation algorithm to construct the equivalent control input of SMC.

The second method of this paper applied an NN to produce the gain of the corrective control of SMC. Furthermore, the output of the switching function the corrective control of

SMC was applied for the learning and training of the NN. There was no equivalent control of SMC used in this second method. The weights of the NN were adjusted using a sliding mode backpropagation algorithm, that was a backpropagation algorithm using the switching function of SMC for its plant sensitivity. Thus, this second method did not use the equivalent control law of SMC, instead it used a variable corrective control gain produced by the NN for the SMC.

Brief stability analysis was carried out for the two methods, and the effectiveness of our control methods was confirmed through computer simulations.

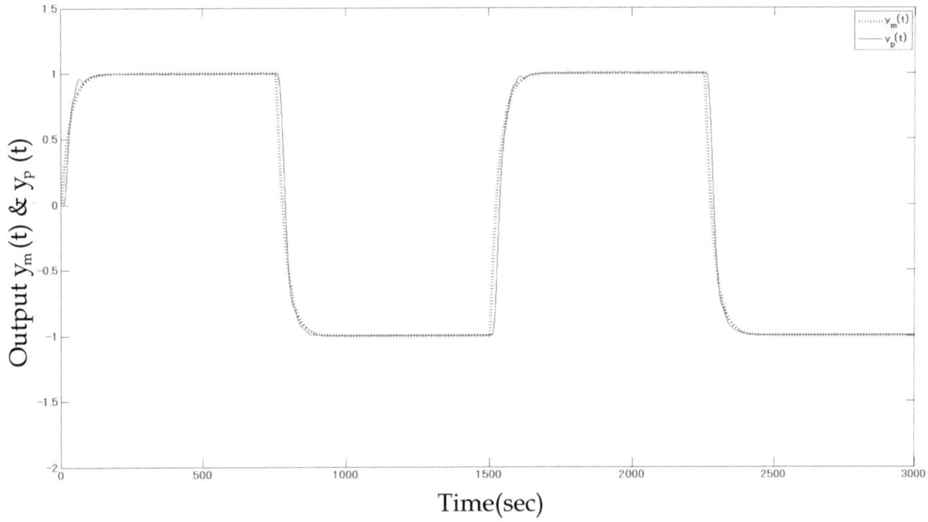

Fig. 5. $y_m(t)$ and $y_p(t)$ using SMC with a variable corrective gain using NN

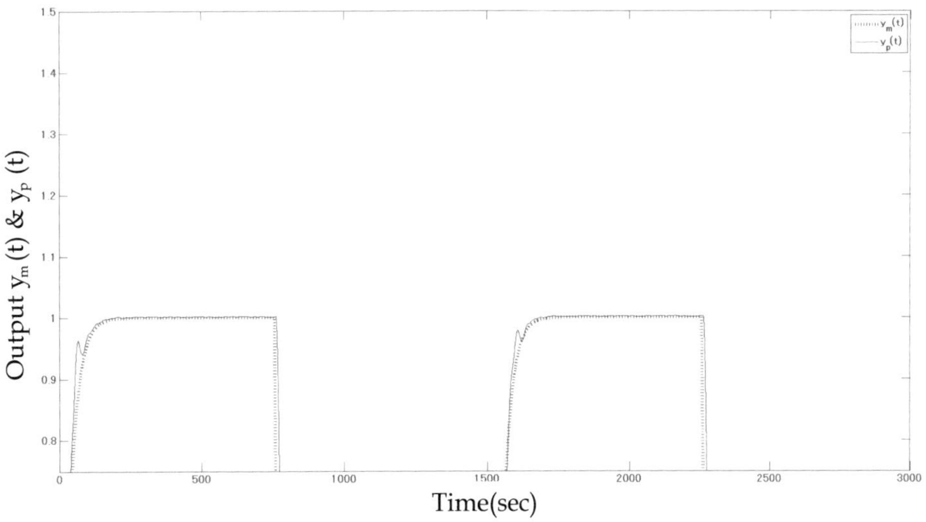

Fig. 6. Magnified upper parts of the curves in Fig. 5

6. References

Ertugrul, M. & Kaynak, O. (2000). Neuro-sliding mode control of robotic manipulators. Mechatronics, Vol. 10, page numbers 239-263, ISSN: 0957-4158

Hussain, M.A. & Ho, P.Y. (2004). Adaptive sliding mode control with neural network based hybrid models. Journal of Process Control, Vol. 14, page numbers 157–176, ISSN: 0959-1524

Phuah, J.; Lu, J. & Yahagi, T. (2005) (a). Chattering free sliding mode control in magnetic levitation system. IEEJ Transactions on Electronics, Information, and Systems, Vol. 125, No. 4, page numbers 600–606, ISSN: 0385-4221

Phuah, J.; Lu, J.; Yasser, M.; & Yahagi, T. (2005) (b). Neuro-sliding mode control for magnetic levitation systems, Proceedings of the 2005 IEEE International Symposium on Circuits and Systems (ISCAS 2005), page numbers 5130–5133, ISBN: 0-7803-8835-6, Kobe, Japan, May 2005, IEEE, USA

Slotine, J.E. & Sastry, S. S. (1983). Tracking control of nonlinear systems using sliding surface with application to robotic manipulators. International Journal of Control, Vol. 38, page numbers 465–492, ISSN: 1366-5820

Topalov, A.V.; Cascella, G.L.; Giordano, V.; Cupertion, F. & Kaynak, O. (2007). Sliding mode neuro-adaptive control of electric drives. IEEE Transactions on Industrial Electronnics, Vol. 54, page numbers 671–679, ISSN: 0278-0046

Utkin, V.I. (1977). Variable structure systems with sliding mode. IEEE Transactions on Automatic Control, Vol. 22, page numbers 212–222 , ISSN: 00189286

Yasser, M.; Trisanto, A.; Lu, J. & Yahagi T.(2006) (a). Adaptive sliding mode control using simple adaptive control for SISO nonlinear systems, Proceedings of the 2006 IEEE International Symposium on Circuits and Systems (ISCAS 2006), page numbers 2153–2156, ISBN: 0-7803-9390-2, Island of Kos, Greece, May 2006, IEEE, USA

Yasser, M.; Trisanto, A.; Haggag, A.; Lu, J.; Sekiya, H. & Yahagi, T. (2006) (c). An adaptive sliding mode control for a class of SISO nonlinear systems with bounded-input bounded-output and bounded nonlinearity, Proceedings of the 45th IEEE Conference on Decision and Control (45th CDC), page numbers 1599-1604, ISBN: 1-4244-0171-2, San Diego, USA, December 2006, IEEE, USA

Yasser, M., Trisanto, A., Lu, J. & Yahagi, T. (2006) (b). A method of simple adaptive control for nonlinear systems using neural networks. IEICE Transactions on Fundamentals, Vol. E89-A, No. 7, page numbers 2009–2018, ISSN: 1745-1337

Yasser, M., Trisanto, A., Haggag, A., Yahagi, T., Sekiya, H. & Lu, J. (2007). Sliding mode control using neural networks for SISO nonlinear systems, Proceedings of The SICE Annual Conference 2007, International Conference on Instrumentation, Control and Information Technology (SICE2007), page numbers 980-984, ISBN: 978-4-907764-27-2, Takamatsu, Japan, September 2007, SICE Japan

Young, K.D.; Utkin, V.I. & Ozguner, U. (1999). A control engineer's guide to sliding mode control. IEEE Transactions On Control System Technology, Vol. 7, No. 3, ISSN: 1063-6536

Sliding Mode Control Approach for Training On-line Neural Networks with Adaptive Learning Rate

Ademir Nied and José de Oliveira
Department of Electrical Engineering, State University of Santa Catarina
Brazil

1. Introduction

This chapter includes contributions to the theory of on-line training of artificial neural networks (ANN), considering the multilayer perceptrons (MLP) topology. By on-line training, we mean that the learning process is conducted while the signal processing is being executed by the system, i.e., the neural network continuously adjusts its free parameters from the variations in the incident signal in real time (Haykin, 1999).

An artificial neural network is a massively parallel distributed processor made up of simple processing units, which have a natural tendency to store experimental knowledge and make it available for use (Haykin, 1999). These units (also called neurons) are non-linear adaptable devices, although very simple in terms of computing power and memory. However, when linked, they have enormous potential for nonlinear mappings. The learning algorithm is the procedure used to do the learning process, whose function is to modify the synaptic weights of the network in an orderly manner to achieve a desired goal of the project (Haykin, 1999).

Although initially used only in problems of pattern recognition and signal processing and image, today, the ANN are used to solve various problems in several areas of human knowledge.

An important feature of ANN is its ability to generalize, i.e., the ability of the network to provide answers in relation to standards unknown or not presented during the training phase. Among the factors that influence the generalization ability of ANN, we cite the network topology and the type of algorithm used to train the network.

The network topology refers to the number of inputs, outputs, number of layers, number of neurons per layer and activation function. From the work of Cybenko (1989), networks with the MLP topology had widespread use, because they possessed the characteristic of universal approximator of continuous functions. Basically, an MLP network is subdivided into the following layers: input layer, intermediate or hidden layer(s) and output layer. The operation of an MLP network is synchronous, i.e., given an input vector, it is propagated to the output by multiplying by the weights of each layer, applying the activation function (the model of each neuron of the network includes a non-linear activation function, being the non-linearity differentiable at any point) and propagating this value to the next layer until the output layer is reached.

Issues such as flexibility of the system to avoid biased solutions (*under□tting*) and, conversely, limiting the complexity of network topology, thus avoiding the variability of solutions (*over□tting*), are inherent aspects to define the best topology for an MLP. This balance between bias and variance is known in the literature as "the dilemma between bias and variance" (German et al., 1992).

Several algorithms that seek to improve the generalization ability of MLP networks are proposed in the literature (Reed, 1993). Some algorithms use construction techniques, changing the network topology. That is, from a super-sized network already trained, methods of *pruning* are applied in order to determine the best topology considering the best balance between bias and variance. Other methods use restriction techniques of the weights values of MLP networks without changing the original topology. However, it is not always possible to measure the complexity of a problem, which makes the choice of network topology an empirical process.

Regarding the type of algorithm used for training MLP networks, the formulation of the backpropagation algorithm (BP) (Rumelhart et al., 1986) enabled the training of fedforward neural networks (FNN). The algorithm is based on the BP learning rule for error correction and can be viewed as a generalization of the least mean square algorithm (LMS) (Widrow & Hoff, 1960), also known as delta rule.

However, because the BP algorithm presents a slow convergence, dependent on initial conditions, and being able to stop the training process in regions of local minima where the gradients are zero, other methods of training appeared to correct or minimize these deficiencies, such as Momentum (Rumelhart et al., 1986), QuickProp (Fahlman, 1988), Rprop (Riedmiller & Braun, 1993), setting the learning rate (Silva & Almeida, 1990; Tollenaere, 1990), the conjugate gradient algorithm (Brent, 1991), the Levenberg-Marquardt algorithm (Hagan & Menhaj, 1994; Parisi et al., 1996), the fast learning algorithm based on the gradient descent in the space of neurons (Zhou & Si, 1998), the learning algorithm in real-time neural networks with exponential rate of convergence (Zhao, 1996), and recently a generalization of the BP algorithm, showing that the most common algorithms based on the BP algorithm are special cases of the presented algorithm (Yu et al., 2002).

However, despite the previously mentioned methods accelerating the convergence of network training, they cannot avoid areas of local minima (Yu et al., 2002), i.e., regions where the gradients are zero because the derivative of the activation function has a value of zero or near zero, even if the difference between the desired output and actual output of the neuron is different from zero.

Besides the problems mentioned above, it can be verified that the learning strategy of training algorithms based on the principle of backpropagation is not protected against external disturbances associated with excitation signals (Efe & Kaynak, 2000; 2001).

The high performance of variable structure system control (Itkis, 1976) in dealing with uncertainties and imprecision have motivated the use of the sliding mode control (SMC) (Utkin, 1978) in training ANN (Parma et al., 1998a). This approach was chosen for three reasons: because it is a well established theory, it allows for the adjustment of parameters (weights) of the network, and it allows an analytical study of the gains involved in training. Thus, the problem of the training of MLP networks is treated and solved as a problem of control, inheriting characteristics of robustness and convergence inherent in systems that use SMC.

The results presented in Efe & Kaynak (2000), Efe et al. (2000) have shown that the convergence properties of gradient-based training strategies widely used in ANN can be improved by utilizing the SMC approach. However, the method presented indirectly uses the Variable Structure Systems (VSS) theory. Some studies on the direct use of SMC strategy are also reported in the literature. In Sira-Ramirez & Colina-Morles (1995) the zero-level set of the learning error variable in Adaline neural networks is regarded as a sliding surface in the space of learning parameters. A sliding mode trajectory can then be induced, in finite time, on such a desired sliding manifold. The proposed method was further extended in Yu et al. (1998) by introducing adaptive uncertainty bound dynamics of signals. In Topalov et al. (2003), Topalov & Kaynak (2003) the sliding mode strategy for the learning of analog Adaline networks, proposed by Sira-Ramirez & Colina-Morles (1995), was extended to a more general class of multilayer networks with a scalar output.

The first SMC learning algorithm for training multilayer perceptron (MLP) networks was proposed by Parma et al. (1998a). Besides the speed up achieved with the proposed algorithm, control theory is actually used to guide neural network learning as a system to be controlled. It also differs from the algorithms in Sira-Ramirez & Colina-Morles (1995), Yu et al. (1998) and Topalov et al. (2003), due to the use of separate sliding surfaces for each network layer. A comprehensive review of VSS and SMC can be seen in Hung et al. (1993), and a survey about the fusion of computationally intelligent methodologies and SMC can be found in Kaynak et al. (2001).

Although the methodology used by Parma et al. (1998a) makes it possible to determine the limits of parameters involved in the training of MLP networks, their complexity still makes it necessary to use heuristic methods to determine the most appropriate gain to be used in order to ensure the best network performance for a particular training.

In this chapter, an algorithm for on-line ANN training based on SMC is presented. The main feature of this procedure is the adaptability of the gain (learning rate), determined iteratively for every weight update, and obtained from only one sliding surface.

To evaluate the algorithm, simulations were performed considering two distinct applications: function approximation and a neural-based stator flux observer of an induction motor (IM). The network topology was defined according to the best possible response with the fewest number of neurons in the hidden layer without compromising the ability of network generalization. The network used in the simulations has only one hidden layer, differing in the number of neurons in this layer and the number of inputs and outputs of the network, which were chosen according to the application for the MLP.

2. The On-line adaptive MLP training algorithm

This section presents the algorithm with adaptive gain for on-line training MLP networks with multiple outputs that operates in quasi-sliding modes. The term "quasi-sliding regime" was introduced by Miloslavjevic (1985) to express the fact that the extension to the case of discrete time under the usual time for the continuous existence of a sliding regime, does not necessarily guarantee chattering around the sliding surface in the same way that it occurs in continuous time systems. Moreover, in Sarpturk et al. (1987) it was shown that the condition proposed by Miloslavjevic (1985) for the existence of a quasi-sliding mode could cause the system to become unstable. Now, let us specify how the quasi-sliding mode and the reaching condition are understood in this paper.

Definition 1. *Let us define a quasi-sliding mode in the* ε *vicinity of a sliding hyperplane* $s(n) = 0$ *for a motion of the system such that*

$$|s(n)| \leq \varepsilon \tag{1}$$

where the positive constant ε *is called the quasi-sliding-mode band width (Bartoszewicz, 1998).*

This definition is different from the one proposed in Gao et al. (1995) since it does not require the system state to cross the sliding plane $s(n) = 0$ in each successive control step.
The convergence of the system state to the sliding surface can be analyzed considering the convergence of the series

$$\sum_{n=1}^{\infty} s(n). \tag{2}$$

If the convergence of the series is guaranteed, then the system state will converge, at least assimptotically, to the sliding surface $s(n) = 0$.
Consider Cauchy's convergence principle (Kreyszig, 1993): The series $s_1 + s_2 + \cdots + s_n$ converges if and only if, for a given value $\varepsilon \in \Re^+$, a value N can be found such that $| s_{n+1} + s_{n+2} + \cdots + s_{n+p} | < \varepsilon$ for all $n > N$ e $p = 1, 2, \cdots$. A series is absolutely convergent if:

$$\sum_{n=1}^{\infty} |s(n)| \tag{3}$$

is convergent. To study the convergence of the series given by (3) the ratio test is used (Butkov, 1968). Thus, it holds that:

$$\left| \frac{s(n+1)}{s(n)} \right| \leq Q < 1. \tag{4}$$

Definition 2. *It is said that the system state converges to a quasi-sliding regime in the vicinity ε of a sliding surface $s(n) = 0$ if the following condition is satisfied:*

$$|s(n+1)| < |s(n)|. \tag{5}$$

Remark: From Definition 2, crossing the plane $s(n) = 0$ is allowed but not required.

Theorem 1. *Let $s(n) : \Re^2 \to \Re$, the sliding surface defined by $s(n) = CX1(n) + X2(n)$, where $\{C, X1(n)\} \in \Re^+$ and $X2(n) \in \Re$. If $X1(n) = E(n)$, being $E(n) = \frac{1}{2}\sum_{k=1}^{m_L} e_k^2(n)$ defined as the instantaneous value of the total energy of the error of all the neurons of the output layer of an MLP, where $e_k(n) = d_k(n) - y_k(n)$ is the error signal between the desired value and actual value at the output of the neuron k of the network output at iteration n, m_L is the number of neurons in the output layer of the network, and $X2(n) = \frac{X1(n) - X1(n-1)}{T}$ is defined as the variation of $X1(n)$ in a sample period of T, then, for the current state $s(n)$ to converge to a vicinity ε of $s(n) = 0$, it is necessary and sufficient that the network meet the following:*

$$sign(s(n))\,[C(X1(n+1) - X1(n)) + X2(n+1) - X2(n)] < 0 \tag{6}$$

$$sign(s(n))\,[C(X1(n+1) + X1(n)) + X2(n+1) + X2(n)] > 0, \tag{7}$$

being $sign(s(n)) = \begin{cases} +1, & s(n) \geq 0 \\ -1, & s(n) < 0 \end{cases}$ *the sign function of $s(n)$.* ◊

Proof: Defining the absolute value of the sliding surface as follows

$$|s(n)| = sign(s(n))s(n), \tag{8}$$

then, from (5) it holds that

$$|s(n+1)| < |s(n)| \Rightarrow sign(s(n+1))s(n+1) < sign(s(n))s(n).$$

As $sign(s(n))sign(s(n)) = 1$, yields

$$sign(s(n))[sign(s(n))sign(s(n+1))s(n+1) - s(n)] < 0.$$

If $sign(s(n+1)) = sign(s(n))$, then $sign(s(n))[s(n+1) - s(n)] < 0$. Replacing the definition of $s(n)$ as given by Theorem 1 yields

$$sign(s(n))\,[CX1(n+1) + X2(n+1) - (CX1(n) + X2(n))] < 0 \Rightarrow (6).$$

If $sign(s(n+1)) = -sign(s(n))$, then $sign(s(n))[-s(n+1) - s(n)] < 0$. Replacing the definition of $s(n)$ as given by Theorem 1 yields

$$sign(s(n))\,[CX1(n+1) + X2(n+1) + CX1(n) + X2(n)] > 0 \Rightarrow (7).$$

To prove that the conditions of Theorem 1 are sufficient, two situations must be established:

- The sliding surface is not crossed during convergence. In this situation, it holds that

$$sign(s(n+1)) = sign(s(n)).$$

  Considering $s(n) = CX1(n) + X2(n)$ and $s(n+1) = CX1(n+1) + X2(n+1)$, one can write (6) as

$$sign(s(n))[s(n+1) - s(n)] < 0 \Rightarrow sign(s(n+1))s(n+1) < sign(s(n))s(n),$$

  and using (8) yields $|s(n+1)| < |s(n)|$. The validity of (7) for this situation is trivial, i.e.:

$$sign(s(n))[s(n+1) + s(n)] = |s(n+1)| + |s(n)| \Rightarrow (7).$$

- The sliding surface is crossed during convergence. Now, for this situation it holds that

$$sign(s(n+1)) = -sign(s(n)).$$

  Considering, again, $s(n) = CX1(n) + X2(n)$ and $s(n+1) = CX1(n+1) + X2(n+1)$, one can write (7) as

$$sign(s(n))[s(n+1) + s(n)] > 0 \Rightarrow sign(s(n+1))s(n+1) < sign(s(n))s(n),$$

  and using (8) yields $|s(n+1)| < |s(n)|$. The validity of (6) for this situation is trivial too, i.e.:

$$sign(s(n))[s(n+1) - s(n)] = -|s(n+1)| - |s(n)| \Rightarrow (6).$$

□

From Theorem 1, it can be verified that (6) is responsible for the existence of a quasi-sliding regime for $s(n) = 0$, while (7) ensures the convergence of the network state trajectories to a vicinity of the sliding surface $s(n) = 0$. One can also observe that the reference term from the sliding surface signal $sign(s(n))$ determines the external and internal limits of the range of convergence for the following expressions:

$$C(X1(n+1) - X1(n)) + X2(n+1) - X2(n) \tag{9}$$

$$C(X1(n+1) + X1(n)) + X2(n+1) + X2(n). \tag{10}$$

To study the convergence of the sliding surface $s(n) = CX1(n) + X2(n)$, the decomposition of (9) and (10), with respect to a gain η, is necessary in order to obtain a set of equations for these variables and, from the conditions defined by Theorem 1, to determine an interval in $\Re$ due to a gain η, that can guarantee the convergence of the proposed method.

Theorem 2. *Let $s(n) : \Re^2 \rightarrow \Re$, the sliding surface defined by $s(n) = CX1(n) + X2(n)$, where $\{C, X1(n)\} \in \Re^+$ and $X2(n) \in \Re$. If $X1(n)$, $X2(n)$ and T are defined as in Theorem 1, then, for the current state $s(n)$ to converge to a vicinity ε of $s(n) = 0$, it is necessary and sufficient that the network meets the following:*

$$sign(s(n)) \left[c_1\eta^2 + c_2\eta - s(n) + CX1(n)\right] < 0 \tag{11}$$

$$sign(s(n)) \left[c_1\eta^2 + c_2\eta + s(n) + CX1(n)\right] > 0, \tag{12}$$

where $\{c_1, c_2\} \in \Re$. If the following restrictions are taken into account:

$$c_1 > 0 \tag{13}$$

$$c_2 < 0 \tag{14}$$

$$\Delta = c_2^2 - 4c_1c_3 > 0, \tag{15}$$

being $c_3 = \begin{cases} -s(n) + CX1(n), \text{ (in (11))} \\ s(n) + CX1(n), \text{ (in (12))} \end{cases}$ *then, the existence of a limited region for the gain η that satisfies both conditions for convergence is guaranteed.* ◊

Proof: Initially, consider that:

$$X1(n) = \frac{1}{2}\sum_{k=1}^{m_L}(d_k(n) - y_k(n))^2 = \frac{1}{2}\sum_{k=1}^{m_L}(d_k^2(n) - 2d_k(n)y_k(n) + y_k^2(n)), \tag{16}$$

$$X1(n+1) = \frac{1}{2}\sum_{k=1}^{m_L}(d_k^2(n+1) - 2d_k(n+1)y_k(n+1) + y_k^2(n+1)), \tag{17}$$

$$X1(n-1) = \frac{1}{2}\sum_{k=1}^{m_L}(d_k^2(n-1) - 2d_k(n-1)y_k(n-1) + y_k^2(n-1)), \tag{18}$$

$$X2(n+1) = \frac{X1(n+1) - X1(n)}{T}. \tag{19}$$

From (16), (17), (18), (19) and considering the definition of $X2(n)$ given by Theorem 1, one can derive the terms of (9) taking into account that $d_k(n-1) = d_k(n) = d_k(n+1) = d_k$. Thus, it holds:

$$\begin{aligned}
&C(X1(n+1) - X1(n)) + X2(n+1) - X2(n) = \\
&C(X1(n+1) - X1(n)) + \left(\frac{X1(n+1) - X1(n)}{T}\right) - \left(\frac{X1(n) - X1(n-1)}{T}\right) \\
&= \frac{1}{T}\left[(TC+1)X1(n+1) - (TC+2)X1(n) + X1(n-1)\right] \\
&= \frac{1}{T}\Bigg[(TC+1)\frac{1}{2}\sum_{k=1}^{m_L}(d_k^2 - 2d_ky_k(n+1) + y_k^2(n+1)) \\
&\qquad -(TC+2)\frac{1}{2}\sum_{k=1}^{m_L}(d_k^2 - 2d_ky_k(n) + y_k^2(n)) + \frac{1}{2}\sum_{k=1}^{m_L}(d_k^2 - 2d_ky_k(n-1) + y_k^2(n-1))\Bigg] \\
&= \frac{1}{T}\frac{1}{2}\sum_{k=1}^{m_L}\Big[TC(-2d_ky_k(n+1) + y_k^2(n+1) + 2d_ky_k(n) - y_k^2(n)) - 2d_ky_k(n+1) \\
&\qquad +y_k^2(n+1) + 4d_ky_k(n) - 2y_k^2(n) - 2d_ky_k(n-1) + y_k^2(n-1)\Big].
\end{aligned} \tag{20}$$

In the same way, it is possible to derive the terms of (10) taking into account the same considerations used to derive (9). Thus:

$$
\begin{aligned}
&C(X1(n+1)+X1(n))+X2(n+1)+X2(n)=\\
&C(X1(n+1)+X1(n))+\left(\frac{X1(n+1)-X1(n)}{T}\right)+\left(\frac{X1(n)-X1(n-1)}{T}\right)\\
&=\frac{1}{T}\left[(TC+1)X1(n+1)+TCX1(n)-X1(n-1)\right]\\
&=\frac{1}{T}\Bigg[(TC+1)\frac{1}{2}\sum_{k=1}^{m_L}(d_k^2-2d_ky_k(n+1)+y_k^2(n+1))\\
&\qquad+TC\frac{1}{2}\sum_{k=1}^{m_L}(d_k^2-2d_ky_k(n)+y_k^2(n))-\frac{1}{2}\sum_{k=1}^{m_L}(d_k^2-2d_ky_k(n-1)+y_k^2(n-1))\Bigg]\\
&=\frac{1}{T}\frac{1}{2}\sum_{k=1}^{m_L}\Big[TC(d_k^2-2d_ky_k(n+1)+y_k^2(n+1)+d_k^2-2d_ky_k(n)+y_k^2(n))\\
&\qquad-2d_ky_k(n+1)+y_k^2(n+1)-2d_ky_k(n-1)-y_k^2(n-1)\Big]. \qquad (21)
\end{aligned}
$$

From (20) and (21) one can identify the term $y_k(n+1)$ as the target variable from which it is possible to obtain the gain η. Then, doing

$$y_k(n+1)=y_k(n)+c\eta, \qquad (22)$$

$$y_k^2(n+1)=y_k^2(n)+2y_k(n)c\eta+(c\eta)^2, \qquad (23)$$

replacing (22), (23) in (20), (21), respectively, and considering $e_k(n)=d_k-y_k(n)$, yields:

$$
\begin{aligned}
&\frac{1}{T}\frac{1}{2}\sum_{k=1}^{m_L}\Big[(TC+1)c^2\eta^2-2(TC+1)ce_k(n)\eta\\
&\qquad+2d_ky_k(n)-y_k^2(n)-2d_ky_k(n-1)+y_k^2(n-1)\Big] \qquad (24)
\end{aligned}
$$

and

$$
\begin{aligned}
&\frac{1}{T}\frac{1}{2}\sum_{k=1}^{m_L}\Big[(TC+1)c^2\eta^2-2(TC+1)ce_k(n)\eta\\
&\qquad+2TC(d_k-y_k(n))^2-2d_ky_k(n)\\
&\qquad+y_k^2(n)+2d_ky_k(n-1)-y_k^2(n-1)\Big]. \qquad (25)
\end{aligned}
$$

Finally, taking into account the result of $X1(n)-X1(n-1)$, one can obtain the conditions (11) and (12) defined in the Theorem 2, with the coefficients given by:

$$
\begin{aligned}
c_1&=\frac{1}{2}\left(C+\frac{1}{T}\right)c^2\\
c_2&=-\left(C+\frac{1}{T}\right)\sum_{k=1}^{m_L}ce_k(n)
\end{aligned}
$$

$$c_3 = \begin{cases} -s(n) + CX1(n), & \text{(in (11))} \\ s(n) + CX1(n), & \text{(in (12)).} \end{cases} \tag{26}$$

To analyze the intervals of convergence limited by the conditions of (11) and (12) it is necessary to determine the limits of these intervals. It can be verified that the intervals of convergence are obtained from a parabola, the concavity of this parabola being determined by the value of c_1 (in this case, positive concavity, since $c_1 > 0$).
The general form for the quadratic equation related to the convergence conditions can be written as:

$$c_1\eta^2 + c_2\eta + c_3 \tag{27}$$

where c_3 is the independent term. Considering the value of $\Delta = c_2^2 - 4c_1c_3$ and taking into account that $c_1 > 0$, the roots of (27) are given by:

$$\Delta = c_2^2 - 4|c_1|c_3. \tag{28}$$

According to (28), the value of Δ is related to the signal and the module of the sliding surface $s(n)$. From these considerations, one can proceed with the following analysis:

- If $s(n) > 0$:
  (a) $c_1\eta^2 + c_2\eta - s(n) + CX1(n) < 0$
  (1) $|s(n)| > CX1(n) \Rightarrow c_3 < 0$.
  Roots: $\Delta = c_2^2 + 4|c_1||c_3| \Rightarrow \Delta > c_2^2$. Considering $\Delta = c_2^2\xi_1^2$, being $\xi_1 > 1$, the roots can be written as:

$$\eta = -\frac{c_2}{2c_1} \pm \left|\frac{c_2\xi_1}{2c_1}\right| \tag{29}$$

  (2) $|s(n)| < CX1(n) \Rightarrow c_3 > 0$
  Roots: $\Delta = c_2^2 - 4|c_1||c_3| \Rightarrow \Delta < c_2^2$. There are two possible variations for Δ:
  1^a) $0 < \Delta < c_2^2$: Considering $\Delta = \frac{c_2^2}{\xi_1^2}$, the roots can be written as:

$$\eta = -\frac{c_2}{2c_1} \pm \left|\frac{c_2}{2c_1\xi_1}\right| \tag{30}$$

  2^a) $\Delta \leq 0$: This condition is not considered because it does not meet the restriction (15).
  (b) $c_1\eta^2 + c_2\eta + s(n) + CX1(n) > 0$ Roots: $\Delta = c_2^2 - 4|c_1||c_3| \Rightarrow \Delta < c_2^2$. There are two possible variations for Δ:
  1^a) $0 < \Delta < c_2^2$: Considering $\Delta = \frac{c_2^2}{\xi_2^2}$, being $\xi_2 > \xi_1$, the roots can be written as:

$$\eta = -\frac{c_2}{2c_1} \pm \left|\frac{c_2}{2c_1\xi_2}\right| \tag{31}$$

  2^a) $\Delta \leq 0$: This condition is not considered because it does not meet the restriction (15).

From (29), (30) and (31) the following relationship can be established:

$$\left|\frac{c_2}{2c_1\xi_2}\right| < \left|\frac{c_2}{2c_1\xi_1}\right| < \left|\frac{c_2\xi_1}{2c_1}\right|. \tag{32}$$

Considering $(-\frac{c_2}{2c_1})$ as the center point of convergence intervals and observing (32), a diagram can be drawn identifying, in bold, the intervals of convergence for $s(n) > 0$ as shown in Figure 1.

- If $s(n) < 0$:

(a) $c_1\eta^2 + c_2\eta - s(n) + CX1(n) > 0 \Rightarrow c_1\eta^2 + c_2\eta + s(n) + CX1(n) > 0$

Roots: $\Delta = c_2^2 - 4|c_1||c_3| \Rightarrow \Delta < c_2^2$. There are two possible variations for Δ:

1^a) $0 < \Delta < c_2^2$: Considering $\Delta = \frac{c_2^2}{\xi_2^2}$, the roots can be written as:

$$\eta = -\frac{c_2}{2c_1} \pm \left|\frac{c_2}{2c_1\xi_2}\right| \tag{33}$$

2^a) $\Delta \leq 0$: This condition is not considered because it does not meet the restriction (15).

(b) $c_1\eta^2 + c_2\eta + s(n) + CX1(n) < 0 \Rightarrow c_1\eta^2 + c_2\eta - s(n) + CX1(n) < 0$

(1) $|s(n)| > CX1(n) \Rightarrow c_3 < 0$.

Roots: $\Delta = c_2^2 + 4|c_1||c_3| \Rightarrow \Delta > c_2^2$. Considering $\Delta = c_2^2\xi_1^2$, the roots can be written as:

$$\eta = -\frac{c_2}{2c_1} \pm \left|\frac{c_2\xi_1}{2c_1}\right| \tag{34}$$

(2) $|s(n)| < CX1(n) \Rightarrow c_3 > 0$

Roots: $\Delta = c_2^2 - 4|c_1||c_3| \Rightarrow \Delta < c_2^2$. There are two possible variations for Δ:

1^a) $0 < \Delta < c_2^2$: Considering $\Delta = \frac{c_2^2}{\xi_1^2}$, the roots can be written as:

$$\eta = -\frac{c_2}{2c_1} \pm \left|\frac{c_2}{2c_1\xi_1}\right| \tag{35}$$

2^a) $\Delta \leq 0$: This condition is not considered because it does not meet the restriction (15).

From (33), (34) and (35), it can be established the same relationship defined in (32) and, therefore, the diagram can be drawn identifying, in bold, the intervals of convergence for $s(n) < 0$, as shown in Figure 1. □

Remark: The Theorem 2 guarantees the existence of real intervals for the gain η to satisfy the convergence conditions. However, the Theorem 2 does not guarantee, *directly*, the existence of a *positive interval* for the gain η. Both for $s(n) > 0$ and $s(n) < 0$, it is assured that at least one positive real root exists, which reinforces the existence of a positive interval for η. In (30), (31), (33) and (35), the existence of positive real roots is conditioned by $-\frac{c_2}{2c_1} > 0$. As $c_1 > 0$, the condition is: $-c_2 > 0 \Rightarrow c_2 < 0$, which can be easily verified from the application of the methodology developed in a two-layer MLP.

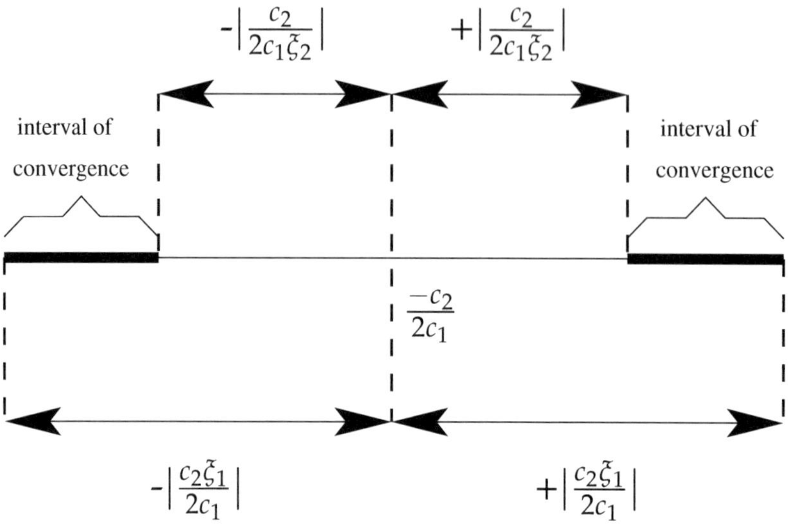

Fig. 1. Intervals of convergence for the algorithm with adaptive gain.

Once $s(n)$ is related to the network topology used, to verify the existence of a positive interval for the gain η, it is necessary to analyze the behavior of convergence conditions for the linear perceptron, the nonlinear perceptron and the two-layer MLP network with linear output. The choice of an MLP network topology was made in order to make the calculations involved in determining the network response to a stimulus simpler, yet still effective.

2.1 Determination of η for the linear perceptron

Let the output, at discrete-time n, of a neuron perceptron with linear activation function be given by:

$$y(n) = \sum_{j=1}^{m_0} w_j(n)x_j(n), \tag{36}$$

where m_0 is the number of inputs of the neuron. The analysis for the determination of the intervals for the gain η is performed for each input pattern of the neuron.
The output of the neuron at time $n+1$ is given by:

$$y(n+1) = y(n) + \Delta y(n) = y(n) + \sum_{j=1}^{m_0} \Delta w_j(n)x_j(n). \tag{37}$$

To calculate (37), it is necessary to determine $\Delta w_j(n)$, which represents the adjustment of the weights of the perceptron at time n. An immediate expression can be obtained from the Delta rule, which gives rise to the LMS algorithm or learning algorithm of gradient descent. Thus, it yields:

$$\Delta w_j(n) = -\eta \frac{\partial E(n)}{\partial w_j(n)} = -\eta\, 2\, \frac{1}{2}\,(d(n) - y(n))(-1)\frac{\partial y(n)}{\partial w_j(n)} = \eta e(n)x_j(n). \tag{38}$$

Once $\Delta w_j(n)$ is set, $y(n+1)$ can then be calculate as follows:

$$y(n+1) = y(n) + e(n)\sum_{j=1}^{m_0} x_j^2(n)\eta = y(n) + c\eta. \tag{39}$$

Therefore, using (39) and considering $c = e(n)\sum_{j=1}^{m_0} x_j^2(n)$, the expressions for the coefficients c_1, c_2 e c_3 of (26) can be obtained:

$$c_1 = \frac{1}{2}\left(C + \frac{1}{T}\right)c^2 = \frac{1}{2}\left(C + \frac{1}{T}\right)e^2(n)\left(\sum_{j=1}^{m_0} x_j^2(n)\right)^2$$

$$c_2 = -\left(C + \frac{1}{T}\right)ce(n) = -\left(C + \frac{1}{T}\right)e^2(n)\sum_{j=1}^{m_0} x_j^2(n)$$

$$c_3 = \begin{cases} -s(n) + CX1(n), & (\text{in } (11)) \\ s(n) + CX1(n), & (\text{in } (12)). \end{cases} \tag{40}$$

After determining the coefficients c_1, c_2 e c_3, the Theorem 2 can be applied to determine the intervals of convergence for the gain η.

2.2 Determination of η for the non-linear perceptron

The output characteristic of this type of neuron is given by:

$$y(n) = \varphi\left(\sum_{j=1}^{m_0} w_j(n)x_j(n)\right), \tag{41}$$

where $\varphi(\cdot)$ is the neuron activation function, continuous and differentiable.

The approach used to determine the neuron output is an approximation of the activation function through its decomposition into a Taylor series, instead of propagating the output signal of the neuron using the inverse of activation function. This approach was chosen because the first terms of the Taylor series provide a significant simplification and mathematical cost reduction for the definition of the intervals of convergence, yet limit the ability of approximating the function to regions close to the point of interest.

Let the output, at time n, of a neuron perceptron with non-linear activation function be given by (41). The output of the neuron at time $n+1$ can be written as:

$$y(n+1) = y(n) + \Delta y(n) = y(n) + \varphi\left(\sum_{j=1}^{m_0} \Delta w_j(n)x_j(n)\right). \tag{42}$$

Applying the decomposition of the first-order Taylor series in (42), yields:

$$y(n+1) = y(n) + \dot{y}(n)\sum_{j=1}^{m_0} \Delta w_j(n)x_j(n), \tag{43}$$

where $\left|\sum_{j=1}^{m_0} \Delta w_j(n) x_j(n)\right| \leq \xi$. Using (38) for the variation of weights at time n, it is possible to define an interval for the gain η related to the first-order Taylor series:

$$\eta \leq \frac{\xi}{\left|e(n) \sum_{j=1}^{m_0} x_j^2(n)\right|}. \tag{44}$$

It can be verified that (44) limits the interval of the gain η in accordance with the desired accuracy (ξ) for the approximation of the activation function of the neuron. Rewriting (43) it follows that:

$$\begin{aligned} y(n+1) &= y(n) + \dot{y}(n) e(n) \sum_{j=1}^{m_0} x_j^2(n) \eta \\ &= y(n) + c\eta. \end{aligned} \tag{45}$$

Therefore, using (45) and considering $c = \dot{y}(n) e(n) \sum_{j=1}^{m_0} x_j^2(n)$, the expressions for the coefficients c_1, c_2 e c_3 of (26) can be obtained:

$$\begin{aligned} c_1 &= \frac{1}{2}\left(C + \frac{1}{T}\right) c^2 = \frac{1}{2}\left(C + \frac{1}{T}\right) \dot{y}^2(n) e^2(n) \left(\sum_{j=1}^{m_0} x_j^2(n)\right)^2 \\ c_2 &= -\left(C + \frac{1}{T}\right) c e(n) = -\left(C + \frac{1}{T}\right) \dot{y}(n) e^2(n) \sum_{j=1}^{m_0} x_j^2(n) \\ c_3 &= \begin{cases} -s(n) + CX1(n), \text{ (in (11))} \\ s(n) + CX1(n), \text{ (in (12))}. \end{cases} \end{aligned} \tag{46}$$

After determining the coefficients c_1, c_2 e c_3, observing the limits imposed by the Taylor series decomposition, the Theorem 2 can be applied to determine the intervals of convergence for the gain η.

2.3 Determination of η for two-layer MLP network

Let the linear output of the k-th neuron of a two-layer MLP network related to an output vector $\mathbf{x}(n)$ be:

$$y2_k(n) = \sum_{j=1}^{m_1+1} w2_{kj}(n) y1_j(n) = \sum_{j=1}^{m_1+1} w2_{kj}(n) \varphi\left(\sum_{i=1}^{m_0} w1_{ji}(n) x_i(n)\right).$$

Due to the existence of two layers, one must do the study of the interval of convergence for the output layer and hidden layer separately. Thus, it follows:

- Output layer: Considering only the weights of the output layer as the parameters of interest, the output k at time n of an MLP network with linear output is given by:

$$y2_k(n) = \sum_{j=1}^{m_1+1} w2_{kj}(n) y1_j(n). \tag{47}$$

Assuming that the adjustment of weights is performed initially only in the weights of the output layer, (47) can be compared to (36) for the linear perceptron. In this case, the inputs of neuron k correspond to the output vector of neurons in the hidden layer (plus the bias term) after the activation function, $\mathbf{y1}(n)$, and the weights, for the vector $\mathbf{w2}_k(n)$. The coefficients c_1, c_2 *e* c_3 are obtained from the use of the equations for the linear neuron *by applying the analysis to the network with multiple outputs*. Thus, the coefficients of the quadratic equation associated with the convergence conditions are defined as:

$$c_1 = \frac{1}{2}\left(C + \frac{1}{T}\right)\sum_{k=1}^{m_2}\left[e_k^2(n)\left(\sum_{j=1}^{m_1+1} y1_j^2(n)\right)^2\right]$$

$$c_2 = -\left(C + \frac{1}{T}\right)\sum_{k=1}^{m_2}\left(e_k^2(n)\sum_{j=1}^{m_1+1} y1_j^2(n)\right)$$

$$c_3 = \begin{cases} -s(n) + CX1(n), \text{ (in (11))} \\ s(n) + CX1(n), \text{ (in (12))}. \end{cases} \tag{48}$$

- Hidden layer: Now, we consider the adjustment of the weights of the hidden layer, $\mathbf{W1}(n)$. For this, the weights of the output layer are kept constant. Therefore, the k-th neuron of the MLP network with two layers with linear output is given by:

$$y2_k(n) = \sum_{j=1}^{m_1+1} w2_{kj}(n)\varphi\left(\sum_{i=1}^{m_0} w1_{ji}(n)x_i(n)\right). \tag{49}$$

The output at time $n+1$ is given by:

$$y2_k(n+1) = y2_k(n) + \Delta y2_k(n) = y2_k(n) + \sum_{j=1}^{m_1+1} w2_{kj}(n)\varphi\left(\sum_{i=1}^{m_0} \Delta w1_{ji}(n)x_i(n)\right). \tag{50}$$

Applying in (50) the decomposition of the first order Taylor series, we obtain:

$$y2_k(n+1) = y2_k(n) + \dot{y2}_k(n)\sum_{j=1}^{m_1+1} w2_{kj}(n)\sum_{i=1}^{m_0}\Delta w1_{ji}(n)x_i(n), \tag{51}$$

where $\left|\sum_{i=1}^{m_0}\Delta w1_{ji}(n)x_i(n)\right| \le \xi$. It is possible to use (38) for the variation of weights at time n. However, for the hidden layer, there is not a desired response specified for the neurons in this layer. Consequently, an error signal for a hidden neuron is determined recursively in terms of the error signals of all neurons for which the hidden neuron is directly connected, i. e., $\Delta w1_{ji}(n) = \eta\sum_{k=1}^{m_2} e_k(n)w2_{kj}(n)x_i(n)$. From the expression of $\Delta w1_{ji}(n)$ it is possible to define an interval for the gain η of the Taylor series decomposition:

$$\eta \le \frac{\xi}{\left|\sum_{k=1}^{m_2} e_k(n)w2_{kj}(n)\sum_{i=1}^{m_0} x_i^2(n)\right|}. \tag{52}$$

Although (52) is assigned to a single neuron, the limit for the gain η must be defined in terms of the whole network, choosing the lower limit associated with a network of neurons. Decomposing (51) yields:

$$y2_k(n+1) = y2_k(n) + \dot{y2}_k(n)\sum_{j=1}^{m_1+1}\sum_{k=1}^{m_2} e_k(n)w2_{kj}^2(n)\sum_{i=1}^{m_0} x_i^2(n)\eta, \tag{53}$$

Therefore, using (53) and considering $c = \dot{y2}_k(n)\sum_{j=1}^{m_1+1}\sum_{k=1}^{m_2} e_k(n) w2_{kj}^2(n)\sum_{i=1}^{m_0} x_i^2(n)$, the coefficients $c_1,\ c_2\ e\ c_3$ can be obtained as follows:

$$c_1 = \frac{1}{2}\left(C+\frac{1}{T}\right)\sum_{k=1}^{m_2}\left[\dot{y2}_k^2(n)\sum_{j=1}^{m_1+1}\sum_{k=1}^{m_2} e_k^2(n)\left(w2_{kj}^2(n)\right)^2\left(\sum_{i=1}^{m_0} x_i^2(n)\right)^2\right]$$

$$c_2 = -\left(C+\frac{1}{T}\right)\sum_{k=1}^{m_2}\left(\dot{y2}_k^2(n)\sum_{j=1}^{m_1+1}\sum_{k=1}^{m_2} e_k^2(n) w2_{kj}^2(n)\sum_{i=1}^{m_0} x_i^2(n)\right)$$

$$c_3 = \begin{cases} -s(n)+CX1(n), \text{ (in (11))} \\ s(n)+CX1(n), \text{ (in (12))}. \end{cases} \qquad (54)$$

Thus, from the coefficients obtained in (48) and (54), the Theorem 2 can be apply, with the final interval for the gain η determined by the intersection of the intervals defined by convergence equations obtained for the hidden layer and the output layer, observing the limit imposed by the Taylor series decomposition. It should be noted also that, in (48) and (54), the coefficients c_1, c_2 e c_3 are dependent on C e T. This implies that, for the determination of C, the sampling period should be taken into account.

3. Simulation results

This section shows the results obtained from simulations of the algorithm presented in Section 2. The simulations are performed considering two distinct applications. In Section 3.1 the proposed algorithm is used in the approximation of a sine function. Then, in Section 3.2, the proposed algorithm is used for observation of the stator flux of the induction motor.

3.1 On-line function approximation

This section presents the simulation results of applying the proposed algorithm for the learning real-time function $f(t) = e^{\left(-\frac{1}{3}\right)} sin(3t)$. The following parameters were considered for the simulations: integration step = 10μs; simulation time = 2s; sampling period = 250μs. The same simulations were also performed considering the standard BP algorithm (Rumelhart et al., 1986), the algorithm proposed by Topalov et al. (2003), and two algorithms for real-time training provided by (Parma et al., 1999a;b). For these algorithms, the training gains (learning rates) were chosen in order to obtain the best result, using the same initial conditions for each of the algorithms simulated.

The network topology used in the simulation of the algorithms was as follows: an input, 5 neurons in the hidden layer and one neuron in the output. The size of the hidden layer of the MLP was defined according to the best possible response with the fewest number of neurons. The hyperbolic tangent function was used as the activation function for the hidden layer neurons. This same function was also used as the activation function for the neuron of the output layer in the standard BP algorithm and on the two algorithms proposed by (Parma et al., 1999a;b). For the algorithm presented in this paper and that proposed by Topalov et al. (2003), the linear output for the neuron of the output layer was used.

The simulation results of the proposed algorithm are shown in Figure 2. For the confidence interval, $\xi = 1.5$ was used to approximate the hyperbolic tangent function using the first-order Taylor series.

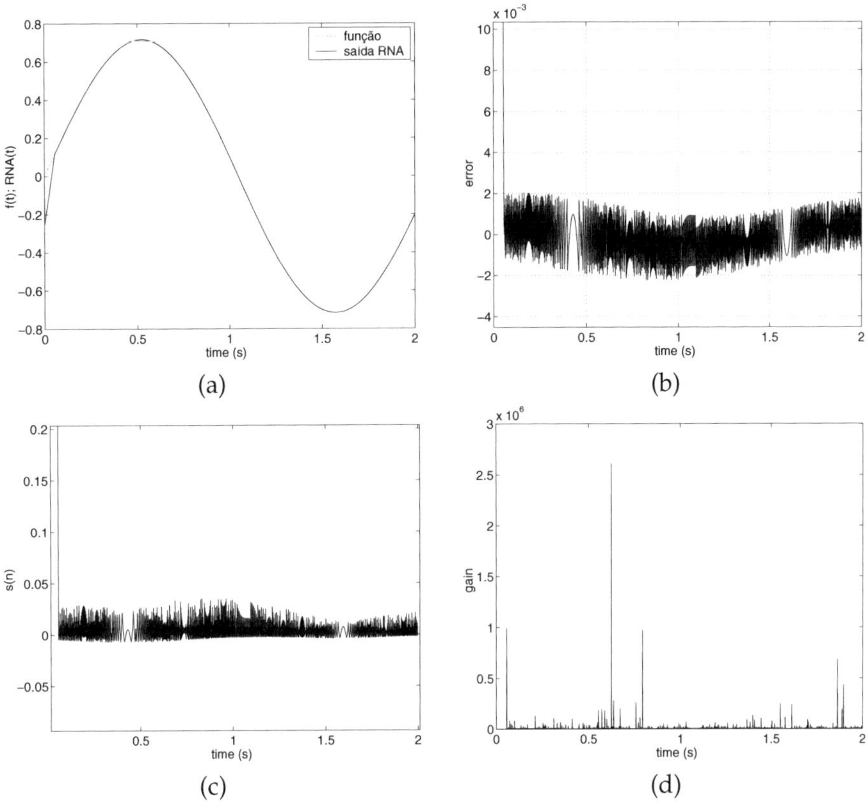

Fig. 2. Simulation results of the approximation of $f(t)$ using the presented algorithm: (a) output $f(t)$ x ANN(t); (b) error between output $f(t)$ and ANN output; (c) behavior of $s(n)$; (d) adaptive gain.

In the simulation, the value of 10,000 was adopted for the parameter C. The function $f(t)$ is shown dashed while the output of ANN is shown in continuous line. The graph of the approximation error for the sine function considered, the behavior of the sliding surface $s(n)$, and the training gains obtained from the proposed algorithm during the simulation time are also presented.

The fact that the proposed algorithm uses the gradient of error function with respect to weights, causes oscillations in the learning process, implying the need for high gains for the network training. These oscillations are also felt in the behavior of the sliding surface, as can be seen in the graph (c) of Figure 2.

Figure 3 shows the simulation results of the algorithms proposed by Parma et al. (1999a;b) and Topalov et al. (2003).

The coefficients and the gains of the algorithms were adjusted by obtaining the following values: 1st Parma algorithm - $C1=C2=10000$, $\eta1=3000$, $\eta2=10$; 2nd Parma algorithm - $C1=C2=10000$, $\eta1=200$, $\eta2=100$; Topalov algorithm - $\eta=10$. These three algorithms presented similar results, especially considering the time needed to reach the sine function, which is much smaller compared with the algorithm proposed in this paper. The proposed algorithm uses a gain adjustment which penalizes the reach time of the function $f(t)$. On the other

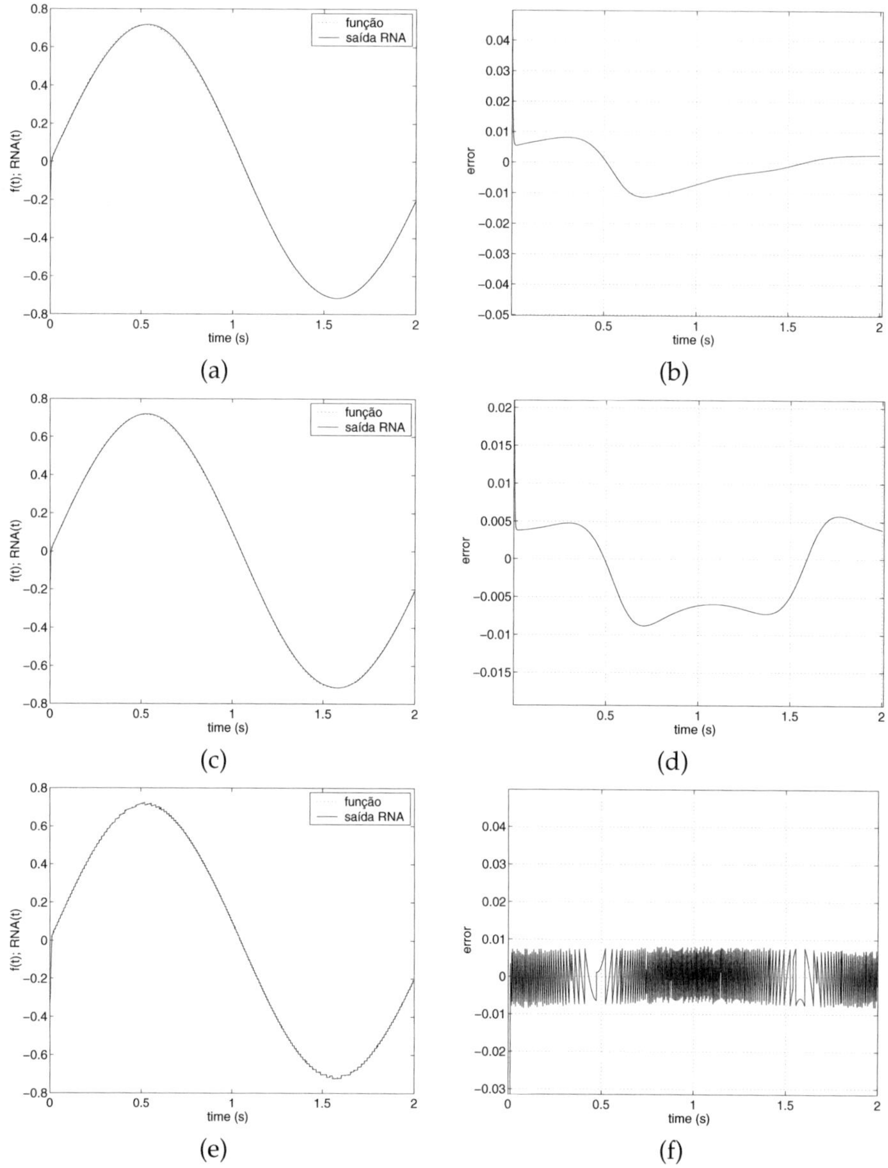

Fig. 3. Simulation results of the approximation of $f(t)$ using the proposed Parma and Topalov: graphs (a) and (b): - 1st Parma algorithm; graphs (c) e (d) - 2nd Parma algorithm; gráficos (e) e (f) - Topalov algorithm.

hand, if the errors of function approximation are compared, the proposed algorithm has better performance.

Finally, Figure 4 shows the results obtained using the standard BP algorithm. The adjusted values of gain for the hidden and output layers were, respectively, $\eta 1$=102 e $\eta 2$=12.

As can be easily verified, the standard algorithm BP had the highest error in the approximation of the considered function. This performance was expected for the various reasons outlined

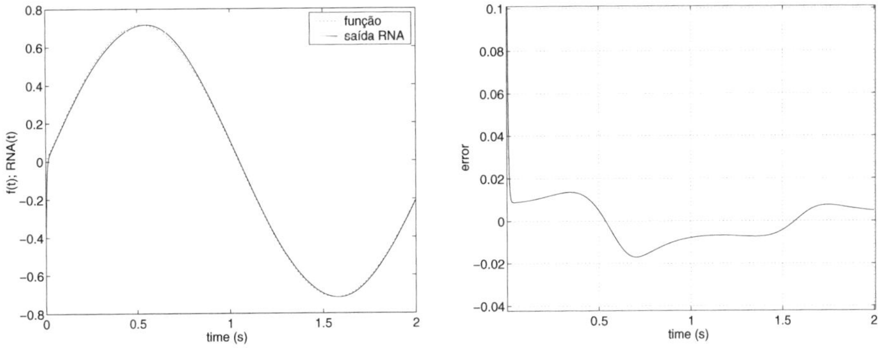

(a) (b)

Fig. 4. Simulation results of the approximation of $f(t)$ using the standard BP algorithm: (a) output $f(t)$ x ANN(t); (b) error between output $f(t)$ and ANN output.

above. The results of this algorithm were presented as a reference, since this algorithm is the oldest of the simulated algorithms.

3.2 Induction motor stator flux neural Observer

Considering the IM drives, the correct estimation of the flux, either the stator, rotor and mutual, is the key to the successful implementation of any vector control strategy (Holtz & Quan, 2003).

The observation, in turn, is a closed loop estimation, which employs, in addition to the input signals, a feedback signal, obtained from the system output signals and the process model.

An important requirement for using an ANN for observing the motor flux, is that training should be done on-line. This approach allows a continuous adjustment of the network weights according to the requirements of the system in which the network operates, in this case, the IM. Figure 5 presents the simulation results of applying the proposed algorithm for training a neural network used as an IM stator flux observer. The following variables were considered: stator flux module (stator flux IM *versus* neural flux observer), electromagnetic torque and motor speed. The IM was submitted to the following transients: 1) start up and speed reversion with no load; 2) loading and unloading (constant torque) the motor at constant speed.

The IM flux can be estimated directly from the voltage equation given by (Novotny & Lipo, 1996):

$$\mathbf{v}_s = R_s \mathbf{i}_s + \frac{d\lambda_s}{dt} \Rightarrow \tag{55}$$

$$\lambda_s = \int (\mathbf{v}_s - R_s \mathbf{i}_s) dt. \tag{56}$$

The main reason for use of (56) is simplicity. The stator flux estimator is independent of the speed measurement if the stationary reference is adopted for the d-q axes (Kovács & Rácz, 1984). This fact makes the approach attractive for use in motor control without speed measurement. Moreover, one can see that the only parametric dependence is the stator resistance, which can be obtained with reasonable accuracy (Novotny & Lipo, 1996). Efficient solutions for the correction of off-set in the integrals of current and voltage can be verified in Holtz & Quan (2003).

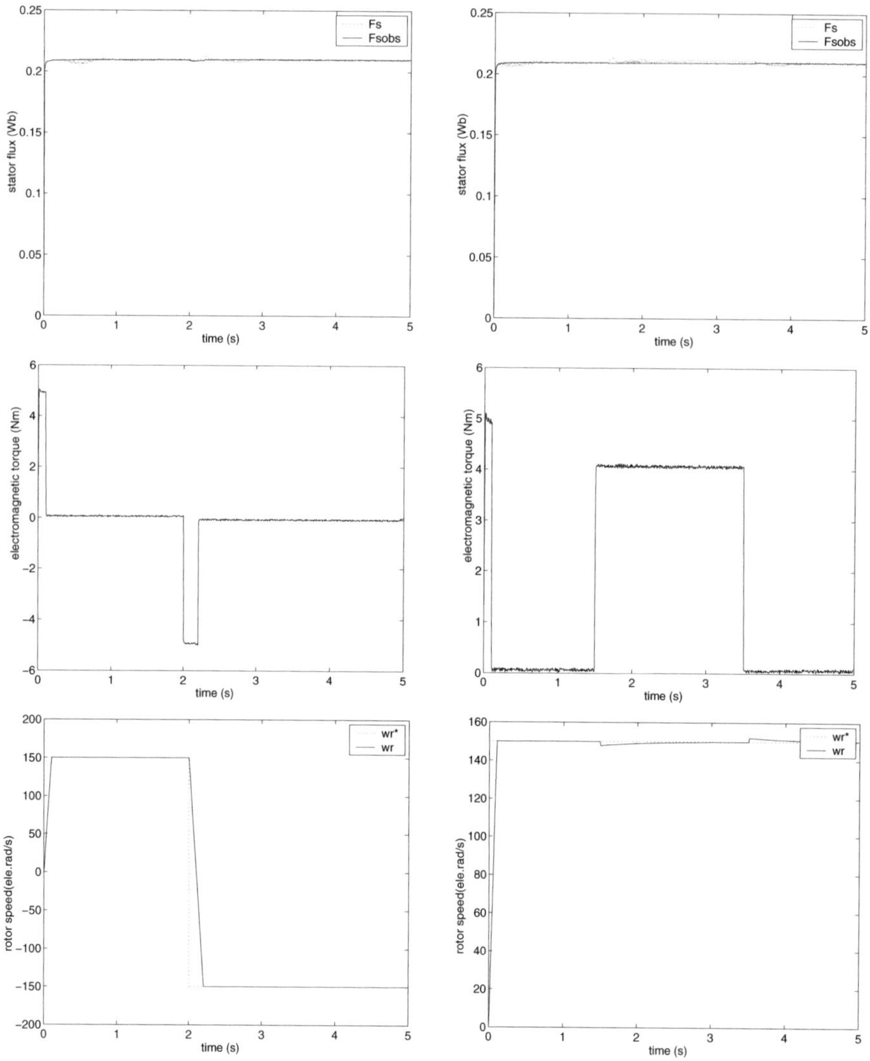

(a) (b)

Fig. 5. Simulation results from neural observer: (a) speed reversal with no load in t=2s; (b) loading and unloading (constant torque) the motor at constant speed of 150 ele.rad/s in t=1.5s and t=3.5s, respectively.

Rewriting (56) considering d-q axes, it follows that:

$$\mathbf{v}_{sd} = R_s \mathbf{i}_{sd} + \frac{d\lambda_{sd}}{dt} \tag{57}$$

$$\mathbf{v}_{sq} = R_s \mathbf{i}_{sq} + \frac{d\lambda_{sq}}{dt}, \tag{58}$$

where R_s is the stator resistance; v_{sd} and v_{sq} are the d-q components of the stator voltage, i_{sd} and i_{sq} are the d-q components of the stator current, λ_{sd} and λ_{sq} are the d-q components of the stator flux, all of them in stator coordinates.

Thus, the d-q components of stator current are used as input of the ANN, and the d-q components of stator flux are the output of the network. The ANN used is the MLP 2-5-2, i.e., 2 inputs, 5 neurons in the hidden layer and 2 outputs. The number of neurons in the hidden layer was determined by analyzing the simulation results, aiming to reduce the computational cost without compromising the results generated by the network. Other studies using a neutral observer can be seen in Nied et al. (2003a) and Nied et al. (2003b).
The IM was submitted to the transients of start up and speed reversion with no load (Figure 5 (a)) and loading-unloading (constant torque) the motor at constant speed (Figure 5 (b)). Both transients are done under the motor speed condition of 150 elec.rad/s. The simulation time was 5 s. A good dynamic performance of the neural observer can be verified since the estimated stator flux tracks the stator reference flux, even during the transients applied to the motor.

4. Conclusion

Using the theory of sliding modes control, the problem of training MLP networks allows the analysis of the network as a system to be controlled, where the control variables are the weights, and the output of the network should follow the reference variable. From this, a methodology was used that allows us to obtain an adaptive gain, determined iteratively at each step of updating the weights, eliminating the need for using heuristics to determine the gain of the network. This methodology was used for on-line training of MLP networks with a linear activation function in the output layer.
The training of the ANN in real time requires a learning process to be performed while the signal processing is being executed by the system, resulting in the continual adjustment of free parameters of the neural network to variations in the incident signal in real time.
From the methodology, an algorithm was developed for on-line training of two-layer MLP networks with linear output. The algorithm presented is general, providing that there are one or more neurons in the output layer of the network.
Regarding the update of network weights, the algorithm updates the weights using the gradient of the error function with respect to the weights (BP algorithm). This weight correction law, despite being widely used for training MLP networks, has its weaknesses, such as the fact that the stability (not asymptotic stability) can only be guaranteed for a set of weights that corresponds to the overll minimum BP algorithm, according to Lyapunov stability theory.
By using the algorithm presented, it is possible to determine a resulting range for the gain η of the network, which is obtained through the intersection of the ranges defined for the hidden layer and output, noting the limit imposed by the Taylor series decomposition. However, the algorithm does not define the final value for the gain η. Thus, it is possible in principle, that any value within a range of positive results be used. Issues are not addressed by the optimization algorithm. However, bearing in mind the necessity of obtaining practical results from the application of the algorithm, we adopted a conservative solution using the gain value η obtained for the limit imposed by the Taylor series decomposition.
Due to the nature of the algorithm, applications that required adjustment of free parameters of the neural network in real time were selected for evaluation.
As a first application, the algorithm was used in the approximation of a sine function. The error of the approximation algorithm presented was the lowest compared with the values of the approximation error made by the other three algorithms simulated.

The other application was related to the use of the algorithm as an observer of the neural stator flux of IM. The results obtained show that the neural observer contributed to the good performance of the variables of flux, speed and torque.
From the simulation results of the algorithm, at least two features of this algorithm can be identify: 1) ease of use, since there is no necessity of determining the gain (or learning rate), which is obtained iteratively by the algorithm, 2) eliminates the need for any information regarding the mathematical model of the system in which the network operates.

5. References

Bartoszewicz, A. (1998). Discrete-time quasi-sliding-mode control strategies, *IEEE Tran. on Industrial Electronics* 45(4): 633–637.

Brent, R. P. (1991). Fast training algorithms for multilayer neural nets, *IEEE Trans. on Neural Networks* 2(3): 346–354.

Butkov, E. (1968). *Mathematical Physics*, Addison-Wesley.

Cybenko, G. (1989). Approximation by superpositions of a sigmoidal function, *Mathematics of Control, Signals and Systems* 2(4): 304–314.

Efe, M. O. & Kaynak, O. (2000). Stabilizing and robustifying the error backpropagation method in neurocontrol applications, *Proceedings of the 2000 IEEE International Conference on Robotics and Automation*, IEEE Press, San Francisco, CA, pp. 1882–1887.

Efe, M. O. & Kaynak, O. (2001). Variable structure systems theory based training strategies for computationally intelligent systems, *Proceedings of the 27th Annual Conference of the IEEE Industrial Electronics Society*, IEEE Press, pp. 1563–1576.

Efe, M. O., Kaynak, O. & Wilamowski, B. M. (2000). Stable training of computationally intelligent systems by using variable structure systems technique, *IEEE Trans. on Industrial Electronics* 47(2): 487–496.

Fahlman, S. E. (1988). Faster-learning variations on backpropagation: an empirical study, *in* D. Touretzky, G. Hinton & T. Sejnowsky (eds), *Proceedings of the 1988 Connectionist Models Summer School*, Morgan Kaufmann, Sao Mateo, CA, pp. 38–51.

Gao, W., Wang, Y. & Homaifa, A. (1995). Discrete-time variable structure control systems, *IEEE Trans. on Industrial Electronics* 42(2): 117–122.

German, S., Bienenstock, E. & Dournsat, R. (1992). Neural networks and the bias/variance dilemma, *Neural Computation* 4(1): 1–58.

Hagan, M. T. & Menhaj, M. B. (1994). Training feedforword networks with the marquardt algorithm, *IEEE Trans. on Neural Networks* 5(6): 989–993.

Haykin, S. (1999). *Neural Networks: A Comprehensive Foundation*, 2nd edn, Prentice-Hall, Englewood Cliffs, NJ.

Holtz, J. & Quan, J. (2003). Drift- and parameter-compensated flux estimator for persistent zero-stator-frequency operation of sensorless-controlled induction motors, *IEEE Trans. on Industry Applications* 39(4): 1052–1060.

Hung, J. Y., Gao, W. & Hung, J. C. (1993). Variable structure control: A survey, *IEEE Trans. on Industrial Electronics* 40(1): 2–22.

Itkis, U. (1976). *Control systems of variable structure*, John Wiley and Sons Inc., New York.

Kaynak, O., Erbatur, K. & Ertugrul, M. (2001). The fusion of computationally intelligent methodologies and sliding-mode control - a survey, *IEEE Trans. on Industrial Electronics* 48(1): 4–17.

Kovács, P. K. & Rácz, E. (1984). *Transient Phenomena in Electrical Machines*, Elsevier, Amsterdam, The Netherlands.
Kreyszig, E. (1993). *Advanced Engineering Mathematics*, 7th edn, John Wiley and Sons Inc.
Miloslavjevic, C. (1985). General conditions for the existence of a quasisliding mode on the switching hyperplane in discrete variable structure systems, *Automation and Remote Control* 46: 307–314.
Nied, A., Junior, S. I. S., Menezes, B. R., Parma, G. G. & Justino, J. C. G. (2003a). Comparative study on flux observers for induction motor drives, *Proceedings of the 7th Brazilien Power Electronics Conference*, Fortaleza, CE.
Nied, A., Junior, S. I. S., Parma, G. G. & Menezes, B. R. (2003b). On-line training algorithms for an induction motor stator flux neural observer, *Proceedings of 29th annual conference of the IEEE Industrial Electronics Society - IECON2003*, IEEE Press, Roanoke, VA, pp. 129–134.
Novotny, D. W. & Lipo, T. A. (1996). *Vector control and dynamics of AC drives*, 1st edn, Cleredon Press.
Parisi, R., Di Claudio, E. D., Orlandi, G. & Rao, B. D. (1996). A generalized learning paradigm exploiting the structure of feedforward neural networks, *IEEE Trans. on Neural Networks* 7(6): 1450–1460.
Parma, G. G., Menezes, B. R. & Braga, A. P. (1998a). Sliding mode algorithm for training multilayer neural network, *IEE Electronics Letters* 38(1): 97–98.
Parma, G. G., Menezes, B. R. & Braga, A. P. (1999a). Neural networks learning with sliding mode control: the sliding mode backpropagation algorithm, *International Journal of Neural Systems* 9(3): 187–193.
Parma, G. G., Menezes, B. R. & Braga, A. P. (1999b). Sliding mode backpropagation: control theory applied to neural networks learning, *Proceedings of the International Joint Conference on Neural Networks*, IEEE Computer Society Press, Washington-DC, pp. 1774–1778.
Reed, R. (1993). Pruning algorithms - a survey, *IEEE Trans. on Neural Networks* 4(5): 740–746.
Riedmiller, M. & Braun, H. (1993). A direct adaptive method for faster backpropagation learning: The rprop algorithm, *Proceedings of the Int. Conf. on Neural Networks*, San Francisco, CA, pp. 586–591.
Rumelhart, D. E., Hinton, G. E. & Williams, R. J. (1986). *Learning internal representation by error propagation*, Vol. 1 of *Parallel distribuited processing: explorations in the microstructure of cognition*, MIT Press, Cambridge, MA, pp. 318–362.
Sarpturk, S., Istefanopulos, Y. & Kaynak, O. (1987). On the stability of discrete-time sliding mode control systems, *IEEE Trans. on Automatic Control* 32(10): 930–932.
Silva, F. M. & Almeida, L. B. (1990). Speeding up backpropagation, *in* R. Eckmiller (ed.), *Advanced Neural Computers*, Amsterdam: Elsevier North Holland, pp. 151–158.
Sira-Ramirez, H. & Colina-Morles, E. (1995). A sliding mode strategy for adaptive learning in adalines, *IEEE Trans. on Circuits and Systems - I: Fundamental Theory and Applications* 42(12): 1001–1012.
Tollenaere, T. (1990). Supersab: Fast adaptive back propagation with good scaling properties, *Neural Networks* 3(5): 561–573.
Topalov, A. V. & Kaynak, O. (2003). A sliding mode strategy for adaptive learning in multilayer feedforward neural networks with a scalar output, *IEEE International Conference on Systems, Man and Cybernetics, 2003*, Vol. 2, IEEE Press, pp. 1636–1641.

Topalov, A. V., Kaynak, O. & Shakev, N. G. (2003). Variable structure systems approach for on-line learning in multilayer artificial neural networks, *Proc. IEEE 29th Annual Conference of the Industrial Electronics Society (2003)*, IEEE Press, Roanoke, VA, pp. 2989–2994.

Utkin, V. I. (1978). *Sliding modes and their application in Variable Structure Systems*, MIR, Moskow.

Widrow, B. & Hoff, M. E. (1960). Adaptive switching circuits, *IRE WESCON Convention Record*, Vol. 4, IRE, New York, pp. 96–104.

Yu, X., Efe, M. O. & Kaynak, O. (2002). A general backpropagation algorithm for feedforward neural networks learning, *IEEE Trans. on Neural Networks* 13(1): 251–254.

Yu, X., Zhihong, M. & Rahman, S. M. M. (1998). Adaptive sliding mode approach for learning in a feedforward neural networks, *Neural Computing and Applications* 7(4): 289–294.

Zhao, Y. (1996). On-line neural network learning algorithm with exponential convergence rate, *IEE Electronics Letters* 32(15): 1381–1382.

Zhou, G. & Si, J. (1998). Advanced neural network training algorithm with reduced complexity based on jacobian deficiency, *IEEE Trans. on Neural Networks* 9(3): 448–453.